SUSTAINABLE HORTICULTURE DEVELOPMENT AND NUTRITION SECURITY

Volume III: Food and Nutrition Security

Compiled by

Dr. Prem Nath

The Technical and Editorial Team

Dr. C.P.A. Iyer

Dr. K.R.M. Swamy

Mr. P.N. Krishna Moorthy

Dr. O.P. Dutta

Dr. B.S. Prabhakar

Mr. P.B. Gaddagimath

Dr. Prem Nath Agricultural Science Foundation (PNASF), Bengaluru, India

Sustainable Horticulture Development and Nutrition Security
Volume III: Food and Nutrition Security

Dr. P.N. Agri. Sci. Foundation (PNASF), Bengaluru-560094

Published by:

Scientific Publishers India

Head Office:
5-A New Pali Road, PO Box 91, Jodhpur
342 001, India
Tel: +91-291-2433323, 2624154
Fax: +91-291-2613449
E-mail: info@scientificpub.com

Branch Office:
4806/24, Ansari Road, Darya Ganj,
New Delhi 110 002, India
Telfax: +91-11- 41511055
Email: delhi@scientificpub.com
Web www.scientificpub.com

ISBN: 978-93-8665-23-86

Citation:

Nath, Prem, (Compiled by) 2018, Sustainable Horticulture Development and Nutrition Security (Volume- III: Food and Nutrition Security), Dr. P.N. Agri. Sci. Foundation (PNASF), Bengaluru and Scientific Publishers India (SP), Jodhpur, Rajasthan. P. 659.

Dedicated to

All stakeholders engaged in horticulture and in attaining nutrition security in the developing world.

—PNASF

PREFACE

In the wake of national and global call on promoting nutrition security, horticultural crops have emerged as promising providers of nutrient-rich food resources. During the last 50 years, the production and consumption of fruits and vegetables have increased significantly, followed by the increase in human population across the globe.

But the fact remains that about 805 million people are suffering from hunger and malnutrition and over three billion people are suffering from micro-nutrient deficiencies, which calls for immediate remedial measures towards healthy life of the people.

In order to promote production and consumption of fruits and vegetables, there is a need for taking concrete steps toward sustainable horticultural development, which will sustain food and nutrition security of the people.

The PNASF has the mission to promote agricultural education, research and development with major focus on food and nutrition security. Between 2009 and 2016, the PNASF had prepared and published 5 volumes of the Basics of Human Civilization- Food, Agriculture and Humanity. During 2017, the PNASF launched another book series on Food and Nutrition Security. The Volume I namely, **Attaining Food and Nutrition Security in the Developing World** was published recently. The Volume-II of this series, namely **Cycle of Nutrients, Nutrition and Nutrition Security in Earth, Plant, Livestock and Human** is in the press. The present **Volume-III, Sustainable Horticultural Development and Nutrition Security** is being published now.

The present volume has 9 chapters, 32 articles by 21 authors/ co-authors of national and international repute. The PNASF acknowledges their valuable contribution with gratitude and thankfulness.

The valuable messages provided by known personalities, Shri. T.B. Jayachandra, Minister for Law, Parlimentary Affairs and Minor Irrigation, Government of Karnataka; Shri M.V. Rajasekharan, former Union Minister of State for Planning, Government of India; and Dr. D.L. Maheswar, Vice-Chancellor, University of Horticultural Sciences, Bagalkot, Karnataka; are very much welcomed and gratefully acknowledged by the Board of Trustees of the PNASF.

We are highly grateful to the Technical Team and Editorial Board consisting of Dr. C.P.A. Iyer, Dr. O.P. Dutta, Dr. K.R.M Swamy, Dr. B.S. Prabhakar, Mr. P.N. Krishnamoorthy and Mr. P.B. Gaddagimath who made unfailing efforts to review and edit the book. All of this was not possible without the support of the secretariat staff, Mrs. Vanita Vinay, who worked hard on the details of the publication.

On behalf of the Board of Trustees, I express my gratitude to PNASF Senior Scientific Councillors for their valuable suggestions. The contribution of the Technical Advisory Committee members during the brain storming sessions on the preparation of the book is appreciated very much.

On behalf of the PNASF, I extend my thanks to Scientific Publishers, Jodhpur, who readily agreed to make this joint publication in this beautiful form and presentation.

May 20, 2017

Prem Nath

Chairman, PNASF, Bengaluru,
Former Asst. Director General,
Food & Agriculture Organization
of the United Nations (FAO-UN)

T.B. JAYACHANDRA

Minister for
Law, Parliamentary Affairs
and Irrigation

Phone: Office: 22254661
22033439
Room No. 327, 3rd Floor
Vidhana Soudha
Bengaluru - 560 001
E-mail: lawministeroffice@gmail.com

DO No. L.P.A. & MI/3171/2017 Date: 21/04/2017

MESSAGE

Horticulture plays an important role in health and welfare of the people. The state of Karnataka, has been one of the leading states of the nation in producing horticultural crops and the state has been the first one to recognize its underlying significance and to establish the Department of Horticulture in the Ministry of Agriculture to support the growers, entrepreneurs and consumers alike.

In the present world, about 805 million people are suffering from hunger and malnutrition and over 3 billion people are suffering from micro-nutrient deficiencies leading to serious human diseases. Horticultural products are the major source of micro-nutrients and hence, play an important role of horticulture in human nutrition and nutrition security. FAO & WHO have already recommended 400g of fruit and vegetables *per capita* per day for improving human health.

The P.N. Agricultural Science Foundation (PNASF), Bengaluru has been doing pioneering work in preparing and publishing valuable books related to Food, Agriculture, Horticulture and Food and Nutrition Security and other important aspects related to Food and Agriculture/Horticulture. The present book entitled, **"Sustainable Horticulture Development and Nutrition Security"**, meets the demand of the information and technology on the subject required by the readers in the present context of Food and Nutrition Crisis. The book attempts to cover all aspects on this subject for the benefit of policy makers, academicians, researchers, teachers and farmers.

I take this opportunity to congratulate PNASF, and thank all authors for bringing out this useful publication.

(T.B. Jayachandra)

M.V. RAJASEKHARAN's Former Union Minister of State For Planning, Government of India		"SRI MALLIKARJUNA NILAYA" # 20 / 1, Kanakapura Road Basavanagudi Bengaluru - 560 004 India Phone: Res. - 080-2657 6763 Fax: Res. - 91-080-2657 8366 Fax: Off. - 91-080-2669 0860 e-mail: mvr1928@gmail.com asiancarebgl@vsnl.net

Date: 11.04.2017

MESSAGE

I am indeed overwhelmed to learn, that Dr. Prem Nath Agricultural Science Foundation, founded and chaired by one of the outstanding and very distinguished agriculture and horticulture scientist of the world repute is preparing a new book entitled, "Sustainable Horticulture Development and Nutrition Security", in response to national and global call by all stakeholders for integrating a very serious, ever increasing problem of hunger and nutrition crisis.

I would like to heartily congratulate and pay my complements to Dr. Prem Nath ji. I know the recently published books, such as, "The Basics of Human Civilization, Food, Agriculture and Humanity Volume I: Present Scenario" 2013. Volume II: Food (2014). Volume III: Agriculture (2014). Humanity in Humans by Humans and for Humans in Securing Food for All, Volume IV: Humanity (2016) and Food Expectations of the People in New Millennium, Volume V: Expectations (2016) and they heartily acknowledged by all.

Dr. Prem Nath at his advanced age keeps everybody inspired by his dedicated and committed work and by his unbelievable contributions in the field of science and technology, especially agricultural and horticultural sciences. I pay my tributes to him.

(M.V. Rajasekharan)

ತೋಟಗಾರಿಕೆ ವಿಜ್ಞಾನಗಳ ವಿಶ್ವವಿದ್ಯಾಲಯ, ಬಾಗಲಕೋಟ
University of Horticultural Sciences, Bagalkot

Dr. D.L. MAHESWAR Vice-Chancellor University of Horticultural Sciences (UHS)	Udyanagiri, Navanagar, Bagalkot-587 104, Karnataka Tel: 91-8354 230351/230364 E-Mail: vc@uhsbagalkot.edu.in

Date: 12/04/2017

MESSAGE

I am delighted to write a message for the book entitled "Sustainable Horticulture Development and Nutrition Security", is a pioneering effort to collate and provide information in one compilation. It is heartening to note that, Horticulture legendry Dr. Prem Nath, Chairman, PNASF, for publishing the book in the wake of national and global call for mitigating hunger and nutritional crisis, for the benefit of stakeholders and society.

Malnutrition is the largest threat to Global Health; Over 900 million people in the world are undernourished. Despite India's 50% increase in GDP since 1991, more than one third of the world's malnourished children live in India. Among these, half of them under 3 are underweight and a third of wealthiest children are over-nourished. The World Bank estimates, that India is one of the highest ranking countries in the world for the number of children suffering from malnutrition.

The relationship between human being and food is permanent, and the resources of food may survive in nature without human intervention, but human being cannot remain alive without food. Thus, addressing issues of food and nutrition security attains its priority among the major issues of food, clothing and shelter for human kind. Moreover, dealing with health issues becomes equally important in the wake of prevailing diseases and health disorders. Recent experimental studies by the WHO has concluded, that adequate consumption of vegetables and fruits prevents certain non-communicable diseases, such as, cancer, blood pressure, diabetes and above all growing obesity among young and old in many regions of the world. In this connection, horticulture sector plays an important role in food security, nutrition and health security and employment generation and its impact on economic growth.

Horticulture is the fastest growing sector, contributes to poverty alleviation, nutritional security and have ample scope for farmers to increase their income and helpful in sustaining large number of agro-based industries, which generate huge employment opportunities.

Rapidly growing demand for horticultural products, especially burgeoning market for processed fruits and vegetables, as well as booming floriculture market is an evidence of the phenomenon which is expected to accelerate horticultural growth in the country. Consequently, horticulture is set to assume a greater role and importance within the agriculture sector and eventually in the national economy. As a result of a number of thoughtful research, technological and policy initiatives and inputs, horticulture in India, today, has become a sustainable and viable venture for the small and marginal farmers. Besides, this sector has also started attracting entrepreneurs for taking up horticulture as a commercial venture. Therefore, there is a great scope for accelerating agricultural development through expansion of horticultural crops.

I compliment the editors for compiling the selected articles from national and international lead sources to bring out a most essential and needful book with a coverage of quite relevant and appropriate subjects, which will contribute in pursuit of ones quest to find solutions for sustainable horticulture development. My best wishes for praiseworthy publication.

Place: UHS, Bagalkot **(D. L. Maheswar)**

THE ARCHITECT OF THE BOOK

Dr. Prem Nath obtained his Ph.D. Degree in Horticulture from the Kansas State University, Manhattan, USA during 1962. Since then, Dr. Nath has been serving in different national and international institutions, in various capacities and in different countries, in food and agriculture sector for more than 50 years.

As Professor and Senior Scientist, he served at Indian Agricultural Research Institute, New Delhi; University of Udaipur, Rajasthan; Indian Institute of Horticultural Research, Hessarghatta, Karnataka and at Nigerian Institute of Horticulture Research, Ibadan, Nigeria during the period 1962-79.

As FAO Research Team Leader, he served at Nigerian Institute for Horticultural Research, Nigeria; National Centre for Horticultural Research and Development Centre, Saudi Arabia; and National Agricultural Research Authority, Yemen Arab Republic during the years 1979-90.

During 1990-94, Dr. Nath served as FAO Regional Team Leader for Asia and Pacific, Middle East, North Africa and Europe Services, Agricultural Field Operations Division, FAO, Rome.

Dr. Prem Nath served as the Special Adviser for Food Security to Director General, FAO, Rome (1994-96). He had a diplomatic position as FAO Resident Representative, Myanmar during 1996-98.

As International Senior Executive, Dr. Prem Nath retired as the FAO Assistant Director General and Regional Representative at the FAO Regional Office for Asia and the Pacific, Bangkok, Thailand by the end of 1999.

Since 2000, Dr. Nath continues to serve as the Chairman of the Dr. P.N. Agricultural Science Foundation (PNASF), Bengaluru. During November, 2002, Dr. Nath was nominated as Convener of the Vegetable Science International Network (VEGINET) by the General Assembly of the International Conference on Vegetables (ICV-2002), and during November, 2009 he was nominated as Chairperson of VEGINET by the International Conference on Horticulture (ICH-2009). During 2012 and 2014, he was nominated as the Chairperson of the SEAVEG Coordination Committee in Thailand.

With the mission to promote agricultural education, research and development with focus on food and nutrition security, the PNASF was launched during the beginning of the new millennium (2000), and has made significant achievements under the Chairmanship of Dr. Prem Nath. He organized successfully two international conferences (International Conference on Vegetables ICV-2002 and International Conference on Horticulture ICH-2009) and also was instrumental in organizing FAO International Workshop on Urban and Peri-urban Agriculture for Asian Countries (UPA-2005); SEAVEG/DOA/AVRDC/AARNET/VEGINET/ HSSI Regional Symposium on High Value Vegetables in Southeast Asia (SEAVEG-2012) in Chiang-Mai, Thailand; FAO/PNASF Regional Workshop on Strengthening Urban and Peri-urban Agriculture towards Resilient Food system in Asia (UPAFSA-2013) at Bangkok, Thailand; and ISEC/PNASF Seminar on Development Trends in Urban and Peri-urban Agriculture, August 2013, Bengaluru. He also assisted other national and international institutions to organize symposia on Horticulture/ Agriculture and actively participated in then.

During his research career, Dr. Nath developed a number of vegetable varieties which were widely adapted by growers and as professor, he guided a number of M.Sc. and Ph.D. students.

During his career, Dr. Nath led and assisted in establishing a number of research institutions in horticulture and agriculture in India, Nigeria, Saudi Arabia and Yemen Arab Republic. He developed and guided a number of national, regional and inter-regional field projects on food and agriculture to its success in different countries.

Dr. Nath is the author/co-author/editor of number of books, hundreds of research and technical papers and thousand of research and technical reports. The books are *Vegetables for the Tropical Region* (ICAR), *Germplasm Resources and Genetic Improvement of Vegetable Crops in the Tropics* (NIHORT, Nigeria), *Food Security and Vegetables-A Global Perspective* (PNASF), *Vegetable Crops-Improvement and Production* (PNASF), *Foods and Nutrition* (PNASF and Studium Press) and *Horticulture and Livelihood Security* (PNASF and Scientific Publishers) and *The Basics of Human Civilization-Food, Agriculture and Humanity; Volume: I- Present Scenario* (PNASF, Bengaluru and New India Publishing Agency, New Delhi) and *The Basics of Human Civilization-Food, Agriculture and Humanity; Volume: II-Food* (PNASF and New India Publishing Agency) *The Basics of Human Civilization-Food, Agriculture and Humanity; Volume: III-Agriculture* (PNASF and Scientific Publishers, Jodhpur); *Humanity in Humans, by Humans and for Humans in Securing Food* For All (The Basics of Human Civilization-Food, Agriculture and Humanity; Volume: IV-Humanity) (PNASF); and *Food Expectations of the People in the New Millennium* (The Basics of Human Civilization-Food, Agriculture and Humanity; Volume: V- Expectations) (PNASF and Westville Publishing House, New Delhi).

The Basics of Human Civilization-Food, Agriculture and Humanity, the brain-child of Dr. Prem Nath was laid out by him in 5 volumes, and he is the author of number of papers in these volumes. Dr. Prem Nath is the author of Volume-IV: Humanity with the title, “Humanity in Humans, by Humans and for Humans in Securing Food for All”.

The latest publication of Dr. Prem Nath is the **Volume-I,** of the **Food and Nutrition** series, namely, **“Attaining Food and Nutrition Security in the Developing World” (2017).**

THE TECHNICAL AND EDITORIAL TEAM

Dr. C.P.A. Iyer obtained M. Sc and Ph.D. degrees in Horticulture from the Post Graduate School, I.A.R.I., New Delhi and did his post doctoral research at the Institute of Horticultural Plant Breeding, Wageningen, the Netherlands, in 1965-66.

Dr. Iyer served as Geneticist (Fruits) at the Indian Institute of Horticultural Research, Bengaluru, in 1968 and later became the Head of the Division of Fruit Crops at the same Institute. He was the All India Project Coordinator for fruits and coordinated research project on fruits on an all India basis. Subsequently appointed as Director, Central Institute of Horticulture, Lucknow, India.

Dr. Iyer was the International Horticulturist of FAO in a Mango Project of UNDP/ FAO in Bangladesh. Subsequently, he was a FAO Consultant for many projects in Nigeria, Myanmar and Bangladesh. Dr. Iyer has traveled extensively worldwide.

Evolved many hybrid varieties in mango, guava and papaya. Dr. Iyer was invited by the CSIRO, Australia to take up the prestigious McMaster Fellowship to organize the Australian National Mango Breeding Project, which he successfully carried out in 1994. Dr. Iyer has published more than 200 papers in scientific journals. Has guided many students for their M. Sc and Ph.D. degrees.

Dr. O.P. Dutta obtained his Ph.D degree in Horticulture from the University of Agricultural Sciences, Bengaluru during 1981.

He has thirty three years of research experience in breeding tropical vegetables for high yield, nutritional qualities and resistance to multiple stresses, twenty five years of experience in hybrid seed production, breeder seed production and production of tropical vegetables and sixteen years of experience in Research Management as Head, Division of Vegetable Crops at IIHR, Bengaluru; planned organized and monitored the activities of 8 Research Laboratories having 18 projects and 17 scientists in the Division of Vegetable Crops at IIHR, Bengaluru.

At Namdhari Seeds Pvt. Ltd., Bengaluru he has developed commercially viable 54 varieties in tropical vegetables, such as, tomato (4), hot pepper (3), watermelon (20), melon (10), gourds (6) and beans (11), gaining 70% market share at the national level.

He has established a sound vegetable seed production system in Bangladesh under the FAO Vegetable Seed Project (GCP/BGD/025 and 028/DEN) jointly funded by DANIDA and Belgium Government 1996-1999. He has provided technical guidance pertaining to vegetable production, vegetable seed production, vegetable breeding to SAARC countries, under SAVERNET programme.

Dr. K.R.M. Swamy is a former Principal Scientist and Head, Division of Vegetable Crops, IIHR, Bengaluru. He obtained his Ph.D. in Horticulture from University of Agricultural Sciences, Bengaluru, in 1984. He has worked in ICAR Research Institutes for 33 years in various capacities. His field of specialisation is Vegetable Breeding, Collection and Conservation of Cashew Genetic Resources, Cashew improvement, Vegetative Propagation of Cashew and Nursery Management. He has handled several research projects and has contributed in the development and release of four high yielding vegetable varieties and three high yielding cashew varieties. He has established the National Cashew Gene Bank (NCGB) at NRC for Cashew, (presently Directorate of Cashew Research - DCR), Puttur, for 451 clonal germplasm accessions of cashew, standardised the "Softwood Grafting Technique for Cashew" and recommended the same for commercial multiplication of cashew varieties; established "Cashew Scion Banks" for 25 released varieties of cashew; established "Cashew Nurseries" both at Shantigodu and Puttur campuses of NRCC. Dr. Swamy has over 200 publications to his credit, which include 39 research papers, 24 seminar/symposia/ conference papers, 28 book chapters, 77 popular articles, 22 technical bulletins, 5 extension folders, 5 teaching manuals, 1 technical report, authored/ co-authored 5 books and edited/co-edited 8 books/souvenirs.

Dr. B.S. Prabhakar obtained Ph.D. in Agronomy, from I.I.T. Kharagpur 1978. He served as Agronomist, Indian Institute of Horticultural Research (IIHR) ICAR, Bengaluru, 1978-2008; Scientist, Directorate of Oilseeds Research, Hyderabad, 1977-78; Research Assistant, AICRP Oilseeds, APAU, Hyderabad, 1970-71; Research Fellow, Rockefeller Foundation, AICRIP, Hyderabad, 1969-70; Principal Investigator, AP Cess Fund/ICAR Project on Technology Assessment and Refinement through Institute-Village linkage programme, 1995-1999; Principal Investigator, NATP/ICAR, Project on Technology Assessment and Refinement in Eastern Dry Zone of Karnataka, 1999-2000; Principal Investigator, NATP/ICAR Project on Protected cultivation of vegetables and flowers in plains and hills, 2000-04.

Dr. Prabhakar took a Study Tour on "Advanced training in protected cultivation of vegetables and flowers, USA and Canada", 1991; As an accomplished agronomist, he contributed significantly in realizing the potential yield of vegetable varieties and hybrids grown in the open field, as well as under protected cultivation using hydroponics organic, inorganic production technologies to a sole crop, and in horticultural based cropping systems.

In his academic career, he guided several M. Sc. & Ph. D. students in Horticulture from University of Agricultural Sciences, Bengaluru and is the author of more than 100 research publications, scientific abstracts, popular articles, technical bulletins, chapters in books and videos.

Mr. P.N. Krishna Moorthy obtained B. Sc (Agri) and M. Sc (Agri. Entomology), in 1972 and 1976, from University of Agricultural Sciences (UAS), Bengaluru.

Served as Research Assistant in All India coordinated research project on sunflower at UAS Bengaluru (1976-1977). Served Indian Institute of Horticultural Research, Bengaluru from 1977 till retirement in 2014. There he served as Scientist S-1 (1977-1984), Scientist S-2 (1984-1986), Senior Scientist (1986-1998) and Principal Scientist (1998-2014).

Associate and PI of the USIF project on "Insect Pest Management in Vegetable crops utilizing thresholds and statistical models" (1992- 1997) funded by USDA. Worked in the "South Asian Vegetable Research Net Work" (SAVERNET) on the IPM of vegetables sponsored by ADB through AVRDC, Taiwan. CC-PI of National Agriculture Technology Project (NATP) "Validation and Promotion of IPM in selected crops in different agro-ecological regions- Vegetables" (2001-2004), Co-PI of the NATP project "Production of pesticide free vegetables" (2000-2004).

Mr. Pulakeshi Basayya Gaddagimath obtained his post graduate in Organic Chemistry from Karnataka University, Dharwad in 1978 and served as a Lecturer in Kittle Science College, Dharwad till 1982. His interest in Mass Communication made him to leave the teaching job and join post graduate course in Mass Communication and Journalism in 1982 as a regular student and graduated from Karnataka University Dharwad in 1984.

During his post graduation in Mass communication and Journalism, he served many national news papers until he joined as Jr. Scientific Officer, at Directorate of Oilseed Research, Hyderabad in 1985. He joined the Indian Institute of Horticultural Research, Bengaluru in 1989, as Information and Publication Officer.

As a media person he has anchored /conducted more than 40 T.V. programmes and live programmes on Doordarshan on various aspects of agriculture. He was also a consultant for developing scripts for documentaries and interactive CDs and has scripted more than 10 documentaries on rural water supply and sanitation. Apart from a good number of official publications at IIHR, Bengaluru and DOR, Hyderabad, he has authored two books, edited four books (two international), published three books and written many technical and popular papers.

CONTENTS

ABBREVIATIONS

AARNET: The ASEAN-AVRDC Regional Network
ACC: 1-Aminocyclopropane-1- Carboxylic Acid
ACOTTA: Autonomous Council of Technology Transfer in Agriculture
ADB: The Asian Development Bank
AEs: Agri-Entrepreneurs
AFTA: The ASEAN Free Trade Area
AICRPs: All India Coordinated Crop Research & Improvement Programmes
AIDS: Acquired Immune Deficiency Syndrome
AIR: All India Radio
AIs: Adequate Intakes
AMUL: Anand District Milk Producers Union Ltd. (Anand Milk Union Limited)
ANBC: Annual Net Bank Credit
AOA: Agreement on Agriculture
APEDA: Agricultural and Processed Food Products Export Development Authority
APMC: Agricultural Produce Market Committee
ARIMA: Auto Regressive Integrated Moving Average
ARMF: Automate Agricultural, Rural & Micro-Finance
ASI: Agribusiness Systems International
ASLA: American Society of Landscape Architects
ATMA: Agriculture Technology Management Agency
ATP: Adenosine Triphosphate
AUP: Administrative Use Permit
AVRDC: Asian Vegetable Research and Development Centre (The World Vegetable Center)
BARC: Bombay Atomic Research Centre
BAS: Biologically Active Substances
BBMP: Bruhat Bengaluru Mahanagara Palike
BDA: Bengaluru Development Authority
BESO: Building Energy Saving Ordinance
BGGCOMS: The Bengaluru Grape Growers' Cooperative Marketing and Processing Society Ltd
BGREI: Bringing a Green Revolution to Eastern India
BHOPCOMS: The Bengaluru Horticultural Producer's Cooperative Marketing and Processing Society Ltd.
BMRCL: Bengaluru Metro Rail Corporation Ltd.
CA: Controlled Atmosphere

CAM: Crassulacean Acid Metabolism
CBOs: Community Based Organizations
CBSE: Central Board of Secondary Education
CCTVs: Closed-Circuit Television
CDCP: Certified Data Centre Professional Training Programme
CDI: Centre for Development Innovation
CITYNET: Regional Network of Local Authorities for the Management of Human Settlements
CMC: Christian Medical College
CN: Cetane Number
COAG: Committee of Agriculture meeting
CODEX: Careful Obliging Dependable Easygoing Xenial
CPB: Colorado Potato Beetle
CPCRI: Central Plantation Crops Research Institute
CPI: Consumer Price Index
CPRI: Central Potato Research Institute
CRS: Corporate Social Responsibility
CSA: Community gardens, Community Supported Agriculture
CSIRO: Commonwealth Scientific and Industrial Research Organisation
CSRI: Central Scientific and Industrial Research
CT: Cholera Toxin
CVDs: Cardiovascular Diseases
CWR: Crop Wild Relatives
DAC: Department of Agriculture and Cooperation
DARE: Department of Agriculture Research and Education
DBM: Diamondback Moth
DDG: Deputy director General
DDT: Dichlorodiphenyltrichloroethane
DI: Dissemination Index
DIHAR: Defense Institute of High Altitude Research
DLHS: District Level Health Surveys
DNA: Deoxyribonucleic Acid
DOA: The Department of Agriculture,Thailand
DRDO: Defence Research and Development Organization
DRIs: Dietary Reference Intakes
DSCL: DCM Shriram Consolidated Ltd.
DSS: Decision Support System
DTH: Direct-To-Home
EARs: Estimated Average Requirements

EBMUD: East Bay Municipal Utility District
ECB: European Corn Borer
ED: Electro-Dialysis
EEI: Extension Education Institutes
EFILWC: European Foundation for the Improvement of Living and Working Conditions
EIU: Economist Intelligence Unit
eNAM: e-National Agriculture Market
EPN: Entomo Pathogenic Nematodes
ESG: Environmental Support Group
EU: European Union
F&V: Fruits and Vegetables
FAO: Food and Agriculture Organization of the United Nations
FELDA: Federal Land Development Authority
FFV: Fresh Fruits and Vegetables
FHFI: Fight Hunger First Initiative
FIFA: Federation International de Football Association
FIGs: Farmer Interest Groups
FLDs: Frontline Demonstrations
FPO: Farmer Produce Organization
FRCs: Food Retail Chains
FVCC: Fruits and Vegetable Consolidation Center
FYM: Farm Yard Manure
GA: Gynaecological Age
GAIN: Global Alliance for Improved Nutrition
GAPs: Good Agricultural Practices
GCMMF: Dairy Cooperatives of Gujarat - Gujarat Cooperative Milk Marketing Federation Ltd.
GDP: Gross Domestic Product
GEAC: Genetic Engineering Approval Committee
GHG: Greenhouse Gas
GIDC: Gujarat Industrial Development Corporation
GIS: Geographic Information Systems
GM: Genetically Modified
GMOs: Genetically Modified Organisms
HACCP: Hazard Analysis and Critical Control Points
HDP: High Density Planting
HOPCOMS: Horticultural Produce Cooperative Marketing Society Ltd.
HPR: Host Plant Resistance
HRD: Human resource development

HSM: Hub and Spoke Model
HSST: Horticultural Science Society of Thailand
HVA: High- Value Agriculture
IARI: Indian Agricultural Research Institute
ICAR: Indian Council of Agricultural Research
ICDS: Integrated Child Development Services
ICH-2009: The International Conference on Horticulture
ICM: Institute of Commercial Management
ICMR: Indian Council of Medical Research
ICN-2: The International Conference on Nutrition
ICT: Information and Communication Technology
ICV-2002: International Conference on Vegetables-2002
IDE: International Development Enterprises
IFOAM: International Federation of Organic Agriculture Movement
IFP: Integrated Fruit Production Guide
IFPRI: International Food Policy Research Institute
IIHR: Indian Institute of Horticultural Research
IIPM: Integrated Production and Protection Management
IISc: Indian Institute of Science
IITs: Indian Institute of Technology
INM: Integrated Nematode Management
IOBC: International organization for Biological and Integrated Control of Noxious Animals and Plants
IoT: Internet of Things
IPM: Integrated Pest Management
ISOCARP: International Society of City and Regional Planners
ISRO: Indian Space Research Organization
IT: Information technology
ITC: Indian Tobacco Company Limited
ITD: Innovations in technology dissemination
ITK: Indigenous Traditional Knowledge
ITPB: International Tech Park Bangalore
IWM: Integrated Weed Management
KAPC: Karnataka Agriculture Price Commission
KCC: Kisan Call Centre
KU: Kasetsart University
KVKs: Krishi Vigyan Kendras
LED: Light-Emitting Diode
LEED: Leadership in Energy and Environmental Design

LGBT: Lesbian, Gay, Bisexual and Transgender
MAHAFPC: MAHA Farmers Producer Company Ltd.
MAP: MapShots AgStudio Mobile Application Platform
MAS: Marker Assisted Selection
MDG: Millennium Development Goal
MDGS: Millennium Development Goals
MFI: Micro Finance Institution
MIDH: Mission for Integrated Development of Horticulture
MIS: Minimum Support Price During Distress Sales Under Government Schemes
MLE: Market-Led Extension
MRTN: Maximum Return To Nitrogen
MSP: Minimum Support Price
MSSRF: M. S. Swaminathan Research Foundation
MT: Metric Ton
MYRADA: A Bengaluru-Based Non-Profit Organization
NABARD: National Bank for Agriculture and Rural Development
NARES: National Agricultural Research and Education system
NATP: National Agricultural Technology Project
NBFC: Non-Bank Finance Company
NCDEX: National Commodity and Derivatives Exchange
NCDs: Non-communicable diseases
NCERT: National Council for Educational Research and Training
NERWA: North East Residents Welfare Associations
NF: M/s Namdhari Fresh
NFSM: National Food Security Mission
NGOs: Non Government Organizations
NGT: National Green Tribunal
NHANES-III: The Third National Health and Nutrition Examination Survey
NHB: National Horticultural Board
NHM: National Horticulture Mission
NMAET: National Mission of Agricultural Extension & Technology
NOx: Mono-Nitrogen Oxides
NPOP: National Programme for Organic Production
NPV: Nuclear Polyhedrosis Virus
NSC: National Seeds Corporation
NRC: Nutritional Rehabilitation Centre
NSC & SSCs: National and State Seed Corporations
NSKE: Neem Seed Kernel Extract
NSP: Neem Seed Powder

NSSO: National Sample Survey Organisation
NSW: New South Wales
OECD: Organisation for Economic Cooperation and Development
OFTs: On-Farm Testing
OHCPIB: Organic Horticultural Crops Production Information Bank
PDAs: Personal Digital Assistants
PHC: Pre-Harvest Contractor
PHT: Post Harvest Technology
PLRV: Potato Leaf Roll Virus
PM: Particulate Matter
PMFBY: Pradhan Mantri Fasal Bima Yojana
PMGSY: Prime Minister's National Rural Roads Program
PNASF: Dr. Prem Nath Agricultural Science Foundation
PPb: Parts Per billion
PPV&FR: Protection of Plant Varieties and Farmer's Rights
PRA: Pest Risk Appraisal
PSL: Priority Sector Lending
PSL: Priority Sector Lending
PTC: Perimeter Trap Cropping
PTM: Potato Tuber Moth
PVP: Plant Variety Protection
QTLs: Quantitative Trait Locis
R&D: Research and Development
RBI: Reserve Bank of India
RDAs: Recommended Dietary Allowances
RDI: Recommended Dietary Intake
RE: Retinol Equivalent
RETA: Regional Technical Assistance
RIDF: Rural Infrastructure Development Fund
RKKs: Rallis Kissan Kendras
RKVY: Rashtriya Krishi Vikas Yojana
RNA: Ribonucleic Acid
RO: Reverse Osmosis
RTF: Alternatively Ready To Fruit
Rubisco SSU: 5-Bisphosphate Carboxylase/Oxygenase Small Subunit
SAFAL: The Largest Organised Retail Network of Fruits and Vegetables in the National Capital Region of India
SAMETIs: State Agricultural Management Extension Training Institutes
SAM: Severe Acute Malnutrition

SAPPL: Shree Ajit Pulp and Paper Ltd.
SAR: Systemic Acquired Resistance
SAUs: State Agricultural Universities
SCERT: State Council for Educational Research and Training
SDGs: Sustainable Development Goals
SEAVEG: Southern East Asia Vegetable Symposium
SEP: School Feeding Programme
SFCI: State Farms Corporation of India
SFI: Syngenta Foundation India
SHG-BL: Self-Help Group Bank Linkage
SHGs: Self Help Groups
SJM: Swadeshi Jagaran Manch
SLA: State level Association
SMS: Spent Mushroom Substrate
SMSP: Sub-Mission on Seeds & Planting Material
SO_2: Sulphur Dioxide
SOC: Soil Organic Carbon
SPFS: Special Programme for Food Security
SSCs: State Seeds Corporation
TCDC: Technical Cooperation among Developing Countries
TDM: Time-Division Multiplexing
TDZ: Thidiazuron
TERI: The Energy Resource Institute
TLCV: Tomato Leaf Curl Virus
TOT: Transfer of Technology
UAV: Unmanned Aerial Vehicle
UILs: Upper Intake Levels
UN DESA: Department of Economic and Social Affairs of the United Nations
UNDP: United Nations Development Programme
UN-DSEA: UN Department of Economics and Social Affairs
UNESCO: United Nations Educational, Scientific and Cultural Organization
UNIBPGR: United Nations International Board for Plant Genetic Resources
UNICEF: United Nations Children's Fund is a United Nations programme
UNO: United Nations Organization
UP: Use Permit
UPA: Urban and Peri-urban Agriculture
UPAFSA-2013: Regional Workshop on Strengthening Urban and Peri-urban Agriculture towards Resilient Food System in Asia
UPH: Urban and Peri-urban Horticulture

URBACT: European Exchange and Learning Programme Promoting Sustainable Urban Development

USAID: United States Agency for International Development

USDA: United States Department of Agriculture

USGBC: U.S. Green Building Council

VCM: Value Chain Model

VEGINET: Vegetable Science International Network

VHT: Vapor Heat Treatment

WFP: World Food Programme

WFS-1996: World Food Summit-1996

WHO: World Health Organization

WST: Wireless Sensor Technologies

WTO: World Trade Organization

CONTRIBUTOR'S LIST

B.S. Prabhakar
Former Principal Scientist,
Indian Institute of Horticultural Research (IIHR),
Hessaraghatta lake post,
Bengaluru
E-mail: bspkar@gmail.com

Bindu Ananth
Chairperson,
IFMR Trust & IFMR Holdings,
10th Floor-Phase 1, IIT-Madras Research Park
Kanagam Village, Taramani
Chennai-600113
E-mail: bindu.ananth@ifmr.co.in

Brahm Singh
Chief Technical Advisor,
MTPL, New Delhi
Former Emeritus Scientist,
DRDO-DIPAS, New Delhi
E-mail: brahma88@gmail.com

C. Aswath
Principal Scientist,
Indian Institute of Horticultural Research (IIHR),
Hessaraghatta lake post,
Bengaluru
E-mail: aswathiihr@gmail.com

C. P. A. Iyer
Former International Horticulturist (FAO) and Former Director,
Central Institute of Sub-tropical Horticulture, Lucknow
E-mail: babu_sivan@yahoo.com

C. Prasad
Ex. DDG (Ag. Ext.), ICAR & President, VARDAN,
EB-106, Maya Enclave,
New Delhi - 110064
E-mail: vardan_ngo@yahoo.com

C.M. Devika
Institute for Social and Economic Change (ISEC),
Nagarabhavi, Bengaluru 560072, India
E-mail: devika.cm@gmail.com

G. Senthil Kumaran
Principal Scientist,
Div. of PHT & Agricultural Engineering
Indian Institute of Horticultural Research (IIHR),
Hessaraghatta lake post, Bengaluru
E-mail: senthil@iihr.res.in

H. Ravishankar
Ex- Director, ICAR- Central Institute for Subtropical Horticulture, Lucknow,
Uttar Pradesh
& presently Principal Scientist (Horticulture)
Indian Institute of Horticultural Research (IIHR),
Hessaraghatta lake post, Bengaluru
E-mail: drhravishankar@gmail.com

K. R. M. Swamy
Former Principal Scientist and Head,
Division of Vegetable Crops,
Indian Institute of Horticultural Research (IIHR),
Hessaraghatta lake post, Bengaluru
E-mail:krmswamy@gmail.com

K. S. Krishna Prasad
Former Principal Scientist and Head,
Central Potato Research Station,
Ootacamund, India
E-mail: karpakal@gmail.com

K. Vijayasarathy
Institute for Social and Economic Change (ISEC),
Nagarabhavi, Bengaluru 560072, India
E-mail: vijsaraec@gmail.com

K.K. Poornima
Former Research Fellow,
Dr. P.N. Agricultural Science Foundation (PNASF), Bengaluru
E-mail: poornima88@gmail.com

M.J. Chandregowda
Principal Scientist(Agril.Extension)
Agricultural Technology Application Research Institute,
MRS, H.A.Farm Post, Hebbal, Bengaluru
E-mail: maravalalu@yahoo.com

Meera Pandey
Principal Scientist, Mushroom Research Laboratory
Indian Institute of Horticultural Research (IIHR),
Hessaraghatta lake post, Bengaluru
E-mail: meera@iihr.res.in

Megha Katte
Assistant Professor,
University of Agricultural Sciences (UAS), GKVK, Bengaluru
E-mail: meghakatte91@gmail.com

N.C. Mamatha
Institute for Social and Economic Change (ISEC),
Nagarabhavi, Bengaluru 560072, India
E-mail: mamatha.nc5@gmail.com

P. Parvatha Reddy
Former Director,
Indian Institute of Horticultural Research (IIHR),
Hessaraghatta lake post, Bengaluru
E-mail: reddy_parvatha@yahoo.com

P. Usha Palaniswamy
Dean of Global Pathways Institute, Mumbai
Arcadia University
450 S. Easton Road, Glenside, PA 19038, USA
E-mail: usha.palaniswamy@gmail.com

P.B. Gaddagimath
Formerly Chief Technical Officer (Information and Publication Officer),
Indian Institute of Horticultural Research (IIHR),
Hessaraghatta lake post, Bengaluru
E-mail:puli.iihr@gmail.com

P.G. Chengappa
Former Vice Chancellor, UAS, Bengaluru
National Professor of ICAR
Institute for Social and Economic Change (ISEC),
Nagarbhavi Post, Bengaluru 560 072
E-mail:chengappapg@gmail.com

P.N. Krishna Moorthy
Former Principal Scientist (Entomology)
Indian Institute of Horticultural Research (IIHR),
Hessaraghatta lake post, Bengaluru
E-mail: pnkpadar@yahoo.com

Pradeep Mehta
Scientist - Impact Assessment,
Institute of Rural Research and Development,
Plot No.34, Sector 44, Institutional Area, Gurgaon-122002 (Haryana), India,
Email: p mehta@irrad.org

Prem Nath
Chairman, P. N. Agricultural Science Foundation (PNASF)
Former Asst. Director General,
Food & Agriculture Organization of the United Nations (FAO-UN)
Bengaluru-560 094, INDIA
Chairperson, Vegetable Science International Network (VEGINET)
E-mail:pnasfoundation@gmail.com

R. S. Deshpande
Visiting Professor ISEC
Former Director, Institute for Social and Economic Change (ISEC) and ICSSR National Fellow
Group Leader, S M Sehgal Fondation, Gurugram and Visiting Professor, Nagarbhavi, Bengaluru
E-mail: rs.kalbandi@gmail.com

S. Bisaliah
Former Vice Chancellor,
University of Agricultural Sciences (UAS), Bengaluru
E-mail:sbisaliah@yahoo.com

Sanjay Kumar Singh
ICAR-National Research Centre on Litchi, Muzaffarpur- 842 002, Bihar
E-mail:sanjayhor@rediffmail.com

Savithri Byadigi
PNASF Fellow,
Department of Food Science and Nutrition, UAS, Bengaluru
E-mail: shreya.byadagi8@gmail.com

Shilpa Yatnatti
PNASF Fellow,
Department of Food Science and Nutrition, UAS, Bengaluru
E-mail: shilpayatnatti@gmail.com

Sudha Mysore
Chairperson, ZTMC, PI, ABI
Principal Scientist, Agril Econ,
Fulbright-Nehru Senior Research Fellow (PDF, MSU),
Indian Institute of Horticultural Research (IIHR),
Hessaraghatta lake post, Bengaluru
E-mail: sudhamys@yahoo.com

V. A. Parthasarathy
Formerly National Project Coordinator (Tropical Fruits),
Bioversity International, Indian institute of Horticultural Research (IIHR),
Hessaraghatta lake post, Bengaluru
E-mail: vapartha@gmail.com

I. Introduction

1

Introduction

Prem Nath

We are all indebted to nature and its resources for providing us food resources and food for the subsistence and survival of all of us. Without food we can hardly sustain ourselves.

In the food domain, cereal and legume grains occupy the front line, whereas, horticultural crops have assumed the second line of defense. For healthy diet cereals and legumes provide us with carbohydrates and protein. Whereas, fruits and vegetables provide us minerals and vitamins. Both macro- and micro-nutrients are essential for human growth and development. The fruits and vegetables are the major source of micro-nutrients.

Let me look into the daily activities of human life. In the morning, I found my grandfather drinking water with fresh lime before he went for a walk. At breakfast, our family will have fruit juice, bread with fruit jam or butter and eggs, alternatively *idli* or *dosa*, with sambar of vegetables, milk or juice and sometimes, poori, cooked vegetables, fruit juice or milk. Likewise, at lunch time the menu included rice or chapathi, dal or sambar, cooked vegetables, dry or wet or both, plus some fruits. At dinner, the family will have mostly chapathi, cooked dry and/or wet vegetable or non-vegetarian food and deserts mostly fruits. A hurried analysis will confirm that there is no menu without horticultural food. The method of preparation and volume of consumption vary from region to region and from family to family depending on the socio-economic status.

It is recognized that modern agriculture must diversify production and achieve sustainable higher output. At the same time safeguard the environment, conserve natural resources and adapt to climate changes and ever-rising population. In order to release the pressure on cereals as well as to improve upon the human nutrition through consumption of the other nutritious crops, diversification in cropping pattern provides better option. The increased production and consumption of horticultural crops, particularly vegetables with its wide adoption and provider of important nutrients offer promise for the future. Vegetables are rich source of nutrients (especially vitamins and minerals), besides its medicinal values. As such, vegetables as food and diet supplement are gaining momentum in most countries in the recent years.

Fruit and vegetables are important components of a healthy diet and, if consumed daily in sufficient amounts, could help prevent major diseases, such as, CVDs and certain cancers. According to *The World Health Report 2002,* low fruit and vegetable intake estimated to cause about 31% of ischemic heart disease and 11% of stroke worldwide. Overall it is estimated that up to 2.7 million lives could potentially be saved each year, if fruit and vegetable consumption was sufficiently increased. Recommendations in this direction tend to complement and reinforce other valid messages based on the long-known health benefits of consuming vegetables and fruit as dietary sources of fibre, vegetable proteins and protective micronutrients.

A daily intake of at least 400 g of fruits and vegetables, within the context of ensuring a better general dietary pattern as a population goal, was reconfirmed as an appropriate basic message by the FAO and WHO (2005).

The International Conference on Horticulture (ICH-2009) with the theme "*Horticulture for Livelihood Security and Economic Growth*" held in Bengaluru during November 9-12, 2009, had highlighted its importance and salient features projecting the future steps to be undertaken by all stakeholders. The conference recognized horticulture as the second line of defense with cereals and legumes serving on the front line of the food domain. It was stated that horticulture (a) supports significantly to food security, (b) is the major contributor to nutrition security (c) promotes good health and (d) is source of employment generation- all leading to livelihood security and economic growth.

Today, we have nutrient crisis like never experienced before. Over three billion people are suffering from micro-nutrient deficiencies, causing a number of human diseases across the globe.

In this regard, it is essential that micro-nutrients are both available and accessible to all. This can be done by growing and producing enough fruits and vegetables to fulfill the demand of all people. To make fruits and vegetables available all the time, it is essential to have a sound system of production management and distribution which is amenable to stakeholders and at the same time ensuring better output. This demands for a sustainable horticultural development. In other words, sustainable horticulture industry is a prelude to attaining nutrition security.

We recognize that fruits and vegetables are reservoirs of macro- and micro-nutrients. The pressing demand of today's world is to attain *nutrition security*, which relates to production and utilization of nutrients in the form of vegetables and fruits. With regard to specific nutrient requirements, grain and legumes cannot replace or exchange with vegetables and fruits and *vice-versa*. The balanced nutrition requires the consumption of grains for carbohydrates, legumes for proteins and vegetables and fruits for macro-and micro-nutrients. Thus the nutrition security cannot be attained without vegetables and fruits.

Further, the migration of people from rural to urban areas has demonstrated that the increase in urban population is mostly made of consumers and not producers. This situation has given rise to urban and pre-urban agriculture, where vegetables and fruits are major contributors in attaining food and nutrition security. Until now, horticultural/agricultural researchers considered only the high yield of food crops/vegetables and fruits and food quality depending mostly on colour, size and shape, but the time has come where we have to focus on nutritional quality.

There is a great variation in crop yield per unit area within a nation and crop yield in the industrialized world has much higher crop yield per unit area. Moreover, the food development plan is based on food yield with focus mostly on grain crops. If we need to attain food and nutrition security, the food planning estimates are required, which should be estimated in terms of 'nutrients' in order to provide balanced nutrition to citizens of a nation/world. Further, it is obvious that ongoing grain biased nutrition, needs to be complimented and supplemented with vegetables and fruits.

The recent trends indicated that the production of horticultural crops has increased many times with the lapse of time and also horticultural production has exceeded the cereal production in some developing countries. But the question is whether all these food products are available throughout the year, which is not; again whether the products available are accessible to all, which is not. These questions need to be answered quickly. We expect appropriate response from the expected sustainable horticultural development.

We understand that horticultural products are perishable unlike cereals and legumes. Moreover upto 50% food losses and wastage have been reported to occur in horticultural produce. This leads to significant loss of micro-nutrients already produced with significant investment of time, money and resources. We expect that sustainable horticulture development should solve this problem.

In order to develop sustainable horticulture, basic pillars of education, research, extension and development need to be reviewed and strengthened to make the horticulture industry sustainable.

In relation to each of the above pillars, specific issues need to be taken up, studied and solution to be tailored suiting the requirements.

During the international meets on agriculture/ horticulture held (ICV-2002, ICH-2009, WFS-1996, MDGs, ICN2-2014, SDG-2015) it was realized that horticulture is the major contributor to nutrition security and provides support to food security. It is also confirmed that fruits and vegetables produce much higher yield of food and nutrients than cereals per unit area per unit time.

In the developing world, common man on street or non-elite families in rural or urban areas, do not know much about the word nutrition or nutrients. What they know about is food items locally available and they choose and buy what they prefer or could afford. There is a serious lapse of knowledge on nutrition resulting in millions suffering from malnutrition.

The common consumer needs to know what to eat? How much to eat? And how many times to eat? Because good health of an individual depends on the proper intake of balanced nutrition consisting of essential macro- and micro-nutrients. The knowledgeable and educated people have some education on nutrition and nutrition security but millions of less knowledgeable people need to be educated on the utility of proper food nutrition.

There is a dire need of making a common man aware of the nutrition and nutrients. In the present digital world, this should not be difficult because each individual, whether literate or illiterate, rich or poor has mobile/cell available. Only the government has to activate appropriate program to make common man aware of the usefulness and essentiality of the nutrition program.

Even farmers lack knowledge on food nutrition. In developing world, each country has millions of agriculture extension workers. In India, making use of the existing setup of agriculture extension system, millions of extension workers spread nationwide could be trained on nutrition and designated to extend both horticulture technology and nutrition security. Alternately, a separate cadre of *Nutrition Extension Workers* nation-wide could be employed.

A review of literature indicates that less work has been undertaken by the researchers to improve the nutrient quality of horticultural plants. Sometimes breeding for higher yield at the cost of nutrient quality, has taken place. Nevertheless, breeding researches must consider nutrient quality as one of the major objectives for improving crop varieties.

It is incumbent on researchers to employ traditional breeding practices, as well as enhanced knowledge and practice on *biofortified crops* and/or *edited genes*, which hold promise in the present context.

With the focus on edible plant parts, it is possible to maximize the nutrients harvest. For example, *Moringa* or drumstick is used mostly as fruits, but other regions use tender leaves and/or flowers also which are nutrient-rich. With regard fruits, matured ones are consumed in Southern States, whereas, in Northern-eastern states tender fruits only are consumed. This gives rise to difference in nutrient level harvest. Further, in north-eastern India there are varieties which fruit 12 months of the year. On other side, Coimbatore developed dwarf kitchen variety with shorter life-cycle. In other words, we can learn a lot from one crop to maximize nutrient harvest and consumption. Likewise, the practice on utilizing tender leaves, flowers, fruit and seeds of pumpkin should continue.

In India, we need to gear up and accelerate researches on indigenous fruits like guava, fig, jamun, litchi, star fruit (carambola), local plum, passion fruit and others, which are nutrient-rich; common fruit like mango, to bear every year, to provide fruit throughout the year and to have dwarf trees; nutrient-rich and high yielding peaches, plums, pears could be made available like apple to everyone; introduced fruit crops like kiwi, dragon fruit, mangostein, persimmon and others so popular among consumers could be produced locally.

Both at national and international levels, food and agriculture/horticulture development plans and estimates are basically developed, framed and implemented and narrowed down to cereal production. In the present context of attaining nutrition security, this mode of thinking on 'food' needs to be changed to 'nutrients', which will include necessarily all those crops including fruit and vegetables which provide all macro- and micro-nutrients to ensure balanced nutrition needed for good human health. This idea has been supported strongly during the Southeast Asia Vegetable Symposium (SEAVEG) held during September, 2016 in Kuala Lampur, Malaysia, by Nath (2016), who made the following recommendations;

- **National Planning and Budget:** The grain-based planning, budgeting and production and consumption data need to be replaced. We cannot continue giving step-motherly treatment to nutritious crops like vegetables as against grain crops. We need to have detailed data on vegetable production and consumption to develop nutrition-based food requirements.
- **Breeding for nutrient values:** While breeding, enhancing nutritive values should get priority besides high yield. Along with nutritive values, the nutraceutical value where applicable should be considered.
- **Bridging yield gap:** There is wide range of variation in yield per unit area in each vegetable crop between and among the countries of the Asian region, which needs to be bridged.
- **More Nutrient Harvest:** Along with the economic advantage, the farmers need to be educated to include nutritious crops like vegetables to obtain more nutrient harvest.
- **Bio-diversification of crops:** In the cropping pattern, the vegetables should find place not only as cash crop like cereals, but also as important supplement to nutritious crops.

- **Nutrient-rich crops:** Developing bio-fortification technology should be accelerated and farming using biofortified crops to be encouraged.

At this point, recommendations of the United Nations Sustainable Development Goals (SDG), should be considered.

Goal 2. End hunger, achieve food security and improved nutrition and promote sustainable agriculture (UN, 2015)

Recommendation 8: Review national policies and investments and integrate nutrition objectives into food and agriculture policy, programme design and implementation, to enhance nutrition sensitive agriculture, ensure food security and enable healthy diets.

Without the essential inputs of humans, there is no flower, fruit and vegetables production and to achieve these all the inputs of humanity is essential to further ensure sustainable nutrition security. With humanity in mind, we can enjoy fresh air under tree shade, enjoy flowers of the garden, pluck vegetables from home garden, cook fresh vegetables, eat fresh fruits and enjoy juices of fruits and vegetables to be human and healthy. Horticulture serves humans and promotes humanity.

The present publication has attempted to reflect and discuss the above views and ideas on the subject of sustainable horticulture development and nutrition security in **9** chapters with **32** articles by **32** authors.

The first chapter introduces the book; the second chapter discusses the present challenges and opportunities in horticulture; the third chapter covers the advances made in horticultural research; the fourth chapter highlights the steps undertaken towards enhancing horticulture production; the fifth chapter details the issue on essential farm inputs; the sixth chapter attempts to link horticulture harvest to markets; the seventh chapter highlights the steps undertaken in securing food and nutrition security; the eighth chapter makes a serious study on sustainable horticulture development; and the ninth, the last chapter concludes the book by establishing relationship between horticulture and humanity in securing sustainable horticulture development and nutrition security.

This publication is intended to attract the attention of policy and program developers, university faculty members and students, researchers and general readers including farmers and consumers.

II. Present Horticulture—Challenges and Opportunities

2

Woes of Farmers and Consumers—Challenges to Sustainable Horticulture Development

Prem Nath

1. INTRODUCTION

In the developing world, as against the positive human forces promoting food and nutrition security of the people, there have been negative forces among the population working against it. These range from socio-economic status of the people to unfavourable policies, to damage implementation of the agriculture and food development programmes. Many of these happenings have not been documented officially, and remained mostly with the common masses through the newspaper media. An effort has been made to collate those for the benefit of the stakeholders.

In the present world, where 805 million are suffering from hunger and malnutrition and over three billion from micro-nutrient deficiency. While the efforts are on to promote food and nutrition security for all, there are some negative human forces working against it. There are plenty of instances and evidences which remind us of the seriousness and extent of loss to agriculture and horticulture, food and human life.

2. THE PRELUDE

If we look into the daily activities of human life; in the morning, I found my grandfather drinking water with fresh lime before he went for a walk. At breakfast, our family will have fruit juice, bread with fruit jam or butter and eggs; alternatively idli or dosa, with sambar of vegetables, milk or juice; and sometimes, poori, cooked vegetables, fruit juice or milk. Likewise, at lunch time the menu include rice or chapathi, dal or sambar, cooked vegetables dry or wet or both plus some fruits. At dinner, the family will have mostly chapathi, cooked dry and/or wet vegetable or non-vegetarian food and desserts mostly fruits. A hurried analysis will confirm that there is no menu without horticultural component. The way of preparation and volume of consumption vary from region to region and from family to family depending on the economic status.

In daily life, people or consumers are facing lot of woes, adverse conditions in procuring and consuming the horticultural food products. In this connection it would be useful to take the views of common people who actually face both the unfavourable and favourable conditions on day to day basis.

One morning (April 02, 2017), on one page (3) of the news paper (Deccan Herald), I noticed three different news appearing directly related to horticulture and again on another page (5) importance and significance of this sub-sector in common life of the people in different ways. They may be illustrated as mentioned below.

- **400-yr-old Big Banyan Tree needs more space to spread branches** (Rohit, 2017): The city founded by Kempegowda was just over a hundred years old when the seed of a banyan tree was sown on its outskirts. Over the years, it witnessed Bengaluru's transformation from a sleepy city nestled in the southern corner of Karnataka, to a bustling metropolis. Situated 28km from the city near Kengeri, the 400-year-old tree, aptly named Dodda Alada Mara (Big Banyan Tree) with its coiling leathery branches spread wide is an imposing sight to behold. What's more, the tree, much like the city itself, is only growing bigger. However, Bengaluru might just not have enough space to house one of its oldest inhabitants in all its towering beauty .The horticulture department, which maintains the site of the tree, has sought five more acres to ensure unfettered growth of the tree. Although a proposal to this effect was sent to the government two years ago, the department is yet to receive a response. Environmentalist L. Prakash said the delay in granting land had led the horticulture department to prune some of the tree's branches. "The road connecting Mysuru Road and Magadi Road runs right across Kethohalli, which is where the tree is located. Following complaints from motorists about the branches disturbing their commute, the department was forced to prune those branches. Moreover, land prices in the vicinity have shot up and layouts have mushroomed around the area in recent times. The government has to expedite the land acquisition process," he said. Although the main trunk and root of the banyan tree contacted a disease a decade ago, the tree grows around 50 adventitious roots that develop from a leaf or stem ensuring its sustenance. A popular tourist attraction, Dodda Alada Mara draws hundreds of tourists every week, besides being home to a large number of birds.

  Senior Assistant Director of Horticulture (Bengaluru Rural), Vijay Kumar told TOI that the land to accommodate the trees perpetually spreading branches would be acquired soon. "I took charge six months ago and I have recently learnt that a committee has been formed for finalizing land acquisition for tree expansion. We will acquire private property contiguous with the site at the price fixed by the Karnataka Industrial Area Development Board. We hope to complete the project soon. Meanwhile, we are ensuring the tree is adequately nourished," he added.

  Times View: Besides the stones and pillars that hold the vestiges of the past together, living creatures too form an indispensable part of a city's history. That Bengaluru has a 400-year-old inhabitant, the Big Banyan Tree, should be a source of immense pride for its citizens. At a time when trees are being felled in large numbers to make way for infrastructure projects, a gigantic tree growing further and sprouting new branches every years brings much joy to the heart. The government must ensure that the tree, which has weathered many storms, gets as much space as it needs and continues to remain a witness to the city's endless transformation.

- **Truckers' stir: Veggie prices may surge** (Anon., 2017e)

  The indefinite strike by truckers is likely to hit vegetable supply and send prices soaring. Having received no positive response to their demands, truckers have decided to intensify the protest.

"The strike will continue. Now, even Agricultural Produce Market Committee (APMC) trucks are not running. No petrol/diesel trucks will play," G. R. Shanmukhappa, president of the Karnataka State Lorry Owners and Agents Association, told STOI from Salem. He was in the Tamil Nadu city to attend a meeting of truck owners. He added there was no positive response from the government *vis-à-vis* their demand. Meanwhile, the public is slowly feeling the effect of the ongoing strike, began on Thursday. The prices of essential commodities, especially perishable goods like vegetables and fruits, are seeing a gradual increase. According to a section of dealers vendors, the situation may turn grim if the strike doesn't end in 2-3 days. "What we are selling now is from the existing stock. Once that is exhausted, the prices will go up due to shortage of supply," said a vendor. The truckers are protesting against the move to scrap old vehicles and levy toll on state highways and are demanding a reduction in third-party premium paid to insurance firms. They say the premium has increased by nearly eight times over the years; it recently went up from Rs 22,000 to Rs 42,000 a month.

- **Resident files police complaint against temple official for axing peepal tree** (Anon., 2017d): A resident of Ulsoor lodged a police complaint against a temple executive on Saturday for allegedly axing a 250-year-old peepal tree. B.R. Chandran has filed a complaint with police, BBMP as well as the forest cell. The resident has alleged that Selvamani, executive with Subramanya Temple on Swami Vivekananda Road, cut the tree for no reason.

  "Residents from Ulsoor and nearby places have been worshipping the tree for centuries. On March 23, we were shocked to see that it had been chopped. There was no need to axe the tree. It was healthy," alleged Chandran.

  "We (residents) asked Selvamani whether he had sought BBMP's nod but he screamed at us, saying it was his temple and he didn't require any permission," he added.

  Police said they will transfer the case to the forest cell. "It looks like Selvamani didn't seek approval from any agency to cut the tree," police said.

- **Boy, 12, dies after tree branch falls** (Anon., 2017b): A 12-year-old boy succumbed to his injuries on Friday after a tree branch fell on him, when he was riding pillion with his elder sister on a two wheeler near Nettakallappa Circle, Basavanagudi, South Bengaluru. In death, he gifted a new lease of life to many others as his parents donated his organs.

  The boy, a resident of Basavanagudi, was reportedly not wearing a helmet when the branch fell on him from a height of 15 feet on Thursday evening. He was undergoing treatment at a private hospital.

  Locals alleged that the BBMP's negligence caused the accident as it had not pruned several weak branches of trees in the area. Basavanagudi corporator Katte Sathyanarayana said they had undertaken a drive to prune trees to avoid such incidents.

  "I lost my only son. No mother should suffer like me. We donated his organs so that he will be alive in someone else's body," the boy's mother said.

- **Blossoms of April: tulip fest opens** (Anon., 2017a):

Springtime Blooms: Asia's largest tulip garden Siraj Bagh was opened for visitors in Jammu and Kashmir on Saturday. Spread over 75 acres of picturesque land, the garden is located in the foothills of the Zabarwan Mountain on the banks of the Dal Lake. It was opened ahead of schedule this year.

3. SOME EVIDENCES OF WOES IN HORTICULTURE

Some of the news of horticulture surveys through the news paper media in the last few years will be mentioned below give an idea about general situation.

Food Accessibility

- *How to chop onion out of your cooking* (Ali, 2013).
  - ♦ Price hike forcing people to find options.
  - ♦ Most of the chefs are looking for dishes which do not require onions for preparation.
  - ♦ The alternatives for onion used in the dishes are melon seeds, nut paste or asafoetida. Even shallots and leeks are used as great replacement of onions.
- *Poverty line: INR32 a day in villages, Rs. 47 in cities* (Singh, 2014a; Anon., 2014b)
  - ♦ 30% Indians are poor, conclude Rangarajan Panel.
  - ♦ Rangarajan Panel has estimated 363 million people below poverty line in the year 2011-12 as against the estimated 270 million people below poverty line based on Tendulkar formula.
  - ♦ 29.5% of the Indian population lives below poverty line as defined by Rangarajan committee as against 21.9% according to Tendulkar.
  - ♦ On July 16, 2013, 23 children died after consuming mid-day meal at a government primary school in Masarkh under Bihar's Saran District.

Food Quality

- *EU bans Alphonsos, 4 vegetables* (Anon., 2014a).
  - ♦ The 28-member European Union (EU) has temporarily banned the import of Alphonso mangoes, the king of fruits and four vegetables from India from May 1, sparking protests from the Indian community, law-makers and traders.

- ♦ The temporary ban – proposed by the European Commission and endorsed by the committee – includes mangoes, eggplant, the taro plant, bitter gourd and snake gourd and prohibits the import to tackle the "significant shortcomings in the phytosanitary certification system of such products exported to the EU."

Food Prices

- *Beans for INR 100 per kg, peas for INR 200* (Patel, 2014).
  - ♦ Inadequate supply and rise in toll on Bellary road contribute to zooming veggie prices. Marriage season hikes demand for beans.
  - ♦ Traders said that lack of rainfall has resulted in shortage of supply which was pushing the price upwards. Added to this, the increase in toll on Bellary Road has also hiked costs.
- *Onion tears* (Anon., 2014a2).

  Artificial shortages should not be allowed.
  - ♦ There was a record harvest of onions this year which, at about 19.5 million tones, was 15% more than last year's yield. Yet the prices are 20-30% more than last year's in many places.
  - ♦ Many reasons are cited like the possibility of a deficient monsoon or a big hike in the prices of seeds which may reduce the area under cultivation. But these are not enough to explain the large spurts.
- *Tomato at Rs. 50 / Kg as veggie prices surge* (Anon., 2014a4).
  - ♦ The government is working on a national food grid to link deficient area with producing regions to fight inflation and reduce wastage, estimated at Rs. 44,000 crore a year, food processing industries minister Harsimrat Kaur Badal Said.
  - ♦ Real time data on availability of stocks would be put up online.
- *Prices of veggies, kitchen ingredients hit roof* (Anon., 2014h).
  - ♦ It is not just the prices of vegetables that have skyrocketed; the prices of other kitchen ingredients like coriander powder, chilli powder, ginger and garlic have also increased, digging a deep hole in the pockets of common man.
  - ♦ The increase in price is attributed to farmers from Mandya, Chintamani and other places in Karnataka who supply tomatoes to other parts of the country. They export the produce too, hence, there is a shortage.

Food Corruption

- *Rotten onions sell hot on Tumkur road* (Patel, 2013).
  - ♦ Hawkers reportedly sneak out rejected onions from the busy Yeshvantpur APMC market, only to dry them, sort them and sell them for a little less than market price on busy roads.
  - ♦ The hawkers take home a net profit of Rs. 1,500 to Rs. 2,000 by selling the rejected onions at APMC yard.
  - ♦ The hawkers have propped-up due to the increase in the onion price in the APMC yard.
  - ♦ Following the hawkers practice of selling rejected onions, even the vegetable vendors are picking the rejected onions for Rs. 200 to Rs. 250 per quintal, sort them out themselves and sell it for a neat profit in their shops.

Food and Rain

- *The upset tomato cart* (Bagchi, 2014).
    - The tomato vendor caught in a vicious cycle where his one-day selling can provides him and his family, three square of meal a day.
    - If there is rain or if there is any disruption, his whole stock of vegetable goes rotten and his life is toppled.

Food and Pesticides

- *There are toxic pesticides in your chai: study* (Anon., 2014q).
    - The tea contains DDT, monocrotophos, imidacloprid, thiacloprid, bifenthrin, fipronil and quinalphos are some of the pesticides found in the tea samples.
    - DDT, highly toxic. Monocrotophos is highly toxic orally which killed 23 children in Bihar who consumed it in a meal in 2013. Imidacloprid, potential hazard to liver, kidney, thyroid, heart and spleen. Thiacloprid, toxic to liver and thyroid. Bifenthrin is toxic to skin and causes tremors. Fipronil is toxic to thyroid, liver and kidney. Quinalphos is fatal if swallowed.
- *Beware! The pesticides in the ginger and carrots could make you very sick* (Anon., 2015a).
    - Random test conducted by Bio Centre in Hulimavu on different vegetables has revealed shocking details about the freshness and nutrition value of vegetables.
    - Fruits like papaya, vegetables including cabbage, green chilli, capsicum, carrot, tomato, bottle gourd and ginger were found to contain considerable amounts of the banned pesticide.
    - Short-term exposure to organochlorine pesticide would result in convulsions, headache, nausea, vomiting, muscle weakness, slurred speech and sweating.
    - Long-term exposure due to consumption of such vegetables may damage the liver, kidney and central nervous system.
    - There are several pesticides which the Government has banned. But farmers still use them.

Food Adulteration

- *Peel to win* (Anon., 2014w).
    - Artificial additives can make produce look more succulent while compromising on nutrition. Our checklist for supermarket shoppers
    - Salmon is made pinker
    - Wax on your fruit
    - Meat minus antibiotics
    - Pasteurized almonds
    - Disguised as organic
    - Chicken plus sodium
    - Fruit is ripened artificially

Food Loss

- *Digest this. Bengaluru junks 68% of its food* (Rohith, 2014a).

- ♦ Food waste (leftovers, fruit, vegetables peels, meat, liquid food *etc*) make-up to 60% of the waste generated every day in HSR Layout, Gottigere and Konanakunte area of Bengaluru.
- ♦ HSR Layout, Gottigere and Konanakunte area of Bengaluru, on an average generates 1027.45 kg of waste every day.
- ♦ There are food banks that aims to collect extra food items and distribute it among the needy.

Food, Trade and War

- *Indo-Pak battle of the bulb & berry* (Divya and Khan, 2011).
  - ♦ Indian traders hit back with a ban on tomato exports after Pakistan stopped sending us onions, propelling cross border tensions into a vegetable phase.
  - ♦ The Pakistani government banned the export of onions to India *via* the land route and if it was not done, onion prices in Pakistan would have shot up to over Rs. 100 a kg.
  - ♦ The Indian traders also announced that they would stop sending tomatoes and chillies across the border. If they continued to send the vegetables then there would be domestic tomato crisis in India.

Food and Satellite Dish

- *Shop offers 1 kg onion free with a satellite dish.* (Vora, 2013).
  - ♦ Cashing in on the skyrocketing onion prices, Dinesh Patel, 42, owner of an electronic goods shop in Ahmedabad, hit upon a brilliant idea to boost sale of satellite dishes.
  - ♦ Subscribing to a Direct-To-Home (DTH) service from this electronic goods shop in Krishnanagr, Ahmedabad, guarantees "tears of joy" – that will last for at least three days – if you are a four-member family.
  - ♦ Retail prices of onions have quadrupled over the past three months and have pushed inflation to a seven-month high of 6.46 per cent.
  - ♦ Taking the advantage of the situation to boost sale of satellite dishes, Dinesh Patel came up with this offer.

Food and Civic Sense

- *Litter sense* (Sharma, 2014b).
  - ♦ Lack of civic sense seems to be a universal problem that is threatening our ecology.
  - ♦ The people throw rubbish, kitchen waste, dried flowers, paper dishes used in functions, all debris left after house renovation finds its way to the boundary wall of the powerhouse.
  - ♦ People also throw pea pods, twigs of fenugreek leaves and radish tops in the surroundings.
  - ♦ Lack of civic sense seems to be a universal and grave problem that is threatening the ecology of our nation as whole. Only social revolution spearheaded by children and the youth can perhaps stem the rot.

Nutrition Insecurity

- *3,800 kids are malnourished in Bengaluru* (Anon., 2014c).
  - A survey conducted by Bruhat Bengaluru Mahanagara Palike (BBMP) between June 2012 and April 2014 has identified 3,837 malnourished children in the city.
  - 3, 371 children have been treated, 466 are undergoing treatment at the Nutritional Rehabilitation Centre (NRC), of whom 12 are severely malnourished.
- *Banishing malnutrition* (Anon., 2014n).
  - A community-led initiative to address child malnutrition in Aurad taluk in Bidar district is creating a roadmap for change. A marked change has been achieved in the health indicators through a slew of positive interventions. The locals have set this process of transformation into motion with assistance from MYRADA, a Bengaluru-based non-profit organization, which is implementing the Fight Hunger First Initiative (FHFI) of the German Development Agency, Deutsche Welthungerhilfe.
  - Aurad is home to a sizeable population of severely marginalized Dalits and Adivasis and has some of the worst social and health indicators in the southern state.

Local women have been taught a recipe which is a quick fix solution to meet the nutritional needs of their children. Called 'My Nutrimix', it is a blend of powdered wheat and sprouted green gram combined with groundnut and jaggery and cooked with milk or water. It provides all the necessary calories and is rich in iron and calcium. This mix is provided free of cost to Severe Acute Malnutrition (SAM) children.

Tree Corruption

- *Villagers cop fallen tree for a quick buck* (Kaggere, 2012)
  - Protests of octogenarian eco-warrior who planted it 60 years ago went unheard. Adding insult to injury were officials who auctioned the remaining trunk without even informing Thimmakka about it.
  - A 60 year old tree was cut into log and transported to other places in tractors and luggage autos. Leaving a log behind, which was sold later by the Forest Officials through auction.
  - Thimmakka pleaded the people not to cut and wait for the Government Officials to come. But the miscreants did not heed to the plea of the old lady, and went on to cut down the tree.
  - Thimmakka asks, "Did I plant and raise these trees only for people to plunder them mercilessly? I pleaded with them not to cut the tree, but they would not listen to me."
- *These gangs take supari to kill trees* (Rohith, 2014b)
  - These gangs are usually hired by the advertizing mafia and are paid around Rs. 2-3 lakh, depending on the size of the tree.
  - Hoardings are a big threat. Greenery in prominent places is under attack by the advertisement hoarding mafia. They are cut either to make hoardings more visible or create space for upcoming ones.

- *Supari killers don't spare saplings too* (Rohit, 2014c)
  - BBMP yet to assess loss, experts say installation of CCTVs at Green Zones can curb menace.
  - A set of healthy saplings planted near Yelahanka main road, North Bengaluru, suddenly vanished one fine day. Following complaints from residents, the BBMP forest division found 15 of them missing.
  - The saplings were cleared to ensure that they don't obstruct the view of advertisement hoardings installed nearby.
  - This is also attributed to the lack of staff appointed by the Government to monitor the greenery in the city.
  - To curb on this mafia, "The Tree Preservation Act" should be amended. Illegal felling of trees should be made a criminal offence. Helpline numbers should be launched to assist residents in reporting violations.
  - Environmentalist says installation of CCTV cameras at prominent green zones in the city will go a long way in addressing the concern.

Tree and Concrete

- *Damaging the soul* (Krishnamurthy, 2014).
  - The sad part is that trees no longer dominate the cityscape, but buildings do.
  - Trees have been brutally chopped and removed for road widening or mutilated for the aerial passage of cables.
  - Buildings taller than the towering rain trees known for their shade put those trees in the shade.
- *Encroachment of space around trees on the rise*(Jain, 2014).
  - The space around many trees in the city are being encroached upon by citizens to build 'mini temple', despite the Supreme Court's orders against putting up idols and constructing any structure in public spaces or properties, including the trees on roads.
  - The problem is not limited to one area, as residents from Banaswadi, Chamarajpet, Bull Temple Road among others too have voiced similar concerns.
  - The General Secretary of the Federation of North East Residents Welfare Associations (NERWA), Jude Rajesh Ferrao, said that the State government should impose a ban on such practices.

Politics vs. Advanced Technology

- *RSS wing opposes approval for GM crops field trial* (Anon., 2014a5).
  - GM foods will not be allowed without full scientific evaluation on the long term effects on soil, production and biological impact on consumers.
  - The clearance for GM crop field trials were provided for crops like rice, sugarcane, chickpea, brinjal and mustard.
  - The Swadeshi Jagaran Manch (SJM) functionary said neither the government nor the Genetic Engineering Approval Committee (GEAC) has disclosed the contents of promised scientific evaluation.
  - The government should not rely on the biased and manipulated reports of vested interests among the industry.

Urbanization

- *More than a Lalbagh has been wiped out of city in the last three years* (Kidiyoor, 2014).
  - ♦ As many as 9,281 trees have been chopped to make way for Metro and road widening.
  - ♦ By any standards, the ecological price the city is paying for developmental projects is huge. The BBMP which undertook extension of roads in busy traffic areas, axed 7,853 trees for the same and Bengaluru Metro Rail Corporation cut down 1,428 trees for Metro works.
  - ♦ BBMP earned Rs. 54.81 lakh and Bengaluru Metro Rail Corporation Limited (BMRCL) Rs. 2.29 lakh by auctioning of wood.

Trade

- *Cashew spells cash: adding more bolts to the nut trade* (Strom, 2014).
  - ♦ Farmers who harvest between late January and April, typically store the nuts and sell them off as they need the income, making cashews virtually the equivalent of cash.
  - ♦ Brazilians are the biggest consumers of the yellow and red apples today; a handful were featured in Federation International de Football Association (FIFA)'s advertising for the World cup. Brazil a major global source of cashew nuts, processes only about 12 % of its crop of cashew apples annually because of the challenges posed by their short shelf life.
  - ♦ India is one of the world's largest cashew producers. Some three-quarters of the roughly 660,000 tonnes of nuts grown come from small farmers with orchards of one to two hectares or 2.5 to 5 acres.
  - ♦ Acceso is replacing some of the network, buying nuts at a higher price as well as creating a market for what until now was essentially garbage, the cashew apples. Roughly 500 small farmers here participated in the Acceso program selling their apples to it for 2.5 to 3 rupees a kilo.

Against Nature

- *The curse of Bt Brinjal* (Sirinathsinghji, 2014).
  - ♦ Unproductive, pest infestations, low yields, unwanted produce and unlicensed cultivation flourish as Bt Brinjal is forced on South Asian citizens.
  - ♦ A pilot scheme for commercial cultivation of Asia's first genetically modified (GM) Bt brinjal (eggplant), brought disappointing results for nine of the 20 farmers.
  - ♦ Kajla, Uttara, Nayantara and ISD006, were planted in different climatic regions. Brinjal is one of Bangladesh's most important crops both for consumption and export, making its cultivation a huge health and economic risk for the population.
  - ♦ Many farmers when contacted, reported problems that included bacterial wilt and drought, despite the crop's ability to repel the target pest, the fruit and shoot borer pest. One region, Gazipur, had a failure rate of 4 out of 5 farms, resulting in huge financial losses for the farmers.

Street Food Vendors

- *Walk street occupied (picture caption)* (Anon., 2014a).
    - Fruit vendors have occupied a footpath on Katurinagar main road near Baiyappanahalli making it difficult for pedestrians to walk.

Flower and Trash

- *Lalbagh turns messy as flower show visitors trash civic sense* (Anon., 2015b).
    - It reminded one of the garbage crisis that the city witnessed a few years ago, at the Lalbagh on Tuesday.
    - Every day over 25,000 people visited Lalbagh during the 10-day show. This time, around 400 dustbins and gunny bags were kept across the park. But nobody used it.
    - Seven tractor loads and 10 garbage auto rickshaw loads have been cleared.

Green Cover

- *Green cover halves as concrete jungle expands 13 times over* (Mukherjee, 2015).
    - Vegetation cover in Bengaluru Urban has shrunk from 47% to 22% between 1973 and 2014.
    - The water bodies have reduced from 2.6% to 0.8% between 1973 and 2014.
    - The built-up area has gone up from 2.3% to 31%.
    - According to the study by The Energy Resource Institute (TERI), the green cover contains the pollution in the air and refreshes the air. With the green cover gone, more diseases would crop up.
    - The possibilities of flash floods would go up with the reduced vegetation and lakes also would dry-up fast.

Negative Horticulture

Tomato Prices Soar, households scurry for puree (Anon., 2016e).

As tomato prices hit the roof, the demand for its puree and ketchup has shot up by 40 per cent within one month, a study has claimed. "Demand of tomato puree and ketchup has shot up by 40 per cent within one month as people have curtailed the use of tomato and prefer dishes which do not require much of tomatoes..," the study by industry body Assocham said.

The rise in tomato prices along with pulses have also affected the budget of about 78 per cent families across various cities. The maximum impact was felt in Delhi-NCR followed by Mumbai and Ahmadabad.

"According to the survey, about 78 per cent of households find it difficult to manage their budget and are squeezing families' finances to the lowest level due to sudden rise in price of tomatoes, pulses..," it said.

Govt.'s penalty puts grape growers in the wine yard (Kaggere, 2016).

A winery in Bengaluru thought it would get away with using water to make wine (no grapes at all!). But, eventually, the government found out and now everyone has to pay the price. From July onwards, get ready to pay much more for your favorite bottle of wine. The government is looking to raise the excise duty on it by 177 per cent. The proposal is aimed at punishing the wineries that violated the Wine Policy and recovering the losses incurred by the exchequer. But the government's decision to act against the violation had not only affected wine producers and consumers, but the growers too, especially those cultivating the Bengaluru Blue variety of grapes. As a result of the panic created by the governments gazette notification, several wineries stopped buying grapes, which had left growers in the lurch. The price of Bengaluru Blue grape has plummeted from Rs 25-30 a kg to Rs 5- 8 a kg, according to farmers. In view of the massive repercussions, the Karnataka Wine Producers' Association and Karnataka Grape Growers' Association have decided to hold protests against the government and move the court against the unilateral decision.

177% tax hike dampens spirits of grape growers, wine producers. (Anon., 2016a).

The government, which had promised support to the wine industry in Karnataka, has dashed all the hopes of grape growers with a gazette notification proposing 177% hike in tax. The notification issued on June 22 provides a seven-day window for wine producers to file objections. The taxation will come into effect from July1. Calling it a unilateral decision, grape growers and wine producers have decided to resort to a statewide agitation. Till now, wine producers were paying Rs.17 per liter to the state government as excise duty and additional excise duty. With the new tax rates, they'll have to pay Rs 47 per liter to the government.

Triple whammy as dal, tomato & potato prices surge together (Varma, 2016)

Sometimes its pricey onions driving consumers to tear, at other times, potatoes or pulses. But this summer is different. A ***triple whammy*** of scorching ***tomato*** and ***potato prices***, alongside pulses — arhar and urad – has wrecked family budgets. Potato production was estimated to dip this year to about 46 million tonnes from the previous year's 48 million tonnes.

Jasmine prices shoot through the roof (Nischith, 2017)

Jasmine (Malige ***in*** Kannada), one of the most popular flowers used daily for worship and at weddings, has become prohibitively costly with retail price shooting up five times. The retail price of Jasmine – sold in units of the length of the forearm, called 'mola' in Kannada – has shot up to Rs 100 from Rs 20, while wholesale prices per kilogram have spirallet from Rs 200-300 to Rs 900-1,000. The incredible price rise has been attributed to the high demand of this wedding season compounded by poor harvest in neighbouring Tamil Nadu, from where major volumes of jasmines are procured in Bengaluru. S.G.M. Divakar, flowers association president, told Bengaluru Mirror: "Jasmine has always been in demand and has been a favourite during auspicious occasions and festivals. "With the wedding season picking up, we are expecting more fluctuations in the prices. The reason is because of the bad harvest in Tamil Nadu. There was no proper result from the first crop of jasmine and now the entire land is dry and no water in any parts. So the supply from Tamil Nadu has dropped."

With trucks off roads, veggies rot, traders suffer major loss (Anon., 2016f)

Bengaluru: Vegetable growers and merchants in Karnataka have incurred a daily loss of Rs 5-7 lakh for the last 10 days, thanks to the bandhs and shut downs resulting from the Cauvery row. And it's the onion farmers who have been left teary eyed in particular.

Despite a bumper crop in Ballari and Pune this year, onion farmers and traders could barely make any profit in the last 10 days. That is because not a single Tamil Nadu registration truck ferrying vegetables from farmers to traders was seen in Karnataka during this period. With excess stock, traders now have no option but to slash prices. Also, the daily -wage labourers from Tamil Nadu who usually load and unload trucks across Bengalurus's markets, fled to their native places fearing they could be targeted here.

Said Mohammad Idrees Choudhury, general secretary of Russell Market Traders' Association, "While prices of onion fell sharply, vegetables like beans, eggplants, cabbages, cauliflowers, carrots, coriander, green chillies and garlic got costlier by Rs 20 to 40 per kg, drilling holes in the pockets of customers.

Park Inspection: Mayor finds saplings bought for Rs 2,500 each (Anon., 2017c)

Mayor G. Padmavathi made a surprise inspection of parks on Monday following complaints from Bruhat Bengaluru Mahanagara Palike (BBMP) councillors and citizens over their poor maintenance. Saplings bought at a whopping Rs 2,500 each and additional amount spent for planting them were among the instances of wasteful expenditure that came to the notice of the mayor.

A great leap backwards (Varma, 2016)

In 1958, Chairman Mao ordered that all sparrows over China should be put to death. It was hailed as a necessary step by a strong leader. Farmers were suffering because sparrows tended to eat their grain seeds. Thus began The Great Sparrow Campaign. A countless number of sparrows were indeed wiped out — but there were unintended consequences. Sparrows ate locusts and once the balance in the ecosystem changed, locusts proliferated and destroyed China's crops. There was famine, hunger, starvation: no less than 45 million people died in the three years following Mao's orders. At the start, Mao exhorted them to bear with the inconvenience. But then the pain piled up.

Apart from all the individual suffering, our economy is being eviscerated. Cash is integral to most of the economy. Farmers are unable to sell perishable produce, to buy grains for the new harvest or to pay labourers. Transporters are unable to transport goods across distances. Commerce has shut down in many places, with small businesses going bust. In some places, the barter system is back, as if we've gone centuries back in time.

Festive leftovers (Anon., 2016c)

Garbage thrown on the roadside on the busy Dispensary Road after Deepavali festival on Monday.

BBMP adopts 'plant it, forget it' policy for roadside saplings (Khanna, 2017)

Waters them just 4 times a year; now plans to rope in citizens. BBMP forest cell has decided to rope in citizen groups to plant and maintain road side saplings. The decision comes following criticism that the forest cell is not adequately watering newly planted saplings. On record, the forest cell has ordered its contractors to water saplings just four times a year. But in reality, even this once-in-a-quarter watering is not properly implemented. As a result, a majority of them wither and die.

Ratna Prabha, Assistant Conservator of forests, BBMP forest cell, told DH that the contract with the contractor says that they have to water saplings four times a year. "Last year, the BBMP planted 1.20 lakh saplings, of which 10% did not survive, which is normal," she explains. Pointing at funds crunch, the officer said the forest cell is not able to pay the contractors. "The BBMP owes us Rs 17 crore over the last four years. Lack of funds could be one of the reasons why the contractors are not able to water saplings," she said.

Gone in 30 minutes (Valsan, 2017)

A rain tree that took tens of years to grow into a massive green umbrella on Whitefield Rd was reduced to blocks of wood in minutes. With a slew of `development' projects in the pipeline, Bengaluru will see more such `murders'; unless Bengaluru rises to save Bengaluru.

A 20-member squad armed with hardened ropes, electric saws and various sharp edged weapons descended upon Whitefield Road, near the ITPB campus, in the wee hours of Saturday. The instructions given to them were simple: cut three trees (all fully grown rain trees) into smaller logs in 30 minutes, toss the pieces into the back of a mini-truck, take them to the contractor (who bought the tree at a recent government auction) and don't talk too much.

While the residents of the area and regular commuters are bound to miss the massive green umbrella, this is probably only a taste of things to come, what with a slew of development projects dooming hundreds of city's fully grown trees.

Eating only fruits? You are harming your body (Sadhu, 2016)

Every now and then, we hear of celebrities, especially actors, dropping several dress sizes after following particular diets. While most factor in just enough calories with decreased amounts of carbohydrates, proteins or fat, as the case may be, some go as far as recommending only freshly-pressed juices or fruits as dietary intake. Hollywood actor Ashton Kutcher followed what is known as the fruitarian diet in his preparation to play Steve Jobs, which the late entrepreneur was also apparently a fan of. While Ashton lost weight for the role, he did land in hospital citing stomach ache and pancreatic issues.

Fruits are often an integral part of weight-loss diets, but eating nothing but fruits is not recommended, say experts. "Going fruitarian is trouble in capital letters, because it is not a complete diet," says fitness expert Wanitha Ashok, adding, "Take, for instance, bananas, which provide only 1.1gm of protein for every 100gms of the fruit. Fats are essential for many bodily functions, including maintaining proper hormone levels and brain functions and fruits do not provide for that. For day-to-day living, people should never follow such mindless diets, because they do not provide the requisite nutrition or energy and can lead to long-term health issues."

Caterpillar invasion forces residents to stay indoors (Anon., 2016b)

Bengaluru: K. Veena, a resident of HSR Layout, woke up to a creepy-crawly feel on Friday- the walls of her house were dotted with 30-40 caterpillars. A dazed Veena was petrified at the sight. "I had no idea what these insects were and how dangerous they could be. Soon, I got to know more people in our locality and neighbouring sectors were witnessing something similar," says Veena.

Swarms of caterpillars appeared overnight in many pockets of this southeast Bengaluru locality, forcing residents to stay indoors. Experts believe the insects which live and feed on the leaves of honge mara (pongam or Indian beech tree) ventured out in search of food and shade. Panicky residents chopped four trees and resorted to spraying pesticides on others to ward off the tiny visitors.

Environmentalists, However, say there's no reason to panic. According to them, what the area is witnessing is a natural process for certain species to survive. Says environmentalist Vijay Nishanth, who checked 40-odd trees located in various sectors of the layout: "Caterpillars feed on young leaves of pongam trees. This process of eating plants is the most important part of nature's nutrition cycle. In this case, the caterpillars (common banded awl butterfly) which host on pongam trees, are completing their life cycle. Once they finish consuming the leaves, they go in search of greenery or shade. That's how they they are ending up in neighbourhood houses," he explained.

Horticulture and Environment

The defaced paths (George, 2016),

Handbills stuck on trees, primarily near IT companies, on a regular basis have become an eyesore. Although, the BBMP claims that it has removed most of them, not enough is being done to prevent this.

Bengalureans point out that except from defacing the city, the advertisements don't serve any purpose. The best way to protect the trees from such illegal advertising, suggests Nivedita Kartik, an employee with Oracle, is to bring back the metal rings, that were once put around the trees.

Measured and holistic growth is the answer (Basu, 2016)

Unprecedented temperatures, vanishing green cover, traffic that seems to stay in constant chaos — this isn't the Bengaluru of the '80s. In light of an IISc survey that warns of doom, Mirror talks to experts to get to the root of issues ailing the city. This one focuses on the economic side of things.

For those who can observe, the signs have been all around. Unprecedented temperatures, dwindling resources, vanishing lakes, chock-a-block roads, crumbling infrastructure—this isn't a city that's fighting for survival.

Experts are of the opinion that the ability of a city to survive and prosper indefinitely involves several major factors, such as, an economy, availability of jobs and services, healthcare and attraction of the urban environment and the availability of resources, as well as space for growth.

"Economic viability at the cost of health and environment is no viability at all. Presently, this city is recklessly headed down a slippery slope, buoyed by the unreal imaginations unfurled by the avaricious real estate lobby and its influence in government policies and planning mechanisms," said Leo Saldanha, coordinator of Environmental Support Group (ESG).

"The best way forward is to invest heavily in bottom-up planning, coordinate city-wide monitoring and management tasks through a representative and transparent system and ensure future planning is undertaken democratically and keeping in view environmental, health and ecological indicators," said Saldanha.

Transportation, Infra Woes (Basu, 2016)

Transportation problems in cities are major roadblocks in long-term development and are very closely related to land development, economic structure, energy policies and environmental issues.

"Despite a number of transport proposals on the anvil, such as, the Metro rail, ring roads, peripheral ring roads, *etc*, Bengaluru is expected to face severe traffic congestion in the coming years. The rapid economic growth will result in significant increase in traffic problems. In the absence of planned mass transit systems, a disproportionately high share of trips will be carried by personalised modes of transport, creating chaos and straining infrastructure," said Patil.

"In Bengaluru, 30 per cent of mobility is by foot, so the government must give highest priority to footpaths. This is followed by public transport and private cars, which contribute 10 to 15 per cent of the transport demand, two-wheelers another 15-20 per cent. Hence, 60 per cent demand is met by public transport and non-motorised transport. The city planning, however, concentrates on roads and flyovers, which support cars and two-wheelers. Other cities with better facilities for cyclers and pedestrians have higher sustainability values," said environmentalist A.N. Yellappa Reddy.

A little attention and a relatively small investment can patch up the city's footpaths that have been neglected for years, he added. Bengaluru is experiencing an unprecedented urban growth of 4.6 per cent. The establishment of the IT parks on the outskirts of the city has converted the city and its surroundings into a Silicon Valley of the country. It has also created a lopsided urban sprawl towards the south and east, which is not sustainable, said Bharath H. Aithal, a researcher at IISc.

Unplanned concentrated growth or intensified developmental activities in a region strain natural resources, increase traffic congestion and pollution levels and also change the local climate.

"Our data indicates that residential areas of Bengaluru constitutes 43 per cent of the land area followed by approximately 21 per cent each by roads/transport and green cover. The central core in Bengaluru appears to be underdeveloped. This means that business and commercial activities that normally provide services to the city and its region have not developed fully. Instead, the city has a mix of uses, such as, industrial, agricultural, residential and waterbodies around their cores. This signals that markets have not operated to allocate land to higher value uses, such as, commerce and services, as would be seen in comparable Asian cities. In effect, the land utilisation pattern is influenced by central planning controls and this introduces rigidities into the system," said Patil.

Rise in Emissions (Basu, 2016)

A multicity analysis on greenhouse gas emissions by Aithal showed that haphazard growth and inadequate public transport system in Bengaluru and Hyderabad led to increased use of private vehicles.

Emissions from the transport sector ranged from 43.83 per cent (Bengaluru) and 56.86 per cent (Hyderabad). Aithal's findings showed that Chennai emitted the highest carbon dioxide equivalent emissions per GDP, followed by Bengaluru. Patil's study said the transport sector was the major contributor of particulate matter (PM) 10 and NOx (mono-nitrogen oxides) in Bengaluru. The industrial sector was the No. 1 emitter of sulphur dioxide (SO_2).

"Transport sector of Bengaluru is a major source for PM10 emissions contributing nearly 45 per cent of the total. The main source of NOx emissions in the city is motor vehicles (75 per cent), while DG sets contribute about 20 per cent. Domestic and commercial sector combustions are minor sources of NOx. The main source of SO_2 emissions is industry. This source contributes 60 per cent of the total SO_2 emissions in Bengaluru. The other major sources are transport and domestic use," the study said.

Water-scarce & unlivable: that's B'luru in 2020 (Menezes, 2016)

"Bengaluru would witness 93% of urban cover and only 7% of open spaces and green cover in 2020. This would make the region GHG-rich, water-scarce, non-resilient and unlivable, depriving the city dwellers of clean air, water and environment," the study says.

Bengaluru saw a whopping 1,005% concretization or increase of paved surface from 1973 to 2016. "There are only 1.5 million trees to support Bengaluru's population of 9.5 million, indicating one tree for every seven persons in the city.

Storm, fallen trees and death (Anon., 2016d)

What started out as a pleasantly overcast Monday morning exploded into a rainstorm by afternoon, which uprooted trees, flooded roads and even claimed the life of a commuter. Experts say the monsoon may only be getting started. Fairoz Pasha, 50, was on his two-wheeler at Malleshwaram, 18th Cross when a tree fell on him. Pasha, a resident of Nagawara, was a food supplier for government schools in the city. The police said he was still breathing when they reached the spot. "We took him to KC General Hospital, but the doctors pronounced him 'brought dead'.

Why Do Trees Fall Every Monsoon? (Basu, 2016)

It's one of the biggest hazards of monsoon in the city, one that claimed a life on Monday. Lack of environmental planning is one big reason for it, says environmentalist Suresh Heblikar. Bengaluru is being reinvented every day, but the trees here were planted 60-70 years ago. As a result of the sinking water table, these trees are deprived of adequate nutrition and moisture. "The trees that fall are mostly along the roads. This is because trees by the roads are often strangulated by bitumen and tar and don't get adequate water and nutrition," he says. A solution to the problem is a study on the kind of trees the new Bengaluru requires. "Roads now are planned by engineers who have a very poor understanding of ecology," he says, adding that all roads can't have all kinds of trees.

Monsoon warning: At least 10L weak trees need urgent attention (Rohit, 2016)

The death of two persons crushed under trees uprooted by gusty winds, that lashed the city for a few minutes last week has jolted the BBMP's forest cell out of its slumber. While BBMP officials feel pruning and cutting of trees are the best solutions, environmentalists feel such surveys should have been completed well before the monsoon and problems addressed so that many trees are spared the axe. "Digging up of roads by civic agencies—whether to lay a pavement or sewage lines or optical fiber cable—has been a threat to trees as it often results in roots getting chopped. Also, asphalt or concrete has been laid around full-grown trees. Unscientific pruning of tree branches is another worry. Non-treatment of infected and diseased trees is a problem and so is the failure to identify old trees," he added.

How law that protects trees got stumped (Ghosh, 2016)

The KPT Act was twisted over the years to make tree-felling easier. As trees keep being marked and numbered for felling to ensure road-widening under the controversial Tender SURE project, a closer look at the entire process digs out one uncomfortable truth. The very law that was enacted and subsequently amended to protect trees in Karnataka and capital Bengaluru is increasingly being either violated or circumvented altogether for the felling of trees.

Greenocide: Confident of NGT nod, BDA seeks permission to kill trees (Chaturvedi, 2017)

The Bengaluru Development Authority (BDA) can't wait to get going on the Rs 1,781-crore steel flyover project, the fate of which will be decided on Monday when the National Green Tribunal (NGT) pronounces its verdict. Meanwhile, the BDA, sure of getting the go-ahead for a project that has seen massive outcry in the city over the environmental damage it will do, has sought permission to axe 812 trees to make way for the metal behemoth.

Hot enough to fry an egg (Kumaran, 2017)

There are things that are so obvious that one doesn't need proof to believe them. The scorching summer heat of Bengaluru these days is one example. And yet, when a statement made as a figure of speech – 'so hot you could fry an egg here!' – actually comes true, one is allowed to make a big deal of it. So, on Saturday afternoon, we went up to the terrace, frying pan in hand and decided to check if it was indeed as hot as it felt. In less than an hour, we managed to make an omelet of three eggs. And there is your proof.

Dry heat: look what tree-felling has done to Bengaluru (Bandyopadhyay, 2107)

This summer, Bengalureans are getting to know the kind of heat they've rarely witnessed here before: the dry, under-the-hand-dryer kind of heat. While those who've lived in Delhi or Hyderabad would empathise, for the rest of the population here – that has grown up in the city's natural air-conditioning – the development may seem unbearable. Take heart, though, it's not the temperature, but the lack of moisture in the air, say experts. The other new development is the worsening air quality because of increased dust. The problem is worse in areas where there is a lot of construction and the roads are bad. Residents living in and around tech parks, such as, Electronics City, HSR Layout and Whitefield, are getting the worst of it.

Making space for biodiversity in urban areas (Kedlaya, 2017)

A few years ago, almost every day, I took my coffee in the balcony, observing the extremely animated and expeditious parakeets that never failed to keep an appointment with the old trees huddled on the side of the road. A year later, there were no trees and more birds went homeless. Richard Mabey once said, "To be without trees would, in the most literal way, to be without our roots." By being a mute witness to this slow and deliberate felling of trees in the city, we are unknowingly parting ways with our much-coveted memories, experiences and with it a part of our culture.

A recent report released by the Azim Premji University, Bengaluru states that we are set to lose 2244 trees and 205 saplings from 71 species and 26 families to the proposed steel flyover and road- widening projects. A city known for trees like fig and neem that clean our air and cool our surroundings are all going to be axed, rendering the sky empty of birds.

Disastrous Impacts (Kedlaya, 2017)

Bengaluru's population has been increasing steadily from 8.7 million in 2011 to over 11.5 million in 2016. We hold the third position for being the most populous and second as the most polluted city, besides coming second for generating the highest amount of CO_2 in India. Our air quality, drinking water pollution, noise and light pollution is appallingly high. Lung cancer accounts for about 10.8% of all new cancers in Bengaluru and pulmonologists hold the increasing air pollution as one of the primary perpetrators.

Now, you might wonder what difference will 2,244 trees make to your life? For starters, all the above mentioned dismal statistics may inch higher and other problems will come up as well. No shade, merciless summers and roads lined with closely hugging endless lines of ugly buildings are just some of the problems.

Aesthetics aside, without trees, where will the pollinators go? Even the snakes and geckos are biological pest control agents. Let's not forget that these complex ecosystems contribute to the bio-geo-chemical cycle in an urban environment. Let's not forget the role of trees as natural sound-proofers and as air purifiers.

The study proves that "street segments with trees had on average lower temperature, humidity and pollution, with afternoon ambient air temperatures lower by as much as 5.6°C, road surface temperatures lower by as much as 27.5°C and SO_2 levels reduced by as much as 65%. Suspended Particulate Matter (SPM) levels were very high on exposed roads, with 50% of the roads showing levels approaching twice the permissible limits, while 80% of the street segments with trees had SPM levels within prescribed limits."

The author observes that trees are being cut either to widen roads or to give way to metro or to layout electric cables or to make drainage or to layout water pipe with no provision to let the roots to standing trees to grow or no space/ soil left to grow new trees.

4. DISCUSSION

The above evidences are not all and do not represents all woes faced by the common people. The evidences limit to last couple of years and further limited to some selected newspapers. Based on this, the following points emerge;

- The above evidences reconfirm that horticulture is importantly impregnated in the human daily life.
- It confirms that common masses are facing lot of woes beginning farmers to consumers challenging the sustainability of horticultural development operations which are affecting the common life of the people. Measures are required to access the issues and to take appropriate steps to mitigate the woes.
- It is recognized that vegetables and fruits are major contributors of micro-nutrients, which cannot be replaced by cereals. Also, knowing the figures that over 3 billion people globally are suffering from micro-nutrient deficiency with serious human maladies, the woes of both the vegetable and fruit production and availability and of consumers accessibility to them, need to be addressed appropriately towards sustainable development and promotion of better human health.
- The horticulture sub-sector cannot stand alone and it requires the assistance and cooperation of other related sub-sectors/ sectors in order to make horticulture development sustainable. The healthy diet cannot be provided by horticultural products alone, it needs full support of cereal and legume products. Likewise, horticulture requires essentially the support of other related disciplines of science and technology for its fullest outcome of sucess.

- The difficulties or woes faced by farmers to consumers be identified and listed with priorities plan and implement sustainable horticultural development.
- It is not only the farmers and consumers but the positive contribution of all stakeholders in the horticulture sub-sector that is required to make the horticulture development sustainable.

5. CONCLUSION

A serious review of the above shows that most above happenings are caused repeatedly by some negative minded people and are recurring and affecting the lives of the majority of the people; in other words, some minority negative human forces are working against majority of positive human forces. The negative human actions are akin to flies in cup of drinking milk.

All above situations or happenings are related to food and work against the force promoting food and nutrition security. The role of humanity should be to arrest or nullify these negative forces at all levels of social and administrative ladder for the benefit of mankind at large. The positive attitude and action should prevail over the negative attitude of the people.

REFERENCES

Ali, Amin. (2013). How to chop onion out of your cooking, The Times of India, Bengaluru, October 28, 2013

Anon. (2014a). EU bans Alphonsos, 4 vegetables, Deccan Herald, Bengaluru, April 29, 2014

Anon. (2014a2). Onion tears Deccan Herald, Bengaluru, June 23, 2014

Anon. (2014a). Walk Street occupied (picture caption). Deccan Herald, Bengaluru, December 16, 2014

Anon. (2014a4). Tomato at Rs. 50 / Kg as veggie prices surge The Times of India, Bengaluru, July 16, 2014

Anon. (2014a5). RSS wing opposes approval for GM crops field trial, Deccan Herald, Bengaluru, July 20, 2014

Anon. (2014b). Poverty line: INR32 a day in villages, Rs. 47/- in cities, Deccan Herald, Bengaluru, July 17, 2014

Anon. (2014c). 3,800 kids are malnourished in Bengaluru, Bengaluru Mirror, Bengaluru, August 12, 2014

Anon. (2014h). Prices of veggies, kitchen ingredients hit roof Deccan Herald, Bengaluru, July 17, 2014

Anon. (2014n). Banishing malnutrition, Deccan Herald, Bengaluru, August 12, 2014

Anon. (2014q). There are toxic pesticides in your chai: study The Times of India, Bengaluru, August 13, 2014

Anon. (2014w). Peel to win. Bengaluru Mirror, Bengaluru, September 22, 2014

Anon. (2015a). Beware! The pesticides in the ginger and carrots could make you very sick, Bengaluru Mirror, Bengaluru, February 04, 2015

Anon. (2015b). Lalbagh turns messy as flower show visitors trash civic sense, Deccan Herald, Bengaluru, January 28, 2015

Anon. (2016a). 177% tax hike dampens spirits of grape growers, wine producers, Deccan Herald, Bengaluru, June 29, 2016

Anon. (2016b). Caterpillar invasion forces residents to stay indoors. Mar 26, 2016

Anon. (2016c). Festive leftovers. Deccan Herald. Bengaluru. November 01, 2016

Anon. (2016d). Storm, fallen trees and death. Bengaluru Mirror. June 28, 2016.

Anon. (2016e). Tomato Prices Soar, households scurry for puree. Deccan Herald. Bengaluru, June 23, 2016

Anon. (2016f). With trucks off roads, veggies rot, traders suffer major loss. Times of India. Bengaluru. September 17, 2016

Anon. (2017a). Blossoms of April: tulip fest opens, Deccan Herald. Bengaluru, April 01, 2017

Anon. (2017b). Boy, 12, dies after tree branch falls. Bengaluru. TIMES NEWS NETWORK. April 2, 2017

Anon. (2017c). Park Inspection : Mayor finds saplings bought for Rs 2,500 each. Deccan Herald. Bengaluru, January 03, 2017.

Anon. (2017d). Resident files police plaint against temple official for axing peepal tree. The Times of India (Bengaluru) April 02, 2017.

Anon. (2017e). Truckers' stir: Veggie prices may surge, The Times of India (Bengaluru) April 02, 2017.

Bagchi, Subroto (2014). The upset tomato cart, The Times of India, Bengaluru, May 25, 2014

Bandyopadhyay, Nithindra. (2107). Dry heat: look what tree-felling has done to Bengaluru. Bengaluru Mirror. Apr 16, 2017.

Basu, Mihika. (2016). Measured and holistic growth is the answer. Bengaluru Times, June 03, 2016.

Bosky, Khanna. (2017). BBMP adopts 'plant it, forget it' policy for roadside saplings. Deccan Herald. Bengaluru. February 15, 2017.

Chaturvedi, Atul. (2017). GREENOCIDE: Confident of NGT nod, BDA seeks permission to kill trees. Bengaluru Mirror. Jan 26, 2017

Divya, A. & Khan, Omer Farooq (2011). Indo-Pak battle of the bulb & berry, The Times of India, Bengaluru, January 31, 2011

Ellen, Barry. (2017). Litchi, mystery fruit behind Bihar deaths. Deccan Herald. Bengaluru. February 03, 2017.

George, Nina C. (2016). The defaced paths. Deccan Herald, Bengaluru, Aug 04, 2016,

Ghosh, Subir. (2016). How law that protects trees got stumped. Bengaluru Mirror. October 02, 2016

Kaggere, Niranjan (2012). Villagers cop fallen tree for a quick buck, Bengaluru Mirror, Bengaluru, November 03, 2012

Kaggere, Niranjan. (2016). Govt's penalty puts grape growers in the whine yard Bengaluru Mirror Jun 28, 2016

Kedlaya, Gana. (2017). Making space for biodiversity in urban areas. Deccan Herald, Benagluru Feb 28, 2017

Kidiyoor, Suchith (2014). More than a Lalbagh has been wiped out of city last three years, Bengaluru Mirror, Bengaluru, July 29, 2014

Krishnamurthy, Vinita (2014). Damaging the soul, Deccan Herald, Bengaluru, April 23, 2014

Kumaran, P. (2017). Hot enough to fry an egg. Bengaluru Mirror. Apr 15, 2017.

Menezes, Naveen. (2016). Water-scarce & unlivable: that's B'luru in 2020. Deccan Herald, Benagluru, June 25, 2016.

Mukherjee, Saswati (2015). Green cover halves as concrete jungle expands 13 times over, The Times of India, Bengaluru, February 10, 2015

Nischith, N. (2017). Jasmine prices shoot through the roof. Bengaluru Mirror. February 08, 2017.

Patel, Bharat A. (2013). Rotten onions sell hot on Tumkur road, Deccan Herald, Bengaluru, December 01, 2013

Patel, Bharat A. (2014). Beans for INR 100 per kg, peas for INR 20, Bengaluru Mirror, Bengaluru, May 19, 2014

Prasad, R. (2017). Scientist says international journal violated ethics on Bihar's 'litchi disease' report. Deccan Herald. Bengaluru. February 03, 2017.

Rohit B. R. (2017). 400-yr-old Big Banyan Tree needs more space to spread branches. The Times of India (Bengaluru) April 02, 2017.

Rohit, B.R. (2016). Moonsoon warning: At least 10L weak trees need urgent attention. Times of India. Bengaluru. July 04, 2016.

Rohith B. R. (2014a). Digest this. Bengaluru junks 68% of its food, The Times of India, Bengaluru, October 17, 2014

Rohith B. R. (2014c). Supari killers don't spare saplings too, The Times of India, Bengaluru, September 24, 2014

Sadhu, Rahul. (2016). Eating only fruits? You are harming your body. Nov 30, 2016.

Sharma, Ruchi (2014b). Litter sense, Deccan Herald, Bengaluru, July 24, 2014

Sirinathsinghji, Eva (2014). The curse of Bt Brinjal, Deccan Herald, Bengaluru, August 12, 2014

Strom, Stephanie (2014). Cashew spells cash: adding more bolts to the nut trade, Deccan Herald, Bengaluru, August 11, 2014

Valsan, Binoy. (2017). Gone in 30 minutes. Bengaluru Mirror Feb 12, 2017.

Varma, Amit. (2016). A great leap backwards. Times of India. Bengaluru. November 20, 2016.

Varma, Subodh. (2016). Triple whammy as dal, tomato & potato prices surge together. Times of India. Bengaluru. June 17, 2016.

Vora, Jignesh (2014). Shop offers 1 kg onion free with a satellite dish, Bengaluru Mirror, Bengaluru, November 04, 2013

3

Challenges and Opportunities in Sustainable Horticulture Development

V. A. Parthasarathy, Sanjay Kumar Singh and C. Aswath

SYNOPSIS

For a sustainable production system, horticulturists must look for ways to re-create the CO_2 fixing effect of natural systems. Selection of plants should be based on these ecological considerations as well as on traditional considerations like form, colour and texture. To move toward a sustainable production system, horticulturists must look for ways to re-create the CO_2 fixing effect of natural systems. We can then make more informed and ecologically appropriate choices and move toward sustainability. We should do everything we can to move toward local production, composting and recycling of nutrients and integrated pest management as economic measures as well as environmental ones. We should use cultural methods that limit the need for pesticide and fertilizer inputs and be aware of the broad consequences of our actions. When we do these things, we will be moving toward economic sustainability. In this paper, the challenges as well as opportunities for sustainable horticulture is dealt with in detail. The Toronto declaration made in 2004 can be used to develop road map. Communicating the health benefits of fruits to consumers is an essential ingredient in sustaining product demand, which is a prerequisite for sustainable fruit production.

1. INTRODUCTION

The ultimate definition of a sustainable system is one that serves people, sustains or improves the environment and enhances the economy on the scale of the entire planet and over the next hundred or several hundred years (Doxon, 1991).One of the classical explanation of the sustainable agriculture was given in the USA in their Farm Bill 1990 by Blumfield (2000), "Sustainable agriculture to mean an integrated system of plant and animal production practices having a site-specific application that will over the long term: satisfy human food and fiber needs; enhance environmental quality and the natural resource base upon which the agricultural economy depends; make the most efficient use of non renewable resources and on-farm resources and integrate, where appropriate, natural biological cycles and controls; sustain the economic viability of farm operations; and enhance the quality of life for farmers and society as a whole (Government Printing Office, 1990). According to Blumflied (2000) sustainable agriculture became the term used in policy areas related to questions of the environment and resource conservation and the term organic was used in the marketplace as a label to differentiate among agricultural commodities (Klonsky and Tourte, 1998). Commercial fruit production can be described as a system with the following elements:

Production – Post Harvest – Marketing – Distribution – Retail – Consumer Demand.

Over the past several decades, Integrated Fruit Production and organic fruit production have emerged as possible two approaches to sustainable fruit production.

In future vast majority of floriculture crops or cut flowers sold in India will be labeled as sustainably or organically produced. However, flower identity can be enhanced through "green label" flower certification programs, establishing standards for the reduction of pesticides and other chemicals, conservation of natural resources, worker safety and labor rights.

2. CHALLENGES

Horticulture helps in maintaining healthy ecosystem, water regulation, pollination, erosion control and climate and wind regulation so also world food security issues could have been solved by paying greater importance to the role of horticulture. It involves rigorous cropping expertise, including the improvement, production, distribution and use of vegetables, fruits, woody landscape and greenhouse plants. Horticulture is now one of the fastest growing industries with striking professional opportunities. Current trends indicate that consumers are looking for increased variety, freshness and healthy options in their eating choices. They are also seeking greater ease and a higher proportion of fresh produce in their diets. Those in metropolitan environments are more and more aware of and dependent on green spaces for their livelihoods and wellbeing.

To move toward a sustainable production system, horticulturists must look for ways to re-create the CO_2 fixing effect of natural systems. Selection of plants should be based on these ecological considerations as well as on traditional considerations like form, colour and texture. We can then make more informed and ecologically appropriate choices and move toward sustainability. We should do everything we can to move toward local production, composting and recycling of nutrients and integrated pest management as economic measures as well as environmental ones. We should use cultural methods that limit the need for pesticide and fertilizer inputs and be aware of the broad consequences of our actions. When we do these things, we will be moving toward economic sustainability.

The increased complexity in the existing consumer demands for affordability, safety and continuity of supply; increasing needs to achieve sustainable practices; and the requirements to deal with challenges originating from a more uneven climate, the loss of productive soils through urban intrusion and the loss of low-cost labour. The production of tropical, subtropical and temperate fruit crops is highly influenced by climate and undesirable physiological phenomena and long cropping cycles. True tropical fruits like rambutan, mangosteen or durian is highly demanding both in temperature and humidity requirements, cannot be economically cultivated in the subtropics, but others like mango, litchi or avocado, have become important commercial crops in many subtropical countries through many years of research and grower's efforts.

Hi tech horticulture and precision farming are two of many production practices that have been implemented in India. Management of pests is generally easier and cheaper in sub-tropical climate as compared with the tropics due to the constraints of low temperatures and humidity on pathogen activity. This has important relevance for organic cultivation, which can add value to the production of tropical fruits in the subtropics. Since tropical trees produce fewer annual flushes per year in subtropical climates, tree growth control is easier and suitable pruning techniques allow their cultivation at high density, facilitating high initial yields with optimal management. Proximity to the market is an additional advantage, not only for cost reasons, but also due to a reduced carbon footprint. Special problems and challenges for tropical fruit cultivation in the subtropics include the growing importance of *Fusarium* wilt and weevil borer in cooler climates and the impact of low temperatures in relation to flowering, pollination, fruit set, fruit quality and yield in many tropical tree crops.

A significant development is the increasing share of fresh compared to processed horticultural products traded, taking advantage of advances in post harvest technology. The major challenges are:

(i) One major challenge is to truly stay sustainable. If an outbreak of insects comes to infect a mango grove, it is a lot easier and less time consuming to do a mass spraying with a chemical that will for sure eliminate the pest with one or just a few applications, than have to manage with IPM practices.

(ii) Among fruits, with the exception of banana and papaya, average productivity is 12.00 t/ha, much lower than potential biological yields which can be attributed chiefly to non-availability of quality planting material, dwindling natural resources, resource-poor farmers, low adoption of modern technologies,

(iii) Do mangoes have a potential to grow and produce a good yield in areas that receive more than an annual 60 inches of rain? Can mangoes be efficiently grown in higher altitudes of above 1200m?

(iv) The consumption of many high priced items (typically fresh fruit and vegetables) has been static or declining, including that in high-end restaurants. This has also impacted on the rate of growth in exports from some countries.

(v) Consumers in many developed countries have become more interested in understanding the sources of the foods that they eat, so promotion of locally produced horticultural produce is needed in order to boost their sales.

(vi) The proportion of produce that these markets are selling is still comparatively small on a global scale and is usually (but not always) limited to provision of products during the current growing season—the exception is where local producers sometimes supplement their own products with imported ones.

(vii) Supermarkets have rapidly displaced smaller marketers as the main outlets for the year-round supply of fruit, vegetables and cut-flowers.

(viii) The urban-based consumers have become increasing vocal about issues, such as, carbon taxes, buy-local campaigns, the use of pesticides, labour conditions for farm workers and the sustainability of production methods.

The changes outlined above are merely some of the factors that are impacting on current production practices for horticultural crops. They also provide a plethora of opportunities for horticultural science to solve many of the current problems facing producers and to offer solutions to those issues that are concerning consumers.

3. OPPORTUNITIES

The opportunities lie in developing methods to tackle the challenges. How do we tackle these challenges and build a robust system.

(i) For arid region, important tree species, khejri (*Prosopis cineraria* L.) can provide alternative means of security to the farmers. Farmers are maintaining 30–60 trees of khejri in a hectare as it improves the growth of companion annual crops. Being a legume, it also fixes nitrogen in the soil. All parts of tree are used in one or other way as fodder or vegetable.

(ii) Other tree of importance for arid region is teak (*Tecomella undulata* D. Don), gum Arabic or kumat (*Senegalia senegal* Britton), Israeli babool (*Acacia tortilis* Hayne), Indian jujube (*Ziziphus mauritiana* Lamk.), kair (*Capparis decidua* (Forsk.) Edgew), Bengal quince or bael (*Aegle marmelos* (L.) Corr. Serr.), henna (*Lawsonia inermis* L.) *etc*.

(iii) The increase yield is possible with mulch with a combination of coconut leaves, either green or dried up.

(iv) A sustainable way of using fertilizers for yield increase is to use a weaker dose of fertilizer than recommended.

(v) A need for year-round supply to allow consumers continuity of access to specific items of produce.

(vi) A wide range of interesting products to challenge the existing product range and to entice and retain consumers is required.

(vii) We must ensure the best cultivars with highest possible quality with regard to flavour, appearance and condition.

(viii) Products that are convenient to eat with competitive pricing are beneficial for buyer and producer.

(ix) Develop appropriate production methods, new cultivars and products and to demonstrate the benefits from fruit and vegetable consumption and from the use of plants in various environments.

Diversification may contribute to a sustainable horticulture but how?

- The higher the spatial diversity of crops in a particular area (different species in the same region) and/or temporal diversity (crop rotation), the lower the number of problems caused by pests and diseases.
- The cultivation of a greater number of crops allows a higher efficiency in the use of production factors. Different crops may show the same requirements at different times and this means a more rational use of infrastructures, machinery and labour and minimises the problem of seasonal labour.
- Physiological and nutritional complementary requirements of different species can be exploited, optimising soil performance and avoiding soil nutrient depletion and proliferation of soil pathogens caused by monocrop.
- Increasing the number of crops allows an optimum exploitation of the crop×environment interaction by using the most suitable crops for each environment, which avoids dramatic environment alteration, reduces inputs and minimises contamination.

- A greater number of crops favours stability of production and growers′ income because the cultivation of a higher number of species decreases risks against unpredictable environmental and market changes.
- The introduction of a new crop contributes to improve ethno-botanical knowledge, which is a substantial part of folk culture.

4. BIODIVERSITY CONSERVATION AND IMPROVEMENT

The fruit crops of most horticultural significance are vegetatively propagated and not seed propagated, restricts the options for their conservation other than through relatively expensive live plant collections (which in most countries are under severe funding pressures). Now a days, major shortage of conventional plant breeders world-wide as programmes have become disproportionately now focused on molecular techniques. Formal training in plant breeding as well as the retention of conventional breeding in both tertiary education and government research institutes has also seen a major decline in the past two to three decades. The production of horticultural crops by genetic manipulations is of great concern and a matter of debates regarding the safety and the wider justification of GMOs. The lack of consciousness about the role that conventional plant breeding plays in the production of new cultivars and the uncertainty that often prevails between conventional breeding and plants produced using gene transfer technologies is a matter of concern regarding horticulture practices.

Some fruit species are dioecious, *e.g.*, date palm and some papaya cultivars require cross pollination. Many fruit species/cultivars do better with cross pollination, *e.g.*, avocado, litchi and some mango cultivars. Some fruit crops as in *Annonaceae* are pollinated by beetles, not by honey bees. If the beetle population is not sufficient then pollination may not occur. Hand pollination is an option to increase the percent fruit set, improve fruit quality and yields. Species with potential pollination problems are Avocado, Carambola, Annona (sugar apple), Atemoya, Papaya, Passion fruit, Litchi and Sapota. Species generally without pollination problems are Mango, Banana, Longan, Mamey sapote.

5. ENHANCING NUTRIENT AND WATER USE EFFICIENCY

A major fruit tree problem is secondary and micronutrient element deficiencies. Lack of essential elements affect plant health and thus decrease plant growth, flowering and fruit production and lack of consistent fertilization with needed plant nutrients results in poor tree health and/or biannual fruit production (*i.e.,* a lot of fruit in one year and little to none the next). Lack of or too much nitrogen reduces flowering and fruit production. Some orchardists do not consistently fertilize their trees or when they do fertilizer (very occasionally) they over apply fertilizer which either causes damage to the leaves (salt-burn) or vigorous leaf or shoot growth and little to no flowering. Simply, it is better to use small amounts of fertilizer frequently during the warmer months of the year (March-April to September). Fertilizer applications during the late fall-winter period are either less effective or undesirable depending upon the fruit crop. For solutions to secondary and micronutrient deficiencies, use a foliar micronutrient mix and apply to the leaves 3-4 times per year (anytime from March-Sept.). Ideally purchase and apply a soil drench of chelated iron 2-4 times per year. Alternatively, farmer can try a mix of iron (iron sulphate or iron oxide) and compost and apply to the soil under the tree canopy.

The nutrient use efficiency of orchards might vary between varieties, environmental conditions (*i.e.,* climate and soil type) and farming systems (*e.g.*, high or low level of fertilizer application; high or low labour use; intercrop or mono-cropping; high or low density). With a long history of excessive fertilizer applications, the N use efficiency will remain low over a long period due to the large historical nutrient inputs and the unbalanced N-P-K applications. On the other hand, in orchards with low nutrient inputs in the past, nutrient use efficiency could be either high or low depending mainly on the current fertilizer application (*e.g.*, site specific application or not), the indigenous soil fertility, the water supply during the growth period and soil properties (*e.g.*, water holding capacity which affects the degree of nutrient leaching (Wang Na *et al.*, 2016).

Pressure on water resources for urban, industrial, recreational, conservation and other uses all appear to have higher priority within societies than the availability of water for horticultural crop (food) production. Hence, research is desirable to decide issues, such as, the development of drought tolerant crops, dealing with increased salinity, the management of crops under managed water deficits, use of low quality/ waste water and the use of more efficient application methods, *viz.* Drip irrigation, sprinklers, alternate furrow methods with enclosed systems, such as, simple greenhouses. The yield losses often result from poor irrigation water quality or improper fertilization practices. The improved efficiency of water and fertilizers has the potential to decrease fertilizer leaching, which will help protect water resources. The most common irrigation method for greenhouse production is overhead watering. In this system, half of the applied water and nutrients typically do not reach plants. The excess water carries nutrients and pesticides that may contaminate ground and surface waters. As water quality and pollution issues become important, it is important that greenhouse producers adopt practices that reduce nutrient and pesticide runoff.

Over watering is especially a problem with automatic sprinkler systems as they causes oxygen starvation in the root zone, leaches plant nutrients beyond the root zone. Watering is a must for young trees just planted, during the first three years or so or mature trees with a lot of fruit (except mango). Young trees need consistent watering during dry periods to survive or grow. Many fruit crops (carambola, longan, litchi and papaya) bearing fruit need consistent watering during dry periods to either hold the fruit to maturity or to produce high quality fruit. In contrast, some fruit crops, such as, established mango trees should rarely be watered. Overwatering may reduce fruit quality and delay fruit maturity.

- Trees must receive good quality water (municipal water source), grow salt tolerant fruit crops and fertilize trees frequently with small amounts rather than periodic large amounts to bring sustainability in fruit production.
- Periodically (every 3-6 months) over-irrigate the soil around your trees to "flush" excess salts from the root zone under salt affected areas.
- Keep water off the leaves when irrigating as the salts in the irrigation water will damage them.
- Additional research can be done to see whether mangoes could still be grown in heavy rainfall areas of India with good drainage.

6. SOIL MANAGEMENT

The improper use of a soil amendments (top soil, muck, sludge, sand and compost or some mixture thereof) may harm the health of trees and the diversity of amendments and confusion about their use, *i.e.,* how to tell a good one from a poor one and how to use it, it may be better not to recommend a soil amendment to inexperienced people. Soil amendments should only be used in a limited amount in the planting hole. Furthermore, these materials should only be mixed with the native soil in no more than a 60% native soil/40 % amendment mixture. The reason for this is, that if too much soil amendment is used movement of water between the soil in the planting hole and the native soil may be impeded leading to root rot and decline of the tree.

Table 1. *Technological options for sustainable management of soils* (Lal, 2008).

Objectives	Technological options
Soil conservation	No-till farming, residue mulch, cover crops
Water conservation	Mulching, fertigation, drip irrigation, water harvesting and recycling
Positive nutrient budget	Nutrient cycling through use of crop residue mulch, compost, manure and fertilizers, precision farming, intercropping with legumes
Energy management	No-till or conservation tillage, precision farming, fertigation

Some technological options for sustainable management of soil and water resources for horticulture crops are outlined below:

Mulch farming: It is an important component of improving soil quality, conserving soil and water, moderating temperature and improving yield of horticultural crops. There is a wide range of mulching materials including plastic sheet (*e.g.*, black, clear, translucent and plant biomass. Rainfall capture by plastic mulch resulted in 108% average yield increase over the conventional dry land techniques. Using biomass (*e.g.*, crop residue or mulch from a specifically grown cover crop (such as, elephant grass or guinea grass) can have ancillary benefit of enhancing soil organic carbon (SOC) concentration and nutrient cycling. A combination of plant biomass with no-till farming has additional benefits of erosion control and SOC sequestration (Lal, 2008).

Cover crops: A wide range of cover crops like oats (*Avena sativa* L.), rye (*Secale cereale*), clover (*Trifolium pratense*) and wheat (*Triticum vulgare*) helps in erosion control, weed suppression, improvement in the SOC pool, increase of soil fertility, reduction in soil compaction, improvement of soil tilth, decrease in pest incidence, conservation of soil moisture, reduction in runoff and increase in crop yield (Lal, 2008).

Some specific sustainable horticultural technologies suggested are:

(i) A very sustainable method of dealing with of fruit-borer is having a trap crop of marigold intercropped among tomatoes.

(ii) Intercropping in field vegetables (*e.g.*, white cabbage inter-cropped with white clover; leeks planted in subterranean clover) to reduce dependence on pesticides and to increase the land equivalent ratio (Theunissen, 1997);

(iii) Improving N and P management to enhance nutrient use efficiency of carrots, cauliflower, Chinese cabbage, *etc.*, by reducing leaching losses by moderate irrigation (Paulin *et al.*, 1995);

(iv) Enhancing water use efficiency under irrigated horticulture by matching the biophysical efficiency with the economic efficiency.

(v) Preventing erosion and soil structural decline on sloping lands through soil management practices. Growing cover crops, mulching and use of herbicides for weed control are extremely effective in erosion control under fruit trees established on steep terrains.

(vi) Using minimum tillage and other improved management practices for horticultural crops on peat soils.

(vii) Using compost as an integrated nutrient management option for vegetable production.

(viii) Enhancing soil fertility through soil organic matter management and

(ix) Managing crop residues and biosolids to improve nutrient cycling and reducing leaching losses.

Approaches to sustainable horticulture and resource management options depend on specific questions, such as, (Brumfield, 1996), what management practices affect sustainability? What crops can be established using no-till farming? How can use efficiency of inputs be enhanced? and, how can degraded soils be restored?

7. GROWING UNDER ENVIRONMENTAL CONSTRAINTS (FREEZING, FLOODING, WIND, LIGHT AND SALINITY)

Major challenges of adverse growing environment viz. freezing temperatures are the most common problem faced by fruit trees. Flooding can be a problem in low lying areas or during/after tropical storms and hurricanes. Wind damage can be a problem especially if trees are allowed to become overgrown. Salinity water and/or soil is a problem in some areas, especially along the coast and in the keys. Constant low light levels leads to loss of tree canopy, poor fruit production and lower fruit quality.

- Banana grown adjacent to southwest wall protects the plants during cold weather, in contrast the shade from the banana plants cool the south west wall in summer.
- Star fruit tree with *Alocasia odora* border in eastern side of landscape protected by wind. Under clean cultivation (allows solar heat to penetrate soil surface and be "stored" for re-radiation at night), mulch and grass prevents penetration of solar radiation into soil. The mulch may be pulled back from November to March to facilitate exposure of the bare ground to sunlight to inhibit soil-pests build up.
- Irrigating the ground under and around the fruit trees several days prior to a cold/ freeze event will increase the soil's capacity to store and re-radiate heat during freezing temperatures.The greater the surface area wetted the greater the soil heat storing capacity. This helps in frost avoidance.
- Irrigating during the daytime prior to the night of a predicted freeze is not recommended, especially if it is already cold and/or windy.
- Clean soil, mound up 2-3 ft, check tree trunk for insects and diseases in low lying areas.
- When trees exposed to excessive soil moisture, shallow sandy soils may require augering through the hard pan to improve soil drainage and/or construction of mounds or beds for safely planting tropical and subtropical fruit trees.

- Construction of mounds made from native soil may increase the amount of roots above flood waters. Mounds should be 1-3 ft high and 3-10 ft in diameter.
- To prevent tree from wind or storm damages, maintain trees at a reasonable height with practicing annual or biannual pruning.
- The lower limbs of mature trees may be lost if shaded for too many hours of the day, resulting in fruit trees fruit far that only have to bear above the ground.
- Managing tree height and spread is key to maintaining sufficient light levels and the lower fruit producing canopy of fruit trees.
- Periodic selective pruning of mature trees to open the canopy and to limit the spread of the canopy will maintain light levels throughout the canopy.
- Where salinity stress problems (marginal leaf chlorosis, leaf drop, poor plant vigor/ growth, tree decline, stem dieback, tree death) are there, tolerant plant like coconut palm, ber, phalsa and to a limited extent guava can be grown.

8. ORGANIC HORTICULTURE

Fruit production has addressed sustainability challenges in the past, including pesticide use, post-harvest quality and changing consumer preferences. Both Integrated Fruit Production and organic production were developed with sustainability goals in mind (Granatstein and Kupferman, 2008). A truly sustainable horticulture would not only be organic, but also produce social, environmental and economic justice, thus contributing to rural development on a global scale and becoming part of what has been referred to as sustainable globalization (Somner, 2008).

Since tree fruit and vine systems are perennial and typically involve little tillage after planting, they can be very conducive to increasing soil organic matter. Use of scab-resistant (*Venturia inaequalis*) apple varieties has expanded the potential for organic apple production in humid regions where this disease is difficult to control. If one were to calculate the average sustainability of all organic farms and compare it to the average of conventional farms (Figure 1), it is likely the organic system would illustrate greater sustainability (Granatstein and Kupferman, 2008).

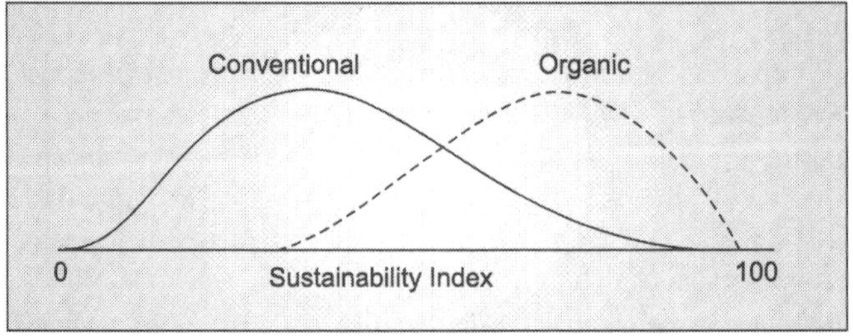

Fig. 1. *Hypothetical distribution of conventional and organic farms along an idealized sustainability index*

9. SUSTAINABLE CROP PROTECTION

The agro-ecosystem design is now getting more attention as a strategy to improve sustainability. A more open canopy helps reduce some diseases and improves pesticide coverage and efficacy, addressing environmental sustainability. The high-density orchards with thin leaf canopies may suffer from increased sunscald of fruit from the reduced protection against the intense solar radiation in the summer. In many regions organic fruit producers must spray more frequently and use more kilograms of pesticide product, often achieving a lower marketable yield (Merwin *et al.*, 2005).

The border plantings of wild rose (*Rosa woodsii*) hedges in Washington State apple orchards augment the bio-control of leafroller by providing alternate habitat for a key parasitoid and add to biodiversity in the system (Pfannenstiel and Unruh, 2003). Establishment of the naturally occurring 'weed' *Ageratum conyzoides* in citrus orchards in China led to near total biocontrol of citrus red mite by natural enemies (*Amblyseius* spp.) encouraged on the weed. Use of scab-resistant (*Venturia inaequalis*) apple varieties has expanded the potential for organic apple production in humid regions where this disease is difficult to control. Verma (1999) reported that combined use of neem cake with *Trichoderma harzianum* and G. virens gave better control of mango wilt pathogen and improved the plant health significantly.

Lal *et al.* (2002) indicated that some plants contained organic substances that act as pest repellents which keep pests away from the crops and avoid potential damage. Diverse nature of plant not only obstructs the adults from egg laying but also the release of volatile allelo-chemicals from a particular crop deters the adult insect from damaging the crops. Sowing coriander (*Coriandrum sativum* L.) or marigold (*Tagetes* sp.) in vegetable fields increases the activity of syrphid flies which predate aphids and other soft-bodied insects. The deep summer ploughing alone has been found be effective for management of several pests, such as, wire worms, cutworms or root grubs, fruit flies in various guards or adjusting the date of planting as observed in rapeseed–mustard(Trivedi and Ahuja, 2011). Mustard can be grown as a trap crop for cabbage to help avoid diamond back moths infecting and destroying cabbage.

Dipping cut rose flowers in either 3 mM of benzoic acid or 3% of H_2O_2 provided high levels of control against botrytis. As we know, continued use of these fungicides is problematical owing to development of fungal resistance and increasing environmental concerns over fungicidal residues. Thidiazuron (TDZ), a compound with potent cytokinin activity, dramatically reduced leaf yellowing in cut flowers, such as, *alstroemeria.* TDZ-treated potted plants of tulip, cyclamen, poinsettia and geranium are dark green for more than a month while leaves on the control plants were yellowing, even abscised.

10. POST HARVEST TECHNOLOGY AND MARKETING

The fruits are relatively cheap and ready source of vitamins and minerals. This is a viable option for diversification from traditional export crops. Horticulture producers have rapidly adopted many such technologies, such as, through the use of near infrared scanning in modern fruit grading equipment which can now sort as many as 1.5 million fruit per hour. The high costs of energy for transport and for storage, changes in consumer use/ preferences and consumer concerns, increasingly mean that these options (only value addition) may no longer be viable for some markets. Together these issues pose a great challenge and would need to be addressed through inter- organizational/ industry talk. Further, with the advent of new technologies, *viz.* remote sensing, environmental monitoring, image analysis and others, there is a growing expansion of the different types of information. Modern trends indicate that consumers are seeking increased variety, novelty and healthy options in their eating choices. They are also seeking greater convenience and a higher proportion of fresh produce in their diets.

11. ROAD AHEAD

The importance of sustainable horticulture has been well recognized and the first road map for the research needs was discussed at Toronto International Symposium on Sustainable Horticulture. The Toronto 2002 Sustainability Declaration (Bertschinger *et al.,* 2004) lists the Research Needs for a Sustainable Horticultural Development in the 21st Century as follows.

1. Horticultural research needs to embody more interdisciplinary and holistic approaches in addition to the traditional reductionist approaches to problem solving.
2. Horticultural research should involve new players from outside the traditional disciplinary area, including more international scope and inclusion of social and cultural expertise.
3. Research regarding sustainability should not end at the farm gate. A dynamic systems approach is needed for planning and conducting research that is more systems oriented.
4. Practical methods are required to bring the concept of sustainability to bear on the horticultural research agenda and to connect with the farm level.
5. We need simple methods for assessing sustainability, both qualitative and quantitative.
6. The choice of research methods for a project needs to be carefully evaluated to maximize sustainability. On-farm, participatory and transdisciplinary methods may be useful and can involve various stakeholders (*e.g.* farmer, consumer). Small farmers and extension workers should not be overlooked and be included in this new research approach.
   (a) Research should address the natural, social, human and physical, capital and finance (according to Pretty's model). If addressing the renewable, natural capital in horticultural systems, research should contribute to its build-up for this purpose, the following topics deserve specific attention:
   (b) Continuous development of new technologies and knowledge for disease and pest management that support sustainability.
   (c) Holistic approach for following fixed nitrogen through the farm system and ecosystem (*e.g.* effect on ground water, NO*x* emissions).
   (d) Development of new disease and pest resistant germplasm that also meets market requirements.
   (e) Enhance soil quality and natural resource management in all horticultural systems (*e.g.* replant diseases, water quality).

One of the strategies to address the concerns of global climate change is to adopt management practices which enhance SOC pool, improve soil structure, reduce risks of soil erosion, moderate soil moisture and temperature regimes, recycle plant nutrients, increase use efficiency of water and nutrients and suppress weeds.

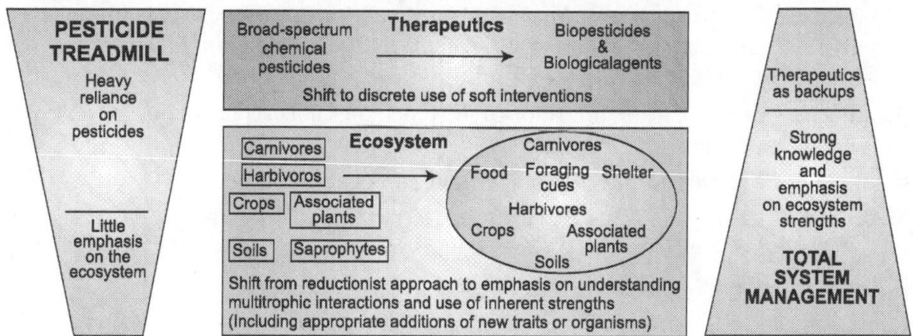

Fig. 2. *Illustration of a shift to a total system approach to pest management through a greater use of inherent strength based on a good understanding of interactions within an ecosystem while using therapeutics as backups. The upsidedown pyramid to the left reflects the unstable conditions under heavy reliance on pesticides and the upright pyramid to the right reflects sustainable qualities of a total system strategy* (Lewis *et al.,* 1997).

12. CONCLUSION

The principles of sustainable horticulture are increasingly being incorporated into conventional agriculture. Integrated crop management, an ecologically based approach to horticulture production that includes management practices, such as, minimum tillage, nutrient balancing and integrated pest management (IPM), have been widely adapted by producers. Many farmers rotate crops, apply fertilizers in bands, mulch, stake and use other practices that reduce their dependence on synthetic chemicals to produce high quality crops. In future, we can expect that the number of environmentally-aware consumers who demand fruits/vegetables/flowers grown without harmful chemicals will continue to grow, but no one should expect that this will happen rapidly. Organic food has been marketed for several decades and remains a small share of food consumption. Growing a large number of crops spreads the economic and production risks and makes the farm less susceptible to changes in supply and demand for a specific crop.

The principles of sustainable agriculture are increasingly being incorporated into conventional agriculture. Producers must be convinced that sustainable systems are profitable before they adopt them. If producers are going to adopt more sustainable systems, more research is needed to develop systems and reduce costs of organic and ICM systems. Methods that include the true costs of production, including family labor and unintended negative side effects, need to be developed. Policy tools that can account for these costs include regulations, liability, information, taxes, risk-benefit models and market-based solutions. Sustainable management of natural resources must be based on enhancing productivity per unit area; time and energy input; maintaining a positive trend in productivity over time; creating SOC sink capacity of natural and managed ecosystems; reducing risks of non-point source pollution and sedimentation; and making managed ecosystems a solution to environmental problems (Lal, 2008).

REFERENCES

Bertschinger,L., J.D. Anderson,, Nico de Groot, D. Granatstein, R.Habib, K. Mulllinix, D. Neilsen, F. P. García, F. P. Weibel and G. Zinati. (2004). Conclusions from the 1st Symposium on Sustainability in Horticulture and a Declaration for the 21st Century. Acta Hort. 638: 509-512

Blumfiled, R.G. (2000). An examination of the economics of sustainable and conventional horticulture. HortTech. 10 (4):687-691.

Brumfield, R.G. (1996). Sustainable horticulture. An overview. Hort. Tech. 6(4):352-354

Doxon, L.E. (1991). Sustainable Horticulture. Hortscience , 26:144-145

Granatstein, D. and Kupferman, E. (2008). Sustainable horticulture in fruit production. Acta Hortic. 767:295-308

Klonsky, K. and L. Tourte. 1998. Organic agricultural production in the United States: Debates and directions. Amer. J. Agr. Econ. 80(5):1119–1124

Lal, O.P., Sinha, S.R. and Srivastava, Y.N. (2002). Role of intercropping on population build up of *Lipaphis erysimi* on cabbage under field conditions. Ann. Pl. Protec. Sci. 10: 377-378

Lal, R. (2008). Sustainable Horticulture and Resource Management. Acta Hort. 767: 19-44

Lewis, W.J., van Leuteren, J.C., Phatak, S.C. and Tumlinson, J.H. (1997). A total system approach to sustainable pest management. Proc. National Acad. Sci. 94:12243-12248.

Merwin, I., G. Peck and Vollmer, E. (2005). Organic orchards in the Northeast: progress, practices and problems. P. 61-62. In: D. Granatstein and A. Azarenko (eds.). Proc. 3rd National Organic Tree Fruit Research Symposium. June 6-8, 2005, Chelan, WA. Washington State University Tree Fruit Research and Extension Center, Wenatchee, WA.

Paulin, B., Lantzka, N., McPharlin, I. and Hegney, M. (1995). Making horticulture sustainable. Western Aust. J. Agric. 36:39-45

Pfannenstiel, R. and T. Unruh. (2003). Conservation of leafroller parasitoids through provision of alternate hosts in near-orchard habitats. P. 256-262. In: Proc. 1st International Symposium on the Biological Control of Arthropods. USDA Forest Service Publ. FHTET-03-05

Sumner, J. (2008). Sustainable horticulture and rural development: more than just organic production. Acta Hortic. 767: 111-122

Theunissen, J. (1997). Inter-cropping in field vegetables as an approach to sustainable horticulture. Outlook on Agric. 26:95-99

Trivedi, T.P. and Ahuja, D.B. (2011). Integrated pest management: approaches and implementation. Indian J. Ag. Sc. 81 (11): 981–93

Verma, V.S. (1999). Studies on Non-chemical Management of Mango Wilt in Nurseries. Ph.D. thesis, Dr. Y.S. Parmar University of Horticulture and Forestry, Nauni, Solan, 119 pp

Wang Na, Joost Wolf and Zhang Fu-Suo. (2016). Towards sustainable intensification of apple production in China— Yield gaps and nutrient use efficiency in apple farming systems. J. Integrative Agriculture. 15(4): 716–725.

❍❍❍

4

Horticulture: International and National Mandate

Prem Nath

1. INTRODUCTION

The author remembers the year 1962 when he joined the Division of Horticulture at IARI, New Delhi. The Division was the offshoot of the Division of Botany created a couple of years ago at that time, but soon became comparable in scientific staff, research programme and infrastructure to other divisions of the institute. The creation of new Division was the first landmark in the history of Indian Horticulture. During that period horticulture was treated as a minor aspect to agronomy in most of the agricultural colleges and states.

The beginning of the land-grant type of agricultural university in each state of India, during 1965 onwards gave lot of impetus, opportunities and status to horticulture in research and education with separate Department of not only horticulture but also fruit, vegetables ornamentals and their preservation.

While at the IARI, in 1960s we horticultural colleagues discussed and dreamt of having separate research institute of Horticulture, Vegetables and Floriculture and Landscape and now in this era we have in some States the University of Horticulture. Now, the IIHR is celebrating successfully its existence of 50 years. The recent data indicate that horticulture production rate has excelled the grain production in India. But the fact remains that with all these advancements, there are about 805 million people suffering from hunger and malnutrition and over 3 billion people suffering from micro-nutrient deficiencies and India shares millions on this account. This shows that we have not done enough and have to work more. This is vital because horticultural products are the major source of micro-nutrients, the deficiency of which causes serious diseases to human being as well as to crop plants.

It is noted that the International Society of Horticultural Sciences, Belgium, continue to organize symposia and conferences on different areas, from time to time which focus more on research and technology. Most of the nations continue to organize meets on horticulture focusing on issues of national interest.

In the present chapter author has attempted to summarize the recommendations and summarize the outcome of selected international meets relevant to our theme of the chapter and indicate the mandate required internationally and nationally in the wake of present food and nutrient crisis and challenging sustainable horticulture development.

2. ICV-2002-THE INTERNATIONAL CONFERENCE ON VEGETABLES (ICV-2002) HELD IN NOVEMBER, 2002 IN BENGALURU

Theme: Vegetables for Sustainable Food and Nutrition Security in the New Millennium

Rising Challenge: The Green Revolution era of 1970's with focus on high inputs and productivity helped the developing nations in tiding over the famine and in overcoming the food shortage but eroded the sustainable agriculture of the post-Green Revolution period. The symptoms of fatigue on the soil because of heavy fertilization and over cropping, the shortage of water because of its indiscriminate use, over use of pesticides and removal of natural vegetation have challenged the progressive agriculture of to-day worldwide. In other words, the harmony which existed before between the natural resources and the food production has resulted into the conflict, which needs to be reviewed.

Vegetables as Supplement: It is recognized that modern agriculture must diversify production and achieve sustainable higher output. At the same time safeguard the environment, conserve natural resources and adapt to climate changes and ever-rising population. In order to release the pressure on cereals as well as to improve upon the human nutrition through consumption of the other nutritious crops, diversification in cropping pattern provides better option. The increased production and consumption of horticultural crops particularly vegetables with its wide adoption and provider of important nutrients offer promise for the future. Vegetables are rich source of nutrients (especially vitamins and minerals), besides its medicinal values. As such, vegetables as food and diet supplement are gaining momentum in most countries in recent years.

Vegetables are higher in productivity than other crops. They provide more food per unit time and area of land. Hybrid cabbage and Chinese cabbage, yield up to 40-60 tonnes per hectare in an average growing period of three months. The more recent advances in technology, particularly hybrids and biotechnology have added value to this group of crops in terms of efficiency and quality (Nath and Dutta, 2002).

Since 1995 the FAO's Special Programme on Food Security has amply impacted the role of vegetable crops among poor communities in the Low Income Food Deficit Countries worldwide (Kobakiwal, 2002).

The International Conference on Vegetables with the theme "Vegetables for Sustainable Food and Nutrition Security in the New Millennium" held during November 11-14, 2002, thoroughly discussed and concluded 'vegetables', as the important and major supplement to the food and nutrition security. Major recommendations are given below:

Value added Germplasm

- The vegetable genetic base should be strengthened through intensive inventorization and characterization of germplasm resources including introduction of exotic and wild germplasm by all the national and state research organizations.
- The participating countries, particularly the developing countries may come out with a proposal for easy exchange or sharing of value added germplasm either by bilateral arrangements or diplomatic relations with other countries within the frame work of safe guarding the IPR issues.
- Research thrust, coordination and the needed support should be streamlined for the utilization of the well documented value added germplasm by private and public sector within the country for developing *"Hallmark"* vegetable varieties having multiple stress resistance or tolerance, sustainable high yield and quality parameters.

Genetic Improvement in Vegetable Crops

- Vegetable varieties with improved shelf life, rich in nutritive value, bio-active phytonutrients, varieties suitable for processing industry and export should be developed for global competitiveness and nutrition security.

Biotechnological Applications in Vegetable Crops

- A policy paper on biotechnological approaches in increasing productivity of vegetables should be brought out by each country.
- Cloning and deployment of genes conferring abiotic stress tolerance and nutrient utilization should be given top priority.
- Work on structural and functional genomics should be initiated to develop capability and utilization in crop improvement.

Production Technology in Vegetable Crops

- Cost effective eco-friendly production technology involving Integrated Production and Protection Management (IPPM) including details regarding selection of the appropriate cultivars, quality seed, adapted cultural practices and production technologies combined with Integrated Pest Management (IPM) are to be reviewed and updated to promote successful production of high quality vegetables. Judicious use of inputs like water, mineral fertilizers and pesticides are to be emphasized. Use of organic manures and IPM technologies are to be promoted for the production of safe vegetables.
- Use of cost effective structures like low and poly tunnels and polynet houses for raising vegetables and planting material which can be adopted as economic activity in vegetable growing areas has been recommended.
- Urban vegetable production systems (urban, periuban and suburban) are to be considered as an emerging opportunity to augment high quality vegetable production to feed the fast growing urban population all over the world.
- Integrated nutrient and water management is an important aspect of vegetable production to minimize the ill effects of excess use and misuse of these inputs and optimize their use to realize the intrinsic genetic potential of high yielding varieties and hybrids of vegetables.

Protection Technology in Vegetable Crops

- Usage of botanicals like neem seed kernel extract (NSKE), neem seed powder (NSP), neem and pongamia soap as effective eco-friendly insecticides for control of insect pests in tomato (fruit borer) and cucurbits (serpentine leaf miner and fruit fly), should be promoted.
- Intensify work on use of bio-control agents (*Paecilomyces lilacinus, Trichoderma harziznum, Verticillium chlamydorporium, Pseudomonas fluorescence*) botanicals (neem and pongamia cakes), endomycorrhizae (*Glomus mosseae* and *G.fasciculatum*) for the integrated manangement of root knot and reniform nematodes in vegetables should be intensified.
- Biotechnological tools for disease resistance breeding work and sensitive diagnostic kits for the identification of new strains of viruses specially in gemini, tospo and llar viruses in vegetables should be developed.

Post Harvest Management in Vegetable Crops

- A coordinated package including vegetable production area and market, as well as processing industry, development of suitable varieties for processing with the matching production technology, creation of pack house facilities at terminal collection centers and training personnel in post harvest management is needed for the speedy development of vegetable processing industry.
- Research on nutrient loss in domestic cooking process, optimization of thermal processing, potential role of dietary fiber and phytonutrients in vegetables, nutritional status of organically grown vegetables, use of biodegradable films in vegetable packing need further intensification.
- Creation of infrastructure facilities, implementation of scientific methods of post-harvest management and transportation to provide assured supply of vegetables from remote areas to the processing sites need to be developed.

Organic Vegetable Production

- Research institutes and agricultural universities should initiate work on organic vegetable production as per the international standards, develop sustainable package for production and take up the validation of the existing practices.
- In order to promote organic farming, each state may identify "Organic Village" where NSOP standards would be applied and compared with conventional farming for soil build up, pest and disease outbreak, quality products and cost-benefit ratio.

Hybrid Seed Production

- National policy support is required for hybrid vegetables R&D and seed production activities for both public and private sector institutions. Collaborative work in R&D, extension and human resources management in vegetable seed production should be taken up.
- Plant variety protection (PVP) systems should be enforced in all developing countries. This will enhance contract hybrid seed production activities and trade.
- Hybrid vegetable seed production is seasonal and a risky business. Crop insurance cover should be extended to all hybrid vegetable seed crops and companies and contract seed growers engaged in seed production.

Vegetables and Nutrition

- Horticulture intervention should be encouraged at the household, community and national level. There is a need to develop and promote food based strategies to combat micronutrient deficiencies and integrate them with other community based nutrition programmes to ensure nutrition security.
- There is a need to establish community based support services (nursery, model garden) to get access to necessary inputs, training and technical information by targeted household garden.
- National agricultural policies should accord emphasis to nutrition linked agriculture development keeping in view the food based dietary guidelines and to ensure nutrition security.

Under-Utilized Vegetables

- Indigenous knowledge on under-utilized vegetable crops should be documented through ethno-botanical studies. The community awareness be strengthened for commercial exploitation of these crops.

- The crops be prioritized to develop under-utilized crops for food and nutrition security and to improve livelihood. The selected crops should fit into the cropping systems.

Marketing and Trade

- There is a need to harmonize the national standards pertaining to "safe vegetables" with that of Codex Alimentarius standards. Awareness among farmers pertaining to international standards need to be generated particularly under WTO regime.
- Guidelincs and rules should be framed to create pest free production sites and places by establishing pest risk appraisal (PRA) unit.

Development Policies and Programmes

- A draft proposal should be prepared by the participating countries for easy exchange / sharing of value added germplasm through bilateral / diplomatic relations with other countries safeguarding the IPR issues.
- Emphasis should be given to develop vegetable varieties suitable for low input technology as well as those providing nutritional security and global competitiveness.
- Integrated production and protection technologies for vegetable crops with easy access to the small farmers should be developed and adopted in each country.

Technical Cooperation among Developing Countries (TCDC)

- Strengthen national, regional and international capabilities to help identify priority areas for Cooperative Development Research and to formulate joint programmes of action that would improve capabilities for production, protection, marketing, distribution and better use of vegetables as dietary supplement for raising nutritional levels and simultaneously improving income levels.
- Help developing countries to formulate policy, design strategy and programmes of action nationally, inter-regionally and globally to increase availability of vegetables in a sustainable manner.
- VEGINET can act as a consultative body for its members to help them in preparation of development projects in vegetable science and technology (PNASF, 2002).

3. JOINT FAO/WHO WORKSHOP ON FRUITS AND VEGETABLES FOR HEALTH (FAO & WHO, 2004)

Non-communicable diseases (NCDs), especially cardiovascular diseases (CVDs), cancer, obesity and type 2 diabetes mellitus, currently kill more people every year than any other cause of death. Four factors in the epidemiology of these diseases – poor diet, physical inactivity, tobacco and alcohol use – are of overwhelming importance to public health.

Fruit and vegetables are an important component of a healthy diet and, if consumed daily in sufficient amounts, could help prevent major diseases, such as, CVDs and certain cancers. According to *The World Health Report 2002,* low fruit and vegetable intake is estimated to cause about 31% of ischaemic heart disease and 11% of stroke worldwide.1 Overall it is estimated that up to 2.7 million lives could potentially be saved each year if fruit and vegetable consumption was sufficiently increased. Recommendations in this direction tend to complement and reinforce other valid messages based on the long-known health benefits of consuming vegetables and fruit as dietary sources of fibre, vegetable proteins and protective micronutrients. The recent Joint FAO/WHO Expert Consultation on diet, nutrition and the prevention of chronic diseases, recommended the intake of a minimum of 400g of fruit and vegetables per day (excluding potatoes and other starchy tubers) for the prevention of chronic diseases, such as, heart disease, cancer, diabetes and obesity, as well as for the prevention and alleviation of several micronutrient deficiencies, especially in less developed countries. The recommendation thus adds to the already strong case the health benefits to be gained from the consumption of fruit and vegetables and paves the way for concrete action advocating increased consumption of these commodities.

WHO has responded to the global rise in NCDs by giving increasing attention to their prevention and control, most recently through the Global Strategy on Diet, Physical Activity and Health endorsed at the Fifty-seventh World Health Assembly on 22 May 2004. Within the framework of this Global Strategy, WHO aims to actively promote an increase in fruit and vegetable intake worldwide. To reach this goal, WHO and FAO have formed a partnership around the theme of fruit and vegetables for health.

WHO and FAO held their first joint workshop – Fruit and Vegetables for Health – at the WHO Centre for Health Development in Kobe, Japan, on 1-3 September 2004. Participants included nutrition-, health- and agricultural scientists, representatives from ministries of health and agriculture, advisers on nutrition from WHO Regional Offices, experts from the World Food Programme, the United Nations Economic Commission for Europe, the Secretariat of the Pacific Community, the International Food Policy Research Institute and the global "5 A Day" community, which promotes fruit and vegetable consumption (FAO & WHO, 2004).

Workshop Conclusions

- While there is strongly suggestive evidence that there is potential for increased consumption of fruits and vegetables to reduce the risk of ischaemic heart disease, stroke and type 2 diabetes mellitus, further research is needed to evaluate specific fruits and vegetables. Further studies are needed on the role of fruits and vegetable consumption in obesity, weight management and cancer prevention.
- A daily intake of at least 400 g of fruit and vegetables, within the context of ensuring a better general dietary pattern as a population goal, was reconfirmed as an appropriate basic message.
- Messages about fruit and vegetable consumption need to be integrated into food-based dietary guidelines, to be country-specific and culturally relevant and coordinated with other messages about healthy diets.
- In the face of ongoing nutrition transition, decreasing cereal consumption and a shift towards excessive fat consumption, there is a risk that the opportunity to promote fruit and vegetables as an integral part of a balanced diet is being lost.
- There are major differences among countries and population groups in the kinds of nutrition-related problems that occur ranging from under-nutrition and micronutrient deficiencies to over-nutrition. Intervention strategies to promote increased fruit and vegetable consumption need to be sensitive to these differences.
- Farmers can be key players in overall strategies aiming to increase fruit and vegetable consumption: they are likely to be motivated by expectations of economic returns. Policies are needed to empower farmers so that diversification can become a reality

along with cost-effectiveness in their production systems and so that their harvests can be scheduled and managed to ensure year-round supplies of a range of safe produce.

- Intervention strategies need to be multidisciplinary and coordinated, should comprise a balance of components to stimulate growth in both demand for and supply of fruit and vegetables. They need equally to recognize and address basic differences that exist between the respective supply chains for fruit and vegetable commodities and to take account of the range of supply and consumption scenarios from subsistence to supermarkets.
- Evaluating projects is essential in order to learn what works and what does not and to avoid wasting time and resources on inappropriate approaches (FAO & WHO, 2004).

4. INTERNATIONAL WORKSHOP ON URBAN AND PERI-URBAN AGRICULTURE FOR ASIAN COUNTRIES, BENGALURU, INDIA (UPA-2005)

The urbanisation process is a reality that drives people to go and live in the cities. It is very pronounced in Latin America where already 85% of the population lives in cities in Asia the average is reaching 60% whereas, in Africa the phenomenon has started later and has now reached 35%. The trend is accelerating and by the year 2010, i.e. in 5 years from now, 50% of the world population will be living in the cities. In this context, besides the increased demand for food, there has been a rapid increase of poverty, unemployment, hunger and malnutrition in the urban and peri-urban environment around the world that are of great concern to central and municipal authorities. The question whether or not urban agriculture could help to address these problems and how was referred to FAO as the Food and Agriculture Organization of the United Nations. As a result, UPA was put on the agenda and debated at the Committee of Agriculture meeting (COAG) in January 1999. The conclusion was "yes", UPA can contribute to feeding the urban population, generate income and employment and thereby reduce poverty. It became clear that in reality many countries had already UPA related activities going on in areas of horticulture, dairy production, small livestock, aquaculture and forestry. However, it was also realised that very few countries had actually investigated the real opportunities and constraints linked to the urban agriculture sector. Usually UPA is going on as an informal activity without due attention for constraints linked to land occupation, water availability and quality, product quality and safety and environmental impact. The COAG session recommended that FAO provides support to its member nations in order to "enable" UPA and give it the recognition it deserves as a component the National Food Security Programmes and as a contribution to meeting the millenium goal to eradicate poverty. Since then, FAO has intervened on the specific request from individual countries through projects of short and medium duration. Lessons learned from these projects indicate that the success and sustainability of UPA interventions require national commitment and municipal ownership with the participation of a series of stakeholders both from the public and private sector. This commitment should ultimately result in the integration of UPA in the city development planning process and into the national agricultural development plans (Baudoin, 2005).

It has been realised that successful implementation of urban and peri-urban agriculture is based on an integrated approach aiming at securing the following three sectors:

(i) securing the basic resources in order to ensure the availability and use of adequate land and water of good quality in sufficient quantity

(ii) securing the production of high quality produce by applying technologies and practices which are environmentally friendly in the urban context, and

(iii) securing the institutional context, monitoring and management of the programme.

Birth and Growth of UPA: During 1995, the FAO realized the growing shift of population from rural to urban area and recognized the future need of augmenting food security for more than 50% urban population globally. Initially, the FAO took the lead with World Bank and UNDP and the birth of peri-urban agriculture took place in support of food security of the growing urban population (Nath, 2013 & 2015). During January 1996, FAO/COAG approved urban and peri-urban agriculture (UPA) as one of its agricultural programs (Baudoin, 2005). Some advances showing overall lead in horticulture are as follows (Nath, 2013 & 2015; Figure 4):

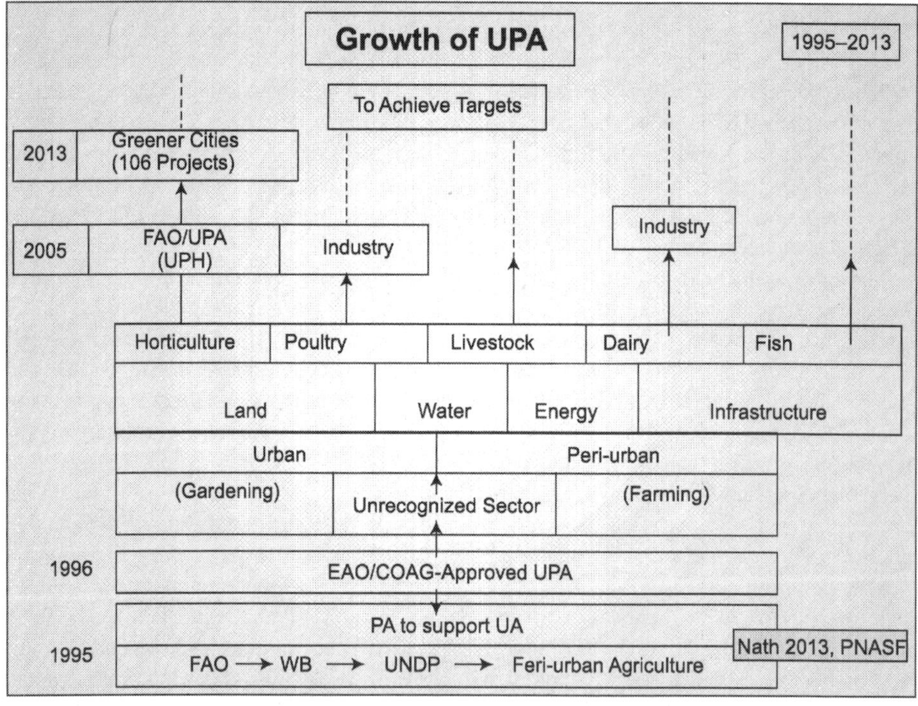

Fig. 4. *Growth of UPA* (Nath, 2013 & 2015)

The Regional Workshop on Strengthening Urban & Peri-Urban Agriculture Towards Resilient Food Systems in Asia (UPAFSA-2013)

Urban and Peri-urban Agriculture (FAO & PNASF, 2013)

During 1995, the FAO realized the growing shift of population from rural to urban area and recognized the future need of augmenting food security for more than 50% urban population globally. Initially, the FAO took the lead with World Bank and UNDP and the birth of Peri-urban Agriculture took place in support of food security of the growing urban population. During January 1996, FAO/COAG approved Urban and Peri-urban Agriculture (UPA) as one of its agricultural program (Nath, 2013).

Context

- Since long urban population is fed by rural production only.
- With the migration of rural young masses urban population is increasing gradually; urban population already increased and expected as follows -1995-50%(-); 2012-50 %(+); 2030-60%; 2050-70% (IFPRI, 1999; UNDP, 2011).
- Major food production still rests with rural area.
- In urban area, the demand for food, job, space, energy, water and other facilities increasing.
- Food security for urban population needs support, also from its own population.
- The migrated labour with farm skills useful to urban and peri-urban agriculture.
- Hence, UPA is becoming a **necessity** rather than **demand** only.
- During 1995-96, while launching the global special programme for Food Security, FAO highlighted the growing trend of shift in population from rural to urban areas.
- **1995-1996:** Realising the need of feeding the teeming rural population, the FAO initiated discussion with UNDP and World Bank which supported the idea to build up Peri-Agriculture to fulfil food demand of growing urban population.
- **1996:** FAO/COAG: Urban & Peri- Urban Agriculture- Definition and scope framed
- **2005:** FAO launched world-wide UPA programme.
  - **Urban and Peri-urban Horticulture (UPH):** Growing Greener Cities
  - **2005-2012:** Peri-urban/Green Zones covering 2000 to 6000 acres around big cities in Asia and Latin America.
- Opportunity for UPA (FAO, 2005)
  - Vicinity to market--reduced transport, packaging, storage expenses/reduced post-harvest losses.
  - Easy access to service-electricity, water, inputs, employment.
  - Develop green zones-improve environment, utilize waste *e.g.* Greener Cities.
- Integrated approach *required*
  - Basic resources- land, water and energy.
  - Production/high quality/technology application.
  - Monitoring and management of the programme.

Conclusion

- The Urban and Peri-Urban Agriculture (UPA) supports food and nutrition security, health improvement, income generation and improved environment.
- It is time to better define national policy and strategy options with relevant resources and technical support mechanisms for the sustainable management of urban and peri-urban agricultural systems, considering crop and fodder production and animal husbandry in conjunction with water use efficiency and food safety criteria.

- Equally important is the need to implement innovative approaches for decentralized marketing, small-scale processing and nutrition education in order to address the integrated production, supply and consumption (Nath, 2013).

5. INTERNATIONAL CONFERENCE ON HORTICULTURE, BENGALURU, INDIA (ICH-2009)

Theme: Horticulture for Livelihood Security and Economic Growth

There is an observable shift in the food habits of people in the last two decades that is reflected as changing cropping pattern, away from cereal base to other commercial crops. During the same period, economic growth in horticulture has far exceeded than that in most agricultural commodities. Annual growth rates for vegetable supplies have surpassed cereals by 200 percent to 800 percent since the 70s, while this has been accelerated in the 1990s. This trend is attributed largely to the changing consumer preferences, powered by increased income and general standard of living led awareness towards diversified and balanced nutrition (Nath & Gaddagimath, 2010).

Growth potential for the production of horticultural commodities is strong in the developing countries and emerging economies. It is reported that China recorded the highest increase in area under horticultural crops with 20 percent of the arable land under this set of crops. The total value of horticultural crops traded at present is more than double that of cereal crops, reaching nearly 21% of total value exported from developing countries.

Experience suggests that development of horticultural crop production promises these countries the ability not only to meet their domestic food needs, but offer diversified income sources to their rural economies. Further, horticultural crop based products are proven supplements of human health, besides being sources of farmer household economic and social advancement. Thus, horticultural commodities on the whole seem to be ideally suited to accomplish the objective of providing 'sustainable livelihood security'.

There is an observable shift in the food habits of people in the last two decades that is reflected as changing cropping pattern, away from cereal base to other commercial crops. During the same period, economic growth in horticulture has far exceeded than that in most agricultural commodities. Annual growth rates for vegetable supplies have surpassed cereals by 200 percent to 800 percent since the 70s, while this has been accelerated in the 1990s. This trend is attributed largely to the changing consumer preferences, powered by increased income and general standard of living led awareness towards diversified and balanced nutrition.

The International Conference on Horticulture (ICH-2009) with the theme "*Horticulture for Livelihood Security and Economic Growth*" held in Bengaluru during November 9-12, 2009, has highlighted its importance and salient features projecting the future steps to be undertaken by all stakeholders. The conference recognized horticulture as the second line of defence with cereals and legumes serving on the front line of the food domain. It was stated that horticulture (a) supports significantly to food security, (b) is the major contributor to nutrition security (c) promotes good health and (d) is source of employment generation- all leading to livelihood security and economic growth. Some of the salient recommendations were (Anon., 2010; Nath & Gaddagimath, 2010):

- Awareness and education programme on food, nutrition and health security be augmented.
- The university syllabus be upgraded; the quality of research should be farmer's/ demand driven and priority and quality of research improved.
- More investment and planning in Transfer of Technology (TOT) were recommended.
- Appropriate policy and programmes formulated to augment food, nutrition and health security and employment and income generation.
- There is need to establish strong linkages among countries and even among states of large countries by networking of Institutes and Organization for exchange of information, seed and plant materials and expertise in horticulture. Awareness and Education programme on International Cooperation among Developing Countries (TCDC) be augmented Nath & Gaddagimath, 2010).

Genetic Resources: Due to large-scale biotic and abiotic activities, to meet the ever increasing demands of human population, the long conserved horticultural biodiversity is getting affected. As a result, the number of vanishing and endangered species is increasing. This situation is further aggravated due to modern monocrop dominant practices. Hence, there is an urgent need to conserve, augment and sustain the use of this valuable horticultural bio-diversity, which is the common heritage of mankind.

Crop Improvement

- To augment livelihood security of urban and peri-urban poor and small and marginal farmers, urban and peri-urban horticulture may be encouraged by the government with participation from various extension machinery and research institutes and the agricultural/ horticultural universities may screen the available varieties / hybrids and technologies for their suitability to urban and periurban situations and, if need be, work may be taken up to evolve or breed varieties / hydrids and develop technologies suitable for urban and peri-urban situations particulary for pot culture, terrace gardening, homested gardens, low cost hydroponics, polyhouse / net home cultivation.
- Climate change and vagaries of climate like droughts and floods, every now and then, cause huge crop losses, leading to food scarcity and economic losses. Since a good number of drought hardy short duration vegetable and tuber crop varieties/ cultivars and deep rooting, wind blocking perennial fruit crop varieties / cultivars are available, contingency plan modules involving vegetables, root / tubers and fruit crops may be developed and standardized for various agroclimatic and crop growing situations to mitigate such situations and also to restrict the soil erosion.
- The WHO has already identified few unexplored areas of research based on the already available scientific evidences of preventive role of fruits, vegetables, spices, tuber crops and medicinal plants, based on nutritional values, phytochemicals, nutraceuticals and antioxidants. Based on the available information and emerging evidence, work may be taken up and strengthened on; identification of wild species, land races and other cultivars, having higher and specific active principles; utilization of such genetic materials in breeding programmes to evolve varieties / hybrids with desired quality traits and health value; and gene marking / gene tagging for specific health value and further work on gene mining and marker aided breeding to evolve varieties / hybrids of higher nutritional values and health care specific traits.

Production Technology

- Attention is needed to develop varieties suitable for processing, export, *etc,* and technologies to manage problems, such as, spongy tissue, alternate bearing and malformation in mango, guava wilt, citrus decline, banana wilt and F_1 hybrids in vegetables. At the same time, cost-effective production technology for fruits, vegetables and ornamental crops is needed, both for optimum as well as suboptimum conditions.
- The currently available production technologies are broad based and are not region and area specific. There is a need to refine these production technologies at microlevels and also specific for small and marginal land holdings, with farmer-participatory approach.
- Climate change is a reality now. Global warming has impacted food security in more than one way. Corrective measures need to be taken to overcome the situation. These include educating the public including policy makers, about the current methods of production and global patterns of food distribution. Aside greenhouse crops, where knowledge about plant responses to environmental factors are generally well defined, our knowledge about the responses of field crops in general and perennial crops in particular is very poorly developed. Improved understanding of the consequences of inadequate winter chilling and the means to counter that, are critical to continued growth and survival of many temperate perennial crops and yet our current knowledge is quite inadequate to deal with the changes that are occurring. The same can be said for knowledge about climatic factors and maturity of different fruit crops. Studies need to be organised and accelerated.

Protection Technology

- Viruses have emerged as serious problems in vegetable and fruit production in various parts of the world. There is an urgent need for their monitoring, using latest molecular diagnostic systems. Further investigations / research is needed with regard to vector specificity and how plant viruses overcome various transmission barriers in their vectors, role played in vector specificity by virus proteins, virus receptors, insect symbionts, barriers to transovarial or vertical transmission of viruses in their vectors, using the new molecular technology to understand and combat plant viruses and their epidemics in horticultural crops.
- Protocols and technologies may be developed for production of botanicals / plant products, biopesticides and biological control agents at the farm gate levels and farmers need to be trained in production of these at the farm gate/village/ community level or at the farm field schools and farm health clinics. This would facilitate generation of more employment to rural educated youth and prevent rural unemployed youth from migration to urban areas in search of employment and livelihood security.

Post harvest Technology

- The post harvest losses to the tune of 20-25% will occur at the harvest stages and at the farm gate levels. It is essential to develop simple, low cost technologies for storage of farm produce at the farm gate levels like zero energy chambers and technologies for removal of field heat from the produce. The technologies already available need to be popularized among the farming community.

- Nutritional security and food safety taking the centre stage in the food consumption and consumer driven food market, labeling of nutrients and their values will have added advantage and standards may be developed for labeling of nutrients and atmost care should be taken for minimizing pesticide residues and chemical preservatives. While Hazard Analysis and Critical Control Points (HACCP) standards need to be strictly imposed for all categories of processing units, investigations on harmful, new and unidentified microorganisms and moulds may be taken up and technologies for destroying / removal of such harmful microorganisms and moulds without harming the food and nutritional value of the value added products may be developed.
- Many of the phytochemicals, nutraceuticals and antioxidants available in fruits, vegetables, tubers, spices, are known to fight many non-communicable diseases. It is most essential to testify and quantify the active principles available in various horticultural crops to understand their role in fighting these diseases and accordingly make dietary recommendations for different groups of populations for health security. This work may be taken up in active collaboration with World Health Organisation -Food and Agriculture Organisation (WHO-FAO) by national horticultural research institutes and nutrition institutes.

Organic Production

- Global awareness about organic products, their health and nutritional benefits and environmental benefits has already been created and in many countries there have been policy interventions to promote organic production of various crops. A wealth of information is available on organic farming of various crops around the world. The need of the hour is to document all such information and various organic production practices and create an 'Organic Horticultural Crops Production Information Bank'(OHCPIB) in network mode, accessible to various organizations and farmers.
- The efficicacy of different organic plant protection practices needs to be evaluated under different cropping situations. Further, these practices need to be validated by conducting on-farm trials in different agro-climatic conditions and seasons.
- Organic farming system needs to be selected based on agro-ecological suitability of crops, season and cultivar, for sustainability and economic viability. Insect pest and diseases forecasting modules have to be integrated with production practices to realise highest returns for the produce.

Underutilized Crops/Crop Diversification

- Under utilized and under exploited plants and some land races of horticultural crops offer best source of livelihood security for tribal and ethnic groups and also would meet the nutritional security of peasants and less advantaged groups, including tribal and rural female populations. Identification of such crops and promotion of cultivation of the same in the native form should be encouraged.
- World over tribal belts have large number of underutilized corps and the tribal communities possess ITK's for value added products and byproducts. Conservation of these crops in the native form and further promotion and scientific validations of the Indigenous Traditional Knowledge (ITK) for the value added products and byproducts is most essential since these in the fresh form and in the value added products form have a niche market and fetch better and higher prices.

- Increase and diversify horticultural crop production, through scheduling and phasing of the production and harvesting in order to achieve cost effectiveness, year –round availability, safety and quality of produce keeping in view the economic and nutritional needs of the targeted population, by developing a production calander of crops by development organizations and made available to farmers through extension functionaries.

Policy Issues

Resources/Funding

- Policy makers and donors have to turn their attention towards enhanced funding for the vegetable and horticultural sector, allowing growers to compete with their products in the global market, increasingly determined by market quality standards and phytosanitary concerns and regulations. Only then will the silent vegetable and horticultural revolution currently underway, benefit a significant portion of the world's poor nations, growers and landless laborers and enable overcome poverty and food insecurity.
- Many current vegetable breeding efforts remain under-funded and disorganized. There is immense need for a more focused, coordinated approach to efficiently utilize funding, share expertise and continue progress in horticultural technologies and programs. Vegetable production can help the poor escape poverty and malnutrition in the 21st century, but only if enough investments are made to improve and sustain the breeding and productivity of vegetable crops.
- Dryland horticulture needs to be promoted on a large scale to improve the livelihood security options of millions of small and marginal farmers. Resource poor, degraded drylands are being diverted to either agroforestry or biofuel crops by poor, small and marginal farmers because of non availability and unaffordability of farmers and also because of attractive terms offered by private / corporate sectors. This needs to be discouraged and the issue is properly addressed by policy interventions, otherwise this would grow as a major barrier in future food availability and supply.

Extension/Training

- Effective partnership can be established between private and public sectors. The public sector can take the lead in generating information on basic research, training manpower and the enforcement of regulatory policies, where as private sector can enhance the flow of technology for the benefit of growers and consumers.
- Most of the horticultural plantations are senile and are under rain–fed condition with poor water management. Prevalence of diseases, pose severe threat to horticulture/ plantation industry. Therefore, such senile orchards have to be rejuvenated and managed. For this purpose there is a need to conduct training programmes for extension personnel and farmers.
- The transfer of technology system in India, after the Training and Visit programme of the World Bank Aid, has been substantially weakened for both funds and manpower. Similar situation has been reported in most of the developing nations. There needs much more investment in TOT, for without extensive and intensive extension work, the research findings may, by and large, remain within the four walls of the research institutions. Because of this weakness, only 32-40% of the improved technologies

have reached the farmers and therefore, the productivity and production have been badly affected. For this purpose, the following measures need to be seriously considered and followed up:

- For aggressive and dynamic extension system for big countries like India and other countries, an Autonomous Council of Technology Transfer in Agriculture (ACOTTA) is most essential within the framework of the Union Ministry of Agriculture. A similar organization needs to be followed at the district level.
- The present SAMETI (State Agricultural Management & Extension Training Institute), established in India and at the district level Agriculture Technology Management Agency (ATMA) and at the village level Farm Schools being promoted by the National and State Extension systems at present are very weak. They need more funding, manpower and administrative responsibilities away from the district administration. In fact, instead of Farm Schools, it could be a better proposition to establish and promote MODEL FARMS and model farmers at strategic locations in villages as a hub of all extension activities and vocational training. This will in fact, could be the Demand Driven Extension and Marketing system, more often talked about in all countries.
- Training both in-service as well as vocational and entrepreneurial persons, is very basic to transfer of technology. We have talked in the past on Teaching, Research and Extension, almost like a slogan. In this chain, now Training should be added as a basic function: we should talk about Teaching, Research, Extension and Training. All Institutions should have similar arrangements and investment for training large number of farmers, farm women and farm youth, again a very big souce for creating jobs for the unemployed.
- Information and Communication Technology (ICT), e-Extension, e-Marketing have started taking shape which needs further strengthening and promotion in India and also in the developing countries. Horticulture being a very potential discipline for multiplying productivity and production as well as for the export, the ICT and e-Extension mechanism including marketing in horticulture can bring about a revolution.

Crop Planning/Area Planning

- Short/medium/long term plan may be developed along with a Master plan of 25/50 years ahead for horticulture development in each state/nation and may be executed in a mission mode on a time scale.
- Increase and diversify horticultural crop production, through scheduling and phasing of the production and harvesting in order to achieve cost effectiveness, year round availability, safety and quality of produce keeping in view the economic and nutritional needs of the targeted population, by developing a production calander of crops by development organizations and make available to farmers through extension functionaries.

Transport/Marketing

- A legislative measure should be enacted for ‘Blue Transport’ for safe transport of horticulture produce / products on the similar lines of ‘Yellow School Vans’.

- Presently, retail chains on the marketing of fresh fruits and vegetables and other horticultural products in the developing countries is emerging faster and takes about 2% of the market share. The main suppliers for this are from large and medium farmers. There is a need to bring majority of small farmers under this network from rural areas through collection centres for high value fruits and vebetables. Supply Chain Management is presently under the clutches of private corporate houses with no farmers representation in their board. There is a need to have two representatives one each from large farm holdings and small and marginal farm holdings to understand the problems of the producers.

Crop Insurance

- Hybrid vegetable seed production is seasonal and risky. Crop insurance cover should be extended to all seed companies and contract seed growers engaged in hybrid seed production and plant propagation.

Infrastructure Development

- Since processing of horticulture produce is mostly concentrated in the urban areas, efforts may be made to extend necessary facility to the rural and periurban areas possibly through PPP Model.
- For vegetatively propagated crops, disease free, certified clonal nurseries should be established to provide basic stocks for multiplication to nurseries and for exchange of plant material.
- In the developing world the number of small farmers with small acreage exceeds that of the large and commercial farmers with large acreage. There is a lack of availability of production technologies, marketing and credit facilities and processing and storage facilities and other infrastructures for the small farmers who form about 75- 85 per cent of the farmers community, whose livelihood is at stake. This is also applicable to India.
- Quality parameters are very important for improving seed health and disease free planting material. Governments of Asian countries should encourage public and private sector companies to set up International standard quality control, molecular biology and seed testing laboratories (ISTA approved). Similarly, there is an urgent need to modernize nurseries with effective quality control system and accreditation for ensuring production and sale of genuine, true to type, disease free, planting material of horticultural crops using micro-propagation, high-tech seedling production, use of electrophoresis and PCR based technologies for assessing genetic purity of seed and planting material.

Nutritional Security

- Among the developing world occupied by the major population of the world, food security, nutrition security, health security and slow employment generation, have challenged the livelihood security. There is a need to accelerate and gear awareness and education programme at all levels including farmers and policy makers and review the existing policy to augment food and nutrition security. Out of one billion people suffering from hunger and malnutrition globally, about 230 million are homed in India. This demands serious consideration in the planning of India.

Technical Cooperation

- Awareness and Education programme on international technical cooperation among the developing countries is needed. In this direction, appropriate steps by Government and Non-Government organizations to promote international cooperation need to be taken up.
- Phytosanitory certification standards for various horticultural crops for imports / exports need to be re-looked into by each country with respect to emergence of new insect pest-disease (resurgent insect pest-diseases), hither to were not important and serious.

Transfer of Technology

- Research-Extension-Farmer Linkage may be established. The linkage achieved in Maharastra in India through ICT, PPP, Integrated cropping system, ATIC, formation of Self Help Groups (SHGs) can be taken as examples and may be replicated in other places as well to reduce the gap between the yields in on- farm and off- farm trials. Sustainable technologies in one agro-climatic condition may not be sustainable or successful or potential in another agro-climatic condition. Hence, it is essential to replicate such technologies with participatory approach and refine them to suit different agro- climatic conditions.
- Farmers should be trained as Integrated Pest Management (IPM) experts and be involved in IPM implementation and should be educated for using appropriate protective gear while applying pesticides.

6. FAO'S SPECIAL PROGRAMME FOR FOOD SECURITY (2004)

Towards achieving the goal, the FAO sponsored Special Programme for Food Security (SPFS) initiated during 1994 needs mention when it was launched in 14 Low Income Food Deficit Countries in different regions of the world which was extended to 71 countries by 2002 (Kobakiwal, 2004). Among the commodities utilized fruits, vegetables and tuber crops figured prominently in increasing the income of the beneficiaries.

There has been significant increase in production of fruits and vegetables worldwide. The world promotion of fruits and vegetables was about 1,158 metric tons in 2000, which had further increased during 1990-2009, export of fresh and procured vegetables had arose from 19.7 million ton to 32 million tons (Fresco and Baudoin, 2004).

In recent report of FAO (2010), the food prices will remain above the pre-crisis level of early 2008 in developing countries, negatively affecting access to food of vulnerable populations (Nath, Gaddagimath, 2010).

Nutrition: Fruits and vegetables play an important role in providing vitamins, minerals, micro-nutrients, as well as proteins to obtain a balanced diet. Starchy vegetables and more specifically the root and tuber crops play an important role as a source of calories to meet the energy requirements.

When farmers have better yields by growing cereals, they are no longer poor and hungry. When they shift to horticulture, by growing fruits and vegetables, they can get more income and this will reflect on a better diet, education and livelihood of the family. In Kenya, the income of $ 1 a day was raised to $3 and $4 a day for those farmers who have been successful in diversifying their agriculture partly or entirely towards horticulture (Fresco and Baudoin, 2004; Nath and Gaddagimath, 2010).

7. SECOND INTERNATIONAL CONFERENCE ON NUTRITION (ICN2) (FAO & WHO, 2014; NATH, 2017)

Recommended actions for sustainable food systems promoting healthy diets

- **Recommendation 8:** Review national policies and investments and integrate nutrition objectives into food and agriculture policy, programme design and implementation, to enhance nutrition sensitive agriculture, ensure food security and enable healthy diets.
- **Recommendation 9:** Strengthen local food production and processing, especially by smallholder (Smallholder farmers include agriculture and food workers, artisanal fisher folk, pastoralists, indigenous peoples and the landless (Committee on World Food Security, Global Strategic Framework for Food Security and Nutrition, 2013). and family farmers, giving special attention to women's empowerment, while recognizing that efficient and effective trade is key to achieving nutrition objectives.
- **Recommendation 10:** Promote the diversification of crops including underutilized traditional crops, more production of fruits and vegetables and appropriate production of animal-source products as needed, applying sustainable food production and natural resource management practices.
- **Recommendation 11:** Improve storage, preservation, transport and distribution technologies and infrastructure to reduce seasonal food insecurity, food and nutrient loss and waste.
- **Recommendation 12:** Establish and strengthen institutions, policies, programmes and services to enhance the resilience of the food supply in crisis-prone areas, including areas affected by climate change.
- **Recommendation 13:** Develop, adopt and adapt, where appropriate, international guidelines on healthy diets.
- **Recommendation 14:** Encourage gradual reduction of saturated fat, sugars and salt/sodium and trans-fat from foods and beverages to prevent excessive intake by consumers and improve nutrient content of foods, as needed.
- **Recommendation 15:** Explore regulatory and voluntary instruments –, such as, marketing, publicity and labelling policies, economic incentives or disincentives in accordance with Codex Alimentarius and World Trade Organization rules – to promote healthy diets.
- **Recommendation 16:** Establish food or nutrient-based standards to make healthy diets and safe drinking water accessible in public facilities, such as, hospitals, childcare facilities, workplaces, universities, schools, food and catering services, government offices and prisons and encourage the establishment of facilities for breastfeeding.

8. SEAVEG-2014: SOUTHEAST ASIA VEGETABLES- REGIONAL SYMPOSIUM ON SUSTAINING SMALL-SCALE VEGETABLE PRODUCTION AND MARKETING SYSTEMS FOR FOOD AND NUTRITION SECURITY

Theme: Families, Farms, Food

Asia is the main hub of vegetable production, contributing around 70% of the global production. However, its productivity is nearly half that of some European countries. A gradual increase in vegetable production (25%) was observed during the past 10 years, but that increase was mainly due to the increased vegetable production in China alone, which accounts for nearly half of the world's vegetable production. Thus there is an ample scope to promote vegetable production in other Asian countries as well. Asia is dominated by smallholders with meagre incomes who cultivate more than 80% of farmland. The vegetable market is fragmented and unorganized. Strategies pertaining to postharvest losses and food wastage, food safety and quality are lacking. The population in the region is undernourished and malnourished. About two-thirds of the global population suffering from nutrient deficiencies live in Asia. Over 70% of the world's malnourished children live in Asia. Emerging environmental challenges, such as, water scarcity, changing weather patterns, pests and diseases, unsustainable use of natural resources, rapid urbanization and pressure on arable land seriously draw the attention of both policy makers and scientists (Hughes *et al.,* 2015).

To discuss the challenges and issues, SEAVEG2014: "Families, Farm, Food, a regional symposium on sustainable small-scale vegetable production and marketing systems for food and nutrition security", was organized by the Department of Agriculture (DOA) Thailand, in collaboration with AVRDC – The World Vegetable Center, Taiwan, the ASEAN-AVRDC Regional Network (AARNET), the Vegetable Science International Network (VEGINET), Kasetsart University (KU), the Food and Agriculture Organization of the United Nations (FAO) and the Horticultural Science Society of Thailand (HSST). More than 200 participants from 25 countries attended the symposium. After discussing the key issues, the following recommendations and policy issues were made (Hughes *et al.,* 2015).

Recommendations

I. Research and Development Issues

1. Vegetable research should be focused on consumers' preferences, nutritive value nutraceutical value, marketing quality and productivity. Research on spices, tuber crops and medicinal plants, rich in phytochemicals, nutraceuticals and antioxidants as identified by FAO/WHO needs further intensification.
2. There is a need to upgrade and scale-up technologies suitable for urban and periurban vegetable production close to the centers of consumption through home gardens, school gardens, rooftop gardens organic gardens, close circuit vertical farming, hydroponics, aeroponics, sprout and microgreen production to mitigate seasonality of vegetables, dependency on off-farm resources and distant marketing. This will facilitate food and nutrition security at household and community levels.
3. Develop and promote environment sensitive management technologies through integrated pest management, integrated water and nutrient management, rain

shelters, low-till farming, use of grafting and low tunnel protected cultivation, use of short duration, drought-tolerant vegetables including tubers and root crops to mitigate biotic and abiotic stresses, thus leading to sustainable vegetable production.

4. Protocols and technologies should be developed for the production of botanicals, plant products, biopesticides and biological agents at the farm level. Rural youths should be trained in the above-mentioned skills, helping them to run farm health clinics for self-employment.
5. Indigenous and exotic genetic resources of vegetables should be collected and conserved in national gene banks and utilized in various breeding programs. Advanced breeding lines resistant/tolerant to biotic and abiotic stresses available at AVRDC should be used in local breeding programs. Collaborative programs with AVRDC in key areas should be encouraged. Skill of vegetable breeders should be improved through specific training programs for effective output.
6. There is a need to modernize nurseries with effective quality control systems, accreditation of true to type, disease-free planting material using micropropagation and improved seedling production methods, including grafted seedlings, to counter losses due to soil-borne diseases. Use of molecular markers should be encouraged for assessing genetic purity of seed and planting material. The private seed industry should be established on a firm footing for the flow of technologies among farmers.
7. Promote nutrition sensitive horticulture through dietary diversification of vegetables, especially traditional vegetables, highlighting their nutritive values, incorporating them in demand-driven breeding programs and channeling the produce to the market mainstream.
8. Consumption of nutritious vegetables alone is not sufficient to address malnutrition. There is a need to focus on multisectorial coordination among various preventive measures like availability of clean water, proper hygiene, good sanitation, health education, early child development and family planning through women empowerment programs and by linking these activities with existing government programs pertaining to health and education at the national level.
9. Pesticide residue limits and safe waiting periods for different chemicals and vegetables should be developed and documented to overcome the non-tariff trade barriers imposed by different countries for export/import of vegetable produce.
10. Market and supply chain infrastructure should be improved to facilitate regional trade among Asian nations.
11. Awareness programmes to prevent food wastage and available low cost postharvest technologies should be popularized through extension services
12. Multi-utility and multi-crop processing machinery/facility should be developed for year-round functioning of the processing industry.
13. Suitability of vegetable crops for organic farming should be identified based upon the validation of various organic farming practices as well as cost-benefit ratios.
14. Asian nations should develop their own quality standards and harmonize them with WTO/UNICEF standards to meet national as well as export/import requirements.
15. WTO/AFTA should review limitations/barriers/laws for transaction of trades among member countries, especially ASEAN (Hughes *et al.,* 2015).

II. Policy Issues

1. Special economic zones for vegetable production should be developed based upon agro climatic, biophysical and socioeconomic parameters.
2. Develop awareness programs pertaining to quality standards and phytosanitary regulations for export market, keeping in view AOA (Agreement on Agriculture) and WTO (World Trade Organization) regulations.
3. In public-private partnerships, the public sector should generate information on basic research, training and enforcement of regulatory policies; the private sector should enhance the flow of technology by putting an efficient distribution system in place.
4. An efficient supply chain management system needs to be established by linking smallholders and producers to the main market. Smallholders and exporters should receive information regularly pertaining to demand and supply of vegetables, consumers' quality preferences and sales and return figures in domestic, national and international markets, enabling them to place their produce effectively in the vibrant market.
5. Facilities pertaining to processing industries should be extended to rural and periurban areas.
6. Investment should be enhanced in upgrading and scaling-up of production technologies suitable for small-scale farmers, in human skill development among rural youths, in urban and peri-urban farming, in dry land farming, in processing and storage facilities, in marketing and credit facilities and in women empowerment programs in the areas of health, hygiene and nutrition.
7. Existing policies pertaining to food and nutrition security should be reviewed, ensuring their sustainability at ecological, economic and social levels.
8. Research-Extension-Farmer linkage should be reviewed to enhance efficiency in transfer of technologies refined for different agroclimatic conditions through participatory approaches involving the public and private sectors.
9. Strong linkages should be established within and among countries through networking of institutes and organizations for exchange of information and technologies, seed and planting material as well as expertise for mutual benefit and development of horticulture. The FAO (Food and Agriculture Organization of the United Nations) under the umbrella of TCDC (Technical Cooperation among Developing Countries) may act as a facilitator for the formation and strengthening of such networks as VEGINET, which was established during the international conference on vegetables (ICV-2002) and other networks, such as, the AVRDC/ASEAN Regional Research and Development Network Steering Committee Meeting (AARNET).
10. Nutrition awareness and training programs for consumers should be intensified (Hughes *et al.,* 2015).

9. SUSTAINABLE DEVELOPMENT GOALS AND TARGETS (UN, 2015b; Nath, 2017)

The Sustainable Development Goals and targets are integrated and indivisible, global in nature and universally applicable, taking into account different national realities, capacities and levels of development and respecting national policies and priorities. Targets are defined as aspirational and global, with each Government setting its own national targets guided by the global level of ambition but taking into account national circumstances. Each Government will also decide how these aspirational and global targets should be incorporated into national planning processes, policies and strategies. It is important to recognize the link between sustainable development and other relevant ongoing processes in the economic, social and environmental fields.

Goal 2. End hunger, achieve food security and improved nutrition and promote sustainable agriculture (UN, 2015)

2.1. By 2030, end hunger and ensure access by all people, in particular the poor and people in vulnerable situations, including infants, to safe, nutritious and sufficient food all year round.

2.2. By 2030, end all forms of malnutrition, including achieving, by 2025, the internationally agreed targets on stunting and wasting in children under 5 years of age and address the nutritional needs of adolescent girls, pregnant and lactating women and older persons.

2.3. By 2030, double the agricultural productivity and incomes of small-scale food producers, in particular women, indigenous peoples, family farmers, pastoralists and fishers, including through secure and equal access to land, other productive resources and inputs, knowledge, financial services, markets and opportunities for value addition and non-farm employment.

2.4. By 2030, ensure sustainable food production systems and implement resilient agricultural practices that increase productivity and production, that help maintain ecosystems, that strengthen capacity for adaptation to climate change, extreme weather, drought, flooding and other disasters and that progressively improve land and soil quality.

2.5. By 2020, maintain the genetic diversity of seeds, cultivated plants and farmed and domesticated animals and their related wild species, including through soundly managed and diversified seed and plant banks at the national, regional and international levels and promote access to and fair and equitable sharing of benefits arising from the utilization of genetic resources and associated traditional knowledge, as internationally agreed.

2.6. Increase investment, including through enhanced international cooperation, in rural infrastructure, agricultural research and extension services, technology development and plant and livestock gene banks in order to enhance agricultural productive capacity in developing countries, in particular least developed countries.

2.7. Correct and prevent trade restrictions and distortions in world agricultural markets, including through the parallel elimination of all forms of agricultural export subsidies and all export measures with equivalent effect, in accordance with the mandate of the Doha Development Round.

2.8. Adopt measures to ensure the proper functioning of food commodity markets and their derivatives and facilitate timely access to market information, including on food reserves, in order to help limit extreme food price volatility.

10. CONCLUSION

The above deliberations and recommendations by the group of participating developing nations have amply demonstrated the role of horticulture in human livelihood and life. The importance of horticulture in humans health is pertinent. The areas of deficiencies have been identified and recommendations made particularly on research and development towards mandatory sustainable horticulture development.

REFERENCES

Anon. (2010). Proceedings and Recommendations of the International Conference on Horticulture (ICH-2009), Bengaluru; ICV-2002 Secretariat, P.N. Agriculture Science Foundation, Bengaluru, India.

Baudoin, W.O. (2005). Urban and Peri-Urban Agriculture: The Concept, Theme and Objectives, Proc. FAO/PNASF Inter. *Workshop on UPA for Asian countries*, 29th June to 2nd July, 2005, Bengaluru, pp. 17-20.

Baudoin, Wilfried O. (2005). Theme Presentation: Scope and Expected Outputs of the Workshop, Proceedings of the International Workshop on Urban and Peri-urban Agriculture for Asian Countries (FNOP/INT/001/NOR).Dr. Prem Nath Agricultural Science Foundation (PNASF), Bengaluru. FAO and PNASF

FAO & WHO. (2004). Fruits and Vegetables for Health Report of a Joint FAO/WHO Workshop, 1–3 September 2004, Kobe, Japan

FAO &WHO. (2014). Framework for Action, Second International Conference on Nutrition, 19-21 November 2014. P.8.

Fresco, Louisce O. and Baudoin, Wilfried O. (2004). Food and Nutrition Security towards Human Security, Food Security and Vegetables-A Global Perspective, P.N. Agricultural Science Foundation, Bengaluru, India, p.7-42.

Hughes J. d'A, Kasemsap, P., Dasgupta, S., Dutta, O.P., Ketsa, S., Chaikiattiyos, S., Linwattana, G., Kosiyachinda, S., Chantrasmi, V. eds. (2015). Proceedings of the Regional Symposium on Sustaining Small-Scale Vegetable Production and Marketing Systems for Food and Nutrition Security (SEAVEG2014), 25-27 February 2014, Bangkok, Thailand. AVRDC – The World Vegetable Center, Publication No. 15-785. AVRDC – The World Vegetable Center, Taiwan.

IFPRI. (1999). World Food Prospects: Critical issues for the Early Twenty First Century; 2020 Vision-Food Policy Report, IFPRI, Washington, USA.

Kobakiwal, A.Q. (2002). Special Programme for Food Security, ICV-2002 Souvenir, International Conference on Vegetables, November 11-14, 2002, Bengaluru, India "Vegetables for Sustainable Food and Nutritional Security in The New Millennium", Dr. Prem Nath Agricultural Science Foundation, Bengaluru.

Kobakiwal, A.Q. (2004). Special Programme for Food Security (SPFS), Food Security and Vegetables-A Global Perspective, P.N. Agricultural Science Foundation, Bengaluru, p. 85-100.

Nath, Prem and Dutta, O.P. (2002). Food and Nutrition Security and Vegetables, ICV-2002 Souvenir, International Conference on Vegetables, November 11-14, 2002,

Bengaluru, India "Vegetables for Sustainable Food and Nutritional Security in The New Millennium", Dr. Prem Nath Agricultural Science Foundation, Bengaluru.

Nath, Prem and Gaddagimath, P.B. (Eds.) (2010). Role of Horticulture in Livelihood Security, Horticulture and Livelihood Security, PNASF & SP, India. p 1-11.

Nath, Prem, Gaddagimath, P.B. and Dutta, O.P. (Eds.) (2005). Introduction, Food Security and Vegetables-A Global Perspective, PNASF, Bengaluru, p 1-6.

Nath, Prem. (2013). Concept and Implementation of Food Security and its Relevance to Urban and Peri-Urban Agriculture. Vil-I. Proceedings of the Regional Workshop on Strengthening Urban & Peri-urban Agriculture towards Resilient Food Systems in Asia (UPAFSA-2013), FAO & PNASF, p 22-25.

Nath, Prem. (2013). Urban and Peri-urban Agriculture, The Basics of Human Civilization — Food, Agriculture and Humanity, Volume-I: Present Scenario, PNASF, Bengaluru and NIPA, New Delhi, pp. 465-475.

Nath, Prem. (2015). Progressive Development of Urban and Peri-urban Agriculture. Development Trends in Urban and Peri-urban Agriculture, Institute for Social and Economic Change (ISEC), Bengaluru; P.N. Agri. Sci. Foundation (PNASF), Bengaluru; Westville Publishing House, New Delhi, p 4-5.

Nath, Prem. (2017). Genesis of Nutrition Security. Attaining Food and Nutrition Security in the Developing World, P.N. Agric. Sci. Foundation, Bengaluru and Westville Publishing House, New Delhi. p 345-346.

Nath, Prem. (2017). United Nations Development Goals. Attaining Food and Nutrition Security in the Developing World, P.N. Agric. Sci. Foundation, Bengaluru and Westville Publishing House, New Delhi. p 482-484.

PNASF. (2002). ICV-2002: Salient Recommendations, Proceedings -International Conference on Vegetables, November 11-14, 2002, Bengaluru, India "Vegetables for Sustainable Food and Nutritional Security in The New Millennium", Dr. Prem Nath Agricultural Science Foundation, Bengaluru, Indian Society of Vegetable Science, New Delhi and Indian Institute of Horticultural Research, Bengaluru. p 13-21.

PNASF. (2009). ICH-20092: Proceedings - The International Conference on Horticulture (ICH-2009), "*Horticulture for Livelihood Security and Economic Growth*" held in Bengaluru during November 9-12, 2009, Bengaluru, India. Dr. Prem Nath Agricultural Science Foundation, Bengaluru, India.

UN. (2015). United Nations General Assembly, A/RES/70/1 Transforming our world: the 2030 Agenda for Sustainable Development, September 25, 2015. p 35.

III. Advancing Horticulture

5. Empowering and Advancing Researches on Vegetables and Fruits
6. Smart Horticulture Towards Sustainable Horticulture Development
7. Green Cities in the 21st Century
8. Elements and Strategic Measures in Maximizing Protected Cultivation of Vegetables Towards Sustainable Horticulture Development in Asian countries
9. Vertical Farming to Supplement Horticultural Food Basket of Urban and Peri-Urban Population in India

5

Empowering and Advancing Researches on Vegetables and Fruits

Prem Nath

1. INTRODUCTION

Over the span of last 100 years, horticulture has grown, developed and achieved substantially but in the present millennium, it is facing serious emerging challenges including nutritional crisis which needs to be addressed in the present food context. In the same vein, research institute like IIHR have achieved significantly on research front, but now new challenges need to be met with new strategies and objectives to achieve targets set in the present context of food and nutrient crisis coupled with climate change.

2. WHY VEGETABLES AND FRIUTS?

- Human nutritional needs are supplied by macronutrients (carbohydrates, proteins and fats) and micronutrients (vitamins and minerals).
- If nutritional requirements are not satisfied, deficiency diseases can result that have widespread effects in the body systems.
- Plants can supply majority of human nutritional requirements and there is evidence that increasing the proportion of vegetarian diet can have positive health benefits.

Nutrients are substances that provide:

(a) energy for activity, growth and all functions of the body, such as, breathing, digesting and keeping body warm.

(b) materials for the growth and repair of the body and for keeping the immune system healthy (McMahon, 2008).

2.1 Nutrients

- Proteins, fats and carbohydrates, also referred as proximate principles, are oxidized in the body to yield *energy*.
- All fresh vegetables, fruits, whole grains, legumes and beans, nuts, seeds and herbs have high nutrient contents.

Vitamins and minerals:

- They do not supply energy but regulate *metabolic activity* and help in utilization of the proximate principles; utilized in formation of body structure and skeleton.

- Human body cannot synthesize vitamins and minerals (few exceptions), must be supplied by food.

Facts of Minerals:

- Minerals play a critical role in maintenance of good health.
- Minerals cannot be created by living beings, such as, plants or animals.
- Plants, animals and humans need minerals for maintaining their health.
- Minerals are elements that originate in the soil and plants absorb minerals from the soil.
- Animals get their mineral requirements from the plants or other animals they eat.
- Most of the minerals in human diet come from plants one eats; some minerals from edible animals.

  In a nutshell, vegetables and fruit provide all essential vitamins and minerals we need and consume for better health.

3. FOOD AND NUTRITION CRISIS

3.1 Hunger

Hunger kills more people every year than ***acquired immune deficiency syndrome*** (AIDS), malaria and tuberculosis combined (WFP, 2014).

Some relevant facts:

1. Over 805 million people in the world do not have enough food to lead a healthy active life. That's about one in nine people on earth.
2. The vast majority of the world's hungry people live in developing countries, where 13.5 percent of the population is undernourished.
3. Asia is the continent with the hungriest people - two thirds of the total. The percentage in Southern Asia has fallen in recent years but in Western Asia, it has increased slightly.
4. Sub-Saharan Africa is another region with the high prevalence (percentage of population) of hunger. One person in four there is undernourished.
5. Poor nutrition causes nearly half (45%) of deaths in children under five, 3.1 million children each year.
6. One out of six children, roughly 100 million in developing countries is underweight.
7. One in four of the world's children are stunted. In developing countries, the proportion can rise to one in three.
8. If women farmers had the same access to resources as men, the number of hungry in the world could be reduced by up to 150 million.
9. 66 million primary school-age children attend classes hungry across the developing world, with 23 million in Africa alone.
10. The ***World Food Programme*** (WFP) calculates that US$ 3.2 billion is needed per year to reach all 66 million hungry school-age children.

3.2 Micronutrient Crisis

Humans require at least 49 nutrients to meet their metabolic needs. Inadequate consumption of even one of these nutrient will result in adverse metabolic disturbances leading to sickness, poor health, impaired development in children and large economic costs to society (Branca, Ferrari, 2002; Golden, 1991; Grantham-McGregor and Ani, 1999; Ramakrishnan, *et al.*, 1999). The required daily amounts for some of these nutrients for adults are reported by the Food and Agricultural Organization, United Nations and the World Health Organization (FAO/WHO, 2000). Importantly, the primary source of all nutrients for people comes from agricultural products. If agricultural systems fail to provide enough products containing adequate quantities of all nutrients during all seasons, the resulting dysfunctional food systems cannot support healthy lives. Unfortunately, this is the case for many agricultural systems in many developing nations in the Global South (Graham, *et al.,* 2001; McGuire, 1993; Schneeman, 2001).

Today, over three billion people are afflicted with micronutrient malnutrition and the numbers are increasing (Mason and Garcia, 1993; Welch *et al.,* 1997; WHO, 1999; World Bank, 1994). Nearly two-thirds of all deaths of children are associated with nutritional deficiencies, many from micronutrients deficiencies (Caballero, 2002).

In the present scenario of food and nutrition-crisis, the lists of major and minor vegetables and fruits have become inadequate. We need to search and identify vegetables and fruits providing essential nutrients required for improving and maintaining human health. Our research should focus on both food and nutrient values of crops.

4. PRESSING DEMAND

We recognize that vegetables and fruits are reservoirs of macro- and micro-nutrients. The pressing demand of today's world is to attain *nutrition security*, which relates to production and utilization of nutrients in the form of vegetables and fruits. With regard to specific nutrient requirements, grain and legumes cannot replace or exchange with vegetables and fruits and *vice-versa.* The balanced nutrition requires the consumption of grains for carbohydrates, legumes for proteins and vegetables and fruits for macro-and micro-nutrients. Thus the nutrition security cannot be attained without vegetables and fruits.

Further, the migration of people from rural to urban areas has demonstrated that the increase in urban population is mostly consumers and not producers. This situation has given rise to *urban and peri-urban* agriculture, where vegetables and fruits are major contributors in attaining food and nutrition security. Uptill now, horticultural/agricultural researchers considered only the high yield of food crops/vegetables and fruits and food quality depending mostly on colour, size and shape, but the time has come where we have to focus on nutritional quality.

There is a great variation in crop yield per unit area within a nation and crop yield in the industrialized world has much higher reduce span per unit area. Moreover, the food development plan is based on food yield with focus mostly on grain crops. If we need to attain food and nutrition security, the food planning estimates are required, which should be estimated in terms of 'nutrients' in order to provide balanced nutrition to citizens of a nation/world. Further, it is obvious that ongoing grain biased nutrition, needs to be complimented and supplemented with vegetables and fruits.

5. RESEARCH MANDATE

In the wake of national and international call on mitigating hunger and malnutrition, the drive on attaining food and nutrition security has gained momentum worldwide. The International Conference on Vegetables (ICV-2002) organized by the PNASF and IIHR in Bengaluru, during 2002 and International Conference on Horticulture organized by the PNASF and UAS (B) in Bengaluru, during 2009, recognized vegetables and fruits as major contributor to nutrition security and extended supplementary role to food security. The first FAO and WHO Joint Workshop on Fruits and Vegetables for Health held in Kobe, Japan during 2004, emphasized that the daily intake of at least 400g *per capita* of fruits and vegetables ensures a better dietary pattern and prevents non-communicable diseases, namely, heart disease, cancer, diabetes and obesity. The International Conference on Nutrition (ICN-2) organized by FAO and WHO in Rome during 2014 has highlighted the significance and importance of nutrition and nutrition security worldwide. The Sustainable Development Goals (SDG) announced by the United Nations has set the target and framework to achieve food and nutrition security in Goal 2 and Goal 13.

Goal 2: states that "End hunger, achieve food security and improve nutrition and promote sustainable agriculture"

Goal 13: states that "Take urgent action to combat climate change and its impacts" It emphasizes on climate change mitigation, adaption and impact reversion.

The role of vegetables and fruits is imminent in contributing to nutrition security. Along with increased promotion of horticulture foods, there is a need to upgrade the nutrient level in these crops. Nutrient production and consumption are the demand of the day by all food consumers. The micronutrients are provided by vegetables and fruits as much required as macronutrients, namely, carbohydrates and proteins provided by cereals and legumes.

6. RESEARCHER AND RESEARCH LEADERSHIP

It was the first day with my designated guide Prof. Charles V. Hall who took me to the Horticulture Farm of the Kansas State University, Manhattan, USA. It was the month of August 1959 and all summer vegetable crops were already harvested. Prof. Hall was taking me round and suddenly I realized we were walking, on plants of squashes, watermelon and muskmelon which were damaged or almost dead and covered almost fully with dirty smelly bold insects all over the field. He said it was squash bug which was very hardy and not easy to control. The pesticides tried did not provide satisfactory control. He added that squash was host of this pest, but once the squash crop was fully damaged, the population moved down to neighbouring cucurbit crops. Thereafter, we forgot about this visit.

We never talked about it for several months or two semesters, but out of curiosity I used to look for some references on this pest and plant (because both plant and pest were unfamiliar) whenever I visited library for my courses. A time came when I completed requisite courses and became eligible for taking up research work. Prof. Hall asked me to think about it and suggest preferred choice. Finally, our discussion ended with my choice on plant resistance to pest. My Master's research ended on resistance of cucurbit seedlings against spotted and striped beetles. Prof. Hall guided me on methodology and did not dictate or prompted me on the topic but he encouraged me all the way. When I observed significant difference in degree of tolerance/resistance between the varieties and among the crops, I was thrilled; and on analysis of data, significant plant ability to resist/ tolerate the pest was observed; it raised confidence in me. It was obvious that I proposed my PhD research on "Genetics of plant resistance to spotted and striped beetles", while Prof. Hall smiled.

This is the live example of a *research leader* who not only guided my research but enthused me to become a young vegetable breeder. We need much of this attitude and personal aptitude in research institution to generate qualitative and productive research. Everybody cannot be a researcher. During 1965, when the author was teaching at the University of Udaipur, Jobner, Rajasthan, the visiting Dr. B.P. Pal, first Direct General of the reconstituted ICAR told me that I was doing a good job helping university and the state in horticulture education and research but my contribution would be more beneficial at the national level. It was a valuable advice of the agriculture leader of the nation.

During 1969, as the founder member of the Indian Institute of Horticultural Research, the author found Dr. G.S. Randhawa, the first Director of the Institute as the favourite Director and good research leader who assisted the author in forming a strong vegetable breeding team of about a ten young energetic and brilliant breeders which marked the first milestone of success of the institute by developing improved vegetable varieties which remained popular among both farmers and consumers for decades and some of them like pumpkin variety Arka Suryamukhi, brinjal variety Arka Kusumaker are popular in markets even after 40 years.

The disease resistant, good quality and high yielding watermelon variety Arka Manik and virus resistant and high yielding okra variety Arka Anamika developed by IIHR remained in commercial cultivation for decades, before they became susceptible to new virus diseases.

7. RESEARCH AREAS

The following areas of research in vegetables and fruits are considered, looking into the emerging issues faced by the nation.

- **Breeding for high nutritional quality:** All this time, the plant breeders have been focusing on enhancing the yield with less or no focus on nutrient increase. The trend is changing and the focus is on nutrients. For example, all these years, breeders have considered only TSS for watermelon but now focus is changing to anthocynin and minerals. Likewise, lycopene and vitamins needs to be upgraded in tomato and so on.
- **Developing adverse climate resistant crops:** Development of crops resistant/ tolerant to biotic and abiotic stress need to be upgraded and accelerated. The drought resistant, flood resistant, heat tolerant, frost tolerant and other improved varieties are very much required. *Basella* sp. grows well in mud soil and *Palak* in saline soils. Hence, emphasis should be to develop of such crops/varieties.
- **Breeding insect pest and disease resistant varieties:** Much breeding researches have been undertaken to develop disease resistant varieties but new strains of diseases are emerging which demand further investigations. Our research efforts on breeding crops resistant to insect pests is relatively behind which could be augmented. The fruit fly resistant variety in mango and other fruit crops are still in demand. Likewise, we are still looking for brinjal plants resistant to borers.

- **Increasing crop yield potential:** We identify that productive yield per unit area and time of vegetable or fruit crops vary from one state to other. For example our tomato yield /ha is much lower than in USA or China; mango yield is relatively lower than in Australia. Furthermore, neighbouring country Thailand produces fresh mangoes almost 12 months of the year, whereas,, we are following traditionally seasonal crop of 3 to 4 months. We are stuck up with alternate bearing in mango. It seems some countries in Latin America have overcome it. While apple is available all over the nation, why good quality national pears, peaches and plums are less seen in the market?
- **Breeding for nutraceutical values:** The trend of developing crop both for nutrient and pharmaceutical values should be encouraged. Bitter gourd and some other vegetables are good examples.
- **Introducing nutrient-rich crops:** The introduced crops like kiwi and dragon fruit have become popular but less used national fruit crops like *ber*, *phalsa*, loquat, *jamoon*, need to be researched and more utilized. Less known fruits like mangosteen, persimmon, rambutan could be produced. These crops will enhance nutrient availability and supply.
- **Integrated pest management:** Building upon the experiences of FAO's Regional IPM project on vegetable crops in Asia during the turn of the century, the strengthening of on-going and development of appropriate research programmes on economic insect pests in vegetables and fruits should yield useful results.
- **Minimum irrigation water utilization:** The researches on minimal requirement of water and infrastructure should continue.
- **Nutrient quality in freshly and processed stored vegetables and fruits:** Ongoing research on both fresh and processed vegetables and fruits should take into account the assessment of the level of nutrients with the lapse of time and the kind of storage provided.
- **Sprouted vegetables:** The ongoing researches and commercialization of sprouted beans with high nutritional value should be encouraged.
- **Upgrading regional vegetables:** Taking into consideration the nutrient providing crops and its limitation to regional availability, it would be useful to introduce them in other regions with similar agro-climatic conditions. As for example, pointed gourd (*parwal*), round melon (*tinda*), longmelon (*kakri*) and others in northern states could be introduced and commercialized in southern states. By now in Bengaluru markets these crops are available. In *Amorphophallus* (*suvarna gedde*, the elephant foot yam), many varieties in southern states do not have itching reaction while handling, unlike varieties in northern states.
- **Maximizing utilization of vegetable plant parts:** With the focus on plant parts, it is possible to maximize the nutrients harvest. As for example, *Moringa* or drumstick is used mostly as fruits, but other regions use tender leaves and/or flowers also which are nutrient-rich. With regard fruits, matured ones are consumed in Southern States, whereas, in Northern-eastern states tender fruits only are consumed. This gives rise to difference in nutrient level at harvest. Further, in north-eastern India these are varieties which fruit 12 months of the year. On other side, Coimbatore developed dwarf kitchen variety with lesser life-cycle. In other words, we can learn a lot from

one crop to maximize nutrient harvest and consumption. Likewise, the practice on utilizing tender leaves, flowers, fruit and seeds of pumpkin should continue. The consumption of fruits of *Coccinia indica* (*tondekayi* or *Kunru*) is common all over the nation but in Thailand tender leaves are also commonly consumed.

The above research areas do not limit the coverage of other areas on emerging issues to be addressed.

- **Establishing harvesting stage:** The harvesting stage of fruits is very much identified based on visual observations. But we need to ascertain the age of fruit when maximum nutrient level is available to be harvested. In case of vegetables, particularly leafy, it is hard but important that the stage/age of crop is ascertained to maximise nutrient harvest.
- **Upgrading nutrient-rich gardens:** The nutrition gardens established in institutions, communities and individuals nationwide do not fulfil all requirement of nutrients required. These gardens end up in having some vegetables and fruits or both without looking into essential nutrient requirements of the community. A precision on selection of vegetables and fruits containing essential nutrients is required to layout the garden to obtain requisite nutrients and continuous nutrient harvest.
- **Strengthening linkages between research and extension:** The popularization and acceptance of new varieties by farmers and consumers is the ultimate milestone in the process of research and extension. The new vegetable crops like round melon (*tinda*), longmelon (*kakri*) and squash with the improved varieties Ark Tinda, Arka Sheetal and Patty Pan respectively which were developed during 1970's by the IIHR could not be popularized. After four decades, Bengaluru markets have started receiving these vegetables with inferior products. These varieties need to be revisited by both researchers and extension specialists.

The research institution should establish priorities depending on the resources available.

8. ADOPTION OF LATEST BREEDING TECHNIQUES

While using the traditional breeding methods the new techniques developed should also be tested and adopted which will address the emerging issues and assist in achieving objectives.

As for example, vegetable breeding for enhancing nutrient quality, the crops can be enriched (*i.e., biofortified*) with micronutrients using plant breeding and/or transgenic strategies, because micro-nutrient enrichment traits exist within their genomes that can be used for substantially increasing micronutrient enrichment traits exist within their genomes that can be used for substantially increasing micro nutrient levels in these food without impacting crop productivity.

In a recent message to the author, Dr. M.S. Swaminathan suggested the establishment of *genetic gardens of biofortified plants* with emphasis on naturally securing biofortified plants like *Moringa* as well as those developed by Mendelian breeding, such as, iron-rich pearl millet and zinc-rich rice.

In order to avoid debate on developing using 'genetically modified' crops, the *edited gene* technique could be employed. In a few years, we would be eating next generation of genetically altered food like potato that do not turn brown or soybeans with healthier mix of fatty acids. A new generation of crops known as gene-edited rather than genetically modified is coming to the market. Created through new tools that snip and tweak DNA at precise locations and as of now fall outside the current regulations.

9. CONCLUSION

The dynamic, forward looking and cohesive thinking with research aptitude of any research institute are pre-requisite to achieving sound research results. The contribution of intelligent and devoted research leader with team spirit provides necessary guidance with utilization of appropriate technologies to young and energetic researchers.

The understanding of the research mandate, areas of research and adoption of appropriate and advanced technologies are keys to ensuring the objectives of research goal.

The vegetable breeding experience confirms that identifying objectives of breeding is prelude and key to successful results of a breeding programme in developing improved varieties. In order to identify the objectives, the views of the choice makers (state, consumer, grower, trader and researcher) which will ensure the acceptance and popularity among the farmers and consumers.

The research areas listed illustrate the current issues related to vegetables and fruit and the research institution may prioritize them depending on the mandate and resources available.

Acknowledgements

The references used and photos cited from different sources are very much acknowledged and utilized for the purposes of education only.

Figure 1. *Some Regional Vegetable Crops to be researched and popularised.*

Parwal, Pointed Gourd (*Trichosanthes dioica*)

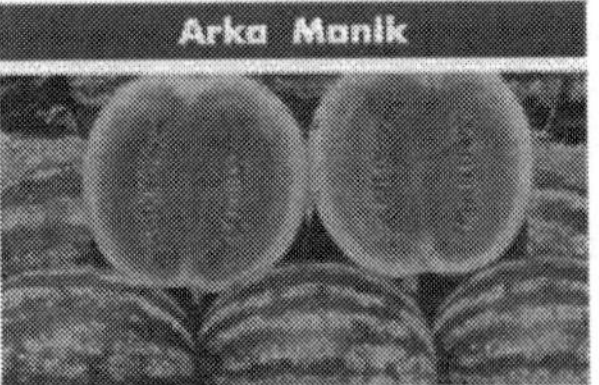

Kalangadi, Watermelon (*Citrullus lanatus* var. *lanatus*)

Tondekayi, Ivy gourd (*Coccinia grandis* (L.) Voigt)

Tinda (*Citrullus vulgaris* var. *fistulosus*) Arka Tinda

Source: *PNASF, Bengaluru*

Cho-cho, Chayote (*Sechium edule* (Jacq.) Sw.)

Hagalkayi, Bitter Guard (*Mormordika charantia*) Arka Harit

Kumbalakayi, Pumbkin (*Cucurbita maxima*) Arka Suryamukhi

Huralikayi, Freanch bean (*Phaseolus vulgaris* L.) Arka Komal

Okra-Arka Anamika (*Abelmoschus esculentus* L.)

Squash- Patty Pan (*Cucurbits pepo*)

Brinjal-Arka Kusumakar (*Solanum melongena* L).

Tomato (*Lycopersicon esculentum*)

Source: *IIHR, Bengaluru*

Moringa-Drumstick (*Moringa oleifera* Lam)

Figure 2. *Some nutrient-rich fruits to be researched and popularized*

Mangosteen (*Garcinia mangostana*)

Passion Fruit (*Passiflora edulis*)

Persimmon (*Diospyros virginiana*)

Phalsa (*Grewia asiatica* L.)

Rambutan (*Nephelium lappaceum*)

Rose Apple (*Syzygium samarangense*)

Source: *Different websites are listed in references.*

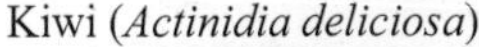

Kiwi (*Actinidia deliciosa*)

Dragon Fruit (*Hylocereus undatus*)

Guava (*Psidium guajava* L.)

REFERENCES

Branca F., Ferrari M. (2002). Impact of micronutrient deficiencies on growth: the stunting syndrome. Annuals of Nutrition and Metabolism 46: 8-17.

Caballero B. (2002). Global patterns of child health: the role of nutrition. Annuals of Nutrition and Metabolism 46: 3-7.

FAO/WHO. (2000). Preliminary report on recommended nutrient intakes. Joint FAO/WHO Expert Consultation on Human Vitamin and Mineral Requirements, FAO, Bangkok, Thailand, September 21-30, 1998, revised July 13, 2000. Food and Agricultural Organization of the United Nations Rome, Italy and World Health Organization, Geneva, Switzerland.

Golden, M.H.N. (1991). The nature of nutritional deficiency in relation to growth failure and poverty. Acta Paediatrica Scandinavica 374, 95-110.

Graham R.D., Welch R.M. and Bouis H.E. (2001). Addressing micronutrient malnutrition through enhancing the nutritional quality of staple foods: principles, perspectives and knowledge gaps. Advances in Agronomy 70, 77-142.

Grantham-McGregor S.M., Ani C.C. (1999). The role of micronutrients in psychomotor sad cognitive development. British Medical Bulletin 55: 511-527.

Mason, J.B., Garcia, M. (1993). Micronutrient deficiency the global situation. SCN News 9, 11-16.

McGuire, J. (1993). Addressing micronutrient malnutrition. SCN News 9, 1-10.

McMahon, Leventin. (2008): "Human Nutrition", Plants as source of Food, The McGraw-Hill Companies.

Nath, Prem. (2017). Empowering and Advancing Researches on Vegetables and Fruits, Special Issue of Indian Horticulture; May-June 2017 on Golden Jubilee Celebrations of ICAR-IIHR, DKMA, ICAR, New Delhi (in press).

Ramakrishnan, U., Manjrekar, R., Rivera, J., Gonzales-Cossio, T., Martorell, R. 1999. Micronutrients and pregnancy outcome: a review of the literature. Nutrition Research 19, 103-159.

Schneeman, B.O. (2001). Linking agricultural production and human nutrition. Journal of the Science of Food and Agriculture 81, 3-9.

Welch, R.M., Combs, Jr G.F., Duxbury JM. (1997). Toward a `Greener' revolution. Issues in Science and Technology 14, 50-58.

WFP. (2012). Two Minutes to Learn About School Meals, WFP, 2012

WFP. (2014). Fighting Hunger World, World Food Programme, http:/www. wfp.org/ hunger/what-is

WHO. (1999). Malnutrition worldwide. Geneva, Switzerland: World Health Organization. http://www.who.int/nut/malnutrition_worldwide.htm. 1-13.

World Bank. (1994). The challenge of dietary deficiencies of vitamins and minerals. In: Enriching lives: overcoming vitamin and mineral malnutrition in developing countries. Washington DC: World Bank, 6-13.

References for vegetables and fruits photographs are listed below:

1. Jamun: https://goqii.com/blog/amazing-health-benefits-and-uses-of-jamun-fruit/
2. Persimmon: https://www.waimeanurseries.co.nz/assets/Uploads/Persimmon-Matsumoto-Wase-Fuyu.jpg
3. Loquat: https://www.aliexpress.com/store/product/3-Loquat-SEEDS-Eriobotrya-japonica-Fruit-seeds-Lovely-Home-Garden-Backyard-Precious-HEIRLOOM-FREE-SHIPPING/822922_938845600.html
4. https://toptropicals.com/pics/garden/11/olymp/6/P6156817.jpg
5. Phalsa: http://fruitlovers.com/Gallery1/PhalsaFlowersSndFruitCloseup.jpg
6. Mangosteen: http://dizonexoticfruittrees.com/mangosteen/img/mangosteen.jpg
7. Mangosteen tree: https://s-media-cache-ak0.pinimg.com/564x/95/1e/74/951e742e21c9f3642518604a44b0d1c3.jpg
8. Rambutan: http://www.fruitsvegetablecinnamonsuppliersrilanka.com/images_cat/Rambutan_Sri_Lanka.jpg
9. Rambutan: http://www.netcostphilly.com/wp-content/uploads/2015/08/tree.jpg
10. Water rose apple Hot pink:http://s262.photobucket.com/user/7_Heads/media/Fruit%20Trees/Fruit_Trees/S_samarangense_005.jpg.html
11. Water rose apple : http://i.ebayimg.com/images/i/201416204031-0-1/s-l1000.jpg
12. Passion Fruit: http://www.handyman.net.au/sites/default/files/article_up/g-passionfruit-main-th522958299BODY.jpg

13. Chow Chow: http://www.agrifarming.in/wp-content/uploads/2017/03/Chow-Chow-Vegetable-Farming-Information..jpg
14. http://towness.co.in/xcart44/images/D/chayote.jpg
15. Tinda: https://images-na.ssl-images-amazon.com/images/I/410JEFT72BL.jpg
16. Parwal: http://agrifarming.in/wp-content/uploads/2015/05/Pointed-Gourd-Cultivation-Parwal.jpg
17. http://www.boldsky.com/img/2013/12/03-parwal.jpg
18. Tondekayi: http://healthyliving.natureloc.com/tindora-ivy-gourd-or-kovakka-nutritive-value-of-vegetables/
19. Vegetables, Indian Institute of Horticultural Research (IIHR), Hesseraghatta, Karnataka, India.
20. Cucurbits, P.N. Agricultural Science Foundation (PNASF), Bengaluru, Karnataka, India.
21. Okra-Arka Anamika: https://kcdn.indiabizclub.com/upload15A/3/2012/axma1o5omagiax01qrt/okra-arka-anamika-1-323.png
22. Squash- Patty Pan: https://usercontent2.hubstatic.com/7061335_f520.jpg
23. Brinjal-Arka Kusumakar: http://www.namdhariseeds.com/sites/default/files/ARKA%20KUSUMAKAR.jpg
24. Tomato: https://www.organicfacts.net/wp-content/uploads/2013/05/Organictomato1.jpg
25. Moringa-Drumstick: http://3.bp.blogspot.com/--v9Al5p_BNU/U3XdFHIAlNI/AAAAAAAAgOw/A3zoWVM0Ph8/s1600/Moringa_Seed_pod_opening_1.jpg
26. Kiwi: http://www.healthyfoodhouse.com/wp-content/uploads/2013/08/kiwi-natural-remedy-for-many-diseases-featured1.jpg
27. Dragon Fruit: https://cdn.authoritynutrition.com/wp-content/uploads/2016/06/dragon-fruit.jpg
28. Guava: http://i.ndtvimg.com/i/2015-11/guava-benefits-25_625x350_71447741866.jpg

6

Smart Horticulture Towards Sustainable Horticulture Development

Usha R. Palaniswamy

1. INTRODUCTION

In realizing sustainable development, "Smart horticulture" is not a label to be merely used as a marketing slogan, but be "smart" in every sense of part and being. The term "smart" is used most often, if not always to describe an approach, item, activity, place or any such thing that combines information technology with its infrastructure and function in terms of input, output and anything in between or thereafter. Horticulture is "smart", if it provides better efficiency for production, distribution and consumption through a variety of information technologies to address social, economic and environmental issues.

Sustainable development is "development that meets the needs of the present without compromising the ability of future generations to meet their own needs" and calls for "building an inclusive, sustainable and resilient future for people and the planet" (United Nations World Commission on Environment and Development, 1987). Since then, substantial interest in sustainability, as applied to all areas of human activity has emerged worldwide.

In achieving human well-being, we consume economic, natural, human and social capital that in turn affect the wellbeing of the overall resources; and the consumption of natural capital in terms of natural resources that directly affects the ecological/ecosystem well being (Figure 1).

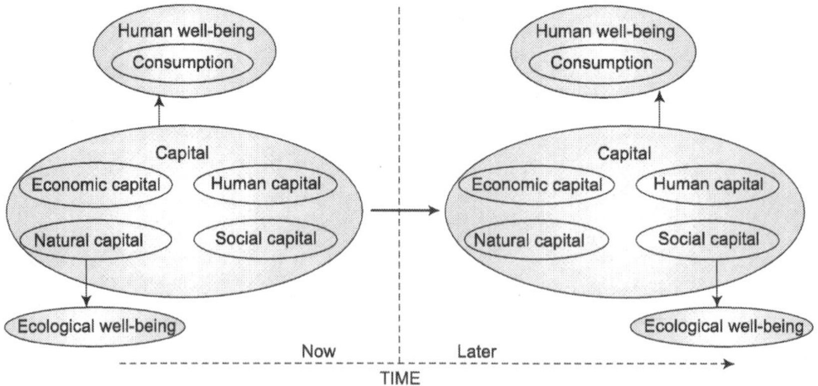

Fig. 1. *Concept of capital and consumption for the human and ecological wellbeing in sustainable development: "now" versus "later" (Source: UN Economic Commission for Europe, 2014).*

To this end, a "sustainable development" agenda will include the three essential dimensions, **economic growth, social inclusion** and **environmental protection** (Figure 2) (http://www.un.org/sustainabledevelopment/development-agenda).

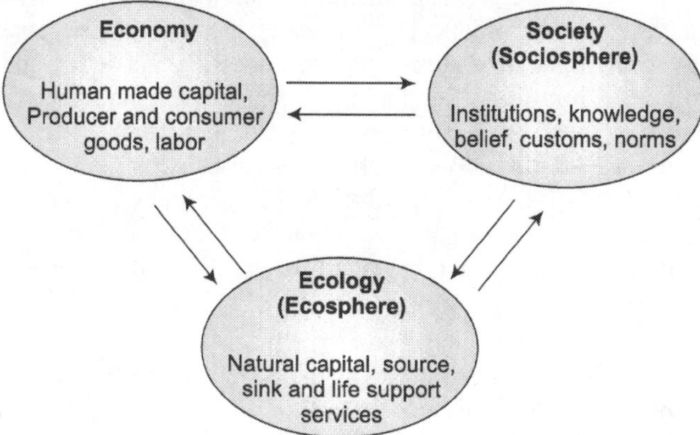

Fig. 2. *The three dimensions of sustainable development (Source: Lawn, 2006).*

As with all areas of sustainable development, horticulture development needs to address these three dimensions for it to be sustainable overtime. Thus, the practice of "smart horticulture" using a complex system of connected intelligent information technologies, will identify, design, develop and adapt a spectrum of horticultural practices for economic growth, social inclusion and best horticultural/agricultural practices in its production, preparation and consumption, ensuring protection of the environment in the process, prudent in the consumption of natural resources which affects the ecological wellbeing of the planet. To this end, agriculture/horticulture must become more: (1) environmentally sustainable and resilient to climate change; (2) efficient, responsive, innovative and flexible in its production and distribution networks; and (3) formulate policies that are responsive to changing social and economic paradigms which simultaneously, optimize effective environmental stewardship, nutrition, economic opportunity and social engagement and empowerment, and inclusive of all segments of populations.

2. HORTICULTURE IN LIFE, LIVELIHOOD AND ECONOMIC GROWTH

The vital role of horticulture in realizing global food and nutrition security and reducing poverty has been well-established. With increasing importance of horticultural crops, especially fruits, vegetables, herbs and spices in the human diets and the wide array of crops classified as horticultural crops including ornamentals and plantation crops, it is perhaps the most profitable venture of all farming activities, in providing prospects of nutrition and food security, economic growth, employment opportunities and scope to raise the income of the farming community.

In India, like most Asian countries where the daily human diet is heavily based on cereals and pulses, efforts to increase the consumption of horticultural crops like fruits and vegetables was only in response to the accumulating scientific evidence of human health promotion and disease preventing benefits of fruits and vegetables in the daily diets.

In India, where historically agriculture has always been a high national priority agenda item for the production of staple crops, horticulture development gained precedence only when the Government of India launched the National Horticulture Mission in 2005-06, with a mandate to promote integrated development in horticulture, to help in coordinating, stimulating and sustaining the production and processing of fruits and vegetables and to establish infrastructure in the field of production, processing, marketing and post-harvest management to reduce post-production losses. This has lead India to become the second largest producer of fruits and vegetables in the world, after China. The annual increase in production area under fruits and vegetables from 1991 to 2005 in India was 2.6% and 3.6% respectively. India is the largest producer of Ginger and Okra amongst vegetables and ranks second in production of potatoes, onions, cauliflowers, eggplant and cabbage. Amongst fruits, India ranks first in production of Bananas (22.94%), Papaya (44.03%) and other fruits (37.57%).

Private and public sectors in India have taken initiatives and with assistance from the Agricultural and Processed Food Products Export Development Authority (APEDA) in starting Centers for Perishable Cargoes and integrated post-harvest handling facilities have been set up in the country. Through this initiative, a few state-of-the-art cold chain infrastructure and quality assurance measures have emerged. Capacity building initiatives at the farmers, processors and exporters levels has also contributed towards this effort. The share of fruits and vegetables in the total value of agricultural exports has increased over the years from 9.5% in 1980-81 to 16.5% in 2002-03. However, in terms of the actual export, India has not reached its potential due to lack of infrastructure for post-harvest handling, storage and transport. For example, although India produces 65% and 11% of world's mango and banana, respectively, ranking first in the world production of both the crops, their exports are nearly negligible of the total agricultural exports from India. India's share in the global market is very low at < 1% only.

Although horticulture has been identified as an area that has plenty of scope for farmers to increase their income and is helpful in sustaining a variety of agro-based industries which can generate large employment opportunities in a community promoting economic growth (http://agriexchange.apeda.gov.in), about 72% of fruits and vegetables produced in rural areas are wasted due to weak marketing, poor transport/roads, inadequate power, insufficient cold storage facilities and lack of infrastructure (Economic Times, 2008). More recent figures also show a decrease in the export of fresh fruits and vegetables from 29,17,405 Metric Ton (MT) in 2013-2014 to 24,04,945 MT in 2015-2016; that of processed fruits and vegetables from 6,19,960 MT in 2013-2014 to 5,20,761 MT in 2015-2016; that of floriculture and seeds from 40,302 MT in 2013-2014 to 33,444 MT in 2015-2016 (APEDA, 2016). The acceptance of India's produce at the world market will depend on India's ability to meet the international standards in terms of quality of the product.

3. HORTICULTURE AND SOCIETY

While food security, including access to and availability of healthy food is a human right (Rideout *et al.*, 2007, UN Committee on Economic, Social and Cultural Rights, 1999) it is also a public responsibility (Beaudry and Delisle, 2005). Food security, defined as a situation when "all community residents obtain a safe, personally acceptable, nutritious diet through a sustainable food system that maximizes healthy choices, community self-reliance and equal access for all" (Dietitians of Canada, 2011, Hamm and Bellows, 2003), is clearly seen as a social movement by growing public interest and activism through several local, national and international membership in organizations (Schiff and Levkoe, 2014).

Horticultural activities bring the community and the neighborhoods together in the cultivation and consumption of fresh produce as a social activity. The act of growing produce locally contributes to the social life of a community where community gardens are planted and managed by the community; the school gardens are laid out and managed by the students and by their parents who participate in the education of their children; the master gardeners programs offered through universities in the United States and Canada, provides horticultural training to individuals who then volunteer as Master Gardeners in their communities by educating the community in horticulture by giving lectures, creating gardens in the neighborhood and participate in many horticultural projects as needed by the community; non-profit organizations as the Herb Society of America in the US, that is dedicated to studying and sharing the knowledge and uses of herbs and promoting knowledge, use and delight of herbs through educational programs by laying out and maintaining herb gardens in the community, sharing recipes and celebrating the use of herbs for culinary and ornamental purposes throughout the year. This kind of social movement serves to underscore alignment of theory and practice in positioning food security as a public right and responsibility (Figure 3).

Study projects have provided evidence of vegetable gardening as a means to the expansion of rural women's capabilities in income generating activities involving agriculture, access to natural resources and participation in community activities with a major impact not only on women's rights but also on the society's well being at large (Agoramoorthy *et al.*, 2012); in schools students plan and grow fresh produce where gardens are integrated into the curriculum as school gardening programs can improve nutrition knowledge, as well as vegetable identification, preference and consumption of fruits and vegetables as part of daily diet (Figure 3).

Fig. 3. *Gardens integrated to local food production in the society/community through vertical, roof top, home, community and school gardens, (Illustration by P. Rajeswari, NJ, USA)*

By 2030, the world's population is projected to be 8.5 billion and increase to 9.7 billion by 2050 and 11.2 billion by 2100. Half of humanity today lives in cities and this is bound to increase with growing employment opportunities in the urban areas and the desired migration of rural populations to the cities. Despite the overall increased production of food, there exists disparities of access to fresh, nutritious and affordable produce and the resulting health implications of their lack of in the so called "food deserts." There is growing concern about the environmental impacts, safety and quality of our food among the larger public and town planning authorities. In addressing this, the urban farming and urban self-sufficiency are emerging issues in the design and transformation of cities (Inayatullah, 2011) in the form of urban horticulture. Urban agriculture/horticulture has taken off well in Chicago (Taylor and Taylor, 2012) as it also creates an alternative use for vacant urban lands (McClintock *et al.*, 2013) wherever available even if for short durations of time. Urban agriculture/horticulture is seen as community farms, community gardens, commercial farms, institutional farms, roof zone top and vertical gardens and private farms/gardens. Integration of agriculture into city zoning and planning by the municipal governments, such as, the GO TO 2040 regional plan of Chicago (with an entire section for sustainable local food systems) (CAM, 2010) or the work of the London Planning and Housing Committee regarding urban food growing systems and planning (London Assembly, 2010). As a result, food production has become an important issue in the field of planning and architecture of "smart cities" for the future. Major professional associations as the American Society of Landscape Architects, has identified urban agriculture as a sustainable element in urban planning (ASLA, 2011) and the American Planning Association has developed a guide for the planning of local food production (Hodgson *et al.*, 2011) to address this issue in the urban "food deserts." The development of sustainable design and planning (*e.g.*, LEED program for buildings) stresses the introduction of innovative systems into buildings, such as, building-integrated agriculture (Komisar *et al.*, 2009). This concept is incorporated in the academics as course offerings for the architecture students studying designing cities for urban agriculture- for example, Carrot City is a research initiative that explores how design can enable the production of food in cities at the Department of Architectural Science, Ryerson University, Canada.

4. HORTICULTURE AND ENVIRONMENTAL PROTECTION

The United Nation's Sustainable Development Goals (SDGs) that came into effect in 2015 call for global action to promote prosperity while "protecting the planet." All actions taken in ending poverty need to include measures to tackle climate change and environmental protection to achieve sustainability. Clearly, environmental protection requires the promotion of integrated and sustainable management of natural resources and ecosystems-sustainable development cannot be achieved without climate action. Of the seventeen sustainable development goals (SDGs), four of the goals are directly related to agriculture/horticulture and the farmers' livelihood: Goal 1. No poverty, Goal 2. Zero hunger; Goal 3. Good health and well being and Goal 12. Responsible consumption and production; and Goal 13. To "promote urgent action to combat climate change and its impacts" applies to agriculture which is a major contributor to enhanced climate change/global warming.

As cultivation of crops continues to contribute the most to a farmer's household's income, farmers will need training and support *via* all possible avenues including the national extension system to increase field productivity with a drive for diversification and post-production for value addition to increase the farmers' livelihood income and quality of life. To achieve this, strengthening the research for development linkages, expanding and increasing the flexibility of the existing extension systems, value chains for fruits and vegetables for grading and packaging and using the modern logistic network for distribution will be necessary rather than an option. Farmers may design a sustainable livelihood, unique to their livelihood needs, that include, means, engaging activities, entitlements and assets by which they can make a living. This may be achieved through an "integrated farming systems" approach in which the outputs, usually the by-products of one component are used as inputs for another component within the farming system creating a recycling mechanism, thus minimizing the input costs and maximizing the return to the farmer from the farming activity. The integrated farming systems may be combinations of agriculture, horticulture, livestock, forestry, dairy, forage cropping, composting, fish/aquaculture and any activity that could minimize the risk, increasing the production and profits by improving the utilization of waste or by-products and crop residues. The benefits of integrated farming systems are plenty: increasing the productivity, profitability, sustainability, food and nutrition base diversity for the farm families, environmental safety, assured income from varied sources throughout the year by minimizing the risk of a "zero income" situation for the farmer, energy savings, increased employment generation for the farming family, as well as the members of the community throughout the year and increased opportunities of value addition to the farm products. Integrated farming can have multiplying effects economically, socially and environmentally–to the farmers, rural community as well as to the rural environment; economically this increases the farmer's net income. The benefits of increased cropping system diversity include decreasing the need for intensive chemical inputs for weed control (Liebman and Gallandt, 1997), thus adding an additional essential dimension to environmental protection.

Prudent utilization of natural resources and protection of the environment include the soil, water and the air. This can be achieved by careful planning and execution of best practices in crop production and protection. This includes crop rotation and intercropping to protect the soil structure and fertility – the benefits of integration of leguminous vegetables in improving the soil fertility, productivity and profitability of the cereal-cereal system are well known. Additional benefits of intercropping include more effective use of water, nutrients and solar energy (Midmore, 1993), soil conservation (Jarenyama *et al.*, 2000), stability of crop yields over time (Horwith, 1985), reduced pest and disease attacks (Theunnissen, 1997) and better land use efficiency as an important component of sustainable farming (Guvenc and Yildirim, 1999). Research evidence also supports that taxonomic diversity of plant species plays an important role in the functioning of the ecosystems (Hooper *et al.*, 2005). As such, compared to other agricultural crops, vegetables utilize water most efficiently in terms of both production and economic efficiency (Ali, 1999) and vegetable production is more inclusive in use of human resources in the form of labor of all groups including women and children (Braun *et al.* 1989; Wann *et al.* 2000).

Developing and ensuring optimal production in the field is only one of the multiple dimensions in attaining sustainable development of horticulture. From the human nutrition point of view, the benefits of traditional diets and increasing the consumption of wild foods to alleviate hunger and malnutrition (Burlingame, 2000) have been reported. Integrating wild vegetables that are good sources of micronutrients into diets has been promoted as the most practical and sustainable way to achieve nutrition security (Chadha and Oluoch,

2003). Since the unit cost of production and per unit area of land of these wild foods are very low (Ali and Tsou, 1997), they are now receiving renewed attention, with the recognition that they could become useful parents in breeding programs, convenient sources of income and vehicles for improved nutrition and increased food supply (Burlingame, 2000; Grivetti and Ogle, 2000; Padulosi *et al.*, 2002; Olorode, 2004). In designing a suitable crop diversification model and making changes to increase the farm income and reduce poverty, the farmers not only consider their economic background and their own needs but also the social parameters like attitudes, values, traditions, peer group pressure and the local culture (Dent, 1990 J.B. Dent, Optimising the mixture of enterprises in a farming system. In: J. Jones and P. Street, Editors, *Systems Theory Applied To the Agriculture and the Food Chain*, Elsevier Applied Science, London (1990), pp. 113–130.Dent, 1990).

5. "SMART" PATHWAYS FOR SUSTAINABLE HORTICULTURAL DEVELOPMENT

Technology in Horticulture:

Information and communication technologies (ICTs), such as, the mobile phones, social media, the "worldwide web," and the big data analytics are useful in data collection, information organization and management and delivering services quickly and efficiently crossing the barriers of time and space. Most of the technology devices we use on a daily basis today are intelligent/smart in sensing the need (diagnostic) and offer a solution to the situation-examples of more popular items include the Amazon's Dash Buttons (https://www.amazon.com/b/?node=10667898011&sort=date-desc-rank&lo=digital-text) that are programmed to alert when one is short on household supplies, to order with a simple click and receive the orders at their doorstep; there are devices that can be fixed in the car that tracks one's driving habits, car's health and its parking location; the cars are able to sense when they are low in gas, oil or issues and alerts with a flash of a light; some devices can sense the tire pressure and alert which of the tires is low on its pressure, give live traffic updates, suggest alternate routes to get the destination; remotely monitor homes for safety, turn on/off the lights and utilities with a simple device connected to their mobile devices and computers.

The advancement in communication technology has evolved into a new concept of the Internet of Things (IoT) that brings "intelligence" to the everyday devices around us. It is a connected world and, using that connectivity, one can bring the internet to digital devices that were once purely analog. IoT is the concept of basically connecting any device with an on and off switch to the Internet and/or to each other. This includes everything from cellphones, computers, home appliances, headphones, lamps, wearable devices and almost anything one can think of. IoT provides an environment where objects, animals/people/things are equipped with unique identifiers capable of data transmission over the Internet without or with minimal need for human-human or human-computer interaction (Gluhak *et al.*, 2011). In 2015, there were 13.4 billion devices connected to the internet and this is expected to increase by 185% to 38.5 billion devices by year 2020 (Juniper Research, 2015). IoT has uses in almost every area of the modern society including Smart Health Care, Smart Cities, Smart TVs and hand held personal electronic devices Smart Industry, Smart automobiles, Smart Agriculture, Precision Agriculture and Smart Homes (Shang *et al.*, 2015; Khan *et al.*, 2012). In the project Farm Bot (FarmBot, 2016), open hardware and software are used in IoT, which focuses on creating open-source automated precision farming machine.

Advances in technology has enabled the collection and processing of "big data" which are extremely large data sets that may be analyzed computationally to reveal patterns, trends and associations, especially relating to human behavior and interactions; this is used widely in all kinds of industries and businesses to provide customized goods and services to individuals. Data collection in large volumes – cross sectional and/or longitudinal has been happening since time immemorial, However, the primary value from big data comes not from the data in its raw form, but from the processing and analysis of it and the insights, products and services that emerge from such analysis and its potential applications. Big data is characterized by its huge (1) volume collecting data from a variety of sources, including business transactions, social media and information from sensor or machine-to-machine data; (2) speed/velocity by which the data streams in at an unprecedented speed that needs to be dealt with in a timely manner in near-real time; and (3) variety by which the data comes in all types of formats – from structured, numeric data in traditional databases to unstructured text documents, E-mail, video, audio, internet searches and financial transactions.

How can this kind of connectivity be applied to horticulture on a large scale? Based on the use of information technology in making decisions, the field of precision agriculture has emerged as a field of agriculture, where information about the variability of soil and crop conditions is gathered and analyzed, in order to maximize the efficiency of crop inputs within small areas of the farm field (Zarco-Tejada *et al.*, 2014). It quantifies the field input and productivity as well as the product quality using applications, such as, mapping, surveillance and exploration using sensors for most part and help make data driven decisions. Remote sensing in the form of smart flying sensors based on an unmanned aerial vehicle (UAV) can cover large farm areas, take measurements, images and process them using an in-build intelligent system (Hernandez *et al.*, 2015).

Taking this one step ahead would involve the use of internet and the worldwide web, tools and services that come with it. The IoT is a concept that not only has the potential to impact how we live but also how we work. But what exactly is the "Internet of things" and what impact is it going to have on horticulture and the farmers and a consumer if any? The IoT is an enormous network of connected "things" between people-people, people-things and things-things. This can connect the farmers, researchers, policy makers and the consumers across the world, give access to most recent research findings and services that can assist in decision making and create networking opportunities. This has become more possible due to the greater availability of broadband Internet, decreasing cost of "connecting", more devices being created and manufactured with Wi-Fi capabilities and sensors built into them, decreasing costs of technology devices and applications and smartphone use and information dissemination is sky-rocketing with new, creative and customizable applications being built to serve the consumers. The major companies involved in the technology services are also ensuring the consumer education as in Apple and Microsoft stores and investing in educational and promotional materials. All of these things are creating perfect platform for the IoT. There is excitement among the populations as they experience the connections and capabilities under the rule of "Anything that can be connected, will be connected," and if you "think of it, you can connect, access and do it with a click of a device," crossing the barriers of time and distance.

Advances in technology has surrounded us by objects that can think, communicate and store information and the IoT has brought magic and what one would consider science fiction to the wonder of real life experiences. IoT allows for virtually endless opportunities and connections to take place creating a new culture of smart people, policies and services based on real time big data intelligence and analytics in a closely knit of environment, many of which we can't even think of or fully understand the impact of today (Figure 4).

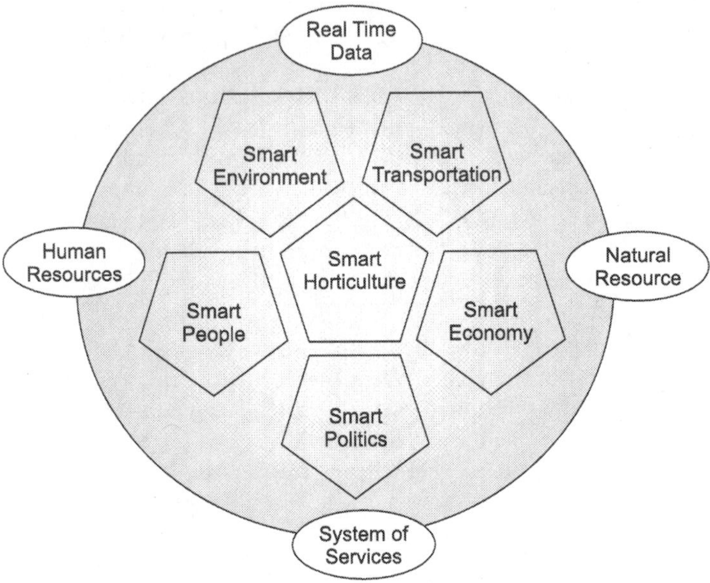

Fig. 4. *Smart horticulture with smart people, policies and services based on real time big data intelligence and analytics in a community.*

The IoT uses machine-produced data, most commonly from sensors, communicated *via* connectivity technologies, such as, mobile phones, to gather real-time information, which can then be analyzed and used to improve processes and decision making. This has taken off in the last few years in developed countries, for example wearable sensors in watches and bracelets to measure vital signs and count steps and progress towards products, such as, self-driving cars and is becoming increasingly common in the developing countries. The number of devices is predicted to grow enormously over the next few years, with 10 billion more internet connected devices coming online by 2019 and out of these 70% will be IoT connections. The internet of things concept is predicted to have an economic impacts of between $3.9 trillion and $11.1 trillion per year by 2025 (McKinsey, 2016, http://www.mckinsey.com).

In global development, IoT can be used in transport, in water, sanitation and hygiene interventions in the developing countries; USAID is currently active in a project that improves water service delivery through the implementation of water flow sensors in East Africa and uses water level sensors to monitor boreholes. On the water, sanitation for hygiene and health front, sensors for behavior change sensors are being used to measure whether behavior change has actually led people to use latrines (sensors on latrine doors) and wash their hands (water flow sensors) in a Mercy Corps Indonesia partnered project with Portland State University's SWEETLab. The concept of horticulture ambient intelligent system using sensors and soil-weather-plant data analytics for the sustainable management of water resources is of significance in the context of climate change as shown in the picture (Vasilenko *et al.*, 2015).

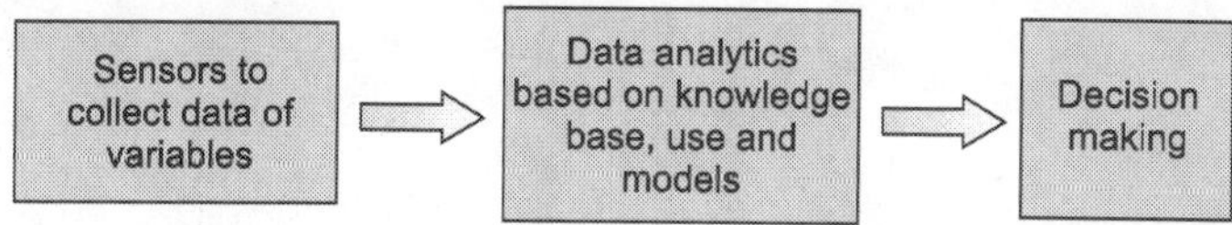

Geographic Information Systems (GIS) have been around for a while now in agriculture and are useful for analysis and spatial data processing. GIS helps in managing spatial data and their associated attributes. It is useful in downloading, storing, analyzing and displaying geographic information, which allows users to create interactive queries, analyze spatial information and edit data. GIS connects precisely located spatial data with tabular databases and thus provides the user to visualize their relationship. GIS connects layers containing various information on the position, which allows a better understanding of the space. Information from the GIS along with water sensors, soil fertility and soil health sensors, weather/climate sensors are already in use in India and Africa to collect and analyze the data to make decisions (Cisco Report, 2016),

Environmental sustainability can be enhanced with the sensors integrated with building designs as has been done in the design of the Gardens by the Bay in Singapore. The conservatories at the Gardens by the Bay are indeed a statement in sustainable engineering that apply a suite of cutting-edge technologies for energy-efficient solutions in cooling by minimizing the solar heat gain by using special glass that allows optimal light in for plants but reducing the heat, roof that is fitted with a sensor-operated retractable sails that opens automatically to provide shade to the plants when it gets too hot. The super tree structures capture the waste heat generated by the running coolers and recycle it to regenerate the liquid desiccant. (http://www.gardensbythebay.com.sg)

While remote sensing and connectivity based systems are in the pipeline and being researched and developed for mainstream large scale agriculture/horticulture/ farming, a more complete system of remote connectivity based on sensing is presently seen for personal home gardening. For example Parrot.com sells a smart garden system that makes gardening easy for home gardens. The system comes with a gadget that measures light, temperature, level of fertilizer, soil humidity and water reservoir level. One can use this device to remain connected in real time to the plant's healthy growth. All the data collected by this device is sent to a smartphone or a tablet with Bluetooth Low Energy wireless technology, an eco-friendly standard. It sends out customized recommendations and alerts to help one better care for your plant. Additionally the device gives access to a database of more than 8,000 plants, for which the individual is given personalized advice.

At a local level, sensors and the internet of things allow communities to collect data for their local contexts. For example, information about local weather, food consumption, produce purchase and marketing. Through this we can see a role for the IoT in global, national, rural and per-urban communities and counties, so it's not just about "smart cities" but also "smart towns," "smart villages" or "smart counties" (Figure 5).

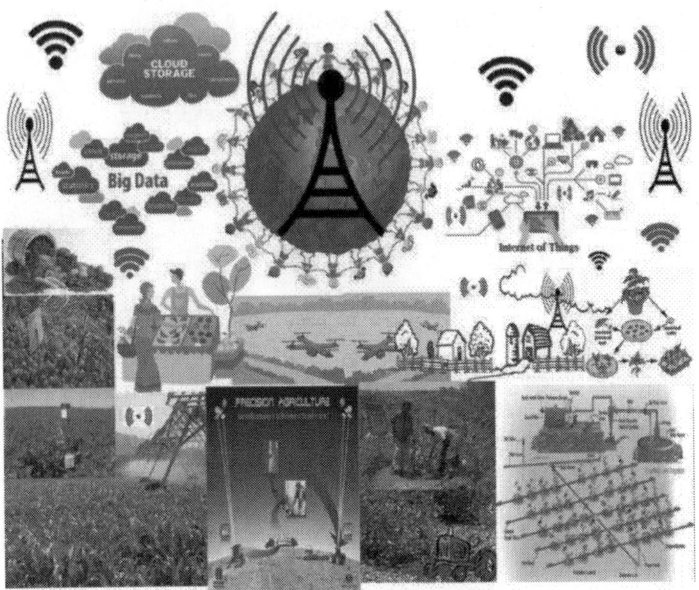

Fig. 5. *Smart agriculture/horticulture: internet of things and big data*

There is huge potential for the IoT to harness the technological innovations, improve research and currently there is a lot of activity from academic institutions around training researchers from developing countries to use sensors to improve their own research activities and funding funneled for research and development in this area. On the policy side, we see examples of where data being collected from the IoT is being used to help policymakers make decisions about allocation and spending.

However, there are technical challenges around bandwidth, connectivity, power supply and scaling sensors at a global level for this to work at this time as countries work out their technology budgets and on the national policy side there exists data flows across the borders – which is experienced due to the countries mandating that data collected has to stay within the country and some countries blocking their devices from access from out of the country. This also creates a security issue that technology companies need to figure out not only a collect, store, track, analyze and make sense of the vast amounts of data that will be generated but also create a safe environment for them as well as the consumers. Every time a device is connected to the internet, another door is added through which to conduct malicious activity- this needs cybersecurity to ensure the safety and security of the network by authenticating and also de-authenticating devices where appropriate. With these challenges on the horizon, it is still very early days in terms of the IoT in the ecosystem that even in advanced developed countries it is not something that one can quickly buy and apply to agriculture/horticulture. However, with the huge surge of excitement, interest and the unlimited possibilities, it may come into fruition sooner than one thinks.

REFERENCES

Agoramoorthy, G., Hsu J.M. and Shieh, P. (2012). India's Women-led Vegetable Cultivation Improves Economic and Environmental Sustainability Scottish Geographical Journal Vol. 128, No. 2, 87–99, June 2012.

Ali, M. (1999). New paradigm of vegetable cultivation in Asia. Asian Vegetable Research and Development Center, Shanhua, Tainan, Taiwan.

Ali, M and Tsou, C.S. (1997). Combating micronutrient deficiencies through vegetables—a neglected food frontier in Asia. Food Policy 22, 17–38.

APEDA. (2016). Fruits and vegetable export data.http://apeda.gov.in/apedawebsite/six_head_product/FFV.htm http://apeda.gov.in/apedawebsite/ accessed Dec. 21, 2016.

ASLA. (2014). Healthy and Livable Communities, Retrieved from http://www.asla.org/livable.aspx. Accessed June 17, 2014.

Beaudry, M., Delisle H. (2005). Public('s) nutrition. Public Health Nutr.8 (6A):743–8.

Braun, J.V., Hotchkiss D and Immink M. (1989). Nontraditional export in Guatemala: effects on production, income and nutrition. Research Report 73. International Food Policy Research Institute (IFPRI) and Institute of Nutrition of Central America and Panama (INCAP), Washington, DC, USA.

Burlingame, B. (2000).Wild nutrition. Journal of Food Composition and Analysis 13, 99–100.

CAM. (2010). Go to 2040: Regional Plan for Chicago (Chicago: Chicago Metropolitan Agency for Planning, 2010).

Chadha, M.L. and Oluoch, M.O. (2003). Home-based vegetable gardens and other strategies to overcome micronutrient malnutrition in developing countries. Food, Nutrition and Agriculture Series, vol. 32. FAO, Rome, Italy.

Cisco Report. (2016). Harnessing the Internet of things for global development. A contribution to the UN broadband commission for sustainable development, 61p.

Dent, J.B. (1990). Optimising the mixture of enterprises in a farming system. In: J. Jones and P. Street, Editors, Systems Theory Applied To the Agriculture and the Food Chain, Elsevier Applied Science, London (1990), pp.113–130.

DeSchutter, O. (2013). Report of the special rapporteur on the right to food, Mission to Canada. Rome: United Nations; 2013.

Dietitians of Canada, (2011). Community food security position of Dietitians of Canada. Toronto, ON: Dietitians of Canada

Economic Times,.(2008). '72 percent of India's fruit, vegetable produce goes waste', Economic Times, 12 May. Available at: http://www.dnaindia.com/india/report_72-percent-of-india-s-fruit-vegetableproduce-goes-waste_1163946.

FarmBot. (2016) "FarmBot genesis", FarmBot [Online]. Available: https://farmbot.io/ [Accessed: 05 February 2016].

Gluhak, A., Krc, S., Nati, M., Pfristerer, D., Mitton, N. and Razafindralambo, T. (2011). A survey on facilities for experimental internet of things research, IEEE Communications Magazine, Vol. 49, No 11, pp. 58-67. ISSN 0163-6804. DOI: 10.1109/MCOM.2011.6069710.

Grivetti L.E. and Ogle B.M. (2000). Value of traditional foods in meeting macro and micronutrient needs: the wild plant connection. Nutrition Research Reviews 13, 31–46.

Guvenc, I., Yildirim, E. (1999). Multiple cropping systems in vegetable production. In: Proceedings of the Organic Agriculture Symposium, 21–23 June, Izmir, Turkey, pp. 288–296.

Hamm, M.W., Bellows, A.C. (2003). Community food security and nutrition educators. J Nutr Educ Behav 35(1):37–43.

Hernandez, A., Murcia, H., Copot, C. and Keyser, R.D. (2015). Towards the Development of a Smart Flying Sensor: Illust. Field of Prec. Agrl. Sensors, 15: 16688-16709.

Hodgson, K, Campbell. M.C. and Bailkey, M. (2011). Urban Agriculture: Growing Healthy, Sustainable Places (Chicago: American Planning Association, 2011).

Hooper, D.U., Chapin, F.S., Ewel, J.J., Hector, A., Inchausti, P., Lavorel, S., Lawton, J.H., Lodge, D.M., Loreau, M., Naeem, S., Schmid, B., Setala, H., Symstad, A.J., Vandermeer, J. and Wardle, D.A. (2005). Effects of biodiversity on ecosystem functioning: a consensus of current knowledge. Ecological Monographs, **75**, 3–35.

Horwith, B. (1985). A role for intercropping in modern agriculture. BioScience 35, 5.

Inayatullah, S. (2011). City Futures in Transformation: Emerging Issues and Case Studies, Futures 43: 7 (2011) 654–661.

Jarenyama, P., Hesterman, O.B., Waddington, S.R. and Harwood, R.R. (2000). Relay-intercropping of sunnhemp and cowpea into a smallholder maize system in Zimbabwe. Agron. J. 92, 239–244

Jarrott, S.E. and Gigliotti, C.M. (2010). Comparing responses to horticultural-based and traditional activities in dementia care programs. Amer. J. Alzheimer's Dis. Dementias, 25(8):657–665.

Juniper Research. (2015). "Internet of Things' Connected Devices to Almost Triple to Over 38 Billion Units by 2020", Juniper Research [Online]. Available: http://www.juniperresearch.com/press/pressreleases/iot-connected-devices-to-triple-to-38-bn-by-2020. [Accessed: 03 February 2016].

Khan, R., Khan, U., Zaheer, R. and Khan, S. (2012). "Future Internet: The Internet of Things Architecture, Possible Applications and Key Challenges. In: 2012 "10th International Conference on Frontiers of Information Technology. IEEE, p. 257-260.

Komisar, J., Nasr, J. and Gorgolewski, M. (2009). Designing for Food and Agriculture: Recent Explorations at Ryerson University, Open house intern. 34:2 61–70.

Lawn, P. (2006). Sustainable Development Indicators in Ecological Economics, Current Issues in Ecological Economics series, Edward Elgar Publishing, Cheltenham Glos, UK.

Liebman, M. and Gallandt, E.R. (1997). Ecological management of crop–weed interactions. Ecology in Agriculture (ed. L. E. Jackson), pp. 291–343. Academic Press, San Diego, CA.

London Assembly. (2010). Cultivating the Capital—Food Growing and the Planning System in London (London: Greater London Authority, 2010).

McClintock, N., Cooper, J. and Khandeshi, S. (2013). Assessing the Potential Contribution of Vacant Land to Urban Vegetable Production and Consumption in Oakland, California, Landscape and Urban Planning 111: 46–58.

Midmore ,D.J. (1993). Agronomic modification of resource use and intercrop productivity. Field Crops Res. 34, 357–380.

Olorode, O. (2004). Conservation of plant genetic resources. African Journal of Traditional, Complementary and Alternative Medicines 1, 4–14.

Padulosi, S., Hodgkin, T., Williams, J.T. and Haq, N. (2002). Underutilized crops: trends, challenges and opportunities in the 21st Century. In: Engels, J.M.M., Ramanatla, R., Brown, A.H.D., Jackson, M.T. (Eds.), Managing genetic diversity. International Plant Genetic Resources Institute, Rome, pp. 323–338.

Ralston, S.J. (2011). It takes a garden project: Dewey and Pudup on the politics of school gardening. Ethics Environ. 16(2):1–24.

Rideout, K., Riches, G., Ostry, A., Buckingham, D., MacRae, R. (2007). Bringing home the right to food in Canada: challenges and possibilities for achieving food security. Public Health Nutr. 10(6):566–73.

Schiff, R. and Levkoe, C.Z. (2014). From disparate action to collective mobilization: collective action frames and the Canadian food movement. Adv Sustain Environ Justice. 15:225–53.

Sempik, J., Rickhuss, C. and Beeston, A. (2014). The effects of social and therapeutic horticulture on aspects of social behaviour. Br J Occup Ther 77(6): 313-319.

Shang, X., Zhang, R., Hu, X. and Zhou, Q. (2015). "Design theory, modelling and the application for the Internet of Things service", Enterprise Information Systems, 10(3):249-267.

Song, M.J., Kim, M.Y., Sim, I.S., Kim, W.S. (2010). Evaluation of horticultural therapy on the emotional improvement of depressed patients by using heart rate variability. Korean J Hort Sci Tech 28(6):1066–1071.

Taylor, J.R. and Taylor, S. (2012). Mapping Public and Private Spaces of Urban Agriculture in Chicago through the Analysis of High-Resolution Aerial Images in Google Earth, Landscape and Urban Planning 108(1): 57–70.

Theunnissen, J. (1997). Intercropping in field vegetables as an approach to sustainable horticulture. Outlook Agric. 26 (2):95–99.

United Nations World Commission on Environment and Development. (1987). *Our Common Future*. Oxford: Oxford University Press. p. 27. ISBN 019282080X. http://www.un.org/sustainabledevelopment/sustainable-development-goals/

UN Committee on Economic, Social and Cultural Rights (CESCR). (1999). General Comment No. 12: The Right to Adequate Food (Art. 11 of the Covenant), 12 May 1999, available at: http://www.refworld.org/docid/4538838c11.html [accessed 27 Dec. 2016].

UN Economic Commission for Europe. (2014). Measuring sustainable Development, Prepared in cooperation with the organization for economic Co-operation and Development and the Statistical Office of the European Communities (Eurostat), https://www.unece.org/fileadmin/DAM/stats/publications/Measuring_sustainabledevelopment.pdf, accessed Nov. 12, 2016.

Vasilenko, A., Ulman, M., Agris. (2015). Concept of Horticulture Ambient Intelligence **System.** On-Line Papers in Eco. Inform, 7(4): 209-216.

Wann, J.W., Peng ,T.K. and Wu, M.H. (2000). Taiwan. In: Ali, M. (ed.) Dynamics of vegetable production, distribution and consumption in Asia. Shanhua, Taiwan, Asian Vegetable Research and Development Center, AVRDC Publication No. 00-498,470 p.

York, M. and Wiseman, T. (2012). Gardening as an occupation: a critical review. Br J Occup Ther 75(2), 76–84.

Zarco-Tejada, P., Hubbard, N. and Loudjani, P. (2011). Precision Agriculture: An Opportunity for EU Farmers—Potential Support with the CAP 2014-2020; Technical Report, Joint Research Centre (JRC) of the European Commission; Monitoring Agriculture Resources (MARS) Unit H04: Brussels, Belgium, 2014.

7

Green Cities in the 21st Century

K.R.M. Swamy

1. INTRODUCTION

Through most of history, the human population has lived a rural lifestyle, dependent on agriculture and hunting for survival. In 1800, only 3 percent of the world's population lived in urban areas. By 1900, almost 14 percent were urbanites, although only 12 cities had 1 million or more inhabitants. In 1950, 30 percent of the world's population resided in urban centers. The number of cities with over 1 million people had grown to 83. The world has experienced unprecedented urban growth in recent decades. In 2008, for the first time, the world's population was evenly split between urban and rural areas. There were more than 400 cities over 1 million and 19 over 10 million. More developed nations were about 74 percent urban, while 44 percent of residents of less developed countries lived in urban areas. However, urbanization is occurring rapidly in many less developed countries. It is expected that 70 percent of the world's population will be urban by 2050 and that most urban growth will occur in less developed countries (PRB, 2016).

The 2014 revision of the "World Urbanization Prospects" by the Population Division of the Department of Economic and Social Affairs of the United Nations (UN DESA) notes that the largest urban growth will take place in India, China and Nigeria. These three countries will account for 37 per cent of the projected growth of the world's urban population between 2014 and 2050. By 2050, India is projected to add 404 million urban dwellers, China 292 million and Nigeria 212 million. The urban population of the world has grown rapidly from 746 million in 1950 to 3.9 billion in 2014. Asia, despite its lower level of urbanization, is home to 53 per cent of the world's urban population, followed by Europe with 14 per cent and Latin America and the Caribbean with 13 per cent. The world's urban population is expected to surpas six billion by 2045. Much of the expected urban growth will take place in countries of the developing regions, particularly Africa. As a result, these countries will face numerous challenges in meeting the needs of their growing urban populations, including for housing, infrastructure, transportation, energy and employment, as well as for basic services, such as, education and health care. "Managing urban areas has become one of the most important development challenges of the 21st century. Our success or failure in building sustainable cities will be a major factor in the success of the post-2015 UN development agenda," said John Wilmoth, Director of UN DESA's Population Division (Anon., 2014d).

The rural population of the world has grown slowly since 1950 and is expected to reach its peak around 2020. The global rural population is now close to 3.4 billion and is expected to decline to 3.1 billion by 2050. While Africa and Asia are urbanizing rapidly, the regions are still home to nearly 90 per cent of the world's rural population. India has the largest rural population with 857 million, followed by China with 635 million. The report notes that a successful urban planning agenda will require to focus on urban settlements of all sizes. If well managed, cities offer important opportunities for economic development and for expanding access to basic services, including health care and education, for large numbers of people. Providing public transportation, as well as housing, electricity, water and sanitation for a densely settled urban population is typically cheaper and less environmentally damaging than providing a similar level of services to a dispersed rural population (Anon., 2014d).

1.1 World of Cities

In 2010, according to the United Nations Population Division, the world crossed a significant demographic threshold for the first time in history, the number of people living in cities exceeded the number living in rural areas. This trend is expected to continue. The U.N. Population Division estimates that between 2010 and 2050 the world's urban population will grow by 80 percent – from 3.5 billion to 6.3 billion. During the same period the number of people living in rural areas is expected to decline by about 18 percent – from 3.4 billion to 2.8 billion. Most of this growth in the world's cities will be concentrated in developing countries. From 2010 to 2050 the urban population of these countries will double – from 2.6 billion to 5.2 billion. In developed countries, the number of people living in urban areas will grow from 900 million to 1.1 billion. Developing countries will account for about 93 percent of all growth in urban populations worldwide. The central role of developing countries in the growth of the world's cities is widely recognized. While urban densities can vary greatly from country to country, it is clear that much of the projected growth in the world's cities will occur in areas – including South and Southeast Asia and parts of Africa – that are generally characterized by very high urban densities (Mitchell *et al.,* 2012).

2. GREEN CITIES

Today, more than half of the world's population lives in cities. As the urban population grows and the effects of climate change worsen, our cities have to adapt. It is time for us to invest in efficiency and renewable energy, rebuild our cities, towns, municipalities and states. Cities need to accelerate their transition to a cleaner, healthier and more economically viable future through improvements in efficiency, investments in renewable technology and regulation reform (Anon., 2016a).

More than 80% of the U.S. population and more than 50% of all people in the world live in urbanized areas. Nature in cities and towns includes parks, gardens, trees, small landscapes and natural areas – all providing many benefits. Recent research has revealed the environmental benefits provided by metro nature, such as, improved air and water quality, energy savings and reduced urban heat island effects. The social sciences provide additional evidence of benefits. Nearly 40 years of research shows that the experience of nature is profoundly important to human functioning, health and well-being. We may intuitively accept that urban nature is important for public health. Given the high population densities of urban areas, every bit of nearby nature has the potential to benefit hundreds and thousands of people daily. People have long recognized that nature in cities and towns provides beauty and respite. However, city green improves public health - ranging from individuals to entire communities. Taken as a whole, metro nature is a "green infrastructure" system that can be planned and integrated with built systems, (such as, transportation and housing) to create more sustainable urban environments (Wolf and Rozance, 2013).

2.1 Green City Initiatives

Green Development initiatives are being implemented throughout the City in private and public projects to incorporate green roofs, energy efficiency, pedestrian and cycling infrastructure, rainwater amenity features, street trees and vibrant streetscapes to encourage a healthy live-work community. Outstanding examples include:

1. City of Burnaby's new public buildings, such as, the Tommy Douglas library and the Edmonds Aquatic and Community Centre, have been developed to LEED (Leadership in Energy and Environmental Design) and/or equivalent green building standards.
2. The City allows for density bonuses in Town Centres through the Community Benefit Bonus Policy, in exchange for community amenities, affordable housing and environmental improvements. This program has resulted in riparian area enhancements and contributions to parkland throughout the City.
3. The new City Public Works Yard has been designed in accordance with LEED principles with an aim to achieve a Gold rating. The facility design includes provision for a green roof, innovative stormwater management and restoration and enhancement of adjacent streamside areas.
4. Through collaboration with the development community, green roofs (eco-roofs) are frequently included in new development, as they improve building energy efficiency, reduce rainwater runoff, provide scenic views from above and contribute habitat. A study carried out in 2002 inventoried 49 green roofs in Burnaby, placing the city among the top two municipalities in the region in overall number of green roofs. Over the past five years about a dozen new green roofs have been constructed on City and private or institutional buildings.
5. Burnaby has partnered with BCIT to investigate opportunities and benefits of green roofs; the program included acquiring $29,000 for monitoring equipment to include the green roof on the Electronic Arts building in the Regional Infrastructure Network.
6. In 2007, the "Green", an aptly-named residential development in the Edmonds Town Centre, was completed under the voluntary Built Green certification program. The 325-unit townhouse development incorporates many of the City's green building guidelines that were initiated at UniverCity, such as, water and energy efficiency measures and a car-sharing program.
7. Burnaby Mountain Secondary School, constructed in 2000, uses an efficient geothermal heating/cooling system. (*Anon.*, 2016b)

Greenspace is vital to the health of cities and citizens alike. It cleans our air and filters our rainwater, it provides shade cover and shelters our wildlife and it creates space for urban communities to live, play and connect with nature.

2.1.1 LEED Standards

LEED (Leadership in Energy and Environmental Design) is an ecology-oriented building certification program run under the auspices of the U.S. Green Building Council (USGBC). LEED concentrates its efforts on improving performance across five key areas of environmental and human health: energy efficiency, indoor environmental quality, materials selection, sustainable site development and water savings. LEED has special rating systems that apply to all kinds of structures, including schools, retail and healthcare facilities. Rating systems are available for new construction and major renovations as well as existing buildings. The program is designed to inform and guide all kinds of professionals who work with structures to create or convert spaces to environmental sustainability, including architects, real estate professionals, facility managers, engineers, interior designers, landscape architects, construction managers, private sector executives and government officials. On its Web site, the USGBC says that LEED defines "a nationally accepted benchmark for the design, construction and operation of high-performance green buildings" and "provides building owners and operators with the tools they need to have an immediate and measurable impact on their buildings' performance." According to the American Institute of Architects, the 69 LEED points that make up the program's specific design points and considerations can be reviewed in a two-hour meeting, during which time the design team and the owner can decide what level of LEED compliance is desirable for their building project. State and local governments around the United States are adopting LEED for public buildings of all kinds and LEED initiatives at the US Departments of Agriculture, Defense, Energy and State drive activity at the federal level. In addition, various types of LEED projects are currently underway in over 40 other countries, including Canada, Brazil, India and Mexico (Margaret, 2016).

According to Matthew (2016) LEED's Rating System is a nationally accepted benchmark for high performance green buildings. Under the LEED rating system, projects can earn points for meeting certain criteria relating to the design, construction and operation of a building. Certified, Silver, Gold and Platinum levels of certification are awarded based upon the total number of points earned. Although there is some regional variation in the Green Building Rating System through Regional Credits, the LEED standard is generally national in scope.

Today, LEED consists of a suite of nine rating systems for the design, construction and operation of buildings, homes and neighborhoods. Five overarching categories correspond to the specialties available under the LEED Accredited Professional program. That suite currently consists of: (1) Green Building Design & Construction, (2) Green Interior Design & Construction, (3) Green Building Operations & Maintenance, (4) Green Neighborhood Development and (5) Green Home Design and Construction (Anon., 2016e).

2.1.2 Green Building Requirements

According to OESD (2016), Building Sustainably Green buildings provide healthy, comfortable building interiors that maximize savings through the efficient use of energy and water and limit construction impacts on the natural environment. Green building requirements for projects requiring a Zoning Use Permit (UP) or Administrative Use Permit (AUP) are listed below. In addition, there are supplemental green building policies that ensure that we continue to divert waste from landfills, reduce energy and water usage in our buildings and help our community meet our environmental and Climate Action Plan goals.

Green building: A green building takes a holistic approach to design, sitting, construction and operation to enhance the well being of its occupants and minimize the negative impacts on the community and natural environment.

1. **Green Building Checklist:** Required for large (10,000 sqft. or more) development projects and projects adding one or more dwelling units.
2. **Green Building Consultations:** Consultations are voluntary. The Planning Department offers free green building consultations to help you improve your project.
3. **LEED Gold (or equivalent) for Downtown Projects:** New buildings and additions in the downtown area require LEED Gold certification or equivalent.

Energy: An energy-efficient building reduces greenhouse gas emissions, lowers utility bills and improves comfort.

1. **Energy Conservation Analysis:** Required for large (10,000 sft. or more) commercial projects and recommended for multifamily and mixed-use projects..
2. **Building Energy Saving Ordinance (BESO)** BESO requires building owners to complete whole-building energy assessments and publicly report the building's energy efficiency information. BESO is required prior to sale of a house or whole-building, except for large buildings over 25,000 square feet and on a phased-in schedule for all buildings, except houses.

Water: Controlling storm water runoff protects the San Francisco Bay and local creeks and watersheds. Construction activities are the largest source of storm water pollution in the Bay. Water conservation helps save water while saving money and energy.

1. **Reduction of Storm water Pollution:** All construction projects must manage storm water pollution. The Storm water Requirements Overview summarizes the onsite storm water treatment and required checklists for projects based on the quantity of impervious surface that they create or replace.
2. **Protecting Health of Creeks:** Requirements vary depending on project scope and distance from the creek.
3. **Fixing Leaks in Sewer Laterals:** Required for remodeling projects and when properties are sold.
4. **East Bay Municipal Utility District (EBMUD) Section 31 Water Efficiency Regulations:** All applicants for new and expanded service are required to comply.
5. **Bay Friendly Basics Landscape Requirements:** Required for permitted projects with new landscape of 500 square feet or greater and renovated landscape area of 2,500 square feet or greater. The Bay-Friendly Basics Landscape Checklist represents the 9 required practices from the Bay-Friendly Landscape Scorecard.

Waste: Conservation and reuse of building materials and recycling construction and demolition debris saves valuable resources and landfill space.

1. **Construction and Waste Demolition Diversion Plan:** Projects must divert construction and demolition debris away from the landfill through reuse, recycle and compost.
2. **On-Site Recycling:** Businesses and multifamily buildings must provide recycling service to comply with the Mandatory Recycling Ordinance for Alameda Country.

2.1.3 UniverCity

UniverCity is Burnaby's flagship sustainable mixed-use community, located adjacent to Simon Fraser University (SFU) on the top of Burnaby Mountain. Planning began with the SFU Official Community Plan in 1996, coupled with a land transfer of 320 hectares of forest land from SFU to the City that were incorporated in the Burnaby Mountain Conservation Area. SFU Community Trust, representing the university and associated stakeholders, implements development in the community in accordance with the adopted Community Plan and approval process. Currently approximately 3,000 residents live at UniverCity; SFU staff, faculty and students make up about 41% of this population. At full build-out, the population of this new urban community will reach approximately 10,000. The community currently includes a mix of residential developments and forms, offices, an elementary school, a child care centre, a grocery store and pharmacy, various businesses and offices and a number of cafes and restaurants. (Anon., 2016b).

UniverCity exemplifies many sustainable urban planning policies, technologies and practices, including:

1. Innovative rainwater management practices and aggressive requirements for on-site infiltration and retention, to protect downstream watercourses, as set out in the 2002 Storm water Management Plan. Monitoring and implementation of the plan is carried out with a multi-stakeholder Adaptive Management Committee, chaired by the City.
2. Requirements for sustainable building performance, water conservation and other green building practices. The Phase 3 Zoning Bylaw for the development requires at least 30% greater energy efficiency compared to the Building Code and offers a density bonus for achieving 45% energy efficiency.
3. Through a partnership with Translink, SFU Community Trust and VanCity, a reduced-cost community transit pass was available to residents between 2002 and 2011, to encourage increased transit ridership. Currently, 36% of UniverCity residents use transit for commuting which is the highest percentage of transit use in the City.
4. A Neighbourhood Energy Utility is currently being developed, to serve the campus and UniverCity, using a highly efficient and virtually non-polluting biomass gasification process that would dramatically reduce CO_2 emissions associated with heating and cooling.
5. The University Highlands Elementary School was built to LEED™ Gold standards and includes outdoor teaching areas and a curriculum strongly focused on the environment and sustainability.
6. A child care facility aims to be one of the first projects in Canada to achieve Living Building certification, an advanced green building rating that exceeds LEED™ standards.
7. Car sharing is available to residents through a car co-operative (Anon., 2016b).

2.1.4 City of Campbell River

In 2008, the City of Campbell River adopted the Green City Strategy, a commitment to move the City towards environmental sustainability. The City has identified key priority areas pertaining to sustainability planning, waste reduction, energy conservation and renewable energy and environmental management and stewardship. One of the key components of the Strategy includes a commitment to develop a greenhouse gas emissions reduction plan for the community and achieve carbon neutrality in municipal operations. The City of Campbell River is a signatory to the British Columbia Climate Action Charter, which requires that the City reduce and where necessary, offset the greenhouse gas emissions from City facilities, fleet and traditional services to zero emissions by 2012 (Anon., 2016c).

1. **Alternative transportation:** These modes of transportation help to reduce greenhouse gas emissions from vehicles and bring many health benefits.
2. **Climate action:** The City of Campbell River is one of 179 BC local governments that have committed to the BC Climate Action Charter. The Charter requires local governments to take action to reduce greenhouse gas emissions, report on community climate initiatives and become carbon neutral in municipal operations by 2012. The City of Campbell River's Community Energy and Emissions Plan and Master Transportation Plan encourage alternative transportation modes, such as, walking, cycling, public transit, carpooling and electric vehicles. These modes of transportation help to reduce greenhouse gas emissions from vehicles. Active forms of transportation also have great health benefits for participants. The City encourages alternative transportation through the installation of cycling and walking paths, way-finding signs, improved transit services and provision of electric vehicle charging stations.
3. **Environmental protection:** Through public consultation, the City has identified goals to ensure that Campbell River protects and conserves aspects of the natural environment for future generations. Environmentally sensitive areas are critical components in maintaining the community's natural attributes and livability. Ecosystem functions minimize risk for hazard lands, particularly those areas at risk of flooding or geo-technical instability. Goals also recognize that the conservation of nature and its ecosystems are a substantial strategy to both mitigate and adapt to climate change. Environmental Protection goals identified in the Sustainable Official Community Plan are to 1) Reduce impacts to Environmentally Sensitive

Areas. 2) Minimize air, land and water pollution. 3) Protect and maintain the urban forest. 4) Minimize the loss of sensitive ecosystem inventory sites.

4. **Food and agriculture:** Food and agriculture have emerged as significant focus areas for sustainable community development, community resiliency and long-term sustainability. Communities are looking at how planning for resilient food and agriculture systems benefit the local economy and overall community health. In 2010, the City of Campbell River developed an Agriculture Plan and incorporated food system targets into the development of the Sustainable Official Community Plan to support public interest in local food programs and the start-up of an agriculture sector in Campbell River. This planning process identified that there is significant opportunity for agriculture in Campbell River.
5. **Green building:** This section offers tips and tricks on how to make buildings more energy efficient. The City of Campbell River's Community Energy and Emissions Plan sets greenhouse gas reduction targets to help reduce community energy use. Green building techniques and energy efficiency and conservation are methods that help the City work toward these goals. To improve your home or office's energy efficiency, you may want to start by having an energy audit or making small energy upgrades, such as, improving insulation, installing efficient lighting and upgrading your heating system. *Rebates:* Many homeowners are making efficiency a priority when undertaking home renovation projects. Improved efficiency can make a renovation more affordable in the long term – and reduce your household environmental impact.
6. **Renewable energy:** The City of Campbell River is a signatory to the British Columbia Climate Action Charter and has a Community Energy and Emissions Plan that targets energy conservation and renewable energy generation.
7. **Waste reduction:** The average household in Campbell River produces 40 kilograms of garbage each month. There is a need to lower this number to reduce our environmental impact and stretch the life of our over-burdened landfill. Reducing consumption is the most effective measure anyone can take. Campbell River can also be made greener by diverting waste into the recycling stream, composting and finding ways to reuse items. In 2009, 38 per cent of waste was diverted through the multi-material drop-off recycling bins, curbside recycling and spring/fall yard waste collection programs available to all households. This is a great start, but can be improved further. "Taking the tons out of trash" (Anon., 2016c)

2.2 Green City Index

The Green City Index series measures cities on approximately 30 indicators across eight to nine categories depending on the region. It covers CO_2 emissions, energy, buildings, land use, transport, water and sanitation, waste management, air quality and environmental governance. About half of the indicators in each Index are quantitative – usually data from official public sources, for example, CO_2 emissions per capita, water consumption per capita, recycling rates and air pollutant concentrations. The remainder are qualitative assessments of the city's environmental policies – for example, the city's commitment to sourcing more renewable energy, traffic-congestion-reduction policies and air quality codes. Measuring quantitative and qualitative indicators together means the Indexes are based on current environmental performance as well as the city's intentions to become greener. The specific indicators differ slightly from Index to Index, taking into account data availability and the unique challenges in each region. For example, the African Index includes indicators measuring access to electricity and potable water and the percentage of people living in informal settlements. Each city receives an overall Index ranking and a separate ranking for each individual category. The results are presented numerically (for the European and the US and Canada Indexes) or in five performance bands from "well above average" to "well below average" (for the Asian, Latin American and African Indexes). Bandings are used in regions where levels of data quality and comparability do not allow for a detailed numerical ranking (Roland Busch, 2016).

More than 20 global experts in urban environmental sustainability from the following organisations advised the Economist Intelligence Unit (EIU) in developing the methodology for the Green City Indexes: (Roland Busch. 2016)

1. African Development Bank
2. Cambridge University
3. CITYNET (Regional Network of Local Authorities for the Management of Human Settlements)
4. European Commission
5. Ford Foundation
6. Harvard University
7. ICLEI (Local Governments for Sustainability)
8. ISOCARP (International Society of City and Regional Planners)
9. Inter-American Development Bank
10. Karlsruhe University
11. Natural Resources Defense Council
12. New York University
13. OECD (Organisation for Economic Cooperation and Development)
14. Regional Plan Association
15. Technical University Munich
16. UN-Habitat
17. University of Pennsylvania
18. URBACT
19. Vienna Institute for Urban Sustainability
20. World Bank.

2.3 Seven Steps to Green City

The following seven steps to green city have been suggested by Roland Busch (2016):

1. **Good governance and leadership at the metropolitan level:** National environmental regulations can have far-reaching impacts and are able to provide overall strategic direction. Yet city-level leadership is just as crucial. Indeed, at their best, national directives incentivize local governments to establish their own

initiatives. For example, the Mexican government has a plan to replace more than 45 million incandescent light bulbs with energy-saving bulbs by the end of 2012. In the US state of Colorado, legislation requires that 30% of all electricity produced should come from renewable sources by 2020. Germany has strong national legislation too, covering many environmental issues including building codes and water management. In addition, there are numerous examples in all parts of the world where national, regional and local jurisdictions follow common goals and set mutually re-enforcing standards. However, the Green City Index series also demonstrates that the national legislation needs to leave enough autonomy to cities to address their most pressing issues and make their own investment decisions. Crucially though, when cities receive more autonomy to set policies they also need the funding to implement them. More must be done to address this problem, especially in the developing world. "Urban government is the crucial level in addressing the urban environment." Xuemi Bai, senior science leader for sustainable ecosystems, CSIRO (Australia's national science agency) "The city is a living organism that needs to be managed as a single entity and just like any living organism, it needs to develop holistically." Nicholas You, Chairman, World Urban Campaign, UN-Habitat.

2. **A holistic approach:** Top-performing cities take a holistic approach to environmental problems, recognizing that performance in one category, such as, transport, is linked to success in others, such as, air quality. These cities often have strong mayors who set an overall strategy. They also have dedicated environmental departments, structured communication and joint target setting between departments with different responsibilities (for example, water, waste management and transport). One of the best examples from the series is Curitiba. As early as the 1960s, city officials implemented several integrated initiatives with the single goal of tackling the effects of rapid population growth. These included policies across a number of departments, including strategies to limit urban sprawl, create pedestrian areas and provide low-cost rapid transit. By the 1980s, one single integrated urban plan addressed issues, such as, the creation of green areas, waste recycling and management and sanitation. Singapore and Copenhagen, both leaders in their regions, also plan holistically. Singapore has the Inter-Ministerial Committee on Sustainable Development, which brings together many different departments to set an integrated strategy on sustainable development. Copenhagen has coordinators in each environmental department who meet regularly to exchange information. In the US and Canada Index, cities like San Francisco, New York, Seattle and Boston integrate their environmental programmes into wider development strategies that simultaneously revitalize their economies and make urban areas more livable. These cities stand out as examples pointing the way forward for urban governments where different departments manage different aspects of sustainable urban development, without following a consistent strategy.

3. **Wealth and Policies:** Wealth is important, but at the early stages of development the right policies matter more. In most regions the Green City Index series shows a clear link between greater wealth and better environmental performance. The reason is obvious: More affluent cities can invest more money in infrastructure and set aside more generous budgets for environmental oversight. However, money is not everything. In each Index, some cities with a below-average income clearly

outperform their peer cities with higher incomes. This was the case, for example, with Berlin and Vilnius in Europe, Bogotá in Latin America, Delhi in Asia, Vancouver in the US and Canada and Accra in Africa. Delhi in particular, with the third lowest average income in Asia (US$2,000 *per capita* versus the average of US$18,600), achieves a relatively good placing. This shows that less-well off cities can adopt policies or low-cost projects to improve environmental sustainability. For example, tree planting is becoming a common environmental activity in Asia, especially for cities with lower incomes. Beijing holds an annual "Voluntary Tree Planting Day" where some two million residents participate, including the president and most senior officials. Belo Horizonte in Brazil has legalized waste picking, which has a major environmental impact but also provides economic and social benefits for waste pickers. Low-income cities can also look to international agencies to finance environmental goals. One example is Vilnius, in Lithuania. The city took advantage of funding from the World Health Organisation's Healthy Cities Project to promote the use of cycling and public transport. It also drew on European Union funds to improve its water supply and treatment network. Wealth can even be counter-productive in the early stages of economic development, when more affluence often correlates with more emissions, more urban sprawl, lower density and more cars. The Asian Index shows that only when GDP *per capita* rises above approximately US$20,000 per person, a "tipping point" occurs and the trend reverses. Cities with a higher income start to consume relatively less water and generate less waste and carbon emissions. A central issue for cities in the developing world is to work towards limiting the environmental impact of rising consumption today, rather than waiting for attitudes to change as incomes grow. This can be done by investing in efficient infrastructure, initiating public education campaigns and setting targets – for example, for more renewable energy, green spaces and air quality as well as addressing the growth of informal settlements.

4. **Civic engagement:** Environmental performance is also a matter of civic engagement. This is seen in the European Green City Index, when comparing the results with an independent report from the European Foundation for the Improvement of Living and Working Conditions (EFILWC). The study measured the level of voluntary participation in organisations. The comparison found that the more volunteerism in the city, the better the score in the Index. In general, involving citizens in environmental decisions is an important element and there are several good examples throughout the world. One of these is the Seoul city government's "Green Seoul Citizen Committee" which is chaired by the mayor and has 100 members from non-governmental organisations and businesses. Another one is Porto Alegre's "participatory budgeting" programme, in which city residents and delegates meet annually to vote on a wide range of municipal spending priorities, including for environmental areas, such as, transport and sanitation. Yet out of more than 120 cities evaluated, just over half receive full marks for involving their citizens in important environmental decisions. It is clear that more needs to be done across the world to engage the public in sustainability issues.
5. **The right technology:** Technology plays an important role in reducing environmental impacts. For example, the "Sustainable Urban Infrastructure:

London Edition", a separate study by Siemens in conjunction with McKinsey and the Economist Intelligence Unit, found that technological levers alone could cut almost 44% of London's total CO_2 emissions. This would go a long way in meeting its overall reduction target of 60% by 2025, without requiring a change in lifestyle or consumption. At the same time technologies can reduce costs for energy, water or waste disposal. The London study referenced above found that about two thirds of the technological solutions it assessed would pay for themselves through energy savings. Collaboration between the private and public sector is also a useful way to pay for the up-front costs. In Berlin, for example, the "Energy Saving Partnership" involves private companies retrofitting public buildings and then benefitting from the cost savings. The scheme has received more than €60 million in private investment and has brought carbon emissions down by more than 600,000 tonnes. Other examples of cutting-edge technologies in the Index series include a state of the art waste-to-energy plant in Amsterdam that provides power to more than three-quarters of the city's households, while sending just 1% of the original waste to landfill. A new landfill in Rio de Janeiro will cut CO_2 emissions by 1.4 million tonnes by capturing methane gas. In Tokyo, technology has allowed the city to use gravity more effectively to deliver water, thereby reduce the need for pumps. It also employs advanced methods to enhance water quality, including ozone and membrane filtration systems. In addition, technology can help cities in the developing world "leap-frog" less sustainable infrastructure in the developed world. One example is Guangzhou's 71-storey Pearl River Tower, called the "world's greenest skyscraper", which includes state-of-the-art technology to utilize solar power and energy-efficient cooling systems.

6. **The green and brown agenda need to go hand in hand:** Developing cities often emphasise the so-called "brown agenda", which focuses on human health and poverty reduction, as distinguished from the "green agenda", which looks to improve the sustainability of eco-systems. The two agendas should go hand in hand, as many experts have pointed out throughout the Green City Index series. Among the benefits, adopting environmentally sound policies reduces municipal waste and sewage (and the spread of disease), improves the efficiency of energy and water provision and creates jobs and wealth through investment in infrastructure. The Johannesburg Development Agency, for example, has transformed decayed inner city areas by integrating urban environmental improvements with social and economic development. Environmentally, Johannesburg's efforts are helping to curb urban sprawl by drawing residents back to the rehabilitated city centre. In these central neighbourhoods, the city has built new mixed-income houses, increased access to municipal services and has extended the public transport network. In addition, the city has boosted security and culture to make these neighbourhoods attractive for businesses as well as for individuals. Unfortunately, in many parts of the developing world, the immediate demands of survival in some cities tend to prevent officials from integrating sustainability into their plans.
7. **Tackle informal settlements:** Experts agree that addressing informal settlements is key to the environmental agenda because they exist outside of formal planning policies and often lead to pollution through inadequate sewerage and waste management. With up to 70% of the African urban population living in slums, the African Green

City Index in particular shows a strong correlation between a city's environmental performance and the percentage of residents living in informal settlements. In brief, the fewer city residents living informally, the better the city performs. The Green City Index series also shows that cities have different approaches, but many experts favour upgrading and integration over removal. As an example, the state water company operating in Curitiba has extended water services and sewerage connections to all of the 1,790 households in the informal settlement of Vila Zumbi dos Palmares, which is located along the banks of one of the city's main water sources. The upgrading has helped to reduce sewage pollution in the river and cut back informal connections to the water system. Likewise, Medellín has connected its cable car system, Metro-cable, to informal sections of the city to provide the people living there with better access to central city areas and job opportunities. Mumbai has what it calls a "slum adoption scheme", which has extended door-to-door waste collection to 550 informal settlements. There are many other examples of attempts to integrate informal settlements that can serve as models for action.

2.4 Green City Principles

A Garden City is a fair, just and harmonious community. It is not restricted to new cities or towns or those built following traditional Garden City town planning, architectural or design principles. A Garden City is about community, not merely about architecture and urban design. It is about building a harmonious community, balancing the best of town and country together to a community where the measure of success is ultimately the happiness of the people who live in it. Below are listed twelve principles that we believe underlie a Garden City. Some are methods and others are objectives. In effect these principles represent doorways into the Garden City; you can enter using one of the many doorways. But contradict or deny any of the principles, then they will also prove themselves to be exits. We declare that any town, city or neighbourhood can be considered as a Garden City if it embraces the following principles (Philip Ross and Yves Cabannes, 2016):

1. **Residents are citizens:** Residents consider themselves to be citizens of the Garden City. This includes people who also work, participate and use the Garden City. They are aware that the town truly belongs to them. There is a culture of rights, duties and responsibilities that comes through citizenship. The town is run for the common good, reflecting and representing the common will with a belief in equality and fraternity as the city is run for the benefit of the many, not the few.
2. **The garden city owns itself:** The Garden City is ultimately owned by its local community and not by a series of landlords. This ownership and governance is derived from the people who live and work in the city and who are its citizens acting for the common good. If the Garden City is its own landlord then it is answerable to and controlled by its citizens, ideally as a Community Land Trust managed by democratic structures that make it both inclusive and accountable.
3. **The garden city is energy efficient and carbon neutral:** A Garden City has a harmonious relationship with nature and is energy efficient. A Garden City is a carbon neutral city and does not pollute; its planning, design and resources are deployed to achieve this goal. Citizens and the government in the Garden City have a collective responsibility in their daily lives to design and implement such policies. This could

be ensuring the provision of clean, safe and efficient public transport, the ability to navigate the Garden City by walking or cycling on one hand and the ability to reduce waste, recycle and reuse resources by citizens on the other.

4. **Provide access to land for living and working to all:** The Garden City promotes urban agriculture, the ability for citizens to grow most of their own food, even in an urban area. There is a right of free and fair access to the land for all residents to grow their own food whether it is through common allotments, common land, farms, productive streets and parks or private gardens. Alongside this is the right to affordable housing and also the right of access to resources in urban areas to build or run their individual or collective businesses or workshops. It is a productive city that aims at its own self-sufficiency providing opportunities for agricultural work, crafts, commerce and industry. Rents are provided to encourage self-sufficiency and regeneration, provided in partnership with tenants, not just for tenants. The goal is for the City to be productive and sustainable in its own right, not as a dormitory settlement or a place of mere consumption.
5. **Fair trade principles are practiced:** The Garden City is committed to the practices and ethics of Fair Trade implementing the credo that its prosperity is not built upon the suffering of others, whether inside its own city limits, inside its own country or internationally.
6. **Prosperity is shared:** The prosperity of the Garden City is shared in practice among all its citizens, not just among the rich, wealthy and establishment. Participatory budgeting through which citizens decide on the priorities for public and community investment is one of the key mechanisms in practice. To secure the wealth and trigger jobs among the community, it can create local or a complementary community currency and set up community banks.
7. **All citizens are equal, all citizens are different:** All citizens in a Garden City are equal regardless of how long they have lived there or how many generations of their family has lived there. There are no special privileges for anyone. A Garden City provides support, treats with dignity those with mental and physical disabilities and values each citizen, irrespective of their religious or sexual orientation.
8. **Fair representation and direct democracy:** There is a right to participate in the Garden City, in what the city does, how it is run and who does what. A Garden City can be made up of many cities and towns but each of these will be comprised of different neighbourhoods and communities, each with differing needs and aspirations. The prosperity of the Garden City is employed to help those in greatest need. Each community and neighbourhood should be empowered and encouraged to form its own free and open association, council or forum to represent and engage the views and needs of that local community. The Garden City will share its decision making. It will devolve some to representatives but also by engage directly and meaningfully with the citizens so all can have an informed say and collective decision making power on the priorities for the Garden City.
9. **Garden cities are produced through participatory planning and design methods:** A Garden City is in harmony with the landscape, water, air, nature and the surrounding countryside. New developments and housing have Garden City space and design characteristics and aim to promote the health and wellbeing of

its citizens, current and future and are developed through participatory methods on fundamental issues, not just cosmetic ones. Public spaces are widely available as an important concept as they provide the means for people to meet and share views and to integrate. These public spaces and facilities bring together young and old, rich and poor, those of different races, religions and backgrounds as a community that celebrates and rejoices in its diversity and exercises tolerance and freedom.

10. **A city of rights that builds and defends the right to the city:** In the Garden City there are universal rights for all citizens, such as, the right to clean air, the right to nutritious food, the right to adequate housing, the right to work and fair wages. There are not only individual rights but collective rights, such as, the collective right to enjoy the city and its majesty as well as collective civic and political rights. In traditional terms, as the City is held in common there is a collective right to these commons. The Right to the City is a superior Right as it is both individual and collective.
11. **Knowledge is held in common, shared and enhanced:** A Garden City is a mutual city that builds a culture of production, sharing and co-operation, not just in terms of its prosperity and governance but also in terms of the knowledge it acquires and generates. It shares and co-operates for the good of the City while still operating competition to create innovation and development.
12. **Wealth and harmony measured by happiness:** The wealth and harmony of the Garden City is measured in the happiness its citizens. It is the only true measurement of a successful city. Their happiness is not based upon the suffering or expense of others.

These are the characteristics of a Garden City. Not all can be present but the guiding principles of a new Garden City will be to: Share, Enjoy and Prosper.

2.5 Expert Advice on Green City

Canada is an urban country. Though the national mythology speaks of canoes and ancient forests, the daily reality for most Canadians is defined by SUVs and traffic jams. Recently, 18 of Canada's city leaders got together in Toronto to demand a new national agenda that reflects this reality. Cities in Canada are undergoing profound change that requires new thinking in terms of architecture, urban design, planning and governance. For the moment, about two-thirds of Canadians live in auto-oriented suburbs, according to research from Queen's University. But younger and older Canadians are showing a desire to go car-free and embrace life in neighbourhoods that are densely built and conducive to walking, cycling and transit. The prominent urbanists, architects and scholars have given their opinions on what things Canada's mayors should be considering: the tools, policies and ideals that will build the city of the 21st century, as given below (Alex Bozikovic, 2015.):

1. **Make people, not cars, happy:** According to Jan Gehl (founding Partner of Gehl Architects; former professor at the Royal Danish Academy of Fine Arts; author, most recently, of How to Study Public in Life), Some of the principles that have worked so well in Copenhagen and now New York could be applied in Canada: a people-oriented, humanistic city-planning model. Copenhagen has worked for 50 years to improve the people landscape. There is an overriding strategy to improve the streetscape and improve conditions for public life; a lot of things are done to

make the streets safe and appealing for people to walk and bicycle. The city is here to make the people happy, not the cars. Plus, Copenhagen has woven in sustainability goals, is working to be carbon-neutral and there are goals around health: People are invited to walk and bicycle in their day-to-day lives. It's about people moving about naturally.

2. **Decrease speed limits:** Gil Penalosa (Chair of the non-profit 8-80 Cities and former commissioner of parks, sport and recreation for Bogota) states that a pedestrian is hit by a car every four hours, on average, in Toronto – and those are the incidents where the police are called. We should lower the speed on neighbourhood streets to 30 km/h or less. When you're walking and the cars are going by at 20 or 30 km/h, you feel at ease; if they are going faster, it's not an enjoyable walk. So the community is much more walkable. It's good for mental health and physical health; it's good for economic development, because if you walk, you buy local. And it doesn't cost anything. It's just a question of changing the laws and changing the culture. This is not just an issue for bigger cities. In a city, such as, Parry Sound, Ont., the population is getting older and driving less. More than half of pedestrian deaths are of people over 60. They're not distracted or talking on their phones; it just isn't safe for them to cross the street.
3. **Empower city governments:** Richard Florida (Director of Cities at the Martin Prosperity Institute at University of Toronto's Rotman School of Management; founder of the Creative Class Group) reported that Canada has 80 per cent of its population in 2 per cent of its land area. The country has to wake up to the fact that cities are our future and they need the substantive power and the fiscal power to build infrastructure and rebuild the urban middle class. Right now, they are hamstrung. If highways were the infrastructure of 20th-century industrial capitalism, then transit, high-speed rail and better airports are the infrastructure of 21st-century capitalism. In Toronto, you see it; we have fallen so far behind on building the infrastructure, we need to move our people around, it's terrifying. Giving cities the power to do that would be a step in the right direction.
4. **Leverage density:** According to Brent Toderian (urbanist and former chief planner of the City of Vancouver), Canadian cities are getting denser and that's a very good thing – supporting sustainability, public health, diversity and creativity. A huge part of doing new density well is ensuring that, on top of great urban and architectural design, cities achieve public amenities and housing diversity to help make them livable and lovable for everyone. Cities like Vancouver have shown how "density bonusing" and "value capture" tools translate to great neighborhoods: new parks and people places; rental and social housing; and many other public benefits. This is how density bonusing works: When more density is proposed by developers, if it is considered reasonable, cities then negotiate additional public benefits as well. In value capture, if a city invests in something like public transit, it can apply a charge on development around that transit, reflecting how public investment has increased nearby land value. Cities around the world have used these techniques.
5. **Embrace the science of big data:** According to Anthony Townsend (researcher at NYU's Rudin Center for Transportation Policy & Management; author of Smart

Cities: Big Data, Civic Hackers and the Quest for A New Utopia) throughout history, cities have either evolved organically, with little oversight; or been planned from the top down, often without evidence [to back up the planning]. In the coming decades, all that will change as the seeds of a new quantitative science of cities begin to bear fruit. Already, mobile-phone-location trails have vastly improved our understanding of daily commuting patterns and the spread of epidemics. In the future, massive new streams of data about urban life, produced by smart infrastructure networks and our own personal devices, will allow us to predict how different designs for neighborhoods and buildings will affect traffic, carbon emissions and quality of life. And as the foundations of a rigorous understanding of cities take shape, we'll understand the fundamental forces that determine why some cities grow and others decline. Cities will still hold many mysteries, but our urban planners will finally be able to shine a light on much of what makes them tick.

6. **Mix residences and workspace:** Vishaan Chakrabarti (principal of Shop Architects; professor at Columbia University) stated that we should be reinventing the nature of the urban building. Think of a model where an individual can rent a private room, but where there are shared amenities, including communal kitchens, work spaces, child care. I think younger single people and also empty-nesters aren't interested in having all that infrastructure for their private residences; and they're also looking for shared experiences. If there's communal space, you begin to interact with people differently. Ideas come out of that and new experiences. This applies also to blocks and neighbourhoods. If you can create living and working environments that are more proximate to each other, you cut back on commute times. People used to walk to work. People used to bicycle. There's nothing new about that. Imagine that the city is a quilt of living and working. Entrepreneurialism comes out of that; a lot of affordability comes out of that. It doesn't work for everyone, but there is now a population of people that forms a market for a different kind of living.
7. **Turn streets into destinations:** According to Janette Sadik-Khan (principal at Bloomberg Associates; Commissioner of New York City's transportation department from 2007 to 2013) to see where smart cities are headed, don't look up at the newest skyscraper; the answer lies beneath your feet. Those that succeed in this urban age will be the ones that design sustainable mobility into the streetscape. Dedicated bus lanes can help speed transit; a safe bike-lane network and a bike-share system located near transit can help expand the reach of buses and trains into more neighbourhoods. City residents are hungry for better streets. Improvements, from plazas and parklets to benches and well-designed sidewalks, help turn them into destinations. These aren't just quality-of-life improvements; they're investments in our economic future. As we saw in New York and now in cities worldwide in our work for Michael Bloomberg, better streets mean better business. I often say, if you want to save the world, move to a city; if you want that city to be a safer, vibrant, more efficient place to live, start by building a bike lane.
8. **Redevelop the inner suburbs:** Ellen Dunham-Jones (architect, author of Retrofitting Suburbia: Urban Design Solutions for Redesigning Suburbs; professor at Georgia Tech in Atlanta) states that The biggest challenge is to connect affordable housing

with affordable transportation. Americans in the lower half of all incomes are paying more on transportation than on housing. We should look at suburbs – and retrofit older commercial corridors with more housing and more access to buses, so that you can drive but you have the alternative of taking transit. Many of our aging suburban properties are underperforming: strip-mall retail, office parks. This is underused property, yet it has terrific infrastructure and a relatively central location in a now-expanded metro area. If we can redirect growth to those areas, in a sustainable pattern, it's a great opportunity to solve the problems of affordability and it always helps the larger community more to redevelop an existing dead property than to tear down trees.

2.6 Green Development

Every other species on this planet produces no "waste", instead, outputs from one organism become resources for another, creating a 'virtuous cycle' and healthy ecosystems. Can human developments also function this way? "Green Development" is a broad concept that addresses this question, with urban development approaches and practices that strive to minimize negative impacts or ideally have a net positive impact, on the environment and nearby ecosystems. Green Development is a rapidly growing field that combines ecological principles with advanced technology, often including social and economic benefits as well. The green development "toolbox" includes:

1. Planning and land use development regulations, policies and practices that aim to create "complete communities" that are vibrant and encourage walking, cycling and transit use;
2. Green building certification and rating systems, such as, LEED™ (Leadership in Energy and Environmental Design) that quantify energy efficiency and other green design goals;
3. Policies and bylaws requiring energy efficiency standards and encouraging green design;
4. Technologies and systems that capture and re-use or re-purpose "waste" products from materials (*e.g.* construction waste) and energy (waste heat) and re-use them as resources; and
5. Engineered "ecosystems", such as, constructed wetlands, green walls, green roofs and rain gardens / bio-swales, that perform specific functions, such as, capture and cleanse storm water runoff (*Anon.*, 2016b).

Bioswales are landscape elements designed to remove silt and pollution from surface runoff water.

Green building and landscape initiatives for public and private developments are frequently considered and advanced by the City through community plan development, bylaw amendments, comprehensive site plans and specific development and building projects. Many developments have incorporated LEED™ standards, high energy efficiency and other significant environmental considerations to help set the framework for broader application of Green Development initiatives in the City (Anon., 2016b).

2.7 Greenest City in Canada

More than 80% of Canadians are living in cities – and this number is expected to increase in the coming years. The Economist Intelligence Unit, sponsored by Siemens, conducted a research project, which resulted in the Green City Index. Based on 9 evaluation criteria (CO2, land use, energy, buildings, air, transport, water, waste and environmental governance), the study ranked 27 major Canadian and American cities from best to worse. Undoubtedly Vancouver is the greenest city in Canada – and by far –, Vancouver is praised for the quality of its air (ranked first in all 27 North American cities, with 4,02 metric tons of CO2 emission per person), its transportation system, its land use (12% of the territory is green, which is remarkable for a city with such high population density), the management of its waste (55% of the municipal waste is recycled – 26% is the average!) and its overall environmental governance (Julia Taylor , 2013).

Vancouver city

2.8 Greening of Canadian Cities

Call it the second wave of urban greening: hot on the heels of urban farming, urban forestry is emerging as the new frontier of sustainable urban planning and land use. Canada's urban forest exists in many forms. Public parks, streetscapes, natural areas and yards form a complex forest ecosystem that can be dramatically different than the vast forest landscapes we typically imagine.

Since, Canada's first Urban Forest Strategy was developed 2006, urban forest policies have been developed by more than 20 municipalities including major urban centres, such as, Toronto, Vancouver and Halifax (Ngaiohotte, 2015).

Like urban farming, urban forestry requires more than simply translating rural land management practices into an urban setting. In particular, operating in an urban environment requires a high degree of engagement and consultation with local people and organizations. Conventional forestry principles are combined with modern information management and communications tools to achieve a broad range of social, ecological and economic benefits.

The Social and Economic Values of Canada's Urban Forests: A National Synthesis unearthed a wealth of information about the benefits of urban forests and identified numerous opportunities for government and public involvement in urban forestry (Ngaiohotte, 2015).

Trees are known to improve air quality; an important service in an urban setting that can improve the health and longevity of residents. They also moderate temperature by providing shade and cooling in the summertime and reducing heat loss by winds in the winter, which translates into lower residential energy bills and lower carbon dioxide emissions associated with energy use. Trees also reduce costs to municipalities by helping rainwater to infiltrate into the soil, thereby lessening wear and tear on storm water infrastructure. And urban forests can produce substantial economic benefits by raising local property values and occupancy rates, attracting shoppers to local business districts and drawing recreational visitors and tourists. To meet the growing demand for urban forestry professionals, colleges and universities across Canada and the USA are developing programs to help students develop skills ranging from planning, geographic data management, communication, outreach and extension in addition to tree selection, planting and maintenance. For example, the University of Toronto hosts an annual urban forestry field course, Fleming College has offered an Urban Forestry Certificate program since 2012 and the University of British Columbia has launched its full undergraduate degree program in Urban Forestry (Ngaiohotte. 2015).

2.9 Create Clean, Green Cities

Municipal solid waste services are fundamental to the quality of life in our communities, our health and our environmental future. The challenge is to continue to reduce the amount of residential waste we create and to capture the value of any waste created as another public resource. We must also extend waste reduction and recycling practices to all commercial and industrial activity. We cannot keep digging and filling up holes with our garbage or releasing toxins from its disposal into our air and water. In order to meet these challenges municipalities must retain accountability, flexibility and control over their solid waste services. Contracting out garbage services means municipalities lose control and flexibility to implement waste diversion programs like recycling and composting. It is critical that municipal governments retain control over waste collection and recycling. Stronger waste reduction, reuse and recycling programs aimed at private industry must support effective, universally-accessible public systems (CUPE, 2016).

2.10 Social Strengths of Green Cities

City parks, gardens and open spaces are often filled with people of all ages and backgrounds looking for a brief respite from their bustling urban lifestyle. Public green spaces draw people for many reasons: to relax, find inspiration, exercise and to gather with friends and family. While there, they inevitably interact with others, be it a smile in passing, joining in informal discussion or play. These seemingly banal and unimportant social interactions can contribute substantially to community enhancement. Researchers and social theorists have looked at the relationship between urban green spaces and community building. From casual interactions to organized events, urban green spaces can provide a neutral space within which people gather, people from different backgrounds come together and interact socially and where relationships and partnerships take form. While personal goals or desires are pursued, community building and increased social capital also emerge, particularly if people share work on a project or goal. A sense of connectedness fosters greater trust and cooperation among individuals, potentially leading to greater community resilience in the face of adversity. These social dynamics help promote community wellness and social cohesion, critical components of a well-functioning society (Wolf and Rozance, 2013).

Green spaces can be the neighborhood places that foster a greater sense of connectedness, trust, cooperation and community resilience

2.11 Landscape Architecture for Green City

When a city does its landscape design correctly, environmental issues, such as, water can be addressed. Because water is so precious. So the cleaning, capturing of water, retaining it, allowing it to go back to replenish aquifers — super important. Of course, planting is extremely important because planting helps to clean water, absorb water. But it also helps to cut down particulate matter, clean the air. It helps to moderate temperature fluctuations. And it makes things more beautiful as well. And then making sure that people can walk safely on sidewalks, that people are connected on foot, that you can ride a bicycle. And then advocating to really cut down on the need for cars. Building better scaled streets so people are in street environments that are more conducive to street life. These are all really important environmental issues (Martha Schwartz, 2012).

Landscape architecture is a lot more than green roofs. The spaces and places that we actually travel on and spend our time on are the streets. People spend maybe eight to 10 times more time on the streets than they do in a park. It's very rare people go to parks, but people are on streets every day. Jan Gehl, a great urbanist from Copenhagen, has promoted the use of streets and the awareness that these are our major public spaces. But they don't just function to beautify. He convinced the city that if they invested in bike lanes the cost of healthcare in general for the population would go down. And that's exactly what has happened. *"Most of our urban environments are not our parks — they are our streets, our sidewalks."* There is also the deployment of the natural environment. Street trees, for example. Rob Adams from Melbourne, Australia has figured out a way in which the street trees are actually being planted in the streets. Each segment of the street actually is a catchment basin for water, which is stored under the street. And then the water's taken out to water the trees during the dry season. But the trees actually bring down the demand for cooling buildings. And all this is an integrated system (Martha Schwartz, 2012).

3. CITY LIVING

3.1 Thoughts on What Cities Must do to Thrive in the 21st Century

According to Julián Castro (2016) the cities must do three things to thrive in the 21st century:

1. **Invest in shared prosperity:** Cities have been among the greatest drivers of human progress throughout history. But it's also true that far too often, that progress was created for some citizens at the expense of others. For example, the same cities to revolutionize public transit or erect new highway systems sometimes did so in ways that destroyed low-income communities and cut off many of their residents from opportunity. Investments to fund new schools, build new parks and create new business districts in wealthier communities too often dwarfed investments in low-income neighborhoods—that's if those investments were made at all. To dismantle entrenched inequality and promote shared prosperity, cities will have to actively correct the mistakes of the past. Brainpower is the key to creating a dynamic, agile workforce and that opportunity has to start early in life. So cities ought to invest in high quality education, particularly early childhood education. Cities must also make smart investments in ports, airports, roads and waterways. They should ensure that all households have access to world-class broadband. And cities need to invest in affordable housing so that every child—no matter where he or she starts in life—has the chance to move up and achieve their dreams.
2. **Embrace regionalism:** Cities must strengthen their connection to surrounding communities—urban, suburban and rural. Here in the United States, the Obama Administration is helping to lead a transformation in how local communities make planning decisions. We've joined forces with mayors and county executives, as well as business leaders and non-profit groups to create unified regional plans that link downtown districts with surrounding counties. As a result, suburbs are now investing in their neighboring cities' downtown areas, helping to create vibrant urban cores that can then boost the entire region's economy. Cities are expanding light rail and other transit so that it connects with rural communities. Rural communities are taking the lead in bringing healthy food options to more families, especially in urban neighborhoods that have long gone without adequate grocery stores. The challenges of tomorrow are too complex for individual communities to try to solve alone. By working together and including every community's voice in our planning decisions, we're helping to create a future where all can thrive.
3. **Foster greater openness and freedom:** There is an intrinsic desire for freedom within each and every one of us. It's part of who we are. So while we invest in housing, infrastructure and schools let's also ensure that every person can get fair legal representation regardless of how much money they have. Let's continue the progress we've seen around the world in the long march for full equality for LGBT people, especially for transgender people who are often at the greatest risk of homelessness or becoming the victim of a hate crime. Let's promote the advancement of women's rights, recognizing that gender equality is an integral part of a just, secure and democratic society.

Free, open societies are those best equipped to foster creativity and entrepreneurship. They are the communities that will prosper most in this Century of Cities. Former Mayor of New York, Ed Koch, once said that this city was "where the future comes to audition." That's still true of New York—and it is also true of cities around the world. The world's cities are often at the vanguard of smart policies and practices to improve the human condition. Mayors of these cities are embracing diversity and fostering greater inclusiveness. Mayors are bringing together the "Three Ps" of partnership—the people, the public sector and private industry—to meet the demands of our rapidly changing world. And they are investing in innovative solutions that will improve how we use our land, water and energy resources to make our planet more sustainable for future generations.

3.2 City Living in the 21st Century

In my opinion living in a 21st-century city presents more disadvantages than advantages, even though nowadays cities are widely thought to be the place to live. One of the first disadvantages I can think of is traffic. Traffic itself is a cause of stress among drivers and citizens in general. A multitude of cars in the same place is more likely to be caught into accidents. Therefore traffic may be considered dangerous. Traffic also affects the environment, causing pollution, noise and possible consequent diseases, such as, asthma. Last winter, Milan, second largest city in Italy, had to face serious problems, due to the high level of pollution and the absence of rain for several weeks. It is almost a matter of fact that cities are over-crowded nowadays. People living in large cities may feel stressed by the general chaos, for example when using public transport. It is true that there are plenty of facilities in big cities, but it might be difficult for them to be of some benefit to all citizens. Over-crowding often affect cities by making them dirty, for instance, from my personal experience, London, Sydney and Paris. Big cities are also believed to be dangerous and more exposed to crimes than towns, but a recent study showed that Melbourne, second largest city in Australia, is the safest and most livable city in the world and I agree with it. Large cities may lack affordable housing and rents might be too high for an average income. Bills, food, transport and the overall cost of living are sometimes extremely expensive. This can lead to an increasing criminality and it might be a reason why some people prefer to live in smaller cities or towns. As stated, cities are sometimes expensive, but on the other hand they offer more job opportunities and sometimes jobs in big cities are better paid than anywhere else. Concluding, I believe today's cities are stressful and chaotic and I agree they offer few advantages (The Mixer, 2016).

3.3 Modern Cities

Around the world, modern cities are centers of economic activity. Their skyscrapers and bustling marketplaces are testament to the development they have driven. Overall, urbanization has brought countries opportunity, prosperity and health. At the same time, modern cities are filled with shadows. Beneath the skyscrapers, behind the marketplaces, the lives of city dwellers are hidden from view. This is especially true for the urban poor living in slums or other informal settlements, which are often excluded from estimates of cities' economic and health development. Relying on city averages, rather than examining differences between neighborhoods and urban subgroups, has further obscured inequalities within cities (WHO, 2016).

The world is urbanized and will become even more so in the future. The rapid increase of people living in cities will be among the most important global health issues of the 21st century. Cities offer unique opportunities for residents to benefit from education, health and social services and to optimize their health and quality of life. At the same time, health hazards, such as, poor housing conditions, lack of access to safe water and sanitation and economic downturns are fuelling a range of health problems. If left unchecked, climate change will multiply these and other urban health risks through heat waves, storms and changing weather patterns. Meanwhile, overwhelmed by the speed of growth, health services in many urban settings are poorly equipped to manage current and emerging public health threats (WHO, 2016).

3.4 Livable Cities

Throughout time people have noted that nearby nature is an important part of places that are livable and have a high quality of life. Having gardens, parks and trees in cities leads to life satisfaction and a positive outlook. Urban green spaces can provide a neutral space within which people come together, social interactions occur (that include people from different backgrounds) and relationships or partnerships take form. While personal goals or desires are achieved, community building and increased social capital also emerge, particularly if people share work on a project or goal. Individual benefits, improved public health and social resilience are potential positive outcomes. A study in urban public housing found that the presence of well managed vegetation was very important feature in promoting the development of social ties within urban public housing. People prefer natural over hard-scape settings and preferences are predictors of the use of environments. Hardscape refers to hard landscape materials in the built environment structures that are incorporated into a landscape. A study found that urban residents dislike and fear treeless, empty common spaces. The addition of trees and grass dramatically changed their perceptions of those spaces. Older adults who have more exposure to green common spaces report a stronger sense of unity among residents within their local neighborhood and experience a stronger sense of belonging to the neighborhood. Strong community relationships may result in individuals being more likely to work together to achieve common goals (*e.g.*, cleaner and safer public spaces), to exchange information and to maintain informal social controls (*e.g.*, discouraging crime or other undesirable behaviors). These situations can influence public health. Communities where residents express high mutual trust and reciprocity have been linked with lower homicide rates. Neighborhoods lacking social cohesion and community wellness conversely, have been related to higher rates of social disorder, anxiety and depression (Wolf and Rozance, 2013).

Urban green can improve livability and value in mixed use residential and commercial settings

Individuals and groups within communities with strong social cohesion and social capital experience many positive benefits

Social and cultural events are often held in parks and gardens, providing opportunities for people to interact and develop social bonds

4. CITIES OF THE FUTURE

4.1 Re-imagining Our Cities for the 21st Century.

Having fallen in love with the structure of the old freight railway from the street, it was what High Line founder Robert Hammond saw when he got up on to the railway itself, that convinced him to try to save it. Here, running through the middle of Manhattan, was a mile and a half of wild flowers. That was in 1999. Now, 12 years on, New York's park in the sky attracted more than 3.7 million visitors last year, has generated $2bn-worth of private investment (£1.3bn) surrounding the park and is predicted to exceed $900m (£562m) in new tax revenues for the city over the next 20 years. Such figures are not to be sniffed at. For what started out as a rescue attempt by two neighbourhood residents (Hammond and Joshua David) with no design background, no plan and no money, has created the city's second most popular tourist attraction after the Museum of Modern Art. It is decisive evidence that it is increasingly the quality of our parks and public spaces, not the towering ambition of our skyline, that make our towns and cities stand out. But it is what the High Line – and to a great extent London's Olympic Park – will make possible, that will define its legacy. By making a virtue of an industrial relic, the High Line offers a new, 21st-century experience of the city – not a sanctuary from it. In fact, the High Line offers views of the city that you'd get nowhere else. So, suddenly, in the middle of Manhattan, is a piece of infrastructure that is not only a green lung, but looks and smells beautiful; is a place that offers a new route to work, space for recreation and contemplation and is regenerating an area. Once we have a park in the sky, the city is ours to re-imagine. Indeed, it must be re-imagined if we are to face the challenges we seem destined to face (GSB, 2012).

London Olympic Park Arial Photo

The winning idea, called "Pop Down" by Fletcher Priest Architects, proposed reinventing the disused tunnels of the Mail Rail that run for 10.5m as a public experience and urban mushroom farm. London with its history of hidden tunnels and lost rivers provided the inspiration for the idea, while the team tried to capitalise on the same sense of drama that the High Line creates by being in between the fabric of the city (GSB, 2012).

Mushroom farm; Underground farm

Between 2009-10, J&L Gibbons and Muf architects delivered 10 projects as part of Making Space in Dalston. The Eastern Curve, once a overgrown and abandoned piece of railway land, was one of them. It is now a popular public space with trees to offer shade, warm weather cooling and improve air quality and vegetable and herb growing areas for local people. In the wake of the Olympic Park's success, it's easy to forget how ambitious a project it was to transform 50 hectares of post-industrial wasteland in East London. More than 2m tonnes of soil was decontaminated and the largest wildflower meadow the UK has ever seen was planted. While we are unlikely to see investment in public space on this scale again for a generation, as with community-led projects, such as, the High Line, if green infrastructure is designed and managed to a high quality, it can give credibility to wider regeneration issues. In straitened times, the creative reuse of our existing landscapes and their architecture, might be our only option (GSB, 2012).

4.2 Building the City of the 21st Century

Cities were and remain the center of innovation, exchange, interests, culture and art. Increasing urbanization is taking place all over the world. In Western countries urbanization is already slowing down and the challenge is to optimize and update existing structures. With the change from an industrial society to a service and knowledge society, cities offer numerous development opportunities for today's highly individualized and fully connected society. The main challenge of cities which want to play a role in the competition for businesses and residents face will be to offer a high level of living quality (Superfuturegroup, 2016).

4.3 Cities of the Future

What lies ahead for our urban world and for the cities that comprise it? Looking to past trends is a useful way of imaging the future, but unforeseen events are inevitable and will certainly shape the future of cities in ways that cannot be predicted fully. Cities will differ from one another based on several factors. Their access to information, technology and the global marketplace will shape them, as will the ways in which they are governed. Migration will continue to influence the size and nature of their populations. Climate change impacts and new disease pandemics could trigger mass migration at an unprecedented scale, altering demographics within countries and cities, changing borders or generating conflicts. Cities without adequate planning or proper governance will find it increasingly difficult to provide affordable land, decent housing, adequate transportation and public services. As a consequence, their political legitimacy will, sooner or later, begin to erode. Non-governmental organizations or the private sector may attempt to fulfill roles previously held by local authorities and fragmentation will ensue. In this scenario, slum dwellers and the urban poor will continue to be overlooked and disparities within cities will continue to grow. At the same time, cities present substantial opportunities for the future. The most prosperous cities will be those that design sustained, comprehensive visions and create new institutions or strengthen existing ones, to implement this vision. This will bring them to look for new methods of close cooperation with regional and central governments and other actors, such as, the private sector, all the while ensuring an equitable distribution of opportunities and sustainable development (WHO, 2016).

4.4 Greenest City in the World by 2020

Vancouver, in British Columbia, Canada, wants to establish itself as "the greenest city in the world by 2020" by demonstrating that economic growth and the welfare of its citizens depends on developing renewable, rapid transit systems and promoting cycling and walking to curb car use. It is one of dozens of cities worldwide working towards improving the life of their citizens while reducing fossil fuel use, but it claims to have the most ambitious targets. Many city administrations in Europe have the support of their governments, but in other countries—particularly Australia and Canada, where governments are actively promoting fossil fuels—cities are having to act alone. Among the "Green Vancouver" targets are doubling the number of green jobs in the city by 2020, from a 2010 baseline of 16,700 and making all new buildings in the city carbon neutral from 2020, while dramatically cutting emissions from existing buildings. Progress towards meeting the city's impressive list of targets includes reductions in air pollution, waste, water use and car journeys. Other aims are to provide a green space within five minutes walk for every citizen, planting thousands of trees and growing food locally. Between 1996 and 2011, while the population in the city center increased by 40 percent, there was a 25 percent decrease in the number of vehicle journeys and a rise in the use of public transport, dedicated cycle routes and walkways. In contrast, Vancouver, which has a population of 600,000, believes that all its power can come from renewables—although getting all heating and cooling and transport without using fossil fuels may take until 2040, depending on whether there is any help from the central government (EcoWatchContributor, 2015).

REFERENCES

Alex Bozikovic. (2015). Expert advice on building the city of the 21st century. The Globe and Mail. Feb. 13, 2015. http://www.theglobeandmail.com/news/toronto/expert-advice-on-building-the-city-of-the-21st-century/article22998067/

Anon. (2016a). Get Your City to Go Green. Earth Day Network, Washington, DC 20036. http://www.earthday.org/campaigns/green-ities/?gclid=CNnf_cm3ls4CFRBrfgod0asGPA

Anon. (2016b). Green Development- City of Burnaby. https://www.burnaby.ca/City-Services/Planning/Environmental-Planning/Green-Development.html?PageMode=Print

Anon. (2016c). Green City- City of Campbell River. http://www.campbellriver.ca/your-city-hall/green-city

Anon. (2014d). World's population increasingly urban with more than half living in urban areas. 10 July 2014, New York. http://www.un.org/en/development/desa/news/population/world-urbanization-prospects-2014.html

Anon. (2016e). The Leadership in Energy and Environmental Design (LEED). Environment and Ecology. *environment-ecology.com › Environment and Architecture.*

CUPE. (2016). Create clean, green cities. Canadian Union of Public Opinion. May 27, 2916.

EcoWatchContributor. (2015). This City Plans to Be the Greenest City in the World by 2020. EcoWatch, May 13, 2015. http://www.ecowatch.com/this-city-plans-to-be-the-greenest-city-in-the-world-by-2020-1882040494.html

GSB. (2012). Re-imagining our cities for the 21st century. Guardian sustainable business Sustainable business blog. 25 October 2012. https://www.theguardian.com/sustainable-business/blog/reimagining-cities-high-line-reinvigorated

Julián Castro. (2016).Three thoughts on how cities can thrive in the 21st century. Sustainable community development: Remarks to global leaders at the United Nations Habitat III *conference in New York City* on *May 16, 2016.* https://medium.com/@SecretaryCastro/three-thoughts-on-how-cities-can-thrive-in-the-21st-century-183469f860f9#.n5a1shl0o

Julia Taylor. (2013). Which Canadian Cities Are the Greenest and Why?. Spacing. June 24, 2013. http://spacing.ca/national/2013/06/24/which-canadian-cities-are-the-greenest-and-why/

Margaret, Rouse. (2016). LEED (Leadership in Energy and Environmental Design). http://searchdatacenter.techtarget.com/definition/LEED-Leadership-in-Energy-and-En

Martha Schwartz. (2012). Interview: Designing Green Cities To Meet 21st Century Challenges. Environment 360 digest. December 5, 2012. Interview by Diane Toomey. https://e360.yale.edu/digest/...green_cities__to...21st_century.../3711/

Matthew, L. Kimball. (2016). Baltimore City Green Building Standards: New Regulations Pose Unanswered Questions. *www.nilesbarton.com/.../baltimore-city-green-building-standards-new-re...*

Mitchell L. Moss, Hugh O'Neill and Appleseed. (2012). Urban Mobility in the 21st Century A Report for the NYU BMW i Project on Cities and Sustainability. https://wagner.nyu.edu/.../NYU-BMWi-Project_Urban_Mobility_Report.

Ngaiohotte. (2015). Urban forestry and the greening of Canadian cities. Spacing, November 02, 2015. http://spacing.ca/national/2015/11/02/urban-forestry-greening-canadian-cities/

OESD. (2016). Green Building Requirements. Energy and sustainable development. Office of Energy and Sustainable Development. http://www.cityofberkeley.info/Green_Building_Requirements/

Philip Ross and Yves Cabannes. (2016). 12 Garden City Principles. 21st Century Garden Cities of Tomorrow'. https://gardencities.info/reference-documents/12-garden-city-principles-from-21st-century-garden-cities-of-to-morrow/

PRB. (2016). Human Population: Urbanization. Population Reference Bureau. http://www.prb.org/Publications/Lesson-Plans/HumanPopulation/Urbanization.aspx

Roland Busch. (2016). The Green City Index A summary of the Green City Index research series A research project conducted by the Economist Intelligence Unit, sponsored by Siemens. https://www.siemens.com/entry/cc/features/greencityindex_international/all/en/pdf/gci_report_summary.pdf

Superfuturegroup. (2016). How do we want our cities to look like in the 21st century? Imagine a city you cannot see on Google Earth!. An article by superfuturegroup. 09 June 2016. http://futurearchitectureplatform.org/news/36/how-do-we-want-our-cities-to-look-like-in-the-21st-century/

The Mixer. (2016). City living in the 21st century – Essay. The Mixer. Language Exchange for Everyone. 27 Feb 2016. http://www.language-exchanges.org/node/120173

WHO. (2016). Modern cities-WHO. www.who.int/...centre/.../hiddencities.../p1_who_un_habitat_hidden_citi

Wolf, K.L. and M.A. Rozance. (2013). Social Strengths - A Literature Review. In: Green Cities: Good Health (www.greenhealth.washington.edu). University of Washington : College of the Environment. depts.washington.edu/hhwb/

8

Elements and Strategic Measures in Maximizing Protected Cultivation of Vegetables Towards Sustainable Horticulture Development in Asian Countries

Brahm Singh

SYNOPSIS

To maximize vegetable production through protected cultivation for sustainable horticulture development in Asian countries, there is an immediate requirement of developing adequate human resource in this area. Another aspect requiring priority is development of suitable varieties/hybrids of vegetable crops for different protected structures for vegetable production in different diverse regions of Asia. Being a different, distinct and potential new area involving vast multi-disciplinary aspects of agriculture, establishing schools of protected horticulture/agriculture in peri-urban and urban areas of the country would be of great help to generate appropriate human resource and technologies, such as, soilless vegetable cultivation, hydroponics, aeroponics, vertical vegetable production, microgreens, mechanization and maintenance of poly houses/structures, pollination devices, robotics, vegetable grafting, specialized disease free seed production, container farming organic protected cultivation, good vegetable production practices, water harvesting, post-harvest handling of produce and use of LED. There is a dearth of empirical knowledge on the technical as well as the economic feasibility of low cost protected-cropping system for vegetables.

1. INTRODUCTION

Asia is one of the most populous and largest continents. It covers 30 per cent of world's land area and 60 per cent of world's population (7.4 billion). As per 2016 estimates Asian countries support 63.37 per cent (4,436,224,473 people) of global human population with a growth rate of 0.94 percent. By 2050 this population is expected to be 75.24 percent of world population (5,266,848,432 people). Therefore, the requirement would increasingly be high of food and more so for nutrition. Horticulture is a major contributor of human nutrition security in which vegetables play significant role. Vegetables are potential crops for urban and peri-urban horticulture. The largest and second largest producers of vegetables, China and India, respectively, are major countries with huge population in Asia. In Asian countries vegetable production has increased 25 per cent or so over the last decade. As per estimates these countries require to increase vegetable production by 47 per cent to meet the growing nutritional needs of growing population by 2050. Besides, more and more people are added to middle income group in these countries increasing the affordability of nutritional foods like vegetables. There are challenges in vegetable production due to rapid urbanization and climate change resulting in host of abiotic and biotic stress. At the same time there are opportunities to enhance production of vegetable crops using new age technology. To deal with these above mentioned stress, protected cultivation of vegetable crops is a potential technology to maximize their production and productivity leading to sustainable horticulture development. China alone has more acreage of horticultural crops production under protection than that of the other countries combined. In 2010, there was more than 4 million hectares under protected vegetable production in China which was over 95 % of all protected production in China and over 80 % of all protected vegetable production in the world. The potential of protected vegetable production have amply been demonstrated in China and rest of the world. World Vegetable Centre, formerly, Asian Vegetable Research and Development Centre, Shanhua, Southern Taiwan with the slogan "prosperity for poor and health for all" is promoting protected vegetable research and development keeping in view its vast potential in context with climate change and augmenting nutrition availability.

2. OBJECTIVE

This article is aimed to describe elements and strategic measures to maximize production of vegetables leading to sustainable development of horticulture in Asia, where approximately 60% of world population is to be provided nutritional security coupled with food security. Protected vegetable production has emerged as a potential technology to enhance vegetable production on sustainable basis both at small holder and large holder farmers. This technology involves innovative methods of plasticulture to smart and hi-tech horticulture. Protected cultivation of vegetables has been found ideal both for urban and peri-urban horticulture.

3. WHAT IS PROTECTED CULTIVATION OF VEGETABLES?

Protected cultivation of vegetables can be defined as a cropping technique wherein the micro climate surrounding the plant is modified /controlled partially or fully as per the requirements of the plant species grown during their period of growth. It enhances returns to the farmers with added advantage of assured remuneration on account of quality produce and in some cases off season production, early production and prolonged production-a desirable trait in vegetables production.

3.1 Advantages of Protected Cultivation of Vegetables

There are several advantages of adopting protected cultivation of vegetables in different agro climatic conditions of Asian countries. Some of the major advantages of protected cultivation are listed below

1. There is large increase in yield, produce quality and revenue under protected cultivation of vegetables (Table-1)

2. Off- season and year-round production of vegetables is possible under protected cultivation practices. This allows farmers to take advantage of market seasonality and off season higher prices. It is well established advantage of protected cultivation.
3. Since, generally, drip irrigation and/ or fertigation practices are followed on large scale vegetable production under protected structures, water evaporation rate is reduced and weed population is minimal resulting in high water use efficiency. There is an added advantage of water harvesting from roof tops of structure.
4. Use of pesticides, generally, is less under protected cultivation resulting in healthier produce.
5. Protected cultivation is an attraction to educated unemployed youths with pride.
6. Affordable low-cost polyhouses, such as, poly tunnel, rain shelter are fabricated locally by small-scale entrepreneurs leading to local employment generation.
7. Hi-tech greenhouse cultivation methods, such as, hydroponics, vertical farming, aeroponics, aquaponics, hi-tech plant nursery production in soilless medium, balcony gardens and others are peri-urban and urban farming of tomorrow which are to be promoted through protected cultivation in the country.

Table 1. *Higher yield under protection in comparison to open field cultivation of vegetables recorded at Pithoragarh (5000ft amsl) in Uttarakhand.*

S.No.	Hybrids/Varieties	Open: Average Yield Q/ha	Protected: Average Yield Q/ha
1	Cucumber Hybrid DARL 102	415	665
2	Cabbage DARL 801	475	650
3	Bitter gourd DARL-41	158	268
4	Tomato Hybrid DARL-405	350	450
5	Capsicum Hybrid DARL- 204	168	240

Source: *DRDO-Defence Institute of Bio-energy Research, Haldwani, Uttarakhand*

3.2 Protective Structures

To modify or control the surrounding climate of the plant, several economical, affordable and effective crop growing structures with sunlight permeable claddings, such as, glass, plastics and screens are used in different Asian countries and places within a country. Most common ones are as under:

3.2.1 Plastic mulches

Plastic mulching is covering the soil around the plant with plastic film (10-15 micron thick). There are different colored plastic films used as mulches, such as, black, silver-black, red, yellow, white-black and other color combinations. Plastic mulching accelerates uptake of micronutrients from the soil by the active root zone, conserves soil moisture thereby reduces the irrigation water requirement of the crop and enhances quantity and quality of the produce with cleaner crop by becoming barrier between soil pathogen and crop canopy.

3.2.2 Plastic low tunnels

These are miniature structures producing greenhouse like effect on crops. Plastic low tunnel facilitates entrapment of carbon dioxide thereby enhancing the photosynthetic activity. It protects plants from harsh climatic conditions, such as, rain, wind, hail, snow, low temperature and other physical factors. These are mainly used for raising nursery for early germination of seeds and production of healthy seedlings of vegetables and other crops. Besides protecting plants from hostile climate these are helpful in round the year cultivation of certain crops. These tunnels are helpful in raising early cucurbits by protecting them from low temperature and frost in winters. Plastic mulch under polytunnels is beneficial to check weeds and conserve water better.

3.2.3 Walk –in tunnels

Walk-in tunnels are structures covered with UV grade film. Tunnels are designed to withstand wind up to 120km/h and trellising loads up to 25 kg/sq. m. Structure in gable configuration can be 8 or 10 meters wide. The height of the tunnels is 2 or 3 meter and length as per availability of land and requirement. Besides raising plant nursery, these are suitable for production of tomato, eggplant, capsicum, cucumbers, leafy vegetables, early production of certain vegetables and other crops. Walk-in plastic tunnels are popular in temperate regions of Asia and helping in protection of vegetables from severe frost and cold.

3.2.4 High roof tunnels with ventilation

These tunnels are similar to walk-in tunnel with suitable size ventilation at the top to facilitate escaping the hot air particularly during summer months. These are suitable for cucumber, capsicum and tomato production.

3.2.5 Insect proof net house

Insect proof net house is a closed crop production system with 2 m height that excludes insect pests via physical barrier of synthetic insect proof 40 mesh, 20% shade net. Double layer net houses are also available which slightly lowers the temperature in summer months. Open top insect proof net houses with more than 2 m height have been found suitable in protecting crops from insects and insect vectors as most of them cannot fly that high.

3.2.6 Shade net houses

Shade net house is considered as one of the major technologies to provide development of nursery of healthy grafts/ seedlings. Shade nets are available with different shade percent, 20, 40, 70 and color. Shade net houses are useful in summer months for rising off season crops.

3.2.7 Naturally ventilated polyhouses

These are naturally ventilated green/polyhouses with single or multiple span and total central height of 5 meters, side ventilation of 3 meters with roll-able poly-cover with or without roof ventilation and double door entry. Such polyhouses are found suitable for round the year production of crops depending on the geographical area. Generally multi-span poly houses are considered more suitable and economical for raising crops. Naturally ventilated polyhouses have been found suitable for cultivation of large number of vegetables besides raising their nurseries in Asian countries.

3.2.8 Polyhouse with pad and fan system

These polyhouses unlike naturally ventilated polyhouse have fan and pad system to regulate temperature and humidity. They have generally height of 4.5 meters with common side and top ventilation. These structures are clubbed with exhaust fans and cellulose cooling pads of 1.8 m in height and 150mm thickness. The cost of construction, running and maintenance is high. Lack of assured power supply becomes a limiting factor for such structures.

3.2.9 Rain shelters

Rain shelter, a low cost naturally ventilated polyhouse, is found ideal for off-season vegetable cultivation in high rainfall areas. Off-season cultivation of vegetables by utilizing low cost protected structures like rain shelter can be considered as a profitable enterprise besides protecting crop nurseries from high rain during monsoon months. High to very high rain fall areas in several Asiatic countries can make use of these during rainy months for raising vegetables.

3.2.10 Hi-tech or climate controlled greenhouses

Hi-tech green houses are climate controlled structures. Green houses have a variety of applications, the majority being, off-season and year-round growing of vegetables, planting material multiplication and hardening/acclimatization of tissue culture plants, crop growing for export market, seed production, plant breeding, bio-technology and new varieties and hybrids development. These are available in different sizes and constructed as per requirement. The sizes vary from as small as 100 sq. m to 10,000 sq m and more. Here most of the parameters temperature, light, humidity, fertilizer, irrigation is sensed and corrected as per program through smart control systems. These can be fully or partially automated.

3.2.11 Innovative structures

There can be several different innovative structures suiting to local environment/climate. Two of such structures i.e soil trench /underground polyhouse, Chinese polyhouse and Ladakhi polyhouse ideally suited for vegetable cultivation in cold deserts have made cultivation of all sort of vegetables possible which was considered in 1980s as impossible due to around six months of very low/sub-zero temperature prevailing atmosphere there. This contribution of DRDO laboratory named as Defense Institute of High Altitude Research (DIHAR), Leh is unique globally.

3.2.11.1 Soil trench

Since cropping period is short in cold arid regions, it was difficult to raise vegetables requiring transplanting, such as, cauliflower, cabbage, knol khol, broccoli, kale, tomato, capsicum, chili and eggplants. Defence Research and Development Organization (DRDO) has standardized soil trench cultivation method of raising early nursery of vegetables in Ladakh cold desert. Seedlings are raised in polythene covered trenches of appropriate sizes, generally, 4.5 m × 1.5 m × 0.9 m during March and April and transplanted as soon as cropping season starts. These trenches during summer are used for cultivation of cucurbits and solanaceous vegetable crops. Trench can be used to rise off-season potato crop (March- May/June).

3.2.11.2 Ladakhi polyhouse

Another structure Ladakhi polyhouse, similar to Chinese polyhouse, is made of unbaked bricks on three sides and polythene on southern side measuring 12 m × 4 m × 2 m size using local poplar and willow wood. It is covered with ultraviolet stabilized white polyethylene film of 200 micron thickness or other films like Rigidex. Crop inside polyhouse is covered over night with 150 gauge black polythene. During frozen winters this structure like soil trench produces leafy vegetables (beet leaf, mustard, fenugreek, celery, parsley and others) and summer months cole crops, solanaceous, root vegetable crops *etc.*

4. SUITABLE VEGETABLES VARIETIES/HYBRIDS

There are number of different vegetable crops one can grow under cover/protected structure, but following vegetables would provide better return besides rich nutrition - Leafy greens/ salad vegetables, microgreens, spinach/palak, cucumbers, tomatoes, /bell peppers, green beans, Swiss chard and squash. The vegetable varieties and hybrids suitable for protected cultivation are different than that of recommended for open field cultivation. Suitable varieties/hybrids under Indian conditions for protected cultivation are as given below.

4.1 Tomato

GS-600, Pant Polyhouse Hybrid Tomato-1 and 2, Rakshita, Naveen 2000, Avinash-2, Snehlata, Heem Sona, VL Tomato-4, Tropic, Dombello, Tuckcross-520, Avtar, Indam Hybrid, All Rounder, Arka Meghali , Kalash 137, Pusa Hybrid 4 and 8, Arka Samarat, Snehlata, VNR F1 (Sakura, Cobra) and several others.

4.2 Cherry Tomato

VR 124, Pusa Cherry tomato, Ole, Sairan, N S Cherry -1, Pant Cherry tomato, Paireso, BSS-834, Rosa, Sheeja (yellow), Nag Moti.

4.3 Capsicum

Bharat (green), Mahbharat (green), Natasha (red), Indra (green),Swarna (yellow) orobelle (yellow), Tanvi (yellow), US26 (yellow), DARL 71 (yellow), Mekong(green), KSP-1070 (green), US 81 (red), Cab-120, Tarquino,1DARL-72 (violet), California Wonder (green), baby capsicum orange-9967422, yellow- 9956434 and red -995559 . Capsicum green varieties become red on maturity/ripening.

4.4 Brinjal/Eggplant

Saropawa

4.5 Cucumber (parthenocarpic)

Isatis, Kian, Green Star2535 (Rizwan), Pant Parthenocarpic Cucumber-2&3, Multi Star (Rizwan), Hilton, Kuk-9, Cucumber Hybrid-1(KPCH-1) from KAAU, KSP-1734, Pusa Parthenocorpic Cucmber-6, Sandhya, Defender.

4.6 Bitter Gourd

Abhisek, Pali, Pusa Rasdar

4.7 Leafy Vegetables (off-season)

Kale, kailon , pok -choi, basil, lettuce, spinach , komatsuna, vegetable mustard (suitable for vertical production)

4.8 Beans

Pole type varieties of different beans. Dwarf beans in vertical system

4.9 Squash

Australian Green, Pusa Alankar, Pusa Pasand, Dhoom-2, Zuccini

4.10 Bottle Gourd

Pusa Naveen and several hybrids

4.11 Watermelon

Sugar Baby, Arka Manik

4.12 Muskmelon

Pusa Sarda

5. NEW AGE PROTECTED VEGETABLE TECHNOLOGIES

In recent years science of protected cultivation has developed several new technologies potential for boosting vegetable production with better nutrition contents. Some of these are mentioned below in brief.

5.1 Soilless Vegetable Production

Production of vegetables in medium other than soil is soilless vegetable production. This is new age crop production technology where plant growing medium is other than soil, such as, rock wool, coco peat, perlite, vermiculite, sawdust, bark chips, sand, gravel, pumice, polyurethane mats *etc*. Water, plant nutrients and mixtures of the above are used as substrates for growing of vegetables indoor or outdoor. It is most suited for vertical gardens under protected conditions or balconies.

5.2 Vegetable Seedling Plugs

Most of the vegetables can be transplanted by raising their seedling plugs in soil less medium under small protected area well in advance (one month or so). The seedlings can be grafts also. Seedling plugs ensure better stand in the field and help in regulating transplanting date reducing their time in main field/structures to avoid competition with weeds and protection from adverse weather.

5.3 Hydroponics

Cultivation of plants in nutrient solutions is known as hydroponic cutlivation. This is generally done under protected structures and in containers arranged horizontally or vertically (vertical garden). Use of nutrient film or fertilizers is made to feed the plants.

5.4 Aeroponics

This is a form of hydroponics. Nutrient and water mixture is sprayed in caged roots under dark at certain intervals with the help of sensors. This conserves water, space and is less cumberome. It demands constant supervision.

5.5 Aquaponics

Involves production of fish and vegetables with same input resource of water and nutrients. Feed is given to the fishes in the pond not to the vegetables or plants grown as hydroponics. Fish pond or tank water is used as hydroponics or areoponics with or without additional fertigation. This results in effective utilization of fish pond water and its nutrients for production of vegatables/plants simultaneously with added returns on input resources.

5.6 Vegetable grafts

Vegetable grafts with resistant rootstock and high as well as quality yield scion are becoming popular to manage biotic and abiotic stress. In Japan 93 % watermelon and 73% cucumber area of total area covered are grafts. World Vegetable Research Centre, Taiwan recommends growing tomato on eggplant accessions EG 195 and EG 203 and sweet pepper on chili accessions PPO273-7502 and PPO 242-62 and Lee B for flood tolerance. In China 20% area under watermelon and cucumber is covered with their grafts. In India sporadic attempts have been made and resistant/tolerant rootstocks have been identified.

5.7 Microgreens

Microgreens are tiny edible plants, rich in flavor and nutrition, that are older than a sprout, but younger than a full-grown plant. Microgreens are harvested after the first "true" leaves have developed. They are rich source of vitamins and minerals. Useful bio-active compounds present in microgreens are mostly phenolics , glucosinolates, betalians, carotenoids, polyacetylene, folic acids, anthocyanins and minerals.

6. SUGGESTIONS

Keeping in view the foregoing aspects and conditions following suggestions are made to promote protected cultivation in Asian countries:

1. Develop climate and farmer friendly suitable low cost structures.
2. There is absolute necessity for breeding of vegetable varieties and F_1 hybrids suitable for protected conditions.
3. Development of protocol for production of particular crop/variety/hybrid for particular structure for particular region to yield desired results.
4. Demonstrate new age protected cultivation technologies, such as, soilless cultivation at farmer's field.
5. Systematic extensive research and development and human resource development are required to meet the growing demands of protected cultivation of vegetables.
6. Use this technology for providing disease free planting material as plug seedlings to enhance crop productivity of vegetables.
7. Review promotional policies in different countries on the subject to make them further farmer friendly.

7. CONCLUSION

The population explosion in Asian countries is a major challenge to meet their vegetable and nutrition requirements. There is a need to popularize and fine tune technologies like protected cultivation that enhance productivity and quality of vegetables in present scenario of climate change, urbanization leading to reduced agriculture land, declining natural resources, increasing biotic and abiotic stress but decreasing availability of working hands in agriculture. Potential of protected cultivation have been amply demonstrated in several Asian countries which needs to be popularized on large scale and continuously improved upon for sustainable horticulture development in Asia.

REFERENCES

Behra, T. K. and Singh, A K. (2016). Plug Plant Production of Horticultural Crops through Hi-Tech Nursery Raising. Doubling Farmers Income through Horticulture. The Horticultural Society of India, New Delhi Publication. 19-25.

Mishra, G.P., Singh, N., Kumar, H. and Singh, S.B. (2010). Protected Cultivation for Food and Nutritional Security at Ladakh, Defense Science Journal, 61(2): 219-225.

Parvatha Reddy, P. (2016). Sustainable crop protection under protected cultivation. Springer E- book. 2016, p.433.

Singh, Brahma. (2016). New Age Protected Cultivation, e- magazine, ISPC, New Delhi publication. 2(1): 1-5.

Singh, B. and Singh, Balraj (2014). Development of protected cultivation in Haryana. Working Group report, Haryana Kisan Ayog, Panchkula, Haryana.. 68p.

Singh, B. Dhaulakhandi, A B. (1998). Application of solar greenhouse for vegetable production in cold desert. In: Proceedings of world renewable energy congress. Florence. Italy. 2311-14- V.

Singh, B. (2016). Protected Cultivation with Special Reference to Vegetable Production. Proceedings XII Agricultural Science Congress, 03-06 Feb. 2015. National Dairy Research Institute, Karnal, Haryana, India 383-394.

Singh, B. (2016). Innovative Protected Horticulture and its Potential for Doubling Farmer's Income. Doubling Farmers Income through Horticulture. The Horticultural Society of India, New Delhi Publication. 543-556.

Singh, B. (2016). New Age Crop Production Technology to Boost Farmer's Income. Indian Horticulture: 61(6), ICAR, Publication: 65-71.

Spehia, R.S. (2015). Effect of poly house on productivity of vegetable crops as compared to open cultivation. Current Science; 2254-2257-108.

Sreedhara, D.S., Kcrutagi, M.G., Basavaraja, I., Kunnaland, H.L.B. and Dodamani, T. (2013). Economics of capsicum production under protected conditions in Northern Karnataka. Karnataka J. Agric. Sci., 217-219-26(2).

Yadav, R. K., Kalia, P., Choudhary, H., Zakir, Husain and Brihama, Dev. (2014). Low-Cost Poly house Technologies for Higher Income and Nutritional Security. International Journal of Agriculture and Food Science Technology: 191-196 -5(3).

9

Vertical Farming to Supplement Horticultural Food Basket of Urban and Peri-Urban Population in India

P.B. Gaddagimath

SYNOPSIS

Vertical farming is growing plants in the space available vertically instead of growing horizontally on ground as being practiced in traditional agriculture. According to Wikipedia, vertical farming is 'the practice of cultivating plant or animal life within sky scraper green houses or vertically inclined surfaces where natural sun light can be augmented with artificial lighting and metal reflectors. Vertical Farming is gaining attention and popularity, which is the dream concept, the images of which are being circulated on the internet is the renderings and conceptual designs from architects and designers. They have borrowed this concept from Prof. Dickson Despommier, an American scientist from Colombia University when he authored the book 'The Vertical Farm: Feeding the World in the 21st Century'. Despommier says, his skyscraper farm has aroused the interest of scientists and investors around the world. Here the growing media used may be soil or soil less cultures like coco peat, hydroponics, aeroponics, aquaponics or pisciponics. In a broader sense Vertical Farming is a potential component of Urban and Peri-urban horticulture towards resilient food system. Vertical farming can be classified into three major groups' *viz.,* 1) Protected cultivation, 2) Cultivation of crops on vertically inclined high rise buildings and 3) Cultivation of crops in Skyscraper green house structures (High-rise urban farms). This paper discusses various aspects of vertical farming like the concept of vertical farming, why vertical farming, the issues involved, urban and peri-urban agriculture/ horticulture, open space urban farms, vertical farming - protected cultivation (green house, poly house, poly tunnels, net house), cultivation of crops on vertically inclined high rise buildings, skyscraper farming (hydroponics and aeroponics) *etc.* and Indian scenario of vertical farming.

1. INTRODUCTION

Vertical farming is growing plants in the space available vertically instead of growing horizontally on ground as being practiced in traditional agriculture. The term 'Vertical Farming' was coined by Mr. Gilbert Ellis Bailey when he authored a book on Vertical Farming in 1915. This concept mentioned by him otherwise as farming underground with the use of explosives, However, today refers to intensive farming of plants or animal life in vacant space that stretch upward (vertical) instead of sprawling outward (horizontal), may it be within high rise buildings/skyscraper green houses /poly houses or green houses/ vertically inclined surfaces like building walls or compound walls or roof tops/balconies/ porticos/vacant space available in different floors of a multi floor building *etc*. According to Wikipedia (2013), vertical farming is 'the practice of cultivating plant or animal life within skyscraper green houses or vertically inclined surfaces where natural sun light can be augmented with artificial lighting and metal reflectors. Vertical Farming that is gaining attention and popularity is the dream concept, the images of which are being circulated on the internet are by and large the renderings and conceptual designs from architects and designers. They have borrowed this concept from Prof. Dickson Despommier, an American scientist from Colombia University when he authored the book 'The Vertical Farm: Feeding the World in the 21st Century,' who became to be known as the father of Vertical Farming. The original vision of Prof. Despommier was a highly scientific technology led tilling high rises vertically in layers inside buildings, instead of horizontally in/on the ground using glow lights (LED lights) and conveyor belts powered by renewal energy sources, in which approximately 100 kinds of fruits and vegetables would be grown on upper floor with lower floors housing birds (chicken) and fish subsisting on the plant waste. Despommier says his skyscraper farm has aroused the interest of scientists and investors around the world. Here the growing media used for farming may be soil or soil less cultures like coco peat, hydroponics, aeroponics, aquaponics or pisciponics. As dreamt by Prof. Despommier, in Japan which is the front runner in such ideas has already implemented many such projects and today it has over 200 vertical farm plant industries already producing and supplying vegetables to its citizens, Recently a new vertical farm plant industry has been established near Kyto delivering 10,000 salads every day to the region. Similarly in US about ten vertical farm plant industries have been established supplying fresh greens, micro greens, vegetables and herbs to the citizens. In a broader sense Vertical Farming is a potential component of Urban and Peri-urban horticulture towards resilient food system. Vertical farming can be classified in to three major groups *viz.,* 1. Protected cultivation 2. Cultivation of crops on vertically inclined high rise buildings and 3.Cultivation of crops in Skyscraper green house structures (High-rise urban farms).

2. WHY VERTICAL FARMING?

Of all the issues of concern affecting the well being of humankind of the world community today, the most important issues grown to higher dimension that have made the task of providing food and nutritional security to world population more difficult are; (1). Population increase, (2). Urbanization - shrinking arable land for food production and (3). Growing poverty.

2.1 Population Increase

In June 2013, the United Nations Department of Economic and Social Affairs, Population Division estimated that as of 2010, the world population was 6.916 billion. The United States Census Bureau estimates that the world population exceeded 7 billion on March 12, 2012. According to a separate estimate by the United Nations Population Fund, it reached this milestone on October 31, 2011.The 2012 UN projections show a continued increase in population in the near future and the global population is expected to reach between 8.3 and 10.9 billion by 2050, while a 2014 estimate forecasts global population between 9.3 and 12.6 billion in 2100 and growing thereafter. Under such situations the world has to provide adequate quantity of quality food to a significantly higher number of human beings in the times to come. This increase in population poses a number of social, political, economic, environmental and ecological and technological problems to achieve future successes on the food and nutritional security front. The world agriculture produces 17% more calories per person than it did 30 years ago despite a 70% increase in population. This is enough to provide food to everyone in the world with at least 2,720 kilo calories per person per day according to an estimate of FAO. In spite of this, at this point of time nearly 870 million people of the 7.1 billion people in the world or one in eight were suffering from chronic under nourishment in 2010 – 2012. Almost all hungry people, 852 million live in developing countries, representing 15% of the population of the developing country (FAO, 2012).

2.2 Urbanization: Shrinking Arable Land for Food Production—A Challenge in Reality

Fifty years ago, two-thirds of the world population lived in rural areas and only one-thirds was in urban areas. Now, a half of them are in urban areas and by 2050, nearly two-thirds of them would live in urban areas which is more than 6 billion. Specifically in Asia, although urban population is just 42.2%, in the last twenty years, Asia's cities have grown at a faster rate than any other region: 13 of the 20 most populated urban areas in the world are now in Asia. Asia is urbanizing at unprecedented speed. Though still predominantly rural, this upward trend is expected to continue for many years to come. In this decade alone, it is anticipated that two- thirds of the growth in the world's cities will occur in Asia, raising its urban population by another 411 million. Most of the urban growth is happening in the developing countries of Asia and Africa. In 2030, of the 41 mega cities, 33 would be from the third world/developing world itself. Migration, primarily rural to urban, is a key driver of this growth. Recently the UN Department of Economics and Social Affairs has come out with the World City Report -2016 which reveals that by 2030 the number of mega cities (urban agglomeration of over 10 million population) would go up to 41 from the present 31and their population would be about 730 million i.e. 8.7% of the world population. The number of mega cities in India would go up to seven by 2030 from the present five mega cities. It is very important to know that Delhi, the capital city of India would continue to be the second most populous city in the world till 2030 and then adding a staggering 9.6 million people to its population–the most in any mega city (UN-DESA 2016). However, together with natural growth, reclassification of rural areas is also an important contributor: every year millions of people become city dwellers by this way even without movement, as their communities transformed into cities because of rapid urbanization. The most striking example as reported by UN-DSEA is that of Delhi. The United Nation's report has not just relied on the administrative boundaries of cities but has instead preferred to use the concept of "Urban Agglomeration" which is the "contiguous urban area or built – up area" as can be seen in case of Delhi agglomeration, satellite towns of Ghaziabad, Noida, Faridabad and Gurgaon are included. People in these contiguous areas are economically and socially integrated with the main city, Delhi. Likewise about 500 million people across the world live in such mega cities - about 6.8% of the world's population or 12% of the world's urban population. Besides Delhi, Mumbai, Kolkata, Chennai and Bengaluru also figure in the list of mega cities and by 2030 Hyderabad and Ahmadabad would also join them as their respective population too would cross 10 million. It is also very interesting to know that only a minority of urban dwellers stay in mega cities and nearly 21% of the world population stays in cities of population between 5,00,000 to 10 million while an even bigger share of 26.8% resides in smaller cities and towns with population less than 5.00.000 says the UN report. Recently the Karnataka Agriculture Price Commission (KAPC) has submitted a study report on effect of urbanization to Govt. of Karnataka which shows that 11% of cultivable land – to the extent of 12.49 lakh hectares in the state has been lost for urbanization and infrastructure development during 2005 to 2015. The data from the report most strikingly reveals that about 60,000 hectares of agriculture land have been converted for non agriculture purpose every year during the period. Nearly 113.54 lakh hectares of land was under cultivation during 2005-06 and most of the land lost towards this is particularly from the mega city Bengaluru and tire II cities like Mysuru and Mangaluru (KAPC study report 2016). As a result, the terminology 'Urban and Peri-urban Agriculture/Horticulture (UPA/H)' has emerged to the forefront and urban planning of states/nations is taking a different dimension, challenging the basic element of food security. Of all the options available for food production in Urban and Peri-urban areas vertical farming holds more promise.

2.3 Growing Poverty—The World Community is Becoming Poorer

Food and nutritional security is the consequence of poverty and growing poverty and reduced exchange power of the population migrating to urban cities from rural areas, price crisis and inflation in food prices. Despite significant progress made in improving living standards of people in almost all developing countries, more than 1.3 billion people in the developing world live in absolute poverty - still struggle to survive on less than a dollar a day. Up to now, poverty has been mainly a rural phenomenon, attributable in part to a vicious circle: a lot of today's degradation of agricultural resources is poverty related and degraded environmental resources contribute to the perpetuation of poverty. Although poverty will continue to characterize the rural landscape, projections show that the number of urban poor will overtake the number in rural areas by early next century. Relative poverty has also increased. Over the past 15 years the world has seen spectacular economic advances for some countries and unprecedented decline for others. The gap in *per capita* income between the industrial and the developing world tripled from \$5,700 in 1960 to \$15,400 in 1993. Disparities have grown within the societies as well. The poorest 20% of the world's people saw their share of global income decline from 2.4% to 1.4% in the past 30 years, while the share of the richest 20% rose from 70% to 85%. That doubled the ratio of the shares of the richest and the poorest from 30:1 to 61:1.

It is very much evident from the foregoing that, in spite of great scientific advancements made across the globe for the wellbeing of human kind, leave apart eradicating poverty, generation to generation, the global community has failed to at least reduce poverty, a cause for food and nutritional insecurity and human suffering. While absolute poverty has direct negative implications for human development, increasing economic disparities against a background of widespread poverty put the social fabric at risk. This disintegrating social fabric will have grave consequences not only for the environment, political stability and the safeguarding of regional and national tranquility but also for food security. In this backdrop it is essential to examine and understand what is UPA/H and what for the UPA/H.

3. URBAN AND PERI-URBAN AGRICULTURE/HORTICULTURE (UPA/H)

A good number of studies have been conducted across the world on population growth, migration of people from rural to urban areas, urbanization and trends of urbanization, food and nutritional security of the urban poor, problems and challenges arising of urbanization and various related issues. With population explosion round the world more particularly in the Asian countries, challenges and problems in the areas of food security, poverty and malnutrition are gaining significance. According to a study, it is expected that by 2020, 85 percent of the poor in Latin America and 40–45 percent of the poor in Africa and Asia will be concentrated in towns and cities. The Food and Agriculture Organization (FAO) estimates that by 2050, global food demand will increase by 70 per cent to feed the estimated global population of 9.3 billion. With the present trend to continue, the vast majority of these people will be living in irregular settlements without access to decent food, shelter, water and sanitation resulting in tremendous pressure on urban food chain leading to chaos in the urban areas. The pace with which the population is growing coupled with rapid urbanization, the planet earth would habitat about 9.00 million population with about 7.00 million population in the urban areas by 2050!

Rapid urbanization has aspired many rural migrants to cities in search of livelihood and better life, has compelled them to be slum dwellers, live in slum areas where such impacts are even higher resulting in extreme poverty and misery paving way for unhygienic city environment, malnutrition among urban poor particularly the urban women and children, increase in unlawful activities and crime rates *etc.* According to an estimate, today more than 571 million people are living in urban slums – around a third of the total urban population in the Asia-Pacific region. This is the rural population who once were dependent on on-farm and off-farm activities for their lively hood, have now low access to agricultural production and the majority of such poor urban households are primarily dependent on purchase of food items: this makes them especially vulnerable to loss of income, price shocks *etc.* resulting in the kind of the misery mentioned in the foregoing. The discussions on strengthening "food environment" in the countries, including safety challenges and concerns, dynamics of agriculture and urban food production, important value chains, trade and market access, regional dimensions, *etc.*, play an important role to achieve food system in a country to support healthy foods.

The most serious consequence of urbanization is shrinking arable land resources and depletion of urban natural resources like water and energy because of over exploitation of resources. Another significant issue that emerges as a serious threat is reduction in the number of human resources actually involved in farming/agriculture which would have a direct bearing on the production of food even in the rural arable land. This situation would further put tremendous pressure on the food supply chain and distribution in the urban areas. The situation poses greater challenges to feed the ever growing population with rapid urbanization-the food and nutritional security of the urban population, particularly the urban poor and disadvantaged would be hard hit.

3.1 Developments in Urban and Peri-urban Agriculture/Horticulture (UPA/H)

Urban and peri-urban agriculture as a recognized economic activity and a means of resilient food system is of recent origin and is trying to attain recognized economic urban food system with its strengths. In countries like Great Britain, USA, Brazil, France, Switzerland, Australia – the developed countries, urban agriculture/horticulture, particularly the vegetable gardening (edible gardens) has been a well-recognized activity. Rapid urbanization and population growth centered in and around the urban areas has created acute shortage and need of food to the urbanites, particularly the urban poor which has created extra pressure on the normal supply chain leading to food deficit and price rise resulting in search for alternate source of food supply in and around the urban areas. There is growing body of literature on UPA since early 1990s recognizing its potential contribution to local economic development. Various studies have brought to the fore front the role of urban and peri-urban agriculture and its significant importance in urban food security and poverty alleviation.

Growing plants in home gardens/kitchen gardens/landscape gardens/in pots or other different containers is not new to civilians. This culture has been in the civic society in various forms for various purposes since ages as humans are part of the nature and desire to be with the nature for their wellbeing. With the advent of urban cities and urbanization, perhaps after industrial revolution, growing of plants particularly the flowers, few fruits, herbs and medicinal plants was in vogue in the elite circles of the society and was considered as a leisure/pleasure, luxury and cost had no consideration as the basic purpose of this activity was for recreation, leisure, hobby, art, aesthetics, status for the elite and rich and was never considered as an economic activity. Contrary to this, growing of fruits, vegetables, herbs and medicinal plants in the back yards/ home gardens/kitchen gardens, rearing of animals and birds in the back yard of the hut/house in the rural areas was purely a subsistence farming activity and what remained after self-consumption was used as a means for bartering purpose which had a component of economics in it. This activity is still in practice in majority of households in the rural areas. But the urban agriculture/ horticulture as a recognized economic activity and a means of resilient food system for the urbanites is of recent origin and is trying to attain recognized economic and urban food system with its strengths. In fact, in countries like Great Britain, USA, Brazil, France, Switzerland, Australia – the developed countries, urban agriculture/horticulture, particularly the vegetable gardening (edible gardens) has been a well-recognized activity and there is a huge market for the vegetable seeds from this sector.

Need is the mother of innovations and development. Rapid urbanization and population growth centered around the urban areas has created acute shortage and need of food to the urbanites, particularly the urban poor which has created extra pressure on the normal supply chain leading to food deficit and price rise resulting in search for alternate source of food supply in and for the urban populace. This gave birth to urban and peri-urban agriculture/horticulture demanding support and recognition from the local as well as national governments.

Urban agriculture received a momentum in the early nineties of the last millennium when various civic bodies and other development organizations started debating about the food crisis in the urban areas and few enthusiasts and entrepreneurs took up farming in and around the peripheries of cities (urban and peri-urban areas); that is how the word 'urban and peri-urban agriculture' came into existance.

There is growing body of literature on UPA since early 1990s recognizing its potential contribution to local economic development (FAO, 2007; FAO, 2009; de Zeeuw, 2011). Various studies have brought to the fore front the role of urban and peri-urban agriculture and its significant importance in urban food security and poverty alleviation and have emphasized the need of a political will and national commitment for policy interventions by the governments in support of and carry forwarding various activities under the ambit of UPA. During 1995, the FAO realized the growing shift of population from rural to urban area and recognized the future need of augmenting food security for more than 50% urban population globally. Initially, the FAO took lead with World Bank and UNDP and the birth of Peri-urban Agriculture took place in support of food security of the growing urban population. During January 1996, FAO/COAG approved Urban and Peri-urban Agriculture (UPA) as one of its agricultural programs (Nath 2013). This was followed by FAO/PNASF International Workshop on UPA in Asia held in Bengaluru, India during 2005, that recognized horticulture as a major component of UPA. Recently organized Regional Workshop on Strengthening Urban and Peri-urban Agriculture towards Resilient Food System in Asia (UPAFSA-2013) during January 28, 2013 by FAO Regional Office, Bangkok, in collaboration with PNASF, Bengaluru, India at Bangkok, Thailand recognized that in urban planning the Greener Cities approach should be on the agenda and UPA should remain as the driving force for the same and further recommended that integration of UPA into development strategies and programs with multi-sectarian and multi-stakeholders participation should be encouraged for sustainability and should be supported by sound policy guidelines and frame work (UPAFSA-2013).

In 1999, during the meeting of the FAO Committee of Agriculture (COAG) in Rome, the member countries reviewed and discussed the role of urban agriculture in the context of urbanization process and recommended that FAO develop an integrated approach to assist the member countries in dealing with the UPA issues at the policy and technical level. Since then, various initiatives have been taken by FAO in this area. These include, implementing a project called Growing Greener Cities to promote urban and peri-urban horticulture (UPH), with a multi stake holders approach to ensure a stable policy and community support; support to establishing a number of school gardens in a number of countries; programs to improve water quality which is also used for urban horticulture and launching the Food for Cities program in 2000. In November 2011 a workshop entitled 'Ensuring resilient food system in Asian cities' was organized by FAO regional office in Bangkok, wherein urban and peri-urban agriculture was covered in various presentations, although this was not the main focus of that workshop (Konuma, 2013). International Seminar on UPA held during Tropical Agriculture Congress in Havana, Cuba (2012) highlighted the progress and development of FAO's Green Cities Program (Nath, 2013). Several countries promote UPH as part of their urban food security strategies.

Today there are a good number of models/methods being practiced to grow a few vegetables, flowers, fruits, few food crops and rear animals and fish under urban and peri-urban conditions for various purposes *viz.,* as subsistence farming, for leisure and pleasure, as a component in landscape gardening *etc.* However, growing food for economic gains and as a means of resilient food system has been of recent origin and importance. Some of the important methods /models of urban and peri-urban agriculture practiced worldwide are; Allotment gardens, Backyard gardens, Beehives, Berry patches, Community gardens, Community Supported Agriculture (CSA), Container gardens, Edible landscapes (landscaping that incorporates food-producing plants), Greenbelt agriculture, Greenhouse agriculture, Hedgerows consisting of edible plants, Herb gardens (culinary and medical), Kitchen gardens, Metro farms, Micro-livestock orchards, Prison farms, Rooftop gardens, Schoolyard gardens, Trellis/fence farms, Vegetable gardens, Vertical agriculture, Vineyards *etc*., are some of the forms of gardens associated with urban agriculture which is found on the website of City Farmer (2014).

However, following are the popularly practiced models/types of gardens most suitable for urban horticulture.

3.1.1 Open space urban farms

Two forms of open space farming *viz.,* 1. Production for urban markets in the urban open areas or in the peri-urban open areas. 2. Backyard/front-yard open areas cultivated mostly, but not only, for home consumption. Public parks, public open space, landscape open space of buildings, private space of residences, roof tops, terraces, balconies, verandas, road sides *etc.* are some of the open space gardens.

3.1.2 Container gardens

This is a simple method of growing/ production of crops in containers. Different types of containers are in used for this purpose and accordingly the following types of gardens.

3.1.3 Wading pool gardens

Crops are grown in plastic wading pool (a shallow plastic tub), a most cost-efficient container, 120-180 cm diameter pool of 30-37.5 cm deep. The pool is filled with growing media like soil, peat moss and manure and the technology for production has been standardized.

3.1.4 Used tire gardens

Used tires are being used as containers for growing crops in urban areas. It is important to isolate the tire surface from the growing medium which could be accomplished by using a plastic liner.

3.1.5 Feed sack gardens

Feed sack containers are the least expensive. Feed sacks come in different forms: burlap bags or woven plastic bags. Due to the pores in them they are most suited as containers to grow vegetables. They are good for one season and need to be replaced every season. Construction of a feed sack container, production technology *etc.* have been standardized.

3.1.6 Wooden crates gardens

Wooden crates of suitable sizes depending upon the extent of area available and growing conditions of the area are used as containers for growing a good number of vegetables, few fruits and flowers in home gardens in the front /back yards.

3.1.7 Wooden box gardens

Wooden boxes designed to accommodate available space in the home gardens are also used for growing the vegetables.

3.1.8 Brick grid gardens

Brick grids of rectangle/square shape in rows are made like containers and used for growing crops. This type of brick grids are very commonly found in many home gardens.

3.1.9 Other containers

There are a number of other containers like pots of different shapes, designs and sizes, grow bags, barrels, plastic pails with holes in them *etc.* are used in urban gardens.

3.2 Some Established Facts And Data

Many studies have been conducted to establish the potentials of urban horticulture. Some of the important studies relevant to the present context are given below.

- It has been estimated that a growing space of only 750 square meters could generate \$425,000 annually and provide employment for up to eight people (Hobbs, 2002).
- One sq. meter of space can produce enough crop to satisfy the need of vegetables per person, in other words can meet the RDA of vegetables per person (Sarpan, 2012).
- A growing space of 120 cm × 120 cm grids produces an average of 126 plants per year. Each grid is divided into 16 units of or 30 cm × 30 cm in which a different crop can be grown (Bartholomew, 1981).
- A growing space of 120 cm × 120 cm can accommodate 60-72 vegetable plants per year.
- Each vegetable plant grown in a pot can give an average yield of 600 g per year.
- A square foot garden for a family of four (14.7 m^2) gives an yield of about 54.00 kg of vegetables.
- A conventional garden for a family of four (73.4 m^2) gives an yield of about 270.00kg of vegetables.

The container gardens may not merit so much for vertical gardens in the sense perceived. However, they can be a component of roof top gardens, terrace gardens, portico gardens hanging gardens, pot culture in vacant space available in different floors of multi floor buildings *etc.*

3.3 Hydroponic Farming

Of late Hydroponic Gardening is being heard a lot which involves a process where plants are grown without using soil. Gravel, sand, perlite and water are used in place of soil. Vegetables, fruits, herbs and flowers can be grown in houses. This technology has been standardized for few crops, However, has many limitations like, higher quantities of macro and micro nutrients, light, huge space, management of temperatures, water management depending upon the type of the plant grown, proper drainage *etc.* Many of the world's largest hydroponic farms are set up in the deserts of the Middle East or unfertile soils in other parts of the world. The advantage of this system is that nearly all of the nutrients poured into the growing medium are absorbed by the plant, making it exponentially more efficient and increasing productivity manifold. Tahir Rana, a nuclear physicist who gave up a job in Canada to set up a vegetable farm in Faisalabad is heard often in the field of hydroponics. He makes extraordinary profits in hydroponic vegetable farming, a new method that dramatically increases productivity and thus farmers' incomes. Rana has set up a small company just outside Faisalabad called Fareed Farmhouse, where he produces three varieties of tomato (cherry tomato, strawberry tomato, beef tomato), cucumber and capsicum. His production capacity is significantly above the norm. Through this technique, farmers can get between 450 and 550 tons of vegetables per acre, compared to the average yield of 15 tons per acre using traditional farming. Rana uses coconut waste imported from Sri Lanka as the solid medium in which he grows his plants. Fareed Farm uses reverse osmosis water purification systems to ensure good water quality. Each plant requires up to two liters of water per day, which needs to be slightly acidic, with a pH of 5.8. The vegetable produced on the farm attract high prices. Customers include large retailers and wholesale chains as well as hotels and restaurants that have historically imported their produce from the European markets. Rana sells the tomatoes for about Rs. 225 ($2.50 US approx) per kilogram, compared to the cost of importing them from the Netherlands, which can run as high as Rs. 800 per kilogram. Hydroponic farming requires an investment of up to Rs1.5 million per acre, though it can yield net profits of up to Rs.3.00 million per acre annually if there is enough capital to set it up initially (Anon., 2012).

3.4 Vertical Farming

Vertical farming is a potential component of urban and peri – Urban agriculture towards Resilient Food System. The meaning or definition of Vertical Farming has already been explained in beginning of the paper. By and large vertical farming can be classified in to three major groups viz.

1. Protected cultivation
2. Cultivation of crops on vertically inclined high rise buildings and
3. Cultivation of crops in Skyscraper green house structures (High-rise urban farms).

3.5 Protected Cultivation of Vegetable/Flowers/Herbs

Protected cultivation of vegetables, herbs and few fruit crops has been a great success in many of the American, European and Australian countries. It is also becoming a popular cultivation alternative in countries like China, Japan, Korea, Vietnam, Thailand and off late in India in the past one decade. World over, nearly 40 countries are involved in protected cultivation of vegetable crops. New production centers are being added constantly, mostly from developing nations.

In simple terms protected cultivation is a technology of growing crops in a protected structures which protects the crop from high intensity of light, high rainfall, winds, insects through structure, polyethylene film/ polycarbonate sheet, shading nets/ thermal nets, insect net, cooling pad, exhaust fan, foggers and drip systems, fertigation equipment *etc.* which controls light, temperature, humidity and irrigation with fertigation and other required growth substances directly into the root zone of the plant. Water and nutrients enter the soil from the emitters, moving into the root zone of the plants through the combined forces of gravity and capillary. In this way, the plant's withdrawal of moisture and nutrients are replenished almost immediately, ensuring that the plant never suffers from water stress, thus enhancing quality, its ability to achieve optimum growth and higher yields. Apart from quantum jump in yield, the protected structures are mechanical barriers to pests and vectors of viral diseases. Partial regulation of microclimate in the green house will minimize the effect of abiotic stresses. It is an eco-friendly production system to minimize the use of harmful pesticides. It has been possible to produce hybrid seeds and seedlings; to produce during off season in tropical plains and breaking the seasonality barrier in vegetables in temperate regions to produce export oriented superior quality vegetables; to combat biotic and abiotic stresses in vegetable production; and to produce vegetables organically. Protected cultivation of vegetables has increased the yields by many folds. Capsicum production in the tropics has increased from 40 t/ha in open field to 100 t/ha in a greenhouse crop of six months. Similarly in tomato grown in green house during summer months, the yield has been increased to about 180 t/ha. The major cost However, is contributed to the structure, which amounts to about 70 per cent of the cost of production of vegetables. (Nath and Swamy, 2013). The different types of protected structures in use are:

3.5.1 Greenhouse

Greenhouses are climate controlled structures. Green houses have a variety of applications, the majority being, off-season growing of vegetables, cut flowers, planting material acclimatization, few fruit crop for export market and plant breeding and varietals improvement.

3.5.2 Polyhouse

Polyhouses are basically naturally ventilated climate controlled structures. Polyhouses have a variety of applications, the majority being, growing of vegetables, cut flowers, production of nursery seedlings, planting material acclimatization, few fruit crops for export market.

3.5.3 Polytunnels

Polytunnels are basically naturally ventilated climate controlled structures with a variety of applications, the majority being, growing of vegetables, flowers, planting material acclimatization, certain fruit crops.

3.5.4 Nethouse

Net houses are basically naturally ventilated climate controlled structures. Net houses have a variety of applications like growing of vegetables, production of vegetable seedlings, cut flowers, certain fruit crops for export market.

Protected cultivation of vegetable crops, cut flowers and certain fruit crops has been a successful venture worldwide and majority of the issues related to structure, growing media, GAP for different crops have been standardized and certain issues like water and nutrient management, plant protection particularly the disease management arising out in the course of production system are being addressed from time to time. As a result today protected cultivation has taken a shape of an industry. However, there are few limitations in this field which needs to be addressed. The most important ones are;

3.6 The Limitations of Protected Cultivation

There are two major limitations. They are:

1. Majority of the technology has been applied to select few crops like capsicum, tomato, English cucumber, melons (water melon and musk melon), bottle brinjal, pole beans, cole crops, lettuce and broccoli in vegetable crops, rose, gerbera, carnation, chrysanthemum, orchids, anthurium in flower crops and very few fruit crops like straw berry orange *etc.* There is a need to come out with more number of vegetable crops and food crops like cereals and pulses.
2. With climate controlled structures it has been possible to increase the yields by two to three folds with quality. However, there has been no increase in the photosynthetic efficiency of the crops. Presently the photosynthetic efficiency of the crops grown in open field conditions has been 9%. If this efficiency of the crops is enhanced to by another 6 to 8 %, there would be at least 3 to 4 fold increase in the yield levels. This can be possible by using LEDs (Light Emitting Diode) in place of the direct sun light. Work in this direction is in progress and already few leads have emerged.

   Over the past decades a small and innovative Dutch horticulture company named PlantLab has developed a revolutionary and high-tech growing concept for cultivation of plants in conditioned production units without sunlight, saving 90% of the water used in normal horticulture, with improved production and reliability. When grown outdoors plant photosynthesis is only about 9% efficient. With the correct balance of colored LED light, PlantLab has increased that efficiency to 15 to 18%. Double the efficiency means increased yield. A ***light-emitting diode (LED)*** is a semiconductor ***light*** source. LEDs are used as indicator ***lamps*** in many devices and are increasingly used for other ***lighting***. By keeping the plants in a contained system, PlantLab can also recycle evaporated water, which helps them grow crops using just one tenth the water as with traditional greenhouses. As PlantLab's crops grow indoor by creating a true "Plant Paradise," plants are not susceptible to pests and diseases; Hence, no pesticides are needed anymore. PlantLab's production facilities can be built almost anywhere: from the Sahara to Upper energy and transportation (Ger Post, 2012).

3.7 Cultivation of Crops on Vertically Inclined High Rise Buildings

While urbanization has become the order of the day, arable land for production of food for the growing population is taking a serious beat. According the UN, the amount of arable land per person decreased from about an acre in 1970 to roughly half an acre in 2000 and is projected to decline to about a third of an acre by 2050.There will simply not be enough farmland to feed us. In other words by the year 2050, nearly 80% of the earth's population will reside in urban centers. The human population will increase by about 3 billion people during the interim. Under such circumstances feeding the billions and billions of bellies would be the greatest challenge if technology is not available to produce food within the available space. Then the question arises, where is the space to grow the food even if the technology is available? Another challenge is the depleting water resources to grow the crops.

3.7.1 Where is the space?

The concept of urban agriculture/horticulture took birth because of rapid urbanization process resulting in depletion of arable land to produce food. Urbanization is contagious and keeps on expanding with time by acquiring land adjacent to it. Today's rural land would be peri-urban and the peri-urban land would be urban over the years as the society progresses. Urbanization is horizontal. Now the question, where is the space to grow? It would be interesting to know that every square meter land acquired for urbanization creates or makes available one fourth square meter area in various forms which could made use of for growing food crops, which otherwise remains unexploited or being used for non-productive purposes. Such areas which merit for the purpose of growing crops/plants are;

- House hold gardens - Kitchen gardens/back yard gardens and land scape gardens in the front of the bungalows/residences/houses.
- Rooftops and terraces of buildings (all buildings in the cities including working places and the residences)
- Porticos, balconies
- Concrete compound walls, vertical building walls *etc.*
- Large private and public buildings and its surrounding space.
- Vacant space in and around corporate houses, public sector and private companies and industries.
- Vacant public places
- Vacant space in Govt. offices, public gardens, educational institutes, banks *etc.*
- Road sides.

However, it may be noted here that no systematic and scientific studies have been carried out in this direction to find out the extent of available unexploited and under exploited space in the urban areas. The fact remains that, there is ample space available to grow crops as an inbuilt space of urbanization. And all this space is most suitable for vertical farming. According to Dickson Despommier, moving farms into buildings might change the way much of humanity grows its food. And while vertical farming is on the cusp of blooming, the main hurdle isn't technology; it's one of engineering and funding. "We know how to grow food inside, that's not the problem. What we don't know yet is how to integrate that technology into a tall building. That appears to be a simple, but perhaps expensive engineering problem." Known as the founder of the modern vertical farming concept, Despommier is a Public Health professor and stresses the urgency of changing the way humanity cultivates its food. Water for growing crops would be a scares commodity in the near future, technology of production of crops with increased yields would be the priority area of work worldwide. In this direction it is important to note that off land farming uses a fraction of the water required for, the land farming. This is why Despommier believes that vertical farming will take root first in "water challenged" areas of the world, like the Middle East (GLOBE NET, 2009).

3.8 Limitations of Vertical Farming

Vertical farming is a recent concept and technologies are being developed. Even though the technologies are available, they are restricted to the crop growing situations of the regions/countries where vertical farming in inclined high rise buildings has been implemented/being practiced and cannot be duplicated in other regions as broad spectrum technologies. Under such circumstances the major limitations of vertical farming in inclined high rise buildings are;

1. Very limited vegetable crops have been identified for the purpose; However, vegetable crops in vogue are being grown without any scientific validation.
2. No varieties/hybrids have been exclusively bred for the purpose of vertical farming as result the yields and quality of the produce cannot be guaranteed.
3. The production technologies and GAPs for these crops have not been standardized.
4. Presently the high rise inclined buildings are not built to suit vertical farming. There is a need to re-engineer the existing buildings by providing additional structures to suit the need.

While the above limitations could be taken up as the researchable issues, it is essential that all the high rise buildings coming up in the smart cities/green cities and tire two cities in the near future should be engineered in such a way that they used for vertical farming.

3.9 Skyscraper Farming-Aeroponics with Vertical Gardens

In the field of vertical farming, 'Aeroponics' is a new concept and a technology. Increasing urbanization have led Columbia's Dickson Despommier and his students to came up with a technology of tilling high-rises using grow lights and conveyor belts powered by renewable energy sources, in which approximately 100 kinds of fruits and vegetables would grow on upper floors with lower floors housing chickens and fish subsisting on the plant waste. This may all sound like pie in the sky, but Despommier says his skyscraper farm has aroused the interest of scientists and investors around the world. And though the project is still in the blueprint phase, if it pans out, not only will large numbers of individuals be able to source their food locally, but ecosystems destroyed by years of exposure to toxic pesticides will also be restored. The farming technology includes hydroponics, which uses 70 percent less water than normal agriculture. They also use aeroponics, which uses 70 percent less water than hydroponics. Indoors, we can control everything including temperature and humidity as well as nutrition (Anon., 2009). Aeroponic grow methods have the ability to efficiently utilize space and resources. Aeroponics doesn't need soil or growing material and the attendant weight and volume, can also grow quite a number of plants without a tremendous amount of weight. So this technology is attractive in spaces like rooftops where weight is an issue where in the growing plants with their roots suspended in a deep air or growth chamber receives excellent aeration and a moist environment. This is the main advantage of aeroponics which uses an aggregate material, such as, rocks or pebbles. The suspended plant roots receive 100% of the available oxygen and carbon dioxide to the roots zone. This will accelerate biomass growth and reduce rooting times. Another distinct advantage of aeroponics is that any species of plant can be grown. In an aeroponic system the micro environment can be finely controlled and uses about 90% less than conventional, soil-based farming (Anon., 2013).

Vertical farming with aeroponics basically derived from Controlled Agriculture Environment initiated by NASA some 15 years ago is highly technology driven involving high capital investments. Very few countries like USA, Japan, Taiwan, Korea, China, Singapore and Australia have been successful in implementing this technology. For European and Asian countries this is quite new and as far as India is concerned, the field is open. With the Govt. of India announcing developing a good number of smart cities in the coming years with huge investments, it is time that vertical farming particularly, Skyscraper Farming with aeroponics has tremendous potentials as resilient food system.

4. INDIAN SCENARIO

In India, growing plants particularly the flowers, fruits, vegetables and herbs and medicinal plants is not new to the urbanites. This has been in practice since ages. If we just go through ancient history and modern history we do find good number of examples of growing fruits, flowers and certain vegetables in princely gardens, palaces and large buildings in princely states. It was also practiced by rich and elite class of families residing in the capital cities and large towns of princely states (urban areas as we classify today). This practice was purely personal, an activity of leisure/pleasure/entertainment or recreation in the form of hobby, art, aesthetics, symbol of status/luxury and never was considered as an economic activity. Here cost had no consideration at all. On the contrary, in rural areas growing of vegetables, fruit trees, flowers, some herbs and medicinal plants in the back yards of houses, home /kitchen gardens, area surrounding the huts was practiced in vogue by the farmers purely for subsistence i. e. for self consumption and what remained as surplus was distributed to surrounding farm families either as good will or was used as a means for bartering purpose. Even though this activity was also personal, it had the component of economics in it. This activity is still in practice in many households in rural India. But the situation today is different.

4.1 Container Gardens, Kitchen/Backyard Gardens, Nutrition Gardens

Karnataka state was the first state to establish a separate department of horticulture followed by other states and today almost all the states in India have established separate department of horticulture offering training/ skill development programs on kitchen gardening/ nutrition gardens/ roof top gardening/ mushroom cultivation to the urbanites particularly for the women/house wives. There is a growing awareness about urbanization and its consequences on the life of the urbanites and the urban population is feeling the pinch of it.

As a consequence of urbanization, urban and peri-urban agriculture/horticulture has emerged as a recognized economic activity and means of resilient food system for the urbanites. Already it has made inroads in mega cities, tier I and tier II cities and many towns. Even though urban and peri-urban horticulture is of very recent origin in India, some of the private seed companies like Pocha seeds of Pune, Maharashtra, Indo American Hybrid seeds company of Bengaluru of Karnataka had made a beginning almost about 40 years back focusing more on ornamental plants and annual flower seeds. The Indian Institute of Horticulture Research, Bengaluru, Karnataka had come out with the concept of 'Nutrition Garden' way back during 1990 itself and was offering training programs also and further during 2013 it introduced the concept of Kitchen Garden Seed Packets also.

A private seed company, Sarpan Seeds Pvt. Ltd. Co. Dharwad, Karnataka has introduced this Kitchen Garden Seed Pack (containing seed packets of 10 vegetable seeds of 10 grams each which can be grown in pots/containers and open space in the front/backyards of residences) during November 2007 and has been in the market in different states since then. The sale figures from 2007 to September, 2013 indicate a gradual increase in the sale. The company also informed that there is an ever growing demand for Kitchen Garden seeds from cities and towns in different states. Another example is Dr. Doshi's City Garden Methods in Mumbai which has become a house name for kitchen gardens and growing vegetables and fruits in containers innovated and promoted (RAUF, 2011).

Today every year the state departments/ state agriculture/horticulture universities, NGOs, private event management companies horticulture shows and krishi melas in which a good number of private seed companies participate in exhibitions and showcase the concept of Edible Gardens by displaying the live pot plants of vegetables and sell kitchen garden seed packets which is gaining more popularity over the years. A good number of societies/grower clubs have also been established in cities and towns by interested growers and enthusiasts involved in urban and peri-urban horticulture.

Recently, some eight months back I came across an interesting incident when I started construction of my house in Bengaluru. The contractor to whom I had awarded the construction work had housed two labour families at the construction site for Mason work of the building. After three months I was surprised to see beautiful crop of tomato, brinjal and some vegetable greens in about 200 square mt. of vacant land just adjacent to the construction site. In fact I was happy and excited and after interacting with these labours, I was told by them that it is very difficult for them to manage the daily living in city like Bengaluru with the wages they get and what is wrong if they grow few vegetables in the vacant unused land for self consumption as vegetables are essential in their daily diet. On further enquiry, I came to know that they are farm labours/agriculturists from Raichur in Karnataka and have migrated to Bengaluru in search of better livelihood. It is clear from this that;

1. the families migrated to cities from rural areas in a hope of better living are in great misery in urban and peri-urban agglomerations.
2. they are well aware about the nutritional aspect of their diet.
3. most important point to be noted here is that majority of such people were once either farmers or farm labours and they are well versed with farming practice.

This makes a very strong human resource for the urban and peri-urban agriculture/ horticulture. What is needed is to recognize this as the need of the hour and take up massive awareness programs well equipped with men, materials and resources involving all the stake holders of urban and peri-urban horticulture like the civic society, corporate houses, private and public establishments, NGOs, agricultural and horticultural universities, central and state agricultural/horticultural research institutes and the government.

To understand the strengths, potentials and scope of urban and peri-urban horticulture a study was conducted in 2013 by the author in Bengaluru metro city to find out how far urban horticulture would be successful if practiced by the urbanites. The results of the study are given below.

- Bengaluru has more than 8,000 Bungalows, can afford a conventional garden of 73.4 m^2which can produce 2160.00 kgs of vegetables per year.
- Bengaluru has more than 20,000 residences of which 15000 can afford a Square Food Garden of 14.7 m^2 that can produce 810.00 tons of vegetables per year.
- More than 20,000 residences in Bengaluru can take up container gardens on rooftops, terraces *etc.* which can produce 720.00 tons of vegetables per year.
- Bengaluru has more than 4,000 government buildings including educational institutes, banks *etc.* which can afford a growing space of 750-800 m^2 that can produce 10,800.00 tons of vegetables per year.
- Bengaluru has more than 1000 private and corporate sectors which can afford 1000 m^2 growing area which can produce 13,500 tons of vegetables.

To sum up, if urban horticulture is implemented with right spirit and in right direction, Bengaluru metro city can produce 27,990.00 tons of vegetables per year! (Gaddagimath, 2013).

4.2 Vertical Farming

As already indicated earlier by and large vertical farming can be classified in to three major groups *viz.,*

1. Protected cultivation
2. Cultivation of crops on vertically inclined high rise buildings and
3. Cultivation of crops in Skyscraper green house structures (High-rise urban farms).

4.3 Protected Cultivation of Vegetable/Flowers/Herbs

Protected cultivation of flowers, vegetables, herbs and few fruit crops is of recent origin in India. In fact commercial protected cultivation of flowers received a big leap when the Govt. of India announced liberalization policy way back in the middle of 1990s during which a good number of private firms and corporate houses entered into this as 100% export oriented business. A good number of public issues were also made to take up this as a promising business venture. However, this did not take off so well to the expected heights as sustainability of these projects was under great threat because the technology and patented planting material were imported and overheads for these were the main bottleneck for the sustainability of the projects. As a result many of the projects did not take off in the beginning stage itself, some became sick units and a very few units modified and refined the technology to suit the growing conditions at the local and regional levels and are still doing a good business cultivating roses, gerbera, carnation orchids, anthuriums and some exotic flowers. Till then cultivation of vegetables in protected environment was not attempted where as it had already taken a leap in majority of developed nations. Realizing the importance and potentials of protected cultivation of vegetables and in order to refine and domesticate the production technology and engineering of the infrastructure of the protected structures for growing vegetables and flower crops under protected environment, it was during the later years of the last century the Indian Council of Agricultural Research (ICAR) took up work on protected cultivation of vegetables and flowers under the National Agriculture Technology Project with World Bank assistance. Work was taken up across the country in Bengaluru, Pune, Himachal Pradesh, NEH states in two major vegetables viz., capsicum (including colour capsicum) and tomato and rose, gerbera and carnation in flower crops. Under this project studies were conducted on the suitability of playhouses, green houses, net houses, poly tunnels and attempts were made to standardize the GAPs for the crops. As result today protected cultivation of vegetables has become a successful enterprise which has been encouraged by the state horticulture departments in the form of back ended subsidies under the National Horticulture Mission (NHM) and National Horticulture Board (NHB). Many commercial banks also have come forward to support this field and NABARD has taken special interest in promoting this field. Along with this production of planting material of fruit crops ornamental and flower crops and production of vegetable seedling production under low cost net houses has also taken a business shape. Today there are more than 800 sustainable and successful projects of poly house production vegetables in the country covering crops like capsicum/colour capsicum, tomato, parthinocarpic cucumber (English cucumber), melons (watermelon, muskmelon, cantaloupes), pole bean, bottle bribjal, broccoli, lettuce *etc.* A good numbers of portals, websites, computer applications, databases and even whats apps groups on protected cultivation have been developed for the benefit of the growers.

4.4 Vertical Farming

Cultivation of crops on vertically inclined high rise buildings and cultivation of crops in Skyscraper green house structures (High-rise urban farms) using hydroponic and aeroponic culture is almost new to India. Attempts have been made by enthusiastic entrepreneurs in successfully growing vegetables and strawberry fruit with hydroponics and aeroponics cultures and in many places we find various groups/associations of hydroponics culture for exchange of ideas and technology.

The word hydroponics is derived from two Greek words 'hydro' meaning water and 'ponos' meaning labour. This word was first used by Dr. Gericke, a California Professor in 1929 who developed this commercial technology of growing plants which had been a technique in his laboratory. The hanging gardens of Babylon, the floating gardens of the Aztecs of Mexico and those of Chinese are examples of hydroponics (Resh, 1997). Growing of vegetables in Dal Lake of Srinagar, India is a standing example of hydroponics culture. In conventional or traditional agriculture the growing media invariable is soil. If the plants are grown in nutrient media or any other substrate (other than soil) containing essential nutrients such culture is known as hydroponics or soilless culture or dirt free culture. In soilless culture or hydroponics the nutrients are directly made available to the plant roots making the plant to fully express its genetic potential. Hydroponic culture can be broadly categorized into three categories *viz.,* 1) Substrate based farming with nutrient solutions using cocopeat, vermiculate, perlite, gravel, sand, haydite, clay pebbles, vydron, rockwool, wood dust, paddy husk *etc*. 2) Nutrient film technique and 3) Deepwater culture. Similarly aeroponics culture is a process or technology of growing plants in a hydro-atomized moisture and nutrient rich environment under controlled conditions that support plant growth. So to say the basis of vertical farming is hydroponic/aeroponic culture in nutrient solution.

In India hydroponic and aeroponic culture or the vertical farming are still not able to adopt the technology because of high initial costs, highly science and technology driven field and suitable market (consumers/buyers) for this enormous and highly perishable quality produce. However, here and there attempts have been made in this direction. To mention a few;

1. Himachal Pradesh Agricultural University, Palampur, Himachal Pradesh, has been working on production technology of vegetables under hydroponic conditions.
2. ICAR-IIHR, Bengaluru has been working on standardization of technology for production of vegetable seedling, protected cultivation of vegetables.
3. Aeroponic potato seeds production technology has been standardized by ICAR-CPRI, Shimla and is being practiced by many farmers.
4. The ICAR, Research Complex, Goa has standardized the technology of gowning green fodder hydroponically.
5. Krishi Vigyan Kendra, Kannur, Kerala – vertical farming integrated with animal husbandry for sustainable livelihood
6. Bidhan Chandra Krishi Viswavidyalaya in Nadia - high rise vertical farming structures
7. Vertical Farming Association of India – awareness programs in the form of conducting conferences, seminars, meetings *etc*. and
8. Some of the private firms/consultants are into hydroponics and aeroponics for selected crops and are popularizing their technologies.

In a message to the inaugural issue of the International Magazine on Vertical Farming-Nov. 2015 the former Director General of ICAR Dr. S. Ayyappan points out "In the recent past there has been spurt in protected cultivation through vertical farming and I gather more than 500 urban commercial vertical farms have come up world over with the exception of India". Similarly the former Deputy Director General (Horticultural Sciences), ICAR Dr. N. K. Krishna Kumar in a message to the same magazine very optimistically puts up "Future horticulture is intertwined with urbanization and peri-urban horticulture. There have been limited attempts in protected cultivation focused on selected crops. While the focus on protected cultivation including low cost shade net cultivation is largely horticulture based and peri-urban driven, it is a fact that the world is rapidly getting urbanized. Vertical growth is the order of urbanization, often called 'concrete jungle' with a disdain for parks, lakes and greenery. There are opportunities and challenges. Whether one likes it or not, it will change and shape the future world we would be living in." Many such views are the expressed facts of scientists, technocrats, entrepreneurs across the country about vertical farming in India. To have or not to have vertical farming in the sense being perceived and understood today is being debated at various corners and levels in the country, the facts remains that we are lagging behind in this field for various obvious reasons. If not today, we essentially need it tomorrow for our own well being and it is time that the authorities at appropriate levels look at this with all seriousness and plan for absorbing the future shock in this direction. Already the Govt. of India has announced development of 98 smart cities, tire1 and tire 2 cities are re-demarked and emergence of new cities and towns being taking a new toll and peri-urban area is being redefined. It is time that all the stake holders including the government to give a new thought and include vertical farming in their developmental plans, particularly in urban development plans.

As already discussed in the foregoing, vertical farming particularly cultivation of crops on vertically inclined high rise buildings holds more promise towards Resilient Food System, the question is who can do this or take the responsibility of growing crops in the vertically high rise buildings and what should be the approach? The coming years would definitely throw an insight in these issues to be addressed to make a sound and sustainable urban and peri-urban resilient food system with horticulture crops aiding the food and nutritional security of the urban and peri-urban populace.

5. CONCLUSION

The practice of growing plants in home gardens/ backyards / containers, roof tops / open spaces *etc.* is in vogue in urban areas since ages. Rapid urbanization has posed many challenges of which food and nutritional security of the urban populace is badly hit in which the urban poor are the most vulnerable. Urban and Peri-urban Agriculture / Horticulture as an alternate sector of food production has more potential to address many issues. In this context vertical farming as a potential component of Urban and Peri-urban Agriculture/Horticulture offers ample scope and opportunities which deserves recognition as an independent entity at various levels of policy makers and the government. It needs a national will and a political commitment from the local, regional and national levels. Given the right recognition with technological support and investments, vertical farming would definitely aid in addressing the ever growing threat of food and nutritional insecurity in the coming years.

REFERENCES

Anon. (2007). Best practices in urban agriculture- A background report for the City of Kamloops to support development of a urban agricultural strategy / True Consulting Group, February 2007,ISBN 978-1-895984-23-1,http://extension.missouri.edu/food systems/urbanagriculture.aspx,

Anon. (2009). A Rooftop and Terrace Vegetable Garden in Brooklyn. Posted July 20, 2009. http://brooklynroofgarden.com/2009/07/20/a-rooftop-and-terrace-vegetable-garden-in-brooklyn/

Anon. (2009). Aeroponics in the news with "vertical gardens". http://www.igrow hydro. com/blog/226/aeroponics-in-the-news-with-vertical-gardens/

Anon. (2012). Urban horticulture, from Wikipedia, the free encyclopedia. http:// en.wikipedia.org/wiki/Urban_horticulture

Anon. (2012). Vertical farming, Wikipedia, the free encyclopedia. http://en.wikipedia.org/ wiki/Vertical_farming

Anon. (2013). Urban agriculture, from Wikipedia, the free encyclopedia.

Anon. (2016). India will have seven mega cities by 2030, The Times of India, Sept.07, 2016.

Anon. (2016). Karnataka Agriculture Price Commission (KAPC), Govt. of Karnataka, 2016.

Anon. (2016). Urbanization effect: Karnataka is losing its farm land at fast pace, The Times of India, Nov. 14, 2016.

Bailkey, M. and J. Nasr (2000). From Brownfield to Greenfields: Producing Food in North American Cities.

Bansal, G. L. (2015). Hydroponics culture-the basis of vertical farming, International magazine on vertical farming (inaugural issue), 2015.pp40-45.

Bartholomew, Mel. (1981). Square-foot gardening. http://www.squarefootgardening.com/ http://www.nepanewsletter.com/square.html

Bruinsma, Wietse, Marielle Dubbeling (Eds.) (1998). Resource Guide on Urban Agriculture. Wageningen, Netherlands. 276 pp.

City Farmer. (2014). Canada's Office of Urban Agriculture (http://www.cityfarmer.org).

Christine Zimmermann-Lossl. (2015). Vertical Farming – Center for Foodopolis, International magazine on vertical farming (inaugural issue), 2015.pp 17-18.

Cooke, R. D. (1998). Foreword. *In*: Resource Guide on Urban Agriculture (Eds. Wietse Bruinsma, Marielle Dubbeling). Wageningen, Netherlands. 276 pp.

Cotthem, Willem Van. (2008). Various containers used in urban gardening (Technology for the poor), http://www.technologyforthe poor.com/Urban Agriculture/Garden.

De Freitas, Susan. (2012). Smart tech for scaled-up urban agriculture.

De Zeeuw, H., Veenhuizen, V. and Dubbeling, M. (2011). The Role of Urban Agriculture in Building Resilient Cities in Developing Countries. *Journal of Agricultural Science*, 1-11.

Dickson, Despommier. (2015). The Vertical Farm – Feeding the Cities,, International magazine on vertical farming (inaugural issue), 2015.pp 11-16.

Ebenezer, Job S. (2008). Urban Agriculture: A Guide to Container Gardens President, Technology for the Poor, 877 Pelham Court, Westerville, Ohio – 43081, technology for the poor@yahoo.com

Food for the Cities- www.fao.org/fcit/en/food-for cities@dgroups.org

Fraser, Evan D.G. (2002). Urban Ecology in Bangkok Thailand: Community Participation, Urban Agriculture and Forestry. *Environments* 30 (1): 37-49.

Gaddagimath P. B. and Aswath C. (2015). Vertical Farming towards Resilient Food System in India: A Potential Component of Urban and Peri-urban Horticulture, International Magazine on Vertical Farming (inaugural issue), 2015.pp 24-39.

Gaddagimath, P. B. (2015). Urban horticulture-Civic and Corporate Social Responsibility. *In* Develpoment trends in Urban and Peri-urban Agriculture. Deshpande, R. S., Nath Prem, Chengappa, P. G., Kannan Elumalai, Swamy, K. R. M. and Iyer, C. P. A.(Eds) Westville Publishing House, New Delhi. Pp 230.

Ger Post. (2012). Urban Horticulture in Smart Cities: Open innovation as a Strategy for the Global Marketplace. *The Ispim Magazine* (Netherlands), 2012.

GLOBE NET. (2009). The High-Rise Urban Farms of the Future. https://www.greenbiz.com/news/2009/11/24/high-rise-urban-farms-future

Goldstein, M., *et al.* (2011). Urban agriculture: a sixteen city survey of urban agriculture practices across the country. Page 4. Retrieved from http://www.georgiaorganics

Growing Greener Cities- www.fao.org/ag/agp/greener cities

Hampwaye, G., Nel, E. and Ingombe, L. (2013). The role of urban agriculture in addressing household poverty and food security: the case of Zambia. Gdnet.org. Retrieved 2013-04-01.

Hendrickson, Mary K. and Mark Porth. (2012). Urban Agriculture -Best Practices and Possibilities-Report developed for the Urban Sustainability Directors in the cities of Columbia, Kansas City and St. Louis, Missouri, Division of Applied Social Sciences.

http://en.wikipedia.org/wiki/Urban_agriculture

http://money.cnn.com/2010/12/13/technology/urban_farming_high_tech/index.htm

http://www.tgdaily.com/sustainability-features/66705-smart-tech-for-scaled-up-urban-agriculture

Konuma, Hiroyuki. (2013). Proceedings of Regional Workshop on Strengthening Urban &Peri-urban Agriculture towards Resilient Food Systems in Asia, Bangkok, Thailand, January 28-30, 2013.

Maria. (2013). DIY Aeroponics. http://www.google.co.in/imgres?imgurl=http: //www.farmxchange.org/wp-

Nath, Prem. (2013). Key note address- Concept and Implementation of Food Security and Its Relevance to Urban and Peri-urban Agriculture, Proceedings of Regional Workshop on Strengthening Urban &Peri-urban Agriculture towards Resilient Food Systems in Asia, Bangkok, Thailand, January 28-30, 2013.

RAUF. (2013). What is urban agriculture? RUAF Resource Centres on Urban Agriculture and Food Security. http://www.ruaf.org/node/512

Resh Howard, M. (1997). Hydroponic Food production, Fifth edition, Woodbridge Press, 1997. pp23.

Smit, J., Ratta. A. and Naqr. J. (1996). C/&n Ag~iwlfrrre: Food. Jobs md Swttrimrh ~eCitier. UNDP. New York.

Taguchi, Makiko. (2013). Global Perspective on strengthening UPA towards resilient food systems, Proceedings of Regional Workshop on Strengthening Urban &Peri-urban

Agriculture towards Resilient Food Systems in Asia, Bangkok, Thailand, January 28-30, 2013.

Tarathep, Chanvit. (2013). Opening remarks, Proceedings of Regional Workshop on Strengthening Urban &Peri-urban Agriculture towards Resilient Food Systems in Asia, Bangkok, Thailand, January 28-30, 2013.

The World's Cities in 2016, United Nations, Department of Economics and Social Affairs

UN-HABITAT. (2008). Land and Sustainable Food Production. UN-HABITAT, Nairobi, Kenya.

United Nations. (2012). World Urbanization Prospects: The 2011 Revision, Department of Economic and Social Affairs, Population Division, United Nations, New York.

Wikipedia. (2013). https://en.wikipedia.org/wiki/Vertical_farming.

IV. Enhancing Horticulture Production

10

Pathways to Enhancing Sustainable Fruit Production

C. P. A. Iyer

SYNOPSIS

There are different definitions for the word 'sustainability' as used in agriculture/horticulture. However, most agree to certain attributes, namely, conserving or protecting the essential agrosystem- base; provide adequate quality and quantity of food to meet the present and future requirements; optimize crop output per unit of capital. labor, land and energy output and they are profitable enough to provide farmers with adequate living standards and support viable rural communities. The key principle for sustainable fruit growing is to take a holistic view of the industry at every level. Short-term and problem-focused approaches that fail to recognize and account for the complexity of inter-relations between activities on- and off- the farm will not increase the sustainability of the enterprise. With regard to individual farm planning for sustainable fruit production, the primary aspect is to recognize that each property is a part of a greater landscape which should help to plan orchards to co-exist with the surrounding environment.

While the organic fruit growing has been welcomed world-wide, particularly with consumers paying a higher price for such fruits, the economic sustainability in the long run has limited the expansion of organic fruit production in certain regions. While the traditional fruit production system appear to provide excellent models of sustainability, they are threatened with demand for higher quantity and a change in quality.

***Keywords*:** Fruit, sustainability, organic fruit growing.

1. INTRODUCTION

The word 'sustainability' has been defined by many workers in various words. However, it all amounts to the ability to meet the needs of the present without compromising the ability of the future generations to meet their own needs. Many people define 'sustainability in agriculture' only in terms of local environmental effects. The definition of sustainability in a crop production system must be much broader than that (Doxon, 1991). The ultimate definition of a sustainable system is one that serves people, sustains or improves the environment and enhances the economy on the scale of the entire planet and over the next several hundred years. As it is difficult to imagine all the repercussions of our present action on a long term basis, it is important to use all scientific tools available with us to project all kinds of reactions possible to assess the long term effects. Most people who are working towards sustainability in horticulture appear to think first in environmental terms, particularly water- and nutrient- use and pest control measures. However, sustainability should include not only the immediate surroundings of an orchard but also how it impacts the overall global atmosphere.

While sustainable farming systems have been defined in various ways, all share certain attributes. First, they conserve and protect the essential agro - ecosystem resource bases (*e.g.*, soils, water and genetic diversity). Second, they provide adequate quantity and quality of the food and fiber to meet present and future requirements. Third, they optimize crop output per unit of capital, labor, land or energy input. Finally, they are profitable enough to provide farmers with adequate living standards and support viable rural communities. Beyond these four basic attributes, there is little agreement about the best definition of sustainability.

2. SUSTAINABLE FRUIT PRODUCTION

With regard to sustainable issues in fruit production, traditional fruit production systems appear to provide excellent models of sustainability. They were developed with minimal external inputs and they have endured for very long periods of time. There are growers in whose fields of avocado, cashew, olive, apple, grapes and other perennial fruits have been harvested for centuries in their farms. These traditional fruit production systems are sustainable on their own merits. However, with regard to forestry areas, rampant deforestation for timber, grazing and conversion to commercial cash crops threaten these forests, where the local people depend on them for food, income and invaluable diversity of flora and fauna they harbor. Besides, they do not meet the food demands of increasing human populations (Merwin and Pritts, 1993).

Agricultural sustainability is closely linked with availability and quality of water and soil. Population affected by water scarcity was 130 million in 1990, 436 million in 1995 and projected to be 800 million in 2025 and 3950 million in 2050. *Per capita* agriculture land availability in 1990 and 2025 respectively is estimated to be 0.05 and 0.03 ha in Egypt, 0.10 and 0.04 ha in Kenya, 0.33 and 0.05 ha in Tanzania, 0.35 and 0.07 ha in Pakistan, 0.20 and 0.12 in India. Land scarcity is exacerbated by the severe problem of soil degradation in developing countries. Soil degradation depletes the soil organic carbon (SOC) pool leading the emission of CO_2 into the atmosphere.

Sustainable horticultural practices, similar to grain crops, include mulching with crop residues and synthetic materials, growing cover crops, water harvesting and conservation tillage or even no-till farming. Sustainable management of natural resources must be based on enhancing productivity per unit area; time and energy input, maintaining a positive trend in productivity over time, creating SOC sink capacity of natural and managed eco-systems; reducing risks of non-point source pollution and sedimentation. Relevant indices of sustainable horticulture and resource management include soil quality and resilience and factors affecting them and temporal changes in productivity and use efficiency of non-renewable or input of limited resources (Lal, 2006).

The SAI platform (Anon., 2009) document provides a set of principles for sustainable fruit production for the mainstream market in all regions of the world. The idea of the fruit growers should be to ensure the safety and quality of their products that will satisfy the highest expectation of the fruit industry and consumers. Besides, on- farm practices should ensure that crops are produced under sustainable economic, social and environmental conditions. A good management of farming system constitutes the grass roots of the system's economic, environmental and social sustainability. Hence, it first pays attention to planning and managing well the overall farm system itself. The initial orchard management action for sustainability can be initiated and planned by the fruit growers or a group of growers themselves.

2.1 Sustainable Farming Systems for Fruit Growing

According to SAI platform the following questions need to be answered and the grower convinced about the sustainable farming system particularly for fruit growing. Some of which are;

- The varieties suited to the local climate, soil and an assessment about the pests and diseases in that locality.
- Do not expand fruit plantations in protected areas nor in any part of high conservation value forests.
- Are there natural habitats on the farm? Are rare species of plants/ animals threatened by growing the crop?
- Preferences should be given to varieties, rootstocks or trees that are adapted to local conditions, resistant or tolerant to common important pests and diseases and meeting customers' requirements.
- In case of genetically modified plants, rootstocks or scion material, one should comply with all the regulations in place for both countries of production and consumption.
- Plant spacing shall be chosen to minimize the use of plant protection inputs and facilitate harvest.
- Are the major energy inputs known? – How can their impact on climate change be reduced?
- In order to maintain or improve the physical and biological characteristics of the soil as well as to reduce mineralization and leaching of nutrients, practices should focus on crop rotation, vegetative cover, maintenance of soil organic matter, avoidance of persistent pesticides and optimization plan of irrigation. A crop nutrient management plan is to be developed so that nutrient loss is minimized and the crops' needs are met based on soil, plant analysis and so on. Chopping and incorporation of crop residues as well as organic manure or compost can be used to increase organic matter content, improve nutrient and water retention and reducing soil erosion.
- Heavy metal build up shall be prevented by the use of good quality fertilizers. Where considered a risk, industrial residue water or sludge shall not be spread on vegetation.
- Trees shall be trained and pruned to achieve a manageable uniform size, a balance between growth and regular yields and allow an optimal distribution of solar radiation and external sprays in the canopy.
- In case of aerial sprays for plant protection measures in orchard, drift minimization techniques must be followed.
- Non- chemical measures are encouraged to prevent and reduce pest, disease or weed infestation. These include indirect measures like choice of disease and pest resistant

crop/ variety, mechanical and physical methods of crop husbandry, good fertilizer- and irrigation- practices, *etc*.

- The principles to reduce, reuse and recycle of the wastes are to be understood well.
- Producers shall look to diversify the sources of income by diversifying their client base and / or growing alternative commodities on the farm, either for on- farm consumption or to be sold externally.
- For ensuring environmental security in fruit growing areas, adequate care has to be taken with regard to soil, water, biodiversity, climatic change, energy and waste management.
- Practices should ensure prevention of soil erosion and loss by biological and mechanical control methods. One should avoid steeply sloping field for cultivation or follow contours with operations for soil preparation, Plant hedges to maintain the soil in steep areas as well as use cover crops or permanent covers.
- To support local economy and contributing to the build-up of sustainable livelihoods, preference should be given to local community in recruitment of staff.

2.2 Sustainability of Production Systems

While discussing sustainable fruit production, it is useful to learn from experiments conducted in various locations. Reganold *et al.,* (2001) did an elaborate study of sustainability of three production systems in apple, namely organic, conventional and integrated apple production systems in Washington State from 1994 to 1999. All three systems gave similar yields. The organic and integrated systems had higher soil quality and potentially lower negative environmental impact than conventional systems. When compared with conventional and integrated systems, the organic system produced sweeter and less tart apples, higher profitability and greater energy efficiency. Their data showed that organic system ranked first in environmental and economic sustainability, the integrated system second and the conventional system last.

Newley and Treverrow (2006) while discussing the challenges facing current and future generations of horticultural producers mentioned that it is most important to increase sustainability by minimizing the impact of various practices on the environment and repair the effects of past practices. It is first necessary to have an economically viable production system if the cost of these investments in environmental enhancement is to be borne by the current generation of farmers. With regard to NSW, Australia, the authors felt that impact that need addressing include those of salinity, raised water tables, soil erosion and degradation, residues of persistent pesticides and off- site impacts of nutrients. These concerns are relevant to many parts of India also. Positive action is needed now to protect, restore and enhance the natural resources on which our agriculture and horticulture industries depend upon. The key principle for sustainable fruit growing is to take a holistic view of the industry at every level from the individual plant – environment interaction though to orchard and landscape level. Short term and problem- focused approaches that fail to recognize and account for the complexity of interrelations between activities on and off the farm will not increase enterprise sustainability. According to Newley and Treverrow (2006), planning needs to be considered at regional, local and individual enterprise level. With regard to NSW, Australia, for regional planning the

NSW government has recently established regional catchment management authorities and the Department of Natural Resources to direct community and government efforts in managing and enhancing the resource base. These reforms aim to keep their river systems in a healthy state while giving irrigators greater certainty in their access to water and to conserve and enhance the natural vegetation of the state. New and existing horticultural enterprises will need ~~to~~ enrolment with the Dept. of Natural Resources for guidance on water harvesting and irrigation licenses. The NSW Department of Planning has embarked on projects to identify land that has regionally significant agricultural potential and to protect this land from development which reduces the productive potential of the region. Consultation and cooperation between industries, local authorities, planning instruments, recognize and protect prime agricultural land and avoid conflicts that may render more of the limited horticultural land unproductive. At an individual farm planning level, farmers are able to better define the limitations of their soils, landscape and climate and locate each farming system on the most suitable plants of their properties while retaining native vegetation on the parts best suited for conservation. Once a whole of farm plan is worked out, sustainable management of the enterprise can be developed.

With regard to individual farm planning for sustainable fruit production, the primary aspect is to recognize that each property is part of a greater landscape which should help to enable planning of crops and cropping practices to co-exist with the surrounding environment. The role of agroforestry for ecosystem resilience cannot be underestimated. Agroforestry helps to conserve and protect natural resources by, for example, mitigating non-point sources pollution (*e.g.*, dust), controlling soil erosion and creating wildlife habitat. It facilitates flexible resources to rapid shift in ecological conditions while at the same time maintaining or restoring soil and water resources. Once a whole of farm plan is worked out, sustainable management of the enterprise can be developed. Firstly, attention should be given to sustainable vegetation management. This is a critical component of sustainable orchard management particularly in its contribution to biodiversity, protection of soils and management of water tables. Opportunities to retain stands of native vegetation should be given as they have a positive effect on the farm in particular and over the wider landscape in general. Benefits include the provision of habitat for birds and other pest control agents, wind protection for crops and associated reduction in irrigation demand (Newley and Treverrow, 2006).

With regard to sustainable water management for fruit orchards, over- irrigation and poor management of drainage water leads to major changes in the hydrological characteristics of the catchment and can have detrimental effects on other producers and the environment many miles downstream. The loss of preciously productive land to raised water tables and the associated salinization of soils are occurring on a serious scale. A well-planned and managed irrigation and drainage system should achieve a healthy vigorous crop with very little run- off above that natural level, only small losses to the water table and minimum discharge of salt, fertilizer, pesticides and organic matter into the river system.

Crop residues can be a good source of organic matter, provided their retention is compatible with the good crop hygiene essential for disease management. Pests and disease guide could indicate where crop hygiene is an issue. The challenge in managing crop nutrition is to supply the nutrient levels required at the right time to meet crop demands while preventing on- site soil degradation and minimizing off-site movement of nutrients. For fruit orchards a feedback on nutrient requirement is the regular use of soil and leaf analysis. A thorough understanding of the fate of the range of fertilizer products will enable producers to choose products that will best achieve their aims without any detrimental after- effects. For example, some forms of nitrogen fertilizers are much less acidifying than others and should be preferred. With regard to insect management, IPM has replaced older approaches to pest control which was based on routine calendar spraying or spraying pests on sight. Biological control agents are important components of modern pest management system. To achieve minimal pest it is essential to monitor insect population to know when a serious problem is developing.

With regard to diseases of fruit crops, utilization of rootstocks that are resistant to key soil- borne diseases or pests is an important step. Disease control strategies have already been developed in a range of crops that use weather data to predict high risk conditions and allow growers to apply fungicides when and where they are most effective. Cultural control by good crop hygiene is another essential element of sustainable disease management.

Granatstein and Kupferman (2008) have made an elegant analysis of factors concerning sustainable agriculture or horticulture. Maintaining the integrity of the agricultural resource base (land, water and biodiversity) is necessary for continued production. But history is replete with examples of countries and cultures that allowed their resource base to degrade over time, undermining their ability to provide for their needs. The desire to avoid this fate is a prime motivation for the emergence of the idea of sustainable agriculture. While earliest sustainability discussions for agriculture revolved around soil erosion and conservation in fruit production, pesticide issues have been more central to the development of sustainability concerns. Sustainable agriculture is more to be considered a goal rather than a specific set of farming practices. Fruit production faces a whole range of sustainability issues today spanning all the economics, environmental and social dimensions.

The economic issues include rising production costs (*e.g.*, labor) with static or declining prices; retail consolidation leading to more sellers than buyers and less economic power for producers, global competition and counter seasonal production in opposing hemispheres. Environmental issues around pesticides, water use and quality, energy, biodiversity and air (*e.g.*, Methyl bromide) all relate to sustainability, but often on a longer time frame than economics. Social sustainability encompasses worker safety and other labor issues, the health imparting benefits of fruit in the diet, urbanization and land use changes.

With regard to perennial fruit trees, sustainable fruit production has many dimensions. When the perception of the consumer with regard to the desirability of certain cultivar changes, many of the well-established orchards will have to replaced with new cultivars. The replacement of the well-known Red Delicious apples with new deeper colored apple cultivars to meet the market- demand and subsequently the waning of demand for the new cultivars owing to their poor flavor and eating- quality in comparison to their predecessors resulted in the price drop. In the meanwhile, thousands of hectares of Red Delicious apples were removed during the late 1990s and early 2000s in Washington State. Such examples are many. Thus, demand can be considered as a key system condition for sustainable fruit production. With regard to fruits, the main factors that influence demand could be cultural preferences, dietary diversity, taste, health attributes (real or perceived), prices and transport qualities. A new challenge is climate change that may undermine the suitability of an existing premier production area for the current fruit product, such as, shifting zones for wine grape production (Jones *et al.*, 2005) or emergence of new pests.

2.3 Elements of Integrated Fruit Prodution

In the 1970s, the International organization for Biological and Integrated Control of Noxious Animals and Plants (IOBC) produced technical guidelines under IP umbrella for Pome fruits (*Prunus* spp.) and vines (*Vitis vinifera*) specific to Europe and this was one of the first attempts to delineate the specifics of sustainable fruit production (IOBC, 1994). These guidelines for pome fruits, detail the following elements for integrated fruit production, namely:

(i) Professionally trained, environmentally- and safety- conscious growers.
(ii) Conserving the orchard environment
(iii) Site, rootstocks, cultivars and planting system for new orchards.
(iv) Soil management and tree nutrition
(v) Alleyways and weed free strip
(vi) Irrigation
(vii) Tree training and management
(viii) Fruit management
(ix) Integrated plant protection
(x) Effective spray application methods and
(xi) Post-harvest chemical treatments.

The level of details provided in the guidelines varies among these elements with the plant protection receiving the most attention. With the publication of IFP guidelines, many fruit tree producing regions in Europe responded by developing their own technical standards and certification systems based on IOBC documents and tailored to the regions. In Europe, government policy often provided direct financial incentives to growers who adopted IFP. By 1994, 40% of the pome fruits areas in Western Europe were under an IFP program. In the 1990s IFP programs were also developing in New Zealand, USA, Canada, Argentina and South Africa.

The recent emphasis on high density planting (HDP) in temperate, sub-tropical and tropical fruit crops has its ramifications on sustainability. For example, early commercial bearing of the crop helps in economic sustainability; low canopy height improves worker safety by eliminating much of ladder work, benefitting social sustainability. A more open canopy helps to reduce diseases and improves pesticide coverage and efficacy, addressing environmental sustainability. But as in most sustainability studies, there are trade- offs. Yuri *et al.*, (2004) reported HDP orchards with their typical leaf canopies may suffer from increased sun scald of fruit from the reduced protection against the intense solar radiation in summer increasing fruit cullage and sometime requiring evaporative cooling of the orchard with irrigation water, particularly in semi-arid region.

According to Granatstein and Kupferman (2008), many practices employed in modern HDP enhance sustainability. Water conservation and efficient usage can be ensured by micro jet sprinklers and drip irrigation. Legumes and organic amendments can provide a significant portion of the fertility needs, offsetting fossil fuel based fertilizers. Mulches in the tree row can control weeds, conserve moisture and build soil organic matter. IPM tools, such as, monitoring, mating disruption and granulosis virus can reduce traditional pesticide inputs for Codling Moth control especially those organo-phosphate materials with a high level of environmental and health concerns. Similarly, border plantings of wild rose (*Rosa woodsii*) hedges in Washington State apple orchards assisted the biocontrol of leaf rollers by providing alternate habitat for a key parasitoid. According to Liang and Huang (1994) establishment of the naturally occurring weed, *Ageratum conyzoides* in citrus orchards in China lead to near total biocontrol of Citrus Red Mite by natural enemies (*Amblyseius* spp.) encouraged on the weed and this practice was adopted on over 135,000 ha of citrus.

Organic farming is one approach to increasing sustainability that is market- driven and growing rapidly. The major guiding principle is the use of natural materials for crop production and the avoidance of long lasting synthetic materials like fertilizers and pesticides which have an adverse effect on the atmosphere. When organic farming expanded in the 1980s, certification programs became necessary to guarantee the consumer who is indeed paying a higher price that the product is extremely safe and free of any adverse chemical effects. However, organic farming and sustainable fruit culture need not to be synonyms. It is true that organic matter, a key consideration in organic farming is the most important aspect of sustainable soil management. Since tree fruit and vine systems are perennial and typically involve little tillage after planting, they can be very conducive to increasing soil organic matter. On the other hand, tree, fruit and vine crops require a high level of pest management to produce marketable crops. Organic growers are greatly restricted in the pest control products they can use. The allowed products are generally less effective and of shorter duration than products that are used under IFP. Organic tree fruit growers in U.S.A. will commonly fumigate the soil prior to replanting an orchard rather than risk the economic devastation from replant disease, for which there are no proven organic controls (Granastein and Kupferman, 2008).

The conflict between the potential for improved environmental sustainability of organic systems and their challenge in maintaining economic sustainability has limited the expansion of organic fruit production in certain regions. In many regions organic fruit producers must spray more frequently and use more kilograms of permitted pesticide product, often achieving a lower marketable yield (Merwin *et al.,* 2005). Use of scab resistant apple varieties has expanded the potential for organic apple production in humid regions where this disease is difficult to control. Reganold *et al.,* (2001) made a comparison of conventional, integrated and organic apple production in side- by -side plots in Washington. The integrated and organic had the best soil quality while the organic had the lower environmental impact based on a pesticide rating system. Tree growth and fruit yields were similar across all systems as were production costs. However, the authors estimated that the organic system would need a 12% price premium to match the financial break-even point in the conventional system.

In the U.S., many food eco- labels other than 'organic' have been developed including the Lodi-Woodbridge sustainable wine grapes program in California, Food Alliance and Salmon Safe in Oregon and Fair Trade and other social sustainability programs, which offer eco-labels that are somewhat broader than organic and more in line with IPG.

The common goal is to achieve increased sustainability on farms through a feedback mechanism (the label) whereby consumers can support positive changes on farms through purchase decisions. However, consumer research (Hartman, 1997) suggests that altruistic motivations (protecting the environment, worker safety) tend to rank lower than self-interest motivations (pesticide residue- free, nutrition, taste, *etc.*) when consumers respond to eco-labels.

Merwin and Pritts (1993) made a very elegant analysis regarding the sustainability of modern fruit production systems particularly with regard to the U.S. situation. According to them, although there are different definitions of 'sustainability', most agree to certain attributes, namely, (i) they conserve and protect the essential agro-ecosystem resource bases namely, soils, water and genetic diversity, (ii) they provide adequate quantity and quality of the food and fiber to meet present and future requirements, (iii) they optimize crop output per unit of capital, labor, land or energy input and (iv) they are profitable enough to provide farmers with adequate living standards and support viable rural communities. At a first glance traditional fruit production systems appear to provide excellent models of sustainability. However, intrinsically sustainable they may be, they are threatened and losing ground since they may not be able to cope up with the present day requirements with regard to quantity and quality. The modern fruit production systems differ in many ways with the traditional systems. They need more capital to establish and must be precocious and productive to be profitable. They are planted at a high density, needs less pesticides per unit area and they are easier to prune and harvest. Intercrops are rare. Many of the local cultivars and races have been replaced by newer cultivars originated from scientific breeding programs. Pest control is accomplished through synthetic chemicals and breeding resistant/tolerant varieties. Tropical fruits are now available in developed nations throughout the year from growers around the world. Changing consumer demands coupled with stringent soil and climatic requirements mandate today's fruit planting be reestablished continuously. In such cases specific replant problems develop in orchards and vineyards and soil fumigation with broad spectrum biocides, such as, methyl bromide is necessary to maintain previous level of production. With the recent loss of registration for most soil fumigants, it is questionable whether some fruit production centers can be sustained unless effective substitutes are developed (Merwin and Stiles, 1989). Since resistance to soil borne diseases in strawberry was not a top priority in breeding program, more than 90% of the U.S. strawberry cultivars rely extensively on pre- plant fumigation for control of soil diseases, sustainability issues of strawberry growing crops up. With regard to apple growing also, in Washington 85% of the apple replant sites require soil fumigation when older orchards are replanted with intensively managed high density system. However, ban of methyl bromide and other soil fumigants will have severe impact on these fruit industries (Smith, 1993).

Some of the fruit crops like blueberries have low agrochemical requirements and can approach sustainability when grown appropriately. Some blueberry plantations have remained productive for more than 60 years with little external input other than fertilization and pruning. Similarly, they thrive on organic resources of nitrogen and have very low demands for other nutrients. Other examples of crops and varieties that can be grown successfully with minimum chemical input are new and old disease- resistant apples, Native American grapes, primo cane fruiting raspberries and native fruits like papaya. In future, growers may rely more on these pest resistant varieties and indigenous fruits.

In France where nearly 200 nations signed a landmark deal to limit climate change, some of the vineyards and people in wine making business are employing green practices to limit climate change. They change to eco -friendly practices which enable their champagne to command higher prices. In one farm in Champagne, one floor of the building make use of natural light to brighten up the space during the day and efficient insulation which keep the building warm. More energy- efficient pressing techniques are used in the winery. All producers are required by French law to recycle 100% of discarded materials from their grapes crushing endeavors including the water used to wash down the press. As much as 90% of plastics and glass from champagne production is recycled. There are grape farmers who increase biodiversity in their field by planting grasses and hedges for insects to nest and feed on grape pests. There are growers who use pheromones to control insect pests and these entire eco-friendly practices saw a dramatic 90% drop in the use of pesticides while the whole Champagne region saw a 50% drop.

Among the fruits crops considered most at risk by UNIBPGR are apple, avocado, mango and pear collections, preservation and utilization of genetic materials on different locations will be essential for long term sustainability of fruit production. Since tropical fruits have recalcitrant seeds without dormant phase, they do not survive drying or freezing. Hence, the only way they can be perpetuated is in tissue culture or as *in situ* conservation in the field in the biosphere reserves or protected national parks at the centers of origin. In comparison with field crops, fruit production is very labor intensive and provides more employment opportunities in rural communities than field crops and helps to sustain the rural employment base.

The increasing needs to protect the environment require a correct management of wastes including wastes from fruit industry. For example, with regard to lemon, in the zones of its production and processing, as in Spain and Italy, the peel wastes are used as feed for livestock and also for the production of pectins. The mucilages that are obtained from citrus wastes are vegetal polysaccharides similar to pectin and can be used in food, dietary, cosmetic and pharmaceutical industries since recent studies have shown the anti-inflammatory effects can be obtained from orange juice centrifugation pulp (Salomon and Saija, 2014).

3. CONCLUSION

There is increased awareness about sustainability issues in many countries now. Consumers express their interest by paying extra for buying foods with organic, IFP or other eco-labels. While sustainability issues in fruit growing through IFP or organic cultivation have been driven by markets in the more affluent nations in a big scale, in developing countries it is yet to catch up due to various reasons. The ultimate impact from changes in fruit production will be the sustainability impact per unit area times the area influenced, while there is enormous awareness for adopting environmentally safe agro-techniques in orchards and agricultural farms. We are yet to develop reliable methods to quantify improvements in sustainability. These agro-techniques may also be challenged by the factors like climate change. As Newly and Treverrone (2006) have pointed out sustainability remains a goal and the most accurate way to evaluate it is with hind sight, looking back from twenty years in the future and determining whether the actions taken today to increase sustainability actually did so.

For conservation of soil and water resources, growing perennial crops offers advantages like excellent protection of soil because of persistent roots, leaf litter and surface cover crops. Besides, woody plants improve soil structure and nutrient and water retention. Fruit plantings also can provide a relatively rich habitat for both plant and animal species.

The threat from heavy industrialization in many countries where fruit orchards are sacrificed for commercial or residential development is looming large in these days of emphasis on infra-structure building and development. Though restrictive agricultural zoning or conservation easements can deprive the farmer of the lucrative option of selling out to developers, the growers themselves in many cases oppose protective measures.

Genetic engineering is a powerful tool for transferring desirable gene(s) form one organism to other conferring traits, such as, herbicide- resistance, virus resistance *etc.* However, many of these genes that determine such traits have not been maped or located in the genome of most fruit crops. It is also not known whether consumers will accept genetically- modified produce especially foods, such as, fruits with special cultural and symbolic importance.

REFERENCES

Anon. (2009). Principles and practices for sustainable fruit production.SAI Platform. www.saiplatform.org/uploads/pps.fruits.

Doxon, L.E. (1991). Sustainable Horticulture. *HortScience*, 26(12): 1454-55.

Granatstein, D. and Kupferman, E. (2008). Sustainable horticulture in fruit production. *Acta Hort.* 767: 295-308.

Hartman, H. (1997). The Hartman Report: food and the environment, a cosumer's perspectives. Phase-2. The Hartman Group, Bellevire, WA, 64 pp.

IOBC. (1993). Integrated production- principles and technical guidelines. *IOBC/WPRS Bulletin* 17(9): 1-8.

Jones, G.V., White, M.A., Cooper, O.R. and Storchmann, K. (2005). Climate change and global wine quality. *Climate Change* 73 (3): 319-343.

Lal, R. (2006). Sustainable horticulture and resource management. *Acta Hort.* 767: 19-44.

Liang, W. and Huang, M. (1994). Influence of citrus orchard ground cover plants on arthropod communities in China: A review. *Agriculture, Ecosystems, Environment* 50: 29-37.

Merwin, I., Peck, G. and Vollmer, E. (2005). Organic orchards in the Northeast: Progress, Practices and Problems. In Granatstein, D. and Azarenko, A. (eds.). *Proc. 3rd National Organic Tree Fruit Research Symposium*, June 6-8, 2005; Washington State University, Tree Fruit Research and Extension Center, Wenatchee, WA. p. 61-62.

Merwin, I.A. and Prittts, M.P. (1993). Are modern fruit production systems sustainable? *HortTechnology* 3(2):128-136.

Merwin, I.A. and Stiles, W.C. (1989). Root lesion nematodes, potassium deficiency and prior cover crops as factors in apple replant disease. *J. Amer. Soc. Hort. Sci.* 114: 724-728.

Newley, P.D. and Treverrow, N.L. (2006). Sustainable horticulture. *Prime Facts*. 144: 1-11.

Reganold, J.P., Glover, J.D. andrews, P.K. and Hinman, H.R. (2001). Sustainability of three apple production system. *Nature* 410(6831): 926-930.

Salomone, R. and Giuseppe, S. (2014). *Pathways to Environmental Sustainability. Methodologies and Experiences*. Springer, New York.

Smith, T.J. (1993). Orchard replant disease. *Good Fruit Grower*, pp. 35-38.

Yuri, J.A., Lepe, V., Moggia, C., Bastias, R., Quilamapu, I. and Bertschenger, L. (2004). Sunburn in apples. *Obst. and Weinbour*. 140(8): 7-10.

❍❍❍

11

Climate Change and Sustainable Horticulture Development

P. Parvatha Reddy

SYNOPSIS

Global warming due to climate change is the greatest concern of mankind in 21st century. Due to climate change, it is projected that rainfall over India will increase by 15–40% and the mean annual temperature will increase by 3–6° C by the end of 21st century. The established commercial varieties of fruits, vegetables and flowers will perform poorly in an unpredictable manner due to aberration of climate. Melting of ice cap in the Himalayan regions will reduce chilling effect required for the flowering of many horticultural crops like apple, saffron, rhododendron orchid, *etc.* Commercial production of horticultural crops particularly grown under open field conditions will be severely affected. Due to high temperature, physiological disorders of horticultural crops will be more pronounced *e.g.* spongy tissue of mango, fruit cracking of litchi, flower and fruit abscission in Solanaceous fruit vegetables, *etc.* Air pollution also significantly decreases the yield of several horticultural crops and increases the intensity of certain physiological disorder like black tip of mango. The number of disease epidemics has dramatically increased in recent years and also the threat of emerging new diseases, such as, thrips-transmitted tospoviruses and white fly-transmitted begomoviruses in chilli, cucurbits, okra and tomato.

Hence, there is a need to protect these valuable crops for sustainability against the climate change scenario. The most effective way is to adopt sustainable agriculture, using renewable energy, forest and water conservation, reforestation *etc.* To sustain the productivity, modification of present horticultural practices and greater use of greenhouse technology are some of the solutions to minimize the effect of climate change. For adaptation strategies the emphasis should be on development of production systems for improved water-use efficiency and to adapt to the hot and dry conditions. Strategies like changing sowing or planting dates, modifying fertilizer application methods, providing irrigation during critical stages of the crop growth and conservation of soil moisture reserves by mulching with crop residues and plastic mulches are the most important interventions. Development of climate-resilient horticultural crops which are tolerant to high temperature, moisture stress, salinity and climate proofing through genomics and biotechnology would be essentially required. Shift in varietal choice may become necessary in case of grapevines, banana, mango and other important horticultural crops. Moisture stress has to be addressed through the development of varieties tolerant to drought and rain-fed conditions. Delaying the date of planting can be an adaptation strategy in most of the states of the Indo-Gangetic plains to minimize the yield reductions while for West Bengal and *kharif* potato-growing regions, development of heat tolerant varieties should be the main adaptation strategy. Dogridge (*Vitis champini*) has been found promising for improvement in vigour, yield and quality of seedless grapes as well as tolerance to drought and salinity.

In order to evaluate appropriate disease management strategies, there is a need to thoroughly understand the potential climate change impacts on host–pathogen interactions. Horticulture-based farming systems have high potential for sequestering carbon for mitigation of climate change. The perennial trees act as carbon sinks by sequestering the atmospheric carbon. Availability and development of good simulation models for fruit and vegetable crops is lacking in India probably with exception of potato and coconut where InfoCrop models have been adopted and validated for different agro-ecological regions.

***Keywords*:** Climate change, physiological disorders, horticultural crops, global warming, precipitation patterns.

1. INTRODUCTION

At present, the total production of horticultural produce is 234.5 million tons from the 11% of the total cultivated area in India, which contributes 30.4% in value term to GDP of agriculture. India with production of more than 28.2 million tons of fruits and 66 million tons of vegetables is the second largest producer of fruits and vegetables in the world next only to Brazil and China. However, *per capita* consumption of fruits and vegetables in India is only around 46 g and 130 g as against a minimum of about 92 g and 300 g recommended by the Indian Council of Medical Research, New Delhi and National Institute of Nutrition, Hyderabad.

Significant variation in either mean state of climate or its variability, persisting for an extended period (typically decades or longer) is referred as climate change, which may be due to natural internal processes or external forcing or to persistent anthropogenic changes in the composition of the atmosphere or in land use. In atmosphere, energy is absorbed by the land, seas, mountains, *etc.* and simultaneously released in the form of infrared waves. All the released heat is not lost in space, but is partly absorbed by some gases present in very small quantities in the atmosphere (greenhouse gases - carbon dioxide, methane, nitrous oxide, water vapour, ozone and a few others). Thus, increased concentration of greenhouse gases leads to increased temperature which in turn has impact on the world climate, leading to the phenomena known as climate change. The earth's climate system constantly adjusts so as to maintain the balance between the energy that reaches it from the sun and the energy that goes from earth back to space. This means that even a small rise in temperature could mean accompanying changes in cloud cover and wind patterns. Some of these changes may enhance the warming, while others may counteract. Cooling effect may result from an increase in the levels of aerosols (small particles of matter or liquid that can be produced by natural or man-made activities). Positive feedback may result from an increase in water vapour due to high evaporation with rise in temperature and can further add to the warming effect. The significant change may impact agriculture/ horticulture/ fish/ livestock and consequently food supply. Climate change *per se* is not necessarily harmful, but the problems arise from extreme events that are difficult to predict, like more erratic rainfall pattern and unpredicted warm spells shall affect productivity. At the same time, more availability of CO_2 would help in improved yield of root crops and increased temperature may shorten the period of crop growth.

Climate change is predicted to cause an increase in average air temperature of between 1.4° C and 5.8° C, increases in atmospheric CO_2 concentration and significant changes in rainfall pattern (Houghton *et al.,* 2001). The threshold value of temperature rise (2° C) can cause devastating, dangerous and irreversible consequences of warming to manifest world over. Global warming is occurring along with shifting pattern of rainfall and increasing incident of extreme weather events like floods, droughts and frosting. Recent studies suggest clear evidence of reduction in light intensity, rapid melting of glaciers and rise in sea level. The estimates of Indian Institute of Tropical Meteorology, Pune, indicate similar trends in India with slightly higher magnitude. It is projected that rainfall over India will increase by 15–40% and the mean annual temperature will increase by 3–6°C by the end of twenty- first century. Warming is likely to be more pronounced over land areas.

The present challenges like global climate change, water and soil pollution, decreased water availability, urbanization *etc.* adds up to the situation. In combination with elevated temperatures, decreased precipitation could cause reduction in availability of irrigation water and increase in evapotranspiration, leading to severe crop water-stress conditions. Vegetable production is threatened by increasing soil salinity particularly in irrigated croplands which provide 40% of the world's food.

2. CONSEQUENCES

The consequences of rapid climate change include unpredictable high temperature spells, seasonal changes, more erratic rainfall patterns, melting of ice cap, flood, rising sea level and salinization, drought, latitudinal and altitudinal shifts in ecological and agro-economic zones, *etc.* leading to extremity of all kinds (FAO, 2004). Decrease in potential yields is likely to be caused by shortening of the growing period, decrease in water availability and poor vernalization. Salinity affects agriculture in coastal regions which are impacted by low-quality and high-saline irrigation water due to contamination of the groundwater and intrusion of saline water due to natural or man-made events. Furthermore, coastal areas are threatened by specific, saline natural disasters which can make agricultural lands unproductive, such as, tsunamis which may inundate low-lying areas with seawater. Although the seawater rapidly recedes, the groundwater contamination and subsequent osmotic stress causes crop losses and affects soil fertility.

Developing countries in the tropics will be particularly vulnerable. Western Ghats and surrounding regions may be deprived of normal precipitation due to abnormal monsoon. Vulnerability, rarity and rapid extinction of plant species will be among the other consequences. Plains of India will face similar kind of problems. The total agricultural land will shrink and the available land may not remain suitable for the present crops for too long. Farmers have to explore options of changing crops suitable to weather. The climatic changes could lead to major food security issues for a country like India. Unless measures are undertaken to mitigate the effects of climate change, food security in developing countries in the tropics will be under threat.

3. IMPACT ON HORTICULTURAL CROPS

Two major parameters of climate change that have far reaching implications on horticulture are more erratic rainfall patterns and unpredictable high temperature spells that will consequently reduce crop productivity. The climate change will have many impacts on horticulture and a few examples are given below.

- Many areas presently suitable for horticultural crops would become unsuitable in another 25 years. There would be new areas which are presently unsuitable, become highly suitable for cultivation of horticultural crops.
- Production timing will change due to rise in temperature. As a result, photosensitive crop will mature faster.
- The winter regime and chilling duration will reduce in temperate regions affecting the temperate crops.
- Pollination will be affected adversely because of higher temperature. Floral abortions flower and fruit drop will occur frequently.
- The requirement of annual irrigation will increase and heat unit requirement will be achieved in much lesser time.
- Higher temperatures will reduce tuber initiation process in potato, reduced quality in tomatoes and pollination in many crops. In case of crucifers, it may lead to bolting; anthocyanin production may be affected in apples and capsicum. Tip burn and blossom end rot will be the common phenomenon in tomatoes.
- Coastal regions can expect much faster percolation of sea water in inland water tables causing more salinity.

3.1 Fruit Crops

Due to rise in temperature, fruits will develop more rapidly and mature earlier. For example, Citrus, grapes, melons *etc.* will mature earlier by about 15 days. Strawberries will produce more runners at the expense of fruits. Specific chilling requirements of pome and stone fruits will be affected and thus dormancy breaking will be earlier. Delay in monsoon, dry spells and untimely rains during water stress period, supra-optimal temperatures during flowering and fruit growth, hail storms are some of the most commonly encountered climatic conditions experienced by the citrus growers over the past decade or so. Banana cultivation may suffer from high temperature, soil moisture stress or flooding/ water logging. High temperature and moisture stress also increase sunburn and cracking in apples, apricot and cherries and increase in temperature at maturity will lead to fruit cracking and burning in litchi (Kumar and Kumar, 2007). Air pollution also significantly reduces the yield of several horticultural crops and increases the intensity of certain physiological disorders like black tip of mango which is induced by coal fume gases, sulphur dioxide, ethylene, carbon monoxide and fluoride. Leaf production increases by one leaf per month for every 3.3 to 3.7°C rise in minimum or mean temperature from 10-20°C or 13.5 to 25°C. In general, higher temperature increases the rate of plant maturity in annual species, thus shortening the growth stages, during which developing fruits and suckers absorb photosynthetic products. The temperature below 10° C leads to impedance of inflorescence and malformations of bunches. Chilling symptoms on leaves are not seen immediately but it may take 2 to 4 days to appear.

3.2 Vegetable Crops

Vegetables are generally sensitive to environmental extremes and thus high temperatures and limited soil moisture are the major causes of low yields in the tropics and will be further magnified by climate change. Increasing temperatures, decreased irrigation water availability, flooding and salinity will be major limiting factors in sustaining and enhancing vegetable productivity.

3.2.1 High temperatures

With a changing climate, vegetable crops in tropics will be subjected to increased temperature stress. High temperatures can cause significant losses in tomato productivity due to reduced fruit set and smaller and lower quality fruits (Stevens and Rudich, 1978). Hazra *et al.,* (2007) summarized the symptoms causing fruit set failure at high temperatures in tomato; which include bud drop, abnormal flower development, poor pollen production, dehiscence and viability, ovule abortion and poor viability, reduced carbohydrate availability and other reproductive abnormalities. In addition, significant inhibition of photosynthesis occurs at temperatures above optimum, resulting in considerable loss of potential productivity. In bell pepper, high post-pollination temperatures inhibited fruit set, suggesting that fertilization is sensitive to high temperature stress (Erickson and Markhart, 2002).

3.2.2 Drought

Water availability is expected to be highly sensitive to climate change and severe water stress conditions will affect crop productivity, particularly that of vegetables. In combination with elevated temperatures, decreased precipitation could cause reduction of irrigation water availability and increase in evapotranspiration, leading to severe crop water-stress conditions (IPCC, 2001). Drought stress causes an increase of solute concentration in the environment (soil), leading to an osmotic flow of water out of plant cells. This leads to an increase of the solute concentration in plant cells, thereby lowering the water potential and disrupting membranes and cell processes, such as, photosynthesis.

3.2.3 Salinity

In hot and dry environments, high evapotranspiration results in substantial water loss, thus leaving salt around the plant roots which interferes with the plant's ability to uptake water. Excessive soil salinity reduces productivity of many agricultural crops, including most vegetables which are particularly sensitive throughout the ontogeny of the plant. According to the United States Department of Agriculture (USDA), onions are sensitive to saline soils, while cucumbers, eggplants, peppers and tomatoes are moderately sensitive. Plant sensitivity to salt stress is reflected in loss of turgor, growth reduction, wilting, leaf curling and epinasty, leaf abscission, decreased photosynthesis, respiratory changes, loss of cellular integrity, tissue necrosis and death of the plant (Jones, 1986; Cheeseman, 1988).

3.2.4 Flooding

Most vegetables are highly sensitive to flooding and genetic variation with respect to this character is limited, particularly in tomato. In general, damage to vegetables by flooding is due to the reduction of oxygen in the root zone which inhibits aerobic processes. Flooded tomato plants accumulate endogenous ethylene that causes damage to the plants (Drew, 1979). Low oxygen levels stimulate an increased production of an ethylene precursor, 1-aminocyclopropane-1-carboxylic acid (ACC), in the roots. The rapid development of epinastic growth of leaves is a characteristic response of tomatoes to water-logged conditions and the role of ethylene accumulation has been implicated (Kawase, 1981). The severity of flooding symptoms increases with rising temperatures; rapid wilting and death of tomato plants is usually observed following a short period of flooding at high temperatures (Kuo *et al.,* 1982).

3.3 Flower Crops

Melting of ice cap in the Himalayan regions will reduce the chilling requirement for the flowering of many of the ornamental plants like Rhododendron orchid, Tulip, Alstromerea, Magnolia, Saussurea, Impatiens, Narcissus *etc.* Some of them will fail to bloom or flower with less abundance while others will be threatened. Plant species requiring high humidity and water may find them under difficult conditions for survival. Plains of India will also have similar kind of problems and will be affected either by drought or excessive rains, floods and seasonal variations. Commercial production of flowers particularly grown under open field conditions will be severely affected leading to poor flowering, improper floral development and colour. Chrysanthemum is a short day plant. So flowering round the year in open field condition is not possible. Low temperatures shut down flowering in Jasmine (<19° C) and lead to reduction in flower size. Flowers do not open up fully in tropical orchids wherever temperatures are below 15° C. High temperature leads to flower bud drop and unmarketable spikes in tropical orchids when temperature remains > 35° C.

3.4 Plantation Crops

Consecutive drought reduced the coconut production by about three lakh nuts/year for four years. Productivity loss was to the tune of about 3500 nuts/hectare/year in India. Apart from drought, other natural calamities like cyclone *etc.* have impacted the crop production and productivity. In coconut, arecanut and cocoa increased CO_2 led to higher biomass production. But a slight decline in biomass production was apparent at elevated air temperature. All three crops responded differentially under elevated air temperature. In coconut, net photosynthesis rate has reduced but increased that of arecanut and cocoa. However, TDM was slightly decreased in all three crops.

The rainfed cashew crop is highly sensitive to changes in climate and weather vagaries, particularly during reproductive phase. Cashew requires relatively dry atmosphere and mild winter (15-20°C) coupled with moderate dew during night for profuse flowering. High temperature (> 34.4°C) and low relative humidity (< 20%) during afternoon causes drying of flowers, resulting in yield reduction. Unseasonal rains and heavy dew during flowering and fruiting period aggravated the incidence of pests and diseases. All these situations resulted in reduction yield up to 50 to 65%.

3.5 Spice Crops

In general, the productivity showed decreasing trend in most of the black pepper growing areas of India due to increase in maximum and minimum day temperature and decreasing the annual rainfall. In black pepper, Accessions 1380 (IC 316801), 1387 (IC 316803),

1410 (IC 316817), 1423 (IC 316825) and 1430 (IC 316832) were identified as relatively tolerant to drought. In cardamom, RR1 (IC 349591), CL-893 (IC 349537), Green Gold (IC 349550) were found relatively tolerant under Kerala, India conditions. Kashmir's prized saffron crop has suffered 40% drop in production due to shortage of water. Some of the saffron farmers who traditionally relied on rainwater are now looking for irrigation measures to save their rare and labour intensive crop.

Seed spices are winter season crops and commonly grown in arid and semi-arid regions of Rajasthan and Gujarat requiring certain period of low temperature for optimum vegetative growth. Heavy losses have been observed due to combined effect of chilling and frost injury. Cumin, coriander, nigella, ajowan crops are very sensitive to frost. The incidence of frost is responsible for serious loss in yield which almost reaches up to 100%. Fennel and fenugreek are also affected by frost but growth stage plays an important role. So far no efforts have been made to identify the source of resistance against low temperature injury in available germplasm of seed spices.

4. ADAPTATION AND MITIGATION

Farmers in developing countries of the tropics need tools to adapt and mitigate the adverse effects of climate change on horticultural productivity and particularly on vegetable production, quality and yield. Current and new, technologies being developed through plant stress physiology research can potentially contribute to mitigate threats from climate change on horticultural crop production. Technologies that are simple, affordable and accessible must be used to increase the resilience of farms in less developed countries. World Vegetable Centre (AVRDC) has identified germplasm of the major vegetable crops which are tolerant of high temperatures, flooding and drought and developed advanced breeding lines. Efforts are also underway to identify nitrogen-use efficient germplasm. Top research and development priorities of AVRDC include the development of production systems geared towards improved water-use efficiency which are likely to mitigate the effects of hot and dry conditions in vegetable production systems.

4.1 Enhancing Horticulture Crop Production Systems

Various management practices have the potential to raise the yield of vegetables grown under hot and wet conditions of the lowland tropics. AVRDC has developed technologies to alleviate production challenges, such as, limited irrigation water and flooding, to mitigate the effects of salinity and also to ensure appropriate availability of nutrients to the plants. Strategies include modifying fertilizer application to enhance nutrient availability to plants, direct delivery of water to roots (drip irrigation), grafting to increase flood and disease tolerance and use of soil amendments to improve soil fertility and enhance nutrient uptake by plants.

4.1.1 Water-saving irrigation management

If water is scarce and supplies are erratic or variable, then timely irrigation and conservation of soil moisture reserves are the most important agronomic interventions to maintain yields during drought stress. In order to generate income and alleviate poverty of the small-holder farmers in developing countries, affordable small-scale drip irrigation technologies have been developed by the International Development Enterprises (IDE). Drip irrigation delivers water directly to plants through small plastic tubes. IDE found that water losses due to run-off and deep percolation are minimized and water savings of 50-80% are achieved through drip irrigation as compared to most traditional surface irrigation methods. Crop production per unit of water consumed by plant evapotranspiration is typically increased by 10-50%. Thus, more plants can be irrigated per unit of water by drip irrigation with less labour. The water-use efficiency by chilli pepper was significantly higher in drip irrigation compared to furrow irrigation, with higher efficiencies observed with high delivery rate drip irrigation regimes (AVRDC, 2005). The incidence of *Fusarium* wilt was reduced when a lower drip irrigation rate was used in drought tolerant crop like watermelon. In general, the use of low-cost drip irrigation is cost effective, labour-saving and allows more plants to be grown per unit of water, thereby both saving water and increasing farmers' incomes at the same time.

4.1.2 Cultural practices that conserve water and protect crops

Various crop management practices, such as, mulching and the use of shelters and raised beds help to conserve soil moisture, prevent soil degradation and protect horticultural crops from heavy rains, high temperatures and flooding. The use protective coverings (mulches) help to reduce evaporation, moderate soil temperature, reduce soil runoff and erosion, protect fruits from direct contact with soil and minimize weed growth. Further, the use of organic mulch can help to enhance soil fertility, structure and other soil properties. Rice straw is abundant in rice-growing areas of the tropics and generally recommended as mulch for summer tomato production. The benefits of rice straw mulch on fruit yield of tomato have been demonstrated in Taiwan (AVRDC, 1981). In India, mulching improved the growth of eggplant, okra, bottle gourd, round melon, ridge gourd and sponge gourd compared to the non-mulched controls (Pandita and Singh, 1992). The highest yields were obtained when polythene and sarkanda (*Saccharum* spp. and *Canna* spp.) were used as mulching materials. In the lowland tropics where temperatures are high, dark-collared plastic mulch is recommended in combination with rice straw (AVRDC, 1990). Dark plastic mulch prevents sunlight from reaching the soil surface and the rice straw insulates the plastic from direct sunlight thereby preventing the soil temperature rising too high during the day.

During the hot rainy season, vegetables, such as, tomatoes suffer from yield losses caused by heavy rains. Simple, clear plastic rain shelters prevent water logging and rain impact damage on developing fruits, with consequent improvement in tomato yields (Midmore *et al.,* 1992), reduction in fruit cracking and the number of unmarketable fruits. Plastic shelters were responsible for elimination of flooding and rain damage, reduced air temperature and higher yields. Temperature stress, direct rain impact and intense sunlight can also be reduced by using shade cloth shelter. Planting vegetables in raised beds can ameliorate the effects of flooding during the rainy season (AVRDC, 1981). Yields of tomatoes increased with bed height, most likely due to improved drainage and reduction of anoxic stress. Additive effects on yield were observed when tomatoes were planted in raised beds with rain shelters.

4.1.3 Enhanced stress tolerance by grafting

Grafting has been used primarily to control soil-borne diseases affecting the production of fruit vegetables, such as, tomato, eggplant and cucurbits. The soil-borne disease (Fusarium wilt) and virus diseases (bud necrosis virus and Zucchini Yellow Vein Necrosis virus) on water melon can be managed by grafting on bottle gourd rootstock. Grafting can also provide tolerance to soil-related environmental stresses, such as, drought, salinity, low soil temperature and flooding if appropriate tolerant rootstocks are used. Grafting of eggplants was started in the 1950s, followed by grafting of cucumbers and tomatoes in the 1960s and 1970s (Edelstein 2004). Romero *et al.,* (1997) reported that melons grafted onto hybrid squash rootstocks were more salt tolerant than the non-grafted melons. The rootstocks from *Cucurbita* spp. are more tolerant of salt than rootstocks from *Lagenaria siceraria.* (Matsubara, 1989). Grafted plants were also more able to tolerate low soil temperatures. *Solanum lycopersicum* x *S. habrochaites* rootstocks provide tolerance of low soil temperatures (10°C to 13°C) for their grafted tomato scions, while eggplants grafted onto *S. integrifolium x S. melongena* rootstocks grew better at lower temperatures (18°C to 21°C) than non-grafted plants (Okimura *et al.,* 1986).

Vegetables, in general, are unable to tolerate excessive soil moisture and tomatoes in particular are considered to be highly sensitive. In the tropics, heavy rainfall with poor drainage induces water-logged conditions that reduce oxygen availability in the soil thereby causing wilting, chlorosis, leaf epinasty and ultimately death of the tomato plants. Genetic variability for tolerance to excess soil moisture is limited or inadequate to prevent losses. Many accessions of eggplant are highly tolerant of flooding (Midmore *et al.,* 1997). The grafting techniques are developed to improve the flood tolerance of tomato using eggplant rootstocks which were identified with good grafting compatibility with tomato and high tolerance to excess soil moisture. Tomato scions grafted onto eggplant rootstock grow well and produce acceptable yields during the rainy season (Midmore *et al.,* 1997). In addition to protection against flooding, some eggplant genotypes are drought tolerant and eggplant rootstocks can therefore provide protection against limited soil moisture stress.

4.2 Developing Climate-Resilient Varieties

Superior varieties adapted to a wider range of climatic conditions could result from the discovery of novel genetic variation for tolerance to different biotic and abiotic stresses. Genotypes with improved attributes conditioned by superior combinations of alleles at multiple loci could be identified and advanced. Improved selection techniques are needed to identify these superior genotypes and associated traits, especially from wild, related species that grow in environments which do not support the growth of their domesticated relatives that are cultivated varieties. Plants native to climates with marked seasonality are able to acclimatize more easily to variable environmental conditions (Pereira and Chaves, 1995) and provide opportunities to identify genes or gene combinations which confer such resilience.

4.2.1 Tolerance to high temperatures

AVRDC has developed tomatoes and Chinese cabbage with general adaptation to hot and humid tropical environments and low-input cropping systems since the early 1970s. This has been achieved by developing heat-tolerant and disease-resistant breeding lines. The Centre has made significant contributions to the development of heat-tolerant tomato and Chinese cabbage lines and the subsequent release of adapted, tropical varieties worldwide. Indeed, AVRDC's heat tolerant hybrids have resulted in the successful cultivation of

Chinese cabbage in the lowland tropics. The key to achieving high yields with heat tolerant cultivars is the broadening of their genetic base through crosses between heat tolerant tropical lines and disease resistant temperate or winter varieties (Opena and Lo, 1981). The heat tolerant tomato lines were developed using heat tolerant breeding lines and landraces from the Philippines (*e.g.* VC11-3-1-8, VC 11-2-5, Divisoria-2) and the United States (*e.g.* Tamu Chico III, PI289309) (Opena *et al.,* 1989). However, lower yields in the heat tolerant lines are still a concern.

More heat tolerant varieties are required to meet the needs of a changing climate and these must be able to match the yields of conventional, non-heat tolerant varieties under non-stress conditions. A wider range of genotypic variation must be explored to identify additional sources of heat tolerance.

4.2.2 Drought tolerance and water-use efficiency

In slowly developing water deficit, plants may escape drought stress by shortening their life cycle (Chaves and Oliveira, 2004). However, the oxidative stress of rapid dehydration is very damaging to the photosynthetic processes and the capacity for energy dissipation and metabolic protection against reactive oxygen species is the key to survival under drought conditions (Ort, 2001, Chaves and Oliveira, 2004). Drought tests show that *Solanum chilense* is five times more tolerant to wilting than cultivated tomato. *S. pennellii* has the ability to increase its water use efficiency under drought conditions unlike the cultivated *S. lycopersicum* (O'Connell *et al.,* 2007). It has thick, round waxy leaves, is known to produce acyl-sugars in its trichomes and its leaves are able to take up dew (Rick, 1973). Transfer and utilization of genes from these drought resistant species will enhance tolerance of tomato cultivars to dry conditions, although wide crosses with *S. pennellii* produce fertile progenies, *S. chilense* is cross-incompatible with *S. lycopersicum* and embryo rescue through tissue culture is required to produce progeny plants. Research work is in progress at AVRDC to identify the genetic factors underlying drought tolerance in *S. chilense* and *S. pennellii* and to transfer these factors into cultivated tomatoes.

4.2.3 Tolerance to salinity

Success in breeding for salt tolerance requires effective screening methods, existence of genetic variability and ability to transfer the genes to the species of interest. A cross between a salt-sensitive tomato line (UCT5) and a salt-tolerant *S. esculentum* accession (PI174263) showed that the ability of tomato seed to germinate rapidly under salt stress is genetically controlled with narrow-sense heritability (h^2) of 0.75 (Foolad and Jones, 1991). Several studies indicate that salt tolerance during seed germination in tomato is controlled by genes with additive effects and could be improved by directional phenotypic selection (Foolad, 2004).

In bell pepper, salt stress significantly decreases germination, shoot height, root length, fresh and dry weight and yield. Yildirim and Guvenc (2006) reported that bell pepper genotypes Demre, Ilica 250, 11-B-14, Bagci Carliston, Mini Aci Sivri, Yalova Carliston and Yaglik 28 can be useful as sources of genes to develop bell pepper cultivars with improved germination under salt stress. In Tunisia, the root system of salt tolerant cultivar 'Beldi' was unaffected by salt stress. In addition, 'Beldi' significantly out-yielded other test cultivars at high salt treatments.

Related wild tomato species have shown strong salinity tolerance and are sources of genes as coastal areas are common habitat of some wild species. Studies have identified potential sources of resistance in the wild tomato species *S. cheesmanii, S. peruvianum, S pennelii, S. pimpinellifolium* and *S. habrochaites* (Flowers, 2004; Foolad, 2004; Cuartero *et al.,* 2006). Attempts to transfer quantitative trait loci (QTLs) and elucidate the genetics of salt tolerance have been conducted using populations involving wild species. Elucidation of mechanism of salt tolerance at different growth periods and the introgression of salinity tolerance genes into vegetables would accelerate development of varieties that are able to withstand high or variable levels of salinity compatible with different production environments.

4.3 Climate-Proofing though Genomics and Biotechnology

In the past decade, genomics has developed from whole genome sequencing to the discovery of novel and high throughput genetic and molecular technologies. This has opened new way to genetic manipulation of genes associated with tolerance to environmental stresses. National and international institutes are retooling for plant molecular genetic research to enhance traditional plant breeding and benefit from the potential of genetic engineering to increase and sustain crop productivity.

4.3.1 Qtls and gene discovery for tolerance to stresses

The use of molecular markers as a selection tool provides the potential for increasing the efficiency of breeding programs by reducing environmental variability, facilitating earlier selection and reducing subsequent population sizes for field testing. Molecular markers facilitate efficient introgression of superior alleles from wild species into the breeding programs and enable the pyramiding of genes controlling quantitative traits; thus enhancing and accelerating the development of stress tolerant and higher yielding cultivars for farmers in developing countries.

Quantitative Trait Loci (QTLs) for drought tolerance have been identified in tomato in *S. pennellii* and *S. pimpinellifolium. S. pimpinellifolium* is also commonly investigated as source of salt tolerance. AVRDC tomato line CL5915 has heat tolerance. Many genes associated with stress tolerance have recently been determined using high throughput expression assays. Integration of QTL analysis with gene discovery and modelling of genetic networks will facilitate a comprehensive understanding of stress tolerance, permit the development of useful and effective markers for marker-assisted selection and identify candidate genes for genetic engineering.

4.3.2 Engineering stress tolerance

Environmental stress tolerance is a complex trait and involves many genes (Wang *et al.,* 2003). Approximately 130 drought-responsive genes have been identified using microarrays (Seki *et al.,* 2001; Reymond *et al.,* 2000). These genes are involved with transcription modulation, ion transport, transpiration control and carbohydrate metabolism. Research on the physiology of stress tolerance has demonstrated that tolerance to a specific stress is determined by several component traits and controlled by corresponding genes. A combination of a genome-wide scan of expression, using DNA arrays and QTL analysis could provide important information in identifying the major genes association with stress tolerance.

Some of the horticultural crop varieties tolerant to abiotic stress have been presented in given in Table 1.

Table 1. *Some horticultural crop varieties tolerant to abiotic stress (Bose and Mitra, 1996; Hazra and Som, 1999; Rai and Yadav, 2005)*

Crop	Variety/rootstock	Tolerance
Pomegranate	Ruby	Drought
Annona	Arka Sahan	Drought
Fig	Deanna, Excel	Drought
Grape (rootstock)	Dogridge	Salinity
Mango	Bappakai	Salinity
Lime	Rangpur lime, Cleopatra mandarin	Salinity
Tomato	Arka Vikas	Drought/rainfed
Onion	Arka Kalyan	Drought/rainfed
Chilli	Arka Lohit	Drought/rainfed
Dolichos	Arka Jay, Arka Vijay, Arka Sambram, Arka Amogh, Arka Soumya	Photo insensitive
Cowpea	Arka garima, Arka Suman, Arka Samrudhi	Photo insensitive

5. RESEARCH PRIORITIES

It is unlikely that a single method to overcome the effects of environmental stresses on horticultural crops will be found. A systems approach, where all available options are considered in an integrated manner, will be the most effective and ultimately the most sustainable, particularly for developing countries in the tropics under a variable climate. This holistic strategy will need global integration of efforts; the resulting synergies will produce impact more quickly than the individual institutions working in isolation could accomplish. For this to succeed, adequate and long-term funding is necessary, scientific results have to be delivered, best approaches utilized and effective methods sustained to deliver global public goods for impact.

Research should be expanded to further address the potential challenges posed by climate change with an objective of reducing malnutrition and alleviating poverty in developing countries through improved production and consumption of safe horticultural products by adaptation of current systems to the potential impact of climate change. Horticultural crops germplasm with tolerance to drought, high temperatures and other environmental stresses and ability to maintain yield in marginal soils must be identified to serve as sources of these traits for both public and private crop breeding programs. This germplasm will include both cultivated and wild accessions possessing genetic variation unavailable in current, widely-grown cultivars. Special populations are being developed to introgress and identify genes conferring tolerance to stresses and at the same time generate tools for gene isolation, characterization and genetic engineering. Furthermore, agronomic practices that conserve water and protect horticultural crops from sub-optimal environmental conditions must be continuously enhanced and made easily accessible to farmers in the developing world. Some technologies have already been developed on integrated crop management and many of these approaches directly provide solutions to mitigate the effects of climate change. For a technology to be successfully adopted, an effective extension strategy must be in place that includes technical, socio-economic and political considerations. Finally, capacity building and education are key components of a sustainable adaptation strategy to climate change.

Enhancing adaptation of tropical production systems to changing climatic conditions is a huge undertaking. It requires the combined efforts of many national and international institutions and an effective and efficient strategy to be able to deliver technologies that can mitigate the effects of climate change on the diverse crops and production systems. The scientific information and technologies developed through these initiatives must be readily accessible, consolidated and utilized in a strategic way. This can only be achieved through collaboration, complementarity and coordinated objectives to address the consequences of climate change on the world's crop production.

6. CONCLUSIONS

In view of climate change, scientists will have to play a significant role in the climate change scenario and proper strategies have to be envisaged for saving horticulture. The most effective way is to adopt sustainable horticulture; using renewable energy, forest and water conservation, reforestation *etc.* to sustain the productivity modification of present horticultural practices and greater use of greenhouse technology are some of the solutions to minimize the effect of climate change. Development of new cultivars of horticultural crops tolerant to high temperature, resistant to pests and diseases, short duration and producing good yield under stress conditions and judicious management of land use resources will be the main strategies to meet these challenges.

REFERENCES

AVRDC. (1981). Annual Report. Asian Vegetable Research and Development Centre, Shanhua, Taiwan, 84 pp.

AVRDC. (1990). *Vegetable Production Training Manual.* Asian Vegetable Research and Training Centre, Shanhua, Tainan, 447 pp.

AVRDC. (2005). Annual Report. AVRDC – The World Vegetable Centre, Shanhua, Taiwan.

Bose, T.K. and Mitra, S.K. (1996). Fruits: Tropical and Subtropical. Naya Prakash, Kolkata, India.

Chaves, M.M. and Oliveira M.M. (2004). Mechanisms underlying plant resilience to water deficits: prospects for water-saving agriculture. J. Exp. Bot. 55:2365-2384.

Cheeseman, J.M. (1988). Mechanisms of salinity tolerance in plants. Plant Physiol. 87:57-550.

Cuartero, J., Bolarin, M.C., Asins, M.J. and Moreno, V. (2006). Increasing salt tolerance in tomato. J. Exp. Bot. 57:1045–1058.

Drew, M.C. (1979). Plant responses to anaerobic conditions in soil and solution culture. Curr. Adv. Plant. Sci. 36:1-14.

Edelstein, M. (2004). Grafting vegetable-crop plants: Pros and Cons. Acta Horticulturae 65:235-238.

Erickson, A.N. and Markhart, A.H. (2002). Flower developmental stage and organ sensitivity of bell pepper (*Capsicum annuum* L) to elevated temperature. Plant Cell Environ. 25:123-130.

FAO. (2004). Impact of climate change on agriculture in Asia and the Pacific. Twenty-seventh FAO Regional Conference for Asia and the Pacific, Beijing, China, 17-21 May 2004.

Flowers, T.J. (2004). Improving crop salt tolerance. J. Exp. Bot. 55:307-319.

Foolad, M.R. (2004). Recent advances in genetics of salt tolerance in tomato. Plant Cell Tissue Organ Culture 76:101-119.

Foolad, M.R. and Jones R.A. (1991). Genetic analysis of salt tolerance during germination in *Lycopersicon*. Theor. Appl. Genet. 81:321–326.

Hazra, P. Samsul, H.A., Sikder, D. and Peter, K.V. (2007). Breeding tomato (*Lycopersicon Esculentum* Mill) resistant to high temperature stress. International Journal of Plant Breeding 1(1):31-40.

Hazra, P. and Som, M.G. (1999). Technology for Vegetable Production and Improvement. Naya Prakash, Kolkata, India.

Houghton, J., Ding, Y., Griggs, D., Noguer, M. and Van der Linden, P. (2001). Climate Change 2001: The Scientific Basis. Published for the Intergovernmental Panel on Climate Change, Cambridge University Press, Cambridge, UK and New York. 881pp.

IPCC. (2001). Climate change 2001: Impacts, adaptation and vulnerability. Intergovermental Panel on Climate Change, New York, USA.

Jones, R.A. (1986). The development of salt-tolerant tomatoes: breeding strategies. Symposium on Tomato Production on Arid Land. Acta Horticulturae 190:101-114. DOI: 10.17660/ActaHortic.1986.190.8.

Kawase, M. (1981). Anatomical and morphological adaptation of plants to waterlogging. HortScience 16:30-34.

Kumar, R. and Kumar, K.K. (2007). Managing physiological disorders in litchi. Indian Horticulture 52 (1):22-24.

Kuo, D.G., Tsay J.S., Chen, B.W. and Lin P.Y. (1982). Screening for flooding tolerance in the genus *Lycopersicon*. HortScience 17(1):76-78.

Matsumara, S. (1989). Studies on salt tolerance of vegetables-3. Salt tolerance of root stocks Agri. Bill Okayoma Univ. 73: 17-25.

Midmore, D.J., Roan, Y.C. and Wu, M.H. (1992). Management of moisture and heat stress for tomato and hot pepper production in the tropics. In: C.G. Kuo (ed.) Adaptation of Food Crops to Temperature and Water Stress. AVRDC, Shanhua, Taiwan, pp. 453-460.

Midmore, D.J., Roan, Y.C. and Wu, D.L. (1997). Management practices to improve lowland subtropical summer tomato production: yields, economic returns and risk. Exptl. Agric. 33:125-137.

O'Connell, M.A., Medina, A.L., Sanchez Pena, P. and Trevino, M.B. (2007). Molecular genetics of drought resistance response in tomato and related species. In: M.K. Razdan and A.K. Mattoo, (eds.) Genetic Improvement of Solanaceous Crops, Vol. 2: Tomato. Science Publishers, Enfield, USA, pp. 261-283.

Okimura, M., Matsou, S., Arai, K. and Okitso, S. (1986). Influence of soil temperature on the growth of fruit vegetable grafted on different stocks. Bull. Veg. Ornam. Crops Res. Stn. Japan C9:3-58.

Opena, R.T., Green, S.K., Talekar, N.S. and Chen, J.T. (1989). Genetic improvement of tomato adaptability to the tropics: Progress and future prospects. In: S.K. Green (ed.) Tomato and Pepper Production in the Tropics. AVRDC, Shanhua, Taiwan, pp. 70-85.

Opena, R.T. and Lo, S.H. (1981). Breeding for heat tolerance in heading Chinese cabbage. In: N.S. Talekar and T.D. Griggs (eds). Proceedings of the 1st International Symposium on Chinese Cabbage. AVRDC, Shanhua, Taiwan.

Ort, D.R. (2001). When there is too much light. Plant Physiol. 125:29-32.

Pandita, M.L. and Singh, N. (1992). Vegetable production under water stress conditions in rainfed areas. *In*: C.G. Kuo (ed.) Adaptation of Food Crops to Temperature and Water Stress. AVRDC, Shanhua, Taiwan, pp. 467-472.

Pereira, J.S. and Chaves, M.M. (1995). Plant responses to drought under climate change in Mediterranean-type ecosystems. In: J.M. Moreno and W.C. Oechel (eds.) Global Change and Mediterranean-type Ecosystems. Springer-Verlag, Berlin, pp. 140-160.

Rai, N. and Yadav, D.S. (2005). Advances in Vegetable Production. Researchco Book Centre, New Delhi, India.

Reymond, P., Weber, H., Damond, M. and Farmer, E.E. (2000). Differential gene expression in response to mechanical wounding and insect feeding in *Arabidopsis*. Plant Cell 12:707-720.

Rick, C.M. (1973). Potential genetic resources in tomato species: clues from observation in native habitats. In: A.M. Srb (ed.) Genes, Enzymes and Populations, Plenum Press, New York, pp. 255-269.

Romero, L., Belakbir, A., Ragala, L. and Ruiz, M.J. (1997). Response of plant yield and leaf pigments to saline conditions: effectiveness of different rootstocks in melon plant (*Cucumis melo* L) Soil Sci. Plant Nutr. 3:855-862.

Seki, M., Narusaka, M., Abe, H., Kasuga, M., Yamaguchi-Shinozaki, K., Carninci, P., Hayashizaki, Y. and Shinozaki, K. (2001). Monitoring the expression pattern of 1300 Arabidopsis genes under drought and cold stresses by using a full-length cDNA microarray. Plant Cell 13:61-72.

Stevens, M.A. and Rudich, J. (1978). Genetic potential for overcoming physiological limitations on adaptability, yield and quality in tomato. HortScience 13:673-678.

Wang, W.X., Vinocur, B. and Altman, A. (2003). Plant responses to drought, salinity and extreme temperatures: towards genetic engineering for stress tolerance. Planta 218:1-14.

Yildirim, E. and Guvenc, I. (2006). Salt tolerance of pepper cultivars during germination and seedling growth. Turk. J. Agric. Forestry 30:347-353.

ооо

12

Challenges and Opportunities in Potato Production Towards Sustainable Horticulture Development

K.S. Krishna Prasad

1. INTRODUCTION

Potato originated in the Andes Mountains of South America and was introduced to Spain in the 16th century by the Spanish conquerors and later spread to many European countries, from where it was distributed throughout the world. Thus, Europe is considered as the secondary centre of origin (Pandey and Sarkar, 2005). Potato is one of the world's most important food crops among all high value agricultural crops, which is grown in about 160 countries and its importance is highlighted by the United Nations by naming the year 2008 as the ***International Year of Potato*** (FAO, 2007). The crop plays a vital role in meeting the food and nutritional requirements of the global population, particularly in the developing countries.

Organized Research and development of the crop in our country started in 1949 when the Central Potato Research Institute (CPRI) was established at Patna in Bihar. Initially there were two major problems; lack of potato varieties suitable for sub-tropical Indian conditions and production of healthy seed potatoes. As Patna conditions were not suitable to tackle both the objectives, the Institute was shifted during1956 to Shimla in Himachal Pradesh. Since then, the CPRI is conducting the research and developmental activities of potato in our country and the Indian potato program passed through different stages and now has become one of the most successful ventures of agriculture in India (Singh, 2014).

2. AREA, PRODUCTION AND CONSUMPTION

2.1 Area and Production

Potato is an annual vegetative propagated crop, generally grown in mild climates, where the night temperatures are ideally between 15 to 20 ^{0} C, followed by bright sunny days. The Indian sub-tropical plains provide such optimum conditions for potato cultivation during October to February. Thus, the crop has become an important agricultural venture in all the states under varying ecological situations throughout the country as detailed below:

1. Short-day conditions of winter months in the Northwestern and Northeastern plains during October to March.
2. Long-day conditions of summer months in the mid-hills of Himalayas during April to September.

3. Equinox conditions of Deccan Plateau regions during rainy season/*Kharif* crop in the months of May to September.
4. Round-the-year crop in equable climate of Tamil Nadu hills as;
   (a) Rainfed summer crop during April to August;
   (b) Rainfed autumn crop during August to December; and
   (c) Irrigated winter crop during January to May.

Thus the crop serves as a major agricultural produce throughout the year, next to rice and wheat in our country.

The present day global potato production is about 382 million tonnes (MT), of which China (96.1 million MT) and India (46.4million MT) are the major producers (Table 1). The other important potato producing countries are Russian Federation, Ukraine and USA. Basically the crop existed in the colder regions of the world and is more suited to the temperate climatic conditions. However, the diversity of this crop in its adaptability to the sub-tropical and tropical conditions has been well exploited by agricultural research personnel throughout the globe and more so in the developing countries.

Maximum potato crop production in India is taken up during winter months of October to March in the North-Indian plains (Table 2). Uttar Pradesh with 14.5 million MT is the major producer, followed by West Bengal 11.6 million MT and Bihar 6.6 million MT. Gujarat and Madhya Pradesh produce about 2.5 million MT each. Other important Potato producing States are Punjab, Karnataka, Maharashtra and Hill tracks of Himachal Pradesh and Tamil Nadu. Riverbed cultivation is also prevalent in Gujarat, along with mixed cropping in parts of Bihar and Orissa. Two potato crops in a year are taken in Maharashtra, Karnataka and parts of Madhya Pradesh while three distinct potato seasons are prevalent in South Indian Nilgiri and Palani hills.

Table 1. *Area & production in top ten potato producing countries in world*

Country	Area in ('000 ha)	Production (Million MT)	Average Yield (MT/ha)	Total seed (MT)	seed potato % to total production
China	5647.0	96.136	17.024	3100.0	3.22
India	2024.0	46.395	22.760	3238.4	6.97
Russia	2101.5	31.501	14.990	6579.2	2.08
Ukraine	1342.8	23.693	17.650	5100.0	2.15
USA	425.4	20.560	47.150	1250.0	6.07
Germany	244.8	11.607	47.410	555.5	4.78
Bangladesh	495.8	9.435	19.030	443.9	4.69
Poland	276.9	7.689	27.770	950.0	12.35
France	168.0	8.054	47.940	350.0	4.34
Belarus	307.9	6.279	20.390	1200.0	19.11

Data based on FAO statistics for 2014

A perusal of production and unit area yields (Table 1) indicates that lot of variation occurs from country to country and region to region depending on the prevalent climatic variability. For example, in Europe, the unit area potato production is 47 to 48 MT /ha, where the potato is taken up as a summer crop for duration of 140 to 180 days under long day conditions of 14 hours sunlight. The crop is exposed to moderate temperatures of 15°C minimum to 25°C maximum. In contrast to this crop growth period, the potato in India is taken up from mere 65 to 110 days and that too in short day conditions of winter months as a sandwich crop between two main seasons or main crops. Further, suiting to local preferences different varieties are grown by adopting cultural practices suitable to their conditions. Thus, the average potato yield in India is about 22.9 MT/ha and it varies from 9.85 MT /ha in Assam to 29.9 MT /ha (West Bengal) and 30.76 MT/ha Gujarat (Table 2).

Table 2. *Area & production in top ten potato producing states in India*

State	Area ('000 ha)	Production ('000 MT)	Yield (MT/ ha)	Main consumption as
Uttar Pradesh	603.76	14430.28	23.90	Vegetable
West Bengal	386.61	11591.30	29.90	Vegetable & processing
Bihar	322.50	6640.60	20.59	Vegetable
Gujarat	81.27	2499.73	30.76	Vegetable & Processing
Madhya Pradesh	108.87	2299.00	21.12	Vegetable & processing
Punjab	85.25	2132.31	25.01	Seed
Assam	99.77	975.27	9.85	Vegetable
Karnataka	44.40	698.30	15.72	Vegetable & processing
Haryana	29.47	676.01	22.94	Vegetable & seed
Jharkhand	47.21	659.61	13.96	Vegetable & processing

Data based on Directorate of Economics & Statistics, Govt. of India (2013).

2.2 Consumption and Nutritional Value

In developed countries potato serves as fresh staple food, processed products as wafers and chips, industrial starch and also as animal feed. However, in developing countries, it is mainly used as a vegetable (Rai, 2003). The Potatoes are highly nutritious containing essential vitamins, minerals and proteins in addition to starch, hence, easily digestible. The tubers contain 80% moisture and 20% dry matter constituting 14% starch, 2% sugar, 2% protein, 1% minerals, 0.6% fiber, 0.1% fat and vitamins B and C in adequate quantities, making them wholesome food. Similar to other developing countries, India also utilizes most of the potato as a vegetable that is available throughout the year. The potato in our country is mainly grown for table purposes and seed purposes, which are grown in specific regions with specific cultivation practices. Also nowadays there is an upward trend in identifying suitable localities for raising potatoes, that are used for processing purposes in large scale food industries that are specific about the taste and texture of the selected potatoes, which in turn is attributed to the soil and other associated parameters of the particular type locality.

3. PROCESSING INDUSTRY SCENARIO

Firstly, potato crop produces more edible energy and protein per unit area and time than many other food crops. Secondly, for the small and marginal farmers, potato fits well into multiple cropping systems prevalent in tropical and subtropical agro-climatic conditions. Thirdly, the high profitability of potato as a cash crop has made it a viable commercial enterprise. And lastly, rapid technological advances in varietal improvement, agro-techniques, plant protection and storage and processing, have led to expansion of potato production even in non-traditional areas. In comparison to United States and other European countries which utilize more than 50% of their potato for processing, India hardly used about 1% of its potato for processing and less than 1% for export a decade ago. However, at present, the demand for processed products in the country has increased due to urbanization; increased *per capita* income and buying capacity of micro-families; about 7% of the potato is currently being utilized by the food processing units. Potato can be processed into value-added fried and non-fried food products having longer shelf-life. The market for such processed products is currently increasing in India, particularly for potato wafers, chips, frozen French fries, flakes *etc*. To meet the increasing demand, potato processing is emerging as a fast growing industry with more participation of entrepreneurs. The existing potato traders are increasing the capacity of their processing units. In this scenario, potato stands in a unique position to meet the demands due to its diversified uses.

4. SYNERGY BETWEEN AGRICULTURE AND SOFTWARE

The changing food consumption habits due to globalization, rising income levels, is increasing the demand for changed food habits. The micro-family concepts and urban living increases the processed food consumption. At the same time, it becomes more difficult to draw additional land into production, while soil degradation and pressure from alternative uses is increasing fast. In several localities the surface and ground water is becoming less available while the urban and industrial water use is increasing at an alarming rate. India being one of the growing economies requires larger amounts of energy, leading to higher energy prices and thus increases prices of many inputs for agricultural production. Further, global climate change that is increasing the temperatures, changing patterns of precipitation and more extreme weather events are likely to negatively affect the crop growth conditions (Gastelo *et al.*, 2014). However, visualizing these problems, specific agro climatic zones for potato have been demarcated indicating the potential productivity zones with specific information on expected dry matter content of the produce. Such vital information on its best visual representation in the form of various suitable agro-ecological zones have been effectively developed with reference to zones of potential productivity, stress prone areas *etc.*, using a well developed software namely Arc GIS 9.1 (Rawat *et al.*, 2008). This detailed information is useful in identifying localities suitable for raising potato for consumption, seed purposes and processing purposes in today's technology oriented growing economy and very useful in fueling further potato research and development in the country.

5. PRODUCTION TECHNOLOGIES

The export potential of Potato, its products and seed from India is being realized and slowly it has increased to about 2% during the past decade and later, expected to increase at a faster phase. As of now, India utilizes about 65% of potato for table purposes, 8-9% as seed and 8% for processing. A huge quantity of 15 to 16% is lost due to post harvest handling due to lack of infrastructure towards proper storage and transport. Now, there is an urgent need to step up these lacunae as the potato production is expected to increase due to improvement in adoption of latest technologies in crop cultivation.

5.1 Varieties

Cultivation of Potato in India started with import of European varieties that were best suited to the long day conditions of Europe but were not favorable for the expected improvement in potato production in our country. So the variety development program was initiated at CPRI, Shimla, in 1958, which became a very great success and now about 50 high yielding, disease and pest resistant varieties suitable for varying agroclimatic conditions of the country have been developed and are being grown throughout the country. Suiting to local market preferences these indigenously developed high yielding varieties have made tremendous impact on potato production, raising the status of our country as the 2nd most important potato producing country in the world. Currently the potato varieties under cultivation in India are discussed briefly:

Short duration potatoes completing their crop cycle in 60 to 70 days are, Kufri Ashoka, Kufri Chandramukhi, Kufri Garima, Kufri Gaurav, Kufri Jawahar, Kufri Khyathi and Kufri Surya. Most of these varieties have ovoid tubers with white cream flesh and are more preferable for urban markets and thus suitable for north Indian plains and plateau regions of the country.

Medium duration Potato varieties complete their crop cycle in about 80 to 90 days and most of them have late blight disease resistance with a built in viral disease resistance. Most of these varieties are preferred and cultivated throughout the country. They are, Kufri Anand, Kufri Arun, Kufri Bahar, Kufri Giridhari, Kufri Himalini, Kufri Jyothi, Kufri Kanchan, Kufri Lalima, Kufri Sadabahar and Kufri Sutlez.

Long duration potato varieties generally complete their crop cycle in about 110 to 120 days and are specifically suited to a particular locality having Late blight disease. Kufri Badshah is preferred in north Indian plains and plateau regions; Kufri Sindhuri with red flesh is suitable for East Indian Indo-Gangetic plains and Kufri Swarna having combined resistance to late blight disease and potato cyst nematodes is suitable for South Indian Hills. This variety also is drought tolerant.

CPRI has developed four varieties for making potato wafers and finger chips during the last decade and they are Kufri Chipsona 1, Kufri Chipsona 2, Kufri Chipsona 3 and Kufri Frysona. In addition, Kufri Chandramukhi, Kufri Jyothi, Kufri Laukar and Kufri Surya are also preferred for potato processing and are extensively grown in plateau regions.

The Indian potato varieties with short duration crop cycle, early bulking, uniform maturity, possessing heat tolerance, late blight resistance and slow degeneration rate are most suitable for cultivation in other countries, such as, Afghanistan, Bangladesh, Bhutan, Bolivia, Madagascar, Mexico, Nepal, Philippines, Sri Lanka and Vietnam. Besides, tolerance to frost, these varieties with excellent market attributes are preferred elsewhere, make them suitable not only in India but also in other countries, increasing the overall global potato export potential.

5.2 Cultivation Practices

Concomitant to the development of varieties, the cultivation practices were also worked out for different agro-ecological zones in India under the All India Coordinated Potato

Improvement Programme with the involvement of local Agricultural Universities and Agricultural development departments throughout the country. Thus specific package of practices were developed for each region depending on the produce utility. The seed plot technique used for seed production under low aphid periods was further refined by incorporating biotechnological approaches for virus elimination, micro-propagation, combined with True Potato Seed technology and use of serology in viral diagnosis made the seed potato production a commercial venture.

The impact of increased potato productivity from a unit area, by adopting specific package of practices, for different ecological zones was realized by the potato growers. These modified agro-techniques involved in such a way that tuber initiation and development coincided with the availability of optimum nutrients and moisture. Similarly, potato based cropping systems were developed for each ecological zone by working out most profitable crop rotations, intercropping systems and utilizing farm machinery wherever necessary to reduce dependence on labor. All these efforts culminated in increased potato production in the country that brought it to the forefront on the potato scene of the world.

CPRI annually produces about 2500 MT of Breeder's seeds that is multiplied twice by the state departments of Horticulture and Agriculture in collaboration with the State Agricultural Universities, to produce about 4,45,000 MT of Certified Potato seeds to be utilized by the potato growers employing the most innovative technological advances. This alone works out to about 500 million US dollars saving to the country's exchequer, as most of our neighboring Asian countries like Bangladesh, Pakistan, Sri Lanka and even China, import seed potatoes from Netherlands and other European countries.

5.3 Management of Biotic Stresses

The potato crop suffers from various diseases and pests. Among them, the most important diseases are; late blight disease (*Phytophthora infestans*), viruses X, S, Y and potato leaf roll virus (PLRV), aphids (*Myzus persicae*), tuber moth (*Pthorimea operculella*), bacterial wilt (*Rolstonia solanocearum),* potato cyst nematodes *(Globodera pallida* and *G.rostochiensis),*wart disease (*Synchytrium endobioticum*). These are universally present and are causing severe economic losses to the tune of about 37% of potato worldwide (Krishna Prasad, 2008). The CPRI has been able to identify sources of resistance to many of these plant protection problems and has developed strategies to mitigate the losses and successfully incorporated them in the potato varieties and cultivation practices.

5.4 Management of Abiotic Stresses

Efforts were made by plant breeders to incorporate tolerance or resistance to most of the abiotic stresses, such as, heat, drought and frost. In addition to the traditional breeding for biotic and abiotic stresses, several biotechnological interventions, such as, meri-stem tip culture*; in-vitro* conservation and rapid multiplication of genotypes, have been effectively and beneficially incorporated for improved seed potato production both qualitatively and quantitatively. All these efforts have borne fruits by the increased potato production in the country.

6. STORAGE AND MARKETING

6.1 Market Arrivals

Although potato is harvested almost throughout the year in our country, maximum potatoes arrive during February and March. The maximum arrivals are in North Indian *mundies* of Uttar Pradesh, Bihar and West Bengal to the tune of about 30 million tonnes in a very short period of about 60 days. Most important potato markets are in Agra, Aligarh, Burdwan, Delhi, Farrukhabad, Meerut, Howrah and Patna,-which puts lot of pressure on transport, storage, post harvest handling and marketing. This arrival during a short period of the time also puts pressure on the processing units causing improper utilization and reducing the quality of potato products. Unlike in the Europe, where the potato is grown in summer under long photoperiod (14 hours) with a crop duration of about 140 to 180 days followed by cold winters, in India the crop is grown in entirely opposite situations. Here, more than 90 % crop is grown in climates having short photoperiods of about 12 hours with a short crop duration of about 85 to 90 days followed by long hot summer which create both transport and storage problems.

6.2 Storage Facilities

The long hot summers followed by the potato arrivals during February and March makes it very difficult to store them for periods more than 3 months after harvest as they start rotting, shrinking, sprouting and increased vulnerability to microbial attacks. There are a number of traditional low-cost and non-refrigerated storage structures which are in use in many localities and these could be adopted easily by the farmers. Short term storage for nearly 4 months can be adopted by storing the produce in passively evaporative cool stores in plains, on- farm storing of potato in heaps, pits with indigenously available staking and covering materials. In the hills, on-farm diffused light storages are better to store potatoes which will help in getting better price for the produce and also preserves seed potatoes for planting in better condition. Cold storage of table potatoes and processing potatoes are not suitable for consumption as they accumulate reducing sugars after removal from stores. They can be best stored at 10 to 12°C after treating with a sprout suppressant. Developing such effective stores would go a long way in preserving consumption potatoes and processable potatoes for longer periods of 7 to 8 months without loss of keeping quality.

Seed potatoes are best stored in cold stores at 2 to 4°C with 90 to 95% relative humidity. The produce can be stored for nearly 7 to 8 months, after which they are in proper condition for planting. There are about 6300 cold stores in the country with a capacity to store about 30 million MT of fresh agricultural produce. The Uttar Pradesh and West Bengal have about 3200 cold stores which accommodate 75 to 80 % space for potato alone. Thus, in our country about 5000 cold stores are exclusively used for potato alone and annually about 24 million MT of potatoes are cold stored in India. Initially, private sector investment in large cold storage facilities greatly increased the availability of ware potatoes during the hot summer and rainy season and kept the seed potatoes in a reasonable physiological state for the following year's planting. Still, there is a need for development of location-specific improved varieties and transmission of production technologies, particularly through the production of adequate quantity of *in-situ* healthy planting material. At present, the Jalandhar potato market in Punjab is the best for seed potatoes that acts as the major seed potato supplier throughout the country.

6.3 Marketing Channel

Efficient marketing of farm produce is necessary from the growers and consumers point of view. This provides an incentive to the grower who gets better pricing for his produce while the consumer gets his produce at a reasonable price. However, the market channel that is in vogue for the potato market in India usually follows the pattern - *The grower/ producer -village trader -regulated market -commission agent -wholesaler- retailer-consumer.* In this system the retail prices are more in urban areas as compared to the rural markets. This system needs a drastic review to make potato marketing more remunerative to the grower/ producer and the consumer.

7. CHALLENGES AND OPPORTUNITIES

7.1 Challenges

Agricultural situation in the country is passing through a critical stage, where on one hand with increasing population that demands diverse products of superior quality while the natural resources are decreasing. The risks associated with the climate change are also increasing both biotic and abiotic stresses that need to be understood at a fast rate and devise technology to mitigate them. On the other hand there is a decline in the preference for farm work due to urbanization that has become a universal problem with decreased farm size. This impact is more so with a crop like potato that is very sensitive to the climate under which it is cultivated. As the crop production increased, the country could visualize its own deficiencies. It was not in a position to properly utilize the increased potato production for want of sufficient facilities for storage, transportation and post harvest handling of bulky produce.

The gradual increase of the potato production in the initial years in one way helped in stabilizing food and nutritional security of the country and also reflected the synergy between the Government and the farming sector. The increased irrigation infrastructure supported by development of high yielding, short duration, disease and pest resistant varieties helped farmers increase crop cultivation in the cool-post rainy season, particularly in Southern India . This has helped the buildup of private sector to invest in potato processing industries during the last decade. Though potato has been the source of value-addition for agricultural sector, still in India it had to transform from a vegetable supplement to food security option. A publicity campaign can be launched through mass media, such as, television, radio and news papers, highlighting its nutritional value. Such awareness campaigns are already being done for eggs and milk.

In addition, the potato processing industry in India is facing the impediments, such as, (1) Large number of varieties grown in different agro climates with varying cultural practices. (2) Maximum arrivals of raw material followed by hot summer months affecting storage. (3) Fewer facilities with specific conditions of storage for processing potatoes.

This situation contrasted with rapid and explosive growth in potato production during the last decade thus not only placing the country in limelight but also making progressive strides in the introduction of various agricultural policies and their effective implementation; thus in the process making the policy makers realize other limitations that were associated with the increased perishable, bulky produce like potato.

Transport infrastructure: Limitations of road-rail connectivity in potato growing localities, with high transport cost for bulky produce that requires increased labor for handling the produce.

Dissemination of improved technology: The information on refined and modified technologies for increased potato productivity has to be transferred to a large number of small and marginal farmers which in itself is a herculean task.

Storage facilities: Most of our cold stores are maintained at 2 to 4°C involving huge energy consumption, while table and processing potatoes can be conveniently stored at 10 to 12^0 C, reducing energy consumption. This change in storage situations for potato is taking more time for adoption.

Processing potatoes: Quality parameters of processing potatoes require higher dry matter content and should not be stored at 2 to 4°C, as the produce accumulates more of reducing sugars, decreasing the quality of processed product.

Marketing channels: The potato marketing hubs lacked proper grading facilities, quality packing coupled with insufficient storage facilities for specific consumptions.

Post harvest management: The post harvest management of the potatoes most unscientific in all the above mentioned items and increased the losses up to the tune of 15 to 16% of the total harvested produce.

7.2 Opportunities

The vision document of CPRI for 2050 envisages potato production in the country to about 125 million MT with an apportionment of about 62% for ware / table potatoes required for about 1620 million Indians at about 45.5 kg /person/year (Singh *et al*, 2014). By that time, it is expected that India will transform potato from a vegetable supplement to a food security option. This is mainly because of fast rate of urbanization, increased number of working couples, higher disposable income levels and realizing the importance of potato in fast food items creating excellent situation for increased potato consumption. Incidentally, the increased cropped area is not going to put pressure on land availability as most of the potato varieties are all short duration crops that fit into cropping systems as a sandwich crop.

It is estimated that about 25 million MT of potato would be utilized by the processing industries while 6 million MT would cater to the seed purposes as more of short duration, high yielding varieties would be grown with high-tech precision farming. The improvement in logistics, storage and packaging technology would bring down the post harvest losses to a great extent while increasing the potato export market to about 5 million MT of both ware and seed potatoes.

Potato being a labor- intensive crop utilizes about 130 to 145 man days including semi-skilled and skilled hands for cultivation of one hectare of land. Further, about 75% of labor employed is constituted by women folk increasing the gender equality. The input intensive potato crop helps overall development of other agricultural related industries, such as, fertilizers, pesticides, farm machineries and other service sectors that cater financial need to the agricultural segment.

Information on efficient and timely use of fertilizers, pesticides and irrigation coupled with very accurate diagnosis of diseases throughout the country will be one of the most important factors to sustain the potato productivity. This will be complimented by involving biotechnological and molecular techniques to develop short duration, thermo-insensitive, high yielding, pest and disease resistant potato varieties in addition to drought and heat tolerant potatoes. Refinement and increased utilization of certain mechanizations for potato harvesters, graders, bag fillers and packaging for easy transport will be an added advantage.

7.2.1 Improving *in situ* seed production

Although Seed potatoes are presently grown in Northwest Indo-Gangetic plains (Punjab and Haryana) meeting the seed needs of the country, the Deccan plateau and southern India faces shortage of quality seed potato. Of late, the surge in the potato cultivation in this part of the country has made the researchers to develop *in-situ* potato seed production by combining True Potato Seed technology, mass propagation of micro tubers to meet quality potato seeds (Krishna Prasad & Shekhawat, 2003).

The present day traditional potato seed is harvested during January and February in North India and about 100,000 MT is transported to South India at a phenomenal cost of about 20 crores of Indian rupees (200 million rupees) and has to be preserved during summer months from May to July for planting in early to late *Kharif* seasons. This can be avoided if South India itself produces quality seed potato. Further, the produce from these states are arriving in the market when no fresh potatoes are available in the country and thus will be a boon for processing industries. The ports like Chennai, Cochin, Mangalore, Tuticorin and Visakhapatnam (Vizag), both in the East–and West coasts strategically provide an excellent opportunity to export ware potatoes and seed potatoes to South Asian and Southeast Asian countries which depend on other European countries for their seed potato requirements, thus improving the export potential of potato.

Contract farming and Crop Insurance are new concepts in agricultural sector in our country. These items are in vogue in potato farming in our country especially preferred by the processing industries to ensure quality of processable potatoes. Once stabilized, these operating systems will certainly get in to the horticultural industry itself improving the production capacities and utilization of other important horticultural crops.

8. STRATEGIES TO INCREASE PRODUCTION

To achieve the expected targets there should be a big shift in our thinking about the crop. Potato should be considered as an important source of food rather than only a vegetable by making progressive changes in its consumption pattern. There should be multi pronged approach to achieve and sustain increased production. On the research front developing high yielding, short duration, stress resistant potatoes suitable to varying agro ecological situations of the country to meet the consumption purposes and processing purposes still needs to be achieved. This can only be met by exploitation of the available native potato genetic resources for variety development by involvement of modern technologies, such as, biotechnological interventions namely marker assisted breeding, tissue culture and implementing micro propagation techniques for seed production. There should be perfect interaction between seed growers, farmers and the research organizations to successfully bring in this sea of change.

As already indicated, value addition by way of processing potato as wafers, finger chips, flakes, starch for Industrial use, are all the more remunerative and hence, the processing Industry should be encouraged to be established nearer to the production areas. Short duration, high yielding, stress resistant and consumption specific potatoes should be catalogued and efforts should be diverted for high quality seed production and its distribution to specific areas.

Other important factor that needs immediate attention is developing proper infrastructures which require changes in the government policies. They are, road- rail link with improved facilities for grading, quick transport and storing the products for specific purposes of seed (2-4°C), consumption (10-12°C) and processing (10 -12°C with a sprout suppressant) is of utmost importance. The marketing yards should have sufficient graders, packing facilities with semi skilled labor. Export potential of potato should be considered as an important area and a policy decision should be taken and implemented at the earliest to streamline the bottlenecks in export of potatoes either for seed or for consumption.

Transfer of refined technology for efficient use of nutrients, pesticides and timely irrigations along with accurate management practices for diseases and pests to be taken up on war footing. Improved cultural practices and information on potato based cropping system involving intercropping to be given priority in all potato growing seasons. Other minor areas that can be worked upon are as follows: village level cottage industries should be encouraged for processing potatoes by utilizing the solar heat after crop harvest. Broader thinking towards crop insurance and contract farming should be done at the earliest. Specific mechanization implements and tools to be devised and made accessible for planting, cultural operations and harvesting in addition to developing *in situ* grading, packaging and transport facilities for the betterment of the farmers.

9. CONCLUSION

Potato is one of the high value horticultural crops that cater to the nutritional and food security of developing countries. Since attaining independence, India has made very significant achievements in potato production, unit area productivity and its utilization. The present day potato varieties are high yielding, region specific, disease and stress resistant and suitable for growing in varying agro-ecological regions of the country. Low cost precision farming techniques have been developed and adopted to improve the quality and quantity of seed potatoes. At present, the country is the second largest potato producer in the world and is able to extend its developmental activities to other South Asian and Southeast Asian countries. Visualizing the need of the country for the next four decades it is expected that a massive production of about 125 million MT is achieved, with main focus on potato processing and quality seed potatoes and reduction in post harvest losses of bulky produce. Technological interventions, such as, biotechnological approaches and tailor made policies that would enhance promotion of international trade, addressing majorly the supply and demand scenarios is the need of the hour. India is seen taking progressive strides in an attempt to put itself as a global leader in potato cultivation in the international map as well as boosting its very own economic scenario at the national level. An attempt has been made in this article to visualize the challenges and the opportunities that Potato could give to sustain the horticultural development in the country.

REFERENCES

Singh, Birpal. (2014). Potato Scenario-Past, Present & Future. Summer School- Current Trends in Quality Potato Production, Processing and Marketing. Central Potato Research Institute, Shimla 8-28 July. p. 4-10.

Singh, Birpal, Rana, R.K. and Govindakrishnan, P.M. (2014). CPRI Vision 2050. p.26.

FAO. (2007). World Potato Statistics.

Krishna Prasad, K.S. (2008). Management of Potato nematodes; An Overview. Journal of Horticultural Science. 3(2):89-106.

Krishna Prasad, K.S. and Shekhawat, G.S. (2003). New millennium strategies for potato improvement in south India in National Symposium on Horticultural Development in Southern India, TNAU, 7-8 June. p.7-13.

Rai, Mangala. (2003). Role of Potato in Food and Nutritional Security in developing Countries with special reference to India. National Symposium on Potato Research towards National Food and Nutritional security. Central Potato Research Institute, Shimla, 2-3 October, p.93.

Manuel Gastelo, Urich Kleinwechter and Meredith Bonierbale. (2014). Global Potato research for a Changing World. International Potato Centre (CIP), Lima, Peru, Working Paper 2014-1. p.43.

Pandey, S.K. and Sarkar, D. (2005). Potato in India: Emerging Trends and Challenges in the New Millenium. Symposium on Current Perspectives in Potato Research. Potato journal. 32 (3-4): 93-101.

Shashi Rawat, Govindakrishnan, P.M, Chauhan, R.K. and Pandey, S.K. (2008). A Compendium of Climate Based potato Zones in India. CPRI Technical Bulletin No.88. Central Potato Research Institute, Shimla, Himachal Pradesh, India. p.23.

13

Sustainable Intensification of Horticultural Crop Protection

P. Parvatha Reddy

SYNOPSIS

Agro-ecologists opine that interaction among healthy soils, healthy plants and crop diversity as the fundamental requirement for sustainable pest management. Besides regulation of pest populations through enhanced activity of biological control agents; biodiversity performs a variety of ecological services including the production of food and recycling of nutrients in the agro-ecosystems. One of the simplest forms of biodiversification, such as, legume based cropping systems can simultaneously optimize soil fertility and pest regulation. Diverse cropping systems enhance yields by interrupting insect, disease and weed life cycles. The microbiological activity of the soil (population of fungal and bacterial antagonists and arbuscular mycorrhizal fungi that better resist pest attack) is also enhanced by crop diversification. Sustainable pest management makes full use of natural and cultural practices and methods, including host resistance and biological control. In order to stabilize the population of pest species throughout the food web, the new designs should concentrate on managing the farm environment through ecosystem enhancements (i.e., landscape ecology), crop attributes or other means.

Keywords: Agro-ecology, biodiversity, ecosystem enhancements, pest management, sustainable intensification.

1. INTRODUCTION

The extreme form and simplification of nature's biodiversity is crop monocultures that results in the production of an artificial ecosystem requiring constant human intervention in the form of agrochemical inputs which temporarily boost yields along with a number of undesirable environmental and social costs (Altieri, 1987). The monoculture has led to a significant loss of genetic diversity of domesticated plants (FAO, 2001). The need to preserve the biological diversity of agro-ecosystems was emphasized by the UN Conference on the Environment and Development held at Rio de Janeiro in 1992. The ecological principles were continuously ignored, as the agricultural intensification progressed that resulted in recurrent pest outbreaks in many cropping systems, salinization, soil erosion, pollution of water systems, *etc.*

The United Nations Food and Agriculture Organization predict that food production needs to increase by 70 percent globally in order to feed expected population of over nine billion by 2050. Since there is no scope to increase the land available for cultivation, the increase in production should come from sustainable intensification of agriculture-getting more crops out of the same amount of farmland. Earlier, the conventional agricultural intensification (the so called "Green Revolution") relied on monoculture along with heavy inputs, such as, pesticides, inorganic fertilizers and fossil fuels to achieve a decline in global hunger. Despite of this production increase, nearly 800 million people still continue to suffer from hunger and malnutrition around the world.

Crop pests, such as, insects, mites, diseases (caused by fungi, bacteria, virus and nematodes) and weeds are one of the major limiting factors in crop production. The harmful pests are responsible for significant losses that are estimated to vary from 26 to 40% of the attainable (uninjured) yield in major food and cash crops (Oerke, 2006). The total global potential loss due to pests varied from about 40% for potatoes. Overall, weeds produced the highest potential loss (34%), with animal pests and pathogens being less important (losses of 18 and 16%). The crops losses due to pests viewed in terms of food security would represent the equivalent of food required to feed over one billion people (Birch *et al.,* 2011). About 3.5 billion kg of pesticide active ingredients with a value of US $45 billion global market per year (herbicides-42% of sales, insecticides-27%, fungicides-22% and other agrochemicals-9%) are being used every year to manage pests on these crops globally (FAO Stat, 2014). One of the possible approaches to increase food production is to reduce crop losses caused by weeds, pests and diseases.

In order to achieve Sustainable Development Goals (SDGs) (which last until 2030) set forth by the UN (to end hunger, achieve food security and improved nutrition and promote sustainable agriculture) the increased food production should be achieved through environment friendly and economically sustainable manner. The innovative methods of increasing yields while reducing chemical inputs (inorganic fertilizers and chemical pesticides), incorporating trees into the farm landscape and capturing carbon (ecological intensification of agriculture); are already being discovered and implemented by the researchers and farmers. The implementation of these potential sustainable intensification practices needs to be scaled up across the globe to increase food production and productivity.

2. SUSTAINABLE CROP PROTECTION

For over last five decades, chemical control has been the prevailing pest control strategy resulting in safety problems and ecological disruptions. Hence, there are renewed appeals for economically acceptable, effective and eco-friendly alternative control methods, which reduce pest damage while avoiding the cost and negative externalities associated with inorganic pesticides.

The sustainable intensification is defined as a knowledge-intensive process that requires optimal management of nature's ecological functions and biodiversity to improve agricultural system performance, efficiency and livelihood of farmers. It is development of more productive, sustainable production of agro-ecosystems that save on inputs, less harmful to the environment, developing varieties better suited to their environment and inventing new pest and disease management strategies. The sustainable intensification is based on understanding how nature functions so as to exploit its resources for higher production and avoiding practices based on excessive use of chemical fertilizers, fossil fuels, pesticides and water. The key tenet of sustainable intensification is using biodiversity, land, nutrients and water efficiently and in ways that are regenerative, minimizing negative impacts.

2.1 Objectives

The objectives of sustainable intensification include:

- Production of more food with less environmental impact
- Input use efficiency of key production inputs, such as, nutrients, water, pesticides, labor, land and energy
- Use of natural biodiversity (cropping systems and agro-forestry) to overcome biotic, abiotic and economic stresses

2.2 Benefits and Limitations

2.2.1 Benefits

The advantages of sustainable intensification include:

- The full use of available resources (nutrient, radiation, water *etc.*) is utilized under increased species diversity for a desirable compensatory growth because if one species succumbs to pest, weather or harvest; another species fills the void.
- The more stable pest–enemy complexes can be established by perennial vegetation that provides more habitat permanence prevents soil erosion, capture nutrients, offsets nutrient losses through leaching.
- Fallowing and growing green manure cover crops during fallowing restores soil fertility through biologically mediated mechanisms to reduce agricultural pest populations by interrupting the life cycles (with forest regrowth or legume-based rotations).
- Enhancing soil organic matter by addition of high biomass-producing plants forms the foundation for complex food webs which in turn may indirectly influence the diversity and abundance of natural enemies.
- Spatial heterogeneity by a mosaic of agro-ecosystems at the landscape level improves pest management.
- The reduction in the use of pesticides has been reported to increase production levels due to the presence of specific biological controllers for a determined pest (Pesticide Action Network North America, 2009).

2.2.2 Limitations

The limitations of sustainable intensification are as follows:

- Biological control agents have not been identified for some very strong pests. In such cases, the farmers resort to use of chemical pesticides.
- Sustainable intensification is knowledge intensive and requires monitoring for the combined components of the system to produce success. Hence, it is not easy to implement.
- Biological control agents are slow to act and take several years to successfully establish a population and to provide significant pest control.

- More than one type of biological control agent may have to be used to achieve uniform pest control across a variety of different situations and land types, since no single bioagent works in every situation; and a bioagent that works well in one soil type may not work in another soil type.

3. SUSTAINABLE CROP PROTECTION—A NEW PARADIGM FOR 21st CENTURY

An improved understanding of the functioning of agro-ecosystems for crop protection is being evolved since the 1970s (Bottrell, 1980). One of the emerging strategies in crop protection, "sustainable intensification," is being recognized in the present day context (Nicholls and Altieri, 2004). The new paradigm emphasizes on the incorporation of ecological principles into pest management while ensuring high productivity and profitable harvests without causing harm to the environment. The restructuring of the crop production system to incorporate preventative ecological measures that keep organisms from reaching pest status is the long-term pest management strategy. The use of biological processes has been given emphasis to regulate pest populations (as an alternative to direct control via synthetic pesticides) through redesign of cropping system via plant species spatial and temporal diversification, while also preserving and improving the soil health (fertility, biological activity, structure, *etc.*) in sustainable intensification of crop protection (Ratnadass *et al.,* 2012).

The sustainable crop protection strategy helps to maintain bio-ecological equilibrium between plant (crops and non-cultivated plants in areas surrounding the field and in the wider agro-ecosystem) and animal (pests, beneficials and pollinators) communities within agro-ecosystems, helps to maintain habitats to attract indigenous beneficial fauna and/or repel pest fauna with the goal of reducing the risks of pest outbreaks and thus contributes to conservation of natural enemies and biological control. Hence, the agro-ecological approaches should focus on harnessing inherent strengths within ecosystems and be directed more toward bringing pest populations into acceptable bounds rather than toward eliminating them (Figure1).

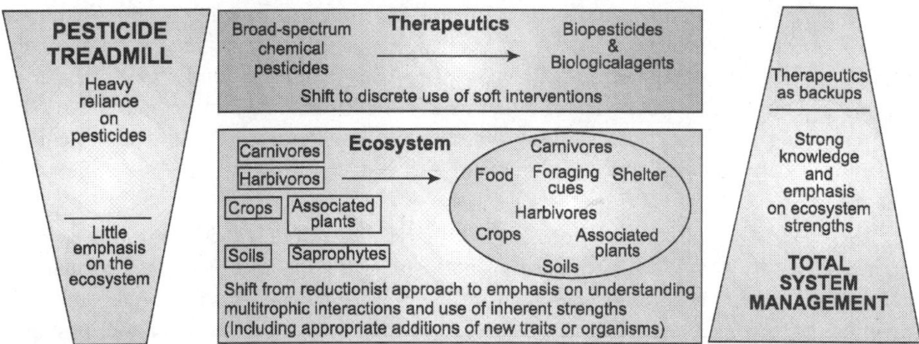

Fig. 1. *Sustainable intensification of pest management. Left-The upside-down pyramid reflects the unstable conditions under heavy reliance on pesticides. Right-The upright pyramid reflects sustainable qualities of agro-ecologically-based crop protection strategy.*

The preventive strategies rather than reactive strategies form the basis of sustainable pest management. The greatest abundance and diversity of above- and below-ground organisms can be enhanced by farmers to bring pests and natural enemies into balance on bio-diverse farms (Figure 2). The benefits of biodiversity can be obtained in farm design by gaining a better understanding of the interactive relationships between soils, microbes, pests and natural enemies. The soil biology, recycling of nutrients, moderating microclimates, detoxifying noxious chemicals and regulating hydrological processes can be enhanced by a highly functioning diversity of crucial organisms.

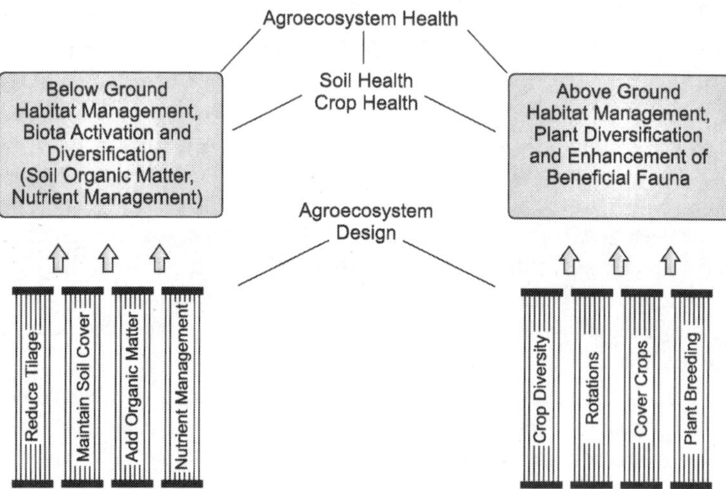

Fig. 2. *The pillars of sustainable pest management*

4. SUSTAINABLE CROP PROTECTION PRACTICES

4.1 Reduced Tillage

The reduced tillage is an innovative and sustainable agricultural production system providing greater economic and environmental advantages. Growing of crops without disturbing the soil through tillage from year to year is called no-tillage farming (zero tillage or direct drilling) or reduced-tillage farming. This practice is helpful to the farmer in many ways including reduction or elimination of soil erosion, recycling of nutrients in the soil, increased amount of water infiltration into the soil, increased organic matter retention in soil and increased amount and variety of life in the soil including pest suppressing organisms.

The well managed conservation agriculture systems through reduced tillage contain high biodiversity below and above the ground which provide a stable habitat for natural enemies of insect pests and pathogens, including antagonistic microorganisms. By adopting good conservation tillage systems especially with crop diversification, pest and disease pressures are significantly reduced over time resulting in saving on pesticide costs.

4.1.1 Weed management

Under reduced tillage, the need for vigilance in weed management is increased as tillage is reduced. Reduced tillage system commonly leads to enhancing grass weeds and volunteer cereals, while reducing broad-leaved weeds (Wrucke and Arnold, 1985). Under conservation tillage, all weed seeds are left on the soil surface under crop residue. The weed growth is suppressed by the release of chemicals (glucosinolates including isothiocyanates) during the decomposition crop residues, which is called allelopathy and can be important with residues from Brassica crops and cereal rye (Haramoto and Gallandt, 2004). Better weed suppression is generally achieved with higher levels of soil surface residues (Chauhan *et al.,* 2006) under reduced tillage.

Awada *et al.,* (2014) suggested the integrated weed management (IWM) systems that include crop rotation, seeding date (early spring), seeding rate (100% or 150% recommended), fertilizer application (fall or spring) and in-crop herbicide rate (50% or 100% of recommended), have the potential to lessen weed populations. The use of IWM systems under conservation tillage systems has been found economical and profitable (Smith *et al.,* 2006).

4.1.2 Insect pest management

The insect pest dynamics is considerably altered by reduced tillage. By following reduced tillage for a number of years, the insect population may decrease because of increased survival of natural enemies, such as, spiders, rove beetles, ground beetles, annelids, parasitic wasps and ants, which can contribute to insect pest management (Chan, 1987). Under reduced tillage systems, thrips and spider mites (in beans) damage is reduced, while slug and cutworm problems are increased.

Stinner and House (1990) reported that 43% of insects and their damage decreased with reduced tillage, compared with 28% where the number or damage increased, while the remaining 29% showed no difference. The pest management should not be an impediment for adopting conservation tillage systems.

4.1.3 Disease management

The effects of reducing tillage on disease pressures are variable, depending on the disease (pathogen virulent or avirulent), the environment (favorable or unfavorable) and the crop (susceptible/tolerant/resistant). In general, the incidence of foliar diseases may increase, while the soil-borne diseases may decrease in reduced tillage systems over time due to increased biological activity and growing numbers of beneficial microorganisms (Bockus and Shroyer, 1998).

For early planted crops like peas, conservation tillage may offer the opportunity to prolong the period before irrigation, which may in turn, reduce soil-borne diseases by keeping the soil drier early in the crop's development. Some virus diseases vectored by aphids are reduced under no-till farming because winged aphids are more often attracted to barren ground (conventional tillage system), compared to ground covered by residue (conservational tillage system).

4.2 Maintaining Soil Cover

Management of insect pests, diseases and weeds by high residue farming system is being adapted by an increasing number of farmers from different regions to suit their conditions and crops. Higher numbers of species that can cause damage to crops and also beneficial organisms that parasitize on eggs of certain pest species occur under high residue farming that is a more natural system (Gonzales and Dave, 1997). The benefits of covering the soil surface with live or dead mulching include reduced weed germination and emergence, controls pests and diseases and shading and physically hindering emerging weeds.

4.2.1 Weed management

The better weed suppression is commonly achieved by higher levels of crop residues covering the soil surface (can also hamper planting) and the effect is species-specific (Chauhan *et al.,* 2006). The chemicals that suppress weed growth are released during the decomposition of crop residues (allelopathy) from Brassica crops and cereal rye (Haramoto and Gallandt, 2004).

High levels of cover crop mulch inhibited weed germination and emergence. The annual weeds with small-seeds (annual bluegrass, black nightshade, chickweed, lambsquarters, purslane, ragweed, hairy galinsoga and pigweeds) are especially susceptible to suppression by mulches (Egley, 1996). The emergence of several weed species with small-seeds was affected to the extent of 20 to 95% and reduced growth of weed seedlings from 8 to 90% by addition of chopped residues of *Brassica hirta*, *B. juncea* and *B. napus* to the soil under greenhouse studies on potato.

The most common annual broadleaf and grassy weeds were suppressed for four to eight weeks in tomatoes by cereal rye cover crop residues on the soil surface, thus eliminating the need for a soil-applied herbicide at transplanting without depressing yield (Smeda and Weller, 1996). Soil surface cover with organic mulches alone cannot prevent 100% of weed emergence. Integration of organic mulches along with good crop rotation gives effective suppression of aggressive weeds, such as, nutsedge and morning glory.

4.2.2 Insect pest management

All organic mulches are useful for suppression of insect pests in comparison to bare soil. The habitat for spiders (which find mulch more habitable) are provided by hay and straw mulches that give 70% reduction in damage due to insect pests on vegetables. Under enough soil cover of organic mulches, even the presence of high numbers (100 larvae per m^2) of the plant-eating insects, like white grubs (*Cyclocephala flavipennis*), cannot cause enough damage to crops.

The increased number and diversity of natural enemies of aphids that predate on them was observed when soil was covered with organic mulches which may also interfere with the visualization mechanism of aphids, when identifying the possible host plants. The aphids may also find it difficult to distinguish the host plants due to reflection of the sunlight in a different way by soil cover as compared to bare soil (Fernandes, 1997).

The straw mulch is responsible for reduction of the early season Colorado potato beetle's (CPB) ability to locate potato plants and for the suppression of its activity in potatoes due to creation of a micro-environment that enhances the population of predators, such as, lacewings, lady beetles and ground beetles. The reduction of defoliation and enhancement of potato yields by one-third compared to plots with no mulch was also noticed in these trials. In apple orchard ecosystems, the use of compost as a mulch should be encouraged for sustainable management of arthropod pests (spotted tentiform leaf miner, *Phyllonorycter blancardella* and migrating woolly apple aphid, *Eriosoma lanigerum* nymphs) by encouraging predator population (Brown and Tworkoski, 2004).

4.2.3 Disease management

The presence of fresh vegetative crop residues on the soil surface provides the most essential medium for growth and survival of pathogens. Hence, special attention is required for solving the pathogen inoculums survival problem by way of biological destruction or decomposition of crop residues by microorganisms until the residues are fully mineralized through proper crop rotations. On the other hand, the presence of more crop residues can form a physical barrier for the completion of the life cycle of certain pathogens, like *Sclerotinia* spp. or prevent pathogens from being spread through soil movement by agricultural equipment, water or wind (Costamilan, 2000).

The incidence of foliar disease of tomatoes planted in a dead mulch of hairy vetch was reduced, gave higher yields and lived longer than tomatoes planted without mulch. Similarly, the Fusarium wilt in watermelon was reduced by 26% when a dead mulch of hairy vetch was used (Stone, 2012). The onset of early blight in tomatoes can be delayed by using cover crops, such as, cereal rye and flowering rapeseed as mulches which reduces soil splash onto leaves.

4.2.4 Nematode management

Bhattacharya and Rao (1984) reported that soil mulching with sugarcane and banana trash reduced the burrowing nematode, *Radopholus similis* and the lesion nematode, *Pratylenchus* sp. in banana roots. The slow wilt/pepper yellows (*Radopholus similis*) and the root-knot nematode (*M. incognita*) population on black pepper (Ichinohe, 1980) were managed by mulching with Guatemala grass.

Sreeja *et al.,* (2001) reported that mulching of kacholam with neem and chromolaena green leaves at 5 kg/m^2 15 days before planting rhizomes, reduced root galling (1 as compared to 5 in control), root-knot nematode population both in soil (80.4 - 81.8%) and roots (86.14 - 89.60%) and increased the rhizome yield (160.7 - 143.3%).

4.3 Addition of Organic Matter

The addition of organic amendments (FYM, green manures, seed meals, oil cakes, saw dust, coir pith, crab shells, dried plant material, good quality compost organic waste and peats) to the soil is responsible for the management of crop diseases, nematode pests (Akhtar and Alam, 1993) and weeds. The use of special composts containing particular antagonistic microorganisms is the recent trend in pest management.

4.3.1 Disease management

The addition of organic amendments to the soil was found to be very effective in the management of diseases caused by pathogens, such as, *Fusarium* spp., *Pythium* spp., *Rhizoctonia solani* and *Sclerotinia* spp). During the decomposition of organic amendments in soil, certain chemicals are released which are toxic to pathogens. The bacterial biocontrol agents, such as, *Bacillus* spp., *Enterobacter* spp., *Flavobacterium balustinum*, *Pseudomonas* spp., *Streptomyces* spp. as well as fungal species including *Gliocladium virens*, *Pencillium* spp. and several *Trichoderma* spp. colonize the compost during the curing phase after heating (Cook and Baker, 1983).

The population of actinomycetes in tomato rhizosphere was higher in organically managed soil than in conventionally managed soil resulting in suppression of corky root severity (Workneh and Van Bruggen, 1994). The organically managed soils contain higher population of antagonistic fungi (*Trichoderma* spp.) and bacteria (fluorescent pseudomonads) (Van Bruggen, 1995). The incidence of Verticillium wilt and common scab of potato was considerably reduced under field conditions by soil application of bone meal, soy meal and poultry manure at 37 t ha^{-1} incorporated to a 15 cm depth (Lazarovits *et al.,* 1999).

Lumsden *et al.,* (1983) reported that the incidence and severity of diseases, such as, *Aphanomyces eutiches* on pea, *Fusarium oxysporum* f. sp. *melonis* on melons, *Phytophthora capsici* on bell pepper, *Rhizoctonia solani* on bean and *Sclerotinia minor* on lettuce was considerably reduced by soil application of composted sewage sludge to soil (10% by volume). Similarly, the incidence of damping-off of pea cultivars (smooth skinned and wrinkled) caused mainly by *Rhizoctonia solani* and *Pythium ultimum* was significantly reduced in field plots by soil amendment of with composted sewage sludge (7-10 t/ha) (Lewis *et al.,* 1992).

4.3.2 Nematode management

A large number of studies have demonstrated the use of organic amendments (like sawdust, oilcakes, sugarcane bagasse, bone meal, sewage sludge, *etc.*) for the management of plant-parasitic nematodes (Oka, 2010). The lesion nematodes, *Pratylenchus* spp on tea bush are satisfactorily controlled by soil application of the Guatemala grass, *Tripsicum laxum* at 20 to 30 tons per ha, incorporation of large quantities of green manure, such as, loppings from dadaps, *Tephrosias* and marigolds (Loos, 1953).

Amendment of soil with sheep and chicken manure reduced the citrus nematode (*Tylenchulus semipenetrans*) population. The infestation level of *Globodera rostochiensis* in potato roots was reduced and development of nematodes in plant roots was slowed down by application of FYM and compost to nematode infested soil that imparted some type of biochemical resistance to the plants by nutrients from these manures.

The release of toxic compounds during the decomposition of organic materials, increase in nematode antagonists and improved crop nutrition and plant growth following soil amendment with organic matter may lead crops to tolerance from plant-parasitic nematodes.

4.4 Nutrient Management

The universal practice in realizing full yield potential in commercial crop production is through properly balanced plant nutrition by integration of organic manures and inorganic fertilizers. The quantity, quality, yield of crops and levels of the incidence of pests (insects, diseases, nematodes and weeds) have been associated with macro- and micro-nutrients. In general, the incidence of pests is increased by the excessive use of nitrogenous fertilizers (Phelan *et al.,* 1995), while application of P or K reduced diseases in 65% of cases and increased diseases or insect pests in 28% of cases (Perrenoud, 1990). Balanced fertilization can provide a proper balance of elements and can stimulate host resistance to pest attack (Luna, 1988).

4.4.1 Insect pest management

About 135 research studies demonstrated more plant damage and/or greater numbers of leaf-chewing insects or mites in nitrogen-fertilized crops, while low pest damage was reported in lesser than 50 studies.

The corn earworm (*Heliothis zea*) infestation levels were reduced by the application of nitrogen fertilizer due to increased husk extension and tightness of husks on sweet corn (Klostermeyer, 1950).

Soil application of 30 and 45 kg P_2O_5 ha^{-1} significantly ($P < 0.05$) reduced the damage by aphids (*Aphis craccivora*), flower thrips (*Megalurothrips sjostedti*) and legume pod borer (*Maruca vitrata*) and consequently higher grain yields were obtained in cowpea cultivars IT91K-180, IT95M-118 and TVU 1890 (Asiwe, 2009). The wireworm populations often tend to increase in soils with low phosphorous.

Sulfur act as an acaricide that can be used for the effective management of mites.

4.4.2 Disease management

The plant nutrition indirectly affects pathogens and diseases, since mineral nutrients are essential metabolic regulators for plant growth. The population of antagonistic microorganisms in the rhizosphere can be enhanced by the quality of root exudates secreted by healthy plants. The antagonistic microorganisms produce toxins that directly affect pathogen development and survival and compete with pathogens for nutrients (Palti, 1981) or produce antibiotics (Tjamos *et al.,* 1992).

The effects of N on disease development are inconsistent mainly due to the developmental stage of N application (Carballo *et al.,* 1994), the type of pathogen: obligate vs. facultative parasites (Marschner, 1995) or the form of N nutrition of the host (Celar, 2003). There is an increase in the severity of infection of obligate parasites, *e.g. Puccinia graminis* and *Erysiphe graminis*, when there is high N supply; while high N supply decreases the severity of the infection of facultative parasites, *e.g. Alternaria, Fusarium* and *Xanthomonas* spp. The incidence of *Fusarium oxysporum*, *Botrytis cinerea*, *Rhizoctonia solani* and *Pythium* spp. decreased at high NO_3, while the incidence of *Pyricularia*, *Thielaviopsis basicola*, *Sclerotium rolfsii* and *Gibberella zeae* decreased at high NH_4 (Celar, 2003).

The reduction in disease incidence and improvement in plant health are generally observed by phosphorus (P) fertilization. Phosphorus is part of many organic molecules of the cell [deoxyribonucleic acid (DNA), ribonucleic acid (RNA), adenosine triphosphate (ATP) and phospholipids] and is also involved in many metabolic processes in the plant as well as in the pathogen. Suppression of the lesions of powdery mildew on the diseased foliage of cucumber, greenhouse-grown roses, field grown mango, nectarine and grapevine was noticed by single application of phosphates (Reuveni and Reuveni, 1998). Likewise, induction of systemic protection against powdery mildew in cucumber caused by *Sphaerotheca fuliginea* was observed by a single phosphate spray of 0.1 M solution (Reuveni and Reuveni, 1998). Control of fungal diseases in seedlings by P fertilization has been achieved due to vigorous root development that permits plants to escape disease (Huber and Graham, 1999).

The incidence of several diseases, such as, *Cercospora* leaf spot in cassava, red rust in tea and seedling rot caused by *Rhizoctonia solani* have frequently been reduced by application of K (Sharma *et al.,* 2005). The development of thicker outer walls in epidermal cells is promoted by application of K, thus preventing disease attack. K also influences tissue hardening and stomatal opening pattern which in turn are closely related to infestation intensity (Marschner, 1995).

4.4.3 Nematode management

Application of high rates of K (2 and 6 kg of K_2SO_4 per tree) significantly reduced the population of *Xiphinema* and *Pratylenchus* in cherry orchards (Kirkpatrick *et al.,* 1959a). The population of these nematodes was negatively correlated with leaf potash (Kirkpatrick *et al.,* 1959b).

The number of galls by *Meloidogyne javanica* in tomato was significantly reduced by increased levels of potash (Gupta and Mukhopadhyaya, 1971). The soil amended with 20% fly ash increased tomato growth (35%) and reduced root galling due to *M. incognita* (root- knot index 1 as against 4 in control), nematode population in soil (80% reduction) and egg mass production (egg mass index 1 as against 4 in control). The reduction in nematode population might be due to toxic compounds (polycyclic aromatic hydrocarbons, dibenzofuran and dibenzo-p-dioxin mixtures) present in fly ash, while the increase in plant growth may be ascribed to increased nutrient availability (N, K, Ca, Mg, Na, B, SO_4).

Sivakumar and Meerazainuddin (1974) reported that the reniform nematode (*Rotylenchulus reniformis*) multiplication on okra was reduced to a great extent by the application of potash alone or potash in combination with phosphorus or nitrogen. The root-knot nematodes can be managed by the application of $CaCN_2$, NaCN and urea cynamide; probably by the release of HCN or NH_3 which are toxic to the nematodes.

4.5 Crop Diversity

The components of agricultural biodiversity include habitat management, intercropping and agro-forestry.

4.5.1 Habitat management

The habitat management is a form of conservation biological control (an ecologically based approach) that provides resources for natural enemies, such as, food and shelter in enhancing biological control in agro-ecosystems. Maintaining different cropped and non-cropped corridors within the agricultural landscape in the form of hedge rows, corridors for natural enemies, flowering weeds to attract beneficials, pollen and nectar resources and beetle banks provide alternative prey resources for pest predators and hosts for parasitoids, as well as shelter from adverse climatic conditions (Altieri and Letourneau, 1982). The conservation and performance of natural enemies can be achieved through high diversity of habitats and vegetation and reduced disturbance.

4.5.1.1 Field margins with biodiversity

The field margins can be converted into reservoirs of natural enemies by providing habitats which supply pollen, nectar and other resources. These natural enemies move from field margins into crops and provide biological control (more in border than in center of fields) of several crop pests. In California, the leaf hopper population in grape vineyards adjacent to French prunes is reduced by the egg parasite *Anagrus epos* (a parasitic wasp), since the prunes harbor an economically insignificant leaf hopper whose eggs provide *Anagrus* with its only winter food and shelter.

4.5.1.2 Creation of natural enemies corridors

The refuge strips of Alyssum flowering plants (increases the diversity and abundance of natural enemies) can be provided amid lettuce crop for multiplication and spread of natural enemies (syrphid flies) of aphids into field centers. The cabbage aphid can be managed by planting strips of buckwheat and tansy leaf in Swiss cabbage fields which increased populations of a small parasitic wasp that attacks the aphid.

Plants from the aster, carrot and buckwheat families are especially useful. Examples of flowering plant mixes might include species from the daisy or sunflower family (Compositae) and from the carrot family (Umbelliferae).

4.5.1.3 Flowering weeds attracting natural enemies

The weeds contributes to the biodiversity of agro-ecosystems, camouflage crops from colonizing pests, making the crops less apparent to their prospective attackers and attracts natural enemies by producing flowers which supply pollen and nectar to the beneficials. The farmers should be educated about the weed management over weed control, since a growing appreciation for the complex relationships among crops, weeds, pests and natural enemies exist.

4.5.1.4 Pollen and nectar resources

The non-prey food, such as, pollen and nectar, is required by most of the natural enemies as part of their diet. Significant increase in the activity, longevity and fecundity of the predators and parasitoids is dependent on the availability of pollen and nectar resources (Hogg *et al.,* 2011). The best nectar sources suitable for small parasitoids and also larger predators include plants having very small flowers. Adult hoverflies, which are aphid predators, use pollen necessary for eggs to mature and produce their young. In order to attract and nourish natural enemies, readymade seed mixes of flowering plants are available.

There is a need to select plants which provide non-prey food to the beneficials without providing resources to pest insects. For example, lacy phacelia (*Phacelia tanacetifolia*) and Nasturtium (*Tropaeolum majus*) provided resources for natural enemies without enhancing moth pests (Baggen *et al.,*1999). Similarly, three types of flowering resource plants sown under the grapevines gave higher egg parasitism of light brown apple moth (*Epiphyas postvittana*) (Begum *et al.,* 2006). The integration of non-prey food along with attraction of natural enemies (Figure 3) in an attract-and-reward approach has potential for enhancing biological control synergy (Simpson *et al.,* 2011).

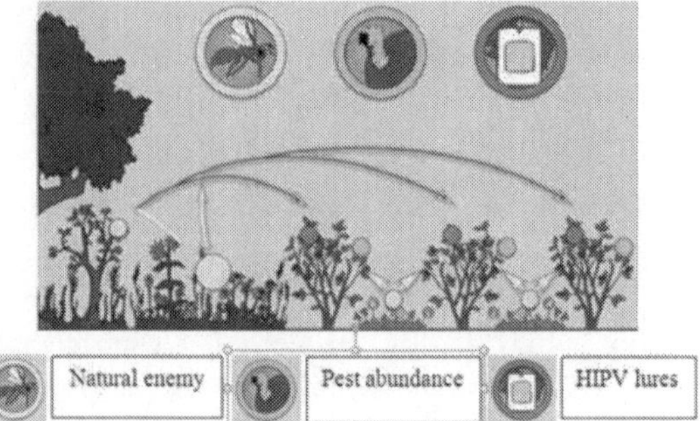

Fig. 3. *Integration habitat manipulation and herbivore-induced plant volatiles (HIPV) to enhance biological control in an agro-ecosystem*

4.5.1.5 Boosting beneficials from beetle banks

Several grasses can provide habitats for high densities of overwintering predators (predaceous ground beetles, *Demetrias atricapillus* and *Tachyporus hypnorum*) with temperature moderation. The "beetle banks" can be established at the centers of fields by sowing orchard grass on earth ridges (Figure 4). The establishment of beetle banks is cost effective, since the natural enemies they harbored were quite effective in preventing aphid outbreaks that pesticide savings outweighed the labor and seed costs required to establish them.

4.5.2 Intercropping

Intercropping systems increases diversity in an agricultural ecosystem, ecological balance, better use of soil nutrients (Igzoburkie, 1971), improves the quantity and quality of products, soil structure, weed control, microclimate manipulation (*e.g.* growing a tall crop to provide a wind barrier), provides habitat for natural enemies and even acts as a trap crop and reduces damage by pests, diseases and weeds. A review of 209 published field studies in which 287 herbivore species were studied showed that 52% of the pest species (149 species) were less abundant in the intercrop. The population of natural enemies of the pests was also higher in 53% of the studies in the intercrop (Andow, 1991).

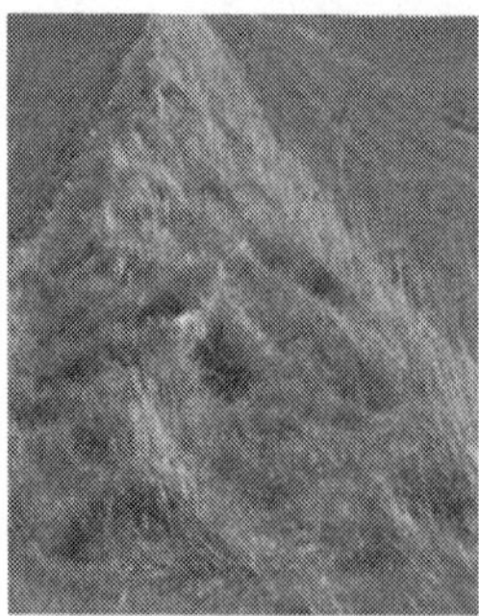

Fig. 4. *The beetle bank*

4.5.2.1 Insect pest management

Intercropping can be employed to manage insect and mite pests by disrupting their ability to find their hosts (Andow, 1991) or the intercrop does not host key pests on the main crop or insects are actually repelled by the odor of non-host plants. The odors released by plants in the cabbage family attract Crucifer flea beetles (*Phyllotreta cruciferae*), but the random movement of flea beetles within that area is obstructed by intercropping, resulting in reduction of flea beetle numbers. The Colorado potato beetles' movement is disrupted by planting strips of rye–hairy vetch cover crop between widely spaced rows of potatoes, since the newly emerged beetles find their hosts primarily by walking. Later, mowing of the cover strips and mulching the soil interferes with beetle movement and provides habitat for beetle predators. Undersowing of clover to deter cabbage root fly (Finch and Edmonds, 1994) and *Medicago litoralis* to deter carrot root fly (Rämert and Ekbom, 1996) effectively prevented the pests.

Trap intercropping is one of the methods of crop diversification that is emerging as a component in integrated pest management (IPM). Trap crops are grown within crops, within farms or within landscapes in a smaller area before or with the main crop in many cases. The efficiency of pest management can be enhanced by concentrating the pests in one location and destroying them by pesticides or cultural/ biological methods. Zalom *et al.,* (2001) recommended this system for the management of Lygus bugs in organic strawberry production. Management of cabbage pests ((*Plutella xylostella, Crocidolomia binotalis, Hellula undalis, Brevicorne brassicae*) and tomato fruit borer (*Helicoverpa armigera*) using Indian mustard and African marigold (*Tagetes erecta*) as trap crops have been reported (Figure 5) (Srinivasan and Krishna Moorthy 1991, Srinivasan *et al.,* 1994).

The dead-end trap crops are very much attractive to insect pests but their offspring cannot survive on trap crops and prevent their movement from the trap crop to the main crop later in the season (Badenes-Perez *et al.,* 2005). Yellow rocket, *Barbarea vulgaris* var. *arcuata*, acts as a dead-end trap crop for the diamondback moth, *Plutella xylostella* (Shelton and Nault, 2004), while sun hemp, *Crotalaria juncea*, as a trap crop for the bean pod borer, *Maruca testulalis* (Jackai and Singh, 1983). Hoy (1999) reported that planting genetically engineered trap crop of Bt (*Bacillus thuringiensis*) potatoes early in the season acts as dead-end trap crops to attract immigrating Colorado potato beetle (*Leptinotarsa decemlineata*) and prevented colonization of the interior of the field that was planted to non-Bt potatoes.

Fig. 5. *IPM in cabbage and tomato using mustard and marigold trap crops*

Perimeter trap cropping (PTC) attract pests from the main crop, concentrating and/or killing the pest in the border area, while reducing pest numbers on the unsprayed cash crop in the center by preserving natural enemies. The diamondback moth (DBM) infestations in Florida were kept below action thresholds in commercial cabbage fields by surrounding them with two rows of collards (Mitchell *et al.,* 2000). The DBM population on the collards was reduced by a naturally occurring parasitic wasp and prevented its spread into the cabbage. The pesticide costs were saved to the extent of $118 to $158 per ha in view of 56% fewer insecticide sprays to manage DBM in cabbage fields protected by the PTC system than conventional fields.

The "push - pull trap cropping strategy" was developed for the management of the Colorado potato beetle (*Leptinotarsa decemlineata*) using neem-based antifeedant (push) to repel and early perimeter planting of potato as trap crops (pull) to attract the pest (Martel *et al.,* 2005). The early-planted potato trap crop was sprayed with a slow-release attractant aggregation pheromone at weekly intervals that significantly attracted more adult beetles, eggs and larvae of the Colorado potato beetle (Martel *et al.,* 2005) where they can be treated with insecticides. Alternatively, the aggregation pheromone can enhance attraction and subsequent biocontrol by predators, such as, the spined soldier bug (*Podisus maculiventris*) (Dickens *et al.,* 2002).

4.5.2.2 Disease management

Intercropping systems have the potential to act as effective disease management tools and reduced the incidence of diseases in 53% of cases, while in 18% instances it increased them due to reduced cultivation and increased shading, favoring some pathogens; associate species serving as alternative hosts; and crop residues serving as a source of pathogen inoculums (Francis, 1989).

The bacterial wilt (*Ralstonia solanacearum*) incidence in potato was most effectively reduced by intercropping with maize and cowpea. Intercropping of tomato with garlic, maize, marigold onion and sorghum decreased bacterial wilt (*R. solanacearum*) and increased crop stand. Fininsa and Yuen (2002) found that the four types of cropping systems (sole cropping with bean, row cropping with maize-bean, mixed cropping with sorghum-bean and broadcast intercropping with maize–bean–sorghum) delayed onset of epidemic, lowered bean bacterial blight (*Xanthomonas campestris* pv. *phaseoli*) incidence, severity and reduced the disease progress rate. Intercropping of tomato with French bean and brinjal reduced the incidence of tomato leaf curl virus (TLCV) as the inter crops act as trap crops for the whiteflies which transmits the TLCV virus.

4.5.2.3 Nematode management

Intercropping systems can also be used for nematode management. Banana nematodes (*Radopholus similis, Pratylenchus coffeae* and *Helicotylenchus multicinctus*) were effectively managed using the intercrops like *Crotalaria juncea*, *Tagetes erecta*, *Sesamum indica* and *Acorus calamus* (Charles and Venkitesan, 1993). The rate of multiplication of the citrus nematode, *Tylenchulus semipenetrans* was considerably reduced by the intercrops, such as, marigold and mustard to 10.03 and 8.59, respectively (Mani, 1988).

Intercropping of tomato with marigold at 1: 4 and 1: 6 ratios and mustard at 1: 2 ratios was found to be effective in reducing root galls, egg masses and nematode population at harvest and gave cost: benefit ratio of 1: 8.36, 1: 7.88 and 1: 3.31, respectively (Rangaswamy *et al.*, 1999). Shivaprasad *et al.*, (2001) reported that the final nematode population of *M. incognita* on brinjal was reduced (14.41%) by intercropping with sweet potato cv. Sree Bhadra. The above treatment also significantly lowered the number of galls, egg masses and root-knot index; and enhanced brinjal fruit yield. Intercropping of lettuce with *Tagetes erecta* reduced root galling by 72% and root-knot nematode population by 94% (Perwez *et al.*, 1988).

Planting the intercrop marigold in tea plantations reduced the lesion nematode *Pratylenchus loosi* population and increased leaf yield by 7%. Medhana *et al.*, (1985) reported that intercropping of betel vine with *Tagetes erecta* was effective in reducing the population of *M. incognita* by 41.4% and root galling by 54%.

4.5.2.4 Weed management

Intercropping systems are considerably effective in suppression of weed population density and biomass production. A review of several studies revealed that the weed biomass in the smother intercrop species was lower in 47 cases and higher in 4 cases than in the main crop grown alone (as a sole crop); a variable response was observed in 3 cases. When intercrops were composed of two or more main crops; the weed biomass in the intercrop was lower than in all of the component sole crops in 12 cases, intermediate between component sole crops in 10 cases and higher than all sole crops in 2 cases. Forage grass and legume species have been extensively used for successful management of weeds in a number of temperate zone cropping systems. There are strong economic incentives for intercropping, particularly where the cost of herbicides is relatively high.

Intercropping of maize/cassava with cowpea (*Vigna unguiculata*) and egusi melon (*Citrullus anatus*) gave adequate weed suppression during the critical period of 4-8 weeks after planting without affecting the maize/cassava yield as compared to reduction in cassava yield by 49% and maize yield by 53% in control (uncontrolled weeds) (Unamma *et al.*, 1986). The above treatment also gave better economic returns than smother crops and herbicides. Similarly, the intercrop egusi melon was found as an effective labor-saving method for weed control in plantain [*Musa* (AAB Group)] (Obiefuna, 1989), yam (*Dioscorea* spp.) (Akobundu, 1980) and yam/ maize/cassava intercrop (Akobundu, 1980), production systems.

4.5.3 Agro-forestry

Agro-forestry systems increase habitat complexity, which generally correlates positively with abundance and diversity of natural enemies due to top-down regulation both at the field (Iverson *et al.*, 2014) and landscape levels (Chaplin-Kramer *et al.*, 2011); and bottom-up factors by providing food resources or improving microclimate or indirectly by enhancing host plant nutritional conditions or water availability (Sileshi *et al.*, 2008b). In general, most aspects of natural pest management are benefitted by agro-forestry. Weed abundance was significantly less under agro-forestry. Agro-forestry significantly enhanced natural enemies of pests more abundantly by the complex landscapes including natural habitat (Chaplin-Kramer *et al.*, 2011). It caused a significant reduction of both pest abundance and plant damage in perennial crops, while it had no effect in annual crops.

The reduction of pest problems in agro-forestry systems (Stamps and Linit, 1998) may be due to the following reasons (Vandermeer, 1989):

- Greater niche diversity and complexity makes it more difficult for pests to find their host plants.
- Plant species may act as trap or repellent crops.
- Higher plant diversity increases natural enemy population and pressure on pests.
- Provision of alternative food sources (*e.g.* flowers supply pollen, nectar) for adult parasitoids.
- Provision of sites for mating, oviposition and over wintering (Stamps and Linit, 1998).

4.5.3.1 Insect pest management

The tri-trophic interactions occur between the plants (trees, crops and weeds), herbivores (pests) and their natural enemies (predators, parasitoids and pathogens) (Sileshi *et al.,* 2001). These tri-trophic interactions are affected in a variety of ways by the plant community (Sileshi *et al.,* 2008).

The trees in multi-strata agro-forestry can increase the abundance of natural enemies (Sileshi *et al.,* 2008). The parasitic wasp *Cephalonomia stephanoideri* and the entomopathogenic fungus *Beauveria bassiana*, which control the coffee berry borer, are favored by the shade trees (Beer *et al.,* 1998). *Helopeltis* damage to cocoa can be managed by planting coconut in cocoa plantations to provide nesting sites for the predatory ants *Dolicoderus* and *Oecophylla* (Way and Khoo, 1991). More diverse traditional agro-forestry systems enhance predator–prey ratio as compared to intensified systems (Klein *et al.,* 2002). The abundance of soil fauna is enhanced when organic material from the improved fallows is incorporated into the soil in annual cropping systems, probably due to an improved living environment for ground and soil-based natural enemies (Barrios *et al.,* 2012). More below-ground natural enemies (particularly ants, carabid beetle larvae and centipedes) have been recorded in maize-based agro-forestry as compared to the corresponding monoculture (Sileshi and Mafongoya, 2006).

4.6 Cover Crops

The cover cropping (growing of pure or mixed stands of legumes and other crops) is responsible for improving soil structure and water infiltration, enhancing soil fertility, preventing soil erosion, modifying the microclimate and suppressing pests including insects, pathogens, nematodes and weeds.

4.6.1 Disease management

Everts (2002) reported that the cover crops decrease populations of bacterial and fungal pathogens by breaking their life cycles. Inclusion of mustards or other species of the Brassicaceae family as cover crops, suppress fungal disease populations through biofumigation (release of naturally occurring toxic chemicals during the degradation of glucosinolate compounds in their plant cell tissues) (Lazzeri and Manici, 2001).

The soil inoculum of bacterial blight (*Ralstonia solanacearum*) in banana was reduced by growing cover crops like kudzu (*Pueraria phaseoloides*) and sorghum. Vegetable crop diseases caused by *Rhizoctonia* sp. can also be reduced by growing oats as cover crop (Sustainable Agriculture Network, 1998).

Abawi and Widmer (2000) reported that the root rot disease complex [caused by several pathogenic fungi (*Fusarium solani* f. sp. *phaseoli*, *R. solani*, *Pythium ultimum* and *T. basicola*) and the plant-parasitic nematode (*Pratylenchus* spp.) individually or in any possible combination] in beans was suppressed by growing rapeseed, wheat and rye cover crops. Similarly, *Sclerotinia* (white mold) in lettuce and *Verticillium* wilt in cauliflower can be managed by brassicas as cover crops (Hartz *et al.,* 2005). Likewise, the biotic factors (fungal and nematode pathogens) in strawberry replant problem were suppressed by utilizing Indian grass (bunch grass) (*Sorghastrum avenaceum*) and brown mustard (*Brassica juncea*) as cover crops which led to 32% and 28% higher strawberry yields, respectively compared to yields from strawberries in fumigated soil (Seigies and Pritts, 2006).

The soil-borne diseases which cause 10-15% losses in vegetable crops can be managed by disease suppressive cover crop rotations with Sudan grass, Brassica, millet cover crops resulting in significant increases in yield. The other examples of green manure cover crops being effective against vegetable diseases include lucerne hay residues against *Sclerotinia sclerotiorum* in lettuce (Asirifi *et al.,* 1994) and common root rot of pea (*Aphanomyces eutieches*) (Williams Woodward *et al.,* 1997); buckwheat against common scab (*Streptomyces scabies*); and Verticillium wilt of potatoes (Wiggins and Kinkel, 2005).

The two ways by which the green manure cover crops suppress diseases include:

- Provide organic substrates for sustenance of microbial communities (bacteria, non-pathogenic *Fusarium* species, Streptomyces and other Actinomycetes) which in turn suppress pathogens through competition, antiobiosis, parasitism or by inducing systemic resistance in plants (Hoitink and Boehm, 1999).
- Have a direct toxic effect on the pathogen (glucosinolates produced during biofumigation) (Larkin and Griffin, 2007).

4.6.2 Insect pest management

The cover crops provide pollen and nectar resources and overwintering habitats for generalist predators through habitat augmentation (Bugg and Waddington, 1994). The leaf hopper pests in grape vineyards can be managed by planting clover and other legume ground covers that attract beneficial wasps and spiders and provide them with habitat and food source (Hanna *et al.,* 2003). The seasonal population of predatory mite, *Euseius tularensis* can be enhanced to reduce pest populations of citrus thrips by planting various leguminous cover crops, such as, bell bean, woolly-pod vetch, New Zealand white clover and Austrian winter pea which provide sufficient pollen as a feeding source (Grafton-Cardwell *et al.,* 1999). Growing of a green manure cover crop (non-host) can significantly decrease infestations of cabbage root fly (*Delia radicum*) (Finch and Edmonds, 1994). The egg-laying by the cabbage root fly was reduced by 36-82% when cauliflowers were planted amongst 24 other non-host plant species (Finch *et al.,* 2004).

A significantly lower incidence of insect pests was observed in orchards with rich floral undergrowth, mainly because of an increased abundance and efficiency of predators and parasitoids, than clean cultivated orchards (Smith *et al.,* 1996). The presence of goldenrod (*Solidago* sp.), lamb's-quarter *(Chenopodium album),* ragweed (*Ambrosia* sp.) and smartweed (*Polygonum* sp.); which provided alternate hosts for the parasite *Macrocentrus ancylivorus*; gave effective management of the oriental fruit moth in peach orchards (Bobb, 1939). Likewise, the parasitism of tent caterpillar eggs and pupae in apple orchards increased four-fold and eighteen-fold, respectively; while parasitism of codling moth larvae increased five-fold by the presence of wild flowers over non-weedy orchards (Leius, 1967). By maintaining floral diversity throughout the growing season in California vineyards, in the form of summer cover crops, such as, buckwheat (*Fagopyrum esculentum*) and sunflower (*Helianthus annus*), supports large numbers of predators and parasitoids, thereby favoring enhanced biological control of leaf hoppers and thrips in grape vineyards (Nicholls *et al.,* 2001).

The commonly used cover crop species and the beneficial insects they attract are presented in Table 1.

Table. *The effect of common cover crop species that attract beneficial insects*

Cover crop	Beneficial insects encouraged
Buckwheat	Parasitic wasps, ladybugs, tachinid and hover flies, lacewings
Clovers	Parasitic wasps, big eyed bugs, minute pirate burgs, ladybugs, tachinid flies and aphid midges
Hairy vetch	Minute pirate bugs, ladybugs, predatory and parasitic wasps
Cereals	Ladybugs

4.6.3 Nematode management

The warm season legume cover crops are effective in reducing populations of certain plant parasitic nematodes by breaking their life cycles (Vargas-Ayala *et al.,* 2000). The root-knot nematode damage in snap bean was reduced by rotation with certain cover/green manure crops like crotalaria, hairy indigo, joint vetch, pigeon pea and velvet bean. Similarly, Loos (1961) reported that the burrowing nematode *Radopholus similis* on banana can be eliminated by planting of sugarcane or cover crops, such as, pangola grass (*Digitaria decumbens*), *Panicum maximum* var. *trichoglume* and *Phaseolus atropurpureus* following the destruction of bananas. Likewise, the root-knot nematode (*M. incognita*) population on black pepper was reduced by growing of non-host cover crop like Siratro (*Macroptilium atropurpureus*) in the interspaces (Ichinohe, 1975).

4.6.4 Weed management

The weed suppression by cover/green manure crops results in saving labor used for hoeing and reducing the use of herbicides, lowering production costs. A weed control of common lambsquarters (*Chenopodium album*) and redroot pigweed (*Amaranthus retroflexus*) level nearly equal to that of a standard herbicide treatment was obtained in lettuce by incorporation of rapeseed cover crop foliage into the soil (Boydston and Hang, 1995).

The allelopathic compounds produced by some cover crops can suppress weeds. The allelopathic effects of cover crops, such as, crimson clover and hairy vetch is enhanced by their incorporation in soil rather than left on the surface in no-till management (Teasdale and Daughtry, 1993). The weed emergence is also delayed by cover crops, giving a competitive edge to the primary crop.

4.7 Crop Rotation

Crop rotation is an important tool in organic agriculture for the management of insect and mite pests, diseases, nematodes and weeds. The inclusion of crops that do not support growth and reproduction of the pathogen in rotation reduces the amount of soil-borne inoculum. The crops in the same family generally should not follow one another in the field. For example, cucumbers should not follow cantaloupes, since both belong to the same family. Similarly, a cucumber-melon-squash rotation obviously invites disease problems. A minimum of at least two years should lapse between the crops from same families. The typical crop rotation may include Brassicaceae (Cole crops), followed by Asteraceae (lettuce, cut flowers), followed by Solanaceae (tomatoes, potatoes, peppers, eggplants) and followed by Cucurbitaceae (squashes, cucumbers, melons).

Some common characteristics of the crop pests that can be controlled by rotations (Flint and Roberts, 1988) include:

- Fairly narrow host range of the pest.
- Source of pest should be within the field.
- Pest's inability to survive long periods without a living host.
- Pest should be as immobile as possible.

4.7.1 Disease management

A study in California showed that when carrots were grown after onions lowered populations of root dieback pathogens (*Pythium* and *Rhizoctonia*) and produced more marketable yield. The club root of cabbage, *Plasmodiophora brassicae* was effectively controlled by growing tomato, cucumber, snap bean and buckwheat in rotation. Growing broccoli immediately before cauliflower resulted in a reduction in Verticillium wilt by producing more specific glucosinolate which stimulates myxobacteria that reduce survival of Verticillium microsclerotia. Verticillium wilt diseases of other crops in California were also suppressed by broccoli.

Both Verticillium wilt and scab diseases of potato were reduced when corn or alfalfa was grown the previous year rather than potato. Verticillium wilt severity also was lower when a buckwheat green manure preceded potato. Broccoli is also a good crop to grow in rotation with lettuce and other crops susceptible to white mold, *Sclerotinia minor*.

Management of different crop diseases using crop rotation is presented in Table 2.

Table 2. *Management of crop diseases using crop rotation*

Crop	Disease/ Causal organism	Crop rotation recommended
Banana	Panama wilt, *Fusarium oxysporum* f. sp. *cubense*	Rice or sugarcane
Potato	Common scab, *Streptomyces scabies*	Wheat, peas, oat, barley, lupin, soybean, sorghum, bajra
	Bacterial wilt, *Ralstonia solanacearum*	Maize, wheat, barley, oat, sun hemp, finger millet, cabbage, onion, garlic
Tomato, brinjal	Bacterial wilt. *Ralstonia solanacearum*	Cowpea-maize-cabbage, okra-cowpea-maize, maize-cowpea-maize, finger millet-brinjal (Pusa Purple Cluster)-French bean

Crucifers	Black rot, *Xanthomonas campestris* pv. *campestris*	Rice
Cucumber	Powdery mildew, *Sphaerotheca fuliginea, Erysiphe cichoracearum*	Rice in low lands

4.7.2 Insect pest management

Insect pests that are less mobile and that feed on specific plants in limited areas are the easiest to control. Farmers try to disrupt insect growth and breeding schedules by including non-susceptible crops in the rotation. When farmers expect problems from particular insects, they may want to physically separate susceptible species with non-susceptible species and thus make it harder for insects to cross over to crops they favor.

Researchers in Georgia reported that growing cover crops in the rotation attract high densities of the beneficial insects like big-eyed bugs, ladybugs and other beneficial insects in v*etc*hes and clovers that help to keep pest populations in check (Bugg and Waddington, 1994). There is anecdotal evidence that beneficial insects have destroyed Colorado potato beetles feeding on eggplant planted into strip-tilled crimson clover.

The carrot fly (*Chamaepsila rosae*) can be managed by crop rotations with non-host crops resulting in reduced chemical pesticide use. Crop rotation with cereals (sorghum, maize) reduced the incidence of field bean pod borer (*Adisura atkinsoni*). Similarly, crop rotation (Paddy – sweet potato – cowpea) can minimize sweet potato weevil (*Cylas* spp.) damage.

4.7.3 Nematode management

Rice or green gram grown after banana suppressed the population of *Radopholus similis, Pratylenchus coffeae* and *Helicotylenchus multicinctus* (Rajendran *et al.,* 1979). Planting of sugarcane or pangola grass (*Digitaria decumbens*) following the destruction of banana infected with *R. similis* eliminates the nematode after 10 weeks. Planting of pineapple after pangola grass reduced the population of *M. incognita, Criconemoides* spp., *Helicotylenchus* spp. and *Rotylenchulus reniformis* and increased the fruit yield (60 tons/ha as against 45 tons/ha in control) and first ratoon crop (48 tons/ha as against 23 tons/ha in control) (Ayala *et al.,* 1967).

In North Western Hills, rotation of potato with French beans, leafy vegetables, maize and wheat helped in minimizing the root-knot nematode damage and increasing the productivity of the crop (Raj and Nirula, 1969). A crop rotation pattern wherein potato is grown once after 3-4 other crops (using French bean, peas, cabbage and carrot), decreased the cyst nematode (*Globodera rostochiensis*) population by 98 to 99% and increased yields of potato by 90%. A decrease in root-knot nematode population on tomato, brinjal, okra, chilies and sponge gourd occurs following marigold, spinach and bottle gourd (Khan *et al.,* 1975). There was sudden decline in stunt nematode (*Tylenchorhynchus brassicae*) population when cabbage and cauliflower were rotated with wheat (Siddiqi *et al.,* 1973).

4.7.4 Weed management

Many short duration pulses viz., green gram and soybean effectively smoother weeds without causing reduction in the yield of main crop. The spring lettuce crops will have less weed competition by following crop rotation with a fast growing, vigorous summer cover crops, such as, Japanese millet, cowpea or soybean and then planting fall broccoli. Planting *Cassia tora* will help to cover and suppress the growth of *Parthenium*. In certain parts of India, crop rotation using marigold (*Tagetes* spp.) during rainy season, instead of the usual crop, is found effective in reducing *Parthenium* infestation in cultivated areas.

4.8 Plant Breeding

Plant breeding is responsible for strengthening numerous plant defenses against pests, such as, insect and mite pests, diseases, nematodes and weeds. The plant defenses may be either structural (tough leathery leaves, leaf hairs and spines) or toxic plant chemicals that deter feeding (allelopathic chemicals that deter weed growth and leaf waxes that form barriers for pests and diseases).

The selection for a single resistant gene (monogenic or oligogenic in vertical resistance) could inadvertently eliminate other genes affecting other pests or genes that play a role in attracting the natural enemies of pests, since the focus of plant breeding for pest resistance is frequently limited to a specific pest/plant interaction. The development of new races of the pest can easily breakdown the resistance, since single gene is responsible for resistance. The initial amount of inoculum is reduced in vertical resistance which can delay the epidemic.

The horizontal resistance (also called polygenic resistance, minor gene resistance or field resistance) is governed by many genes and it is effective against many races of the pathogen. It slows down the rate at which disease increases in the field.

Our knowledge is steadily expanding on the role of plants' in their own defense that can be used to breed and engineer plants whose defenses work harmoniously with natural systems. There is a need for more research as well as plant breeding programs which focus on enhancing natural defenses. More emphasis should be placed on open-pollinated crops rather than hybrids for their adaptability to local environments and greater genetic diversity.

4.8.1 Disease resistance

Host plant resistance is an important tool to control diseases of major horticultural crops in developing countries, especially potato, cassava, chickpea and other horticultural crops (Table 3). The use of resistant varieties is very much welcomed by resource poor farmers because it does not require additional cost and is environment-friendly.

Table 3. *Horticultural crop varieties resistant to diseases.*

Horticultural crop	Pest/Disease	Resistant varieties
Papaya	Ring spot virus	SunUp, Rainbow.
Potato	Late blight	Kufri Sutlej, Kufri Badshah, Kufri Jawahar (in plains), Kufri Jyothi, Kufri Giriraj, Kufri Kanchan, Kufri Meghad (in hills).
Tomato	Powdery mildew (PM)	Arka Asish.
	Leaf curl virus	Avinash-2, Hisar Anmol.

Brinjal	*Phomopsis* blight	Pusa Bhairav.
	Little leaf	Pusa Purple Long, Pusa Purple Cluster (Field resistant).
Chili	PM	Arka Suphala (T).
	Mosaic, leaf curl	Pant C-1, Arka Harita, Arka Meghana.
French bean	Angular leaf spot, mosaic	Pant Anupama.
	Rust, bacterial blight	Arka Anoop.
	Rust	Arka Bold, Swarna Priya, Swarna Latha, Arka Anoop.
Pea	PM	Pusa Pragati, Jawahar Matar 5, Jawahar Peas 83.
	PM, Rust	Arka Ajit, Arka Karthik, Arka Sampoorna.
Okra	Yellow vein mosaic virus (YVMV)	Pusa Sawani, Arka Abhay, Arka Anamika, Hisar Unnat, Varsha Upkar, Pusa A-4, Parbhani Kranti (T), Punjab Kesari, Punjab Padmini, Sun-40, Makhmali.
Cucumber	PM	Swarna Poorna.
	PM, downy mildew (DM), angular leaf spot, anthracnose	Poinsette.
Cabbage	Black rot	Pusa Mukta.
Cauliflower	Black rot	Pusa Snowball K-1, Pusa Shubhra.
	Curd blight	Pusa Synthetic, Pusa Shubhra..
	DM	Pusa Hybrid-2.
Onion	Purple blotch, basal rot	Arka Pitamber, Arka Kirtiman, Arka Lalima.
	Purple blotch, *Alternaria porri*	Arka Kalyan.
Garlic	Purple blotch, *Stemphylium* disease	Agri-found White.
Muskmelon	PM	Arka Rajhans, Pusa Madhuras (MR).
	PM, DM	Punjab Rasila.
Watermelon	PM, DM, anthracnose	Arka Manik
Carrot	PM	Arka Suraj.
Cardamom	Mosaic	IISR Vijetha
Ginger	Soft rot	Maran

PM - Powdery mildew, DM - Downy mildew, MR - Moderately Resistant, T – Tolerant

4.8.2 Nematode resistance

Host plant resistance achieved by traditional breeding programs can provide a valuable protection against some nematodes (Table 4).

Table 4. *Horticultural crop varieties resistant to nematodes.*

Horticultural crop	Pest/Disease	Resistant varieties
Banana	Burrowing nematode, *Radopholus similis*	Kadali, Pedalimoongil, Ayiramkapoovan, Peykunnan, Kunnan, Pisang Seribu, Tongat, Vennettu Kunnan, Anaikomban.
Citrus	Citrus nematode, Tylenchulus semipenetrans	Trifoliate Orange, Swingle Citrumelo.
Grapevine	Root-knot nematode, *Meloidogyne incognita*	Rootstocks Dogridge, 1613, Salt Creek.
Tomato	Root-knot nematode	Hisar Lalit, Pusa Hybrid-2, Arka Vardan.
Carrot	PM, Root-knot nematode	Arka Suraj.
China aster	Root-knot nematode, *M. incognita*	Shashank, Poornima (MR).
Tuberose	Root-knot nematode, *M. incognita*	Sringar, Suvasini (T).
Mentha	Root-knot nematode, *M. incognita*	Kukrail, Arka Neera.
Black pepper	Root-knot nematode, *M. incognita*	IISR Pournami (T).
Ginger	Root-knot nematodes	IISR Mahima

MR – Moderately Resistant T – Tolerant

4.8.3 Insect resistance

Host plant resistance (HPR) to insects is an effective, economical and environment friendly method of insect pest control. Considerable progress has been made in identification and development of crop cultivars with resistance to the major pests in different crops (Table 5).

Table 5. *Horticultural crop varieties resistant to insect pests*

Crops	Insect pests	Resistant varieties
Chilli	Thrips	NP 46 (T)*
Field bean	Jassid, aphid & pod borer	Pusa Sem-2, Pusa Sem-3.
Okra	Fruit borer	Pusa A-4.
Onion	Thrips	Arka Pitamber, Arka Kirtiman, Arka Lalima.
Pumpkin	Fruit fly	Arka Suryamukhi.

*T – Tolerant

4.8.4 Biotechnological approaches for pest management

Some scientists believe that genetically modified crops fit into the gambit of sustainable intensification to fulfill the objective of food security (Birch *et al.,* 2011). The long-term theory and genetics for improved durable resistance is through molecular biology and genetic engineering. Genetically modified crops (transgenics) can be developed by introducing one (or a few) foreign "good" gene/s into the best accepted cultivars. The next 20 years will likely to see a substantial increase in the use of genetically engineered plants. Genetically engineered plants have been designed to resist pests without the need for pesticides.

4.8.4.1 Papaya

Transgenic ring-spot virus-resistant papaya (cvs. Rainbow, SunUp) have been genetically engineered to contain a virus gene that encodes for the production of the coat protein of the virus. As a major component of viruses, the coat protein's primary function is to protect viral genetic information. Expression of this gene in the resulting papaya lines renders the plants resistant to the virus (Figure 6).

Fig. 6. *Transgenic papaya test in field. Yellow plants on the left are non-transgenic papaya; plants on right are transgenic 'Rainbow' papaya.*

4.8.4.2 Potato

Colorado potato beetle (CPB) resistance in potato has been achieved through the incorporation of a gene from the *B. thuringiensis* (Bt) protein into potatoes (Russet Burbank, Atlantic). Several developing countries have transformed potato varieties with the Bt gene to express resistance to the potato tuber moth.

Desiree potato plants were genetically modified to resist attack by insect species belonging to the orders Coleoptera (Colorado potato beetle - CPB) and Lepidoptera (potato tuber moth-PTM and European corn borer-ECB), through the insertion of a hybrid gene, SN19 under the control of a chrysanthemum ribulose-1, 5-bisphosphate carboxylase/oxygenase small subunit (Rubisco SSU) promoter and terminator (Naimov *et al.,* 2003). Transgenic plants were shown to be resistant against CPB larvae and adults, PTM larvae and ECB larvae. These are the first transgenic plants resistant to pests belonging to two different insect orders.

4.8.4.3 Brinjal

Bt Brinjal has been developed by inserting a gene cry1Ac from a soil bacterium, *B. thuringiensis* through an *Agrobacterium*-mediated gene transfer. The Event EE1 was introgressed by plant breeding into various local varieties, such as, Malpur Local, Manjari Gota, Kudachi Local, Udupi Local, 112 GO and Pabkavi Local (Figure 7).

Fig. 7. *Damage caused by brinjal fruit and shoot borer (left) and damage-free Bt brinjal developed by MAHYCO (right)*

5. CONCLUSIONS

The exploitation of synergy and complementarity that result from the various combinations of crops, trees and animals in agro-ecosystems that feature spatial and temporal arrangements, such as, polycultures and agro-forestry systems is the key strategy in sustainable intensification. The best practices that will encourage the desired biodiversity components can be determined by identifying the type of biodiversity that is desirable to conserve and/or enhance in order to carry out ecological services, such as, biological pest control, nutrient cycling and soil and water conservation. The key ecological services to agro-ecosystems can be provided by encouraging those agricultural practices that enhance the diversity and abundance of below- and above -ground organisms (Altieri and Nicholls, 2000).

It is necessary to develop an understanding of the ecology of natural enemies within and outside of the cultivated habitat and identifying those resources that are necessary for their survival and reproduction, in order to implement the diversification strategy more rapidly (Gurr *et al.,* 1998). If within the crop contribution to the overall natural enemy population in subsequent years is minor, then habitat management should be given more importance to specifically increase the source populations from outside the crop to ensure a larger number of immigrants by addition of plant species to provide alternate hosts and/ or food sources, addition of habitats as suitable overwintering sites or the provision of corridors within the cropping system to facilitate movement of natural enemy population.

REFERENCES

Abawi, G.S. and Widmer, T.L. (2000). Impact of soil health management practices on soil-borne pathogens, nematodes and root diseases of vegetable crops. Appl. Soil Ecol. 15:37–47.

Akhtar, M. and Alam, M.M. (1993). Utilization of waste materials in nematode control: A review. Bioresource Technology 45:1–7.

Akobundu, I.O. (1980). Live mulch: a new approach for weed control and crop production in the tropics. *Proceedings of the 1980 British Crop Protection Conference – Weeds*, 2:377-382. British Crop Protection Counci, Thornton Heath, UK.

Altieri, M.A. (1987). *Agroecology: The Scientific Basis of Alternative Agriculture.* Westview Press, Boulder, CO.

Altieri, M.A. and Letourneau, D.K. (1982). Vegetation management and biological control in agro-ecosystems. Crop Protection 1:405-430.

Altieri, M.A. and Nicholls, C.I. (2000). Applying agro-ecological concepts to development of ecologically based pest management systems. In *Proc. Workshop Professional Societies and Ecological Based Pest Management Systems*, pp 14–19. National Research Council, Washington, DC.

Andow, D. (1991). Vegetational diversity and arthropod population response. Annual Review of Entomology 36:561–586.

Asirifi, K.N., Morgan, W.C. and Parbery, D.G. (1994). Suppression of Sclerotinia soft rot of lettuce with organic soil amendments. Australian Journal of Experimental Agriculture 34:131-136.

Asiwe, JA.N. (2009). The impact of phosphate fertilizer as a pest management tactic in four cowpea varieties. African Journal of Biotechnology 8(24):7182-7186.

Awada, L., Lindwall, C.W. and Sonntag, B. (2014). The development and adoption of conservation tillage systems on the Canadian Prairies. International Soil and Water Conservation Research 2(1):47–65.

Ayala, A., Roman, J. and Gonzalez, J. (1967). Pangola grass as a rotation crop for pineapple nematode control. Journal of Agriculture Science, University of Puerto Rico **51**: 94-96.

Badenes-Perez, F.R., Nault, B.A. and Shelton, A.M. (2005). Manipulating the attractiveness and suitability of hosts for diamondback moth (Lepidoptera: Plutellidae). Journal of Economic Entomology 98: 836–844.

Baggen, L.R., Gurr, G.M. and Meats, A. (1999). Flowers in tri-trophic systems: mechanisms allowing selective exploitation by insect natural enemies for conservation biological control. Entomologia Experimentalis et Applicata 91:155-161.

Barrios, E, Sileshi, GW, Shepherd, K and Sinclair, F. (2012). Agroforestry and soil health: linking trees, soil biota and ecosystem services. In DH Wall, RD Bardgett, V. Behan-Pelletier, J.E., Herrick, T.H. Jones, K. Ritz, J. Six, D.R. Strong, W.H. and van der Putten (Eds.) *Soil Ecology and Ecosystem Services,* pp. 315-330. Oxford University Press, Oxford.

Beer, J., Muschler, R., Kass, D. and Somarriba, E. (1998). Shade management in coffee and cacao plantations. Agroforestry Systems 38:139-164.

Begum, M., Gurr, G.M., Wratten, S.D., Hedburg, P.R. and Nichol, H.I. (2006). Using selective food plants to maximize biological control of vineyard pests. Journal of Applied Ecology 43:547–554.

Bhattacharya, R.K. and Rao, V.N.M. (1984). Effect of soil covers and soil moisture regimes on nematode population in soil and in root of banana. J. Res. Assam Agri. Univ. 5:206-209.

Birch, A.N.E., Begg, G.S. and Squire, G.R. (2011). How agro-ecological research helps to address food security issues under new IPM and pesticide reduction policies for global crop production systems. *J. Exp. Biol.* 62:3251–3261.

Bobb, M.L. (1939). Parasites of the oriental fruit moth in Virginia. Journal of Economic Entomology 32:605.

Bockus, W.W. and Shroyer, J.P. (1998). The impact of reduced tillage on soilborne plant pathogens. Annual Review of Phytopathology 36(1):485–500.

Bottrell, D.R. (1980). *Integrated Pest Management.* Council on Environmental Quality, US Government Printing Office, Washington, DC.

Boydston R.A. and Hang, A. (1995). Rapeseed (*Brassica napus*) green manure suppresses weeds in potato (Solanum tuberosum). Weed Technology 9:669-675

Brown, M.W. and Tworkoski, T. (2004). Pest management benefits of compost mulch in apple orchards. Agri. Ecosystems Envt. 103:465-472.

Bugg, R.L. and Waddington, C. (1994). Using cover crops to manage arthropod pests of orchards: A review. Agriculture, Ecosystems and Environment 50:11-28.

Carballo, S.J., Blankenship, S.M. and Sanders, D.C. (1994). Drip fertigation with nitrogen and potassium and postharvest susceptibility to bacterial soft rot of bell peppers. J. Plant Nutr. 17:1175–1191.

Celar, F. (2003). Competition for ammonium and nitrate forms of nitrogen between some phytopathogenic and antagonistic soil fungi. Biol. Control 28:19–24.

Chan, K.Y. (1987). An overview of some tillage impacts on earthworm population abundance and diversity - implications for functioning in soils. Soil and Tillage Research 57(4):179–191.

Chaplin-Kramer, R., O'Rourke, M.E., Blitzer, E.J. and Kremen, C. (2011). A meta-analysis of crop pest and natural enemy response to landscape complexity. Ecology Letters 14:922-932.

Charles, J.S.K. and Venkitesan, T.S. (1993). *Status Report on the Nematological Investigations on Banana in Kerala Agricultural University*. Kerala Agri. Univ., Vellanikkara, 43 pp.

Chauhan, B.S., Gill, G.S. and Preston, C. (2006). Tillage system effects on weed ecology, herbicide activity and persistence: A review. Animal Production Science 46(12):1557–1570.

Cook, R.J. and Baker, K.F. (1983). *The Nature and Practice of Biological Control of Plant Pathogens.* American Phytopathological Society, St. Paul, Minnesota.

Costamilan, L.M. (2000). Doenças de plantas e alternativas de controle em plantio direto. Costamilan, L.M. In: *7o Encontro Nacional de Plantio Direto na Palha,* pp 103-104. FEBRAPDP.

Cunha Fernandes, J.M. (1997). As doenças das plantas e o sistema plantio direto. In: JM Cunha Fernandes (ed.) *Seminário Internacional do Sistema Plantio Direto. Anais,* pp 43-80. EMPRABA, Passo Fundo.

Dickens, J.C., Oliver, J.E., Hollister, B., Davis, J.C. and Klun, J.A. (2002). Breaking a paradigm: male-produced aggregation pheromone for the Colorado potato beetle. J. Exp. Biol. 205:1925–1933.

Egley, G.H. (1996). Stimulation of weed seed germination in soil. Reviews of Weed Science 2: 67–89.

Everts, K. (2002). Reduced fungicide applications and host resistance for managing three diseases in pumpkin grown on a no-till cover crop. Plant Disease 86(10):1134-1141.

FAO. (2001). *Improving Nutrition through Home Gardening*. FAO, Rome.

FAOStat. (2014). FAO, Rome, Italy, Available online: faostat.fao.org (accessed on 1 December 2014).

Finch, S., Billiald, H. and Collier, R.H. (2003). Companion planting – do aromatic plants disrupt host-plant finding by the cabbage root fly and the onion fly more effectively than non-aromatic plants? Entomologia Experimentalis et Applicata 109:183-195.

Finch, S., Edmonds, G.H. (1994). Undersowing cabbage crops with clover – effects on pest insects, ground beetles and crop yield. IOBC/WPRS Bulletin 17(8):59-167.

Fininsa, C. and Yuen, J. (2002). Temporal progression of bean bacterial blight (*Xanthomonas campestris* pv. *phaseoli*) in sole and intercropping systems. European J. Pl. Path. 108:485-495.

Flint, M.L. and Roberts, P.A. (1988). Using crop diversity to manage pest problems: Some California examples. American Journal of Alternative Agriculture 3:163-167.

Francis, C.A. (1989). Biolgical efficiencies in multiple cropping systems. Advance in Agronomy 42:1-42.

Gonzales, A. and Dave, R.D. (1997). Comparison of egg parasitism of *Empoaska kr*aemeri by *Anagrus* sp. in common beans under no-tillage and conventional tillage. CEIBA 38(1):49-54.

Grafton-Cardwell, E.E., Ouyang, Y.L. and Bugg R.L. (1999). Leguminous cover crops to enhance population development of *Euseius tularensis* (Acari: Phytoseiidae) in citrus. Biological Control 16:73-80.

Gupta, D.C. and Mukhopadhyaya, M.C. (1971). Effect of N, P and K on root-knot nematode *Meloidogyne javanica* (Treub) Chitwood. Sci. Cult. 31:246–327.

Gurr, G.M. van Emden, H.F. and Wratten, S.D. (1998). Habitat manipulation and natural enemy efficiency: Implications for the control of pests- Chapter 9. In B Pedro (ed.) *Conservation Biological Control,* p. 155-183. Academic Press, San Diego.

Hanna, R., Zalom, F.G. and Roltsch, J. (2003). Relative impact of spider predation and cover crop on population dynamics of *Erythoneura variablis* in a raisin grape vineyard. Entomologia Experimentalis et Applicata107:177-191.

Haramoto, E.R. and Gallandt, E.R. (2004). Brassica cover cropping for weed management: A review. Renewable Ag. and Food Sys 19(4):187–198.

Hartz, T.K., Johnstone, P.R., Miyao, E.M. and Davis, R.M. (2005). Mustard cover crops are ineffective in suppressing soil-borne disease or improving processing tomato yield. HortSci. 40(7):2016–2019.

Hogg, B.N., Bugg, R.L. and Daane, K.M. (2011). Attractiveness of common insectary and harvestable floral resources to beneficial insects. Biological Control 56:76-84.

Hoitink, H.A.J. and Boehm, M.J. (1999). Biocontrol within the context of soil microbial communities: a substrate dependent phenomenon. Annu. Rev. Phytopathol. 37:427-446.

Hoy, C.W. (1999). Colorado potato beetle resistance management strategies for transgenic potatoes. Am. J. Potato Res. 76:215–219.

Huber, D.M. and Graham, R.D. (1999). The role of nutrition in crop resistance and tolerance to disease. In: Z Rengel (Ed.) *Mineral Nutrition of Crops Fundamental Mechanisms and Implications,* pp. 205–226. Food Product Press, New York.

Ichniohe, M. (1975). Infestation of black pepper vines by the root knot nematode, *Meloidogyne incognita* at Tome-Acu, Para, Brazil. Japanese Journal of Nematology 5:36-40.

Igzoburkie, M.U. (1971). Ecological balance in tropical agriculture. Geographical Review 61:519-529.

Iverson, A.L., Marin, L.E., Ennis, K.K., Gonthier, D.J., Connor-Barrie, B.T., Remfert, J.L. *et al.,* (2014). Do polycultures promote win-wins or trade-offs in agricultural ecosystem services? A meta-analysis. Journal of Applied Ecology 51:1593-1602.

Jackai, L.E.N. and Singh, S.R. (1983). Suitability of selected leguminous plants for development of *Maruca testulalis* larvae. Entomol. Exp. Appl. 34:174–178.

Khan, A.M., Saxena, S.K., Siddiqi, Z.A. and Upadhyay, R.S. (1975). Control of nematodes by crop rotation. Indian Journal of Nematology **5**: 214-221.

Kirkpatrik, J.D., Mai, W.F., Fisher, E.C. and Parker, K.G. (1959a). Population levels of *Pratylenchus penetrans* and *Xiphinema americanum* in relation to potassium fertilization on Montmorency sour cherries on Mazzard rootstock (Abstr). Phytopathology 49:543.

Kirkpatrik, J.D., Mai, W.F., Fisher, E.C. and Parker, K.G. (1959b). Relation of nematode population in nutrition of sour cherries (Abstr). Phytopathology 49:543.

Klein, E., Katan J. and Gamliel A. (2011). Soil suppressiveness to Fusarium disease following organic amendments and solarization. Plant Disease. 95: 1116–1123.

Klostermeyer, E.C. (1950). Effect of soil fertility on corn earworm damage. Journal of Economic Entomology 43:427-429.

Larkin, P. and Griffin, T.S. (2007). Control of soil-borne diseases of potato using Brassica green manure. Crop Prot. 26:1067–1077.

Lazarovits, G., Conn, K.L. and Potter, J. (1999). Reduction of potato scab, Verticillium wilt and nematodes by soymeal and meat and bone meal in two Ontario potato fields. Can J. Plant Pathol. 21:345–353.

Lazzeri, L. and Manici, L.M. (2001). *Allelopathic effect of glucosinolate-containing plant green manure on Pythium sp and total fungal population in soil.* Hortscience. 36:1283–1289.

Leius, K. (1967). Influence of wild flowers on parasitism of tent caterpillar and codling moth. Canadian Entomologist 99:444-446.

Lewis, J.A., Lumsden, R.D., Miller, P.D. and Keinath, A.P. (1992). Suppression of damping-off of peas and cotton in the field with composted sewage sludge. Crop Prot. 11:260-266.

Loos, C.A. (1953). *Meloidogyne brevicauda* n. sp. a cause of root-knot of mature tea in Ceylon. J. Sci. (B) 23:119-124.

Loos, C.A. (1961). Eradication of the burrowing nematode, *Radopholus similis*, from bananas. Plant Disease Reporter 45:457-461.

Lumsden, R.D., Lewis, J.A. and Miller, P.D. (1983). Effect of composted sewage sludge on several soil-borne pathogens and diseases. Phytopathology 73:1543-1548.

Luna, J.M.(1988). In Global perspectives on Agroecology and Sustainable Agricultural Systems. *Proc. Sixth. Int. Sci. Conference of IFOAM.* Santa Cruz, CA. p. 589-600.

Mani, A. (1988). Effect of interculture of marigold and mustard with acid lime on citrus nematode, *Tylenchulus semipenetrans.* International Nematology Network Newsletter 5(4):14-15.

Marschner, H. (1995). *Mineral Nutrition of Higher Plants*, 2nd ed., Academic Press, London, 889 p.

Martel, J.W., Alford, A.R. and Dickens, J.C. (2005). Synthetic host volatiles increase efficacy of trap cropping for management of Colorado potato beetle, *Leptinotarsa decemlineata* (Say). Agric. For. Entomol. 7:79–86.

Medhane, N.S., Jagdale, G.B., Pawar, A.B. and Darekar, K.S. (1985). Effect of *Tagetes erecta* on root-knot nematodes infecting betel vine. Int. Nematol. Network Newsl. 2(3):11-12.

Mitchell, E.R., Hu, G. and Johanowicz, D. (2000). Management of diamondback moth (Lepidoptera: Plutellidae) in cabbage using collard as a trap crop. HortScience 35:875-879.

Naimov, S., Dukiandjiev, S. and de Maagd, R.A. (2003). A hybrid *Bacillus thuringiensis* delta-endotoxin gives resistance against a coleopteran and a lepidopteran pest in transgenic potato. Plant Biotechnol. J. 1:51–57.

Nicholls, C.I. and Altieri, M.A. (2004). Agroecological bases of ecological engineering for pest management, In: G.M. Gurr, S.D. Wratten, M.A. Altieri (eds.) *Ecological Engineering for Pest Management. Advances in habitat manipulation for arthropods*, p. 33-54. CSIRO, Collingwood (Australia) & CABI, Walingford (UK).

Nicholls, C.I., Parrella, M. and Altieri, M.A. (2001). The effects of a vegetational corridor on the abundance and dispersal of insect biodiversity within a northern California organic vineyard. Landscape Ecology 16(2):133-146.

Obiefuna, J.C. (1989). Biological weed control in plantains (I2lusa AAB) with egusi melon (*Colocynrhis citrulus* L.). Biological Agriculture and Horticulture 6:221-227.

Oerke, E.C. (2006). Crop losses to pests. Journal of Agricultural Science 144:31-43.

Oka, Y. (2010). Mechanisms of nematode suppression by organic soil amendments – A review. Applied Soil Ecology 44:101–115.

Palti, J. (1981). Cultural practices and infectious crop diseases. Springer, New York, 243 p.

Perrenoud, S. (1990). *Potussium and Plant Health*, 2nd ed. International Potash Institute, Bern, Switzerland (IPI Research Topics No. 3).

Phelan, P.L., Mason, J.F. and Stinner, B.R. (1995). Soil fertility management and host preference by European corn borer, *Ostrinia nubilalis*, on *Zea mays*: a comparison of organic and conventional chemical farming. Agric. Ecosyst. Environ. 56:1–8.

Perwez, M.S., Rehman, M.F. and Haider, S.R. (1988). Effect of *Tagetes erecta* on *Meloidogyne javanica* infecting lettuce. Int. Nematol. Network Newsl. 5(3):18-19.

Pesticide Action Network North America. (2009). La agroecología aporta un conjunto de soluciones para las crisis y presiones ambientales que enfrenta la agricultura en el siglo. In *Agroecología y Desarrollo Sostenible.* Conclusiones de la Evaluación Internacional de las Ciencias y Tecnologías Agrícolas para el Desarrollo, dirigida por la ONU.

Raj, B.T. and Nirula, K.K. (1969). Trials on the eradication of root-knot nematodes from potato fields with DD (Abstr.). *All India Nematology Symposium*, Indian Agri. Res. Inst., New Delhi, India.

Rajendran, G., Naganathan, T.G. and Sivagami, V. (1979). Studies on banana nematodes. Indian Journal of Nematology **9**:54.

Rämert, B. and Ekbom, B. (1996). Intercropping as a management strategy against carrot rust fly (Diptera: Psilidae): A test of enemies and resource concentration hypotheses. Environmental Entomology 25:1092-1100.

Rangaswamy, S.D., Parvatha Reddy, P. and Nanje Gowda, D.N. (1999). Management of root-knot nematode, *Meloidogyne incognita* in tomato by intercropping with marigold and mustard. Pest Mangmt. Hort. Ecosystems 5:118-121.

Ratnadass, A., Fernandes, P., Avelino, J. and Habib, R. (2012). Plant species diversity for sustainable management of crop pests and diseases in agro-ecosystems: a review, Agronomy for Sustainable Development 32:273–303.

Reuveni, R. and Reuveni, M. (1998). Foliar-fertilizer therapy - a concept in integrated pest management. Crop Protection 17(2):111-118.

Seigies, A.T. and Pritts, M. (2006). Cover crop rotations alter soil microbiology and reduce replant disorders in strawberry. HortSci. 41(5):1303–1308.

Sharma, S., Duveiller, E., Basnet, R., Karki, C.B and Sharma, R.C. (2005). Effect of potash fertilization on Helminthosporium leaf blight severity in wheat and associated increases in grain yield and kernel weight. Field Crop Res. 93:142–150.

Shelton, A.M. and Nault, B.A. (2004). Dead-end trap cropping: a technique to improve management of the diamondback moth, *Plutella xylostella* (Lepidoptera: Plutellidae). Crop Protection 23: 497–503.

Shivaprasad, B., Reddy, B.M.R. and Krishnappa, K. (2001). Effect of trap cropping in management of root-knot nematode in brinjal. *Natl. Cong. on Centenary of Nematol. in India – Appraisal & Future Plans*, Indian Agri. Res. Inst., New Delhi, p. 94-95.

Siddiqui, Z.A., Khan, A.M. and Saxena, S.K. (1973). Studies on Tylenchorhynchus brassicae II - Effect of temperature and moisture on multiplication of nematode. Indian Journal of Entomology 26: 139-147.

Sileshi, G., Kenis, M., Ogol, C.K.P.O. and Sithanantham, S. (2001). Predators of *Mesoplatys ochroptera* Stål in Sebania-planted fallows in eastern Zambia. BioControl 46:289–310.

Sileshi, G. and Mafongoya, P.L. (2006). Long-term effects of improved legume fallows on soil invertebrate macrofauna and maize yield in eastern Zambia. Agriculture, Ecosystems and Environment 115:69-78.

Sileshi, G., Schroth, G., Rao, M.R. and Girma, H. (2008). Weeds, diseases, insect pests and tritrophic interactions in tropical agroforestry. In D. R. Batish, R. K. Kohli, S. Jose, & H. P. Singh (Eds.), *Ecological Basis of Agroforestry* (p. 73-94). London: CRC Press.

Simpson, M., Gurr, G.M., Simmons, A.T., Wratten, S.D., James, D.G., Leeson, G., Nicol, H.I. and Orre, G.U.S. (2011). Field evaluation of the 'attract and reward' biological control approach in vineyards. Annals of Applied Biology 159:69-78.

Sivakumar, C.V. and Meerazainuddin, M. (1974). Influence of N, P and K on the reniform nematode, *Rotylenchulus reniformis* and its effect on the yield of okra. Indian J Nematol 4:243-244.

Smeda, R.J. and Weller, S.C. (1996). Potential of rye (*Secale cereale*) for weed management in transplant tomatoes (*Lycopersicum esculentum*). Weed Science 44:596-602.

Smith, M.W., Arnold, D.C., Eikenbary, R.D., Rice, N.R., Shiferaw, A. Cheary, B.S. and Carroll, B.L. (1996). Influence of groundcover on beneficial arthropods in pecan. Biological Control 6:164-176.

Smith, *E.G.*, Upadhyay, B.M., Blackshaw, R.E., Beckie, H.J., Harker, K.N. and Clayton, G.W. (2006). Economic benefits of integrated weed management systems in field crops in the dark brown and black soil zones of Western Canada. Canadian Journal of Plant Science 86:1271–1279.

Sreeja, P., Nageswari, S. and Mohandas, C. (1998). Reaction of cassava cultivars to root-knot nematode, *Meloidogyne incognita* (Kofoid and White, 1919) Chitwood, 1949. In *Nematology: Challenges and Opportunities in 21st Century* (Usha K. Mehta, ed.), p. 137-139. Sugarcane Breeding Institute, Coimbatore.

Srinivasan, K. and Krishna Moorthy, P.N. (1991). Indian mustard as a trap crop for management of major lepidopterous pests on cabbage. Trop. Pest Manag. 37:26–32.

Srinivasan, K., Krishna Moorthy, P.N. and Raviprasad, T.N. (1994). African marigold as a trap crop for the management of the fruit borer *Helicoverpa armigera* on tomato. Int J Pest Mangmt 40:56-63.

Stamps, W.T. and Linit, M.J. (1998). Plant diversity and arthropod communities: Implications for temperate agroforestry. Agrofor Syst 39:73–89.

Stinner, B.R. and House, G.J. (1990). Arthropods and other invertebrates in conservation-tillage agriculture. Annual Review of Entomology 35(1):299–318.

Stone, A. (2012). *Cover Crops for Disease Suppression.* eXtension Foundation webinar.

Sustainable Agriculture Network (SAN). (1998). *Managing Cover Crops Profitably*, 2nd ed. Sustainable Agriculture Network (SAN), Beltsville, MD.

Teasdale, J.R. and Daughtry, C.S.T. (1993). Weed suppression by live and desiccated hairy vetch. Weed Science 41:207-212.

Tjamos, E.C., Papavizas, G.C. and Cook, R.J. (1992). Biological control of plant diseases: progress and challenges for the future. NATO ASI Series, Plenum Press, New York, 462 p.

Unamma, R.P.A., Ene, L.S.O., Odurukwe, S.O. and Enyinnia, T. (1986). Integrated weed management for cassava intercropped with maize. Weed Research 26:9-17.

van Bruggen, A.H.C. and Grunwald, N.J. (1996). Tests for risk assessment of root infection by plant pathogens. In: Doran, J.W., Jones, A.J. (Eds.), Methods for Assessing Soil Quality. Soil Science Society of America, Madison, WI, p. 293–310.

Vandermeer, J. (1989). *The Ecology of Intercropping*. Cambridge University Press, Cambridge.

Vargas-Ayala, R., Rodriguez-Kabana, R., Morgan-Jones, G., McInroy, J.A. and Kloeppe,r J.W. 2000. Shifts in soil microflora induced by velvet bean (*Mucuna deeringiana*) in cropping systems to control root-knot nematodes. Biological Control 17:11-22.

Way, M.J. and Khoo, C.K. (1990). Colony dispersion and nesting habits of the ants, *Dolichoderus thoracicus* and *Oecophyllas maragdina* (Hymenoptera: Formicidae), in relation to their success as biological control agents on cocoa. Bulletin of Entomological Research 81: 341-350.

Wiggins, B.E. and Kinkel, L.L. (2005). Green manures and crop sequences influence potato diseases and pathogen inhibitory activity of indigenous *Streptomycetes*. Phytopathology 95:178-195.

Williams-Woodward, J.L., Pfleger, F.L., Fritz, V.A. and Allmaras, R.R. (1997). Green manures of oat, rape and sweet corn for reducing common root rot in pea (*Pisum sativum*) caused by *Aphanomyces euteiches*. Plant and Soil 188:43-48.

Workneh, F. and van Bruggen, A.H.C. (1994). Microbial density, composition and diversity in organically and conventionally managed rhizosphere soil in relation to suppression of corky root of tomatoes. Appl. Soil Ecol. 1:219–230.

Wrucke, M.A. and Arnold, W.E. (1985). Weed species distribution as influenced by tillage and herbicides. Weed Science 33:853–856.

Zalom, F.G., Phillips, P.A., Toscano, N.C. and Udayagiri, S. (2001). *UC Pest Management Guidelines: Strawberry: Lygus Bug*. University of California Department of Agriculture and Natural Resources, Berkeley, CA.

ooo

14

Underutilized Vegetables Towards Sustainable Horticulture

Poornima K.K.

1. INTRODUCTION

According to Food and Agriculture Organization of the United Nations (FAO) estimates, a total of 805 million people did not have access to a sufficient supply of dietary energy to live a healthy life, of which 526 million, nearly two-thirds, are living in the Asia-Pacific region. FAO projects that, globally, the value of agricultural production must increase 60 per cent by 2050 in order to meet the demands imposed by population growth, income increase and urbanization (Konuma and Dawe, 2016). This could be attributed to lot of pressure on cultivating major vegetables crops which require intensive care and lot of inputs to get better yields. In contrast to this, there are vegetables which are less known but consumed locally which are grown with less efforts and require lesser inputs as compared to the commercial vegetable crops. These are termed as "Underutilized Vegetables". They play an important role at the village level, as farmers grow them in smaller spaces requiring lesser inputs, lesser efforts and used as nutritional supplements. Underutilized vegetables are the vegetable crops which are not exploited by the scientific community, not promoted by the extension officers and thereby lack of knowledge among the consumers. Promoting these crops would promote crop biodiversity, environment protection and ease the pressure on the environment and on the farmers in growing commercial vegetable. This chapter focuses on how underutilized vegetables could be promoted to bring about sustainable horticulture production and thereby reducing the pressure on intensive cultivation by providing consumers with larger array of vegetables crops. Most importantly these species are very well adapted to the local environment. These crop species contribute for food security, nutrition supplement, additional income, ecosystem stability and cultural stability.

2. UNDERUTILIZED VEGETABLES

Neglected or underutilized vegetables are those to which little attention is paid or which are entirely ignored by agricultural researchers, plant breeders and policymakers. Typically, underutilized vegetables are not traded as commodities. They are wild or semi-domesticated varieties and non-timber forest species adapted to particular, often quite local environments. Many of these varieties and species, along with a wealth of traditional knowledge about their cultivation and use, are being lost at an alarming rate (Padulosi, Thompson and Rudebjer, 2013). There are 17 mega-biodiversity holding countries falling under the following five major geographical regions:

- **South America:** Brazil, Peru, Colombia, Ecuador, Venezuela
- **Central America:** Mexico
- **North America:** USA
- **Africa:** South Africa, Madagascar, Democratic Republic of Congo.
- **Asia and the Pacific:** Indonesia, China, India, Australia, Papua New Guinea, Malaysia, Philippines.

Of the 17 mega-biodiversity holding countries, seven countries fall in Asia and the Pacific region. This indicates that major species diversity is found in this region and these countries are ecologically and ethnically more diverse. The natives here have been and are still dependent on indigenous wild, semi-domesticated and domesticated of edible less-known, underutilized plants and other economic plants for their daily needs of food, feed, shelter and health care. It is found that the tribes in India and Thailand used about 59% and 85% of the dietary energy respectively from the traditional food supplies from local food production. This rich array of food is available in two countries in Asia alone. It is estimated that humans directly modify and use more than 40% of the earth's terrestrial ecosystems, harnessing their productivity for human benefit and modifying their composition, so as to meet their needs. Thus, the traditional agro-ecosystems have been sustained by native farming societies to their benefit under subsistence farming with a wide variety of major and minor food species. Understandably, the role of underutilized species has evolved over time and as it is today, it adds to the quality of life besides meeting needs of the rural poor in particular (Arora, 2014). It is estimated that about 80,000 plant species have been used by humans either as food or for other needs like fibre, medicinal use and industrial uses (Kermali *et al.,* 1997). As against this vast collection of plant species only 30 species of crops are known to feed the world (Harlan, 1975). According to Prescott-Allen and Prescott (1990), 90 per cent of the food plant supplies are provided by only 103 plant species. However, limits this estimate to about 80 crop plants. This provides an ample opportunity to explore/utilize crop resources to meet our food demand.

3. SUPPLEMENTARY NUTRIENT RESOURCE

The major emphasis was on increasing the production of cereals, namely, wheat and rice in most of the Asian countries. This was because majority of the population depend on these cereals for their nutritional needs. The pulses, vegetables and fruits which add-up to the quality of diet along with these cereals were neglected or eaten in small quantities. These pulses, vegetables and fruits are considered as secondary nutrient providers. While taking the example of cow pea, which is cultivated in Sub-Saharan Africa, it is the major source of protein and a staple food for the people. Whereas, the same crop is considered as underutilized crop in the Mediterranean countries. Few crop species which are staple in certain part of the world may be underutilized in other countries vowing to various reasons.

In many villages in India, the concept of kitchen gardens exists, wherein the farmers grow vegetables, flowers and fruits for their daily needs. Some of the examples are, moringa, gourds, chillies, marigold, green leafy vegetables and others. Moringa or drumstick leaves are a rich source of many micronutrients like beta carotene, folic acid, calcium, iron and vitamin C and also provide good quality protein. They are often used to mutually supplement and enhance the quality of cereal based diets (Tontisirin, 2014). Further pumpkin leaves are also a good source of vitamins and minerals. The leaves are rich in vitamin A, B6, C, E, folic acid and iron. Eating pumpkin leaves enhances immunity, helps during lactation and keeps skin and bones healthy. Leaves are eaten in Africa, India and other Asian countries. But this is not commercialised. The leaves are locally consumed.

Wild plant species are also used as vegetables in Mangalore and Gulbarga districts in Karnataka. These plant species are high in nutrition with medicinal values. Few plant species used in Gulbarga are *Basella alba* Linn., (juice of leaf is used for constipation, particularly in children and pregnant women and in urinary disorders. It is a good source of Vitamin C & B, cures mouth ulcers), *Lactuca scariola* Linn., (it is a weed and leaves are used as vegetables and is a good source of iron and calcium), *Momordica cymbalaria* Fenzl, *Amaranthus spinosus* Linn., *Amaranthus viridis* Linn., *Celosia argentea* Linn., *Digera muricata* (L.,), *Portulaca oleracea* Linn., *Portulaca quadrifida* L., *Sesbania grandiflora* (L.), *Coccinia indica and Cocculus hirsutus* (L.) *(*Rajasab and Isaq, 2004). The wild plant species used as vegetable in Mangalore are; *Madhuca neriifolia, Oxalis corniculata* L., *Sauropus androgynus* (L.) Merr., *Syzygium caryophyllum* (L.), *Basella alba* L., *Talinum triangulare, Vigna unguiculata* and *Cucurbita moschata.* These plants are used by Mangalore people to prepare curries, *chutneys*, *patrode* and also for pickling purpose. Most of them have medicinal values (Pavithra, Sridhar and Keshavachandra, 2016).

4. ADDITIONAL FIELD GROWING CAPACITY/ADAPTABILITY

Farmers maintained kitchen gardens to grow crops which they were not able to grow on the main field due to many reasons. The other purpose of kitchen garden was to meet the daily vegetable needs of the family. Every household has a piece of land devoted for kitchen garden near to their home. Kitchen garden area is also used to store their dry straw of finger millet or paddy or maize. These gardens are watered with the waste water from the house, thus conserving water. It is the integral part of households in the rural areas. Kitchen gardens play an important role in providing nutrition to the farmers. Maintaining kitchen garden is not-so labour intensive nor does it require lot of care as compared to the crops cultivated in the field. The kitchen garden is always less than an acre area. It is very handy for the farmers, as with less input they could grow plants which provide them with the common vegetables needed for the daily use. Crops grown in the kitchen gardens are adapted to the environment and very hardy. They have the ability to withstand adverse conditions. The farmers save the seeds and conserve them for sowing. The daily vegetable needs are met, which secures the nutrition of the family. Using these underutilized vegetables, the farmers prepare unique cuisine which is very local.

5. CROP DIVERSIFICATION

In olden days, farmers used to follow mixed cropping system wherein along with the main crop they used to grow pulses, oil seed crops and vegetables. Gradually, more emphasis was given to produce more and more cereal crops, such as, paddy, wheat, maize etc. Farmers started to grow them in their field neglecting the other crops. Millets are one such example wherein the area under these crops decreased giving way to cereals and vegetables. There are traditional food systems that maximize use of diversity. They could help to alleviate hidden hunger issues and provide nutrients, texture and flavour often missing from monotonous energy-dense diets characteristic of transition diets. Many of these traditional food systems integrate the use of local crops, animal source foods and other components that are now considered underutilized. These include small-scale community-based systems like home gardens and smallholder farms that are still practised by many smallholder farmers in the Asia-Pacific region (Talukder, 2010).

Notable in these local and traditional food systems is the key role of women. In addition to being heavily involved in producing the food, women manage the knowledge on seed types and uses, food preparation and storage and crop management. Women also take charge of marketing the produce and decide what their children consume (Sebastian *et.al.*, 2014). The importance of harnessing agro-biodiversity in our food systems and enhancing traditional food systems that use it is amplified by the fact that only 30 crops provide 90 percent of the world's food energy intake and four – rice, wheat, maize and potatoes – supply 50 percent of the world's food energy needs. This is despite the fact that of the estimated 3,00,000 plant species worldwide, 10,000 have been used for human food since agriculture started approximately 7,000 years ago (FAO, 2010). The Asia-Pacific region is the centre of diversity for many important but underutilized crops, such as, small millets (finger millet, kodo millet, barnyard millet, foxtail millet and little millet), minor but locally important legumes (black gram, rice bean, lablab bean, horse gram), cultivated minor and wild tropical fruits and indigenous vegetables. In addition, the diversity in foods, flavourings and other components generated by aquaculture and other production systems is being studied and promoted (FAO, 2010).

6. CONTRIBUTION TO NUTRITION AND FOOD SECURITY

Worldwide, 1,500 to 2,000 plant species have been used as vegetables; for Southeast Asia, close to 1,000 species with uses as vegetables are known. Of the 1,000 species, about 120 species are cultivated commercially or used for home consumption. The term 'indigenous vegetables' primarily refers to plants grown in their centers of origin or diversity, but also encompasses plant species introduced from other geographical areas that have adapted well to the new environment (Lin *et al.*, 2009).

Major cereals, vegetables and fruits are bred for higher yields, pest and disease resistance, longer shelf life, size and shape rather than for their nutritive value. Asian countries have attained self sufficiency in production of cereal crops. But the production and supply of legumes is low or meagre. Major legumes cultivated in this region are soybean, groundnut, chickpea, lentil, common bean, field pea and pigeon pea. The unexploited legumes are broad bean, horse gram, lablab bean, sweet pea (*Lathyrus* spp.), moth bean, rice bean and winged bean. These crops provide the protein needed and their production and supply should be given more emphasis. Many of these indigenous food legumes play a vital role in protein nutrition to poor farm families, especially to women and children. The reason why these crops are not exploited is because of the preference of the consumers. Majority of the people depend on livestock supply to meet their protein needs. But to secure the nutrition and food security of the people living in remote areas, it becomes evident to commercialize these locally cultivated crops, which provide vitamins, minerals and proteins. They would also provide the rural people the economic stability.

For example, the indigenous crops have higher nutritive value than the hybrids or genetically developed crops. Indian gooseberry is one of the richest sources of vitamin C. Moringa leaves are a rich source of beta-carotene, folic acid, calcium, iron and vitamin C and they also provide good quality protein. Horsegram is another crop which belongs to pulse group, provides protein, iron, calcium and polyphenols. The other group of crops are millets namely, foxtail millet, kodo millet, little millet, barnyard millet, proso millet are forgotten crops and finger millet is common among rural people, but their nutritional value is much higher than the major crops which are cultivated extensively. For instance compared to rice, finger millet grain is eight times richer in calcium, four times richer in minerals and two times richer in phosphorus-protein components and has a well-balanced amino-acid profile. It is a good source of methionine, cystine and lysine and is also rich in important vitamins, such as, thiamin, riboflavine, folin and niacin (MSSRF, 2002).

Among vegetables, amaranthus, okra, cowpea, coccinia are important in the South Asian countries. The reason why these are not well-known or not preferred is due to lack of information on the use of the vegetables and also due to consumer preference and lower production. These minor vegetables are no doubt are rich in nutrients but their utility is only limited to the local consumption. Locally these are produced in small areas.

7. ECONOMICS OF UNDERUTILIZED VEGETABLES

Vegetables, in general, but also many traditional vegetables, such as, amaranth (*Amaranthus* spp.), jute mallow (*Corchorus olitorius*), African nightshade (*Solanum scabrum*), Asian eggplant (*Solanum melongena*) and African eggplant (*Solanum aethiopicum*), drumstick (*Moringa oleifera*), bitter gourd (*Momordica charantia*), water spinach (*Ipomoea aquatica*), Chinese kale (*Brassica oleracea* var. *alboglabra*), edible rape (*Brassica napus*), roselle (*Hibiscus sabdariffa*), Malabar spinach (*Basella alba)*, slippery cabbage (*Abelmoschus manihot*), winged bean (*Psophocarpus tetragonolobus*) and many gourd species are of considerable commercial value and thus can make a significant contribution to household income. A number of examples of profitable cultivation of traditional vegetables in East and West Africa are, worowo (*Solanacio biafrae*), cockscomb (*Celosia argentea*), African eggplant (*Solanum macrocarpon*) and amaranth (Hughes and Ebert, 2013).

Value addition by applying appropriate production and postharvest techniques ensures that high quality produce reaches the market and satisfies consumer expectations. Consumer studies with regard to the purchase and consumption of kale (*B. oleracea*) in Nairobi, Kenya revealed that urban kale consumers care most for nutritional, sensory and safety attributes of the produce (Ngigi *et al.*, 2011). In Eastern Africa and Southeast Asia selected traditional vegetables are becoming an increasingly attractive food group for the wealthier segments of the population and are slowly moving out of the underutilized category into the commercial mainstream (Weinberger, 2007). Attracted by the strong market demand, seed companies are beginning to explore and develop these popular crops, thus strengthening the formal seed sector (AVRDC, 2008).

Few underutilized vegetables could be bred for their nutritional value and commercially developed to meet the nutrition needs of the consumers. Amaranth is one such leafy vegetable which can be cultivated with lower inputs and its yield ranges from 32-40 t/ha. The nutritive value of this vegetable is almost same as spinach and more than that of cabbage (Elbert *et al.,* 2011).

Moringa or drumstick is another vegetable which has the potential to be commercialized. The Moringaceae family comprises 13 species that fit into three broad life forms with distinct geographic origins (Olson, 2013). Among them *Moringa oleifera* is the predominantly cultivated species. It is widely grown in the tropics of Asia, Latin America, the Caribbean Islands and sub-Saharan Africa and can also be found in Florida and the Pacific Islands. It is a perennial softwood tree native to the sub-Himalayan ranges of India, Pakistan, Bangladesh and Afghanistan (Fahey, 2005). Moringa has already reached the status of an economically important crop in India, the Philippines, Ethiopia and Sudan (Fahey, 2005). Moringa is used in industries also. The seed contains about 40% oil, known as ben oil. The oil is non-drying, resists rancidity and is used for cooking, lubrication and in the cosmetic industry (Duke, 2013). The leftover pressed cake or ground moringa seeds are used to purify drinking water and to flocculate contaminants (Fahey, 2005 and Bosch, 2004). The wood can be used for dying (blue color) (Duke, 2013). The coarse fiber of the trunk is suitable for making mats, cordage and paper (Polprasid, 1996).

Of the 75 traditional plant-derived oils tested in India for biofuel production, the oil derived from *M. oleifera* showed good potential (Azam *et.al.,* 2005). The aptness of moringa oil for biofuel production was confirmed in a specific study using *M. oleifera* seeds from Pakistan (Rashid, 2008). Biodiesel obtained from moringa oil is characterized by a high cetane number (CN) of 67, one of the highest among biodiesel fuels. Moringa is a fast-growing tree that adapts well to hot, semi-arid regions with as little as 500 mm annual rainfall (Bosch, 2004). It also tolerates occasional wet or waterlogged conditions for a short period of time (Polprasid, 1996), but prolonged flooding leads to a significant loss of plants (Patricio, 2013). Greater use of moringa has good potential in the fight against hunger and malnutrition in the developing world by improving nutrition and health of the rural and urban poor, increasing incomes of smallholder farmers and enhancing environmental services by controlling soil and wind erosion and providing shade and clean water (Ebert, 2014).

8. FUTURE PERSPECTIVE

The underutilized vegetables are vegetables grown in limited areas and are locally adapted. These crops are hardy species which could grow under the adverse conditions. These vegetables also are found to have high nutrition content which could be best exploited to mitigate the nutritional insecurity of the people. Promoting these vegetables could mean, to produce additional food with varied nutrition content, which is required for the well-being of the humans. Scientists could exploit these varieties for their disease, pest and drought resistance. As stated above, different parts of moringa plants could be exploited for industrial uses. These vegetables not only provide nutrition but also produce different products which are useful in textile, processing, medicine industries. To commercialise these crops, awareness about these crops among the city dwellers particularly, children should be effectively made. Making them to know about these species, we are inculcating the knowledge of plant species, which are available and used as food. Consumers should be made aware of these food species with their nutritional benefits. We all know that woman play an important role regarding food intake of most families whether they are in cities or in rural areas. Woman should be made to know about these species, so as to make them popular for sustainable production.

Further;

- It calls for greater investment in long-term research and breeding programs and improved seed supply sources for these crops to ensure they can be competitive in the marketplace.
- Research and breeding of underutilized fruit and vegetable crops are not sufficiently funded compared with the few main staple crops. Funding should be made to attract more private sector breeders.
- Promote intellectual property rights through better documentation and validation.
- Developing and disseminating sustainable management plans for underutilized foods.
- Documentation of traditional knowledge and practices.
- To involve policy-makers in promoting and popularizing indigenous foods and integrate food biodiversity in government policies and programmes.
- To frame code of conduct (legal framework) on wild indigenous foods.
- To include underutilized foods in GAPs, CODEX and other standards.
- Forest food collection should be made legal among local community; at the same time promote sustainable conservation use: domestication of some wild plants (introducing to farming systems).
- To encourage those who are maintaining plants *in situ* (monetary, ownership, recognition).
- Promote processing and post-harvest technology.
- Including syllabus on underutilized vegetables to make them understand about our food diversity in the school/ colleges/universities.
- To develop seed banks containing germplasm of underutilized vegetables; (Durst and Bayasgalanbat, 2014).

9. CONCLUSION

Underutilized vegetables are rich in nutrition, but cultivated in small quantities in kitchen gardens. These could be commercially exploited for their unique characteristics like pest- and disease- resistance and drought resistance. This valued germplasm could be commercially exploited for adding more nutritive value to the human diet. This could bring about well-being of the humans. The micro nutrients with the additional nutrition provided by these crops could enhance the immune system. In addition to exploiting these crops for their nutritional value, scientists could also explore on the other products which could be extracted from different parts of the plants. More research can be focussed on value addition of these products as per consumer preference.

REFERENCES

Arora, R. K. (2014). "Diversity in Underutilized Plant Species – An Asia-Pacific Perspective", https://www.bioversityinternational.org/fileadmin/user_upload/online_library/publications/pdfs/Diversity_in_Underutilized_Plant_Species_An_Asia-Pacific_Prespective_1938.pdf

AVRDC. (2008). Point of Impact: Healthy Urban Fast Food: A New Maasai Enterprise; AVRDC - The World Vegetable Center: Tainan, Taiwan, 2008.

Azam, M.M.; Waris, A.; Nahar, N.M. 2005. Properties and potential of fatty acid methyl esters of some non-traditional seed oils for use as biodiesel in India. *Biomass Bioenerg*. **29**: 293–302.

Bosch, C.H. (2004). *Moringa Oleifera* Lam. *In*: Plant Resources of Tropical Africa, Volume 2: Vegetables; Grubben, G.J.H., Denton, O.A., Eds.; Backhuys Publishers: Kerkwerve, The Netherlands, 2004; p. 392–395.

Duke, J.A. (2013). *Moringa Oleifera* Lam. Available online: http://www.hort.purdue.edu/ newcrop/ duke_energy/Moringa_oleifera.html (accessed on 21 November 2013).

Durst, Patrick and Bayasgalanbat, Nomindelger. (2014). "Promotion of Underutilized Indigenous Food Resources for Food Security and Nutrition in Asia and the Pacific", RAP Publication 2014/07. FAO Regional office for Asia & Pacific, Bangkok. p.3.

Ebert, A.W., Wu, T.-H., Wang, S.T. (2011). International Cooperators' Guide—Vegetable Amaranth (*Amaranthus* spp. L.); AVRDC - The World Vegetable Center: Tainan, Taiwan, 2011; p. 8.

Ebert andreas W. (2014). "Potential of Underutilized Traditional Vegetables and Legume Crops to Contribute to Food and Nutritional Security, Income and More Sustainable Production Systems", Sustainability, 6, 319-335; doi:10.3390/su6010319. Source: www.mdpi.com/journal/sustainability

Fahey, J.W. (2005). *Moringa oleifera*: A review of the medical evidence for its nutritional, therapeutic and prophylactic properties. Part 1. *Phytochemistry*, **47**: 123–157.

FAO. (2010). The Commission on Genetic Resources for Food and Agriculture (CGRFA), second report on the state of the world's plant genetic resources for food and agriculture. Food and Agriculture Organization of the United Nations (FAO). Rome, Italy.

Harlan, J.R. (1975). Crops and Man. Amer. Soc of Agronomy and Crop. Sci. Soc. of America, Madison Wisconsin, USA, p.292.

Hughes, J.d'A., Ebert, A.W. (2013). Research and Development of Underutilized Plant Species: The Role of Vegetables in Assuring Food and Nutritional Security. In Proceedings of the 2nd International Symposium on Underutilized Plant Species: Crops for the Future—Beyond Food Security; Massawe, F., Mayes, S., Alderson, P., Eds.; International Society for Horticultural Sciences (ISHS): Korbeek-Lo, Belgium, 2013; Volume 2, p. 79–91.

Kermali, S.R., Anishetty N.M. and Cooper H.D. (1997). Promoting development and commercialization of underutilized crops and species in the FAO Global Plan of Action. In: Domestication, Production and Utilization of New Crops (eds. J. Smartt and N. Haq), ICUC, Southampton, UK, p. 19-27.

Konuma, Hiroyuki and Dawe, David. (2016). "Attaining Food and Nutrition Security for the Masses in Asia through 2015: Food Expectations of the People in the New Millennium": Vol.V, Dr. P.N. Agricultural Science Foundation, Bengaluru and Westville Publishing House, New Delhi (p. 10).

Lin, L.J., Hsiao, Y.Y. and Kuo, C.G. (2009). Discovering indigenous treasures: Promising indigenous vegetables from around the world. AVRDC – The World Vegetable Center Publication No. 09-720. AVRDC – The World Vegetable Center, Shanhua, Taiwan. p. 317.

MSSRF. (2002). Enlarging the contributions of nutritious neglected crops to food security and to income of the rural poor. Asia Component. Nutritious Millets (IFAD, IPGRI, MSSRF). Summary Proceedings of the Inception Workshop, MSSRF, 24-25 January, 2002.

Ngigi, M.W.; Okello, J.J.; Lagerkvist, C.L.; Karanja, N.K.; Mburu, J. (2011). Urban consumer's willingness to pay for quality of leafy vegetables along the value chain: The case of Nairobi Kale consumers, Kenya. *Int. J. Bus. Soc. Sci.*, **2**:209–216.

Olson, M. (2013). The Home Page of the Plant Family Moringaceae. Available online: http://www.mobot.org/gradstudents/olson/moringahome.html (accessed on 21 November 2013).

Padulosi, Stefano; Thompson, Judith and Rudebjer. (2013). Neglected and Underutilized Species,http://www.bioversityinternational.org/uploads/tx_news/Fighting_poverty__hunger_and_malnutrition_with_neglected_and_underutilized_species__NUS__1671_03.pdf.

Patricio, H.G., Palada, M.C., Ebert, A.W. (2012). Adaptability and Horticultural Characterization of Moringa Accessions under Central Philippines Conditions. In High Value Vegetables in Southeast Asia: Production, Supply and Demand; Proceedings of the SEAVEG 2012 Regional Symposium; Holmer, R., Linwattana, G., Nath, P., Keatinge, J.D.H., Eds.; AVRDC - The World Vegetable Center: Shanhua, Tainan, Taiwan, 2013; p. 61–70.

Pavithra, M., Sridhar, K.R. and Keshavachandra, K., (2016). Traditional nutritional attributes of native flora of Southwest coast of India, *Journal of Agricultural Technology.* **12**(4):777-796. Online http://www.ijat-aatsea.com.

Polprasid, P. (1996). *Moringa oleifera* Lamk. In: PROSEA—Plant Resources of South-East Asia No. 8: Vegetables; Siemonsma, J.S., Piluek, K., Eds.; Prosea Foundation: Bogor, Indonesia, 1996; p. 213–215.

Prescott-Allen and A. Prescott. (1990). How many plants feed the world. Conservation Biology **4:** 365-374.

Rajasab, A. H. and Isaq, Mahamad. (2004). Documentation of folk knowledge on edible wild plants of North Karnataka, Department of P G Studies and Research in Botany, Gulbarga University, Gulbarga 585 106, *Indian Journal of Traditional Knowledge* (4): 419-429.

Rashid, U., Anwar, F., Moser, B.R., Knothe, G. (2008). *Moringa oleifera* oil: A possible source of biodiesel. *Bioresour. Technol.* **99**: 8175–8179.

Sebastian, L. S., Chandrabalan, D., Mattei, F., Cogill, B. and Frison, E. (2014). Promoting Traditional Food Systems for better Nutrition and the Biodiversity International Nutrition Strategy, Promotion of Underutilized Indigenous Food Resources for

Food Security and Nutrition in Asia and the Pacific, RAP Publication, FAO, Nations Region Office for Asia and the Pacific, Bangkok, 2014.

Talukder, A., Haselow, N.J., Osei, A.K., Villate, E., Reario, D., Kroeun, H., SokHoing, L., Uddin, A., Dhunge, S. & Quinn, V. (2010). Homestead food production model contributes to improved household food security and nutrition status of young children and women in poor populations. FACTS [online], Special Issue 1. Available at, http://factsreports.revues.org/index404.html, accessed on 17 April 2012.

Tontisirin, T. (2014). Promotion of Underutilized Indigenous Food Resources for Food Security and Nutrition in Asia and the Pacific, RAP Publication 2014/07. p. 21, 22.

Weinberger, K. (2007). Are indigenous vegetables underutilized crops? Some evidence from Eastern Africa and Southeast Asia. *Acta Hortic.*, **752**: 29–34.

❍❍❍

15

Strengthening Integrated Pest Management (IPM) for Sustainable Organic Horticulture Production

P.N. Krishna Moorthy

SYNOPSIS

In recent times organic farming is gaining lot of importance worldwide. In India, the National Programme for Organic Production (NPOP) was framed only in the year 2000. Since then, many central government and state government departments and agencies are promoting organic farming in a big way by giving subsidies. Farmers are also expecting premium prices for organic produce. Many malls/shops have come up to sell organic products at a premium price. These factors have prompted many farmers to switch to organic farming. Many certifying agencies have also been established for organic crop production (for inputs and products) and for exports. Many farmers soon realized that organic farming is not easy. Often, the yield obtained by organic farming is low. This is mainly due to insect pests and diseases. Many times weather factors contribute for pest and disease problems. These have to be managed effectively by organic methods only by following organically permissible Integrated Pest Management (IPM) tools. IPM gives maximum benefit to the farmers. Hence, there is a need to strengthen the IPM under organic farming. In this paper an attempt is made to bring all the organically permitted IPM strategies for insect pest and disease management, which can be used under organic farming of horticultural crops. Some of the methods that can be utilized in organic farming are cultural practices (use of tolerant/resistant varieties, seed treatment, uprooting infested plants, destruction of affected plant parts, increasing spacing between plants), use of pheromone traps, light traps *etc*. In addition, use of bicarbonates, sulphur, copper products, kaolin clay, botanicals like neem and pongamia, biological control methods like releasing of parasitoids like *Trichogramma,* use of antagonists like *Trichoderma,* biopesticides like *Bacillus thuringiensis*, nuclear polyhedrosis virus (NPV), *etc*., have to be integrated wherever possible to strengthen the IPM programmes for sustainable organic horticulture. Future lines of work have also been suggested.

1. INTRODUCTION

Organic farming is a system which largely excludes the use of synthetic inputs like fertilizers; pesticides *etc.* and rely heavily on farm yard manure, green manure, crop rotation, composting, biological methods of nutrient mobilization and plant protection. In 2005 International Federation of Organic Agriculture Movement (IFOAM) defined organic agriculture as a production system that sustains the health of soils, ecosystems and people. It relies on ecological processes, biodiversity and cycles adapted to local conditions, rather than the use of inputs with adverse effects. Organic agriculture combines tradition, innovation and science, to benefit the shared environment and promote fair relationships and a good quality of life for all involved (IFOAM, 2016).

Indian National policy on organic farming called as National Programme for Organic Production (NPOP) was framed in 2000 by setting up a task force for organic farming at the national level and by setting up a National Centre for Organic Farming at Ghaziabad, Uttar Pradesh. Many government agencies like Indian Council of Agricultural Research, State Agricultural Universities, Agriculture and Processed food Products Export Development Authority (APEDA), Spices Board, Tea Board, Coconut Development Board, Cocoa and Cashewnut Board have also been giving more emphasis for organic farming. The Coffee Board has also published a book on 'Organic Coffee Production' (Raghuramulu and Kamala Bai, 2006). In addition, all state governments have been promoting organic farming since then. However, it is only the Government of Sikkim which stopped importing chemical fertilizers into the state in 2003 and in December 2015 Sikkim was declared as the first and only organic state in India.

As per APEDA (2016), India produced around 1.35 million MT of certified organic products including all varieties of food products namely sugarcane, oil seeds, cereals and millets, cotton, pulses, medicinal plants, tea, fruits, spices, dry fruits, vegetables, coffee *etc*., in 2015-16. Madhya Pradesh had the highest area under organic certification. When exports are considered, oil seeds (50%) and processed foods (25%), account for major share as compared to other products like cereals and millets (17%), tea (2%), pulses (2%), spices (1%) and dry fruits (1%). This indicates that export of organically produced fresh fruits and vegetables is very low and there is a great potential to increase the exports of organically produced fruits and vegetables in this country.

Farmers are also expecting premium prices for organic produce. Many malls/shops have come up to sell organic products at a premium price. These factors have prompted many farmers to switch to organic farming. Many certifying agencies have been established for organic crop production (for inputs and products) and for exports. Many farmers soon realized that organic farming is not easy, but difficult. More often due to poor crop growth, insect pests and diseases, many farmers' have reverted to the use of synthetic fertilizers and pesticides, This is mainly due to difficulty in controlling insect pests and diseases. Many times weather factors contribute for pest and disease problems. These have to be managed effectively by organic methods only.

Most of the organic farmers also feel that there is a need to have organically permissible pest management technologies (also called as Integrated Pest Management). The importance of Integrated Pest Management (IPM) or shifting away from synthetic pesticides was realized way back in 1960s with the publication of "Silent Spring" by Carson (1962). This book showed the dark side of synthetic pesticides to the world. The use of synthetic chemicals not only proved hazardous to environment, health and also became ineffective in due course of time as the insect pests and pathogens developed resistance. They also caused pest outbreaks, pest resurgence *etc*. Overall, they proved to be harmful to the ecosystem. This fact was initially observed in cotton and was later in almost all crops with the emergence of resistant pests, pathogens and emergence of new pests. In view of these problems associated with chemical method of pest control, scientists have developed many alternative technologies that can be used under organic farming.

When we look at the pest control practices given in text books of insect and disease management, use of synthetic chemicals is one of the suggested methods and there are many other alternate methods. Chemical method is one of the options or the last resort (Ayyar, 1963; Hill, 1975; Metcalf and Luckman, 1994; Ciancio and Mukerji, 2007).

When we talk of IPM in organic farming, it will exclude the use of synthetic pesticides. In scientific literature review on the management of insect pests and diseases of any crop, the major emphasis has been given for the development of alternative strategies of pest control, particularly the use of botanicals, biocontrol agents (parasites, predators and pathogens), pheromone traps and many cultural practices. However, legislative measures like banning the use of synthetics (as seen from the example of Sikkim state) is the only way to make an area or state completely organic. It was possible in Sikkim because farmers were not relying much on fertilizers and pesticides and the productivity was very low.

At present we do not have any foolproof protocols for IPM to completely control many pest problems by non-chemical means. Lots of information on organic methods of pest control is available in the internet. Recent reviews on organic pest management include that of Krishna Moorthy *et al.,* (2004), Krishna Moorthy (2005), Reddy (2008) and Caldwell *et al.,* (2013).

In this paper an attempt is made to give an account of IPM strategies that can be used to strengthen pest and disease management in organic horticulture. IPM requires understanding the crop, pest problems and all possible management tactics. We have many organic IPM strategies which have been validated in farmers' fields under conventional farming. Further, many of the non-chemical methods described for insect pest and disease management in text books can be adopted easily. Some of the methods that have been successfully utilized in farmers' field include uprooting infested plants, removal and destruction of affected plant parts (basal insect/disease infested leaves, infested fruits or twigs), following wider spacing between plants, use of pheromone traps, light traps, use of botanicals like neem and pongamia, biological control methods like releasing of parasitoids like *Trichogramma,* sprays of bio-pesticides like *Bacillus thuringiensis, Beauveria bassiana*, nuclear polyhedrosis virus (NPV) formulations. In addition, use of bicarbonates, sulphur, copper products, kaolin clay, can be used to manage diseases. All these tools have to be integrated wherever possible to strengthen the IPM programmes for sustainable organic horticulture. The author has personally demonstrated the IPM practices in cabbage, cauliflower and tomato in more than 200 farmers' fields under conventional farming systems which were found to be highly effective and economical.

The methods of IPM can be classified as cultural, mechanical or physical, elemental, biological and botanical control. In this paper, an attempt is made to give an overview of effective strategies for managing insect pests and diseases of horticulture crops.

2. CULTURAL METHODS

Healthy soil is most essential to get healthy crop and good yield in organic farming. Poor plant growth predisposes plant to diseases and pest problems. Good quality farm yard manure (FYM), green manure, vermi-compost *etc*., have to be produced within the farm itself to reduce the cost of production. These help in the sustainability of organic farming. Cultural methods have to be followed to get good crop and also for pest control. These are the first and foremost in organic method of pest control. Commonly used cultural operations can be very effectively used to manage many pests. Many of these methods are general and some are very specific for a pest. These generally target the weak points and habits of the pests. These are also the cheapest of all control methods. However, basic principles of good agronomic practices have to be followed. For example, stagnation of water is harmful for most of the crops. Hence, crop should be grown only under well drained soils. Crop rotation has to be compulsorily followed with a legume crop like cowpea or green manure crops like *Dhaincha* (*Sesbania aculeata*) or sun hemp (*Crotalaria juncea*) Crop rotation improves soil health and reduces soil borne diseases and nematodes. Application of FYM also has to be done to get proper growth of plants. Though summer ploughing and hoeing help to expose soil inhabiting insects to sunshine and also to the predatory birds, it is always better have some crop cover in the soil as exposure of the soil to sunshine reduces its organic content.

Some cultural practices that can be successfully used in pest management for strengthening organic farming are given below.

2.1 Adjusting Sowing and Planting Time

The IPM starts right from selecting the proper crop for any season. Many crops come up well during rainy season when the weather is cool. In summer, draught tolerant varieties or hybrids have to be planted. French beans come up very well during June-January when the temperature is not high. However, it is better not to sow them in June-July as stem fly will be very serious if there is a continuous dry spell. When the weather cools down, stem fly incidence also reduces. In summer, tomato leaf curl virus is very serious and fruit set is also very low due to high temperature. Hence, it will be better to avoid planting tomato in summer or it can be planted in such a way that fruiting should start before the temperature reaches >35^{0}C. Diamondback moth is a serious pest of cabbage and cauliflower during summer months. Hence, better to avoid these during summer months. However, growing mustard as trap crop and spraying of neem seed kernel extract or neem seed powder extract gives very good control of this pest even during summer. Weekly Spraying of neem soap/ pongamia soap along with mustard as trap crop gives moderate control of this pest. Hence, if neem seeds or neem soap/pongamia soap are not available, it is better to avoid taking up cabbage/cauliflower during summer). Fruit fly (melon fly) (*Bactrocera* spp.) infestation is very high during rainy season. Hence, growing cucurbits like bitter gourd during rainy season should be avoided as far as possible.

2.2 Selection of a Suitable Variety

Locally available traditional varieties are generally more tolerant to pests and diseases. However, the yield may be low compared to commercial varieties which perform well under high fertilizer and pesticide umbrella. Grow pest resistant varieties or hybrids wherever available as the risk of crop failure due to these problems get reduced drastically. In recent times, more emphasis is given by research workers and seed companies to develop pest resistant varieties/hybrids coupled with high yields as compared to only high yield (Singh and Malhotra, 2013).

Recently some heat tolerant varieties and photo insensitive varieties have been released in some vegetables. Hybrids are generally more tolerant to heat than local varieties. Such varieties/hybrids can be grown in summer or in off season also. For example, cabbage and tomato hybrids can be grown even in summer which can perform better than local or old varieties. Photo insensitive dolichos (hyacinth bean) or pea varieties can also be grown throughout the year.

In okra, yellow vein mosaic disease is a serious disease. It is transmitted by a whitefly vector *Bemicia tabaci*. Many tolerant/resistant varieties and hybrids are commercially available. Similarly, for tomato leaf curl disease, another virus disease transmitted by the same whitefly, many resistant varieties and hybrids are available. Bacterial wilt is another major pest problem in tomato and brinjal. Many tolerant varieties/hybrids are commercially available and these should be chosen for organic farming. A list of varieties tolerant to pests of horticultural crops is given by Reddy (2008). Such, specifically bred tolerant varieties or hybrids can also be planted as they give better yield if the soil fertility is high. In tomato and many other vegetables, multiple disease tolerant varieties/hybrids are available. Growing these will help in reducing the pest incidence. An organic farmer has to decide on the crop and varieties based on all the above factors.

However, insect tolerant varieties or hybrids are very rare in any crop. Hence, for strengthening IPM, there is a need to breed insect resistant varieties or hybrids.

2.3 Seed Treatment

Treatment of seeds with beneficial antagonistic formulations of *Trichoderma viride, T. harzianum* and *Pseudomonas fluorescens* has been found to be beneficial in enhancing germination or reducing many seedling diseases. This also reduces nematodes. Treating the seeds with butter milk, cow urine *etc.* has been found to be beneficial in enhancing germination or reducing damping off diseases (Sridhar *et al.,* 2013). Soaking seeds of cowpea, beans, coriander, *etc*., for 24 hours in diluted curd/butter milk before sowing in soil or moist coco-peat and covering them with plastic enhances germination (unpublished observation of the author). More research work in this line is required.

2.4 Hot Water Treatment

Incidence of seed borne bacterial, fungal and viral diseases of small seeded vegetables like tomato, pepper and brassicas can be reduced by hot water treatment (CAFÉ, 2016). A list of crops and the diseases that can be reduced by such hot water treatments is available at its website. The temperature of water for treating seed varies from 45 to 50°C and the treatment period varies from 10 to 60 minutes depending on the crop.

Treating matured harvested mango fruits in hot water at 52^0C for 45 minutes or water vapour also reduces fruit fly damage in mango. This is specifically done before exporting mangoes to meet import requirements of some countries. Hot water treatment of mango also reduces anthracnose and stem end rot diseases (Aveno *et al.,* 2004).

2.5 Salt Water Treatment

Soaking freshly harvested fully matured mango fruits in salt water for 45 minutes to one hour reduces fruit fly damaged fruits (unpublished report of the author). This will kill the eggs and early instars maggots in the skin of the fruit. As this is a more sustainable technology than hot water treatment, which requires special equipment for maintaining the temperature, more research is required in this line.

2.6 Substrate/media for Raising Seedlings

Generally, seedlings are grown in raised seed beds. Presently many farmers of traditional agriculture have abandoned this and are growing them in coco peat (also called coir pith, a byproduct of coir industry), plastic trays (protrays) in South India. This enhances germination. Raising seedlings in coco peat instead of soil reduces many seedling diseases and also enhances germination. It is almost sterile and is having many essential nutrients. It absorbs water very easily (about 5 times of its dry weight) (Mapa and Kumaru, 1995; Prabhu and Thomas, 2002). Under this technology, 8-10 plastic trays sown with the seeds in moist coco peat are kept one over the other for a few days to get uniform and higher germination. Coco peat can also be mixed with beneficial microorganisms like *T. viride, T. harzianum* and *P. fluorescens* which are supposed to reduce seedling diseases.

2.7 Creating Good Habitat or Shelters for Beneficial Organisms

Plants beneficial for parasitoids of insect pests can be planted in and around the plot. Birds prey on different stages of insects. To promote such preying activity, bird perches slightly above the crop canopy should be erected to facilitate resting and watching place for birds (Rana, 2016).

Flowering plants have to be planted around the main crop to give shelter and nectar to natural enemies. Such multi-cropping system creates diversified ecosystem that encourages beneficial organisms in the environment.

2.8 Staking

Many vegetables are creepers and always keep in contact with soil. As a result they are more prone for diseases, particularly during rainy season and also by flood irrigation. Hence, they should be supported by staking and training. It is very good to grow tomato, cucurbits and leguminous crops by staking as this reduces disease incidence and facilitates harvesting. Staking exposes the leaves to more sunshine and aeration. This also reduces incidence of insects like cucurbit borer, *Diaphania indica.*

2.9 Provide more Spacing between the Plants and Rows

Sufficient spacing has to be given according to the crop growth and its spread to give enough room for air movement and sunshine. This also facilitates better plant growth and higher yield. In tomato, for many indeterminate varieties, the spacing of more than 1 m between rows and 75 cm between plants reduces disease incidence and its spread. In cabbage and cauliflower, more spacing than recommended for traditional agriculture will encourage good crop growth with less disease or no disease incidence. Crop rows can also be oriented to get good sunshine and good air movement.

2.10 Irrigation Methods

Drip irrigation and mulching with plastic, live plants or dry leaves reduce water splashing or leaves and plant parts touching the soil. Flood irrigation should be completely avoided. Watering in the morning at the time of sunshine also helps in reducing the spread of diseases. Watering should not be done when the RH is high.

Sprinkler irrigation should be avoided under organic farming as water drops are likely to promote buildup of humidity and thereby disease incidence. However, a strong jet of water can be used to reduce diamondback moth, aphids, thrips incidence in some crops. This may also cause flower drop. Hence, this should be done judiciously.

2.11 Cropping Systems (Growing of Barrier/Companion Crops and Trap Crop/Multiple Cropping)

Barrier crops have been used as a management practice since 1950s for reducing the spread of non-persistently transmitted aphid-borne viruses (Fereres, 2000). The height of the barrier crop is important to camouflage the main crop and thus reduce the risk of virus transmission. But the competition between the barrier and the protected crop should be minimal. Barrier crop of maize can be used to reduce the incidence of insect pests like brinjal shoot- and fruit- borer if grown around the crop properly to suppress the visibility of the main crop.

Growing friendly plant species (companion plants) together with the main crop enhances growth, flavor and protect plants from pests. Companion plants assist in the growth by attracting beneficial insects, repelling pests or providing nutrients, shade or support to beneficials. For example, growing marigold reduces nematodes in between crops like banana, tomato, melons *etc.* (Natarajan *et al.*, 2006). Planting *Borage* (*Borago* sp.) with *squash/strawberries/ tomatoes* reduces horn worm (Wikipedia, 2017).

Growing more than one crop at a time also ensures some returns even when one crop fails. For example, in banana, many vegetable crops like French beans, dolichos, radish, *etc*., can be grown initially when the crop canopy is low. In many fruit crops like mango, guava, citrus, where wide spacing is given; non-competing intercrops can be grown. Coconut, arecanut, banana, cardamom *etc.* are extensively intercropped in coastal Karnataka and Kerala. It is also interesting to note that when coconut is grown along with arecanut (main crop) it is not attacked by dreaded black headed caterpillar, *Opisina arenosella* (unpublished observation by the author).

Growing of susceptible or preferred hosts as trap crops along with the main crop will attract the pests in large numbers. This will minimize the damage on main crop. For example, marigold is grown as trap crop for tomato fruit borer management in tomato (Srinivasan *et al.,* 1994). While planting, for every 16 rows of tomato one row of 45-50 days old marigold seedlings (var. Tall African Marigold bearing yellow or orange flowers) is planted. This trap crop was also found effective in chilli to reduce the same borer (Gundannavar *et al.,* 2007).

In cabbage and cauliflower, mustard is sown as trap crop for all the pests of cabbage (Srinivasan and Krishna Moorthy, 1991). Mustard is planted for every 25 rows of cabbage/ cauliflower. Mustard foliage has to be protected from pests by 3-4 sprays of neem soap or pongamia soap (Krishna Moorthy *et al*., 2003). Otherwise, the pests attacking mustard may migrate to the main crop. Chinese cabbage can be used as a trap crop for diamondback moth and other insect pests in cabbage (Satpathy *et al.,* 2010). Yellow rocket has been used as a trap crop for diamondback moth (Shelton and Nault, 2004). In greenhouse trials, the egg laying preference for yellow rocket varied between 24-66 fold over cabbage, but no larvae were able to develop on yellow rocket.

3. MECHANICAL AND PHYSICAL METHODS

3.1 Soil Solarization

Covering wet soil with transparent polythene or polyvinyl chloride mulch (preferably UV stabilized) and exposing to sunshine for 4-6 weeks increases soil temperature and thereby many soil borne disease, nematodes, weeds *etc.* can be controlled. This technique was discovered in Israel and was found to control diseases caused by *Verticillium*, *Rhizoctonia solani*, *Fusarium oxysporum* f. sp. *vasinfectum* or *melonis* or*obanche*, *Sclerotium rolfsii*, *Pratylenchus* as well as weeds in many tropical countries (Katan, 1984). Under polyhouses the temperature can reach even up to 70^0C (Elmore *et al.,* 1997). Therefore, this can also be used for nematode management by integrating with other methods of control.

3.2 Light Trap

In cabbage and cauliflower, light traps are useful for reducing high diamondback moth incidence. Light traps are being used by many farmers in and around Bengaluru to trap and kill diamondback moth (DBM) adults. A 100 watts bulb is hung in front of a thin gunny bag or polythene sheet (90 cm x 120 cm) smeared with oil or grease. The gunny bag or polythene sheet are hung slanting or vertically with the help of sticks. Some farmers also keep a bucket with water below the bulb. The bulbs are kept burning throughout night. Adults of DBM get attracted to light and get trapped in the oil or grease or fall to the water in the bucket. Daily the trapped adults have to be removed and oil/ grease have to be smeared afresh. About 6-8 light traps are erected per ha. Within 3-4 days most of the DBM adults get killed and facilitate effective control of the pest. It will be impossible to control the DBM from a heavily infested crop where adults are hovering in large numbers without erecting light traps. When adults are found in large numbers, spraying of *Bt* or neem soap or pongamia soap has to be combined with the use of light traps.

3.3 Pheromone Trap/Sex Pheromones

These are chemicals released by an organism to attract an individual of the opposite or same sex and encourage them to mate or perform some other function closely related with sexual reproduction. Sex pheromones are highly specific for each species and specific pheromone has to be used to attract specific pest. Generally they are synthesized based on the original pheromone produced by the insect and used to monitor the presence of the pest and take appropriate control measures if required. When excess pheromone is released, it causes confusion and may disrupt mating. Some chemicals like methyl eugenol and cue-lures which attract male oriental fruit fly and melon fly are called as para pheromones. They mimic the sex pheromones and attract males. These can also be used to suppress fruit flies as a component of IPM.

Some lepidopteron like brinjal shoot and fruit borer (*Leucinodes orbonalis*) and melon fly *Bactrocera cucurbitae* (attacking cucurbits) can be managed by pheromone traps. Oriental fruit fly *Bactrocera dorsalis* (attacking mango, guava, *etc.*) can be attracted by pheromone traps in large numbers and killed (Verghese *et al.,* 2006). Melon fly males get attracted to a para pheromone cu-lure (Dhillon *et al.,* 2005). The management of IPM of brinjal shoot and fruit borer with pheromone traps (without use of any synthetic insecticide) was demonstrated in Bangladesh (Alam *et al.,* 2003). However, the pheromone technology alone has not been that effective in India. There is a great potential to use this technology to manage this pest in India by combining with other IPM components like destroying infested shoots and fruits, release of egg parasitoid, *Trichogramma chilonis* and application of botanical or *B. thuringiensis.*

3.4 Growing Crops Under Nylon Net

Raising tomato nursery under nylon net is a standard recommendation for reducing tomato leaf curl disease in the main field. Cultivation of vegetables and other crops under polyhouse or nylon net reduces many lepidopteran pests like brinjal fruit borer, tomato fruit borer, tobacco caterpillar *etc.* But this may aggravate some other pest problems like red spider mite, white fly, leaf miner *etc.*, which may become very difficult to control. Hence, such net houses should be insect proof; otherwise, they may increase pest problems.

3.5 Destruction of Infested Plants and Plant Parts

This is a very important and most effective plant protection method for the control of leaf miner and borers. Leaf miner, *Liriomyza trifolii* is the dreaded pest in many crops like tomato, beans and cucurbits. In tomato, this pest can be managed by removing leaf miner infested leaves or part of the leaf miner infested leaves just before planting and after planting at weekly interval. In cucurbits leaf miner first comes to cotyledons. Hence, these cotyledons (or part of the infested leaf) with leaf miner have to be destroyed after the emergence of two more leaves, so as not to affect plant growth. Destruction of leaf miner infested leaves have to be repeated at weekly interval for 3-4 times. In beans, leaf miner comes to unifoliate leaves immediately after germination. These infested leaves have to be removed at about 10 days after planting.

Caterpillars of tomato fruit borer *H. armigera,* brinjal borer *L. orbonalis,* dolichos pod borer *Adisura atkinsoni etc*., migrate and cause damage to more than one fruit or pod. Hence, it is advised to periodically remove and destroy infested fruits and pods. Many pests like webber and shoot borer of many crops can also be managed by clipping and destroying the infested shoots (*e.g.* mango shoot borer (*Chlumetia transversa* and webber *Orthaga* spp). Infestation of lemon butterfly (*Papileo demoleus*) can be reduced by collection and destruction of infested leaves and plant parts, particularly in nursery.

Many fungal and bacterial diseases infecting plants spread by splashing rainfall or water drops or by irrigation water. Generally lower leaves get infested first. Hence, bottom leaves touching the irrigation water and any leaf showing disease should be plucked and destroyed to reduce disease incidence and spread.

Virus diseases like leaf curl, bud necrosis are spread by whiteflies in tomato. Plants showing these symptoms have to be destroyed by regular monitoring of the plot at least up to first harvest. Plants showing these symptoms should not be left as such, since they act as source of inoculum for further spread of disease. Tomato spotted wilt disease is another important virus disease. The plants infested with this disease stop growing. The infested plants have to be immediately uprooted and destroyed. This method can also be adopted to manage bunchy top of banana, another dreaded mycoplasma disease.

When branches of perennial fruit trees, plantation crops *etc.* get infested with diseases, affected parts should be pruned at periodical intervals and destroyed wherever possible.

Severely infested plants have to be cut and destroyed to prevent the spread of the problem further.

3.6 Destruction of Egg Masses and Gregarious Larvae

In cabbage, scouting and mechanical destruction of gregarious and early instar larvae of tobacco caterpillar, *Spodoptera litura* and leaf webber, *Crcidolomia binotalis* is highly effective (Krishna Moorthy *et al.*, 2003).

Vegetables are grown as intercrops in between young coconut plants. Adults of tobacco caterpillar (*S. litura*) lays eggs on the coconut leaves and hatched first in star larvae move to the intercrops grown below and develop on them. Hence, scouting and destruction of egg masses of the insect pest on the coconut leaves will help in reducing the pest incidence.

3.7 Destruction of Alternate/Collateral Hosts

Alternate hosts of fruit sucking moth, *Ophideres (Othreis) fullonica* like Gulvel, Vasanvel, Chadelvel growing in citrus orchard should be destroyed. Many weeds serve as alternate hosts for the whitefly, *B. tabaci,* transmitting virus diseases in tomato and okra. Hence, such weeds should be removed all-round the main crops.

3.8 Field Sanitation

Collection and destruction of dead wood, rotting plants or plant parts, fallen fruits, *etc.* which act as reservoir for many pests is important. Hence, sanitation, i.e. removing such plant materials and destroying them plays a very important role in IPM. For example, fallen or left over mango and guava fruits attract fruit flies. Therefore, collecting them and destroying them is very important part of IPM.

3.9 Specific Products (other than Biological Control) for Disease Management in Organic Farming

Beckermann (2008) and Andrews and Baker (2015) have reported some products like bicarbonates of sodium and potassium, copper based fungicides, sulphur, kaolin clay, can be used in organic farming for disease control.

3.9.1 Bicarbonates

Sodium bicarbonate (baking soda) and potassium bicarbonate can be used under organic farming to manage powdery mildew. Potassium bicarbonate spray (10g/l) was found to reduce powdery mildew in grapes (Sawant and Sawant, 2008). This chemical is much safer than sodium bicarbonate and hence, can be used as a component of organic IPM. It can be used as a contact fungicide applied @ 2 to 8 kg/ha, with a maximum of 8 applications per year. Ammonium bicarbonate can also be used for managing powdery mildew.

3.9.2 Copper based products

Copper, fixed copper hydroxide, copper oxide, copper oxychloride, copper octonate and copper sulphate are permitted under organic farming (Caldwell *et al.,* 2013). Copper hydroxide and Bordeaux mixture are derived from copper sulphate and hydrated lime should be applied at neutral pH to avoid phytotoxicity. These are well known as fungicides and also can be used to reduce bacterial diseases to a certain extent. Bordeaux spray is very commonly used in grapes and arecanut in India. However, some crops are sensitive. Further, copper gets accumulated in environment with repeated use. Hence, it should be used judiciously.

3.9.3 Sulphur

Sulphur is one of the oldest fungicides used. Paulsen (2005) has given a list of plant diseases where Sulphur can be used. It is extensively used to control powdery mildew and mites. Lime Sulphur is a product derived by boiling lime and Sulphur together. Often it is used as alternate to Sulphur formulations. Sulphur is also applied to soil as ammonium sulphate or potassium sulphate or as elemental Sulphur to control scab (*Streptomyces scabis*) and black scurf (*Rhizoctonia solani*) in potato and black scurf in sweet potato and *Rhizoctonia* rot in beet root. It is generally applied as a protectant while lime Sulphur can be used even after pest incidence. However, many crops are Sulphur shy, hence, they should be used carefully.

3.9.4 Kaolin clay

Edible kaolin clay, generally used in toothpaste and processing industry is considered to be very safe as organic input. When sprayed it forms a particle barrier for egg laying, feeding and also against a few diseases. It is available as WP formulations. These are sprayed at weekly intervals to get good control. Strong and persistent coating of kaolin is required for successful control of pest problems in apple (Hinman and Ames, 2011).

For strengthening IPM, further studies with kaolin, bicarbonates and lime Sulphur have to be undertaken in vegetables and other crops. The formulations of these products are not generally available in India. Hence, this problem also should be addressed.

4. BIOLOGICAL CONTROL

Many natural enemies of pests are present in nature. They have to be conserved by good cultural practices. For this, flowering plants, multi-cropping systems are congenial. Monoculture is not good. Good cultural practices help in maintaining ecological balance. Intervention by chemical spray or soil application should be minimal. This approach helps in maintaining the pest incidence to the minimum. This is called as Natural Control.

Many times pests have been introduced to new areas along with plant materials or by other meas. To keep such introduced pests under check, there may not be any natural enemy of such introduced pests. To suppress such introduced pests, natural enemies from the native habitat of the pest are introduced. This is called classical biological control. To control cottony cushion scale of citrus, *Icerya purchasi*, a lady bird beetle, *Rodolia cardinalis* was introduced to California from Australia. This was highly successful. This was also repeated in India and in many other countries. Recently, parasitoids of the papaya mealy bug, *Paracoccus marginatus,* were also introduced to India which brought down the pest substantially in Tamil Nadu and Karnataka. Such successes of biological control in India are compiled by Singh (2004).

Mass multiplication and inundative releases of the egg parasitoid, *Trichogramma* is recommended for the control of many lepidopteran borers (Krishnamoorthy, 2012). For controlling fruit borer, *H. armigera* in tomato, six releases at weekly interval @ 40,000/

ha with the first release coinciding with 50% flowering in tomato is recommended. In brinjal, clipping damaged shoots and releasing egg parasitoid, *Trichogramma chilonis* @ 50000/week starting from shoot damage combined with 2 sprays of *Bt* formulation (at peak flowering) significantly reduced borer incidence.

In tomato, egg laying starts even before flowering i.e. at 20-26 days after planting. There is a need to control initial incidence itself to get better control. So, further studies are required in this direction.

Nuclear Polyhedrosis Virus (NPV) is highly species specific and does not affect other organisms. The most commonly used NPV is *Helicoverpa* NPV and is highly effective against tomato fruit borer. The NPV @ 250 larval equivalents/ha (2 x 10^9 ployhedral bodies/ml) has to be sprayed 3 times after flowering at weekly intervals (28, 35 and 42 days after planting) along with soil application of neem cake, collection and destruction of bored fruits at periodic intervals to get effective control (Krishna Moorthy *et al.,* 2002, 2003). NPV has to be sprayed in the evening by mixing with 2% jiggery (gur) to prevent its degradation due to ultraviolet rays. Jaggery also acts as a phagostimulant. However, the quality of NPV formulation is very important to get good control. Hence, there is a need to regulate the quality of the NPV supplied.

Many species of *Bacillus* naturally occur in soil. They can be effectively used in pest management. Among them, *B. thuringiensis* (commonly called as *Bt*), *B. subtilis* and *B. pumilus* are important. *Bt* is effective against lepidopteran caterpillars. In India, it was extensively used to control diamondback moth (DBM), a notorious pest of cabbage in late 1980s. However, DBM developed resistance very fast and became ineffective and it was withdrawn from the market in 1990s. However, due the withdrawal, the formulations are now effective. This can be judiciously used to reduce DBM in cabbage and cauliflower. *Bt* was also found to reduce brinjal shoot and fruit borer and is currently recommended along with pheromone traps and release of *T. chilonis* as components of IPM.

B. subtilis can be used to manage bacterial diseases. This inhibits spore germination of pathogens and interferes with the attachment of the pathogen to the plant. *B. subtilis* is also reported to induce systemic acquired resistance (SAR) against bacterial pathogens. *B. pumilus* can be used against rust and mildews. *B. pumilus* strains and *B. subtilis*, trigger plant's natural defenses.

Fungal agents like *Metarrhizium* sp., *Beauveria* sp., *Nomuraea rileyi*, *Paecilomyces farinosus* can be used to bring down many pests belonging to Lepidoptera, Coleoptera *etc. Verticillium lecanii* can be used against Homopteran pests *viz.,* scales and aphids. Entomopathogenic nematodes (EPN) have also been found to reduce many lepidopterans and root grubs.

4.1 Antagonistic Microorganisms

Antagonistic fungi are widely used in biocontrol of diseases and also nematodes. They colonise and grow on the seed and root surface and prevent the development of pathogens. Important antagonistic fungi include, *Trichoderma harzianum, T. viride, T. virens, Pochonia chlamydosporia* and *Paecilomyces lilacinus* (Reddy, 2008). Among all, *Trichoderma* is very important and extensively used in organic farming. Reddy (2008) has given an exhaustive list of horticultural crops where these can be used.

4.1.1 *Trichoderma* spp.

Trichoderma spp. have been known since 1920s as biocontrol agents against plant pathogens and nematodes. Important species are *T. harzianum, T. viride* and *T. hamatum.* These colonize the root epidermis and outer cortical layers and release bioactive molecule. The biocontrol agent generally grows in its natural habitat on the root surface and so affects root disease in particular, but can also be effective against foliar diseases. The suspected mechanisms of these agents include antibiosis, parasitism, inducing host-plant resistance and competition (Harman, 2006). *Trichoderma* is used for seed treatment, applied to nursery beds, coco peat and also to soil along with FYM or neem cake. It is also applied to perennial plants like coconut, arecanut, black pepper *etc*. Many commercial formulations are marketed. To reduce the cost of application, FYM or neem cake are enriched with *Trichoderma*. For this purpose, 1 kg of the formulation is mixed with 1 ton of moist FYM, covered with polythene or gunny bags and incubated for 3-4 weeks. This is applied as spot application to the plants or ridges or to soil while preparing the land along with the regular FYM. It is also applied @ 6 g/kg of potted soil.

Siddiqui and Shaukat (2004) observed the synergistic effect of *T. harzianum* on the production of nematicidal compound(s). They opined that in view of the inconsistent performance of the biocontrol agents under field conditions, application of a mixture of compatible *T. harzianum* and *P. fluorescens* would enhance the efficacy and reliability of control.

Bacterial wilt infects many crops. Though several bacterial wilt resistant brinjal varieties and hybrids are available in India, there is a need to develop resistant varieties for different types of brinjal preferred in different regions. Another alternative will be to use biocontrol agents for managing this disease with region specific varieties. To improve the performance of *Trichoderma* and other antagonistic fungi, compatibility studies have to be undertaken and more efficient formulations have to be developed.

5. BOTANICALS

Use of any plant part or by-product or extract or even formulation for pest management can be termed as botanical. More than 2000 plant species reportedly exhibit insecticidal properties. Considering the disadvantages of synthetic chemicals, lot of focus is given on the bio-efficacy of botanicals in pest management in recent times. Isman and Grieneisen (2013) have reviewed the current status of global publications on botanicals. In spite of voluminous information on the bio-efficacy of botanicals, only a handful of findings have been pursued to logical end. Among the success stories of the use of botanicals, neem is in the forefront. Many World Conferences on neem have been conducted so far. They have contributed immensely for the success of IPM. Other important plants having insecticidal properties include Pongamia, Pagoda tree and custard apple. Another new area of interest in IPM is essential oils. Pyrethrin, another botanical derived from *Tancetum* spp., is mainly used indoors tobacco and rotenone (derived from Derris) are banned for organic farming.

5.1 Neem

Use of neem in IPM is a success story. Neem (*Azadirachta indica*) is native to India and can grow in most arid, sub-tropical and tropical areas of the world. It is also known as a traditional remedy in controlling many diseases in humans and animals and is called

Sarva Roga Nivarini (cure for all ailments) in Sanskrit. Extensive work has been done on its chemistry and many patents and products have been developed based on neem and are used in plant, human and animal health. However, of all the properties of neem, its effect on insect pests is well documented. On insect pests, neem acts as a repellant, antifeedant, insect growth regulator (IGR) and also as an insecticide. Many books giving information on neem in IPM are available (Schmutterer, 2002; Gahukar, 1995; Vijyalakshmi *et al.,* 1995). Krishna Moorthy and Krishna Kumar (2002) have reviewed the status of the use of neem in IPM of vegetables. Use of neem products, i.e., neem formulations, neem cake, neem oil and neem soap are given below.

5.1.1 Neem formulations

All neem formulations are usually derived from neem seeds. Other products derived from neem seeds are neem oil and neem cake. Neem cake is the residual seed meal remaining after extraction of oil from seeds. This is sold and used extensively as an organic fertilizer. Neem seed contains more than 100 bioactive compounds (Schmutterer, 2002) and among them, azadirachtin, a triterpenoid, is considered to be the most important one. Azadirachtin mainly acts as insect growth regulator, anti-feedant, oviposition deterrent and also as insecticide. Many neem formulations are available in India and world over. These products are broad-spectrum insecticides and act by contact or ingestion. These are generally based on azadirachtin, the main principle in neem seeds. This is having very strong insecticidal property at very low doses. In 1980 and 1990s, the formulations developed were not very stable and had low azadirachtin content. Nowadays, many neem formulations with high azadirachtin (>10000 ppm) are available in market. They are much more effective and also can be used to manage many insect pests like thrips, leaf miner, DBM in cabbage, tomato fruit borer *etc.* and can be used as a component of IPM.

5.1.2 Oiled neem cake

There are two types of neem cake, one is oiled neem cake and another is de-oiled neem cake. Oiled neem cake contains 8-10% oil. It is extracted by cold expeller process, whereas, de-oiled neem cake has very low oil (<4%) and is a byproduct of solvent extraction. Most of the active ingredients present in neem are lost in the solvent extraction process. Many farmers in India generally apply neem cake to many crops without knowing its beneficial effect. Neem cake is a byproduct of oil extraction from neem seeds. Neem cake is very well known for its nematicidal property. It can be applied @ 250-1000 kg/ha to reduce nematode problems. Soil application of neem cake was also found to reduce many insect pests in many crops. Soil application of oiled neem cake at planting @ 250 kg/ha and repeated at flowering reduces leaf miner and fruit borer in tomato. In cucumber, bitter gourd and gherkins, soil application of oiled neem cake immediately after germination and repeated at flowering reduces leaf miner, fruit fly and borer which is far superior to conventional pesticides. In brinjal and okra (ladies finger) also, soil application at flowering and repeated 2-3 times at 30-45 days interval reduces shoot and fruit borer, hoppers *etc.* In vegetable pigeon pea and dolichos lab lab, soil application of neem cake at flowering and repeated after 20-25 days gives excellent control of pod borers. Several studies on the benefits of neem cake on insect pests of many crops have been carried out. However, the type of neem cake is not mentioned in most of the reports. It is observed that the smell of neem cake plays a very important role in its efficacy (Krishna Moorthy and Krishna Kumar, 2002). The smell is governed by the volatiles of neem cake.

Oiled neem cake has more than 50 volatile compounds (Shivashankar *et al.,* 2011). Neem volatiles (odour) act as repellants and also have ovicidal properties. They also kill young caterpillar when exposed to the neem cake volatiles. Hence, its smell is very important. Hence, only good quality oiled neem cake with 8-10 % oil should be used. The main limitation of neem cake is that its efficacy gets reduced at high temperatures and high wind velocity. Now soil application of oiled neem cake is a standard recommendation for many vegetables (IIHR, 2007; 2010). This has to be used as a component of IPM at 250 kg/ha or 8-12 g/plant. Recently a controlled release formulation of neem cake was developed (Krishna Moorthy *et al.,* 2014a) to make it persist for longer time. However, it may take long time to make it commercially feasible.

5.1.3 Pulverised neem seed powder extract

Neem seed kernel extract has been a standard recommendation since early 1980s for many insect pests. Under this technology, the kernel has to be separated from the seed coat, powdered, soaked in water overnight and filtered to get 5% strong solution; or the kernel had to be wet ground, kept overnight and filtered to get 5% strong solution. However, this formulation did not become popular with farmers due to the difficulty in getting seeds, separating kernels, powdering and wet grinding. Bad smell was another deterrent. Every time fresh solution had to be prepared. Now pulverized whole neem seed powder extract (PNSPE) is recommended. Under this technology, whole neem seeds are pulverized and can be stored up to three months. The powder is soaked whenever required and kept overnight and filtered through a double layered nylon net to get 4% solution for spraying. Pulverised neem seed powder is effective because while powdering the temperature of the powder does not go beyond room temperature and as a consequence, all the active ingredients responsible for insecticidal action are retained. The powder also can be stored for 3-5 months without losing its effect. The bad smell encountered while filtering is also reduced. One hectare requires 40 kg of powder for one spray. It is very cheap and effective on all the pests of cabbage and cauliflower. It is also effective for fruit fly control in cucurbits, stem fly in beans and leaf miners in all crops. It can be used as a general spray given at 10 days interval for managing all insect pests of vegetables. Further work is required on its efficacy on other horticultural crops.

5.1.4 Neem oil

Sprays of neem oil at 1-2% against insect pests is very common. This is found moderately effective against many insect pests like chilli thrips, brinjal shoot and fruit borer, mites *etc.* Oil is shaken with soap nut powder (2 gm/litre of water) or soap solution or liquid soap to make an emulsion and sprayed. Spraying has to be done immediately after making the emulsion otherwise oil may separate from water. However, oil sprays may be toxic at high temperatures and also to some plants. Hence, it has to be sprayed selectively in the evening only. Oil sprays are also known to reduce powdery mildew in vegetables.

5.1.5 Neem soap

Neem soap for agriculture use is prepared by using potassium hydroxide. They are also called as soft soaps or insecticidal soaps. Soft soaps are universally permitted inputs for organic farming (Caldwell *et al.,* 2013). Neem soap for agriculture use was developed by the Indian Institute of Horticultural Research, Bengaluru. This was found effective on a number of insect pests (Krishna Moorthy and Krishna Kumar, 2002) and can be used in organic vegetable production. They are in the form of concentrates and are sprayed at 1% concentration at 7-10 days interval on diamondback moth in cabbage and cauliflower, leaf hopper, leaf miners and thrips in many vegetable crops. Their spray also reduces tomato fruit borer, okra fruit borer when given at critical time of pest incidence. They were also found to reduce the thrips under polyhouse conditions. Soap has to be applied as component of IPM rather than sole control measure in order to get best results.

There are two main limitations while using soft soaps - thorough coverage of the leaf area where insect or mites are feeding is essential and the efficacy may get reduced due to rapid loss of residual toxicity at high temperatures. However, there is great potential to use them in the IPM of many vegetables like cabbage, tomato, okra, chilli and cucurbits. Soaps also reduce jassids in mango and leaf miner in citrus. They can be used as a component in IPM of many other crops like cotton, pigeon pea *etc.* The sprays have to be done regularly before the buildup of the pest incidence to get good control.

More work is required on its efficacy in other horticultural crops. Studies on the effect of soil application of soap solution on the incidence of diseases and nematodes have to be undertaken.

5.2 Pongamia

Pongamia pinnata is another potential plant useful in insect pest management. The active principle in its seeds is karanjin. This chemical is supposed to be effective on aphids and mites. Some commercial formulations with 40,000 ppm of karanjin are marketed in India. Pongmia oil, cake and seed powder can be used just like neem. Pongamia oil also reduces powdery mildew. However, it is not as effective as neem in controlling insect pests. Pongamia soap is prepared like neem soap. This was again developed at the Indian Institute of Horticultural Research, Bengaluru. This gives immediate knock down effect of diamondback moth larvae, but it is less effective in summer on diamondback moth as compared to neem soap. Pongamia cake can be used as an alternative to neem cake. But more research is required on pongamia and karanjin and their efficacy on insect pests and diseases in different crops.

5.3 Essential Oils

Use of essential oils in pest management is of recent origin. Isman (2000), Koul (2008) and Tripathi *et al.* (2009) have reviewed the current status of essential oils in insect pest management. Aromatic characteristics of essential oils attract or repel insects, protect plant from heat or cold and also utilize them for defense. Essential oils get evaporated easily at room temperatures or at slightly higher temperatures. These are generally used in preparing perfumes and incense sticks and also in aromatherapy. They are also used as additives in food, cosmetics, plastics and resins to get distinct flavor. These are permitted to be used under organic farming (Caldwell *et al.,* 2013).

Commercially available essential oil preparations of Basil (*Ocimum tenuiflorum* syn. *O. sanctum*) (also called as Tulsi) and mint (*Mentha arvensis*) were found promising against chilli thrips (*Scirtothrips dorsalis*) in capsicum (Krishna Moorthy *et al.,* 2013a) and onion thrips (*Thrips tabaci*) (Krishna Moorthy *et al.,* 2013b). To reduce the cost and make it economically viable, Indian Institute of Horticultural Research has developed formulations of essential oils (Krishna Moorthy *et al.,* 2014b). Presently several essential oils are used in aromatherapy and additional work needs to be undertaken to study their efficacy on insect pests and to develop cheap and effective formulations.

5.4 Pyrethrins

Pyrethrins are derived from dried flowers of *Tancetum sinerarifolium* and *T. coccineum.* These plants have originated in Balkan (Rumania, Italy and Greece). In olden days in the Middle East, dried flowers were used to get rid of lice. Presently grown in Kenya, Tanzania, Equador, Australia, Tasmania and is generally used indoor to control house flies and mosquitoes. This causes immediate knock-down of the insects but they revive after some time. To overcome this problem, pyrethrin (active ingredient in the dried flowers) is added with piperonyl butoxide (non-organic). Though synthetic pyrethriods are synthesized based on pyrethrins, they are not permitted under organic farming. Formulations of pyrethrins marketed in advanced countries for organic farming are added with oils and soaps to enhance their efficacy.

5.5 Garlic

Garlic is another botanical, which has potential in IPM. Garlic is supposed to be antifeedant, insecticidal, nematicidal and repellent. Garlic extract has been reported to be effective on coconut mite.

5.6 Custard Apple

It is commonly known as 'Seetha Phal' or 'anona' (*Anona squamosa*). The active principle in this plant includes acetogenins - annonin I or squamocin. Acetogenins are 10,000 times more effective against cancer than standard chemotherapy chemicals. Hence, more research work should be undertaken to strengthen IPM.

5.7 Pagoda Tree

Commonly known as Japanese Pagoda tree (*Sophora* spp., *Styphnolobium japonicum)* is actually a native of China. This is used in traditional Chinese medicine against cancer, tumours *etc*. It was introduced to many western countries and is a popular ornamental plant in South Africa and North America. It is also known as Chinese scholar tree and is cultivated as an ornamental plant in Kashmir. The alkaloid components (matrine and oxymatrine) are found in the roots of S*ophora japonica, S. subprostrata and* from the above ground portion of *S. alopecuroides*. Formulations containing these alkaloids are effective on insects also, particularly on red spider mites and diamondback moth. These are more effective than synthetic chemicals on red spider mites. Not much information is available on this tree. Perhaps they are effective on plant diseases and aphids also. More work is required on this plant and its extracts.

5.8 Other Botanicals

Rotenone extracted from *Derris* spp. and nicotine from tobacco are other important botanicals. Rotenone is used against unwanted fish in water ways. Tobacco or nicotine (extract from tobacco) can be used against soft bodied insects. However, presently they are banned in organic farming considering their adverse effect on health and safety of environment. Hence, they should not be used in organic farming.

5.9 Others

5.9.1 Mineral oils

These are refined petroleum products which can also be used in organic farming. They act like neem oil and cause mortality of the insects by dehydration and suffocation. They also have ovicidal affect. These are also available as horticultural oils in some countries. These should not be applied when temperature is too high or too low. They also should not be applied when the humidity is high to avoid phytotoxicity.

5.9.2 Fish oil rosin soap

This is a product of fish oil and fish industry. This can also be used like neem soap and pongamia soap. However, it gives a bad odour. This can be used to control mealy bugs in grapes and other plants to a certain extent.

5.9.3 Indigenous technical knowledge

We have lot of Indigenous Technical Knowledge (ITK), which have not been validated scientifically to recommend them universally. For example, many organic farmers claim that plant decoction prepared from tulsi (*Ocimum spp.*), castor (*Ricinus communis*), *Calotropis*, *Adathoda vasica* etc. can control most of the insect pests and diseases. We also have many more indigenous plants which are used in Ayurveda and traditional medicine systems. Though we have got stray reports about their efficacy on pests, they are not pursued to their logical end. Therefore, we must concentrate on these indigenous technologies to come out with sustainable IPM for all crops. This will not only reduce the cost of production but also strengthen the IPM.

6. CONCLUSION

In India, commercial crop production is not adopted in many areas. Most of the backward areas are still following subsistence farming. These should be made into commercial organic farming hubs, particularly in horticultural crops, which are more remunerative. People who are used to apply fertilizers and pesticides have different set of mind. They always want good crop at any cost. Hence, we should concentrate more in backward areas and try to improve their livelihood by adopting sustainable organic farming, particularly in horticulture crops. It is heartening to note that none of the organic farmers have committed suicide so far in this country.

For a sustainable organic farming, all necessary cultural practices have to be followed first. Developing more and more multiple pest resistant varieties is another prerequisite for organic farming. This will reduce the risk in organic crop production. Biological control is one of the cheapest methods of control. However, timely availability and quality of the biocontrol agents and other inputs are also important. More research is required to develop combination formulations of antagonistic organisms to make them more effective. Locally available plants have to be identified for their pesticidal properties and simple technologies should be developed to use them in IPM programmes. More research is required on the use of salt water treatment for controlling mango fruit fly as this can be adopted by any farmer. Many available ITKs have to be validated in different areas and promoted. These technologies have to be taken up for demonstrations by extension agencies and NGOs. By following the IPM, cost of production can be reduced and thus the profitability can be increased substantially. This will strengthen IPM and make organic crop production more sustainable.

REFERENCES

Alam, S.N., Rashid, M.A., Rouf, F.M.A., Jhala, R.C., Patel, J.R. Satpathy, S., Shivalingaswamy, T.M., Rai, S., Wahundeniya, I., Cork, A., Ammaranan, C. and Talekar, N.S. (2003). Development of an integrated pest management strategy for eggplant fruit and shoot borer in South Asia. Shanhua, Technical Bulletin No. 28. AVRDC Publication No. 03-548. AVRDC—the World Vegetable Center. Taiwan.

Andrews, Nick and Brain, Baker. (2015). Can I use this product for disease management in my organic farm? Organic agriculture, retrieved from http://articles.extension.org/pages/18360/can-i-use-this-product-for-disease-management-on-my-organic-farm on October 16 2016

APEDA. (2016). Agriculture and Processed food Products Export Development Authority (http://apeda.gov.in/apedawebsite/organic/Organic_Products.htm#).

Aveno, J. L. and Orden, M.E.M. (2004). Hot water treatment of mango. A study of four export corporations in the Philippines. P. 4 retrieved from http://www.thaiscience.info/journals/Article/KLST/10424429.pdf in October 2016.

Ayyar, Ramakrishna T.V. (1963). Handbook of Economic Entomology for South India. Controller of Stationery and Print. Madras, India.

Beckermann, J. (2008). Disease Management Strategies. Using Organic Fungicides. BP-69-W Department of Extension, University of Purdue. retrieved from https://www.extension.purdue.edu/extmedia/BP/BP-69-W.pdf in October 2016.

Caldwell, B., Sideman E., Seaman A., Shelton, A., Smart, C. (2013). Resource guide for organic insect and disease management. New York State Agricultural Experiment Station, New York.

Carson, Rachel. (1962). Silent Spring. The Riverside Press, Cambridge, MA.

CAFÉ. (2016) Centre for Agriculture Food and Environment. https://ag.umass.edu/services/hot-water-seed-treatment.

Ciancio, A. and Mukerji, K.G. (2007). General Concepts in Integrated Pest and Disease Management. Springer Science & Business Media, AA Dordlretch, Netherlands.

Dhillon, M.K., Singh R., Naresh J.S., Sharma H.C. (2005). The melon fruit fly, *Bactrocera cucurbitae*: A review of its biology and management. Journal of Insect Science. 5:1-16.

Elmore, Clyde, l., Stapleton, James J. and Bell, Carl E. and Devay, James, E. (1997). Soil Solarization- A Non pesticidal Method for Controlling Diseases, Nematodes and Weeds. Division of Agriculture and Natural Resources. University of California.

Fereres, A. (2000). Barrier crops as a cultural control measure of non-persistently transmitted aphid-borne viruses. Virus Res.:71:221-31.

Gahukar, R.T. (1995). Neem in Plant Protection. Agri-Horticultural Publishing House, Nagpur, India.

Gundannavar, K.P. Giraddi, R. S., Kulakarni, K. A. and Awaknavar, J. S. (2007). Development of integrated pest management modules for chilli pests. Karnataka J. Agric. Sci., 20: 757-760)

Harman, G.E. (2006). Overview of mechanisms and uses of Trichoderma spp. Phytopathology. 96 (2): 190–194.

Hill, Dennis. (1975). Agricultural Insect Pests and Their Control. Cambridge University Press. London.

Hinman, Tammy and Ames, Guy. (2011). Apples: Organic Production guide. National Sustainable Agriculture Information Service ATTRA, 40 p. retrieved from www. attra.ncat.org in October 2016.

IFOAM. (2016). International Federation of Organic Agriculture Movement (IFAOM). (http://www.ifoam.bio/en/organic-landmarks/definition-organic-agriculture).

I.I.H.R. (2007). Pest Management in Horticulture Crops: Fruits, Vegetables and Ornamentals, Technical Bulletin 30, Indian Institute of Horticultural Research Bengaluru. India.

I.I.H.R. (2010). Production Technology of Vegetables, A Hand book. Indian Institute of Horticultural Research, Hessraghatta Lake Post, Bengaluru, India.

Isman, Murray B. (2000). Plant essential oils for pest and disease management. Crop Protection. 19: 603-607.

Isman, Murray B. and Grieneisen, Michel L. (2013). Botanical insecticide research: many publications, limited useful data. Trends in Plant Science, 19: 140-145.

Katan, J. (1984). Soil solarization. Acta Hortic. 152: 227-236.

Koul, Opender, Walia, Suresh and Dhaliwal, G. S. (2008). Essential oils as green pesticides: Potential and constraints. Biopesticide International. 4: 63-84.

Krishna Moorthy, P.N. (2005). Principles of Insect Pest and disease Management for organic Crop Production. p. 65-70. Paper presented in the National Seminar on Organic Farming-Current Scenario and Future thrust Hyderabad. April 27-28, 2005.

Krishna Moorthy, P.N., Girija Ganeshan., Sadashiva, A.T. and Hebbar, S.S. (2003). Integrated Pest Management in Tomato Cultivation. IIHR Extension Bulletin (NATP) - 02 Indian Institute of Horticultural Research, Bengaluru. India.

Krishna Moorthy, P.N., Girija Ganshan, Varalakshmi, B. and Prabhakar, M. (2003). Integrated Pest Management in Cabbage Cultivation. Extension Bulletin (NATP)-01. Indian Institute of Horticultural Research, Bengaluru, India.

Krishna Moorthy, P. N. and Krishna Kumar, N. K. (2002). Advances in the use of botanicals for the IPM of major vegetable pests. p 272-282. In: Proceedings of the international conference of vegetables (CD ROM publication). Bengaluru, India. November 11-14, 2002.

Krishna Moorthy, P.N., Krishna Kumar, N.K., Prabhukumar, S., Raghunatha, R. and Prasanna Kumar, G.T. (2002). Validation of the IPM of tomato fruit borer by using NPV and trap crop. p. 261-265. In: Proceedings of the symposium on Biological control of Lepidoptera. Bengaluru, India July 7-12, 2002.

Krishna Moorthy, P . N., Saroja, S. and Shivaramu, K. (2013a). Bio-efficacy of neem products and essential oils against thrips. (*Scirtopthrips dorssalis* Hood) in Capsicum. Pest Management in Horticultural Ecosystems. 19(2): 191-193.

Krishna Moorthy, P. N., Shivaramu, K., Krishna Kumar, N. K., Ranganath, H. R. and Saroja, S. (2013 b). Comparative Efficacy of Neem Products, Essential oils and Synthetic Insecticides in the Management of Onion thrips (*Thrips tabaci* Lindeman). Pest Management in Horticultural Ecosystems. 19: 23-26.

Krishna Moorthy, P.N., Soroja, S., Rangnath, H.R., Shivaramu, K. and Achala Paripoorna, K. (2014a). Controlled release formulation of oiled neem cake for insect pest management. Pest Management in Horticultural Ecosystems, 20: 133-136.

Krishna Moorthy, P.N. , Saroja, S., Shivaramu, K. and Achala Paripoorna K. (2014). Bio-efficacy of essential oil formulations of mint, Basil and Geranium against onion thrips, Thrips tabaci Lindeman and chilli thrips, *Scirtothrips dorsalis* Hood under field conditions Pest Management in Horticultural Ecosystems. 20: 137-140.

Krishna Moorthy, P.N., Sannaveerappanavar, V.T. and Rao, N.S. (2004). Integrated Pest management p 36 - 44. In: (Ed: G.K.Veeresh) Operational Methodologies and Package of Practices in Organic Horticulture. Association for Promotion of Organic Farming, Bengaluru, India.

Krishnamoorthy, A. (2012). Exploitation of egg parasitoids for control of potential pests in vegetable ecosystems in India. Comunicata Scientiae 3(1): 1-15.

Mapa, R. B. and Kumaru, G. K. K. P. (1995). Potential of coir dust for agricultural use. Sri Lankan J. Agric. Sci. 32:161–164.

Metcalf, R.L. and Luckman, W.H. (1994). Introduction to Insect Pest Management, John Wiley & Sons, Inc. New York.

Natarajan, N., Cork, A., Boomathi, N., Pandi, R., Velavan, S. and Dhakshnamoorthy, G. (2006). Cold aqueous extracts of African marigold, *Tagetes erecta* for control of tomato root knot nematode, *Meloidogyne incognita*. Crop Protection 25: 1210–1213.

Paulsen Hans-Marten. (2005). Sulphur in organic farming. p. 105-110. In L.J. De Kok and E. Schnug (eds.), Sino-German Workshop on Aspects of Sulphur nutrition of Plants. Braunschweig Federal Agricultural Research Centre (FAL) Germany.

Prabhu, S. R. and Thomas, G. V. (2002). Biological conversion of coir pith into value added organic resource and its application in agri-horticulture: Current status. Prospects and Perspective. J. Plantation Crops. , 30:1–17.

Raghuramulu, Y. and Kamala Bai, S. (2006). Package of Practices for Organic Coffee. Central Coffee Research Institute, Chikmagalur District, Karnataka, India.

Rana, S.S. (2016). Organic Farming (Revised). Department of Agronomy, College of Agriculture, CSK Himachal Pradesh Krishi Vishvavidyalaya, Palampur.

Reddy, P.P. (2008). Organic farming for sustainable Horticulture. Scientific Publishers (India), Jodhpur.

Sawant, S.D. and Sawant, I.S. (2008). Use of potassium bi-carbonate for the control of powdery mildew in grapes. Acta Hortic. 785: 285-292.

Schumutterer, H. (2002). The Neem Tree, *Azadirachta indica* A. Juss and other Meliaceous Plants. II Edition. Neem Foundation, Mumbai, India.

Shelton, A. M. & Nault, B. A. (2004). Dead-end trap cropping: a technique to improve management of the diamondback moth. Crop Protection. 23: 497-503.

Shivashankar, S. Krishna Moorthy, P.N. and Roy, T.K. (2011). Volatile constituents of neem (*Azadirachta indica*) seed cake. J. Medicinal and Aromatic Plant Science. 3:139-143.

Siddiqui, I.A. and Shaukat, S.S. (2004). *Trichoderma harzianum* enhances the production of nematicidal compounds in vitro and improves biocontrol of *Meloidogyne javanica* by *Pseudomonas fluorescens* in tomato. Letters in Applied Microbiology. 38: 169-175.

Singh, H.P. and Malhotra, S.K. (2013). Trend of Horticultural Research, particularly in Vegetables in India and its regional prospects. p 321-343. In: SEAVEG 2012 Proceedings of the Regional Symposium on High Value Vegetables in Southeast Asia: Production, Supply and Demand (SEAVEG 2012), Bangkok January 24-26 2012.

Singh, S.P. (2004). Some success stories in classical biological control of Agricultural Pests in India Asia-Pacific Association of Agricultural Research Institutions (APAARI), Bangkok.

Sridhar Subhashini, Ashok Kumar, S., Abarna Thoyavathy, R. and Vijayalikshmi, K. (2013). Seed Treatment techniques. Centre for Indian Knowledge Systems. Chennai. India.

Srinivasan, K. and Krishna Moorthy, P.N. (1991). Indian Mustard as a trap crop for management of major lepidopterous pests of cabbage. Tropical Pest Management. 37: 26-32.

Srinivasan, K., Krishna Moorthy, P.N. and Raviprasad, T.N. (1994). African Marigold as a trap crop for the fruit borer, *Helicoverpa armigera* on tomato. International J. of Pest Management 40:56-63.

Tripathi, Arun, K., Shikha Upadhyaya, Mnatu Bhuiyan and Bhattacharya, P.R. (2009). A review on prospects of essential oils as biopesticide in insect-pest management. Journal of Pharmacognosy and Phytotherapy, 1: 52-63.

Verghese, Abraham., Sreedevi, K. and Nagaraju. D. K. (2006). Pre and Post harvest IPM for the mango fruit fly *Bactrocera dorsalis* (Hendel). Fruit Flies of Economic Importance: From Basic to Applied Knowledge. In: Proceedings of the 7th International Symposium on Fruit Flies of Economic Importance. Salvador, Brazil September (10-15). p. 179-182.

Vijayalakshmi, K., Radha, K.S. and Vandana Shiva. (1995). Neem: A users manual, Centre for Indian Knowledge systems, Chennai. India.

Wikipedia. (2017). List of companion plants (https//in.wikipedia.org/wiki/list-of-compinionplants(Retrived on 2 Sept 2017).

❍❍❍

V. Essential Farm Inputs

16

Challenges and Opportunities in Supplying Essential Farm Inputs for Sustainable Horticultural Development

Sudha Mysore

1. INTRODUCTION

Horticulture sector in India has emerged as a key sector in the last two decades. With an annual production of over 230 million tones, this sector has also surpassed the food production in the recent times. The average annual growth rate of production is estimated to be more than 4% in the coming few years. These figures only add to the country's global rank in agricultural production, such as, being first in milk, pulses, spices, largest area under rice, wheat or cotton or being the second highest producer next to China and Brazil w.r.t horticultural crops. However, with more than 60% of the Indian rural households continue to be dependent on agriculture as a main source of livelihood on one hand and the ever increasing demand for fruits vegetables and milk as a supplement for nutritious food that is arising out of raising incomes on the other, India needs a more productive competitive and sustainable agriculture sector.

Out of the total arable land about 195 m ha is gross cropped area with 141 m ha net sown area, hardly 65.3 m ha is with assured irrigation, while the rest continues to be rainfed (Anon., 2015) making it even difficult for India to attain its goal of a competitive agriculture. While Ensuring that agricultural growth responds to the growing food security needs is an impeding goal for Indian agriculture (FAO, 2014), increasing agricultural/horticultural productivity per unit area is one of the most significant challenges that needs immediate attention given the unsurmountable constraints in balancing the input supply mechanisms that Indian agriculture faces today. This paper attempts to provide an account of the changing status and prospects of input supply mechanism with specific reference to horticulture.

Ensuring the supply of quality inputs that ensure assured production has been a major challenge in agriculture, more so in case of horticulture. The specific input supply constraints revolve around, lack of access; inadequate availability; assured quality; non-availability in affordable packages; timeliness of supply; lack of knowledge; and finally absence of client specific and farmer friendly technologies/inputs.

The input supply mechanisms that are presently in place can be categorized into, those relating to seed and planting material, irrigation and water related ones, those relating to fertilizers, pesticides or plant health related products and inputs in the form of technological packages.

2. SEED AND PLANTING MATERIAL SUPPLY MECHANISMS

Seed and planting material are the most critical inputs as the response of all other inputs also depend on their quality for sustainable production. Estimates indicate that quality seed alone can contribute to nearly 20% of the total production while in combination with efficient management can be increased to nearly 45%. (Yes Bank, 2014).

While supply and maintenance of quality of seed and planting material in India is largely based in the public sector organisations, the role and contribution of private sector cannot be undermined. Several policy instruments have been enacted by the Government of India to ensure the supply of quality in seed and planting material. Starting from the Seed act of 1966 to the new seed policy of 1988 and further to the Enactment of seed bill 2004, Indian seed production and supply mechanism has come a long way. Enactment of the New Seed development policy (1988-1989) could be termed as a turning point which transformed the seed industry through opening up the sector for private investments and permitting import of quality seeds from across the world especially in high values crops like vegetables and cotton.

Though late than never, the seed bill and the national seed policy since the millennium have been emphasising on the issues of (i) plant variety development & protection;(ii) seed production & quality assurance, (iii) seed distribution and marketing, (iv) infrastructure facilities, (v) transgenic plant varieties, (vi) import & export of seeds and planting materials & (vii) strengthening and promotion of domestic private sector seed industry. Seed bill 2004 has been in focus for a long time with special provisions of registration of varieties, compensation to farmers in case of supply of spurious seeds and seed certification regulation of seed sales.

Seed and planting material multiplication forms an important economic activity that ensures additional income and employment for all the stakeholders involved in this activity, including the farmers and nursery men who are endowed with the requisite agro climate and infrastructure. On-farm multiplication of breeders' seed and certified seed in food grains and vegetables by the National and State Seed Corporations (NSC& SSCs) has been one of the oldest partnership model that ensured price stability in addition to ensuring forward and backward linkages.

Till 2006, when ICAR adopted its Intellectual property rights policy, research institutes under NARS system supplied breeders' seed to seed multiplication centres under the purview of state governments and private seed companies or nursery men at pre- fixed nominal prices as per their requirement indicated through the indent system. The state agencies in turn multiplied the seeds and planting material on their own facility or at selected farmers' fields for a fixed rate and distributed to different states as per the indents received. The private seed companies and nurserymen also multiplied the seeds for sale. This model ensured the primary objective of seed supply, the system did not provide quality assurance nor did the breeders have any claim / right over the intellectual property over their innovations. The system thus does not address the issue of sustainability nor quality assurance (Sidhu *et al.,* 2011).

Germplasm collection, conservation and utilization form a major mandate of research for a number of ICAR institutions. It is beyond doubt that such collections made by ICAR researchers provide valuable base material for crop improvement, which gets into the main stream breeding programme and to the private sector research through various research programmes, under the All India Coordinated Crop Research & Improvement Programmes (AICRPs). Germplasm collection, evaluation, characterization, screening for biotic and abiotic stress factors and utilization has been a major contributor towards crop improvement and disease resistance breeding irrespective of the crop involved. The value of the germplasm thus collected or the actual costs involved in the whole process of evaluation and selection have seldom been assessed and quantified.

The Indian seed programme adopts a three generation system of seed multiplication, *viz.*, the breeders seed, the foundation seed and the certified seed and thrives to provide the required protection and quality assurance along with maintaining purity.

2.1 Breeder Seed

Breeder seed is the progeny of nucleus seed of a variety and is produced by the originating breeder or by a sponsored breeder. Breeder seed production is the mandate of the Indian Council of Agricultural Research (ICAR) and is being undertaken with the help of;

(i) ICAR Research Institutions, National Research Centres and All India Coordinated Research Project of different crops;

(ii) State Agricultural Universities (SAUs) with 14 centres established in different States;

(iii) Sponsored breeders recognized by selected State Seed Corporations and

(iv) Non-Governmental Organizations.

ICAR also promotes sponsored breeder seed production programme through the National Seeds Corporation (NSC) / State Farms Corporation of India (SFCI), State Seeds Corporation (SSCs), Krishi Vigyan Kendras (KVKs) *etc.*

The requirement of seeds in the form of indents are collected from various seed producing agencies by the State Departments of Agriculture and submitted to the Department of Agriculture and Cooperation (DAC), Ministry of Agriculture, Government of India. This information gets sent to the Project Coordinator/Project Director of the different crops in ICAR, which in turn is diverted different SAUs/ICAR institutions for final allocation of production responsibility. These indents and the responsibilities get thoroughly discussed and each centre is then allotted the production target based on their capability and infrastructure available with them, at least a year and half ahead of the season. The seeds so produced also get monitored by the respective breeders and project directors and the National Seeds Corporation as well.

The actual production of breeder seed by different centres is intimated to DAC by ICAR. On receipt of information from ICAR, the available breeder seed is allocated to all the indenters in an equitable manner. Care is also taken to ensure the relevance of a specific variety to a region or a state. The breeder seed produced is lifted directly by the Director of Agriculture or foundation seed producing agencies authorized by him.

2.2 Foundation Seed

Foundation seed is the progeny of breeder seed and is required to be produced from breeder seed or from foundation seed which can be clearly traced to breeder seed. The state run organisations like NSC, SFCI, State Seeds Corporation, State Departments of Agriculture and few private seed producers, who have the necessary infrastructure facilities are entrusted with the responsibility of producing foundation seed. Foundation seed is required to meet the standards of seed certification prescribed in the Indian Minimum Seeds Certification Standards, both at the field and laboratory testing.

2.3 Certified Seed

Certified seed is the progeny of foundation seed and must meet the standards of seed certification prescribed in the Indian Minimum Seeds Certification Standards, 1988. In case of self pollinated crops, certified seeds can also be produced from certified seeds provided it does not go beyond three generations from foundation seed stage-I (Anon., 2017).

The production and distribution of quality/certified seeds is primarily the responsibility of the State Governments. Certified seed production is organized through State Seed Corporation, Departmental Agricultural Farms, Cooperatives *etc*. The distribution of seeds is undertaken through a number of channels i.e. departmental outlets at block and village level, cooperatives, outlets of seed corporations, private dealers *etc*. The efforts of the State Governments are being supplemented by NSC and SFCI which produce varieties of national importance. NSC markets its seeds through its own marketing network and also through its dealer network. SFCI markets its seeds mainly through the State Departments of Agriculture and the State Seed Corporations. The production of certified seed by NSC and State Seed Corporations is mainly organized through contract growing arrangements with progressive farmers. SFCI undertakes seed production on its own farms. The private sector has also started to play an important role in the supply of quality seeds of vegetables and crops like hybrid maize, sorghum, Bajra, cotton, castor, sunflower, paddy *etc*.

However, the scenario gets very different when F1 hybrids are involved. Since, vegetables include F1 hybrids, the system has been shifting. The private seed companies play a very important role in the supply of seeds for vegetables. It is to be noted that more than 400 or 500 small and medium sized private seed companies are operating in India. However, the main focus of private seed companies has been on the high value low volume seeds. The low value high volume seeds of cereals, pulses and oilseeds is still dominated by the public sector seed corporations. Atpresent 15 State Seeds Corporation and two National level seeds Corporations (National Seeds Corporation of India & State Farms Corporation of India) are functioning in the country. Besides, significant quantities of seeds are also produced by the State Departments of Agriculture, where the State Seeds corporations are not in existence. The contribution of private sector seed companies in total seed production of the country is substantial as seen in the Fig 1.

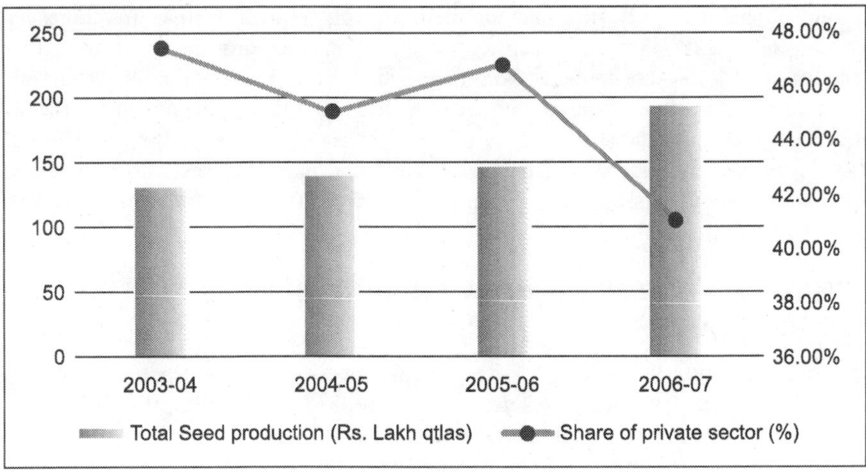

Fig. 1. *Total Seed production & Share of Private sector(%)*
(**Source:** *compiled by Seeds Division of DAC*)

Constraints in the existing seed production and distribution system that need attention include,

1. Labeling/certification and phyto sanitary certification of the seeds produced,
2. Disease-free seeds, bud woods of elite varieties with focus on solarization/sterilization, microbial amendments, grafting techniques, need for their popularization,
3. Tissue culture protocols need to be developed and commercialized in all crops. This activity needs to be taken up under PPP mode.
4. High-tech nurseries for quality planting material are needed in large numbers and Standards (Anon., 2010).

3. FERTILIZERS AND PESTICIDES

Private companies with or without foreign ancillaries have been the primary sources of fertilizers and pesticides to the producers irrespective of the crop. These companies through their established retailer network not only service the producers for their fertilizer needs but also help promote the sale of branded seeds of some of the reputed private seed companies. It is not an exaggeration, if it is said that these retail trade shops of fertilizer and pesticides at the village/town level have over the years become the primary hubs that have tilted Indian agriculture towards a high cost input use mechanism with very high and indiscriminate usage of fertilizers and pesticides. The large scale soil test data from different agro climatic zones in India indicated a total shift in the soil C:N ratio to alarming levels suggesting an over use of inorganic fertilizers. Similar is the case with the pesticide residues in soil/water in India.

In view of the importance attached to the use of balanced nutrition for crop productivity enhancement through fertilization, a number of fertilizers are subsidized under Indian conditions. These subsidies though helped in demonstrating and popularizing new agricultural technologies and modern inputs, but continuation of these subsidies beyond and forever needs to be revisited and curtailed. Though public as well as private research over the years has developed and demonstrated usefulness of organic forms of fertilisers and biopesticides that are more environment friendly and help reduce the over use of inorganic forms, these new types are not been able to reach the target audience at the level that is required due to the strong lobby of fertilizer and pesticide companies who don't want to lose their margins. There is thus a strong and immediate need to evolve better mechanisms of input supply to help reach large number of farmers and also to ensure the supply of safe and useful products.

4. CHANGING FORMS OF INPUT SUPPLY SOURCES FOR PRODUCERS

Efforts have been on to modify the input supply mechanisms for producers and make sure that the farmers have one stop solutions to their needs at the village level itself. Both public and private sector institutions have been making these new forms of input supply mechanism that are making dent in the farmers lives. Some of the notable ones include,

4.1 Setting up Cooperatives

Cooperatives have been the oldest and time tested form of farmer/producer organization that could provide them a bargaining power through collective action. However, the movement as such is said to have failed due to the red tapism that has crept into the system through the dominating role played by the increased interference by the government institutions. There is a need for introducing useful structural reforms to improve functioning of the cooperatives with greater efficiency and viability.

Several new models of cooperatives in the form of community based/ similar interest groups have emerged that have shown reasonable success over the co-operatives, yet adopting the useful points from collective action. Through effective participation and collective action, it has been demonstrated that critical and scarce resources like land, water, credit *etc* can be better managed. Farmer Interest Groups (FIGs), user associations and cooperatives, Self Help Groups (SHGs) and NGOs have shown useful examples of good and effective governance in managing to provide collective action and bargaining power but also in efficient dispensation of input supply and productive use of inputs.

4.2 Farmer Interest Groups Federated into District Associations

Because of resource constraints and meager holdings small farmers' capacity building is a major strategy which may lead to their empowerment for taking decisions on key farm issues. In India, under the NATP, FIGs operating at grassroots level on commodity/activity base are being strengthened. These village level groups have started federating at the block level which is the next level in a three-tier system, which in turn will further federate upwards into district unions. Representatives of district level body would be members on the Governing Board of the ATMA whereas, representatives of the block unions would function as member of the Farmer Advisory Committees (FACs) operating at the Farm Information and Advisory Centre established under ATMA at block level. This kind of federation would enable the farmers to take up variety of entrepreneurial activities at the farm level.

4.3 Strengthening Storage Facilities at Village Level

India has an organized network of procurement, storage and distribution of food grains through public agencies like Food Corporation of India, warehousing corporations and cooperative sector. Despite the massive infrastructure, small farmers need to store a sizeable quantity at village level for food, feed and seed purposes in addition to the stock meant for sale during need. Capacity of small farmers in terms of farm level storage infrastructure and technical know-how needs to be built up. Quality preservation before warehousing may be useful, as the farmers are entitled for 75 percent of the value of the deposited goods as loans from banks against pledging of the warehouse receipts.

4.4 One-stop Agricultural Support Centres: TATA Kisan Kendra

A venture of the renowned TATA Business house of India, TATA Kisan Kendra aims to support farmers through services, such as, (i) agro input service, (ii) farm equipment leasing, (iii) agronomy services, (iv)bulk blending, (v) training and information and allied services. In addition, these centres also help farmers procure seeds, pesticides and fertilizers at affordable prices. The 'crop clinics' established under the same banner is a variant to provide the information regarding which crop to grow, when to grow and the specific amount of nutrients required *etc.,* for the producers.

4.5 Hariyali Kisaan Baazaar (HKB)

DCM Shriram Consolidated Ltd. (DSCL), a leading diversified Indian corporate house with over 35 years of experience in the agri-input markets, has introduced an innovative business venture, Hariyali Kisaan Bazaar, which is an example of a commercially sustainable business model becoming the vehicle to bring about agricultural transformation and inclusive growth. This model encompasses a chain of agri-input retail stores and promoted by DCM Shriram Consolidated Ltd. (DSCL) offering services, such as, (i) Last-mile delivery of relevant Agri-technology, (ii) Agri-Inputs : Provide a range of inputs, such as, seeds, fertilizers, nutrients, pesticides, (iii) Financial Services : Access to credit, insurance & banking services; (iv) Market Linkages: Access to new markets/ buyers for farmers produce, (v) Warehousing & Commodity exchanges : Providing farmers with quality warehousing facilities & access to commodity exchanges, (vi) Household Goods: Provides a range of good quality household goods for the farmer's family. The primary objective of this initiative has been to promote inclusive growth through Investment in Rural Infrastructure, providing need based need-based infrastructure to the farmers. These facilities include, warehouse for agri-inputs or farm produce, an agri-advisory centre equipped with qualified agronomists, a veterinary centre with qualified veterinary doctors & availability of medicines, an information centre with IT linkage to a team of scientists. Availability of diesel and petrol. Addition of bank branches and ATMs to these centres will help in achieving the government's objective of taking credit to the micro level.

These centres also aim at Improving farmer's productivity and profitability by providing 24X7 technology support through a team of dedicated qualified agronomists, Critical last mile delivery of agri-advice to the farmers to ensure adoption of appropriate modern agri-practices. Focus is on shifting farmers from subsistence to technology- led commercial farming. Aggregation of Farm Produce and providing access to information and Use of IT are the other important features of this support.

5. CONTRACT FARMING

The model introduced by Pepsi Co. in 1998 as an Informal Contracts is the most accepted form of contract farming in India. In this model, the farmer is contracted to plant the contractor's crop on his land, expected to harvest and deliver to the contractor, a quantum of produce, based upon anticipated yield and contracted acreage, at a pre-determined price. The contractor in turn supplies the farmer with selected inputs that help enhance productivity of the crop. Contract Farming can be considered as a viable method to enhance agricultural productivity along with the assurance to provide the pre-determined prices to the farmers. Traditionally contract farming was confined to production of sugarcane, tea, coffee, cotton, milk, *etc*. Contract farming has been gaining ground as a solution for assured supplies of uniform quality for processors and traders. Corporatisation of contract farming has potential to boost farm incomes and to increase global trade of Indian agri-produce. The farmers, especially small farmers also benefit in the process from diversification, technological up-gradation and assured market. The pace of contract farming is on the rise after the amendment to APMC Act by the state governments.

5.1 Agri-business Franchising

An agribusiness franchise can be defined as — a right, permission or license (often established by contract) granted by an agribusiness firm (called the franchisor or franchising agency) to another agribusiness firm (called the franchisee) for the latter to distribute, manufacture and/or use the trade name of the farmer's products and services usually in a specified territory assigned to the latter firm by the former firm. (Fosu, 1989). There are several variations of Agri-business Franchising in India and one of them is a model adopted by SAPPL.

SAPPL is a private potato supply chain company- works through franchise model with a network of 36 franchisees in collaboration with a development project in Uttar Pradesh that provides farm input supply and produces buyback services to smallholders. The development project, Sunhara India, is funded by Bill and Melinda Gates Foundation and Implemented by Agribusiness Systems International (ASI), a US-based non-profit consulting firm. The franchisees are the hubs from where, farmers seek and obtain various services like input supply, extension advice and disposal of their output of potato on a pre-agreed price and market outlet. The franchisees are appointed by SAPPL based on their extensive experience with farmers and the potato crop and works in many states of India in potato seed supply and output procurement and in turn supplies to various potato processors.

The SAPPL franchise contract is a detailed one specifying the categories of the products to be supplied, all crop protection chemicals including bio-control agents, packaging materials, seeds, potato seeds, irrigation equipment and farm equipment and controls retail price of the products supplied by it to the franchisees. A franchisee is expected to offer a certain minimum investment for the business of franchising. The SAPPL trains the franchisees from time to time as per its contract and even to farmers who were clients of the franchisee. Through stringent processes, the company ensures that the product ownership is transferred to the franchisees on delivery and payment for the same. Important conditions for becoming a franchisee with SAPPL includes, a good farmer base, accounting knowledge, no criminal background and some investment capacity.

5.2 Other Innovative Approaches

In a unique system, the World Bank has been involved in an R&D in Agricultural Technology through two national level projects with pan-India implementation (the National Agriculture Technology Project and the National Agriculture Innovation Project) coordinated by the Government of India's Indian Council for Agricultural Research (ICAR), since 2006-07.

Notable features of this include, new approaches towards the dissemination of agricultural technology, such as, the Agriculture Technology Management Agency (ATMA) model, that helped in diversification of agricultural production in various states.

Sustainable agricultural practices through watershed and rainfed agriculture development (Karnataka, Himachal Pradesh, Uttarakhand), soil reclamation efforts (Uttar Pradesh) and, more recently, improved groundwater management practices (for instance, in the upcoming Rajasthan Agriculture Competitiveness Project).

Improved access to rural credit and greater gender involvement in rural economic activities through rural livelihood initiatives undertaken by a number of states (Andhra Pradesh, Bihar, Madhya Pradesh, Orissa, Rajasthan, Tamil Nadu) and soon to be scaled up by GOI with Bank support through a National Rural Livelihood Mission.

Agricultural insurance by advising GOI on how to improve the actuarial design and implementation of the insurance program (*e.g.* rating methodology and product design, index insurance, use of mobile and remote sensing technology to measure yields, *etc*.).

Better rural connectivity through IDA support to the Prime Minister's National Rural Roads Program (PMGSY) and by connecting rural poor and smallholder farmers through collective action to public services through Self-Help Groups (and SHG federations), Water User Associations and Farmer Producer Organizations. Recently the Bank's Board of Executive Directors approved the National Rural Livelihood Mission, which supports SHG approaches through a pan-India approach.

5.3 Initiatives Under the Indian Council of Agricultural Research

Setting up of Institute technology management Units & Business Planning and development units across different ICAR research institutes has been one of the most successful new approach adopted through the NAIP funding. These ITMUs have been commercialsing viable technologies through Licensing to the private entrepreneurs so as to enable access of these technologies across different farmers at reasonable price without losing the quality. Under this initiatives, institutions like the Indian Institute of Horticulture Research (IIHR), Bengaluru have commercialized several technologies under Five theme areas, *viz,* (i) Seed and Planting material, (ii) Plant health management products, (iii) Post harvest technologies , (iv) farm equipment and machinery and (v) biotechnological products.

Under the theme area, seed and planting material, the institute licenses the parental lines of improved varieties and F1 hybrids to seed companies for an up front value, under an MoU. The companies produce the seed and supply to the producers. Further, the institute itself undertakes the quality seed production of F1 hybrids under the Seed Village concept benefiting the farmers for seed production and also ensuring regular uninterrupted supply of seed to the growers.

More than 300 biopesticide companies have licensed Biopesticides like *Trichoderma Viridae*, *Trichoderma Harzianum* and others on license and mass produce the same and supply to needy farmers. Other useful bio and microbe based products and crop specific nutritional formulations also have been licensed for ensuring mass production and supply to farmers.

It is imperative that such novel and innovative models need to be initiated as to ensure uninterrupted supply of quality inputs to farmers for enhancing production and productivity of horticultural crops.

REFERENCES

Anon. (2010). Proceedings of the National Conference on Production of Quality Seeds and Planting Material – Health Management in Horticultural Crops, March 11-14, 2010, New Delhi – A Detailed report.

Anon. (2015). http://www.gktoday.in/blog/gross-cropped-area-and-net-cropped-area-in-india/

Anon. (2017). seednet.gov.in/Material/IndianSeedSector.htm.

Parwez, Sazzad. (2014). Value creation, sustainability & integration of agricultural supply chain of India, AIMA Journal of Management & Research, August 2014, Volume 8 Issue 3/4, ISSN 0974 – 497.

Sharma, Rita, Swanson Burton, S. and Sadamate, V. V. S. (2001). Field-Testing New Methodologies for Planning and Implementing Extension Programs: A Comparison of Innovative Extension Projects in India and the United States, paper presented at the AAIEE, Baton Rouge, Louisiana.

Sharma, Rita. (1999). Agricultural Research and Extension Interface – Indian Experience, Agricultural Research and Extension Interface in Asia, APO, Tokyo.

Sidhu, Amrik Singh, Krishna Kumar, N.K., Hegde, Krishna, M.R. Moorthy, A., Ganeshan, S. & Mysore, Sudha .(2011). Ensuring commercial production of quality planting material in Horticulture, Public Private Partnership Models and success stories, National Seed Association , April-June 2011, p. 35-41.

YES BANK. (2014). Food and Agribusiness Strategic Advisory and Research (FASAR) Group, Agri-Infrastructure in India - The Value Chain Perspective. p. 5.

❍❍❍

VI. Linking Horticulture Produce to Marketing

17. Linking Horticulture Producers to Markets
18. Linking Horticulture Produce to Market in achieving Sustainable Horticulture
19. Meeting the Crisis and Challenges in Prices: In Fruits and Vegetables
20. Skilled and Smart Farmers Organizations Driving Farm and Rural Development

17

Linking Horticulture Producers to Markets

P.G. Chengappa, N.C. Mamatha, C.M. Devika and K. Vijayasarathy

SYNOPSIS

Linking horticultural producers, like other agricultural producers to markets has been a major challenge in India. Fruits and vegetables being perishable, markets being fragmented and unorganised with superfluous intermediaries pose multiplicity of problems in linking growers to markets.The supermarkets or food retail chains (FRCs) are making efforts to link vegetable farmers to markets by establishing backward and forward linkages. The concept of ready to retail adopted by the FRC is to the benefit of the farmers as they realized higher returns by adopting improved technology and growing high value horticulture crops. Direct supply by farmers allowed the retail chain to simultaneously increase control over quality, supply reliability and price stability. Institutional approach, such as, cooperatives, associations, self help groups and producer companies have proved successful in terms of procuring required quality inputs at competitive prices and sale of the produce in an efficient manner. Contract farming is touted as a strong viable alternate channel for linking horticultural farmers to markets. Contracting institutions provide quality inputs and technical support in production and marketing. The emerging electronic marketing of horticulture produce are virtual market place where farmers transact directly with a processor or exporter and realize better price for their produce. These emerging institutional arrangements have helped in providing alternate market channels for horticultural farmers to break away from the clutches of traditional players. However, the sustainability, replicablity and up-scalability of these new interventions are a still a challenge as they have succeeded only in few regions and crops.

Keywords: Food retail chains, intermediaries, contract farming, producer collectives, e-marketing

1. INTRODUCTION

Linking farmers to markets can seize a whole range of activities, from the very small and localized to the very large.The concept does, however, assume the development of long-term business relationships rather than support for ad hoc sales (Shepherd, 2007). Linkages between farmers and market call for priority attention to issues on access to technology, information on institutional arrangements, support services, policies, capacity building, identification and development of markets (MANAGE, 2013). The linkage between perishable products is rather weak under Indian conditions attributable to reasons like small scale and dispersed nature of producers, inadequate arrangements for supply of quality inputs at right time, unorganized and fragmented markets, superfluous middlemen in the supply chain, low infrastructure contributing to higher logistic costs and absence of corporate involvement in establishing forward and backward linkages. Absence of efforts at percolating market information has been a major constrain in the efficient functioning of the markets and horticultural produces are no exception. In the recent years there have been some concerted efforts to link farmers to the markets. These efforts include creation of alternate marketing channels which provide better pricing and reducing the transaction costs.

Horticulture, comprising of number of fruits, vegetables, flowers, spices, medicinal and aromatic crops is a key sector in the Indian economy. More than 60 types of vegetables are grown in India in tropical, subtropical and temperate agro-climates. The Fruits and Vegetables (F&V) sector has been a driving force in stimulating a healthy growth trend in Indian agriculture. India is the second largest producer of fruits and vegetables (Horticultural Statistics, 2015). This sector provides ample opportunities to enhance farm incomes as generally horticultural crops are more profitable as compared to most farming activities. Besides, it provides ample employment opportunities and scope to raise the income of the farming community. In fact, horticultural production provides the best opportunity for the estimated 1.5 to two billion people living in smallholder households to escape poverty as the returns per unit area are much higher as compared to food grains. F&V together constitute about 92 % of the total horticultural production in India. During 2013-14, India's contribution in the world production of fruits was 13.6 % and vegetables 14 % respectively (NHB, 2015).

Vegetables being perishable, markets being fragmented and unorganised with superfluous intermediaries pose multiplicity of problems in linking horticultural growers to markets. The major marketing channels in horticultural crop involves a number of intermediaries like the pre harvest contractor, assemblers, transporters, commission agent, wholesaler and retailer operating between the producer and the final consumer. In developed country market, there has been a rapid shift from sales through open market to direct sales that involve linkages and alliances from production to consumption. Whereas, in developing countries the farmers are linked to buyers by identifying traders/commission agents who in turn link producers to final buyers. Technological developments in horticultural crops have facilitated some diversification. Varieties of horticultural crops have been developed for cultivation in off season, under diverse climatic conditions and with various attributes to attract consumers. Adopting different models for collective marketing by the small and marginal farmers will help to realise better access to input and output market and increased share in the consumer rupee. This can be through institutional approaches, such as, cooperatives, growers associations and self help groups. The group approach is much needed for getting benefits from marketing. Farmers can benefit from the emerging super markets and value chains if linked effectively. In this paper we have attempted on linkages witnessed in India *viz;* Linkage through retail food chains (super markets); Contract Farming; Producer collectives and e-marketing for fruits and vegetables.

1.1 Need for Linking Farmers to Market

Food basket is undergoing a significant change in India. There has been a perceptible consumption shift towards horticultural produce. F&V are rich source of vitamins, minerals, proteins and carbohydrates which are essential in human nutrition. As the population is increasing, the demand for such food is also increasing. To meet the demand of proper nutrition, supply chain plays a very vital role. Owing to the very short shelf life and perishable nature, these items require proper transportation, handling and storage facilities in order to reach in fresh state to a customer (Veena *et al.,* 2011). The movement of produce from farm to ultimate retail outlets or consumers through traditional supply chain faces a number of constraints and farmers are at the mercy of the middlemen particularly the commission agents. In India, the fruits and vegetables farmers receive only one-third to one-half of the final price (Chakrabarti and Mukhopadhyay, 2009). This indicates that the existing supply chains are long and crowded and for that marketing costs and margins are high. Fresh Fruits and Vegetables (FFV) in India is marketed mostly either through regulated Agricultural Produce Marketing Committee (APMC) markets or totally unregulated local F&V markets. Marketing through these traditional channels is characterized by very little attention to grading, sorting and storage, poor handling during loading, unloading and transport resulting in loss of 30-40 percent of production (Singh and Singla, 2011). Thus it is high risk business and requires good market linkage for viability. Moreover, F&V crops are more suitable for small holders as they are more labour intensive, provide recurring income and a mechanism of risk management against field crop failure. But, they are more input intensive, require more post-harvest handling, more perishable and their profitability is dependent on market (ibid).They suffer from high wastage or rejection, there is no Minimum Support Price (MSP) or alternative protection against price risk. Fluctuations in the output price are one of the most important problems for the small farmers (Praveen, 2014). Horticulture commodities are highly seasonal that results in fluctuation of prices from month to month. These fluctuations ultimately affect the return to the growers. Imperfect coordination between supplies and demand and lack of infrastructure facilities plays an important role in their marketing. Other factors that contribute to rise in prices of horticulture products are: increase in the demand for protein-enriched food and green vegetables; spoilage due to inadequate storage; hoarding; increased cost of transportation and less importance given for cultivation as compared to cereals and pulses. The strong surging demand for horticultural commodities to meet the nutrional security and reducing market inefficiency to improve the returns of horticultural grower necessitate linking horticultural farmers to the markets.

2. LINKAGE THROUGH FOOD RETAIL CHAINS (FRC)

Retailing in India is highly unorganized, fragmented and predominantly small due to family owned business and poor access to capital, technology and regulations. Retailing has undergone transition with the rise of FRCs or super markets. During the process of evolution there was gradual shift from fragmented local markets to larger centralized wholesale markets. Emergence of organized FRCs were witnessed after liberalization and entry of corporate entities. Emergence of FRCs is due to changing political, demographic and economic features like rise in economic growth and income, proliferation of brands and consumer awareness. In India organized retail evolution initiated during pre 1990s. The retail industry is the second largest employer in India, next to the farming sector. The retail industry provide employment to about 40 million people of which about 0.5 million are employed in the organized retail sector all in urban areas (Sing and Singla, 2011). As e-commerce offerings continue to evolve, nowhere is the impact of their efforts felt more than in India's traditional retail industry.

2.1 Farmers' Access to Markets

During recent years organized retail sector has adopted various formats to source food commodities including fruits and vegetable. These mainly include backward integration with fruits and vegetable growers for procurement, either directly or indirectly and selling through the outlets of the FRCs with in-house brands like: Spencer's, Reliance Fresh, More, Namdhari Fresh Limited, ITC's Choupal Fresh and few others. These FRCs have attempted many changes in the supply chain management and logistics of F&V through the use of quasi-formal and formal contracts to ensure timely delivery of products with desired quality attributes. Even farmers preferred to supply their produce to the fruits and vegetable consolidation center (FVCC) operated by the FRCs, as it provided them stable prices and assured market, compared to the highly volatile prices at the wholesale market (Chengappa *et al.,* 2016). Therefore, they can be viewed as new institutional mechanisms for linking farmers with modern markets and improving supply chain efficiency and farmer livelihoods. Since the Foreign Direct Investment in retailing is allowed only in single brand chains, most of the supermarket growth in India has been driven by the domestic players unlike in many other developing countries of Asia and Latin America (Singh and Singla, 2011).

The organised retailers pursue two different business models: Hub and Spoke Model (HSM) and Value Chain Model (VCM). Organized retailers like Spencer's Retail, More and Food Bazaar are adopting 'Hub and Spoke' business model of retailing fruit and vegetables. In this model, farmers organised retailers, wholesalers and customers form the main actors in the chain. Small farmers and contract farmers who executed a trade contract with the organised retailers are the primary source of supply of vegetables to the organized retailers. The Reliance Fresh follows VCM procuring the vegetable directly from farmers and sell to customers by avoiding intermediaries. This model is based on its core growth strategy of backward integration and progressing towards building an entire value chain starting from the farmers to the end consumers. Here farmers organised retailers and customers form the main actors in the value chain. Small farmers, contract farmers and lease farmers are the primary source of supply of vegetables to the organised retailers. Independent farmers supply their produces to the consolidation centres managed by the FRCs; contract farmers and lease farmer's produce are picked up by consolidation centres. The consolidated centre supplies vegetables to multiple hubs, depending upon the requirement of the product.

The different traditional and modern marketing models or channels for different horticultural crops are presented in the Table 1. Most common marketing channel for all vegetable crops involves: P-W-R-C and in some of the cases middlemen will come in the place of wholesaler followed by P-R-C and P-C. In different states for the same crops different marketing channels exist because of the situations and convenience. Only few states like Punjab and Karnataka are practicing direct marketing models. The direct market models are the best because the producer share in consumer rupee is 100% and will completely eliminate middlemen (Dastagiri and Immanuelraj, 2013).

Table 1. *Preferred marketing channels or models for horticultural crop in different states in India*

Andhra Pradesh	Crops	Potato	Tomato	Rose	Grape
	Preferred channel	P-W-R-C	P-M-W-R-C	P-W-R-C	P-M-W-R-C
Tamil Nadu	Crops	Brinjal	Potato		
	Preferred channel	P-W-R-C	P-W-R-C		
West Bengal	Crops	Brinjal	Bhindi	Guava	Marigold
	Preferred channel	P-M-W-R-C	P-M-W-R-C	P-M-W-R-C	P-M-W-CA-R-C
Manipur	Crops	Tomato	Cabbage	Anthurium	
	Preferred channel	P-R-C	Cabbage	P-R-C	
Rajasthan	Crops	Tomato	Carrot	Kinnow	
	Preferred channel	P-CON-C.A-R-C	P-CON-C.A-R-C	P-CON-W-R-C	
Punjab	Crops	Potato	Tomato	Brinjal	Okra
	Preferred channel	P-C	P-C	P-C	P-C
Karnataka	Crops	Banana	Tomato		
	Preferred channel	P-HOPCOMS-C	P-HOPCOMS-C		

P = producer, W = wholesaler, M = middlemen, CA = commission agent, R = retailer, C = consumer, CON = contractor; HOPCOMS-horticultural produce cooperative marketing society Ltd.

Source: *Dastagiri and Immanuelraj, 2013.*

2.2 Profitability of FRC and Non FRC Farmers

Profitability and transaction costs of FRC and non FRC farmers for four major vegetables, namely, cabbage, cauliflower, carrot and tomato are depicted in table 2. Market price for FRC farmers are comparatively higher than non FRC farmers. Further, FRC farmers had considerably lower transaction costs as compared to non FRC farmers. For example, the transaction cost worked out Rs 180/ton in Cabbage, Rs 189/ton in cauliflower, Rs 775/ton

Table 2. *Net returns in vegetables production under FRC and non FRC farmers.*

Particulars	Cabbage		Cauliflower		Carrot		Tomato	
	FRC farmers	Non FRC farmers	FRC farmers	Non FRC farmers	FRC farmers	Non FRC farmers	FRC farmers	Non FRC farmers
Price received (Rs/tonne)	3490	3000	8430	7000	15500	14000	6540	5500
Transaction cost/ tonne (Rs)	180	700	189	1200	775	1905	640	1000
Total cost/ tonne (Rs)	1077	1739	2060	3219	3364	4093	2036	2550
Net returns/ tonne (Rs)	2413	1261	6370	3781	12136	9908	4504	2950

Source: *Chengappa et al., 2016*

in carrot and Rs 640/ton in tomato as compared to Rs 700/ton, Rs 1200/ton, Rs1905/ton and Rs1000/ton for non FRC farmers respectively. Net returns was higher in all crops for FRC farmers compared to non FRC farmers. This clearly indicated that the FRC linkage has been beneficial to farmers as they received higher prices and lower transaction costs.

3. MARKET LINKAGE THROUGH CONTRACT FARMING

Contract farming is an agreement between farmers and processing and/or marketing firms for the production and supply of agricultural products under forward agreements, frequently at predetermined prices (Eaton & Shepherd, 2001). Basically, contract farming is aimed at assured buy back and hence, assured price and supply chains for creating and sustaining value addition for the produce (Dileep, *et al.,* 2002). The basic elements of contract farming include pre-agreed price, quality, quantity and timely delivery of produce. Eaton and Shepherd (2001) classified five major models in which contractual arrangements occur between the firm and farmers. A sponsor (contracting firm or corporation) decides to follow a model depending on the market demand, production, processing requirements and economic and social viability of the farmers.

1. **Centralized model:** Here a sponsor buys from large number of small farmers. The firm provides inputs, such as, seeds, fertilizers, pesticides, credit and machines. This model was used for annual crops and crops which often require a high degree of processing. Tomato processing by PepsiCo in Punjab taking control of most of the production involving a group of farmers, processing and marketing is a typical example of centralized model.
2. **Nucleus Estate model:** It is a variation of the centralized model where the sponsor also manages a central estate or plantation. It is mainly found in case of fresh vegetables and fruits for export (Badodiya *et al.,* 2012). Namdhari's Fresh, a unit of Namdhari Seeds started in year 2000 at Bengaluru practices contract farming for export fresh vegetables and provide quality produce to domestic customers. Presently, more than 40 different vegetables and fruits are being cultivated by Namdhari's Fresh at various productions centers across the country. The different agro-climatic zones enable them to produce most vegetables all round the year. European standards of Good Agriculture Practices (GAP) are strictly followed at production level. In addition to their own captive production, Namdhari's Fresh has contract production involving farmers. The company supplies quality input, technical backstop and buys back the vegetable produced by the contract farmers.
3. **Multi-partite Model:** The multipartite model usually involves the government, statutory bodies and private companies jointly participating with the local farmers. Each entity may be responsible for providing different goods and services, such as, credit, inputs, equipments, transportation, processing and marketing facilities. An example is the firm, Rallis which provides all inputs, technical support and finance to registered growers for a specific crop and facilitates the sale of produce at reasonable prices. The company follows a consortium approach. It has tied up with banks and buyers. The system is operated through a network of 10 Rallis Kissan Kendras (RKKs) across the country.
4. **Informal model:** It is characterized by individual entrepreneurs and small companies. The crops usually require only a minimal amount of processing or packaging for

resale to the retail trade or local markets like as vegetables. Typically, the firm's involvement in actual production and input procurement is very minimal. Many supermarket chains in India usually follow this model to ensure a steady supply of produce (Manjunatha *et al.,* 2016).

5. **Intermediary model:** This involves sponsors providing contracting linkages to farmers with intermediaries like collectors, farmer groups and NGOs. The intermediaries have informal arrangements with farmers, thus it disconnects the link between company and farmer which is the main disadvantage of this model.

3.1 Positive and Negative Impacts of Contract Farming

The proponents of contract farming argue that there is an increase in income of participating farmers due to improved access to market, employment in farm, better use of resources and better management practices. Firms enable the farmers to reduce their price risk through predetermined prices, production risk through provision of technology and extension services. Further, farmers use the contract agreement as collateral to arrange credit with a commercial bank in order to fund inputs (Badodiya *et al.,* 2012). In addition, the potential advantage for sponsors could be overcoming land constraints, political acceptability, production reliability and shared risk and quality consistency. On the other hand, they also face problems from social and cultural constraints, land availability, farmers' discontent, extra-contractual marketing and input diversion. Most contract farming schemes occur only in the regions where commercial agriculture are already established and offering reliable infrastructural access to markets (Manjunatha *et al.,* 2016). Due to unequal power in terms of bargaining, negotiating or enforcement of agreements contracts between agribusiness firms and farmers tend to be unfair and lead to exploitation.

4. LINKING TO MARKET THROUGH PRODUCER COLLECTIVES

Producer collectives are a new avenue for the smallholder producers by bridging the gap between productivity and market accessibility through a guaranteed market for produce and access to machinery and modern technologies (Raju and Singh, 2014). Producer collectives include rural cooperatives, agricultural /horticultural cooperatives, Farmer Produce Organization (FPO), Non−Government Organizations (NGOs), Civil society organizations, private sector promoted POs, Self-help Groups and producer companies. Collectives are consistent, reliable and generally pay higher price to their farmer suppliers than local middlemen through signing forward purchase contracts several months in advance of the harvest to supply products to buyers.

Cooperatives: Cooperatives are formal, legal and registered organizations, whose main objective is to meet common economic, social and cultural needs. They are jointly-owned and democratically controlled enterprises. Most rural cooperatives are engaged in supply of inputs with only few in marketing. To tackle farmers' weak bargaining power and high commission charges, the Horticulture Producers' Co-operative Marketing and Processing Society Ltd. (HOPCOMS) was founded in 1959 in Bengaluru. The Board of HOPCOMS consists of 20 members, of which 11 members are elected from among the producers and the rest are the Govt. nominees. As of 2012 the HOPCOMS has 16221 growers as members. Main objectives of HOPCOMS is to promote and encourage the development of horticultural produce by training and providing inputs, technical advice and marketing facilities. Based on the rates released by the Agricultural Produce Market Committee (APMC), HOPCOMS fixes the procurement and sale price for different fruits and vegetables brought for sale by its members. HOPCOMS offers multiple facilities to the farmers. It purchases the horticulture produces directly from farmers at competitive prices; provide transport facilities from collection centers to HOPCOMS main procurement centre; ensures correct weighment and immediate cash payment; arrange for essential input supply, such as, fertilizers, plant protection chemicals, garden implements and seeds at reasonable prices. Arrange subsidized supply of plastic crates to its farmer members for handling fruits and vegetables. Conducts farmers tours and technical guidance on production and processing of Horticultural crops.

Farmer Producer Organisations: FPOs is to collectivize small farmers for backward linkage for inputs like seeds, fertilizers, credit, insurance, extension services and forward linkages, such as, collective marketing, processing and market-led agriculture production (Mondal, 2010). These are perceived to empower farmers through collective bargaining along with instilling an entrepreneurial quality to farming. Some are formally registered, others are not. Scaling and sustainability is a main problem in FPOs.

Non Governmental Organisations: NGOs and Civil society organizations are largely organized on a broad farmers welfare goals. Their activities include capacity-building, agricultural extension, community projects and participation in local government decision-making. It strengthens and influence to create new market segments like fair trade organic markets and environment friendly markets. However, long run sustainability is a major problem.

Corporate Social Responsibility (CRS): Private sector organizations under their CSR program promote POs by providing input, credit and delivery of extension services. It provide link to POs for supply of agricultural products or raw materials. They are engaged in out-grower schemes or contracts for the production and delivery of specific products like seeds and vegetables. These POs do not face the challenge of sustainability as a part of the corporate profit is earmarked for such activities. Syngenta Foundation India (SFI) is in a strong and unique position to address the increasing complexity of the challenges farmers face. During 2009-13, SFI embarked the theme –'market-led extension' (MLE), where the emphasis was to 'produce together and sell together'. Essential features of this approach included linking vegetable producers' groups with markets through fewer intermediaries. The major components of MLE in vegetables include facilitating purchase of agri-inputs, providing knowledge of production and marketing of vegetables and linking farmers to markets. In addition SFI diversified using agri-entrepreneurs (AEs) to anchor market-led extension in vegetables. SFI supports the entrepreneurs at business start-up. Thereafter, they work independently with neighboring farmers on a commercial basis. SFI connects the entrepreneurs to bank, which provides credit both to the AE and their farmer clients.

Producer companies: The name 'Producer Company' is provided in order to indicate that the members have to necessarily be a primary producer. It is a hybrid between a Private Limited Company and a Cooperative Society as it minimizes interference from other external bodies thus ensuring that the real power rests with members of the company itself. Such a company functions like any other private limited company with only difference being that the profits earned flows back to farmers who own the company. MAHA Farmers Producer Company Ltd., (MAHAFPC) in Maharashtra is one such example of a successfully operating FPO. These State level Association (SLA) are involved in Agriculture and allied value chain. The SLA are both a 'for-profit' or a 'notforprofit' entity, working directly with Farmer Organizations to help to mitigate the difficulty of farmers through institutional development services, capacity building, system and process integration, technical training, dissemination of information, financial intermediation, market linkage development, policy advocacy and facilitating social security and economic development.

5. ELECTRONIC MARKETING OF HORTICULTURAL PRODUCE

Lack of information and access to markets are the biggest impediments to crop planning and higher farm gate realization. Agriwatch is an information distribution service provider for farmers and traders with access to market and technical information on crops and other commodities which they disseminate through internet, emails, newspapers, magazines, SMS and web services. It has around 25000 paid subscribers spread throughout the nation who benefited from the provision of market rates through the web, magazines, newspapers and SMS services (Singh, 2005).

Indian government launched a unified pan India trading portal eNAM (e-National Agriculture Market) for farmers. This was designed to create a unified national market for agricultural commodities. The NAM portal provides a single window service for all APMC related information and services. This includes commodity arrivals and prices, buy and sell trade offers, provision to respond to trade offers and among other services.While material flow (agriculture produce) continue to happen through mandis, an online market reduces transaction costs and information asymmetry (NAM, 2016). NAM addresses diverse state multiple APMC marketing regulations by creating unified market through online trading platform. It promotes uniformity, streamlining of procedures across the integrated markets, removes information asymmetry between buyers and sellers and promotes real time price discovery which is based on actual demand and supply. NAM promotes access to a nationwide market for the farmer, with prices commensurating with quality of the produce. Provides online payment and availability of better quality produce and at more reasonable prices to the consumers.

The e-choupal initiative of the Indian Tobacco Company Limited (ITC) is one such effort. E-choupal is a virtual market place where farmers can transact directly with a processor and can realize better price for their produce.The ITC provides information on market prices, agronomic practices, inputs and weather through internet kiosks free of charge. A farmer can sell his produce to ITC at the market price and can also avail inputs from them (Birthal *et al.,* 2007).

6. CONCLUSION

Horticulture sector has been a thrust area for increasing income and employment in India. However, the linkage between horticultural farmers to markets in India has been poor. Domination of pre harvest contractors and commission agents still persists in the marketing chain of fruits and vegetables. On other side, new value chains are emerging due to new institutional arrangements but are highly regional in nature limiting the market access and development of markets. Farmers in most of the states are still stuck to traditional marketing channels as the new formats are yet to make a dent. Adopting different institutional approaches for collectivization, such as, horticultural producers cooperatives, growers associations, self help groups and contract farming will help the small and marginal farmers to realise better access to input and output markets. These types of group approaches should be well supported and encouraged to increase the bargaining power of farmers. Electronic marketing improves competitiveness in market through larger participation of buyers with all the market information.

REFERENCES

Badodiya, S. K., Gour, C. L., Sharma, P. and Ram, H. (2012). Contract Farming: An Approach for Agriculture Development. Indian Research Journal of Extension Education. 2: 145-147.

Birthal, P. S., Jha, A. K. and Singh, H. (2007). Linking farmers to markets for high-value agricultural commodities. Agricultural Economics Research Review, 20.

Chengappa, P. G., Mangala, K. P. and Dega, V. (2016). Is Farmer–Food Retail Chain Linkage Feasible? In Organised Retailing and Agri-Business, Springer India. 241-250.

Dastagiri, M. B. and Immanuelraj, T. K. (2013). Supermarket Supply Chains in Horticulture in India: The Novel Marketing Models, Effects and Policies. Agrotechnology.

Dileep, B.K., Grover, R.K. and Rai, K.N. (2002). Contract farming in tomato: An economic analysis. Indian Journal of Agricultural Economics. 57(2): 197-210.

Eaton, C. and Shepherd. A. (2001). Contract farming: Partnerships for growth. Agricultural Services Bulletin 165. Food and Agriculture Organization. Rome.

Horticultural Statistics. (2015). Horticultural Statistics at a Glance. Department of Agriculture, Cooperation & Farmers Welfare, Ministry of Agriculture & Farmers Welfare Government of India. Accessed from http://agricoop.nic.in/imagedefault/hortstat_glance.pdf.

MANAGE. (2013). Training Programme on Linking Farmers to Market. Accessed from http://www.manage.gov.in/studymaterial/LinkingFarmers-E.pdf

Manjunatha, A. V., Ramappa, K.B., Lavanya, B. T. and Mamatha, N.C. (2016). Present Status and Prospects of Contract Farming In India. International Journal of Agriculture Sciences. 8(7): 1072-1075.

Mondal, A. (2010). Farmers' Producer Company (FPC) Concept, Practices and Learning: A Case from Action for Social Advancement. Financing Agriculture. 42(7): 29-33.

NAM. (2016). Small Farmers' Agribusiness Consortium Department of Agriculture, Cooperation & Farmers' Welfare Ministry of Agriculture and Farmer's Welfare, Government of India, Accessed from http://www.enam.gov.in/NAM/home/about_nam.html.

National Horticulture Board. (2015). Ministry of Agriculture and Farmers Welfare Government of India.

Praveen, K. V. (2014). Linking small farmers to market. Markets, Trade and Institutions for Agricultural Development, 1(1.38), 98. Accessed from http://iari.res.in/files/Divisions/CAFT_Training_Manual_2014.pdf#page=98.

Raju, K.V. and Singh, P.K. (2014). Producers' Collectives in Sustainable Agri-value Chain, Rural Entrepreneurship and Enterprise for Inclusive Growth, 3rd National Rural Management Symposium, 20-21November, KIIT University Bhubaneswar.

Singh, A. (2005). Case study: Institution based information systems, India indian agribusiness systems pvt.Ltd. AGRIWATCH, Accessed from http://www.eagriculture.org/sites/default/files/uploads/media/Agriwatch.pdf.

Singh, S., and Singla, N. (2011). Fresh Food Retail Chains in India: Organisation and Impacts CMA Publication No. 238 (1). Allied Publishers.

Shepherd, A. (2007). Approaches to linking producers to markets. Food & Agriculture Org.13.

Veena, Babu, K. N., and Venkatesha, H. R. (2011). Supply Chain: A Differentiator in Marketing Fresh Produce. The IUP Journal of Supply Chain Management, 8(1), 23-36.

❍❍❍

18

Linking Horticultural Farmers to Market for Increased Sustainability—Concepts and Cases

Sudha Mysore

1. INTRODUCTION

In the rapidly changing trade order that is driven by technology, distances seem to be shrinking leading to what is typically termed 'an integrated production space' where all essentials including food cross-over miles to meet the demand. Integrated production and supply arrangements hold promise for ensuring efficient supply chains operating across regions. Growing living standards lead enhanced awareness towards balanced nutrition is increasing the demand for safe and nutritious food despite the increased food miles, there by suggesting that an efficient production system alone may not suffice unless they are also 'well linked to the markets'.

The paradigm shifts in the new economic order have shown shifting focus and investment into sectors hitherto have remained unnoticed. Increased focus is being laid on integrated production and supply chains operated by investment rich international players through modernized retail supply chains. The pressure is on to increase investment into the agriculture based supply chains which appear to promise modernizing the agriculture/ horticulture supply chains for enhanced value addition and reduced post-harvest losses in handling and transit. One of the major contributors to this change is also the growing revolution in means and mode of communication. These proposed changes have an unparalleled bearing on the agrarian sector of our economy as well, with more than 60% of the population dependent on farming as a livelihood option, for rapid modernization and set to get integrated into such supply chains.

Horticulture, comprising a number of fruits, vegetables, flowers, spices, medicinal and aromatic crops has emerged a key sector in the Indian economy, surpassing even the food grain production in the last one decade. The sector has been recognized as a potential contributor and crops of future, in view of their sheer higher profitability both in domestic and export markets, higher employment generation potential as also providing the much needed balanced nutrition for increased societal welfare. The ever increasing budgetary allocation by the Government of India to this sector from a mere Rs. 38 crores during the IV plan period (1975-80) to over Rs. 4,500 crores in the X plan, to over Rs. 69,500 crores in XII plan highlights the growing significance of this sector (Sudha *et al.,* 2013).

The sector's performance in terms of increased production and productivity, expanding export earnings, besides the other micro level indicators, such as, the unit value realization of different commodities in export earnings, proportionate share of producer in consumer's rupee and changing cropping patterns are all suggestive of growth and expansion. The last decade in fact has witnessed change at every stage at a pace which has hither too not been envisaged as a remote possibility. Each one of these changes have impacted the horticultural production ecosystem on the whole and have had specific impact on the overall production and productivity of these set of crops but have very specific and yet significant mark on horticultural produce marketing, especially with the issue of 'Linking farmers to market'.

The concept of linking farmers to market is not new, as various forms and types have been attempted by various organizations with varying degrees of success. The advantages of an assured linkages between producers and consumers are enormous. Besides the fact that producers are assured of a potential buyer at an assured price, all other stakeholders along the market chain also are assured of a pre specified quality and quantity, thereby improving the overall efficiency of the market chain. Attempts at linking producers to markets help in bulking up the produce there by reducing the transaction costs and bring in economies of scale. While such arrangements may increase the requirement of man power in providing extension support, increasing the employment opportunities as well as the transaction costs.

The concept of linking farmers to market has been put into effective implementation not only by the government organisations but also the non-government or the producers themselves. Each one of such attempts have had varying degree of impact based on their scale of implementation, while much needs to be accomplished in this sector as yet. This article attempts to put together some of these efforts and their impact and deliberate on the effectiveness for inferring the way forward.

2. STATUS OF MARKETING OF HORTICULTURAL CROPS

Horticultural crops being highly seasonal and perishable are capital and labour intensive and need care in handling and transportation. Their bulkiness makes the handling and transportation a difficult task, leading to huge post-harvest loss which is estimated at around Rs. 23,000 crores or nearly 35 percent of the total annual production (CII, Mckinsey, 1997). Their seasonal production pattern results in frequent market gluts and associated price risk, thereby forcing the farmers into distress sale to pre harvest contractors and commission agents. A typical marketing channel of horticultural crop thus involves a number of intermediaries like the pre harvest contractor, commission agent, wholesaler, retailer operating between the producer and the final consumer. Each of these market intermediaries performs a specific market function of assembling or distribution that involves a cost to them, thereby making them claim a share in the market margin. Though, it is said that an efficient market provides for the distribution of market margins in proportion to the task performed by each market intermediary, it is seldom so. With each one of the market intermediary adding his/her cost and margin, huge disparity is seen in the price paid by the ultimate consumer and that received by the producer, which in economic terms is called the price spread. As is obvious, the price spread along a given marketing channel is directly proportional to the number of market intermediaries involved along the channel (Gupta and Rathode, 1998), lowest being in the direct marketing involving the producer and the consumer. Given their highly perishable and bulkiness and seasonal production pattern, horticultural crops are the most vulnerable commodities for high price spread.

It is well known a fact that in a market function, the physical movement of the produce is along the market chain, while the monetary and information flow is in the reverse direction. In other words, the producer only produces while it is the marketing network including the trader, commission agent or the wholesaler, who hold the key information about the demand for a specific commodity. It is this access to 'information about the impending demand' that empowers a market intermediary to bargain or take away a larger share in the marketing margin (Crawford, 1997).

The diverse group of horticultural crops differs in their marketing needs as well. Most of the fruits being bulky and highly seasonal are sold through the pre-harvest contractor (PHC) at the field much before they come to harvest. Very often, the PHC takes most of the production risks due to pests and diseases and also the cost of maintenance, while he makes his margin through bulking (Sudha and Froukje, 2006). Vegetables, barring cabbage and cauliflower, are mainly sold through the commission agents at the market, who intern transports the produce to the distant markets and makes his margin, traditional flowers are self marketed at the wholesale auction centres (Subrahmanyam, 1989).

Absence of efforts at percolating market information has been a major constraining factor in the efficient functioning of the market. In a typical market operation, each one of the market intermediary contributes by way of either a transfer function or an assembly function. In this process, each one tries to optimize their risk and maximize their margin. The distribution of market margin depends purely on the access to information and the strategy that is being adopted in forwarding the trade. The primary focus therefore is to increase the accessibility to information by all the market intermediaries so as to enable a level playing ground for them. Such an increased access to market information is expected to bring about the market regulation automatically there by reducing the monopoly power of any single market intermediary. Therefore, the main effort so far in the market regulation activity has been to create marketing infrastructure, such that it enhances the access to market information, besides the enactment of market regulation acts and market interventions whenever required.

3. STATUS AND PROSPECTS OF MARKET REGULATION, A TIME LINE ANALYSIS

Though agricultural marketing is inclusive of horticultural commodities, they differ from the agricultural commodities by way of being highly perishable, seasonal and bulky. This group of commodities was recognized as the major contributors of farm income much later than the other crop groups. Hence, they have been included into the standard marketing system much later than their counter parts. The subject of horticulture marketing and its development attracted attention of policy makers only during the 3rd Five Year Plan.

Most agricultural commodity markets generally operate under the normal forces of demand and supply. However, with a view to protecting farmers' interest and to encourage them to increase production, the Government also fixes minimum support/statutory prices for some crops and makes arrangements for their purchase on state account whenever their price falls below the support level. The role of Government is promoting organized marketing of agricultural commodities in the country through a network of regulated markets. To achieve an efficient system of buying and selling of agricultural commodities, most of the state Governments and Union Territories have enacted legislations (APMC

Act) to provide for regulation of agricultural produce markets. The basic objective of setting up of network of physical markets has been to ensure reasonable gain to the farmers by creating environment in markets for fair play of supply and demand forces, regulate market practices and attain transparency in transactions.

The purpose of regulation of agricultural markets was to protect farmers from the exploitation of intermediaries and traders and also to ensure better prices and timely payment for his produce. Over a period of time these markets have, however, acquired the status of restrictive and monopolistic markets, providing no help in direct and free marketing organized retailing and smooth raw material supplies to agro–processing, competitive trading, information exchange and adoption of innovative marketing systems and technologies. Farmer cannot sell his produce directly in bulk except on retail basis to the consumers. Farmers have to bring their produce to the Market yard. Exporters, processors and retail chain operators cannot get desired quality and quantity of produce for their business due to restrictions on direct marketing. The processor cannot buy the produce at the processing plant or at the warehouse. The produce is required to be transported from the farm to the market yard and then only it can be purchased and taken to the plant. There is thus an enormous increase in the cost of marketing and the farmer end up getting a low price for his produce.

Under the Agricultural Produce Market Committee (APMC) Act, only State Governments are permitted to set up markets. Monopolistic practices and modalities of the State-controlled markets have prevented private investment in the sector. The licensing of traders in the regulated markets has led to the monopoly of the licensed traders acting as a major entry barrier for a new entrepreneur. The traders, commission agents and other functionaries organize themselves into associations, which generally do not allow easy entry of new persons, stifling the very spirit of competitive functioning.

Agriculture sector needs well functioning markets to drive growth, employment and economic prosperity in rural areas of the country. In order to provide dynamism and efficiency into the marketing system, large investments are required for the development of post harvest and cold chain infrastructure nearer to the farmers' field. Keeping in view the need for market regulation, a Task Force was set up by the Ministry of Agriculture which estimated the investment need of Rs.12,230 crores by the 10th Plan for infrastructure development for agricultural marketing.

Accordingly, major reforms have been brought forth in market regulation and the model Act of Agricultural Produce Marketing Committee (APMC) has been adopted in 2003, that provides for establishment of Private Markets/Yards, Direct Purchase Centres, Consumer/Farmers Markets for direct sale and promotion of Public Private Partnership in the management and development of agricultural markets in the country. It also provides for separate constitution for Special Markets for Commodities like onions, fruits, vegetables, flowers *etc.* A separate Chapter has been included in the legislation to regulate and promote contract-farming arrangements in the country. It provides for prohibition of commission agency in any transaction of agricultural commodities with the producers. It redefines the role of present Agricultural Produce Market Committee to promote alternative marketing system, contract farming and direct marketing and farmers/consumers markets. It also redefines the role of State Agricultural Marketing Boards to promote standardization, grading, quality certification, market led extension and training of farmers and market functionaries in marketing related areas. Provision has also been made in the Act for constitution of State Agricultural Produce Marketing Standards Bureau for promotion of Grading, Standardization and Quality Certification of agricultural produce. This would facilitate pledge financing, E-trading, direct purchasing, export, forward/future trading and introduction of negotiable warehousing receipt system in respect of agricultural commodities.

There has been conspicuous attempts at altering the market functions, the main objective being shortening the market/supply chain by creating alternate channels of marketing involving both producers and consumers; initiating collective action by different stakeholders; ensuring dissemination market information to the benefit of the producers and consumers. Several policy instruments have been utilized by different state governments to ensure effective market regulation and augment farmers from such violent daily price fluctuations in vegetables. Notable among these being creation of alternate marketing networks both by public and private sector organizations with an objective of ensuring better bargaining power to the producers and reduce the length of the marketing channels and improve the farmers share in consumers rupee.

These efforts over the years could be classified into two major categories as those that promote 'comparative advantage through better input use and those that promote competitive advantage or ensure better output price (Table 1) and those that have been initiated by the government and those by non-government organisations.

Table 1. *Categorisation of initiatives towards linking horticultural farmers to markets* *

I. Focused on Comparative Advantage Better input use	**II. Focused on Competitive Advantage Better Output value**
A. *Government Initiatives*	A. *Government Initiatives*
I a. Improved Input use efficiency	II a. Creating alternate marketing channels
Precision farming for vegetables in Tamil Nadu	Farmers markets (Raitu bazars)
	Co-operative marketing networks (HOPCOMS)
	Price Forecasting
B. *Non-Government initiatives*	B. *Non-Government Initiatives*
Export oriented production of grapes in Maharashtra	III a. Specialized markets (Public- private) - SAFAL
Integrated Cut flower production in Ooty, TN; Kodagu, Karnataka	b. NAMDHARI

Source: *Own compilation based on personal experience.*

3.1 Initiatives Aimed at Improving the Input Use Efficiency

Several examples of producer initiatives can be identified that aimed at enhancing the crop productivity through better input use efficiency besides also reducing the length of the market channel or facilitating a bargaining power through collective action. Each one of these efforts have resulted in reducing the length of the market channel as also bringing down the price spread and improving the production and marketing efficiency irrespective of the agency responsible for the initiative (Sudha *et al.,* 2010).

3.1.1 Precision farming for vegetables in Tamil Nadu—A Government initiative

A government initiative, this model was in operation in the Dharmapuri district of Tamil Nadu state. The model involves a budget of over Rs. 7.2 crores, spread over a period of three years in Dharmapuri and Krishnagiri districts of Tamil Nadu state covering an area of 400 ha. *Crop diversification* with vegetables is the primary focus of the project. Each one of the willing farmer needs to cultivate at least one hectare under this scheme. The model provides for fertigation system involving a cost of Rs. 75,000/ha along with cultivation expenses up to Rs. 40,000/ha. The beneficiary is required to incur only 10 percent of this expenditure as equity, besides the land and labour contribution towards cultivation. The farmer is expected to choose from a set of vegetables to be cultivated under fertigation conditions. The technical guidance for crop production is provided by periodic visits by an expert team. The team also advises and helps farmer market his produce by providing market information. Tomato, cabbage, cauliflower and chili were the vegetables included under this model. Farmers observed a 20 percent saving in cost due to the use of fertigation and a yield increase to the tune of over 25 to 60 percent in different crops.

3.1.2 Initiatives by non-government organizations or producer based initiatives

Export oriented production of grapes in Maharashtra: Initiated during the mid 90s, this model depicts spatial and vertical integration along the supply chain for grapes with the cooperative and public sector partnership between the farmers and the state government. The three tier structure of the model constitutes farmer groups or association at the base (village) level, district cooperative at the second and federation of cooperatives at the state level. Each farmer needs to enroll as a member of the association closer to his production base with a nominal payment (in the range of Rs. 5000/farmer). All such associations then become a part of the district level cooperative, which in turn is grouped under the state level federation. State government has helped set up cold storage units at district level so as to enable all the members coming under its purview can assemble and store their produce at a nominal fee. The federation then helps in accessing distant domestic markets and the export markets so as to maximize returns. The model, thus empowers the farmer a better bargaining power by creating infrastructure to come out of the clutches of the distress sale to PHC, prolonging the marketing season, access to alternate and distant markets to reap the advantage of better and higher price. Through collective action farmers also have access to import better technical know-how for value addition on farm. This model also helped farmers plan the area under grape for different purposes, *i.e.,* for exports, for raisin making or for distant domestic markets.

Integrated cut flower production: Due to frequent fluctuations in the international tea prices, the farmers from Nilgiris, who are tea planters, have diversified into cut flower production in the recent times. Formed as a small farmer's consortium, the Nilgiris Integrated Floritech Co. Ltd., (NIFCO) pools the flower produce and markets it under a common brand. This initiative by small farmers and cut-flower growers is neither a mass movement nor a co-operative, but has a unique production model envisaged as being operative on a cluster basis. With a total project outlay of over Rs. 10 crores, the project promotes export-oriented floriculture cluster development zones in the 12 watershed areas of Nilgiris district. Each of these clusters would comprise a minimum of 40 units, taking the total number to 500 units in the 12 clusters. A small farmer willing to take part in the consortium could take 500 m^2 polyhouse on lease for three years from the company and choose to cultivate among the 12 exotic flowers identified. The consortium organizes to provide a tripartite agreement with the bank for the necessary financial assistance and also provides the farmer with the centralized facilities for packing, transport and export sales. As per the projected production plan, by 2005-06, the annual production from the 200-odd integrated units is expected to cross 300 lakhs of cut-flowers. The project implementation team envisages the integration of 500-plus units by 2007-08 and the production rising commensurately to touch a high of 825 lakhs of cut flowers.

3.2 Initiatives That Aimed at Increased Competitive Advantage—Output Price/Market Oriented

Marketing of horticultural crop produce being a state subject, most of the state governments have initiated efforts at regulating the markets, market information dissemination and creating alternate marketing channels that help reduce the 'price spread'. These include:

3.2.1 Farmers Markets or Raitha Santhe

A Farmers market or Raita Santhe is a market that facilitates direct sale of produce from farmers to the consumers, where the middlemen have been eliminated. Initiated by the Government in the early 90s, the concept was initially introdued and popularised in Punjab as '*Apni mandies*', has gained momentum across the nation. In Karnataka, there are a round 730 Rural Primary Markets (raitha santhe) functioning, where in food grains and fruits and vegetables are sold mainly by small and marginal farmers directly to the consumers. The main attraction of these markets is the total exemption of market fee or commission charges. These markets are managed by the local authorities/ Gram Panchayats, 'Raitha Santhe' can be described as an example of "a beautifully managed direct marketing system". On an average Raitha Santhes transact up to 30-40 tons of vegetables per day valued at Rs/- 6 to 7 lakhs, benefiting up to 160-220 farmers from direct transactions involving no commission.

Infrastructure facilities at Raita Santhes include platforms, accommodating 180 farmers, storage godown, canteen, administrative office building, water facilities, internal concrete roads, street lights, vehicle parking and display boards.

3.2.2 Co-operative marketing networks (HOPCOMS)

The cooperative producers marketing network popular as HOPCOMS, was established as 'The Bengaluru Grape Growers' Cooperative Marketing and Processing Society Ltd., (BGGCOMS) in 1959 with the main objective of encouraging grape cultivation by providing the required inputs, technical know-how, marketing facilities *etc*. The society started handling fruits and vegetables apart from grapes since 1965. In 1983, the name of the society was changed as 'The Bengaluru Horticultural Producers' Cooperative Marketing and Processing Society Ltd. (BHOPCOMS) and subsequently in 1987 it became HOPCOMS.

HOPCOMS has an approved policy of fixing the procurement price slightly higher than the prevailing wholesale price in the market and the stall (outlet) price at a slightly lower level than the ruling retail price so as to maintain a margin of 25 per cent. At present area of operation is extended to Bengaluru urban, Bengaluru rural, Kolar, Chikkaballapur, Channapatna and Ramanagar. It purchases the horticultural produce directly from farmers at remunerative prices, provides transport facilities from collection centers, provide weighment and immediate cash payment; infrastructure facilities include, accommodation for night stay; canteen and banking facilities, minimum support price during distress sales under government schemes (MIS) and assistance for preliminary processing activities. Estimates indicate that more than 10,000 producers from different parts of the state benefit from this alternate marketing network facility. Similar facilities can also be seen in other states, but the success of HOPCOMS is exemplary.

3.3 Establishment of Alternate Marketing Network by Non-Government Organisations

3.3.1 SAFAL market, Bengaluru

SAFAL is the largest organized retail network for fruits and vegetables in the national capital region of India. Currently operating over 400 retail outlets in and around New Delhi, SAFAL is a subsidiary of National Dairy Development Board. SAFAL's supply chain (Figure1) covers 16 states, about 50,000 farmers and over 200 farmer associations. With a special amendment to the regulated markets act, Government of Karnataka in collaboration with National Dairy Development Board (NDDB), started a SAFAL Market near Bengaluru in 2004 with an investment of over Rs 150 crores.

SAFAL market aims at providing the consumers fresh fruits and vegetables, quality produce, products and services through their own retail network outlets. They practice and promote use of chemical-free, safely ripened fruits. Fruits and vegetables are sourced predominantly from known procurement sources, especially from SAFAL Farmers Associations. They ensure safety by following good agricultural practices (GAP) and provide analysis for presence of heavy metal and pesticide residues. With a 1600 tonne capacity, SAFAL market aimed at influencing the market in a big way. SAFAL also has been entering into contract production of tomato and other high value vegetables through their farmers network for procurement and export of processed products (Fig. 1).

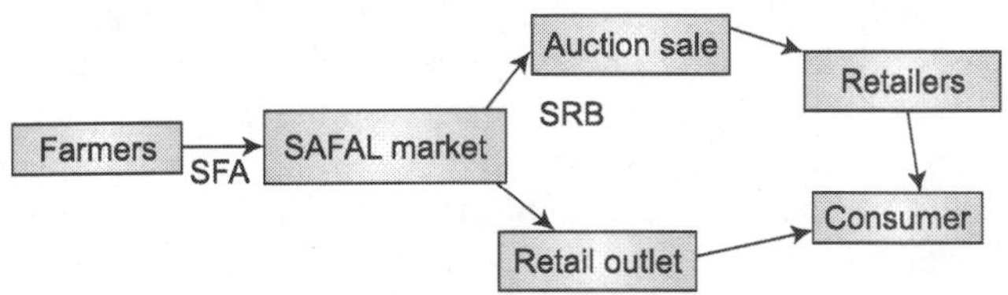

Fig. 1. *SAFAL's Supply chain (Operational Model of SAFAL)*

SAFAL encouraged farmers to form associations (FA) across different production regions such that they would ensure regular supply to SAFAL and get higher share in consumers rupee. The FA supports the farmers in backward and forward linkages with the support of SAFAL Market. 200 such associations have been initiated by SAFAL with appropriate field staff to oversee the operations. On an average each of the Farmers Association transacts business of Rs.50-60 lakhs annually.

The farmers give their indent of supply to the FA indicating how much they will supply on the day. The SAFAL also sends indents to FA of its requirements based on the need of the buyers. The farmers themselves are encouraged to grade their produce and bring in to the FA. The produces are labeled as A or B/ market quality grade. The Transport facilities are arranged to transport the produces from FA or from the farm itself, depend upon the quantity.

This market uses the Dutch auction system (with an auction clock on which prices are reduced) for fruit and vegetables and the English system (with the auction clock counting upwards) is used for onion, potato and garlic. With two auctions per day, the reference price for the morning auction is the wholesale price of the day before, while the afternoon auction uses the morning wholesale price of that day. The produce is procured from 200 farmer associations through local collection centres. The produce is transported to the auction hall, graded, shown in lots and sold the day after the farmers deliver it to the collection centre. Farmers receive payment based on a weighted average of the auction price the next day. Any produce that is not sold through the auction is sold in small outlets at the SAFAL premises.

Three types marketing systems:

1. Electronic auction hall: The selling and buying take place at the press of a button. The retail and whole sale buyers participate in the auction and bid through the electronic button system.
2. SAFAL daily fresh outlets: There are seven outlets in Bengaluru city for the consumers to buy them directly from the shop.
3. Outstation sale: some portion of the produce is also supplied to other networks under the SAFAL brand.

Namdhari Fresh, Bengaluru: A unit of M/s Namdhari Fresh (NF) is located on the Bengaluru and Mysore national highway near Bidadi village. Namdari supply chain (Figure 2) practices both captive farming (by leasing-in land) and contract farming in the ratio of 20:80 on 1500 acres. The contract is informal oral and non-registered and NF contracts individual farmers with bipartite agreement. The contracts are both acreage and quantity contracts. These farmers, primarily small landholders and are reluctant to undertake high price fluctuations in the market, enter with NF contract for the supply of 'A' grade quality for a pre–decided price at the farm gate. NF in turn provides agri-inputs like seeds, along with supplies of fertilizers, pesticide as and when it is required. NF makes it mandatory to use the seed provided by them, thus ensuring the requisite quality of the end product. Over 1500 farmers are under contract for the production and supply of tomato, corn and other selected high value vegetables to NF. The company has over 19 Namdhari fresh retail outlets across Bengaluru which ensures the supply of fresh fruits and vegetables at reasonable prices. The contract production and sale has been highly successful especially for tomato and corn (Fig. 2).

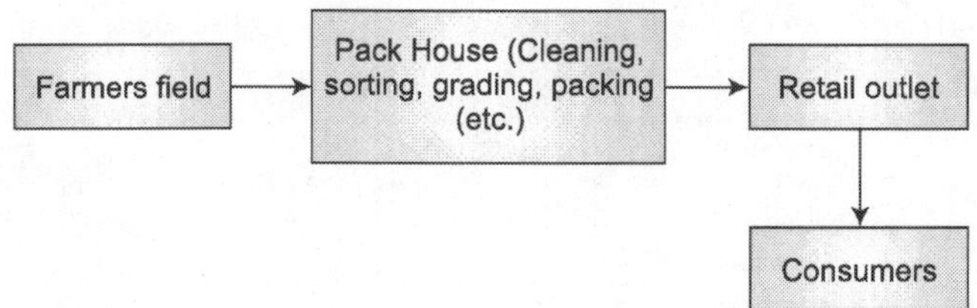

Fig. 2. *NAMDARI Supply chain(Operational Model of NAMDARI)*

4. COMPARISON OF PROVISIONS AND BENEFITS TO THE FARMERS

A comparison is drawn in this section w.r.t the efforts and the benefits that accrued to the producers. It could be seen from the Table 2 & Table 3, that each one of the alternative initiatives had different focus and have shown success in their own way, though limited to a specific time, space or region. As can be seen from the tables, some of these initiatives though were very useful to the growers did not have a direct market linkage, while others had limited access benefiting only a group of growers who were members. Some of the initiatives like SAFAL though have been started through special amendment of the marketing legislation to allow creation of market infrastructure by private parties, the facility is hardly used and benefits only a portion of the growers. The main limitation has been the failure of the facility and its organisers in breaking the barriers of the traditional marketing networks. On the other hand, Namdhari Fresh has been very successful in effectively establishing efficient linkages of farmers to the market, though to a limited extent. The facility also helps in tracing the produce and builds consumer confidence as well.

It is of utmost importance that these initiatives benefited the growers in terms of reduction in the transaction cost or reducing the length of the marketing network or otherwise. Table 3, provides the details of the benefits that accrued to the producer from the use of the described market linking initiatives either by the government of the non-government. Measured in terms of parameters like the cost of transaction, commission paid, distance travelled, quantity disposed off, price realised, mean gain or loss from continued use of the facility over longer periods and efficiency in marketing, it is seen that the effort put forth by the Namdharis and the Raitu Bazar facilities fulfill most of these criteria and could be considered efforts worthwhile for the producers at large in linking them to the markets. Though the others also have some positive points, the uncertainty associated with them in day to day operations.

5. MOBILE TELEPHONE BASED NETWORKS

Several other localised efforts are also in place by private operators through mobile based applications that attempt to link producers to markets. Though successful, these efforts appear to be localised and short lived ones.

Table 2. *Comparison of different initiatives that attempted to link producers to markets.*

S.No	Name/Type	Focus	Organized by	Scope	Limitation
1	Precision farming; collective action	Input use efficiency through Crop diversification	Government , TN	Large scale (> 400 ha)	No direct market linkage
2	Export oriented production of grapes	Complete package	Producers group, supported by Government, MS	Limited to a group of grape growers in a region (110)	Limited access
3	Integrated cut flower production	Crop specific, input use and market linkage	Private initiative, TN	Limited to a group	Limited success
4	*Raitha Santhe*	Direct marketing	Government, KAR	Large no.	Focus on small and marginal farmers with limited produce
5	HOPCOMs	Alternate marketing network	Government, KAR	Over 10,000 growers	Linked to wholesale markets for price realization
6	SAFAL Netwrok	Alternate marketing network	Private initiative, with Government support, KAR	Large scale	could not break the barrier of traditional wholesale market
7	Namdari	Alternate marketing network	Private initiative	Limited to specific region & crops	Limited success

Table 3. *Comparison of advantage to the producer from different alternatives*

	HOPCOMS	NAMDHARI	Raitu Bazar	SAFAL
Average Commission Charged	1.5-2%	0	0	5%
Transaction Cost/Kg (Rs.)	0.60	0.30	0.58	1.25
Average Distance of farm to the market	35km	7km	22km	65km
Average Qty sold/ day/member	5t	2-3t	4.5 t	10
Price /kg	Rs.10-14	Rs.7	Rs.16.8	Rs.10-16
Proportion of total produce sold	25%	90%	60-80%	15%
Mean gain/loss in income from sale	+ Rs. 12-15000	+ Rs. 18000	+Rs. 22000	+/- Rs. 5000-12000
Efficiency	0.68	0.71	0.75	0.64
Constraints	Uncertainty of getting the indent regularly	No uncertainty	No uncertainty	Uncertainty of supplying the produce due to non functional SPAs and lack of demand

6. WAY FORWARD

6.1 Price Forecasts as a Tool for Linking Farmers to Market

Though several efforts have been putforth for linking producers to markets, the most important one is a focus on addressing the issue of high volatility of market prices of horticultural crops, especially that of fruits and vegetables. While each one of these policy instruments has helped augment producers in getting higher returns for their produce, none of them actually attempted to provide a solution to the problem of day to day market price volatility. The prosperity of cultivators depends not only on the increased rate of production but also on the method and efficiency with which they disposed of their produce to their advantage. The efforts of larger production are unfruitful until the produce is marketed efficiently. The market prices are ruled by demand and supply conditions in the market and price received by producer mainly depends on time of sale and terms and conditions of the sale. If the producer has the prior information or 'intelligence' (information in advance) about the pattern and magnitude of the 'price in the specific market' then he can take an informed decision about harvesting the produce and marketing the produce to benefit from this information.

The availability of market intelligence on aspects like the potential markets, the quantum of market arrivals and expected prices in different markets during different months of the year shall go a long way in mitigating the problems of producers due to market price volatility (Sexena *et. al.,* 2015). Market intelligence provides diversified avenues to examine the market behavior of agricultural commodities. 'Market integration' and 'price forecast' are two important components of market intelligence.

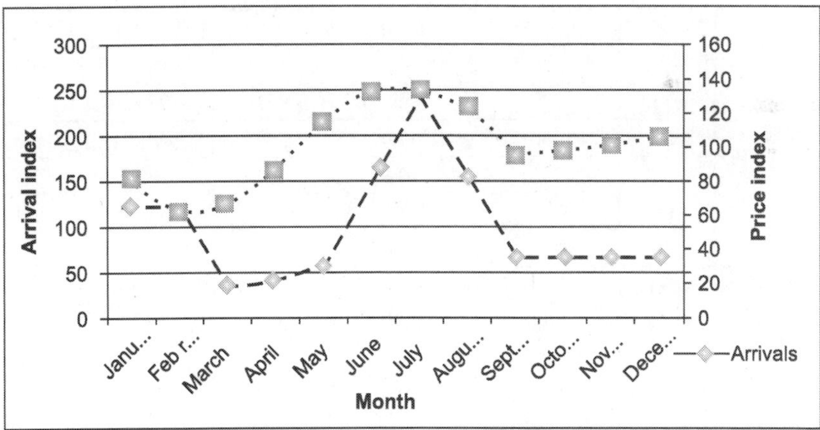

Fig. 3. *Seasonal indices of market arrivals and prices of tomato in Kolar market*

Market intelligence deals with analysing the market signals, such as, the trend in arrivals and prices over a period of time and forecasting the price trends to benefit the producers. This analysis includes, (i) assessing the variability in arrivals and prices, (ii) developing seasonality indices (iii) price prediction through the use of econometric methods and (iv) disseminating these prices in advance to benefit the producers to enable them to make informed decisions on area allotment and harvest time and selecting an appropriate market. Such an attempt has been in operation since three years by a network of economists under a network project on Market intelligence. Under this programme, the variability in arrivals and prices of a commodity in a market are analysed using mean and coefficient of variance. Further, Seasonal indices of arrivals and prices get estimated using ratio moving average method for assessing the seasonality pattern of arrivals and prices.

Considering these fluctuations in trend analysis over the years, 'market intelligence' is provided in the form of forecasted price to the traders and farmers prior to sowing and prior to harvest so as to enable them to take informed decisions on the area allotment and time of harvest and place of sale. Such an exercise is made using standard econometric methods, such as, Auto Regressive Integrated Moving Average (ARIMA) models for every season and forecasted the predicted prices, which were disseminated to traders and farmers through periodic visit to the market and through SMS or voice mail and through regular personal contact. The forecasted prices are then compared with the actual prices prevailed in a market for validation. It was observed that the predicted prices were over 90% accurate and in synch with the actual prices.

Table 4. *Validation of the forecasts and it's comparison with farmers expectations (2014-15)*

Season/ Month/Year/ week		Farmers expectations (Rs./qtl.)	Traders' expectations (Rs./qtl.)	Actual prices (Rs./qtl)	Forecasted prices (Rs./qtl.)	% Deviation
Rabi (Aug 2014)	1st week	2700	2800	2874	2923.50	1.72
	2nd week	2200	2100	2139	2211.55	3.39
	3rd week	2300	2000	2212	2040.60	7.74
	4th week	2100	2000	2177	2107.51	3.19
Kharif (Oct 2014)	1st week	900	900	773	916.03	5.56
	2nd week	700	800	801	794.82	0.77
	3rd week	600	500	569	620.51	9.05
	4th week	700	700	651	628.97	3.38
Rabi (Dec 2014)	1st week	1300	1500	1486	1370.25	7.78
	2nd week	1000	1100	980	1046.48	6.78
	3rd week	800	700	685	784.94	0.08
	4th week	1200	1000	1121	1015.92	9.37
Kharif (May 2015)	1st week	1100	1000	1162	1131.60	2.61
	2nd week	1000	1200	1060	1153.63	8.83
	3rd week	1700	1600	1567	1460.71	6.78
	4th week	1400	1300	1290	1323.44	2.59
Rabi (August 2015)	1st week	800	750	675	699.9	3.68
	2nd week	700	600	623	700.2	3.65
	3rd week	800	750	786	757.44	3.63
	4th week	1100	1000	1082	997.25	7.83

As can be seen from the Table 3, the forecasted prices in almost all the cases were observed to be close to the actual prices prevailed in the market. With less than 10 % deviation the forecasted prices provided farmers a very good opportunity to benefit from this information. As it has been observed that, the forecasted prices, in comparison to the farmers and traders' expectations fared well as well. Discussion with the traders and farmers indicated that an attempt of this nature where the prices (weekly) have been forecasted and disseminated to farmers was the first of its kind and the traders and farmers were highly appreciative of the effort. A total of around 2000 farmers and 30 traders were provided with this forecasted information on prices regularly to benefit them (Table 4).

7. CONCLUSIONS AND POLICY IMPLICATIONS

Linking horticultural producers to markets is of utmost significance given the growing importance of horticultural crops to the nutritional security as well as the economy. There have been several attempts made both by the government and non-government organisations to link horticultural crop producers to markets. Each one of them with a varied focus either in improving the input use efficiency of production or through marketing efficiency have been successful either in limited scale or on large scale, benefiting the producers who have been involved in these efforts. A comparison of these efforts indicated that focused efforts, such as, the Raitha bazars or Namdhari Fresh seemed to have achieved the objective of effective linkage between farmers and markets. Others though useful appear to be engulfed in constraints of uncertainty and high transaction costs.

There appears to be an urgent need to evolve other methods of providing linkages that actually provide information of market prices to the growers and empower them to take informed decisions on the acreage to be planted under a specific crop or when to harvest and where to sell, through 'market intelligence' using econometric methods of price forecast. There is a need to impart skill to young professionals in such estimation procedures and ensure the availability of their services to producers and traders at the market yard itself for efficient use and effective utilisation.

REFERENCES

Agarwal, N. L. and Satya, P. K. (1994). Price Behaviour of groundnut in markets of Rajasthan. Bihar Journal of Agricultural Marketing, 2(4): 309-18.

Chand, Ramesh and Parappurathu, Shinoj. (2012). Temporal and Spatial Variations in Agricultural Growth and Its Determinants, Economic & Political weekly Supplement.

Darekar, A.S. *et. al.,* (2015). An Economic Analysis of Behaviour of Market arrivals and prices of Onion in selected Markets of Western Maharastra. IJTA, National Academy of Agricultural Science, 33(2): 1087-1093.

Sexena, Raka *et. al.,* (2015). A manual on price forecasting techniques, Network Project on Market Intelligence, ICAR, National Institute of Agricultural Economics and Policy Research, new Delhi.

Sudha, M. (2010). Regional pattern of economic growth and horticultural development, In: Proceedings of International Conference on Horticulture, November 9-12, 2009,

PNASF, Bengaluru, p. 2016-2025 ; also adopted as a Chapter in a book entitled, 'Horticulture and livelihood Security', Ed: Premnath and P.B.Gaddagimath Page: 80-93.

Sudha, M., Gajanana, T.M. and Murthy, Sreenivasa, D. (2010). Linking farmers to market, agro industry models partnerships, lessons learnt and perspective – with special reference to horticulture crops in India, Indian Journal of Agricultural marketing, 24(3) 58-67 www.agmarknet.com

Virender Kumar, H. R. *et. al.,* (2005). Behaviour of Market Arrivals and Prices of Selected Vegetable Crops: A Study of Four Metropolitan Markets. Agricultural Economic Research Review,18 (1): 271-290.

19

Meeting the Crisis and Challenges in Prices: In Fruits and Vegetables

Pradeep Mehta and R. S. Deshpande

1. INTRODUCTION

Food inflation has been one of the important policy problems on the agenda of the government. Usually, food is considered as almost synonymous to cereals and pulses. While the prices of pulses are increasing that of cereals are not increasing at the same rate. We forget that in the Indian consumption basket, the importance of fruits and vegetables is equally important. Surprisingly, from the entire debate on inflation, the prices of fruits and vegetables (F & V) do not get any mention. Daily consumption of fruits and vegetable has increased substantially in the recent decades. This has assumed significant proportion in the household budgets. Surely that has several benefits; higher F&V consumption is linked with reduction in incidence of diseases, increase in immunity levels and reduction in mortality rates. In India, F&V are consumed by most of the population: Out of every 1000 households in India, 983 (rural) and 932 (urban) consume good amount of vegetables, whereas, 608 (rural) and 777 (urban) consume fruits (NSSO, 2012). Even the low−income families in India, in general, depend on vegetables to make up their main meals daily. It is estimated that more than 90% of the population has at least 10 pure vegetable meals per week (Chawla, 2008). Fruits are also an important source of nutrients and are consumed regularly by majority of the population, especially in the urban and semi-urban areas. But the crisis of increasing price and other associated issues can change the situation. Naturally, the price rise in the F&V will impact the all income and more severely the medium & low income groups.

India is the second largest producer of F&V in the world, ranking next to China and accounts for about 16% of global fruits production and 10% and vegetable production. But, despite one of the global leader in horticultural sector, F&V covers less than one tenth of the total gross cropped area in India. This is complemented with low intake of F&V in India; the *per capita* consumption of F&V is far below the recommended intake of minimum of five serving. A survey, conducted by Indian Council for research on International Economic Relations (ICRIER) reveals that the average intake of F&V by the young people in the age group of 18-35 is 2.97 servings whereas, the national average is of 3.5 servings per day (Mukherjee, *et al.,* 2016). Overall, there is 74% low F&V consumption amongst adults in India (Hall *et. al.,* 2009). There is a need to increase both area under F&V and consumption of F&V in India, in which price plays a critical role which influences allocation of scarce resources between competing uses (land by farmers; and expenditure on different food and non-food items by consumers).

Increase in prices is directly and negatively influenced by the consumption of products. Due to relatively high income elasticity of F&V as compared to other food items in India, persistent increase in price will surely result in decline in consumption and that would lead to lower nutrition. Incidence of price rise of F&V, in the recent times, has been a matter of concern also due to several reasons. Price rise lowers the purchasing power of every citizen and it is the poor who are most adversely affected due to limited resources available to hedge against inflation. Secondly, the proportion of expenditure of rural and urban people in India on F&V has been increasing at an alarming rate and price rise would impact the total cost of living; For instance, between 1983 and 2004, annual *per capita* consumption of F&V has grown by 150% and 30% respectively and the budgetary shares of both F&V in total food expenditure have increased by around 50% (Kumar *et. al.,* 2011). It is important to note that the share of F&V in total food expenditure, as compared to all other food products, such as, pulses, cereals, meat *etc.* has depicted maximum change in total food expenditure across all income groups including poor and rich *(ibid)*. Third, there is also a general agreement that price rises and volatility leads to inefficient resource allocations and adjustment costs and high prices are bad both for urban and rural poor (Dorward, 2012). Fourth, increase in the price of F&V crops is also a significant factor pushing up the cost of food that also raises concern about problems of food insecurity and hunger. Thus the broad category of food inflation hides a lot below its crust and that is the price rise in F&V.

F&V are seasonal in nature and are more perishable as compared to other food items which results in huge variations in price during different seasons and months of the year. Given the high significance of F&V for nutritional security and low prevalence of its consumption in India, there is a need to understand the dynamics of price change, the challenges brought by price rise and steps to overcome crisis. Supply shocks could be the major factor for price rise but the same could be transient and not a permanent feature.

Therefore, it is vital to explore the extent of variations in prices both in the short and long run for different F&V. We examine here the trend in annual and monthly prices of selected F&V in the last 14 years and 1 year respectively and examine correlation between arrivals of selected F&V in market with their price behaviour. This is followed by exploring the difference between the wholesale and monthly prices. The study then looks into the challenges of price rise of F&V and outlined the measures to overcome crisis. A total of 12 F&V (6 each) that contribute more than sixty percent to production of horticultural crops in India, are selected for the study. Annual and monthly data is obtained from National Horticultural Board (NHB) for the selected F&V. The price and arrival behaviour analysis is done by using data from Delhi market.

2. PRICE RISE IN SELECTED FRUITS AND VEGETABLES IN INDIA

This section explores the dynamics of price changes in selected F&V in India. We have selected 6 fruits and six vegetables for the study. Fruits include apple[1], banana, grapes, litchi, mango and *mosambi* (Sweet Lemon) which cover more than sixty percent of total fruits production. The selected vegetables include brinjal (egg plant), cauliflower, onion,

---

[1] Apple (J&K).

peas, potato[2] and tomato[3] which also cover more than sixty percent of the total vegetable production in the country and are significant in terms of world vegetable production. The annual price data (wholesale price) is collected for last fourteen years[4] from Delhi market and whereas, the monthly price data (wholesale prices and retail prices) is collected for last one year (August 2015 to July 2016) from Delhi market.

The price data for all selected F&V is converted into the real prices by deflating with Consumer Price Index (CPI) series. The trends in the change in real prices of F&V show that the last fourteen years have seen consistent upward trend in the real prices of all selected F&V (Figure 1 and 2). It is important to note that for the selected F&V, the pace of price rise is rather slow until 2008-09, but after that it has grown at faster pace. According to the trends in the price series of all F&V, it seems that the decline in production during 2008-09 and 2009-10 was primarily responsible for high prices of F&V during that time (Nair and Eapen, 2012). The data for 12 F&V also reveals that high price rise in 2008 onwards was linked with decline in production. The production of selected fruits has increased by 8% and 4% during 2002-08 and 2009-15 respectively, whereas, the real prices of fruits have increased by 3% and 12% during 2002-08 and 2009-15 respectively. The production of selected vegetables which has risen by 7% during 2002-08, increased at a slower rate of 5% during 2009-15 whereas, the real price of vegetables increased by 6.9% and 8.1% during 2002-08 and 2009-15 respectively.

As price of F&V fluctuate widely, the growth in price rise is calculated by using three year average. The results reveal that price rise is found to be higher in case of vegetables as compared to fruits. Among the fruits, except for litchi, the growth rate in price increase of all fruits is lower than all selected vegetables. Vegetable crops are part of daily meal and thus higher price rise of vegetable is a cause of concern for consumers. Among fruits, price of mango, litchi, apple and grapes were almost same in 2002 but the growth in real price of litchi was higher than all other fruits resulting in very high relative price for the same. In case of vegetables, except for cauliflower, prices of all other vegetables have grown at compound annual growth rate of around 6-8% in real prices (adjusted for inflation). The growth rate in price of onion is highest among all the selected vegetables.

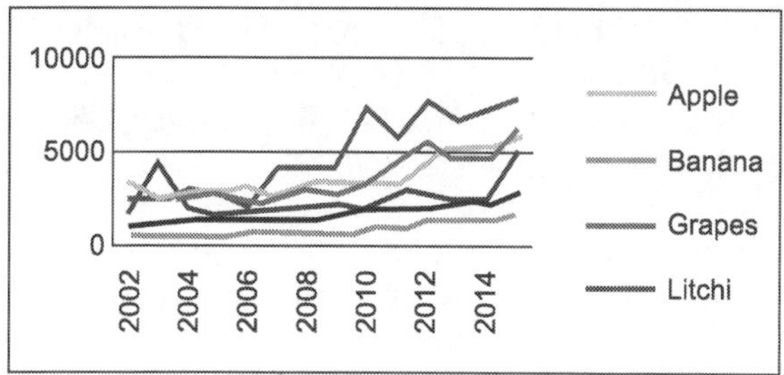

Fig. 1. *Change in Real Price in Major Fruits*

2 Potato (fresh)

3 Tomato (hybrid)

4 Data for few F&V was not available for 2001 at the time of this work.

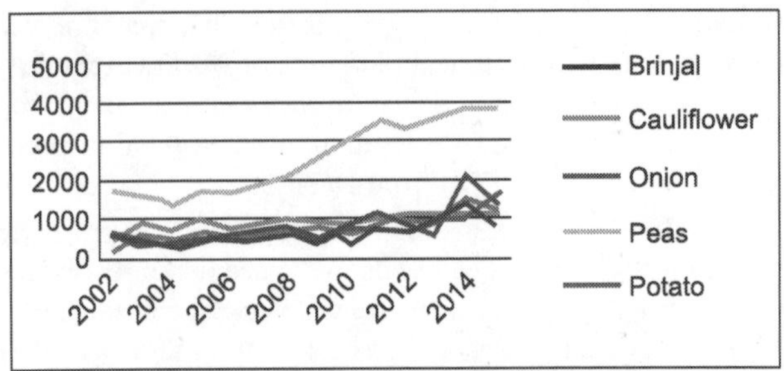

Fig. 2. *Change in Real Price in Major Vegetables*

3. SEASONALITY IN PRICES OF F&V: IMPACTED BY ARRIVALS

The major difference of F&V and other food crops is that F&V are highly seasonal and their production is also very limited to selected geographic or climatic areas. Higher rate of seasonality in the production of F&V results in large variations in their availability in the markets over the year. As most of the F&V are highly perishable and have low shelf life, there is wide oscillation of price of F&V in any given year. The patterns of arrivals of selected F&V in Delhi market and price volatility over the last one year (August 2015-July 2016) show very interesting results. Here we examine the link between arrivals and price variations among different F&V.

The pattern of arrivals of selected F&V reveal that fruit production is more seasonal and their availability restricts to few months in a year, whereas, in case vegetables, not only they are available for most of the months in the year but also their arrivals in Delhi market is more evenly spread over 12 months in a year. For example, litchi and mango are available only for four months in a year i.e., April to July in any given year, whereas, for all other selected fruits, more than 80% arrivals are concentrated for four to five months in a year (Appendix 1). This is also reflected in higher coefficient of variation (CV) of arrival of fruits as compared to vegetables in Delhi market. The CV of arrival from August 2015 to July 2016 is 102% for apple, 31% for banana, 138% for grapes, 233% for litchi, 181% for mango and 76% for Mosambi. On the other hand, vegetables are available evenly throughout the year and their CV in arrival from August 2015-July 2016 is 26% for Brinjal, 25% for Cauliflower, 7% for Onion, 91% for peas, 16% for Potato and 44% for Tomato.

The price pattern of the selected F&V reveal that there is marked fluctuation of price among F&V and it is more acute in case of vegetables (Table 1). The daily price data for F&V provided by National Horticultural Board includes daily minimum price, daily maximum price, daily average wholesale price and daily average retail price. Using the daily price data, we have calculated the highest, lowest and mean price for selected F&V in the last one year i.e., from August 2015 to July 2016. The month-wise pattern is given in Appendix 1. It can be seen that there are marked fluctuations in the prices of vegetables as compared to fruits. The highest price of fruits in the last one year is found to be two times its average price and three to four times its lowest price during 2015-16. In case of vegetables, the highest price is nearly two to three times its average price and even ten to eleven times the lowest price. Even within a single month, the highest price of F&V

was three to four times the lowest price (Table 1). The variations in the prices of F&V are interlinked with arrivals in the market as there is negative correlation between prices and arrivals in the market (Figure 3). This shows that as the arrivals in the market increases, the prices of both F&V go down. This is a normal economic phenomenon but we observed that there are pairs of arrivals and prices that do not show the similar trends. These are the cases where straight economic rule does not apply. The price-arrival correlation is found to be higher in case of vegetables as compared to fruits. The correlation is very high for litchi and peas whereas, it is low for banana, cauliflower and onion.

Table 1. *Price fluctuations between August 2015 to July 2016*

Fruits						
August 2015-July2016	**Apple**	**Banana**	**Grapes**	**Litchi**	**Mango**	**Mosambi**
Seasonal High price (Rs.)	13000	2300	17000	16000	10000	7200
Seasonal low price (Rs.)	2812	600	4000	6000	1500	1900
Average mean price (Rs.)	6200	1258	8532	10860	5321	3637
Vegetables						
August 2015-July2016	**Brinjal**	**Cauliflower**	**Onion**	**Peas**	**Potato**	**Tomato**
Seasonal High price (Rs.)	2700	5333	5500	12875	2375	4464
Seasonal low price (Rs.)	400	250	500	1250	300	400
Average mean price (Rs.)	1054	1601	1762	5384	1227	1777

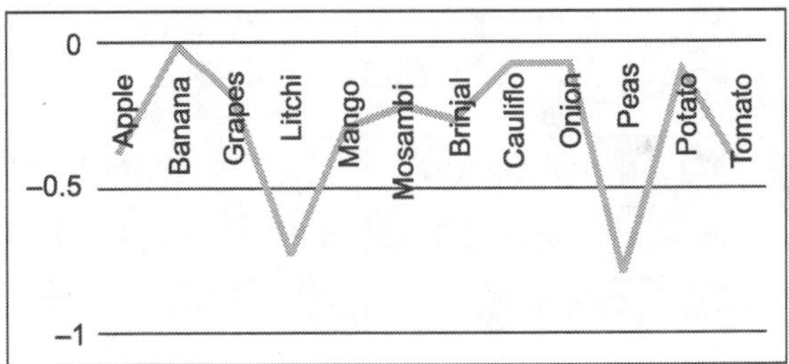

Fig. 3. *Correlation between price and arrival*

4. DIFFERENCES BETWEEN WHOLESALE AND RETAIL PRICES

Understanding the market behavior of horticultural crops is complex especially because of perishability, seasonality and bulkiness (Gandhi and Namoondiri, 2004). Due to its nature, market efficiency plays a prime role in the development of the horticultural sector; market efficiency influences not only the price fluctuations in any given day or month but also the share of the consumer reaching the farmers. For F&V in India, the share of the farmer in the consumer rupee is frequently very low and varies in the range of about 33 to 75 per cent (Dastagiri *et al.,* 2010). Some of the shares are seen in the case of cauliflower at 28.6 *per cent* and onion at 39.1 per cent (Gandhi and Namoodiri, 2004). But there are also incidences especially in case of Onions when farmers have to sell the product at Rs. 1/-

per kg which does not cover even the variable cost of producing the crop (The Huffington Post, 2016).

Low share of farmer in consumer payment reflects high price margins between the wholesale price and retail price. The situation of difference in wholesale and retail price in the last one year (August 2015- July 2016) reveals that in case of fruits, the spread is less as compared to vegetables. In case of fruits, the difference between wholesale and retail price ranges between 23% to 69%, whereas, in case of vegetables it ranges between 47% to 137% (Table 2 and 3).

Table 2. *Percentage difference in wholesale and retail prices of fruits*

	Apple (%)	Banana (%)	Grapes (%)	Litchi (%)	Mango (%)	Mosambi (%)
August 2015	NA	54	86	NA	NA	34
September 2015	51	27	157	NA	NA	21
October 2015	37	25	71	NA	NA	23
November 2015	44	25	57	NA	NA	16
December 2015	64	32	28	NA	NA	23
January 2016	81	34	53	NA	NA	20
February 2016	72	28	47	NA	NA	23
March 2016	72	32	57	NA	NA	19
April 2016	73	35	57	26	31	23
May 2016	68	21	57	37	46	22
June 2016	52	57	86	33	97	21
July 2016	58	17	NA	46	75	27
Yearly average	61	32	69	35	62	23

Table 3. *Percentage difference in Wholesale and Retail Prices of Vegetables*

	Brinjal	Cauliflower	Onion	Peas	Potato	Tomato
August 2015	147	99	70	43	27	134
September 2015	167	148	58	24	31	80
October 2015	108	170	68	30	27	92
November 2015	146	165	72	44	34	77
December 2015	156	138	56	60	68	76
January 2016	128	184	67	57	90	117
February 2016	158	103	95	94	76	119
March 2016	117	90	101	61	62	88
April 2016	136	68	80	58	40	130
May 2016	134	98	89	43	37	NA
June 2016	102	93	91	55	36	88
July 2016	117	105	105	30	39	95
Yearly average	135	122	79	50	47	100

This is due to higher level of perishability of vegetables as compared to fruits. In addition, the marketing of several fruits also follow some standard procedure that also increases their shelf life. For instance, several fruits are graded and generally sold in boxes whereas, vegetables are generally sold in gunny bags. Higher perishability and low shelf life results in higher margin for selling vegetables by different middleman in the market.

Given the short shelf life and high perishability in case of most of the F&V, it is most unlikely that manipulative hoarding could be possible at large scale. Efficiency of the market is very important aspect to reduce the difference in the wholesale and retail prices which could be due to factors, such as, commission agents' fees, cost of transportation, poor storage infrastructure that lead to higher overall costs and handling losses. Due to the presence of numerous middlemen and inability of most farmers to preserve F&V, they are mostly forced to sells the produce at throw away prices, especially in cases of production glut in the market. In addition, the mark*et al*so charges higher fees for F&V due to the nature of these products. In specific case of Delhi, the government collects Rs.80 crore per annum through fees. Between the local *mandi* and the small city *mandi*, another 15 per cent hike in rates comes about and about 3 to 4 per cent of this is due to transportation and labour costs and about five *per cent* is due to packaging and grading. The rest is commission of the agents (The Hindu, 2012). Higher spread of price between farmers and consumers add to the woes of both producers and consumers that require steps to reduce the gap.

5. CHALLENGES OF PRICE RISE

Price rise of fruits and vegetables is a very tricky issue, as unlike cereals, pulses and oilseeds, price increases manifold even within a month. This is primarily due to seasonality in production and high amount of perishability of fruits and vegetable crops which also prevents hoarding for longer time. At times, this results in mismatch between demand and supply leading to sporadic instances of price rises. Due to oscillatory nature of price changes in a season, on the one hand, farmers who sell produce in the markets get lesser share of price rise and on the other hand, consumers end up in paying high prices for products. Even if we take last three years inflation, it was found to be highest among the F&V as compared to other food groups (Table 4).

Table 4. *Inflation in CPI (Combined) Food Groups (in per cent)*

	Weights	**Weights**	**2012-13**	**2013-14**	**2014-15**	**2015-16 (Apr-Dec)***
Base	**2010**	**2012**	**2010**	**2010**	**2012**	**2012**
cereals	14.6	9.7	10.5	13.2	5.2	1.7
meat and fish	2.9	3.6	12.2	12.1	6.3	5.9
milk and products	7.7	6.6	10.4	8.7	10.3	5.7
oil and fats	3.9	3.6	16.6	1.8	2.4	3.9
fruits	1.9	2.9	7.5	11.2	13.9	2.3
vegetables	5.4	6	20.6	26	3.4	1
pulses and products	2.7	2.4	12.3	4.2	7.9	29.6
sugar	1.9	1.4	12.5	-0.7	-0.4	-9.5
spices	1.7	2.5	3.1	7.2	8.6	9.5

* Provisional **Source:** *Economic Survey 2015-16*

The causes of this price spike in fruits and vegetables are complex and are due to a combination of factors, such as, production levels, seasonality, quality *etc.* At the micro and macro level, it raises several challenges both for the producers, consumers and for the economy as a whole.

6. CHALLENGES FOR PRODUCERS (INFLUENCE ON AREA AND INPUT ALLOCATION BY PRODUCERS)

In the market, price of F&V rises to two to three times than the average mean price and even ten to eleven times the lowest price observed in a month. There are two prime issues; first, the price rise is temporary as it depends on the aggregate arrivals in the market which fluctuates widely over a month or year. Second, the price increases several fold after the farmers sell their produce in the wholesale market. For instance, the average difference between the retail and whole price ranges between 23% to 135% for selected F&V. This means the benefits of price rise generally fail to get delivered to the farmers or producers and it is the middleman who gains from sporadic instances of price rise. In case of F&V, an average difference of price which farmer receives and consumers pay ranges from 33-75%, whereas, the share of producers varies from 56-83% in food grains and 79-95% in pulses, 65-96% in oilseeds. In general, farmer growing F&V gets only one rupee out of every Rs 3.50 paid by the consumer, the retailers is getting Rs 0.75, the wholesaler is getting Rs 0.50 and rest of the amount Rs 1.25 is going to commission agents and traders (Dastagiri *et al.,* 2010). One of the biggest challenges is to reduce the gap between the price consumers pay and what farmers get for the produce, so that the benefits of price rise through increased demand can be realised at farm level.

Less proportion of consumer price potentially influences the area and input allocation decisions of the farmers who are guided by the economic rationale in land and other input allocation across different crops. So far, F&V covers around 9% of gross cropped area despite higher returns of the crops. It is estimated that F&V gives four to ten times the return from other crop groups namely cereals, pulses and oilseeds. A one per cent shift in area from non–horticultural crops to horticultural crops adds 0.46 percentage points to growth rate of agriculture sector (Chand, 2016). Despite higher relative profitability, Indian farmers are slow in diversifying to these high value crops. There is still huge scope to increase area allocation to F&V due to their high relative profitability. But due to high leakages and low prices at farm gate, farmers are not able to realise the actual profit from growing these crops.

Increase in price realised at the producer or farmer level can lead to several positive impact. It could potentially increase (with lag) allocation of land and other inputs for F&V relative to other crops that could potentially increase the productivity and total production of F&V. Though it is important to highlight that increased trend of price rise realised at farm level should be consistent over the years to ensure increased allocation of inputs and land to F&V. Especially in case of several fruits, there is a gestation period of three to five years between sowing and ripening of the fruits and increased price should be realised over the period of time to motivate farmers to change area and input to these crops.

7. CHALLENGES FOR CONSUMERS (INFLUENCE ON REAL INCOME AND CONSUMPTION)

Increase in price of F&V has detrimental effects on consumers due to its impact on their purchasing power. Price rise have two immediate effects; 'substitution effect' and income effect. The substitution effect includes changes in relative products available for consumption including both F&V and other food and non-food crops; and the income effect results in increase in the aggregate cost of consumption which results in fall in real income and welfare. Price rise become more burdensome, especially, the last few decades have seen huge jump in expenditure of households on F&V. In between 1983 to 2004, the budgetary shares of both F&V in total food expenditure have increased by around 50% (Kumar *et. al.,* 2011). As the share of F&V in total food expenditure has depicted maximum change in total food expenditure, as compared to food products, such as, pulses, cereals, meat *etc*., price rise in F&V had declined the real income of the people.

The actual consumption of F&V depends on various supply and demand factors. At the individual level, consumers can respond differently to price change rise of F&V, depending on the price elasticity. According to empirical studies, in India, both F&V are more elastic than cereals and across these two categories; vegetables are more inelastic as compared to fruits (Kumar *et al.,* 2011; Pons, 2011). Price rise in F&V vegetables result in decline in real income of the consumers, increased cost of living and less money available for other food and non-food products, reduced consumption of F&V or increased consumption of alternative less nutritious cheap food.

8. CHALLENGES FOR ECONOMY (INFLUENCE ON NUTRITIONAL STATUS AND INCOME)

Increase in the price of F&V raises several challenges for the economy and the government. F&V entails several nutritional benefits but despite India being the second largest producer of F&V in the world, average consumption stand below daily requirement. The *per capita* consumption of F&V is 3.5 serving a day which is far below the recommended intake of minimum of five serving. In the event of higher prices, poor consumers can opt for different strategies like either to skip eating F&V or switch to other less nutrition cheap products. Higher prices of F&V further dent not only the target to increase the consumption of F&V and improve the nutritional status of population but it also reduces the probability to maintain the current levels which are far below than the minimum required levels.

Less intake of nutrition is correlated with deaths and malnourishment among people in the country. Overall it is estimated that low F&V intake is attributable to approximately 2.7 million (4.9%) annual deaths in the world and is considered as the sixth main risk factor for mortality in the world (Sachdeva *et al.,* 2013). The results of Indian Experiment of Infarct Survival (IEIS) showed that consumption of low-fat diet enriched with F&V, compared with a standard low-fat diet, was associated with about 45% reduction in mortality after one-year (Singh RB *et al.,*1992). India is the home of most of the malnourished children of the world and increase prices of F&V could potentially further reduce the nutritional intake. The major influence will be on poor people who still constitute 26% of all people in India. Higher incidence of malnourishment reduces not only the availability of workforce but also reduce the quality of workforce available for work in the economy that adversely affects the gross domestic product. As low intake of F&V is also linked with increased incidence of disease, which means price rise also increases total cost of health care.

The macroeconomic impacts of high F&V prices differ between importing and exporting countries depend upon prior taxes and subsidies and upon any changes in taxes and subsidies made in response to high prices (Dorward, 2012). For importers, high F&V prices in the global market increase import bills, adversely affect the balance of payments and put downward pressure on the domestic currency, encouraging inflation. The opposite effects are experienced by F&V exporters *(ibid)*. High prices of F&V would adversely affect the export trend of the country. During 2015-16, India exported F&V worth Rs. 8,391 crores which comprised of fruits worth Rs. 3,524 crores and vegetables worth Rs. 4,866 crores. Despite being the second largest producer of F&V, India is also a net importer of F&V and currently India's share in the global market is still nearly 1% only. Increase in prices of F&V would further deteriorate India's chance of being one of the major contributors to trade.

9. MEASURES TO OVERCOME THE CRISIS

The demand of F&V has been increasing at a faster rate and is expected to grow at the same pace in the future as well. According to the recent estimates, during the period from 2000 to 2050, while the *per capita* consumption of cereals will decrease by 9%, 47% and 60%, with respect to rice, coarse cereals and maize, the *per capita* consumption of sugar, fruits and vegetables will increase by 32%, 65% and 78% respectively (Hegde, 2012). There is a need to check increase in prices of F&V for the welfare of masses in India who consume F&V daily.

Post 2008 has seen higher rate of growth of price of F&V in India. Also the last three year trend has shown that prices of F&V have grown faster than other food item which is a cause of concern for the consumers of F&V. In addition, trend of huge price variability of F&V within a month further hurts not only the producers but also the consumers who, at time, generally end up paying higher prices for F&V. Price margins between F&V producers and consumer is very high in India owning to fragmented markets, huge post-harvest losses and lack of storage facilities. Undoubtedly, price rise deprive millions of consumers, especially poor people, adequate access to F&V, but in addition to affordability, availability of F&V in different places and access to fresh and quality product is also essential. In nutshell, not only the price but also accessibility, quality and diversity are the facets that require attention to improve consumer's welfare.

Consumer welfare can be addressed by undertaking efforts to increase F&V production, reduce post-harvest losses, improve market efficiency, increase processing *etc.* These will help to improve affordability, accessibility, diverse options and quality of produce. The production of F&V can be increased through increase area allocation and through introduction of smart varieties for better productivity. Increase in production through higher land allocation can be done by increasing farm gate prices and reduction in price and production volatility through seed varieties and market efficiency. In India, there are several instances where the farmers have to sell the produce at throw away prices due to glut in the production in the market. Thus, efforts are required to ensure better farm prices for produce. Smart varieties can be introduced so that productivity of F&V that has been growing at the pace of around 1% could be increased. So far, though, productivity of fruits in India is higher than world's average and productivity of vegetables is below world's average, if compared to individual countries, it is almost half than the most competitive countries. For instance, F&V productivity of India (12 and 17 tonnes per hectare respectively) is about half of the productivity of USA (23 and 32 tonnes per hectare). Also as small and marginal farmers dominate Indian farming, there is lack of mechanization which results in poor productivity and poor quality of crops. Indian farmers are mostly dependent on rainfall which results in huge variability in product availability. Contract farming here can play a pivotal role to induce collective farming, against the risk of weather by the farmers and ensuring better farm gate prices to the farmers.

High prices and seasonality of production are crucial barriers affecting consumption (Mukherjee *et al.,* 2016). Seasonality in production leads to non-availability of F&V in different seasons. But due to high perishability and huge geographic spread of population in India, fresh F&V are not available in all the places even during the season of production of F&V. Consumers find it difficult to buy fresh F&V and in remote areas, they travel to far off places to buy which also increases their cost of accessing F&V. There is a need to ensure F&V availability in different places including the remote areas in the country. F&V stores can be introduced in remote areas through financial incentives and by improving public transportation to these stores. The accessibility can also be increased by introducing small F&V shops in schools, colleges, community centres, hospitals and even workplace cafeterias.

Accessibility and availability of fresh and quality F&V to consumers also can be improved by undertaking measures to reduce post harvest losses, which can be addressed through a supply chain approach involving improvement of collection centres, better storage and communication along the supply chain making use of information and communication technologies (FAO, 2015). Higher amount of wastage in marketing of the crop results in wide difference between the total production and net availability for consumption. In India, between 20 and 30% of total fruit and vegetable production goes to waste owing to spoilage at various steps of the postharvest chain, reducing *per capita* availability to around 50 gm and 110 gm per day for F&V respectively. This results in high losses after the production of the crop, increase in prices of F&V and reduces availability of sufficient F&V. According to the estimated by ASSOCHAM, post-harvest F&V losses in India are over Rs 2 lakh crore annually (The Economic Times, 2013), owing to inadequate cold storage facilities and lack of proper food processing units. In India, where F&V can also be made available in different forms; fresh, dried, juices *etc*. In India so far, only 2.2% of F&V are processed into products, such as, juices and concentrates, pulp, canned and dehydrated products, *etc* while the processing ranges from 30% to 80% in USA, China, South Africa, Brazil, Chile, Philippines, Malaysia, Thailand (Sachdeva *et al.,* 2013 and Mukherjee *et al.,* 2016). In addition, so far, Indian food processing industry is export-oriented. Processing and value addition is a better solution not only for reducing wastage of fruits and vegetables but it also provides diverse options to the consumers to consume F&V in different forms. As the price of processed food also does not vary much, consumers can hedge against sporadic price rise of fresh F&V partially by consuming processed products.

Due to short shelf life and high perishability, mostly F&V get spoiled in the transport which results in lack of availability of fresh and quality produce that is not fit for consumption. It is estimated that between 10 and 60% of the fresh F&V marketed and purchased by consumers in India are rejected as inedible. In villages or small towns the inedible portions of F&V are either fed to animals or are discarded as garbage by consumers in metro cities (APO, 2006). Thus, efforts are warranted to improve the quality of produce by introducing more efficient transportation system, establishment of cold storage facilities and increase processing of F&V. This will help in consumers to have access to more fresh, quality and affordable F&V.

10. CONCLUSION

India is the second largest producer of F&V in the world. F&V covers less than one tenth of the total gross cropped area but their role in the Indian diet is quite crucial. Even with good availability, the average *per capita* consumption of F&V in India is far below the recommended intake. Due to high relative profitability of F&V as compared to many other crops and high nutrient benefits of its consumption, there is a need to increase both area under F&V and consumption of F&V in India. Here, prices play a critical role in influencing the allocation of scarce resources between competing uses (land by farmers and expenditure on different food and non-food items of consumption).

The trends in the change in real prices of F&V show that the last fourteen years have seen an upward trend in the real prices of all selected F&V and growth of real prices is higher in case of vegetables, the demand of which is more inelastic as compared to fruits. Within a given season and month, there is also a marked fluctuation of prices among F&V that increases even by nine to ten times. High difference between the wholesale and retail price that ranges between 23% to 69% in case of fruits and ranges between 47% and 137% for vegetables reflects that most of the benefits of price rise are not realised by the producers or farmers and it is the middlemen who make fortune out of the sporadic spurts in price rise.

Price rise and fluctuations raise several challenges both for the producers, consumers and for the economy as a whole. Due to high marketing cost and market inefficiencies, benefits of price rise generally fail to reach the farmers or producers and it is the middlemen who reap the advantage. The challenge is to reduce the gap between the price consumer pays and what the farmers get for their produce, so that the benefits of price rise through increased demand could reach the purse of the farmer. Increase in price of F&V has detrimental effects on consumers due to its impact on their purchasing power. As F&V are more price elastic than cereals, an increase in price results in decline in real income of the consumers, increased cost of living and reduced consumption of F&V or increased consumption of alternative less nutritious cheap food. Higher prices of F&V result not only in decreasing the consumption of F&V and deprive high nutritional status of population but it also reduces the probability to maintain the current levels which are far below than the minimum required levels. As low intake of F&V is also linked with increased incidence of disease, which means price rise also increases total cost of health care. Despite being the second largest producer of F&V, India is also a net importer of F&V and currently India's share in the global market is still nearly 1% only. Increase in prices of F&V also deteriorates India's chance of being one of the major contributors to trade.

Price rise deprives millions of consumers, especially middle income and poor people. Adequate access to F&V, in addition to affordability, availability of F&V in different places and access to fresh and quality product is essential for better consumer welfare. The challenges emerged from high prices can be addressed by undertaking efforts to increase F&V production, reduce post-harvest losses, improve market efficiency and increase processing. These will help to improve affordability, accessibility, diverse option and quality of produce.

REFERENCES

Asian Productivity Organization (APO). (2006). Postharvest Management of Fruit and Vegetables in the Asia-Pacific Region. Reports of the seminar on Reduction of Postharvest Losses of Fruit and Vegetables 2004 and Marketing and Food Safety: Challenges in Postharvest Management of Agricultural/Horticultural Products in 2005.

Chand, R. (2016). Market Liberalization for Horticulture Revolution at Small Farms. http://niti.gov.in/content/market-liberalization-horticulture-revolution-small-farms

Chawla, A. and Kandhari, P. (2008). Fruits & Vegetables: Imports into and Exports from India. Hindustan Studies and Services Limited and Infolitics.

Dastagiri, M.B., Kumar, B.G. and Diana, S. (2010). Innovative Models in Horticulture Marketing in India. National Centre for Agricultural Economics and Policy Research New Delhi. . http://crida.in/agrl_martng/ISAM/PDF%20FILES/T-II/Dastagiri.pdf.

Deaton, A. and Dreze, J. (2009). "Food and Nutrition in India: Facts and Interpretations." Economic and Political Weekly 44(7): 42-64.

Dorward, A. (2012). The short and medium term impacts of rises in staple food prices. Policy Brief No. 52, March 2012. Future Agricultures Consortium.

Economic Survey. (2016). Economic Survey, Government of India, 2016.

FAO. (2015). Promotion of Fruit and Vegetables for Health. Report of the Pacific Regional Workshop. Food and Agriculture Organization of the United Nations, Rome, 2015.

Gandhi, V.P. and Namboodiri, N. V. (2004). Marketing of Fruits and Vegetables in India: A Study Covering the Ahmedabad, Chennai and Kolkata Markets. IIMA Working Paper No. 2004-06-09, Indian Institute of Management, Ahmedabad.

Hall, J. N., Moore, S., Harper, S.B. and Lynch, J.W. (2009). Global Variability in Fruit and Vegetable Consumption. American Journal of Preventive Medicine. 36:402.

Hegde, N.G. (2012).Water Scarcity and Security in India BAIF Development Research Foundation, Pune. http://www.indiawaterportal.org/articles/water-scarcity-and-security-India.

Kumar, P., Kumar, A., Parappurathu, S. and Raju, S.S. (2011). Estimation of Demand Elasticity for Food Commodities in India. Agricultural Economics Research Review Vol. 24 January-June 2011 p. 1-14.

Mukherjee, A., Dutta, S. and Goyal, T.M. (2016). India's Phytonutrient Report: A Snapshot of Fruits and Vegetables Consumption, Availability and Implications for Phytonutrient Intake. Academic Foundation.

Nair, S.R and Eapen, L.M. (2012). Food Price Inflation in India (2008 to 2010): A Commodity-wise Analysis of the Causal Factors. Economic & Political Weekly, Vol. XIVII, No. 20.

National Council of Applied Economic Research (NCAER). (2014). An Analysis of Changing Food Consumption Pattern in India: A research paper prepared under the project Agricultural Outlook and Situation Analysis Reports, New Delhi.

National Sample Survey Office (NSSO). (2012). Household consumption of various goods and services in India: 66th Round (2009-10), Ministry of Statistics and Programme Implementation: Government of India; 2012.

Pons, N. (2011). Food and Prices in India: Impact of Rising Food Prices on Welfare. Centre de Sciences Humaines, New Delhi. http://www.isid.ac.in/~pu/conference/dec_11_conf/Papers/NathaliePons.pdf.

Sachdeva, S., Sachdev, T.R. and Sachdeva, R. (2013). Increasing Fruit and Vegetable Consumption: Challenges and Opportunities. Indian Journal of Community Medicine. Oct-Dec, 2013, 38 (4) pp. 192-197.

Singh, R.B., Rastogi, S.S., Verma, R., Laxmi, B., Singh, R. and Ghosh, S. (1992). Randomised controlled trial of cardio-protective diet in patients with recent acute myocardial infarction: Results of one year follow up. British Medical Journal. 304, p. 1015–19.

The Economic Times. (2013). India's post-harvest losses over Rs 2 lakh crore annually: Assocham. http://economictimes.indiatimes.com/news/economy/agriculture/indias-post-harvest-losses-over-rs-2-lakh-crore-annually-assocham/articleshow/21652094.cms.

The Hindu. (2012). From farm gate to your plate. December 2012. http://www.thehindu.com/todays-paper/tp-national/tp-newdelhi/from-farm-gate-to-your-plate/article4205641.ece.

The Huffington Post. (2016). Farmer Earns ₹1 After Selling One Tonne Of Onions. May 2016. http://www.huffingtonpost.in/2016/05/24/farmer-earns-1-after-selling-one-tonne-of-onions/.

❍❍❍

20

Skilled and Smart Farmers' Organizations Driving Farm and Rural Development

M.J. Chandregowda

SYNOPSIS

Skill development programs in rural areas can push the farmers to be more adept in handling the challenges and also make their profession rewarding financially. Skill building is necessary to enable a section of the rural population to take up non-farm activities and reduce underemployment in rural areas. Alternative avenues that create jobs outside agriculture in rural areas boost rural entrepreneurship and provide better connectivity with urban centres and infrastructure to make business ideas viable. Considering the importance of skills for both self-employment and getting employed elsewhere, the national skill development mission has initiated a massive exercise to skill rural youth. Improved agricultural practices are skill intensive and imparting new skills to practicing farmers and farm women is an integral part of the capacity building initiatives of Krishi Vigyan Kendras (KVKs). Imparting specific and relevant skills on technological interventions in agriculture and allied sectors by KVKs has helped farmers, farm women, rural youth and other stakeholders through creation of self-employment as well as generating additional income to farmers and farming community. Some of the successful initiatives presented in this paper provide a glimpse of the entrepreneurial models created by KVKs for farmers and farmers organizations.

Keywords: smart farmers, skill development, rural development, KVK.

1. INTRODUCTION

Rural India was home for 833.75 million of its total population of 1210.86 million in 2011, thus reiterating the reality that India continues to live in villages. About 40% of the total population categorized as working population, includes 41.81% rural and 35.31 % urban working population. The fact that the majority of working population lives in villages, mostly engaged in the agriculture sector, signifies the importance of agriculture as the main source of livelihood for majority of the people. Of the total 118.69 million cultivators, 96.87% cultivators lived in rural areas (GOI, 2015). Despite supporting majority of India's total population, the contribution of agriculture to the GDP has been consistently declining. Although Indian economy is growing around 7 percent in recent years, growth rate in agriculture has more or less remained stagnant. A sector that provides work to the most of the country's population, but contributing less than $1/6^{th}$ to its GDP is a reminder of severe imbalance. The statistics demand for a major reorientation and focus towards rural economy. Declining or stagnant agricultural growth is attributed to the challenges facing the farming sector.

The Steering Committee report on agriculture for XI Plan had identified the possible reasons for deceleration in agriculture since mid-1990s. These include, degradation of natural resources, failure in conservation and improvement of rainfed land, knowledge gap with existing technology, low market infrastructure, institutions not geared to help women farmers and plight of small farmers (Dev, 2008). These challenges are not just in India, but it is a global phenomenon. There are growing expectations to maintain and improve environment and face the challenges of climate change; as nowhere are the effects of severe weather more obvious than in farming. We can respond to these opportunities and challenges with better research, technology and policy support. But nothing is more critical to farming's success than ensuring that those who work in it are enthusiastic and skilled. This means attracting the best people to work in farming and then supporting them to build their skills and experience and have successful and rewarding careers (Anon., 2013).

It is estimated that only 4.69 percent of India's total workforce has undergone formal skill training, compared with 52 percent in the USA, 68 percent in the UK, 75 percent in Germany, 80 percent in Japan and 96 percent in South Korea. Based on the Census 2011 and NSSO (68th Round) data, it is estimated that 104 million fresh entrants to the workforce will require skill training by 2022 and 298 million of the existing workforce will require additional skill training over the same time period. Acknowledging the formidable scale of this challenge, the government created a dedicated Department of Skill Development and Entrepreneurship, which became a full-fledged Ministry in 2014. (http://www.pmindia.gov.in/en/news_updates/national-skill-development-mission/). The agriculture sector skill council has identified 142 job areas and developed qualification packs for skilling the rural population. Many more are going to be added to the list depending on the skill requirements.

Skill development programs in rural areas can push the farmers to be more adept in handling the challenges and also make their profession rewarding financially. Skill building can help a section of the population towards non-farm activities and reduce overcrowding of the agriculture sector. Creating alternative income avenues is essential to create jobs outside agriculture in rural areas, boost rural entrepreneurship, develop tourism, cottage and handicraft industries, horticulture, fisheries and better connectivity with urban centres and infrastructure to make business ideas viable (Rajesh, 2016).

Besides skill development, another initiative that has targeted improvement in skill-based farm development is the national policy of farmer producer organizations. Bulk of Indian farmers own marginal (67.1%) and small holdings (17.9%), together accounting for 85% of the total holdings. This poses a huge challenge for the research and development sectors to deliver various services to these farmers. Keeping this in view, Farmer Producer Organisation (FPO) registered under the special provisions of the Companies Act, 1956, has been identified as the most appropriate institutional form around which farmers are organized to build their capacity and collectively leverage their production and marketing strength. Realizing the need for hastening the formation and mobilization of FPOs, the Government of India has come out with a National Policy for the Promotion of Farmer Producer Organisations in 2013. As per the policy, the formation and development of FPOs will be actively encouraged and supported by the Central and State Governments and their agencies, using financial resources from various centrally sponsored and State-funded schemes in the agriculture sector agencies. Detailed guidelines for the promotion of FPOs, including methodology of selecting Resource Institutes, budgets and registration processes have been prepared and made available for the benefit of all stakeholders. These guidelines are meant to guide the Central and State Government agencies engaged in FPO promotion and provide a detailed roadmap to achieve the vision of building FPOs. The seven guiding principles identified in the policy document for sustaining farmer producer organisation development are (i) voluntary and open membership, (ii) democratic farmer member control, (iii) farmer member economic participation, (iv) autonomy and independence, (v) education training and information, (vi) cooperation among FPOs and (vii) concern for the community. The policy document envisages both economic impact and social impact. The economic impact includes increase in productivity, net returns, enhanced availability of inputs, forward and backward integration, employment generation, increase in food and nutrition security and reduced migration. Among social impacts building up of social capital, improved gender relations, increased bargaining power, reduced social complex, leadership role of producers and improved health and nutrition of its members (GOI, 2013).

Krishi Vigyan Kendras (KVKs) are supporting the farmers' organizations by providing the technical back-up through capacity building of members as well as producer organizations promoting agencies. KVKs are fully funded by Indian Council of Agricultural Research (ICAR) and have the required multidisciplinary subject matter expertise to support the district agricultural development. Farm and laboratory facilities available with the KVKs are also used for incubation of farmer entrepreneurs and groups to promote agri-enterprises. KVKs established over period of four decades since 1974 are functioning under the administrative control of State Agricultural Universities, ICAR Institutes and Non-Government Organizations. Horticulture and livestock are the two important sectors that are maintaining higher growth rate and have larger contribution to agricultural GDP of the country. A study of successful smallholders promoted by KVKs in eleven districts of South India confirmed the choice of farmers towards high income activities. The results revealed that horticulture crops were adopted by 80.7% of the farmers. Dairy, sheep and goat, poultry and fisheries constituted the livestock activities and were part of majority of the households. (Fig. 1)

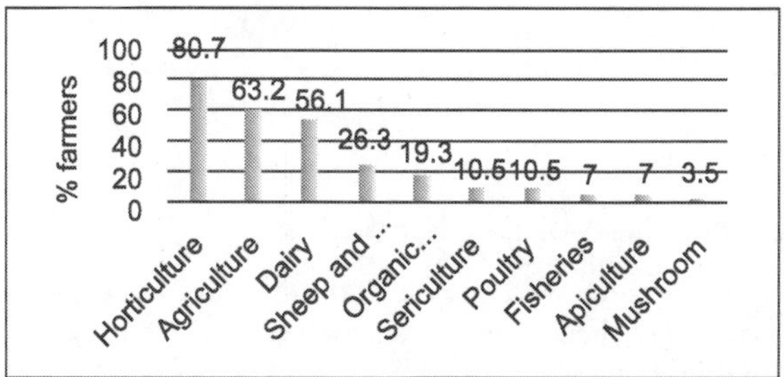

Fig. 1. *Choice of crops/enterprises of successful smallholders*

2. SKILLED FARMERS ARE SMART FARMERS: SUCCESSFUL MODELS PROMOTED BY KVKs

Keeping in view the demand for new skills in promoting intensive agriculture, skill-based technological interventions related to on-farm production of organic manure, bio-products, seeds and planting material, bee keeping and small-scale mechanization were upscaled by the KVKs. The approach adopted by the KVKs and the successful models created in selected areas are briefly explained below:

2.1 Better use of Organic Manure - Vermiculture

Most of the KVKs are training farmers and rural youth in vermiculture as demand for organic manure is increasing. KVK Gadag, Karnataka organised 111 courses during 2005-2015 for building skills of 3221 farmers. Training was followed by post-training handholding and advisory services. It resulted in establishment of 100 large scale production units, besides several farm-level production units. Out of 100 large scale units, 37 units are producing more than 100 t per year. These units have also provided earthworms (2088 kg) to 761 farmers to establish their own units. Study of these 100 vermicompost producers revealed that 44% were small farmers (less than 2 ha farm size) and 42% were big farmers (more than 2 ha farm size). During 2014-15, on an average, each farmer produced 9.54 t out of which used 7.54 t of vermicompost on their farms and sold about 2 t surplus to other farmers. Each farmer supported another 3 farmers in use of vermicompost in their farms. Chilli and onion were the major horticulture crops in the area wherein about 2.6 to 3 t/ha of vermicompost was applied and saved upto Rs.4708 / ha on organic manure. Yields of vermicompost applied fields were better than the control fields and earned a net income of Rs.58076/ha in onion and Rs.37500/ha in chilli (Mudlapur, *et.al.* 2016) (Fig. 2)

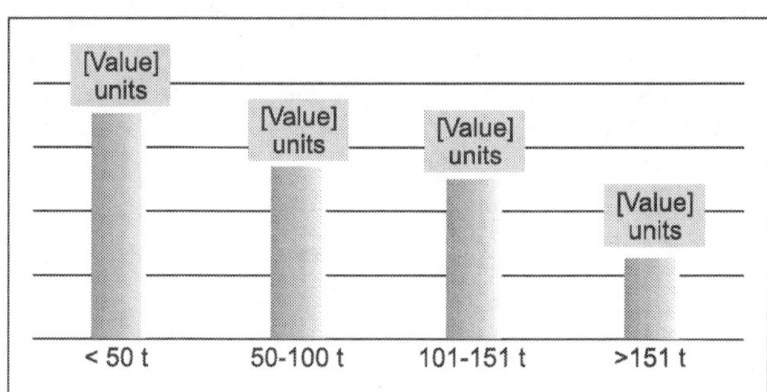

Fig. 2. *Vermiculture units (No.) and annual production capacity (t) in Gadag district*

2.2 Production and Use of Bio-Products

KVKs are producing bio-products and helping farmers / farmers' organizations to produce and use in improving soils health and manages pests in better was. KVKs adopt the following *modus operandi*:

- Asses the performance of various bio-products and bio-agents to different crop-specific problems and situations.

- Skill training on production of bio-products/agents.
- Group formation and motivation.
- Share the KVK lab and farm resources to groups to enable them to produce bio-products in partnership mode.
- Skill perfection in operating lab equipment and production of bio-products.
- Branding, publicity and marketing.

Although the production of bio-products is happening in most of the KVKs, the partnership approach adopted by KVK Wynad, Kerala with a Self-Help Group (SHG) named 'Sabari' for promoting *Trichoderma viride, Pseudomonas flourescens,* Entomopathogens and biofertilizers is mentioned here. KVK provides the technical support in the form of capacity building and shares its laboratory, equipment and other resources. The steps involved in skill development are given below:

Steps in skill training of rural youth and women in preparation of *Trichoderma*

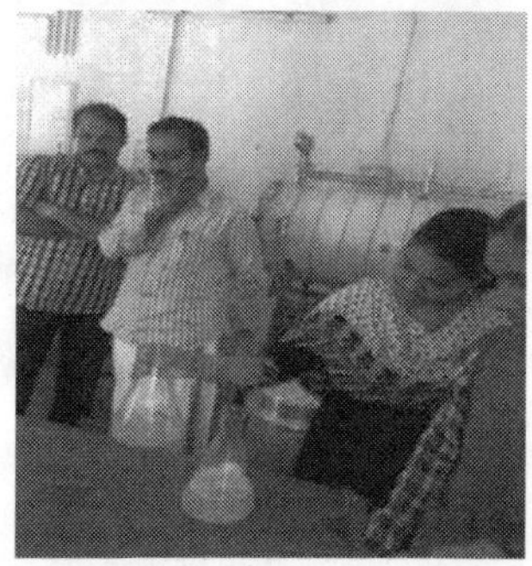

Photo 1: Potato extract

Photo 2: Mixing of dextrose and agar to potato

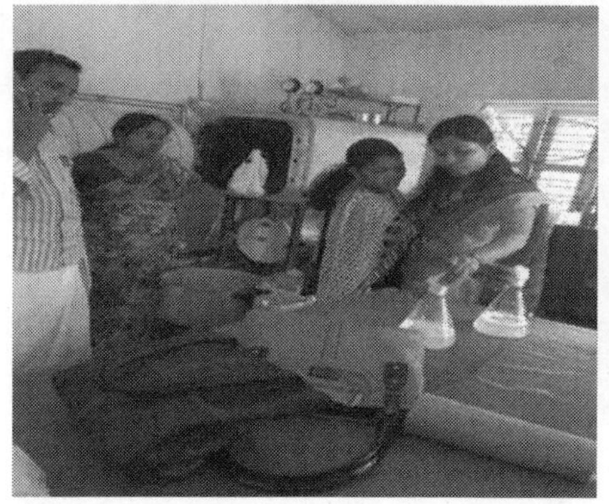

Photo 3: Cotton plug to avoid contamination of dextrose - agar -potato

Photo 4: Sterilization using pressure

Photo 5: Sterilization using auto clave

Photo 6. Sterilization of coffee husk mixture

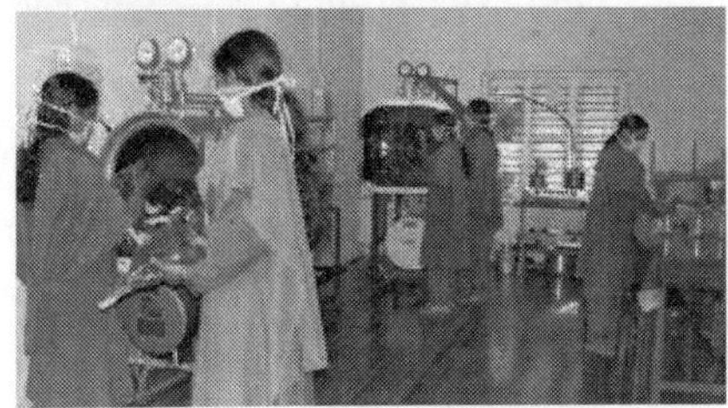

Photo 7. Inoculation with mother culture of Trichoderma in laminar air flow

Photos 7-8. Sterilization, inoculation and incubation

The partnership has been beneficial to KVK as well as the SHG. The quantity of bio-products and bio-agents produced by the KVK in partnership with the women group are given in Table 1. Quantity of *Trichoderma viride* and *Pseudomonas flourescens* produced by the group is increasing every year and used by farmers, mostly for plantation crops like pepper (Table 1).

Table 1. *Production of Biocontrol agents and Biofertilizers in Krishi Vigyan Kendra, Wayanad, Kerala during 2014-17 in partnership with Self Help Group*

Bioproduct	Quantity produced (Kg)			Major skill involved (Steps)
	2014-15	2015-16	2016-17	
Trichoderma viride	10,000	13,270	32,747	Medium: Potato- dextrose- agar Inoculation from mother culture to PDA plates Culture in PDA broth Mix with talc
Pseudomonas flourescens	12,000	10,000	50,000	Medium: King's B agar Inoculation from mother culture to King's B agar plates Culture in King's B broth Mix with talc

Entomo pathogens *(Beauveriabassiana, Metarhizium anisopliae, Lecanicilliumlecani, Pochonia, Paecilomy-ceslilacinus)*	795	1100	1525	Medium preparation: Potato- dextrose-agar Inoculation from mother culture to PDA plates Culture in PDA broth Mix with talc
Biofertilizers (Biopotash, *Azospirillum, Azotobacter,* Arbuscular mycorrhiza (VAM/AMF), *Rhizobium*)	240	1100	6500	Medium: nutrient broth, Nitrogen free semi solid malate, Genson's medium, vermiculite- perlite – cowdung – sand – soil (18: 25: 2: 3: 1), Yeast extract mannitol broth

2.3 Production and Multiplication of Seeds and Planting Material

Participatory seed production is a two-way beneficial approach, benefitting the producers in getting quality seed in addition to higher value for their farm produce on one hand and benefitting the seed producing agency in meeting the quality seed requirement of huge number of farmers.

KVK, Thrissur, Kerala has adopted a partnership approach in vegetable seed production. The homestead production system prevailing in Kerala allows the households to operate small area for cultivation requiring only a small quantity of vegetable seeds each time. Homestead production of vegetables has also become important for growing vegetables with minimum pesticide residues. The KVK has partnered with two groups of women from peri-urban localities of Thrissur in meeting the demand of vegetable seeds from peri-urban farmers/ households. Two groups of women were formed and were trained on seed to seed skills. Once trained skills were not adequate enough to produce quality seeds and Hence, were re-skilled during a season-long hands-on production experience. Quality standards were checked and then the groups members were allotted farm, machinery and laboratory facilities to produce, process, pack and sell seeds under the technical supervision of KVK experts. The quantity of seeds and planting material produced and sold are available in Table 2 and 3.

Table 2. *Quantity of Vegetable Seeds produced and sold by the Seed Growers Groups in partnership with KVK Thrissur, Kerala*

	Quantity (Kg) of seeds produced				Value (Rs.) of seeds sold			
	2013-14	2014-15	2015-16	Total	2013-14	2014-15	2015-16	Total
Cowpea	139	266.9	251.81	657.71	83400	243340	245241	571981
Pumpkin	213	81.26	90	295.16	319650	121890	1620	443160
Bottlegourd	55	11.95	13.5	80.45	44000	9560	10800	64360
Snakegourd	68.2	48.69	70.71	187.60	122760	87633	141410	351803
Amaranthus	58	111.25	55	224.25	72500	195167	68750	336417
Bitter gourd	75.3	86.56	102	263.86	150600	173120	204000	527720

Brinjal	5	16.76	11.86	33.62	7500	25134	21351.6	53985.6
Cucumber	30	22.4	0	52.4	45000	33600		78600
Tomato	1	0.01	0.6	1.61	5000	5000	3000	13000
Bhindi	192	47	42.32	281.32	192000	47000	63480	302480
Ash gourd	321.1	22.8	3.5	347.4	561925	39900	7000	608825
Water melon	20	0	0	20	40000	0	0	40000
Chilly	15	0	13	28	42000	0	82240	124240
Palak	0	0	20	20	0	0	276500	276500
Drumstick	0	0	88	88	0	0	39000	39000
Bush Dolichos	0	0	150	150	0	0	900	900
Dolichos Bean	0	0	7	7	0	0	560	560
Salad Cucumber	0	0	2	2	0	0	40000	40000
Ginger	40	0	0	40	2600	0	0	2600
Turmeric	634	0	0	634	29205	0	0	29205
Mother spawn	517.2	524.8	411.2	1453.2	64620	78720	48150	277890
Total	**2383.8**	1240.375	1332.501	4867.576	1782760	1060064	1254003	4183227

Table 3. *Quantity of Vegetable Seedlings produced and sold by the Seed Growers Groups in partnership with KVK Thrissur, Kerala*

Crop	Seedlings (No.) produced				Value (Rs.)of seedlings sold			
	2013-14	2014-15	2015-16	Total	2013-14	2014-15	2015-16	Total
Tomato	1296	465	0	1761	1296	698	0	1994
Onion	2808	0	13050	15858	2808	0	3068	5876
Chilly	818	117030	0	117848	818	175545		176363
Cauliflower	28681	0	5686	34367	57362	0	10698	68060
Cabbage	30000	0	5349	35349	60000	0	556	60556
Marigold	3625	0	0	3625	3625	0	0	3625
Gomphrena	250	0	0	250	250	0	0	250
Mango graft	36	0	1423	1459	5400	0	213450	218850
Pepper cuttings	8360	27650	27650	63660	50160	276500	42330	368990
Nutmeg	172	0	0	172	8600	0	0	8600
TC Banana	3174	2305	0	5479	63480	34575	0	98055
Other Vegetable seedlings	0	0	93472	88894	0	0	215460	215460
	79220	147450	146630	368722	253799	487318	485562	1226679

Growing vegetable nurseries is an emerging area of rural entrepreneurship. KVK, Mysuru, Karnataka has been training farmers and entrepreneurs on nurseries since 2007-08. The first batch of 15 participants were able to establish 10 nurseries independently and a few worked in other nurseries. The success achieved by the initial nurseries motivated more youth to undergo training either at KVK (about 50) or through hands-on experiential training in the existing nurseries. By 2016, there were about 215 nurseries, on an average earning a net income of Rs.20000 per month. There are nurseries having a turnover of Rs 3 lakh per month. The case of Shri Rajabuddi, Village Duddagere, Taluk of T. Narasipur District is truly inspiring. He is not only doing business, but also educating farmers on choice of crops and varieties keeping markets and seasons in mind. His own experience of successful vegetable cultivation from technical support from KVK, Mysuru is enabling him to provide agronomic and plant protection advises to his client farmers.

2.4 Micro Irrigation

Irrigation uses 65% of all water diverted for various uses worldwide. Irrigation method is a complex subject, not always well understood by farmers and the public alike. Decreasing water availability, higher costs and heightened environmental concerns of agricultural water diversions are issues that growers will not be able to ignore in the future. (Elias *et al.*, 2003). Making the agriculture sector less dependent on rainfall by effective and innovative irrigation methods has been an area long discussed. The skill required in irrigation methods is enormous and several KVKs are promoting micro-irrigation technicians to support the farmers in adoption of micro-irrigation methods in their farms (Table 4). (Photo 9)

Table 4: *Skill training of micro-irrigation technicians at KVK Sivagangai, Tamil Nadu, India*

S.No	Important skills in micro-irrigation	Detailed learning structures
1	Head units installation	consists of a pump or over head tank.
2	Mainline installation and maintenance	50 mm, 63 mm and 75 mm HDPE or PVC pipes are used
3	Submain installation and maintenance	45 mm, 50 mm HDPE pipes are used. Submain 45 mm, 50 mm HDPE pipes are used.
4	Laterals erections and field flow	12 mm and 16 mm LLDPE pipes are used
5	Drippers and its maintenance	Pressure compensating type 2 lph, 4 lph, 8 lph are more suitable
6	Filter unit and its maintenance	Sand filter for coarse particle detention. Mesh filter (120 micron) for fine particle detention.
7	Fertiliser unit and its maintenance	A fertiliser tank with ventury setup is used for applying liquid or dissolved fertilisers. Pressure gauges 0.5 to 2.5 Ksc and gate valve depending on pipe size are also essential for the system to control and monitor the flow. (Photo 12)
8	Water soluble fertilizers application	Water soluble fertilizers and its usage for supply of nutrients through fertigation systems

Photo 9: *Skill training on micro-irrigation at KVK Sivagangai, Tamil Nadu*

2.5 Small Scale Mechanization

The coconut tree climbing device developed and further fine-tuned in Kerala was promoted in a collaborative effort between Coconut Development Board and KVKs in Kerala, Karnataka and Tamil Nadu. Week-long training programmes were organised by the KVKs for groups of 20-25 participants on climbing and safety measures. The Coconut Development Board supplied the climbing device, free of cost to the trained persons. The skilled youth and women trained by KVKs are practising the climbing as a vocation and earning up to Rs.1500/day. It has become a source of livelihood for entrepreneurs on one end and helped the farmers on the other by reducing the cost on harvesting coconuts and cleaning of coconut trees. This has overcome the skilled labour constraint faced by farmers in Kerala and Coastal Karnataka.

Photo 10: *Skilled training of women SHG members on coconut tree climbing*

KVK, Tumakuru has trained 340 persons on coconut climbing since 2012 with the support of Coconut Development Board and Karnataka State Rural Livelihood Programme (Fig 3). These trained persons are now known as Coconut Technicians, by being self employed as well as providing consultancy services. Two Coconut Climbers Associations have been formed with 12 members each and are climbing up to 80 trees per day.

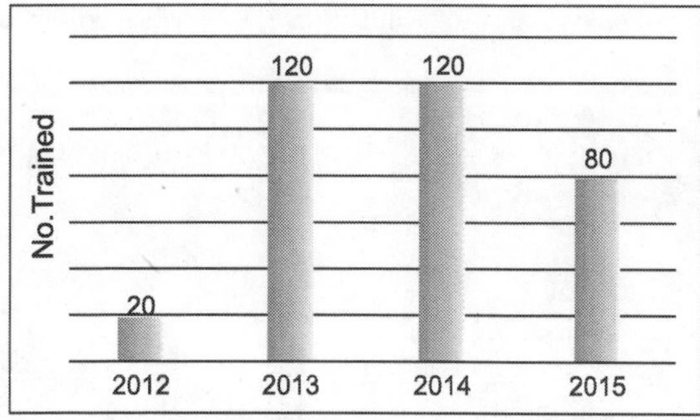

Fig. 3. *Coconut technicians trained by KVK Tumakuru*

On an average, each member earns about Rs.1000 to 1200 per day. They are also providing services to CPCRI and Coconut Development Board on survey, taking up plant protection measures on community basis *etc.*

2.6 Bee Keeping as an Income Generating and Crop Yield Boosting Tool

Krishi Vigyan Kendra, Chamarajanagara, Karnataka has been organizing bee keeping training since 2006 to the interested youth as a vocational training programme. Among those who got trained, Raju, an entrepreneur from Naguvally, Chamarajanagar Taluk got trained again at three different places to make sure that the skills are adequate to start an agri-business of his own. He started production and selling of honey in Karnataka through setting up his own enterprise with the brand name Aparna honey having Agmark and TIN No. The unit has employed 15 members in production and five members in marketing. He is now resource person for training on honey production at KVK and at other places. He is maintaining *Serena indica, Mellipona and Mellifera* species. Under National Horticulture Mission Scheme of Government of India, he is supplying bee colonies, bee-stands and its intermediaries to the trainees. He is marketing honey comb for five star hotels and is marketing 1 to 1.5 t honey per year. He has started a novel idea of marketing honey in sachets. Regularly attends Krishi Melas and other exhibitions to attract the customers and also to promote this enterprise as more remunerative. He got best farmer award from UAS, Bengaluru for his innovative ideas. Motivated by his success, several farmers are getting trained and are making efforts to start the enterprise. The technical support provided by KVK, Chamarajanagara, high returns with small investment and minimum labour requirement have popularized bee keeping with improved management practices in the District.

On the other hand, KVK, Koppal, Karnataka a predominantly paddy growing belt under command area irrigation is also promoting bee keeping for a different purpose. The major purpose is to boost crop yields through bee-aided pollination. The KVK has trained 35 farmers and formed an association to promote bee-keeping. The major skills involved are establishment of honey colony, use of honey bee in improving pollination in field crops, maintenance of honey bee colony in summer, honey bee colony capture, honey bee colony multiplication.

Photo 11-12. *Farmer with his Honeybee hives and extracting honey*

2.7 High Density Planting: An emerging approach to enhance fruit production

Several KVKs have promoted high density planting (HDP) in fruit crops like mango, banana, papaya and have helped farmers to increase production and profits. KVK, Tuticorin, Tamil Nadu is promoting high density planting in Guava. The components of HDP in guava practiced by farmers are;

Season: June-July/Sep-Oct to avoid mortality of young plants due to heat and moisture stress

Varieties: Lalit (Red flesh), Shweta (white flesh) and Lucknow-49

Spacing: Three different densities being practiced by farmers are (i) 3m (row to row) × 1.5m (plant to plant) accommodating 2222 plants/ ha, (ii) 3m × 3m accommodating 1111plants/ha ad (iii) 6m (row to row) × 3m (plant to plant) accommodating 555plants /ha.

Planting: Farmers apply 10-15kg of organic manure with handful (roughly 400-500gm) of superphosphate mix with top soil and plant the seedlings in the center of the pit and press the soil firmly. Staking the plants avoid air pockets near the root zone and reduce mortality.

Canopy Management: Pruning begins at an early stage of plant growth to develop single trunk. Trees are topped to a uniform height of 60 - 70 cm from the ground level. Three to four equally spaced shoots are retained around the stem, allowed to grow for 4-5months until they attain a length of about 40 - 50 cm. The selected shoots are further pruned to 50 *per cent* of their length for inducing multiple shoots from the buds below the cut end. Newly emerged shoots are allowed to grow up to 40-50 cm and are pruned again for the emergence of new shoots. This is done until the desired shape is achieved. Initial canopy management and pruning operations continue during the second year after planting. After two years, the short branches within the tree canopy produce a compact and strong structure. All the plants are confined to a hedge shape of two m inter row width and 2.5m height for which pruning is performed in January-February and May-June every year. In Tuticorin district, 12 farmers are having HDP Guava and three of them are getting fruits regularly (more than 3-4 years old). Advisory services are being provided to all the farmers on regular basis.

3. CONCLUSION

Improved agricultural practices are skill intensive and imparting new skills to practicing farmers and farm women is an integral part of the capacity building initiatives of KVKs. The sill intensive agricultural technology interventions described here are only a few examples of rural entrepreneurship promoted through KVKs. Imparting specific and relevant skills on technological interventions in agriculture and allied sectors by KVKs has helped farmers, farm women, rural youth and other stakeholders through creation of self-employment as well as generating additional income to farmers and farming community. Considering the importance of skills for both self-employment and getting employed elsewhere, the national skill development mission has initiated a massive exercise to skill rural youth with the help of Krishi Vigyan Kendras. The rural India will surely witness a revamped scenario with the help of skilled and smart farmers and farmers' organizations.

REFERENCES

Anon. (2013). Future of Farming Review Report. www.gov.uk/defra.

Dev, Mahendra S. (2008). Challenges for Revival of Indian Agriculture, National Centre for Agricultural Economics and Policy Research, New Delhi http://krishi.icar.gov.in:8080/jspui/bitstream/123456789/804/1/jha%20memorial%20lec_1.pdf.

Elias Fereres, David, A. Goldhamer and Larry, R. Parsons. (2003). Irrigation water management of horticulture crops. HortScience 38(50):1036-1042.http://www.crec.ifas.ufl.edu/academics/faculty/parsons/PDF/280503hs.pdf.

Mudlapur, S.K., Adapur S.H. and Hiregoudar, L.G. (2016). KVK interventions in spread of vermicompost technology: An impact study. Impact study series No.11., KVK Gadag, Karnataka.

GOI. (2013). Policy & Process Guidelines for Farmer Producer Organisations, Government of India,file:///C:/User)/admin/Desktop/Final%20Edited%2007.07.2016/Reviews/FPO-Policy&Process-GuidelinesDAC2013.pdf.

GOI. (2015). Agricultural Statistics at a Glance 2014.Government of India, Ministry of Agriculture Department of Agriculture & Cooperation, Directorate of Economics & Statistics, New Delhi.

GOI. (2017). http://www.pmindia.gov.in/en/news_updates/national-skill-development-mission/.

Rajesh Agarwal. (2016). Future of agriculture in India: Is the nation ready? Business standard,http://www.business-standard.com/content/b2b-chemicals/future-of-agriculture-in-india-is-the-nation-ready-116053000558_1.html.

Acknowledgement

The author thanks KVK, Gadag, Tumakuru, Koppal and Chamarajanagara of Karnataka, KVK, Thrissur and Wynad of Kerala and KVK, Sivagangai and Tuticorin of Tamil Nadu for sharing and providing the details that formed the basis for this chapter. The support provided by Director, ICAR Agricultural Technology Application Research Institute, Bengaluru is also gratefully acknowledged.

VII. Horticulture in Securing Food and Nutrition Security

21. Vegetables: Food and Nutrition Security
22. Nutrition Security of a Girl Child is the Foundation to Ensure Healthy Generations
23. Plantation Crops – Enhancing Production and its Strategic Role in Promoting Nutritional Security
24. Edible Mushrooms towards Achieving Nutritional Security of Small and Marginal Farm Families
25. Nutrition Awareness Through School Children: Experiences and Lessons Learnt
26. School Garden- The Harbinger of Nutrition Education

21

Vegetables: Food and Nutrition Security

Prem Nath

PREAMBLE

- Today the world produces adequate food for everyone but the unequal distribution has created a gap between the countries who produce food more than they consume and those countries with deficit production.
- At present some 805 million people in the world do not have enough food to lead healthy active life. That's about one in nine people on earth.
- "Give me neither poverty nor riches but give me only my daily bread", the hungry said.
- The vast majority of the world's hungry people live in developing countries, where 13.5 *percent* of the population is undernourished.
- Asia is the continent with the most hungry people - two thirds of the total.
- With the above scenario before us it would be appropriate to revisit ourselves and dig into the root causes and adopt measures to mitigate food and nutrition insecurity. Vegetable is considered as one essential among all food commodities providing nutrients.

1. BASICS OF HUMAN CIVILIZATION

Food is basic to human being and agriculture is the catalyst between the two which is basically supported by natural resources. The human being is entirely dependent on food and food is fully dependent on agriculture and agriculture is dependent on human resources with humanity. Even though food, agriculture and humanity have their own entity; they are inseparable for the survival and existence of each other. Hence, the three pillars, namely, food, agriculture and humanity forming a triangle symbolize the roots of human civilization, which needs to be revisited in the present context as illustrated by Nath (2016) in the Figure 1.

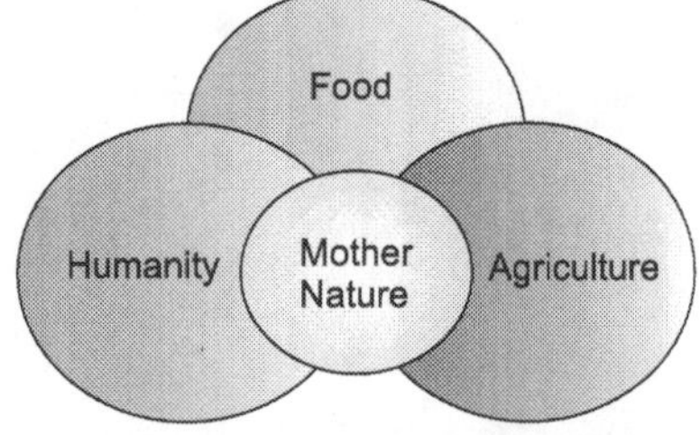

Fig. 1. *Basics of Human Civilization-Food, Agriculture and Humanity (Nath, 2016)*

There is a lot to learn about food itself which is not easy. The *first* thing to learn is, whether the nation has enough food for all - locally produced or imported. *Secondly*, whether we have all kinds of food to make healthy diet *e.g.* carbohydrates, proteins, fats, fibres, vitamins, minerals *etc*? We have varieties of food available, but we forget that we have to eat healthy food for maintenance and growth of the body. We must know what to eat, how much to eat or when to eat towards better human health? *Thirdly*, do we have the knowledge on the sources of food? *Fourthly*, what essential elements they contain to consider them as healthy food. And *Fifthly*, we need to learn the composition which will make healthy diet.

2. HUMAN FOOD, NUTRIENTS AND SOURCES

As a matter of fact, the visible food is a capsule of nutrients which essentially influences our body metabolism and health and consequently its growth and development.

The seven major classes of food/nutrients required by human beings are carbohydrates, proteins, fats, fiber, minerals, vitamins and water.

The origin, production and consumption of food are illustrated in Figure 2. The food or nutrition cycle is a continuous process which repeats itself for the growth, development and maintenance of human body and health.

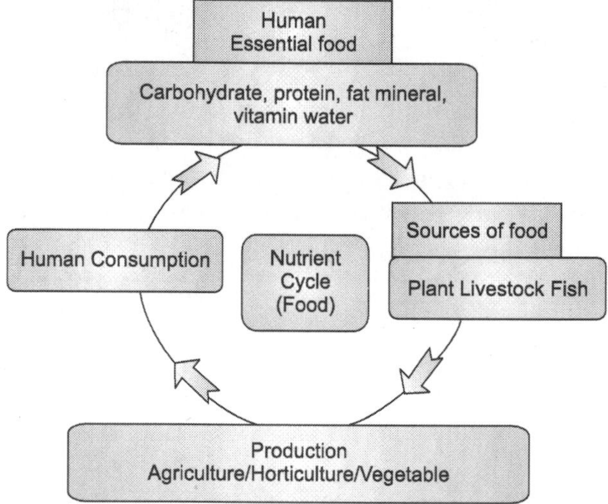

Fig. 2. *Food and Nutrition Cycle*

Definition: Nourishing is the process by which an organism assimilates food and uses it for its growth and maintenance. In another words, food is something which provides nutrients

Properties: Nutrients are substances that provide:

(a) energy for activity, growth and all functions of the body, such as, breathing, digesting food and keeping warm.
(b) materials for the growth and repair of the body and for keeping the immune system healthy.

2.1 Nutrients

- Protein, fat and carbohydrates also referred as proximate principles are oxidized in the body to yield *energy.*
- All fresh vegetables, fruits, whole grains, legumes and beans, nuts, seeds and herbs have high nutrients.

Vitamins and minerals:

- Do not supply energy but regulate *metabolic activity* and help in utilization of the proximate principles; utilized in formation of body structure and skeleton.
- Human body cannot synthesize vitamins and minerals (few exceptions), must be supplied by food.

Facts of Minerals

- Minerals play a critical role in maintenance of good health.
- Minerals cannot be created by living things, such as, plants or animals.
- Plants, animals and humans need minerals for maintaining their health.
- Minerals are elements that originate in the soil and plants absorb minerals from the soil.
- Animals meet their mineral requirements from the plants or other animals they eat.
- Most of the minerals in human diet come from plants one eats; some minerals from edible animals.

2.2 Nutrients, Properties and Vegetable Sources

Since the present discussion is limited to vegetable only, the Table 1 below lists the nutrients and its properties and vegetable sources.

Table 1. *Nutrients, properties and vegetable sources*

Nutrients	Properties	Sources
Carbohydrates/Starch	Energy giving	Roots and Tubers
Fats	Source of energy containing essential fatty acids and have vitamin (A, D, E, K), like functions.	Vegetable Oils
Phytochemicals	Ailment healing properties	Garlic, onion, turmeric, tea, soybean, chilli, *etc.*
Antioxidants (phytochemicals in human diet)	Prevents cancer, cardio-vascular diseases and degenerative diseases with oxidative stress	Abundant in fruits and vegetables
Vitamin C/ Ascorbic acid / Anti-oxidant	Superstant anti-oxidant	Cantaloupe, cabbage, broccoli, spinach, baked potato, sweet potato, tomato and every green vegetable.

Vitamin E / alpha-tocopherol / Anti-oxidant	Low level in blood increases risk of cancers in gastro-intestinal tract, breast, prostrate, lung and colon	Soybean oils, broccoli.
Vitamin A / Retinol	Deficiency include poor appetite, lack of vigor, frequent colds, eye infection, night blinders, skin disorders and poor susceptibility to infections	Beta-carotene- precursor of vitamin A in carrot, sweet potato leafy vegetables, gourds and pumpkins, *etc.*
Vitamin C / Ascorbic Acid	Essential for normal growth, maintenance of body issues; protection against infections; promotes healing.	Cantaloupe, brussels sprouts, broccoli, cabbage, baked potatoes, tomatoes, sweet potato, spinach and cauliflower
Vitamin K / Chlorophyll	Prevents bleeding, promotes clotting in injury, makes bone dense.	Leafy green vegetables
Biotin / B-Complex	Required for carbohydrate and protein metabolism and fat synthesis.	Plentiful in green vegetables
Nutrient/ Mineral	**Properties**	**Sources**
Boron	Regulates use of calcium, phosphorus and magnesium by the body: deficiency may cause abnormal growth cystises or tumors.	Vegetables
Calcium	Essential for formation and maintenance of strong bones and teeth; involved in regulation of heart beat muscle action, transmission of nerve impulses; deficiency causes cramps in muscles.	Dark green leafy vegetables, squash, artichoke, broccoli, cabbage, spinach, green beans, watermelon *etc.*
Iron	Essential for production of hemoglobin; increases resistance to stress, diseases and prevents fatigue; deficiency causes anemic.	Green leafy vegetables watermelon, legumes
Zinc	Essential for normal immune function; activates enzymes needed to breakdown proteins and transports CO_2 by red blood cells; low level may impair senses of smell and taste	Legumes
Phosphorus	Essential for calcification of teeth and bones in combination with calcium; deficiency causes bone diseases, nervous disorders, regarded growth, weakness; part of DNA and RNA	Legumes
Magnesium	Essential for cellular respiration-proper nerve and muscle functions; in metabolism of fat and protein; deficiency includes heart failure, convulsions, kidney damage, depression *etc.*	Beans, dark green vegetable, watermelon

Potassium	Essential for proper functioning of muscles; involved in release of energy and production of glycogen and protein; deficiency causes irregular heartbeat, excessive perspiration of hands and fat, weakness, tiredness.	Parsley, carrot, cabbage, broccoli, brussels sprouts, corn, spinach, radish, tomato, squash and pumpkin
Manganese	Coordinates with enzymes to help nerve, brain and muscles for cell processes; helpful in metabolism of carbohydrates, fats and protein; deficiency causes impaired fertility, growth retardation and birth defects	Spinach, tea leaves
Sulphur	Provides immunity against infections; aids digestion of fats and controls metabolism; deficiency seldom noticed.	Leafy vegetables, black gram

(**Source:** *Ahuja, Nath and Swamy, 2010; USDA, 2016; Nandi & Bhattacharjee, 2004*)

Considering the extent of quality and quantity of nutrients in different vegetables, it is recognized that vegetables are essential sources necessarily required by human beings.

3. EXPRESSION OF INAPPROPRIATE NUTRITION

3.1 Definition (Tontisirin, 2000)

- **Malnutrition** is a broad term for a range of conditions that hinder good health, caused by eating too little, too much or not the right food.
- **Under nutrition** results from chronic food insecurity, in which people do not have enough to eat satisfy their basic energy requirements.
- **Protein-energy malnutrition** occurs when the body does not receive enough protein and calories. The effects of protein-energy malnutrition include **underweight** (low weight for age), **stunting** (low height for age) **and** wasting (low weight for height.)
- In **micronutrient deficiency,** certain essential vitamins and minerals are lacking in the diet, leading to impaired growth and poor health.
- Eating too much, **over-nutrition,** an ever- growing threat as economies expand and lifestyles become sedentary, leads to health problems ranging from diabetes to heart disease.
- **Health improvement:** The contributions of fruits and vegetables to prevention of deficiencies are well-understood by scientists, the obstacles to resolving these problems are usually related to the availability, affordability and consumer awareness. The role of these foods in preventing several diet-related chronic diseases, such as, cardiovascular diseases, type 2 diabetes and cancers is very significant.
- **Aging populations:** Chronic diseases, such as, cardiovascular diseases, type 2. diabetes and cancer generally affect the middle aged and elderly segments of the population. Globally, the number of persons aged 60 years or over is expected nearly to triple, increasing from 673 million in 2005 to two billion by 2050 (UN, 2007).

3.2 Non- Communicable Diseases and Micronutrient Deficiencies

- **Cardiovascular diseases:** Although cardiovascular diseases (CVD) usually affect middle age and elderly adults, the diets consumed early in life can influence the risk of stroke and coronary heart disease. Children's consumption of a heart healthy diet can reduce the prevalence of obesity, high cholesterol and high blood pressure in adults. When fruits and vegetables replace foods that are high in animal fat, salt and sugar this can lower health risks.
- In adolescents and adults, high consumption of fruits and vegetables has been associated with lower risk of cardiovascular diseases.
- **Type 2 diabetes**
  - ♦ It is estimated that there are 180 million cases of diabetes mellitus (DM) worldwide (over 90% are type 2 DM); without strong preventive actions, this number may double by 2030 (Barnard *et al.,* 2009 and Wylie-Rosett and Vinicor, 2007). Observational and clinical studies have shown that diets that are comprised of an appropriate diversity of plant foods (whole grains, legumes, vegetables and fruits) can improve the management of type 2 diabetes. (Barnard *et al.,* 2009).
  - ♦ **Cancers:** The International Agency for Research on Cancer, the World Cancer Research Fund and American Institute for Cancer Research believe that it is probable that high fruit and vegetable intakes may reduce a number of cancers (namely, oesophageal, colo-rectal oral, pharynx, stomach, larynx, lung, ovary and kidney cancers). More research is needed to identify the specific types of fruits and vegetables and their particular components, that have an impact on disease.
  - ♦ Based on current knowledge, it appears that dark green leafy vegetables, yellow and red vegetables, cruciferous vegetables, as well as red and purple fruits and vegetables may be protective against cancer. It is plausible that the vitamins C, E and folate have an effect on cancer. The phytochemicals, such as, carotenoids, phenolics, isoflavonoids, isothiocyannates (ITC) and indoles also have anticarcinogenic effects (Kim and Park, 2009).
  - ♦ Cruciferous vegetables, such as, broccoli, cauliflower, cabbage, kale and mustard are common worldwide and they are rich in carotenoids, vitamin C, folate and soluble fibre. There is evidence from animal studies that these foods contain phytochemicals that have anticarcinogenic effects (Kim and Park, 2009). The latest research suggests that these vegetables, especially Brassicas, may lower the risk of gastric and lung cancers.
- **Chronic energy deficiency:** The Asia and Pacific region experienced continued improvement in alleviating chronic energy deficiency during the last 20 years, with the proportion of the population affected falling from 41 to 16 per cent in East and South East Asia and from 33 to 22 per cent in South Asia. Total numbers declined in East and South East Asia from 289 million to 258 million people, but South Asia experienced an increase from 237 million to 254 million. To support an active health life, dietary energy must come from diverse food sources (FAO, 1996).

3.3 Compounding Factors

The severity of the problems of micronutrient deficiencies is a compelling reason to call for greater production and consumption of vegetable and fruits. It is well- documented that these foods can contribute to the prevention of deficiencies which can be debilitating and can lead to premature deaths.

- **Population rise:** Current trends showed that another 1.5 billion people will be added to the population of developing Asian countries by 2025. More than half of the global increase towards 2025 was expected to take place in South Asia (FAO, 1996).
- **Nutrition issues**
  - **Protein energy malnutrition (PEM):** Currently over three fourths (79%) of the world's malnourished children live in Asia (FAO, 1996).
  - **Stunting and wasting** are both highly prevalent in Asia and this is reflected in the number of children who are underweight. According to ACC/SCN (2000), over a third of the world's stunted children reside in the Asian region as a whole.
  - **Low birth weight:** Regional data show that the incidence of intra- uterine growth retardation (IUGR) and LBW is highest in Asia (12.3%). Almost a third of all babies in South Asia are born with low birth weight which results from intrauterine malnutrition and from the cumulative effects of malnutrition among mothers during their infancy, childhood, adolescence and pregnancy.
  - **Major micronutrient deficiencies:** Micronutrient deficiencies of nutritional concern in South East Asia are iron deficiency anaemia (IDA), iodine deficiency disorders (IDD) and vitamin A deficiency (VAD). Other micronutrient deficiencies, such as, rickets reportedly are a problem in certain parts of the Near East and Southern and East Asia. Zinc and selenium deficiencies are reported in China and some parts of Central Asia (Nandi and Bhattacharjee, 2004).
  - **Iron deficiency anaemia (IDA):** South East Asian countries have the highest prevalence of anaemia, with almost 80% of pregnant women in the region being anaemic, of which countries of South Asia, contribute to the bulk of anaemia (Criteria for determining the presence of anaemia in pregnant women and preschool children below Five years of age is Hb < 11g/dl (WHO, 1998).
  - **Iodine deficiency disorders (IDD):** The extent of goiter in the world is estimated to be as more than 200 million people, added to which should be six million overt cretinism. Around half of these are reportedly in South East Asia, including India. Iodine is absent from the soil in many Himalayan and alluvial regions and high land pressure has eroded top soil and with it iodine, that would be consumed with foods grown there.

3.4 Other Factors Influencing Nutrition

- **Challenges of urbanization:** Urbanization also profoundly affects dietary and food demand patterns. It gives rise to growing concern about the quality and safety of food, particularly foods processed and purchased outside the home and to modification in food preferences caused by changing life styles (Gopalan, 1994; Seidler, 2000). New approaches are needed to promote locally produced foods, including research into the aspects of supply, processing and convenience which make food commodities preferable in the urban setting.

- **Low consumption of fruits and vegetables:** The low consumption of fruits and vegetables in the vast Asian regions is a persistent phenomenon, confirmed by the findings of food consumption surveys. Nationally representative surveys in India (India Nutrition Profile, 1998) indicate a steady level of consumption of only 120-140 g/d, plus about other 100 grams of roots and tubers and some 40 grams of pulses. In China, on the other hand, a country that is undergoing a strikingly fast transition, the amount of fruit and vegetables has gone up from 344 g/d in 1982 to 369 g/d in 1992 (WHO & FAO, 2003), although it is still below the recommended goal.

4. MAJOR RISING CHALLENGES

There are number of persisting and emerging challenges aggravating food and nutrition insecurity. Some of the major challenges are discussed below.

- **Migration of people:** The migration of people from rural to urban area is one the emerging challenges where urban population is getting higher than the rural population and will demand more food and produce nil or much less food. United Nations Population Division (UNPD) reported the projection of population as follows: 2030: Urban 60%+Rural 40%; 2050: Urban 70%+Rural 30%;
- **Hunger and malnutrition:** The greatest challenge of today is the number of hungry and malnourished people to be fed. During 2009, 1020 million hungry and malnourished people were reported (FAO, 2009), whereas, in 2015, 805 million people were hungry and malnourished with Asia-having two-third of total. The reduction in the total is reported to be influenced by the implementation of the Millennium Development Goals.
- **Price rise:** The emerging challenge of price rise of food commodities in developing countries has caused crisis among the common people and political circles in developing countries. Vegetable price rise of 30–80% and inputs price rise of 20-40% have become important cause of food insecurity.
- **Job opportunities:** This is one of the major challenges of the developing countries where food and nutrition security is directly related to job security. According to FAO report, 770 million were unemployed or working poor below poverty lines (FAO, 2010).

5. FOOD AND NUTRITION SECURITY

Food and Nutrition Security is attained when all people, at all times, have the physical and economic access to sufficiently safe and nutritious food to be healthy and active (FAO, 2004, 2006, 2014).

5.1 Essential Elements of Food and Nutrition Security

As indicated by FAO, the essential elements of food and nutrition security are identified as follows;

- **Availability:** The availability of food is dependent on its production which is governed by agriculture which is supported by research and development of efficient and cost-effective technology.

- **Stability:** This specifically relates to continuous supply of food every season and every year.
- **Accessibility:** The availability of food could be met by efficient production system, but it may not be accessible to all as it is happening today where 800 million people are hungry and malnourished. The change in policy must ensure that every individual has economic capacity to buy or access food.
- **Quality and safety:** It is important and essential that the food available and consumed are nutritious and safe. Food safety laws need to be developed and enforced for the benefit of all consumers.

5.2 Operational Mechanism

Keeping in mind the above essentials of FNS, the operational mechanism needs to be developed to ensure achieving the targets on specific areas. It is of serious concern to the operation management to employ these essential elements in developing operation strategy to achieve the aims and objectives of the FNS. The Figure 3 illustrates clearly the essential elements of food and nutrition security and lists ingredients or factors involved in fulfilling the objectives.

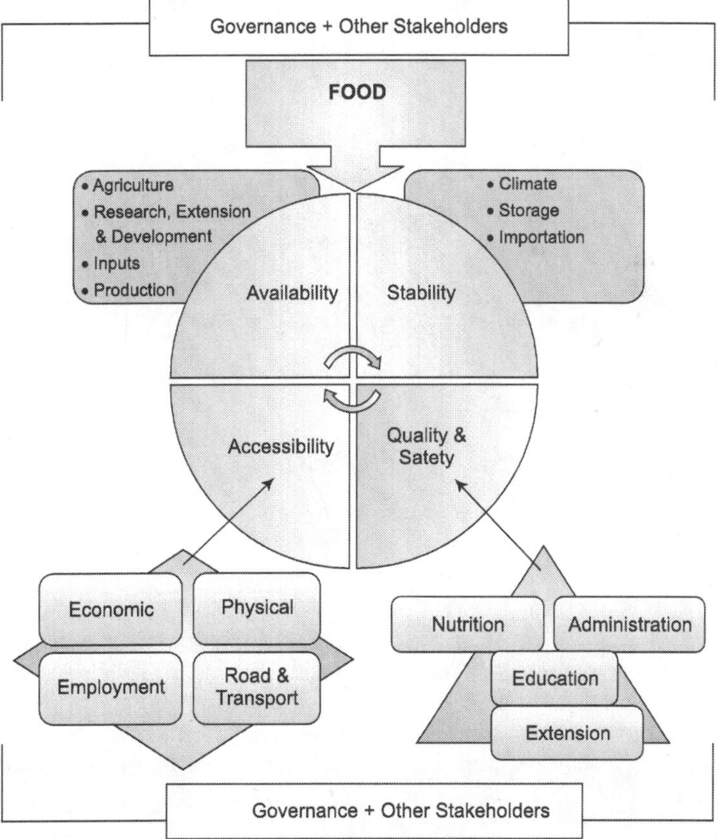

Fig. 3. *Essential Elements of Food and Nutrition Security. (Nath, 2016)*

The Figure. 4 is self-explanatory to five areas of field operation and the related activities to meet the targets of food and nutrition security.

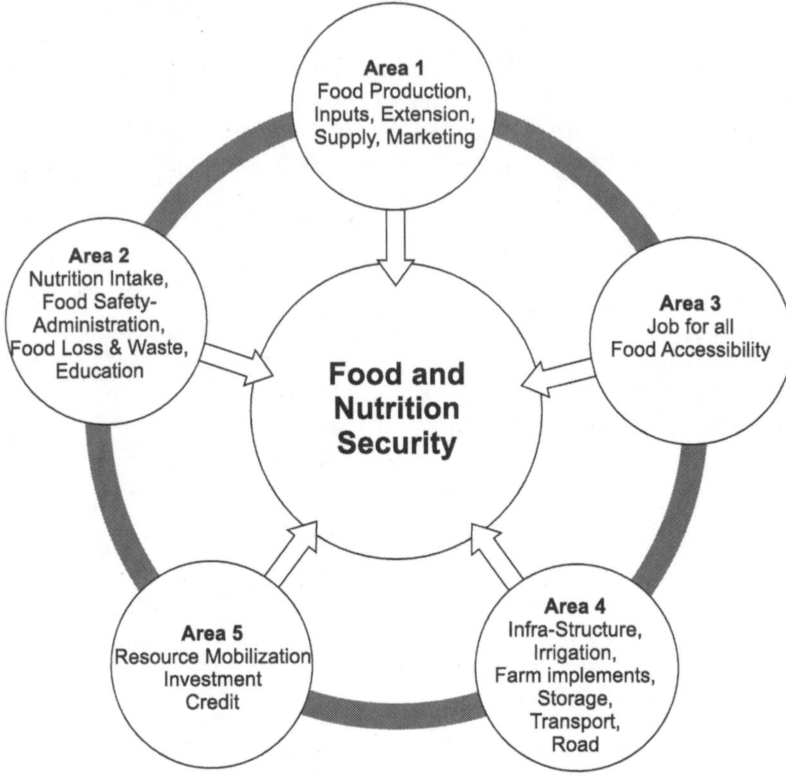

Fig. 4. *The Areas of Operational Focus. (Nath, 2017)*

The Figure 3 and 4, signify and identify the mode of field operation to ensure food and nutrition security objectives and targets.

6. FOOD UTILIZATION AND MINIMIZING WASTE

Apart from making the food available, it is equally important how the food is being used by the consumers. The target should be to maximize its use in food consumption and minimizing its waste.

Proper nutrition is essential to start people on the right path to a long, healthy life. All people should have sufficient quantities of nutrient-rich, safe foods in order to meet their energy and nutrient requirements. Without adequate nutrition, children start life at risk for a range of health problems; adults are unable to function fully in society. *But nutrition is not merely an input that determines human well-being, it is also a reflection, quality of life improves and human potential is maximized, leading to increased productivity and economic strength.*

- **Preventive action is needed:** Public health agencies have stressed that at least 400 grams of fruits and vegetables should be eaten per day (WHO, 2003). Some recommend that the amount should be raised to 600 grams per day (World Cancer Research Fund /AICR 2009).
- **Research on fruits and vegetables:** Fruits and vegetables are very complex and recent research focuses on a few components. The particular components are beta carotene (precursor for vitamin A), lycopene, alpha- tocopherol (Vitamin E), foliate and flavonoids. Current research suggests that each of these compounds may have great public health significance in terms of preventing disease, disability and premature mortality.
- *One-third of food produced globally every year* does not reach human mouths it is either lost in transit or wasted by the consumers themselves. This amounts to a staggering *1.3 billion tonnes every year*. These chilling figures are part of a report — Avoiding Future Famines — released by the UNEP at the Rio+20 summit on sustainable development.
- Consumers *throw away nearly 222 million tonnes of food in edible condition every year* in the rich countries of North America and Europe. This is very close to the total food production of Sub-Saharan Africa which is 230 million tonnes per year.
- In low-income countries, food is mainly lost during the early and middle stages of the food supply chain; much less food is wasted at the consumer level. In Europe and North America, 280-300kg of food is wasted per head every year. In Sub-Saharan Africa and south and Southeast Asia, *per capita* food wastage is 120-170kg per year.
- According to a United Nations report, *globally 33% of food produced goes waste.* And each person wastes around 70kg of food every year.
- Approximately 11 kg of food per person per year is wasted in developing Asian countries (FAO, 2010).
- Approximately 80 kg of food per person per year is wasted in developed Asian countries, such as, China, Japan and Republic of Korea (FAO, 2010).

6.1 Unused Food

The philosophy and practical approach are to utilize as much edible food material as possible after it is produced, whether plant, animal or fish. As for example, mostly the root of radish or turnip is consumed but some people utilize also the tender leaves which are nutritious equally. In case of pumpkin, mostly fruit is utilized but some people consume flowers, tender leaves and seeds which are nutritious. Again, commonly drumstick fruit is consumed, but flowers and tender leaves are also nutritious and edible. In India in *Coccinia* fruits only are consumed whereas, in Thailand tender leaves are commonly used. The essence is that some edible parts are not being consumed or wasted in absence of knowledge or because of food habit.

6.2 Underutilized Food

There are several plants and animals which are being utilized well as food in one country and region but not in others. It is probably because of the fact that these fruits and vegetables were not produced in the region, Hence, not known to consumers. As for example, spinach beet *(Beta vulgaris var. bengalensis)* is common nutritious spinach commonly grown and consumed in northern India is now getting popular in south India too, morning glory *(Ipomoea aquatica)* is so common in Southeast Asia is not commonly known in India; roundmelon (*Praecitrullus vulgaris*), So common in northern India, can be grown in other Asian countries; Mangalore cucumber (*Cucumis sativus sp.),* so popular in south India, can be made popular in other parts of India.

7. INTERNATIONAL INTERVENTIONS

Within three decades, there has been a number of international interventious on food and nutrition security which may be listed below.

7.1 Right to Food

- The right to food was formally recognized in the very first international human rights document, the Universal Declaration of Human Rights, adopted by the United Nations in 1948.
- The Preamble to the FAO Constitution sets "ensuring humanity's freedom from hunger" as one of its basic purposes (Diouf, 1998).

7.2 International Conference on Nutrition-1992 (FAO & WHO, 1992)

During International Conference on Nutrition-1992 (ICN-1992), the Global Plan of Action was designed to provide guidelines for governments, acting in partnership with non-governmental organizations, the private sector, local communities, families and households and the international community, including international organizations, multilateral financing institutions and bilateral agencies, to achieve the objectives of the World Declaration on Nutrition adopted by the International Conference on Nutrition.

7.3 Special Programme for Food Security (SPFS), FAO, Rome (1994)

The first ever field oriented global programme on food security was launched for low-income- food deficit-countries by FAO during 2004. The project was implemented and succeeded in more than 100 countries over decades and included importantly vegetables among other food commodities.

7.4 World Food Summit 1996, Rome (FAO & WHO, 1996)

During World Food Summit 1996, the Rome Declaration on World Food Security and the World Food Summit Plan of Action lays the foundations for diverse paths to a common objective - food security, at the individual, household, national, regional and global levels. Food security exists when all people, at all times, have physical and economic access to sufficient, safe and nutritious food to meet their dietary needs and food preferences for an active and healthy life. In this regard, concerted action at all levels is required.

7.5 The International Conference on Vegetables (ICV-2002), November 11-14, 2002, Bengaluru, India

The International Conference on Vegetables with the theme "Vegetables for Sustainable Food and Nutrition Security in the New Millennium" held during November 11-14, 2002 in Bengaluru, India marked the first milestone in the history of vegetable science and development by recognizing the role of vegetables as major supplement to food and nutrition security (Nath, Gaddagimath and Dutta, 2004).

7.6 Fruit and Vegetables for Health; Joint FAO/WHO Workshop, 1-3 September 2004, Kobe, Japan

According to The World Health Report 2002, low fruit and vegetable intake is estimated to cause about 31% of is chaemic heart disease and 11% of stroke worldwide. *Overall it is estimated that up to 2.7 million lives could potentially be saved each year if fruit and vegetable consumption was sufficiently increased.*

The Joint FAO/WHO Expert Consultation (2004) on diet, nutrition and the prevention of chronic diseases, recommended the intake of a minimum of 400g of fruit and vegetables per day (excluding potatoes and other starchy tubers) for the prevention of chronic diseases, such as, heart disease, cancer, diabetes and obesity, as well as for the prevention and alleviation of several micronutrient deficiencies, especially in less developed countries.

7.7 International Workshop on Urban and Peri-urban Agriculture for Asian countries, June 29 to July 02, 2005, Bengaluru, India (FAO and PNASF, 2005)

While emphasizing the role of Urban and Peri-urban Agriculture, the FAO/PNASF workshop participated by Asian countries highlighting on policy and strategy issues, recognized the contribution of different food commodities in augmenting food and nutrition security. Fruit and vegetables played an important role in providing nutrition security.

7.8 Regional Workshop on Strengthening Urban and Peri-urban Agriculture toward Resilient Food Systems in Asia (UPAFSA-2013), FAO and PNASF, January 28-30, 2013 Bangkok, Thailand

It was recommended that in urban planning, the greener cities approach should be on the agenda and UPA should remain the driving force for the same. The fruit and vegetables figured as important commodities to supplement food and nutrition security.

7.9 International Conference on Nutrition-2014

During ICN-2, the Framework for Action was to guide the implementation of the commitments of the Rome Declaration on Nutrition adopted by the Second International Conference on Nutrition held in Rome, Italy, on 19-21 November 2014. Building on existing commitments, goals and targets, this Framework for Action provided a set of policy options and strategies for governments to undertake.

7.10 The Millennium Development Goals (2000-2015) (UN, 2000)

Goal 1. Eradicate extreme poverty and hunger: Target for 2015: Halve the proportion of people living on less than a dollar a day and those who suffer from hunger.

7.11 Sustainable Development Goals (SDG)(2016-2030)

Sustainable development goal and targets (UN, 2015) relating food and nutrition security are mentions below:

Goal 2. End hunger, achieve food security and improved nutrition and promote sustainable agriculture

- By 2030, end hunger and ensure access to all people, in particular, the poor and people in vulnerable situations, including infants, to safe, nutritious and sufficient food all year round.
- By 2030, end all forms of malnutrition, including achieving, by 2025, the internationally agreed targets on stunting and wasting in children under 5 years of age and address the nutritional needs of adolescent girls, pregnant and lactating women and older persons.
- By 2030, double the agricultural productivity and incomes of small-scale food producers, in particular women, indigenous peoples, family farmers, pastoralists and fishers, including through secure and equal access to land, other productive resources and inputs, knowledge, financial services, markets and opportunities for value addition and non-farm employment.
- By 2030, ensure sustainable food production systems and implement resilient agricultural practices that increase productivity and production, that help maintain ecosystems, that strengthen capacity for adaptation to climate change, extreme weather, drought, flooding and other disasters and that progressively improve land and soil quality.
- By 2020, maintain the genetic diversity of seeds, cultivated plants and farmed and domesticated animals and their related wild species, including through soundly managed and diversified seed and plant banks at the national, regional and international levels and promote access to and fair and equitable sharing of benefits arising from the utilization of genetic resources and associated traditional knowledge, as internationally agreed.

The international interventions have brought about significant awareness, education, guidelines and targets on food and nutrition activities for the benefit of the nations worldwide.

8. IN PROSPECTIVE

National Planning and Budget: The grain-based planning, budgeting and production and consumption data need to be replaced. We cannot continue giving step-motherly treatment to nutritious crops like vegetables as against grain crops. We need to have detailed data on vegetable production and consumption to develop nutrition-based food requirements.

Breeding for nutritive values: While breeding, enhancing nutritive values should get priority besides high yield. Along with nutritive value, the nutraceutical value where applicable should be considered.

Bridging yield gap: There is wide range of variation in yield per unit area in each vegetable crop between and among the countries of the Asian region which needs to be bridged.

More Nutrient Harvest: Along with the economic advantage, the farmers need to be educated to include nutritious crops like vegetables to obtain more nutrient harvest.

Bio-diversification of crops: In the cropping pattern the vegetables should find place not only as cash crop like cereals but also as important supplement to nutritious crops.

Awareness and Education on Nutrition: Traditionally extension workers in rural areas are engaged in extending technology and inputs but they never train rural people on consumption/nutrition of food. On the whole, the education on nutrition and consumption of food is almost minimal or absent surprisingly in both rural and urban areas. The national cadre of extension workers on nutrition is much needed if we need to attain food and nutrition security.

Price Rise: We are all concerned about the price rise of vegetables which is a local or national or international concern. The governance has to address this major issue. Besides, the price rise in farm inputs should also be examined and considered. The vegetable crop production needs relatively more inputs.

Save food: Among the emerging issues, the loss and wastage of edible food has gained much significance wherein vegetables particularly leafy vegetables account for loss upto 50 percent. This can be arrested and utilized.

Increase in Urban Population: As mentioned earlier, the vegetable is a major contributor to nutrition security. With growing urban population, horticulture particularly vegetables has assumed greater importance in Urban and Peri-urban agriculture.

9. MESSAGE

- Food and Nutrition Security for all cannot be attained unless everyone has food accessibility and food accessibility cannot be attained unless everyone has economic accessibility and economic accessibility can be attained only if every one has access to earning (Nath, 2015).
- It is recognized that horticulture crops including vegetables are the second line of defense with cereals and legumes as the first in the front line of food domain (Nath, 2009).
- We need to recognize that we can not ensure nutrition security followed by good health without the consumption of vegetables (Nath, 2016).
- During UN Day Celebrations 2014, the United Nations Secretary General, Mr. Ban Ki-moon, stated "In a single world of plenty, no one, not a single person should go hungry. But almost one billion people in the world do not have enough to eat. I want to see an end to hunger everywhere within my life time".

REFERENCES

ACC/SCN (2000). 4th Report on the World Nutrition Situation, Geneva, Switzerland.

Ahuja, K.G. Nath, Prem and Swamy, K.R.M. (2010). Foods and Nutrition, P. N. Agricultural Science Foundation (PNASF), Bengaluru and Studium Press, New Delhi; p. 485.

Barnard, N.D., Katcher, H.I., Jenkins, D. JA., Cohen, J. and Turner-McGrievy, G. (2009). "Vegetarian and vegan diets in type 2 diabetes management" Nutrition Reviews, 67 (5): 255-263.

Diouf, Jacques. (1998). News and Highlights-Fighting for the right to food, http://www.fao.org/News/1998/981205-e.htm, 10 December 1998, FAO, Rome.

FAO. (1996). The Sixth World Food Survey, FAO, Rome.

FAO. (1997). Special Programme for Food Security- Technical Fact Sheets, FAO Activities and Prospects, FAO, Rome.

FAO. (2004). The State of Food Insecurity in the World, FAO, Rome, Italy.

FAO. (2006). Policy Brief -Food Security, June 2006, Issue 2. Published by FAO's Agriculture and Development Economics Division (ESA) with support from the FAO Netherlands Partnership Programme (FNPP) and the EC-FAO Food Security Programme FAO Agricultural and Development Economics Division: http://www.fao.org/es/esa/.

FAO. (2009). More People than Ever are Victims of Hunger, Background Note for Press Conference 19th June, 2009, FAO, Rome.

FAO. (2010). United Against Hunger-World Food Day 2010, FAO/RAP Publ. 2010/23 Bangkok, Thailand.

FAO. (2014). Food is Life-Save Food-Leaflet, FAO/RAP, Bangkok, Thailand, E-mail: Save-Food-AP@fao.org.

FAO, IFAD and WFP. (2013). The State of Food Insecurity in the World 2013. The multiple dimensions of food security. Rome, FAO. p. 13.

FAO and PNASF. (2005). Proceedings of the International Workshop on Urban and Peri-urban Agriculture for Asian Countries, June 29 to July 02, 2015, Food and Agriculture Organization the United Nations (FAO), Rome and P. N. Agricultural Science Foundation (PNASF), Bengaluru, p. 257.

FAO and PNASF. (2013). Proceedings of Regional Workshop on Strengthening Urban and Peri-urban Agriculture towards Resilient Food Systems in Asia (UPAFSA-2013), FAO/RAP, Bangkok and PNASF, Bengaluru, January 28-30, 2013, Bangkok, Thailand; p. Vol. I 100- Vol. II 100.

FAO & WHO. (1992). Action on Nutrition, International Conference on Nutrition, World Declaration and Plan of Action for Nutrition. Rome, December, 1992. Food and Agriculture Organization of the United Nations and World Health Organization. p. 53.

FAO & WHO. (2004). Fruit and Vegetables for Health, 1-3 September 2004: Kobe, Japan, Report of a joint FAO/WHO workshop, World Health Organization. p. 46.

FAO &WHO. (2014). Framework for Action, Second International Conference on Nutrition, 19-21 November 2014 FAO and WHO; p. 8.

Gopalan, C. (1994). Trends in food consumption patterns : Impact of developmental transition: Nutrition in the nineties – Policy India Nutrition Profile. (1998). National Family Health Survey-2 (NFHS-2), Govt. of India, New Delhi.

Kim, M.K. and Yoon Park, J. H. (2009). Cruciferous vegetable intake and the risk of human cancer: epidemiological evidence, Conference on Multidisciplinary approaches to nutritional problems Symposium on 'Nutrition and Health, Proceedings of the Nutrition Society. 68: 103-110.

Kobakiwal, Abdul Q. (2004). Special Programme for Food Security (SPFS), Food Security and Vegetables-A Global Perspective, P. N. Agriculture Science Foundation, Bengaluru; p. 85-100.

Nandi, Biplab K. and Bhattacharjee, Lalita. (2004). Emerging issues in nutritional security, Food Security and Vegetables - A Global Perspective, P.N. Agriculture Science Foundation (PNASF), Bengaluru.

Nath, Prem. (2009). Role of Horticulture in Livelihood Security, Horticulture and Livelihood Security, Scientific Publishers (India), Jodhpur & P.N. Agricultural Science Foundation (PNASF), Bengaluru, p. 1-11.

Nath, Prem. (2013). Concept and Implementation of Food Security and its Relevance to Urban and Peri-urban Agriculture, Proc. FAO/PNASF Regional Workshop on Strengthening Urban and Peri-urban Agriculture towards resilient food systems in Asia, January 28-30, 2013 in Bangkok, Thailand, p. 20-25.

Nath, Prem. (2015). Introduction, Food Expectations of the People in the New Millennium, Dr. P.N. Agri. Sci. Foundation (PNASF), Bengaluru; Westville Publishing House, New Delhi, p. 570.

Nath, Prem. (2016a). Genesis, Messages and Considerations of the Basics of Human Civilization- Food, Agriculture and Humanity: Humanity in Humans, by Humans and for Humans in Securing Food for All, The Basics of Human Civilization – Food, Agriculture and Humanity; Vol. IV- Humanity. P.N. Agric. Sci. Foundation, Bengaluru; p. 250-258.

Nath, Prem. (2016b). Attaining Food and Nutrition Security in the Developing World, Dr. P.N. Agri. Sci. Foundation (PNASF), Bengaluru; and Westville Publishing House, New Delhi, (In press).

Nath, Prem, Gaddagimath, P. B. and Dutta, O. P. (Eds.). (2004), Food Security and Vegetables-A Global Perspective, P. N. Agricultural Science Foundation (PNASF), Bengaluru; p. 412.

Tontisirin, Kraisid. (2000). A Millennium Free from Hunger, World Food Day-2000, FAO, Rome.

UN. (2000). The Millennium Development Goals, http://www.un.org/millenniumgoals/poverty.shtml. © United Nations. 2016.

UN. (2007). World Population Prospects: The 2006 Revision, Highlights, Working Paper No. ESA/P/WP.202. United Nations Department of Economic and Social Affairs, Population Division, New York.

UN. (2012). Zero Hunger Challenge - UN Secretary-General Message, June 19, 2012.

UN. (2015a).United Nations General Assembly, A/RES/70/1 Transforming our world: the 2030 Agenda for Sustainable Development, September 25, 2015. p. 35.

UN. (2015b). Sustainable Development Goals (SDGs), http://www.un.org/sustainable development/hunger/© United Nations. 2016.

USDA. (2016). United States Department of Agriculture (USDA), Agricultural Research Service. 2011. USDA National Nutrient Database for Standard Reference, Release 24. Nutrient Data Laboratory Home Page, https://www.ars.usda.gov/northeast-area/beltsville-md/beltsville-human-nutrition-research-center/food-surveys-research-group/docs/main-service-page.

WHO. (1998). The World Health Report-Life in the 21st century: A vision for All, WHO.

WHO. (2003). WHO and FAO launched in 2003, a joint Initiative to promote fruit and vegetables for health worldwide, WHO, Geneva and FAO, Rome.

WHO & FAO. (2003). Diet, nutrition and the prevention of chronic diseases. Report of the joint WHO/FAO expert consultation, WHO Technical Report Series, No. 916 (TRS 916). Geneva, 28th January-1st February. 2002; p. 160.

❍❍❍

22

Nutrition Security of a Girl Child is the Foundation to Ensure Healthy Generations

B.S. Prabhakar, Shilpa Yatnatti and Savithri Byadigi

SYNOPSIS

A mother is born with the birth of a baby girl and her beautiful journey through infancy–childhood– adolescence– pregancy and motherhood, should be comfortable and well nourished to give birth to a healthy and robust child of next generation. Nutrition of a girl child varies with age though it is similar to a male child during infancy through childhood and pre–teen age. Nutritional requirement of adolescent girls will be specific for those planning for a child, pregnant woman and mothers feeding the child.

Nutrient needs throughout women's life cycle are well documented. Natural source of nutrients depends mainly on locally produced food crops. Socio-economic conditions play an important role in supply of adequate nutrients to the growing female child. Research efforts are on to find out affordable sources of nutrients through locally available food sources. However, specific vitamins and minerals need to be supplemented to meet the special requirement of woman of child bearing age.

It is important to promote the consumption of fruits and vegetables that takes care of a large part of requirement of vitamins and minerals apart from other nutrient elements.

Keywords: Maternal and child nutrition, infant nutrition, teenage nutrition, Nutrition needs during pregnancy.

1. INTRODUCTION

A female child is the most beautiful Divine gift. Therefore, nurture, nourish, educate and safe guard every girl child for better tomorrow. Caring for a girl child is the fundamental human responsibility, because the well being of future generation depends on her health and wellness. A mother is born with the birth of a baby girl and her beautiful journey through infancy– childhood– adolescence– pregancy and motherhood, should be comfortable and well nourished to give birth to a healthy and robust child of next generation.

On the contrary, female children are being eliminated even before birth. Deliberate ill treatment and neglect of female child is deplorable. Ideally the male and female proportion should be equal and any attempt to tilt this balance affects the society adversely.India is already experiencing the turbulance and againising trends in our society. Female child has equal right to live with dignity as that of a male child. Female infanticide, discrimination and exploitation are disturbing developments in our society today.

2. FEMALE INFANTICIDE IN INDIA

Female infanticide in India has a history spanning centuries (Visaria, 1983; Daniel Grey, 2011). Poverty, dowry system, births to unmarried women, deformed infants, lack of support services and maternal illnesses, such as, postpartum depression have been cited as reasons for female infanticide in India (Oberman, 2005; Giriraj, 2004; Chandran *et al.,* 2002). Female children in India are not only at risk at the time of birth, but are also at risk during infancy. There is a significant decrease in the sex ratio between birth and up to the age of four. Between 1978 and 1983 out of the twelve million girls born each year, only nine million will live to be fifteen (Dube *et al.,* 1999). A more recent estimate by UNICEF, for 2013, reports that under-five child mortality rate was 55 per 1,000 female births in India, The children's rights group CRY has estimated that of 12 million females born yearly in India 1 million die of various causes within their first year of life (Krishan, 2012). Infants die from natural causes, malnutrition, lack of clinical services, diseases and homicide, Extreme poverty with inability to afford raising a child is one of the reasons given for female infanticide in India (Giriraj, 2004; Chandran *et al.,* (2002) Prabha Unnithan (2013). The dowry system in India is another reason that is given for female infanticide. Other major reasons given for female infanticide include unwanted child, such as, those conceived after rape, deformed child in impoverished family and those born to unmarried mothers lacking reliable, safe and affordable birth control (Giriraj, 2004; Christine Alder and Ken polk, 2015; Hausfaster, 2008) Relationship difficulties, low income, lack of support coupled with mental illness, such as, postpartum depression have also been reported as reasons for female infanticide in India (Chandran *et al.,* 2002; Chandra *et al.,* 2002; Friedman and Resnick, 2007). The effect of malnutrition, particularly micronutrient and vitamin deficiency, depends on sex and it adversely impacts female infant mortality (Ian Darnton-Hill *et al.,* 2005).

According to the decennial Indian census, the sex ratio in the 0 to 6 age group in India has risen from 102.4 males per 100 females in 1961 (life expectancy of 41; GOI, 2001) to 104.1 in 1981, to 107.8 in 2001, to 108.8 in 2011 (life expectancy of 66 years (GOI, 2013, Bhaskar and Gupta, 2007). The female to male sex ratio for children in 0-6 age group is within the normal natural range in all eastern and southern states of India, but significantly higher in certain western and particularly north-western states, such as, Punjab, Haryana and Jammu & Kashmir (118, 120 and 116, as of 2011, respectively; Chandramouli, 2011) In the age-group 0-1, the state of Jammu & Kashmir has reported India's highest male to female sex ratio in 2011 census (GOI, 2011).

For 50 years, from 1960-2011, fewer and fewer girls were allowed to be born in India. This situation, what is referred to as gendered arrangement, is old and commonly misunderstood. In India, there are "about 7.1 million fewer girls than boys aged 0-6 years." This is due to "prenatal sex determination with subsequent selective abortion of female fetuses" (Jha *et al.,* 2011). Data from the 2011 population census, provided by the Office of the Registrar General and Census Commissioner of India, Ministry of Home Affairs, also proves this. There are 914 girls per 1000 boys between the ages of 0 and 6. Future estimates paint a grim picture of the imbalanced demographic slide, the effects of which, research estimates, will spill beyond India (Fig. 1).

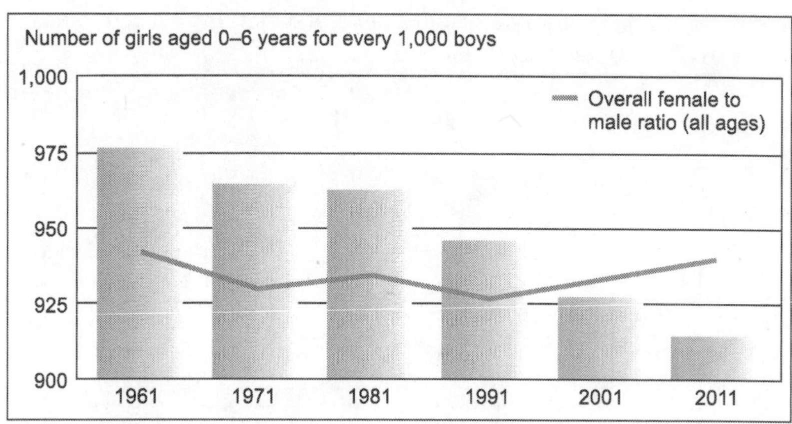

Source: *Census of India, 2011*

Fig. 1. *Falling number of girls aged 0-6 years in India since 1961*

In 1990, Nobel Laureate and economist, Amartya Sen, alerted the world to the phenomenon of "missing women." He said that more than 100 million women were missing from the world, which challenged the commonly held belief that women make up fifty percent of the world's population. Such a view is a generalization of the fact that there are a relatively higher number of women compared to men in Europe and North America. For a majority of the world, However, and especially in South Asia, the gender ratio favors men over women (Sen, 1990). Women in these parts of the world are considered "missing" in the sense that they are not allowed to be born.

It is clear from the foregoing that there are several reasons for the ill fate of a female child. If survives, even then the plight of a girl child is far from satisfactory and is of great concern. The basic requirement is the nutrition of the girl child right from infant to toddler to an adolescent to a pregnant woman and to a mother neglected deliberately or due to ignorance. Nutrition of a woman is not being taken care of throughout her life cycle. This call for immediate action to make sure that female child is taken care of throughout her life to ensure healthy generation in future.

3. MALNUTRITION: A SILENT EMERGENCY

Malnutrition in India is called 'The Silent Emergency'. The proportion of under-nutrition among children and women in India is one of the highest in the world. India is committed to halving the prevalence of underweight children by 2015 as one of the key indicators of progress towards the Millennium Development Goal (MGD). In spite of unprecedented economic growth, improvement in childhood nutritional status in India over the last decade has been slow. Suboptimal infant and young child feeding practices in particular continue to be a serious challenge to reducing malnutrition among children. High rates of maternal under-nutrition measured by low body mass index and anaemia adversely affect the health and survival of mothers and newborns. In contrast to other development indicators, several indicators of malnutrition are on the increase, including anaemia and wasting. Furthermore, significant disparities in nutrition levels exist across regional and societal groups (USAID, 2008) (Fig. 2).

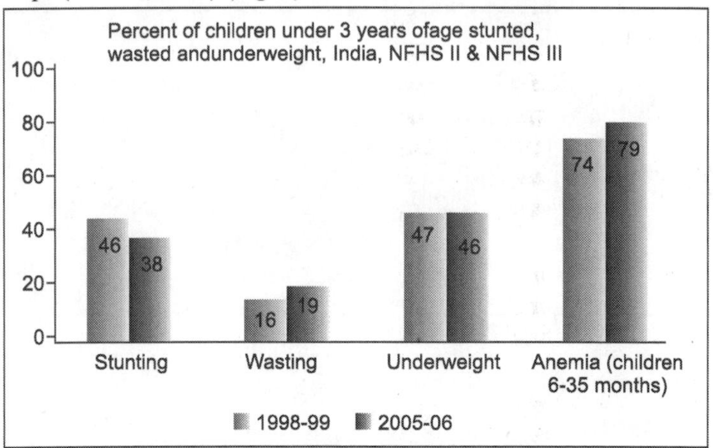

Fig. 2. *Malnutrition affecting children below 3 years of age (USAID, INDIA, 2008)*

Malnutrition is the major underlying cause of 50% of deaths among children. 8.3 million babies born are of low birth weight (less than 2.500 grams). Nearly half (46%) of children under 3 years of age or 31 million children are underweight (low weight for age). This number of malnourished children is equivalent to the entire population of Canada. 25.5 million or 1 in 3, children less than three years of age are stunted (low height for age) .only 1 in f5 children 6 to 23 months of age receive the recommended appropriate feeding (Fig. 3).

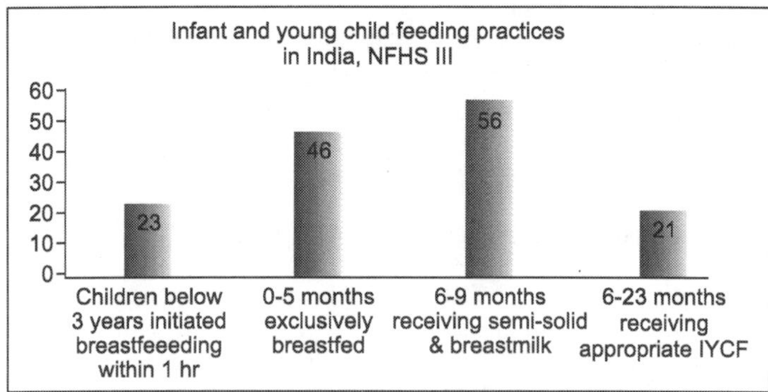

Fig. 3. *Feeding practices in India (USAID, INDIA, 2008)*

4. MATERNAL AND CHILD NUTRITION

The complex relationship between maternal nutrition and child–birth outcomes emphasizes the need for consistent and thorough assessments of women's diet throughout pregnancy and individualized nutrition education to promote positive birth outcomes. Every infant and child has the right to good nutrition according to the Convention on the Rights of the Child. Under-nutrition is associated with 45% of child deaths globally in 2012, 162 million children less than 5 were estimated to be stunted and 51 million have low weight-for-height, mostly as a consequence of poor feeding and repeated infections; 44 million were overweight or obese. About 38% of infants 0 to 6 months old are exclusively breastfed. Few children receive nutritionally adequate and safe complementary foods; in many countries only a third of breastfed infants 6–23 months of age meet the criteria of dietary diversity and feeding frequency that are appropriate for their age. About 800 000 children's lives could be saved every year among children under five, if all children 0–23 months were optimally breastfed (UN Report, 2008).

The UN estimates that 2.1 million Indian children die before reaching the age of five every year – four every minute–mostly from preventable illnesses, such as, diarrhoea, typhoid, malaria, measles and pneumonia. Every day, 1,000 Indian children die because of diarrhoea alone (UN Report, 2008).

Research in India has shown that low birth weight children have increased CVD risk factors and a unique birth cohort in Delhi has shown that low infant weight and rapid childhood weight gain, increases the risk of type two diabetes. Progress has been made in the role of specific nutrients in the maternal diet. In the Pune Maternal Nutrition Study, low maternal vitamin B12 status predicted increased adiposity and insulin resistance in the children, especially if the mother was foliate replete. It is not only maternal undernutrition that causes problems; gestational diabetes, a form of foetal over nutrition (glucose excess), is associated with increased adiposity and insulin resistance in the children, highlighting the adverse effects of the 'double burden' of malnutrition in developing countries, where undernutrition and over nutrition co-exist.

Recent intervention studies in several developing countries have shown that CVD risk factors in the offspring can be improved by supplementing undernourished mothers during pregnancy. Results differ according to the population, the intervention and the post-natal. Ongoing studies in India and elsewhere seek to understand the long-term effects of nutrition in early life and how best to translate this knowledge into policies to improve health in future generations (Anon., 2009).

Nutrition of mother and infant up to 6 months (till the baby is exclusively breast fed) is interlinked. Mother's nutrition during preconception, antenatal and post natal and throughout her reproductive years and foetal, neonatal and child nutrition are fundamental for a healthy start of next generation.

Globally, food insecurity is associated with micronutrient deficiencies and an estimated 4.5 billion people worldwide are affected by some form of nutrient deficiency. Of this 4.5 million, the most vulnerable groups are young children and women of childbearing age. (Dickinson *et al.,* 2009) The micronutrients that are most important for maternal and child health outcomes include iron, vitamin B12, folic acid and vitamin D (de Benoist, 2008; Kovacs, 2008; Black *et al.,* 2008)

One in three women, 15 – 49 years of age, has a Body Mass Index (BMI) below 18.5 indicating severe nutritional deficiency and under-nutrition. Among pregnant women, 58% are anaemic. Anaemia increases the risk of maternal death (Fig. 4).

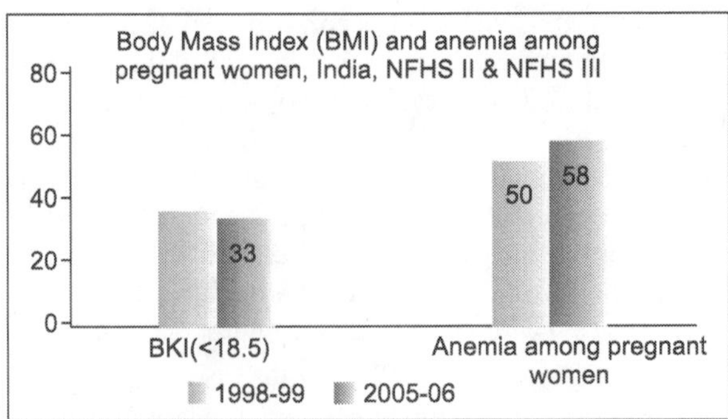

Fig. 4. *Body Mass Index and anemia among pregnant women in India (USAID, INDIA, 2008)*

Deficiencies in micronutrients affect many women of reproductive age and are associated with adverse health effects. A recent review of the maternal malnutrition concluded that 10–19% of women of reproductive age are seriously undernourished. With a body mass index of less than 18.5, these women are at increased risk of delivering low-birth-weight infant (Black *et al.,* 2008). In the developing world, the social and health problems that come along with such deficiencies are often compounded by high fertility rates, repeated pregnancies and short intervals between pregnancies.

The causal association between poor nutrition and ill-health can be seen in the case of anaemia. Anaemia before and during pregnancy is associated with increased maternal and prenatal mortality and morbidity and affects almost two-thirds of pregnant women in developing countries (Royston and Armstrong, 1989). The WHO Quarterly Statistics showed that two-thirds of pregnant women in India have a haemoglobin count so low that it increases the risk of maternal mortality rate to 17–20%. (WHO, 1998) Others have reported that the incidence of anaemia varies from 40% to 90% in India and contributes to 10–15% of direct maternal deaths (Mudaliar and Menon, 2005).

Anaemia in pregnancy generally results from nutritional deficiencies; almost 75% of anaemia is due to iron deficiency followed by folate and vitamin B12 deficiencies (Allen and Casterline –Sabel, 2000). Furthermore, physiological changes in pregnancy lead to an increase in the demand of many nutrients; especially iron, folic acid and zinc. (Mohammed and Hyten, 1989) Since most of the women in the developing countries start pregnancy with depleted body stores of these nutrients, their extra requirement of iron and folic acid is even higher than usual. If these needs are not met, the state of anemia may worsen. Fortunately, anaemia during pregnancy is for the most part, easily preventable and treatable with nutrition, supplementation or fortification.

5. NUTRITION, SUPPLEMENTATION AND FORTIFICATION

The approaches available for increasing micronutrient intake include dietary modification, supplementation and food fortification. Dietary modification involves changing the type and kind of food consumed, while supplementation involves the addition of certain essential micronutrients and vitamins usually in pill form. Food fortification is the public health policy of adding micronutrients (essential trace elements and vitamins) to foodstuffs to ensure that minimum dietary requirements are met.

While each approach has its inherent advantages and disadvantages, these different approaches for improving micronutrient status of populations are more often complementary options rather than competing ones (Shimpton and Schultink, 2002).

6. NUTRITION NEEDS OF INFANTS

Appropriate strategies for the prevention and management of maternal and child under-nutrition in developing countries must be a global priority. Evidence-based interventions that can improve maternal nutritional status and foetal growth include: "fortified food supplements, conditional cash transfers to address household food insecurity and micronutrient interventions (Bhutta and Haide, 2009).

Good nutrition is essential for the growth and development that occurs during an infant's first year of life. When developing infants are fed with the appropriate types and amounts of foods, their health is promoted. Positive and supportive feeding attitudes and techniques demonstrated by the caregiver help infants develop healthy attitudes toward foods, themselves and others throughout the first year. Many physiological changes occur that allow infants to consume foods of varying composition and texture. As an infant's mouth, tongue and digestive tract mature, the infant shifts from being able to only suckle, swallow and take in liquid foods, such as, breast milk or infant formula, to being able to chew and receive a wide variety of complementary foods. At the same time, infants progress from needing to be fed to feeding themselves. As infants mature, their food and feeding patterns must continually change. For proper growth and development, an infant must obtain an adequate amount of essential nutrients by consuming appropriate quantities and types of foods. During infancy, a period of rapid growth, nutrient requirements per pound of body weight are proportionally higher than at any other time in the life cycle (USDA, 2009).

Infant's nutrition needs are met with nutrient-rich breast milk formula and specialized baby foods. Infancy signifies a time of rapid growth and development between birth and the transition to becoming a wobbling toddler. Children reach the toddler years at different rates, but infancy typically ends around 12 months as the baby begins to walk and babble words. Meeting adequate nutrition needs during infancy is critical to normal brain and organ growth. Male and female infants generally require the same amounts of nutrients, but the estimated energy requirements of males are slightly higher.

6.1 Carbohydrates, Protein and Fat

The macronutrients, consisting of carbohydrates, protein and fat, supply the infant with calories to meet the daily estimated energy requirement. Carbohydrates are the primary fuel for cell energy and aid in the proper absorption of protein and fat, which serve as the basic building blocks for organ maturation. All infants from 0 to 6 months need 60 g of daily carbohydrates; from 7 to 12 months, the need increases to 95 g per day. Protein needs from 0 to 6 months are 9.1 g; from 7 to 12 months, 11 g per day. Infants from 0 to 6 months need 31 g of fat per day. From 7 to 12 months, their intake decreases slightly to 30 g per day.

6.2 Vitamins A, C, D, E and K

Infant easily gets vitamins from breast milk or formula. Do not give the infant supplements due to the risk of toxicity, which can be fatal. All infants from 0 to 6 months need 400 mcg of daily vitamin A; from 7 to 12 months, this need increases to 500 mcg. The vitamin D requirement for infants 0 to 12 months is 5 mcg per day. They need vitamin E at 4 mg per day for the first 6 months and 5 mg from 7 to 12 months. Intake of vitamin K is 2.0 mcg up to 6 months; from 7 to 12 months, it is 2.5 mcg. Vitamin C daily intake is 40 mg for the first 6 months and increases to 50 mg from 7 to 12 months.

6.3 B Vitamins

The B vitamins comprise a complex group of eight specific vitamins. Infants from 0 to 12 months need daily intake of thiamine at 0.2 to 0.3 mg, riboflavin at 0.3 to 0.4 mg, niacin at 2 to 4 mg and vitamin B-6 at 0.1 to 0.3 mg. Daily folate is needed at 65 to 80 mcg, vitamin B-12 at 0.4 to 0.5 mcg, pantothenic acid at 1.7 to 1.8 mg and biotin at 5 to 6 mcg.

6.4 Essential Minerals

An infant needs a variety of essential minerals, but the most significant micronutrients of focus include calcium for bone development, iron for healthy blood cell formation, zinc to protect the immune system and sodium to sustain blood volume, fluid balance and proper cell function. Breast milk and infant formula are plentiful sources of these minerals.

6.5 Water

Water makes up more than half of kids' body weight and is needed to keep all parts of the body functioning properly. There's no specific amount of water recommended for children, but it's a good idea to give them water throughout the day — not just when they're thirsty.

Babies generally don't need water during the first year of life. If the child doesn't like the taste of water, add a bit of lemon or lime for flavor. Fruits and veggies are also good sources of water. Kids should drink more water when ill, when it's hot out or when engaged in physical activity.

7. TODDLERS AND PRE-SCHOOLERS

At about six months most babies are ready to start solid foods like iron-fortified infant cereal and strained fruits, vegetables and pureed meats. Because breast milk may not provide enough iron and zinc when babies are around six to nine months, fortified cereals and meats can help breastfed babies in particular.

Once you do start adding foods, don't go low-fat crazy. Although the AAP recently released updated guidelines stating fat restriction in some babies is appropriate, in general, you don't want to restrict fats under age two because a healthy amount of fat is important for babies' brain and nerve development (Shu, 2008).

Toddlers and preschoolers grow in spurts and their appetites come and go in spurts, so they may eat a whole lot one day and then hardly anything the next It's normal and as long as you offer them a healthful selection, they will get what they need (Stern, 2008).

One area parents should probably keep under watch is calcium. Calcium, the body's building block, is needed to develop strong, healthy bones and teeth. Children may not believe or care that milk "does a body good," but it is the best source of much-needed calcium.

Still, there's hope for the milk-allergic, lactose-intolerant or those who are just impartial to milk. Lactose-free milk, soy milk, tofu, sardines and calcium-fortified orange juices, cereals, waffles and oatmeal are some calcium-filled options. In some cases the doctor may recommend calcium supplements.

Fibre is another important focus. Toddlers start to say "no" more and preschoolers can be especially opinionated about what they eat. The kids may want to stick to the bland, beige, starchy diet (think chicken nuggets, fries, macaroni), but this is really the time to encourage fruits, vegetables, whole grains and beans, which all provide fibre. Not only does fibre prevent heart disease and other conditions, but it also helps aid digestion and prevents constipation, something you and your child will be thankful for (Shu, 2008).

8. ELEMENTARY SCHOOL CHILDREN

Don't be surprised if your child quits meat. "It isn't uncommon for a 6- or 7-year-old to suddenly decide to be a vegetarian once they understand animals and where food comes,. This doesn't mean she won't get enough protein. Animal tissue isn't the only place we get protein, rice, beans, eggs, milk, peanut butter, all have protein. So whether your child goes "no-meat" for a week or for life, she'll likely still get sufficient amounts of protein (Stern, 2008).

Areas that might be a little too sufficient are sugars, fats and sodium. This time is when kids first go to school and have a little bit more choice in what they eat, especially if they're getting it in the cafeteria themselves, Cakes, candy, chips and other snacks might become lunchtime staples. The body needs carbohydrates (sugars), fats and sodium, but should be eaten in moderation, as too much can lead to unneeded weight gain and other health problems. Packing your child's lunch or going over the lunch menu and encouraging her to select healthier choices can help keep things on track (Shu, 2008).

9. PRE–TEENS AND TEENS

As puberty kicks in, young people need more calories to support the many changes they will experience. Unfortunately, for some, those extra calories come from fast food or "junk" foods with little nutritional value. Some adolescents go the opposite way and restrict calories, fats or carbohydrates. Adolescence is the time kids start to become conscious of their weight and body image, which, for some, can lead to eating disorders or other unhealthy behaviours, it is recommended that parents keep an eye on things by being aware of changes in eating patterns and making family dinners a priority once or twice a week (Stern, 2008).

Like calories, calcium requirements are higher. Calcium is more important than ever during the tween and teen years because the majority of bone mass is built during this time. However, fewer than 10 percent of girls and just more than 25 percent of boys ages 9 to 13 are at or above their adequate intake of calcium. Encouraging kids to have milk, milk products or calcium-rich alternatives, even if it means they have to add chocolate syrup to their milk to make it more palatable, should help them get more calcium.

In addition to more calories and calcium, your child's gender may play a role in whether he or she needs more of a particular nutrient. For instance, teen girls need more iron than their male counterparts to replace what's lost during menstruation and males need slightly more protein than girls. Although getting your child to eat healthy — regardless of his or her age — can be a constant battle, it is one well worth fighting. A healthy child becomes a healthy adult and only with your support and guidance will your child be both (Table 1).

Table 1. *Healthy diet of children and teens (Childhood Nutrition - HealthyChildren.org)*

Age	Fruits (Cups)	Vegetables (Cups)	Grains (ounces)	Meats & Beans (ounces)	Milk (Cups)	Oil (tsp)
2-3	1	2	3	2	2	3
4-8	1-1½	1½	4-5	3-4	2	4
9-13(females)	1½	2	5	5	3	5
9-13 (males)	1½	2½	6	5	3	5
14-18 females)	1½	2½	6	5	3	5
14-18 (males)	2	3	7	6	3	6

10. NUTRITIONAL NEEDS OF TEENAGERS

Both teenage girls and boys need proper nutrition to optimize growth and development and maintain healthy body weights. With the exception of calorie requirements, nutrition needs are generally the same for boys and girls ages 1 to 13, according to the Institute of Medicine. However, teen girls and boys ages 14 and older are developing into women and men and there are some distinct differences in their nutritional needs. Adolescence is a period of rapid physical, emotional, intellectual and social maturation. To support this growth, teenagers need extra calories, calcium and iron and sufficient protein. Teenage eating habits are less by parents and family and more by peers, media messages and body image issues.

Following a period of slow growth during late childhood, the teenage growth rate is as rapid as that of early childhood. By the end of adolescence, teens attain most of their adult height and weight. Although adolescence spans a period of five to seven years, teens do most of their growing during an 18-24 month period called the "growth spurt." To support this rapid growth, teenagers need to consume lots of calories and other nutrients. During adolescence, teens go through puberty, a process that involves total body maturation and the development of adult sexual function. The body composition of both females and males changes during adolescence. Before puberty, females have approximately 19% body fat, which increases to about 22% after puberty. Males maintain body fat percentage of approximately 15%, but during adolescence, males gain two times more muscle mass than females.

Eating well-balanced meals helps male and female teens maintain healthy body weights, excel in sports and stay focused at school. Getting the right number of calories and choosing a variety of foods from each food group will help ensure the teen meets his/her daily nutritional needs.

10.1 Calorie Counts

Male teens generally need more calories daily than female teenagers because they usually have larger frames and more muscle mass. Still, individualized calorie needs are based on your teen's activity level. The Dietary Guidelines for Americans 2010 reports that teen girls ages 13 to 18 need about 1,600 to 2,400 calories a day, while teenage boys within the same age range often require 2,000 to 3,200 calories daily.

10.2 Protein Foods

According to the Dietary Guidelines for Americans 2010, teens eating 2,000 calories a day need about 5.5 ounces of protein foods daily, while teenagers consuming 2,600 calories daily require about 6.5 ounces from the protein foods group each day. A one ounce equivalent from the protein foods group includes one egg, a half ounce of nuts or seeds, one-fourth ounce of legumes or tofu or 1 ounce of meat, poultry or seafood (1 ounce 28.3 g).

10.3 Dairy Foods

Regardless of whether they are eating 1,600 calories or 3,200 calories a day, teens should aim to consume 3 cups from the dairy foods group daily, according to the Dietary Guidelines for Americans 2010. A 1-cup portion from the dairy group equals two cups of cottage cheese, one cup of milk or yogurt or 1.5 ounces of cheese, Calcium, abundant in dairy foods, is essential for bone strength and development in teens.

10.4 Whole Grains

Because whole grains are rich in healthy carbohydrates, fiber, vitamins and minerals, teens should choose them instead of refined grains when possible. The Dietary Guidelines for Americans 2010 suggest teens eating 2,000 calories a day consume six ounces of grains and teenagers eating 2,600 calories daily aim for nine ounces of grains per day. A 1-ounce portion from the grains group equals a half cup of brown rice, quinoa, oatmeal or whole-wheat pasta, one cup of whole-grain cereal or one slice of whole-grain bread.

10.5 Fruits and Vegetables

The USDA suggests teens eating 2,000 to 2,600 calories a day need two cups of fruits and 2.5 to 3.5 cups of vegetables each day. A one cup equivalent of fruits equals one cup of fresh fruit, a half cup of dried fruit and one cup of 100-percent fruit juice and one cup of raw or cooked veggies, one cup of vegetable juice or two cups of leafy greens counts as a one cup portion from the vegetables food group.

10.6 High-Performance Oils

Teens should aim to obtain 25 percent to 35 percent of their daily calorie intake from healthy fats, notes the Institute of Medicine. The USDA recommends teens eating 2,000 calories a day consume the equivalent of 6 teaspoons of oils and teenagers eating 2,600 calories daily get eight teaspoons of oils daily. A one teaspoon portion from the oils group includes 1.5 teaspoons of peanut butter, one teaspoon of vegetable oil, eight large olives, one tablespoon of Italian salad dressing, one-third ounce of nuts and slightly less than one-fourth of an avocado.

10.7 Dietary Choices

The desire for independence is a defining characteristic of the teenage years. One of the ways that teenagers assert their independence is through their dietary choices. They eat more meals away from home and often have money to purchase meals and snacks on their own. The eating habits of the family take on less importance, while the influence of peers and media messages increases. Adolescence is also a time of dietary experimentation. For example, many teens follow vegetarian or vegan diets. Issues related to body image also affect dietary choices. According to the Youth Risk Behaviour Study of 1990, few adolescents are satisfied with their appearance. 34% of the females surveyed reported feeling "too fat" and 44% of female students said they were trying to lose weight. The most popular methods for losing weight are exercise and skipping meals. In addition, teens often use vomiting as a way to control their weight. More alarming, this survey found that many of the students who feel fat are well within a normal weight range for their height. The dietary choices of teens may also be impacted by various health conditions, most notably anorexia nervosa, bulimia nervosa and obesity.

11. IMPLICATIONS FOR PREGNANCY

The greater the amount of uncompleted growth at conception, the greater the energy and nutrient needs above those normally required during pregnancy (Story, 1992; Story and Alton, 1995) Gynaecological Age (GA), the difference between chronological age and the age at menarche, can be used as an indirect measure of physiological maturity and growth potential (Story, 1992). A pregnant adolescent with a GA of two years or less may still be in a period of growth and will have increased nutrient requirements compared to an adolescent who has finished her growth. Growth is generally completed four years after menarche (Story and Hermanson, 1990).

12. PRECONCEPTION NUTRITIONAL STATUS

Recently, attention has been placed on the importance of nutrition during preconception and the early months of pregnancy for healthy infant outcomes. For example, approximately 50% of cases of neural tube defects may be prevented with adequate intakes of folic acid from the time of conception and throughout the early months of pregnancy (Rayburn *et al.,* 1996). Some dietary conditions that increase nutritional risk at the time of pregnancy are: adherence to a vegan diet; substantial under- or overweight; participation in highly competitive sports; and rigid dieting/eating disorders. Conditions, such as, these may require additional nutrients to replenish nutrient stores. Pregnant adolescents may enter pregnancy with inadequate nutrient reserves. However, pregnancy can motivate adolescents to improve their diets to have a healthy baby. Pregnancy provides a window of opportunity for practitioners to educate teens about the importance of healthy eating during pregnancy and to encourage healthy eating practices. A number of national reports have recommended that more emphasis be placed on preconception and early pregnancy care (Rosen *et al.,* 1991). While preconception care should be standard for all women, it may not be the most effective approach for reaching adolescents, who often have unintended pregnancies and have the lowest rates of health care utilization of any age group. A more effective approach may be primary prevention strategies aimed at improving the nutritional health of all adolescent females. Efforts to improve eating patterns and achieve healthy weights can be implemented, in part, through school-based nutrition programs and integration of nutrition and health promotion counselling into primary care services.

13. NUTRITION NEEDS OF THE PREGNANT ADOLESCENT

Research indicates that among young, growing pregnant adolescents, there may be competition for nutrients between the mother and foetus (Naeye, 1981; Frisancho *et al.*, 1984) This hypothesis was first put forth (Naeye,1981), who reported that foetuses grow more slowly in 10– to16–year olds than in older women and speculated that there may be maternal– foetal competition for a limited supply of nutrients due to diminished placental blood flow.6 Frisancho7 in a study of Peruvian adolescents also found that growing adolescents gave birth to smaller babies and speculated on maternal foetal competition. The recent findings from Scholl, *et al.*, 8, 9 suggest the possibility that, among young, still growing pregnant adolescents, there may be for nutrients between mothers and their foetuses. The area of maternal growth and maternal-foetal competition remains controversial (Stevens *et al.*, 1993).

Limited information is available regarding the nutrient needs of pregnant adolescents. Dietary Reference Intakes (DRIs) represent the new approach adopted by the Food and Nutrition Board to provide quantitative estimates of nutrient needs to help individuals optimize their health and prevent disease. The DRIs replace and expand upon the Recommended Dietary Allowances (RDAs) which focused on reducing classic deficiency diseases. The DRIs contain four categories of recommendations for nutrient reference values: RDAs, Adequate Intakes (AIs), Estimated Average Requirements (EARs) and Tolerable Upper Intake Levels (Uls).

13.1 Energy

Energy needs are difficult to determine because they vary greatly among adolescents. The following factors influence total energy needs: Growth status – Pregravid weight– Physical activity – Stage of pregnancy– Body composition. Sufficient energy is a primary dietary requirement of pregnancy. If energy needs are not met, available protein, vitamins and minerals cannot be used effectively for various metabolic functions. Energy requirements are greater for pregnant adolescents than for their no pregnant peers. Current recommendations suggest pregnant women should increase their average energy intake by 300 calories per day during the second and third trimesters of pregnancy (Inst. Med, 1990). Younger adolescents may require higher energy intakes during pregnancy than older women. In general, pregnant adolescents should not consume less than 2000 calories/day and in many cases, higher intakes are needed. Because caloric needs are so variable (38-50 Kcal/kg/day), the best assurance of adequate energy intake is a satisfactory weight gain. Adolescents who begin pregnancy underweight, are still growing or who are physically active may need additional energy intake during pregnancy.

13.2 Carbohydrates

The majority of calories eaten during pregnancy (at least 50%) should come from carbohydrate-containing foods, such as, grain products, fruits and vegetables. One study found that excessive sugar intake may Increase the risk of delivering a small-for-gestational age infant (Lenders *et al.*, 1997). Fibber can help alleviate constipation and haemorrhoids, which often occur during pregnancy whether pregnant or not, 25-35 g of fibber is needed.

13.3 Protein

Protein needs increase during pregnancy. Adequate protein intake is required to support continued growth and development in both the fetes and the young pregnant adolescent. The RDA for protein for adolescent females is approximately 45 g per day. An additional allowance of 10 g per day is recommended throughout pregnancy (NAS, 1989). High-protein powders and specially formulated high-protein supplements or beverages may be potentially harmful (increasing risk of preterm birth) and should be avoided during pregnancy (Inst. Med., 1990). Diets of American adolescent females, both pregnant and non pregnant show that the majority meet the RDAs for protein intake. Factors that may increase the risk for inadequate protein intake include:– Low socioeconomic status (protein foods tend to be expensive)– Low energy intake (when energy needs are not met, protein is catabolised for energy)– Exclusion of all animal products, such as, eggs, dairy products and meats (most commonly occurs in vegan or macrobiotic vegetarians or adolescents with eating disorders). A careful dietary assessment can determine the protein intake of a pregnant adolescent. The quality of the protein, as well as adequacy of energy intake should be considered. About two-thirds of total protein should be of high biologic quality, such as, eggs, milk, meat or other animal sources.

13.4 Fat

Fat is an essential nutrient which provides essential fatty acids necessary for the development of feta brain and central nervous tissues. Fat also plays a role in transporting the fat-soluble vitamins and is a concentrated source of calories, helping to meet the energy needs of pregnancy. One study found that infants born prematurely or of low birth weight were more likely to have had mothers who were inadequately nourished and had dietary deficits of essential fatty acids (Doyle, 1989). The U.S. Dietary Guidelines recommend that Americans limit their fat intake to no more than 30% of calories. These recommendations are generally accepted during pregnancy, although meeting calorie needs is the primary concern. With caloric needs of 2000 per day, 30% of calories from fat would equal 66 g of fat per day. Monounsaturated and polyunsaturated fatty acids should be the predominant types of fat consumed during pregnancy. Sources of these fats include olive, canola, safflower and sunflower oils, tub margarine, peanuts, peanut butter, walnuts and fish (such as, salmon, tuna and bluefish). Vegan and macrobiotic vegetarians need to pay close attention to the types of fats that they consume to assure adequate essential fatty acid intake for foetal development.

13.5 Fluids

Water helps the body to maintain proper temperature, transports nutrients and waste products, moistens the digestive tract and tissues and cushions and protects the developing foetus. Fluid needs increase during pregnancy, primarily because of increased blood volume. At least 8 cups of non-caffeinated fluids should be consumed each day. Water is the best choice, as it is rapidly absorbed by the body. Milk, juice and no caffeinated beverages, such as, herbal teas contribute to fluid needs. Caffeinated beverages increase urinary output and should be consumed in limited amounts.

13.6 Iron

Iron deficiency is one of the most common nutritional problems among both pregnant and non-pregnant adolescent females and occurs in all socioeconomic groups. Iron deficiency anaemia during the first two trimesters is associated with a twofold increased risk for preterm delivery and a threefold increase risk for delivering a low birth weight baby (Scholl *et al.,* 1992). The need for iron increases during adolescence due to increases in lean body mass, red blood cell mass and onset of menses. Iron requirements are further increased during pregnancy due to expansion of maternal plasma volume and growth of the foetus and placenta. The need for iron increases as pregnancy progresses Iron needs increase throughout the second trimester reaching a peak requirement in the third trimester when foetal demands are greatest. The pregnancy RDA for iron is 30 mg per day, a level twice that for non pregnant adolescents. Despite improved absorption of iron during pregnancy, this level is not likely to be achieved through diet. National data suggest that only one fourth of adolescent girls meet the RDA for iron. Data from NHANES-III indicate that 9% of non-pregnant 12-15 year old females and 11% of 16-19 year old females had iron deficiency anemia.22 The prevalence of anaemia in low-income pregnant adolescents enrolled in the WIC program in 1990 was 11%, 16% and 37% in the first, second and third trimesters, respectively (Beard, 1999).

To meet the iron requirements of pregnancy and prevention deficiency anaemia, the CDCP recommends low dose (30 mg/day) supplements of iron at the first prenatal visit (CDCP, 1998). Low-dose iron can be given alone or as part of a multivitamin, mineral supplement of appropriate composition for pregnancy. Liquid and chewable forms of iron are available if have trouble swallowing tablets or capsules. To ensure adequate iron absorption, iron supplements should be taken at bedtime or between meals with water or juice, not with milk, tea or coffee. Pregnant adolescents should be encouraged to eat iron rich foods and foods that enhance iron absorption (CDCP, 1998). Food sources: The best sources of iron are meats, dried beans and peas, dark green leafy vegetables, whole grain or enriched breads and fortified cereals Iron from animal sources is well absorbed. Iron from plant sources is poorly absorbed, although its absorption is enhanced by vitamin C, meat, fish and poultry.

13.7 Calcium

Adolescents grow at the greatest rate of any age group after infancy and accumulate 37% of their total bone mass during this growth spurt. Because maximum bone mass is acquired during adolescence, the calcium deposited during adolescence determines the risk of osteoporosis and fracture in adulthood. Bone mass is dependent on calcium intake, growth and pubertal development, exercise and genetic and racial factors. Unfortunately, during this time of tremendous calcium need, most adolescents eat a diet that is very deficient in calcium. Girls are twice as likely to be deficient as boys (85% and 43% respectively). Other factors contributing to poor bone mineralization include adolescent pregnancy, anorexia nervosa, excessive exercise and various chronic medical conditions. To avoid osteoporosis in later life, adequate dietary calcium intake should be recommended and calcium supplementation considered in all adolescent patients.

During pregnancy, approximately 25-30 g of calcium is transferred to the foetus (Inst. Med., 1997). Most of this is acquired during the third trimester, when calcium is deposited into the foetal skeleton at a rate of about 330 mg per day (Inst. Med., 1990). In an adult woman this represents about 2.5% of typical maternal bone calcium stores (no data is available for pregnant adolescents). The DRIs for calcium during pregnancy are not increased above non pregnancy levels, as calcium absorption increases during pregnancy (Yate *et al.,* 1998). The pregnant adolescent, theoretically, could have an increased need for calcium because of her need to support her own bone consolidations as well as that of the foetus (Inst. Med., 1997). Adequate calcium intake is especially important for pregnant adolescents who are still increasing their bone mass. A higher peak bone mass during adolescent years is protective against postmenopausal osteoporosis. Approximately 40% of peak bone mass in girls is accumulated during the adolescent years (Weaver *et al.,* 1999). For pregnant and lactating women, the DRI is 1300 mg calcium per day for those age 18 or younger and1000 mg calcium for those older than 18. A few studies indicate that pregnant adolescents may benefit from a high calcium intake (Inst. Med., 1997). A lower incidence of preterm delivery and low birth weight was observed in pregnant adolescents randomized to receive 2000 mg per day supplemental calcium (Villar and Rapke, 1990). Low calcium intakes are well-documented in adolescent females. Only 18% of females 9-19 years of age meet dietary recommendations for calcium (DRIs are1300 mg calcium/ day for adolescents 9-18 years). In NHANES- III, the mean calcium intake of adolescent girls 12-15 years old was 796 mg per day and 822 mg per day among 16-19 year old girls. Soft drinks appear to displace milk in the diets of adolescents. Adolescents who consume soft drinks daily have lower calcium intakes (Hamack *et al.,* 1999). To achieve the recommendation of 1300 mg calcium per day, the pregnant adolescent requires at least four glasses of milk each day or the equivalent, such as, one cup of yogurt or 1.5 oz of cheese (Villar and Rapke, 1990).

Fortified foods: There are several calcium fortified foods on the market, including orange juice, bottled water, cereal, breakfast bars and bread. Some are fortified to the same level of calcium as milk (300mg/serving). The amount of calcium is listed on the label.

Supplements: Supplemental calcium is indicated when adolescents are unable (*e.g.*, because of lactose intolerance) or unwilling to achieve the recommended dietary calcium intake from food sources. Multi-nutrient supplements do not provide this much calcium so a separate calcium supplement is needed. Unlike iron, calcium absorption is enhanced when the supplement is taken with meals rather than between meals. Most chemical forms of calcium carbonate, citrate, lactate or phosphate, have the same absorption rate. On average, 35% of the calcium in calcium citrate malate is absorbed compared to about 30% of the calcium in calcium carbonate and other supplements. Of the calcium supplements commercially available, calcium carbonate contains the highest proportion (40%) of elemental calcium by weight and is the least expensive. Certain calcium preparations, such as, bone meal calcium, dolomite and calcium carbonate from oyster shells, have been found to contain lead and other heavy metal contaminants. Calcium supplements should be labelled "lead-free" (NDC, 1997).

13.8 Zinc

Zinc affects protein synthesis and is essential for growth. The recommended intake during pregnancy is 15 mg of zinc per day. Young pregnant teenagers who are still growing may have higher needs. In one recent study, low intakes of zinc were associated with an increased risk of low birth weight an preterm delivery in adolescents (Scholl and Hediger, 1995). Zinc requirements are highest in the third trimester when the foetus acquires two-thirds of its zinc stores. Zinc intake has been shown to be below two-thirds of the RDA for both pregnant and non pregnant adolescents.

Diets that are very high in fibre or phytate, such as, vegan diets may decrease zinc absorption. Routine iron and folate supplementation may also impair zinc absorption. Therefore, good dietary sources of zinc should be emphasized. Food sources: The best sources are seafood, meat and eggs. Less rich sources are legumes and whole grains.

13.9 Folate

Folate, essential for nucleic acid synthesis, is required in greater amounts during pregnancy as a result of maternal and fetal tissue growth and red blood cell formation. Folate deficiency during pregnancy may result in intrauterine growth retardation, congenital anomalies or spontaneous abortion (CDCP, 1992). The DRI for folate during pregnancy is 600 mcg per day. Supplementation with folic acid during early gestation has been related to a 50% protective effect against first occurrence and a 72% protective affect against recurrent neural tube defects (i.e., spina bifida, anencephaly, encephalocele). Because these birth defects occur early in gestation, before pregnancy is apparent and because many pregnancies, especially those of adolescents, are unplanned, it is advised that all women of childbearing age who are capable of becoming pregnant consume 400 mcg of folic acid daily (CDCP, 1992). The diets of adolescents are often limited in folate. Teens who skip breakfast or who have limited amounts of fruits and vegetables in their diets are at greater risk for inadequate folate intakes. Although the United States began fortification of grain products with folic acid in 1998, there is still the possibility that females will not consume adequate amounts of folate. Pregnant adolescents need to be counselled on foods that are good sources of folate, as well as on the role of folic acid in preventing neural tube defects. Food sources: Folic acid, the synthetic form of folate, is found in enriched bread, pasta, crackers, breakfast cereal and rice. Good sources of folate include: dry beans, lentils, chickpeas, peanuts orange juice oranges, strawberries, pineapple juice, kiwi fruit, leafy greens, okra, sweet corn, beets and broccoli.

13.10 Vitamin D

Vitamin D facilitates the absorption of dietary calcium and helps deposit calcium into bones. Maternal vitamin D deficiency can lead to neonatal hypocalcemia and delayed growth, as well as maternal osteomalacia. The recommended intake of vitamin D for women of childbearing age is 200 IU (5 mcg), whether pregnant or not. The major source of vitamin D for humans is the exposure of the skin to sunlight. Very few foods naturally contain vitamin D. In the US almost the entire intake of vitamin D comes from fortified milk products and other fortified foods, such as, breakfast cereals (Inst. Med., 1997).

13.11 Vitamin A

A maternal deficiency of vitamin A can impair feta growth. The daily requirement, whether pregnant or not, is 5000 International Units (IU). Vitamin A deficiency–is rare in American women of childbearing years. However, if adolescents have low consumption of fruits, vegetables, dairy products and fortified cereals, there is the potential for having an inadequate intake of vitamin A. Vitamin A at high levels is a documented teratogen. Pregnant adolescents should avoid unprescribed supplements that contain vitamin A, especially at dosages exceeding 800 RE (~4,000 IU). (Inst. Med., 1992). Congenital malformations have been documented in mothers using either isotretinoin or etretinate, vitamin A analogues for treatment of cystic acne, during pregnancy. Topical applications of retinoid do not appear to be harmful. Food sources: The richest sources of preformed vitamin A are liver, egg yolks, dairy products, margarine, butter and fortified cereals.. Sources of beta carotene, which are converted to vitamin A include carrots, broccoli, winter squash, greens (spinach, kale, mustard, turnip), tomatoes, apricots, sweet potatoes, red peppers, peaches, pumpkin, mango and cantaloupe.

13.12 Vitamin C

The RDA for vitamin C during pregnancy is 70 mg per day. It is recommended that regular cigarette smokers consume at least 100 mg of vitamin C per day since smoking increases metabolic turnover of the vitamin, leading to lower concentrations in the blood (NAS. 1999). Vitamin C can have a beneficial effect on iron status Usual daily dietary intakes of vitamin C of 25-75 mg can enhance the intestinal absorption of dietary non-heme iron by two to four fold when the two nutrients are contained together. Food sources: oranges and other citrus fruit, greens and red peppers, collard greens, broccoli, spinach, tomatoes, potatoes, strawberries, fortified juices.

13.13 Vitamin B6

The recommended level of vitamin B6 is 1.9 mg for pregnant adolescents. It is not uncommon for adolescent females to have intakes below this level. Food sources: The richest sources are chicken, fish, liver, pork and eggs. Other good sources include whole wheat grains, oats, peanuts and walnuts.

13.14 Vitamin B12

To help prevent megaloblastic anaemia, pregnant vegetarians, especially vegans, should be sure their diets are adequate in vitamin B12, which is primarily found in animal foods. Food sources: beef, ham, pork, turkey, chicken, herring, salmon, trout, tuna, clams, oysters, lobster, shrimp, milk, cheese, yogurt and egg yolk. Fortified breakfast cereals, fortified soy milk and nutritional yeast (not brewer's yeast) also provide vitamin B12.

13.15 Prenatal Supplement Use

Obtaining nutrients from food reduces the likelihood of nutrient imbalances, nutrient interactions and toxicities. A low-dose vitamin-mineral supplement is recommended for pregnant adolescents who do not regularly consume a diet that meets the nutrient needs of pregnancy, are strict vegetarians, have a multiple gestation, who smoke more than 20 cigarettes per day or abuse alcohol or drugs (Inst. Med., 1990). A calcium supplement (600 mg/day) is recommended for adolescents who do not drink milk or do not eat calcium-rich foods daily (Inst. Med., 1990). To prevent iron deficiency anaemia, iron supplementation at a low dose, about 30 mg of elemental iron per day for non anaemic pregnant adolescents is recommended (CDCP, 1998). A study by Scholl *et al.,* (1997) found that the use of prenatal supplements by low income, urban women in the Camden Study reduced the risk of pre–term delivery and low birth weight. Since prenatal supplements are often large, adolescents may be more likely to comply with a standard or chewable multivitamin supplement.

13.16 Lactose Intolerance

About 25% of US adults have lactose intolerance and develop symptoms of diarrhoea and bloating after ingestion of a large dose of lactose, such as, is contained in a quart of milk. Primary lactose deficiency begins in childhood and may become clinically apparent in adolescence (Inst. Med., 1997). Many lactose intolerant individuals can tolerate a glass of milk or its lactose equivalent, especially when consumed with meals. Suggestions for improving calcium intake among adolescents with symptoms of lactose intolerance include - Try smaller amounts of milk with meals – Eat aged cheeses, such as, cheddar and Swiss. They contain only small amounts of lactose.– Eat yogurt with active cultures. The bacterial cultures "auto digest" lactose within the intestinal tract– Use lactose-reduced or lactose-free milk and milk products available in the stores. Use lactose enzyme preparations to treat the milk purchased at the grocery or swallow an enzyme tablet before consuming dairy food (NDC, 1997).

14. FRUITS AND VEGETABLES: THE NUTRI-NUGGETS TO PROMOTE HEALTH

Eating fruits and vegetables is one of the most positive health habits that adults, youth and children can adopt. During the past decade, vegetarianism has risen in popularity. Well-planned vegetarian diets can satisfy the nutritional needs and promote normal growth of infants and children. Research has highlighted nutritional advantages to vegetarian diets and has indicated that this style of eating can lead to lifelong healthy eating habits when adopted at a young age. Several vitamins, minerals and macronutrients may be deficient within a vegetarian diet. Careful nutrition assessment and counseling will allow nurse practitioners to play a key role in encouraging families to adopt healthy eating habits to assist in disease prevention (Dunham and Kollar, 2006).

The Expert Committee of the Indian Council of Medical Research, taking into consideration the nutrient requirements, has recommended that every individual should consume at least 300 g of vegetables (GLV : 50 g; Other vegetables: 200 g; Roots &Tubers : 50 g) in a day. In addition, fresh fruits (100 g), should be consumed regularly. Since requirements of iron and folic acid are higher for pregnant women they should consume 100g of leafy vegetables daily. High-calorie vegetables and fruits are to be restricted for overweight/ obese subjects.

ICMR has recommended the following dietary allowances (RDA) for different age groups in India (Table 2).

Table 2. *Nutrient requirement of different age groups*

Summary of RDA for Indians 2010

Group	Particulars	Body Wt. kg	Net energy kcal/d	Protein g/d	Visible Fat g/d	Calcium mg/d	Iron mg/d	Vit A µg/d Retinol	Vit A µg/d β-carotene	Thiamine mg/d	Riboflavin mg/d	Niacin equivalent mg/d	Pyridoxine mg/d	Ascorbic Acid mg/d	Dietary folate µg/d	Vit B_{12} µg/d	Magnesium mg/d	Zinc mg/d
	Sedentary work		2320		25					1.2	1.4	16						
Man	Moderate work	60	2730	60.0	30	600	17	600	4800	1.4	1.6	18	2.0	40	200	1.0	340	12
	Heavy work		3490		40					1.7	2.1	21						
	Sedentary work		1900		20					1.0	1.1	12						
	Moderate work		2230	55.0	25	600	21	600	4800	1.1	1.3	14	2.0	40	200	1.0		10
	Heavy Work		2850		30					1.4	1.7	16						
Woman	Pregnant woman	55	+350	82.2	30	1200	35	800	6400	+0.2	+0.3	+2	2.5	60	500	1.2	310	
	Lactation 0-6 m		+600	77.9	30					+0.3	+0.4	+4	2.5					12
	6-12 m		+520	70.2	30	1200	25	950	7600	+0.2	+0.3	+3	2.5	80	300	1.5		
	0 - 6 months	5.4	92 kcal/kg/d	1.16 g/kg/d	--		46µg/kg/d			0.2	0.3	710 µg/kg	0.1				30	---
Infants	6 - 12 months	8.4	80 kcal/kg/d	1.69 g/kg/d	19	500	05	350	--- 2800	0.3	0.4	650 µg/kg	0.4	25	25	0.2	45	---
	1-3 years	12.9	1060	16.7	27		09	400	3200	0.5	0.6	8	0.9		80		50	5
Children	4-6 years	18.0	1350	20.1	25	600	13			0.7	0.8	11	0.9	40	100	0.2-1.0	70	7
	7-9 years	25.1	1690	29.5	30		16	600	4800	0.8	1.0	13	1.6		120		100	8
Boys	10-12 years	34.3	2190	39.9	35	800	21			1.1	1.3	15	1.6	40	140	0.2-1.0	120	9
Girls	10-12 years	35.0	2010	40.4	35	800	27			1.0	1.2	13	1.6				160	9
Boys	13-15 years	47.6	2750	54.3	45	800	32	600	4800	1.4	1.6	16	2.0	40	150	0.2-1.0	165	11
Girls	13-15 years	46.6	2330	51.9	40	800	27			1.2	1.4	14	2.0				210	11
Boys	16-17 years	55.4	3020	61.5	50	800	28			1.5	1.8	17	2.0	40	200	0.2-1.0	195	12
Girls	16-17 years	52.1	2440	55.5	35	800	26			1.0	1.2	14	2.0				235	12

332

Please see appendices13.1 to 13.8 for a wide range of vegetables rich in nutrients

15. CONCLUSION

It is important to save and nourish the girl child to build a healthy society. The nutrition requirement of a girl child is similar to that of a male child during infancy, childhood and preteen ages and it changes for a girl during adolescence, preconception and child bearing periods.

Eating fruits and vegetables along with standard meals promotes good health by providing vitamins and essential minerals. Additional supplements are required by some individuals based on the medical conditions that are prescribed by the physician. The above information should not be used as a substitute for the medical care and advice of the paediatrician. There may be variations in treatment that paediatrician may recommend based on individual facts and circumstances. The nutrient requirement may vary among the populations of different nations. However, the ideal range of nutrient requirement from infancy to adulthood may be comparable. Further there is intermixing of communities in the present day world. The minute details of dietary nutrients requirement for individuals always depend on medical advice. General recommendations, However, help the governments to plan food and nutrition security measures for their populations.

The findings reiterate the need to take care of the nutrition of a girl child from her infancy to her motherhood. It is bounden duty of our society to rake care of mothers to be for ensuring healthy generations forever.

Acknowledgement

The authors are thankful to Dr. Prem Nath Agricultural Science Foundation, Bengaluru, for providing fellowships and considering the paper for publication.

REFERENCES

Allen, L, Casterline-sabel, J. (2000). Prevalence and causes of nutritional anemias. *In*: Ramakrishnan U, editor. Nutritional anaemias. Boca Raton: CRC Press. p. 17–21. (Cited from Nutrition in Maternal and Child Health - Unite For Sight.

www.uniteforsight.org/women-children-course/nutrition-women-childre)n).

American Academy of Paediatrics. (2015). Childhood Nutrition - HealthyChildren.org:// *www.healthychildren.org/.../nutrition/.../Childhood-Nutrition.asp*2015.

Anon. (2009). Indian J. Medical Research 130 . 2009.

Aubri John. (2001). Nutritional Needs for Male & Female Infants - Livestrong.com *www.livestrong.com › Parenting(2001).*

Beard, J.L. (1994). Iron deficiency: assessment during pregnancy and its importance in pregnant adolescents. Am J Clin Nutr 1994;59(2 Suppl):502S-508S discussion 508S-510S. (Cited from NUTRITION NEEDS OF ADOLESCENTS

www.epi.umn.edu/let/pubs/img/adol_ch3.pdf).

Bhaskar and Gupta (2007). India's missing girls: biology, customs and economic development Oxford Review of Economic Policy, Volume 23, Number 2, p. 221–238.

Black, R.E., Allen, L.H., Bhutta, Z.A., *et al.* (2008). Maternal and Child Under nutrition Study Group. Maternal and child under nutrition: global and regional exposures and health consequences. *Lancet* 2008; 371:243–60. (Cited from Nutrition in Maternal and Child Health - Unite For Sight *www.uniteforsight.org/womens-children-course/nutrition-women-childre)n).* Bhutta ZA, Haider BA.(2009) Prenatal micronutrient supplementation: are we there yet? CMAJ 2009;180(12): 1188–1189 (Cited from Nutrition in Maternal and Child Health - Unite For Sight *www.uniteforsight.org/women-children-course/nutrition-women-childre)n).*

CDCP. (1992). Recommendations for the use of folic acid to reduce the number of cases of spina bifida and other neural tube defects. MMWR 1992;41(RR-14):1-7. (Cited from NUTRITION NEEDS OF ADOLESCENTSwww.epi.umn.edu/let/pubs/img/adol_ch3.pdf).

CDCP. (1998). Recommendations to prevent and control iron deficiency in the United States. Centers for Disease Control and Prevention. MMWR 1998;47(RR-3):1-29. (Cited from NUTRITION NEEDS OF ADOLESCENTS.www.epi.umn.edu/let/pubs/img/adol_ch3.pdf).

Chandran *et al.* (2002). Post-partum depression in a cohort of women from a rural area of Tamil Nadu, India: Incidence and risk factors, The British Journal of Psychiatry, 181(6), p. 499-504(cited from Female infanticide in India - Wikipedia, the free encyclopedia.

http:// www. en.wikipedia.org/wiki/Female_infanticide_in_India2015.

Chandra *et al.* (2002). Infanticidal ideas and infanticidal behavior in Indian women with severe postpartum psychiatric disorders, J Nerv Ment Dis. 2002 Jul, 190(7), p. 457-61 (cited from Female infanticide in India - Wikipedia, the free encyclopedia.

http:// www. en.wikipedia.org/wiki/Female_infanticide_in_India2015.

Chandramouli, C. (2011). Child Sex Ratio in India, Registrar General & Census Commissioner, India (2011).

Christine, Alder and Ken, Polk (cited from Female infanticide in India - Wikipedia, the free encyclopedia*http:// www. en.wikipedia.org/wiki/ Female_infanticide_in_India2015.*

Daniel, Grey. (2011). Gender, Religion and Infanticide in Colonial India, 1870—1906, Victorian Review, Vol. 37, No. 2, p. 107-120 (cited from Female infanticide in India - Wikipedia, the free encyclopedia*http:// www. en.wikipedia.org/wiki/ Female_ infanticide_in_India2015.*

Dickinson, N. *et al.* (2009). A framework to explore micronutrient deficiency in maternal and child health in Malawi, Southern Africa. *Environmental Health* 2009, 8 (Suppl 1): S13. (2) (Cited from Nutrition in Maternal and Child Health - Unite For Sight

www.uniteforsight.org/women-children-course/nutrition-women-childre)n).

de Benoist B. (2008). Conclusions of a WHO technical consultation on folate and vitamin B12 deficiencies. *Food Nutr Bull* 2008; 29 (Suppl 2): S238–44 (Cited from Nutrition in Maternal and Child Health - Unite For Sight

www.uniteforsight.org/women-children-course/nutrition-women-childre)n).

Doyle, W. (1989). Maternal nutrient intake and birth weight. J Hum Nutr Diet 1989;2:407.*)* (Cited from NUTRITION NEEDS OF ADOLESCENTS

www.epi.umn.edu/let/pubs/img/adol_ch3.pdf).

Dube *et al.* (1999). (cited from Female infanticide in India - Wikipedia, the free encyclopedia*http:// www. en.wikipedia.org/wiki/ Female_infanticide_in_India2015.*

Dunham, L., Kollar, L.M. (2006). -Vegetarian eating for children and adolescents Journal of Pediatric Health Care, 2006 – Elsevier.

Erin Coleman, R.D., L.D. (2013). Nutritional Needs of Male & Female Teenagers - Livestrong . *www.livestrong.com › Parenting (2013).*

Friedman and Resnick. (2007). Child murder by mothers: patterns and prevention, World Psychiatry, 2007 Oct; 6(3), p. 137–141, PMCID: PMC2174580 (cited from Female infanticide in India - Wikipedia, the free encyclopedia*http:// www. en.wikipedia.org/ wiki/ Female_infanticide_in_India2015.*

Frisancho, A.R., Matos, J., Bollettino, L.A. (1984). Influence of growthstatus and placental function on birth weight of infants born toyoung still-growing teenagers. Am J Clin Nutr 1984;40(4):801-807. (Cited from NUTRITION NEEDS OF ADOLESCENTS www.epi.umn.edu/let/pubs/img/adol_ch3.pdf).

Giriraj, R. (2004) .Changing Attitude to Female Infanticide in Salem, Journal of Social Welfare, Vol. 50, No. 11, February 2004, p. 13-14 & 34-35 (cited from Female infanticide in India - Wikipedia, the free encyclopedia *http:// www. en.wikipedia. org/wiki/ Female_infanticide_in_India2015.*

GOI. (2001). Data Highlights - 2001 Census Census Bureau, Government of India.

GOI. (2011). Age Data - Single Year Age Data - C13 Table (India/States/UTs)Population Enumeration Data (Final Population) - 2011, Census of India, Ministry of Home Affairs, Government of India.

GOI. (2013). India at Glance - Population Census 2011 - Final Census of India, Government of India (2013).

Harnack, L., Stang, J., Story, M. (1999). Soft drink consumption among US children and adolescents nutritional consequences. J Am Diet Assoc 1999;99(4):436-441. (Cited from NUTRITION NEEDS OF ADOLESCENTS. www.epi.umn.edu/let/pubs/img/ adol_ch3.pdf).

Hausfater. (2008). (cited from Female infanticide in India - Wikipedia, the free encyclopediab*http:// www. en.wikipedia.org/wiki/ Female_infanticide_in_ India2015.*

Ian Darnton-Hill, Patrick Webb, Philip, W.J. Harvey, Joseph, M. Hunt, Nita Dalmiya, Mickey Chopra, Madeleine, J. Ball, Martin, W. Bloem and Bruno de Benoist. (2005). Micronutrient deficiencies and gender: social and economic costs, American Journal of Clinical Nutrition, May 2005, vol. 81, no. 5, p. 1198S-1205S (cited from Female infanticide in India - Wikipedia, the free encyclopedia.http:// www. en.wikipedia. org/wiki/Female_infanticide_in_India2015.

ICMR. (2009). Maternal nutrition: Effects on health in the next generation*icmr.nic.in/i jmr/2009/november/1119.pdf.*

Institute of Medicine. (1990). Nutrition during pregnancy: part, weight gain: part II, nutrient supplements. Washingt1993)on DCNational Academy Press, 1990. (Cited from NUTRITION NEEDS OF ADOLESCENTS.www.epi.umn.edu/let/pubs/img/ adol_ch3.pdf).

Institute of Medicine. (1992). Nutrition during pregnancy and lactation: an implementation guide. Washington, DC: National Academy Press, 1992. (Cited from NUTRITION NEEDS OF ADOLESCENTS.www.epi.umn.edu/let/pubs/img/adol_ch3.pdf).

Institute of Medicine. (1997). Food and Nutrition Board. Dietary reference intakes for calcium, phosphorus, magnesium, vitamin D and fluoride. Washington, DC: National Academy of Sciences, 1997. (Cited from NUTRITION NEEDS OF ADOLESCENTS www.epi.umn.edu/let/pubs/img/adol_ch3.pdf).

Jha Prabhat *et. al.* (2011). doi: http://cghr.org/wordpress/wp-content/uploads/Trends-in-selective-abortions-of-girls-in-India-2011.pdf.

Key, Janice D.; Key, L. Lyndon Jr. (1994). Calcium needs of adolescents Current Opinion in Paediatrics, Adolescent medicine: PDF Only Lippincott-Raven Publishers.

Krishnan. (2012). (cited from Female infanticide in India - Wikipedia, the free encyclopedia *http:// www. en.wikipedia.org/wiki/ Female_infanticide_in_India2015.*

Kovacs, C.S. (2008). Vitamin D in pregnancy and lactation: maternal, fetal and neonatal outcomes from human and animal studies. *Am J Clin Nutr* 2008;88:520S–8S. (Cited from Nutrition in Maternal and Child Health - Unite For Sight.

www.uniteforsight.org/women-children-course/nutrition-women-childre)n).

Lenders, C.M., Hediger M.L., Scholl, T.O., Khoo, C.S., Slap, G.B., Stallings, V.A. (1997). Gestational age and infant size at birth are associatedwith dietary sugar intake among pregnant adolescents. J Nutr1997;127(6):1113-1117 (Cited from NUTRITION NEEDS OF ADOLESCENTS. www.epi.umn.edu/let/pubs/img/adol_ch3.pdf).

Mohammed, K., Hyten F. (1989). Iron and folate supplementation in pregnancy. In: Chalmers, H, Enkin, M.J., editors. Effective care in pregnancy and childbirth. Oxford: Oxford University Press. (Cited from Nutrition in Maternal and Child Health - Unite For Sight *www.uniteforsight.org/women-children-course/nutrition-women-childre)n).*

Mudaliar, A.L. and Menon, M.K. (2005). Clinical obstetrics. Hyderabad: Orient Longman. p. 147.

National Academy of Sciences, National Research Council. (1989). Recommended Dietary Allowances. (10th ed.) Washington, DC: National Academy Press, 1989. (Cited from NUTRITION NEEDS OF ADOLESCENTS.www.epi.umn.edu/let/pubs/img/adol_ch3.pdf).

National Dairy Council. (1997). Meeting calcium requirements during pregnancy. Rosemont, IL: Dairy Management, Inc., 1997. (Cited from NUTRITION NEEDS OF ADOLESCENTS.www.epi.umn.edu/let/pubs/img/adol_ch3.pdf).

Naeye, R.L. (1981). Teenaged and pre-teenaged pregnancies: consequences of the fetal-maternal competition for nutrients. Pediatrics1981;67(1):146-150. (Cited from NUTRITION NEEDS OF ADOLESCENTS.www.epi.umn.edu/let/pubs/img/adol_ch3.pdf).

Oberman. (2005). (cited from Female infanticide in India - Wikipedia, the free encyclopedia*http:// www. en.wikipedia.org/wiki /Female_infanticide_in_India2015).*

Prabha, Unnithan, N., (2013). Crime and Justice in India, SAGE Publications, p. 257, ISBN 978-81-321-09778(cited from Female infanticide in India - Wikipedia, the free encyclopedia

http:// www. en.wikipedia.org/wiki/Female_infanticide_in_India2015.

Rayburn, W.F., Stanley, J.R., Garrett, M.E. (1996). Periconceptional folate intake and neural tube defects. J Am Coll Nutr 1996;15(2): 121-125. (Cited from NUTRITION NEEDS OF ADOLESCENTS. www.epi.umn.edu/let/pubs/img/adol_ch3.pdf).

Rosen, M.G., Merkatz, I.R., Hill, J.G. (1991). Caring for our future: a report by the expert panel on the content of prenatal care. Obstet Gynecol 1991;77(5):782-787. (Cited from NUTRITION NEEDS OF ADOLESCENTS

www.epi.umn.edu/let/pubs/img/adol_ch3.pdf).

Royston, E., Armstrong, S. (1989). Preventing maternal deaths. Geneva: WHO. p. 85–147(Cited from Nutrition in Maternal and Child Health - Unite For Sight *www.uniteforsight.org/women-children-course/nutrition-women-children).*

Sen, Amartya. (1990). More Than 100 Million Women are Missing. *The New York Review of Books*. doi: http://www.nybooks.com/articles/archives/1990/dec/20/more-than-100-million-women-are-missing/?pagination=false.

Scholl, T.O., Hediger, M.L., I.G. Ances. (1990). Maternal growth during pregnancy and decreased infant birth weight. - The American journal of ..., 1990 - Am Soc Nutrition.

Scholl, T.O., Hediger, M.L., Cronk, C.E., Schall, J.I. (1993). Maternal growth during pregnancy and lactation. Horm Res.,1993 ;39(Suppl 3):59-67. (Cited from NUTRITION NEEDS OF ADOLESCENTS.www.epi.umn.edu/let/pubs/img/adol_ch3.pdf).

Scholl, T.O., Hediger, M.L. (1993). A review of the epidemiology of nutrition and adolescent pregnancy: maternal growth during pregnancy and its effect on the fetus. J Am Coll Nutr 1993; 12(2): 101-107. (Cited from NUTRITION NEEDS OF ADOLESCENTS www.epi.umn.edu/let/pubs/img/adol_ch3.pdf).

School, T.O., Hediger, M.L., Fischer, R.L., Shearer, J.W. (1992). Anemia vs iron deficiency: increased risk of preterm delivery in a prospective study. Am J Clin Nutr 1992;55(5):985-988. (Cited from NUTRITION NEEDS OF ADOLESCENTS. www.epi.umn.edu/let/pubs/img/adol_ch3.pdf).

School, T.O., Hediger, M.L. (1995). Weight gain, nutrition and pregnancy outcome: findings from the Camden study of teenage and minority gravidas. Semin Perinatol 1995;19(3):171-181(Cited from NUTRITION NEEDS OF ADOLESCENTS. www.epi.umn.edu/let/pubs/img/adol_ch3.pdf).

School, T.O., Hediger, M.L., Bendich, A., Schall, J.I., Smith, W.K., Krueger, P.M. (1997). Use of multivitamin/mineral prenatal supplements: influence on the outcome of pregnancy. Am J Epidemiol 1997; 146(2):134-141.

Shrimpton, R., Schultink, W.(2002). Can supplements help meet the micronutrient needs of the developing world? Proc Nutr Soc 2002;61:223–9(Cited from NUTRITION NEEDS OF ADOLESCENTS. www.epi.umn.edu/let/pubs/img/adol_ch3.pdf).

Shu, Jennifer. (2008). Childhood Nutrition - HealthyChildren.org *https://www.healthychildren.org/.../nutrition/.../Childhood-Nutrition.aspx*.

Stern, Loraine. (2008). Childhood Nutrition - HealthyChildren.org *https://www.healthychildren.org/.../nutrition/.../Childhood-Nutrition.aspx*.

Stevens-Simon, C., McAnarney, E.R., Roghmann, K.J. (1993). Adolescent gestational weight gain and birth weight. Pediatrics;92(6):805-809. (Cited from NUTRITION NEEDS OF ADOLESCENTS. www.epi.umn.edu/let/pubs/img/adol_ch3.pdf).

Story, M. (1992). Nutritional requirements during adolescence. In: McAnarney E.R., Kreipe R.E. orr D.E., Comerci, G.D., eds. Textbook ofadolescent medicine. Philadelphia: W.B. Saunders, 1992:75-84 (Cited from NUTRITION NEEDS OF ADOLESCENTS. www.epi.umn.edu/let/pubs/img/adol_ch3.pdf).

Story, M., Alton I. (1995). Nutrition issues and adolescent pregnancy. Nutr Today 1995;30:142-151 (Cited from NUTRITION NEEDS OF ADOLESCENTS. www.epi.umn.edu/let/pubs/img/adol_ch3.pdf).

Story, M. J., Hermanson. (1990) -Nutrient needs during adolescence and pregnancy-management of the pregnant adolescent, 1990 - epi.umn.edu.

UN Report. (2008). Nutrition & Malnutrition Resources specifically for India. The UN ... Tracking Progress in Maternal, Newborn & Child Survival - The 2008 Report

USDA. (2009). Infant Nutrition and Feeding - National Agricultural Library*www.nal.usda.gov/wicworks/Topics/FG/CompleteIFG.pdf.*

Visaria. (1983). Female infanticide in India - Wikipedia, the free encyclopedia*http:// www.en.wikipedia.org/wiki/ Female_infanticide_in_India2015.*

Weaver, C.M., Peacock, M., Johnston, C.C. (1999). Adolescent nutrition in the prevention of postmenopausal osteoporosis. J Clin Endocrinol Metab 1999;84(6):1839-1843. (Cited from NUTRITION NEEDS OF ADOLESCENTS www.epi.umn.edu/let/pubs/img/adol_ch3.pdf).

Villar, J., Repke, J.T. (1990). Calcium supplementation during pregnancy may reduce preterm delivery in high-risk populations. Am Jobstet Gynecol 1990;163(4 Pt 1):1124-1131. (Cited from NUTRITION NEEDS OF ADOLESCENTS. www.epi.umn.edu/let/pubs/img/adol_ch3.pdf).

WHO. (1998). WHO/UNICEF and UNU Iron deficiency: Indicators for assessment and strategies for prevention. Geneva: WHO. (Cited from Nutrition in Maternal and Child Health - Unite For Sight. *www.uniteforsight.org/women-children-course/nutrition-women-children).*

WHO. (2014). Key facts - WHO Infant and young child feeding Fact sheet –pdated February 2014.

Yates, A.A., Schlicker, S.A., Suitor, C.W. (1998). Dietary Reference Intakes: the new basis for recommendations for calcium and related nutrients, B vitamins and choline. J Am Diet Assoc 1998; 98(6):699-706. (Cited from NUTRITION NEEDS OF ADOLESCENTS. www.epi.umn.edu/let/pubs/img/adol_ch3.pdf); *http://www.bbc.co.uk/news/world-south-asia-13264301.*

Nutrients in Vegetables

Appendix 13.1. *Protein rich vegetables*

Vegetables	Protein per 100 g
Paapdi (Double Beans)	8.3 g
Arvi Ka Saag (Colocasia Leaves)	6.8 g
Phool Gobhi Patta (Cauliflower Leaves)	5.9 g
Methi Sabji (Fenugreek Leaves)	4.4 g
Dhania Patta (Coriander Leaves)	3.3 g
Gawar (Cluster Beans)	3.2 g
Shepu Bhaji	3 g

Phool gobhi (Cauliflower)	2.6 g
Saijan ki Phalli (Drumsticks)	2.5 g
Palak Sabji (Spinach)	2 g
Bhindi (Lady Finger)	1.9 g
Tamatar (Tomato)(Green)	1.9 g
Gobhi (Cabbage)	1.8 g
Chukandar (Beet Root)	1.7 g
Bakla (French Beans) Green	1.7 g
Karela (Bitter Gourd)	1.6 g
Aalu (Potato)	1.6 g
Kaddu (Pumpkin)	1.4 g
Kela (Banana) (Raw)	1.4 g
Baingan (Brinjal)	1.4 g
Simlamirchi (Capsicum)	1.3 g
Pyaj (Onion)	1.2 g
Shakarkand (Sweet Potato)	1.2 g
Tondli, Konduri (Kovai)	1.2 g
Pyaj ka patta (Onion Leaves)	0.9 g

Appendix 13.2. *Fat in vegetables*

Vegetables	Fat per 100 g.% of RDA
Arvi Ka Saag (Colocasia Leaves)	2 g
Phool Gobhi Patta (Cauliflower Leaves)	1.3 g
Methi Sabji (Fenugreek Leaves)	0.9 g
Palak Sabji (Spinach)	0.7 g
Dhania Patta (Coriander Leaves)	0.6 g
Shepu Bhaji	0.5 g
Phool gobhi (Cauliflower)	0.4 g
Gawar (Cluster Beans)	0.4 g
Paapdi (Double Beans)	0.3 g
Baingan (Brinjal)	0.3 g
Shakarkand (Sweet Potato)	0.3 g
Simla mirchi (Capsicum)	0.3 g
Chachinda (Snake Gourd)	0.3 g
Kela (Banana) (Raw)	0.2 g
Karela (Bitter Gourd)	0.2 g
Pyaj ka patta (Onion Leaves)	0.2 g
Bhindi (Lady Finger)	0.2 g

Gajar (Carrot)	0.2 g
Tondli, Konduri (Kovai)	0.1 g
Toria (Ridge Gourd)	0.1 g
Kaddu (Pumpkin)	0.1 g
Bakla (French Beans) Green	0.1 g
Saijan ki Phalli (Drumsticks)	0.1 g
Kakdi (Cucumber)	0.1 g
Gobhi (Cabbage)	0.1 g

Appendix 13.3. *Carbohydrate rich vegetables*

Vegetables	Carbohydrate per 100g
Bhindi (Lady Finger)	64 g
Shakarkand (Sweet Potato)	28.2 g
Aalu (Potato)	22.6 g
Kela (Banana) (Raw)	14 g
Paapdi (Double Beans)	12.3 g
Pyaj (Onion)	11.1 g
Gawar (Cluster Beans)	10.8 g
Gajar (Carrot)	10.6 g
Pyaj ka patta (Onion Leaves)	8.9 g
Chukandar (Beet Root)	8.8 g
Arvi Ka Saag (Colocasia Leaves)	8.1 g
Phool Gobhi Patta (Cauliflower Leaves)	7.6 g
Dhania Patta (Coriander Leaves)	6.3 g
Methi Sabji (Fenugreek Leaves)	6 g
Shepu Bhaji	5.2 g
Kaddu (Pumpkin)	4.6 g
Gobhi (Cabbage)	4.6 g
Bakla (French Beans) Green	4.5 g
Simlamirchi (Capsicum)	4.3 g
Karela (Bitter Gourd)	4.2 g
Phool gobhi (Cauliflower)	4 g
Baingan (Brinjal)	4 g
Saijan ki Phalli (Drumsticks)	3.7 g
Tamatar (Tomato)(Green)	3.6 g
Mooli (Radish)	3.4 g

Appendix 13.4. *Calorie rich vegetables*

Vegetables	Calories per 100 g
Shakarkand (Sweet Potato)	120
Aalu (Potato)	97
Paapdi (Double Beans)	85
Arvi Ka Saag (Colocasia Leaves)	77
Phool Gobhi Patta (Cauliflower Leaves)	66
Kela (Banana) (Raw)	64
Pyaj (Onion)	50
Methi Sabji (Fenugreek Leaves)	49
Gajar (Carrot)	48
Dhania Patta (Coriander Leaves)	44
Chukandar (Beet Root)	43
Pyaj ka patta (Onion Leaves)	41
Shepu Bhaji	37
Bhindi (Lady Finger)	35
Phool gobhi (Cauliflower)	30
Gobhi (Cabbage)	27
Saijan ki Phalli (Drumsticks)	26
Bakla (French Beans) Green	26
Palak Sabji (Spinach)	26
Karela (Bitter Gourd)	25
Kaddu (Pumpkin)	25
Simlamirchi (Capsicum)	24
Baingan (Brinjal)	24
Tamatar (Tomato)(Green)	23
Tondli, Konduri (Kovai)	18

Appendix 13.5. *Fibre rich vegetables*

Vegetables	Fibre per 100 g.% of RDA
Saijan ki Phalli (Drumsticks)	4.8 g
Paapdi (Double Beans)	4.3 g
Gawar (Cluster Beans)	3.2 g
Phool Gobhi Patta (Cauliflower Leaves)	2 g
Arvi Ka Saag (Colocasia Leaves)	1.8 g
Bakla (French Beans) Green	1.8 g
Pyaj ka patta (Onion Leaves)	1.6 g
Tondli, Konduri (Kovai)	1.6 g

Baingan (Brinjal)	1.3 g
Gajar (Carrot)	1.2 g
Bhindi (Lady Finger)	1.2 g
Dhania Patta (Coriander Leaves)	1.2 g
Phool gobhi (Cauliflower)	1.2 g
Shepu Bhaji	1.1 g
Methi Sabji (Fenugreek Leaves)	1.1 g
Simlamirchi (Capsicum)	1 g
Gobhi (Cabbage)	1 g
Chukandar (Beet Root)	0.9 g
Chachinda (Snake Gourd)	0.8 g
Karela (Bitter Gourd)	0.8 g
Shakarkand (Sweet Potato)	0.8 g
Mooli (Radish)	0.8 g
Kaddu (Pumpkin)	0.7 g
Kela (Banana) (Raw)	0.7 g
Tamatar (Tomato)(Green)	0.7 g

Appendix 13.6. *Iron rich vegetables*

Vegetables	Iron per 100 g.% of RDA
Phool Gobhi Patta (Cauliflower Leaves)	40 mg
Shepu Bhaji	17.4 mg
Pyaj ka patta (Onion Leaves)	7.4 mg
Kela (Banana) (Raw)	6.3 mg
Paapdi (Double Beans)	2.3 mg
Methi Sabji (Fenugreek Leaves)	1.93 mg
Tamatar (Tomato)(Green)	1.8 mg
Chachinda (Snake Gourd)	1.51 mg
Dhania Patta (Coriander Leaves)	1.4 mg
Phool gobhi (Cauliflower)	1.23 mg
Chukandar (Beet Root)	1.2 mg
Palak Sabji (Spinach)	1.1 mg
Gawar (Cluster Beans)	1.1 mg
Gajar (Carrot)	1.03 mg
Arvi Ka Saag (Colocasia Leaves)	0.98 mg
Gobhi (Cabbage)	0.8 mg
Bakla (French Beans) Green	0.6 mg
Karela (Bitter Gourd)	0.6 mg

Kakdi (Cucumber)	0.6 mg
Pyaj (Onion)	0.6 mg
Simlamirchi (Capsicum)	0.57 mg
Aalu (Potato)	0.5 mg
Dudhi, Louki (Bottle Gourd)	0.46 mg
Kaddu (Pumpkin)	0.44 mg
Toria (Ridge Gourd)	0.4 mg

Appendix 13.7. *VitaminA rich foods*

Vegetables	Vitamin A RAE per 100g
Arvi Ka Saag (Colocasia Leaves)	1000 μg
Shepu Bhaji	598.5 μg
Dhania Patta (Coriander Leaves)	576.5 μg
Palak Sabji (Spinach)	465 μg
Methi Sabji (Fenugreek Leaves)	195 μg
Gajar (Carrot)	157.5 μg
Pyaj ka patta (Onion Leaves)	49.58 μg
Simlamirchi (Capsicum)	35.58 μg
Gawar (Cluster Beans)	16.5 μg
Tamatar (Tomato)(Green)	16 μg
Tondli, Konduri (Kovai)	13 μg
Bakla (French Beans) Green	11 μg
Karela (Bitter Gourd)	10.5 μg
Gobhi (Cabbage)	10 μg
Saijan ki Phalli (Drumsticks)	9.17 μg
Chachinda (Snake Gourd)	8 μg
Baingan (Brinjal)	6.17 μg
Bhindi (Lady Finger)	4.33 μg
Kaddu (Pumpkin)	4.17 μg
Toria (Ridge Gourd)	2.75 μg
Kela (Banana) (Raw)	2.5 μg
Phool gobhi (Cauliflower)	2.5 μg
Aalu (Potato)	2 μg
Shakarkand (Sweet Potato)	0.5 μg
Mooli (Radish)	0.25 μg

Appendix 13.8. *Vitamin C rich vegetables*

Vegetables	Vitamin C per 100 g
Simla mirchi (Capsicum)	137 mg
Dhania Patta (Coriander Leaves)	135 mg
Gobhi (Cabbage)	124 mg
Saijan ki Phalli (Drumsticks)	120 mg
Karela (Bitter Gourd)	88 mg
Arvi Ka Saag (Colocasia Leaves)	63 mg
Phool gobhi (Cauliflower)	56 mg
Methi Sabji (Fenugreek Leaves)	52 mg
Gawar (Cluster Beans)	49 mg
Tamatar (Tomato)(Green)	31 mg
Palak Sabji (Spinach)	28 mg
Bakla (French Beans) Green	24 mg
Kela (Banana) (Raw)	24 mg
Shakarkand (Sweet Potato)	24 mg
Paapdi (Double Beans)	22 mg
Aalu (Potato)	17 mg
Pyaj ka patta (Onion Leaves)	17 mg
Mooli (Radish)	15 mg
Tondli, Konduri (Kovai)	15 mg
Bhindi (Lady Finger)	13 mg

23

Plantation Crops—Enhancing Production and its Strategic Role in Promoting Nutritional Security

K.R. M. Swamy

SYNOPSIS

Plantation crops are the cash crops. Among them, coconut, cocoa, oil palm, palmyrah, arecanut and cashew are predominantly grown by small holders in India. These cash crops supplement and complement nutrition security of people. By themselves, these cash crops and the related value chains have no direct effect on nutrition security of the farmers and plantation workers; they only create an income that is very low for most of the farmers and plantation workers. Under-nutrition in regions that primarily produce cash crops is among the worst in most developing countries, resulting in poor productivity levels, low school performance of children and a poor health situation for farming families. Problems of under-nutrition are intergenerational; when under-nutrition is not adequately addressed, children will grow into undernourished adults and their children will experience the same problems. This affects not only families, but also local economic development. It is estimated that under-nutrition causes losses in GDP of 2 to 3%. Promoting nutrition security offers several opportunities to stimulate economic and pro-poor development in cost-efficient ways. In regions that produce cash crops like tea, cocoa and coffee, nutrition security is rarely on the agenda of decision makers. However, nutrition is the key to trigger a positive effect on the health of farmers and plantation workers and on economic development and productivity.

A rapidly growing global population, accelerating consumption, dietary shifts, climate change and other factors are driving unprecedented price volatility, resource shortages and other risks in supply chains. Assuming the projected population of India by the year 2050 as 1.62 billion, there is likely to be an increasing demand for plantation crops. (coconut, cocoa, oil palm, palmyrah, arecanut, cashew, tea and coffee) products in the domestic and international market, owing to climate related production dips in many parts of the world. Further, cashew, coconut and cocoa demand in the domestic market is also predicted to increase due to life style changes coupled with increase in post harvest processing and product development activities. The strategies to be adopted for increasing the production of plantation crops have been discussed.

An effect on nutrition security can however, be created by diversifying the crops that farmers and plantation workers grow. Major plantation crops, production, nutritional value and health benefits, under-nutrition in major plantation crops growing areas, strategies to stimulate production of plantation crops and nutrition security are discussed in this paper.

Keywords: Plantation crops, nutritional value, health benefits, nutritional security, strategies.

1. INTRODUCTION

The term 'Plantation crops' refers to those crops which are cultivated on an extensive scale in a large contiguous area, owned and managed by an individual or a company. Plantation crops constitute a large group of crops. The major plantation crops, include, coconut, cocoa, oil palm, palmyrah/toddy palm, tea, coffee, cashew, arecanut and rubber. Rubber has no food value; the other plantation crops are high value commercial crops of greater economic importance and play a vital role in our Indian economy. They are the cash crops. Among them, cashew, coconut, cocoa, oil palm, palmyrah and arecanut are predominantly grown by small holders in India. These cash crops supplement and complement nutrition security of people. By themselves, these cash crops and the related value chains have no direct effect on nutrition security of the farmers and plantation workers; they only create an income that is very low for most of the farmers and plantation workers.

India is the leading country in the total production of certain plantation crops in the world. For instance, our production meets the share of 47 *per cent* in tea and 66 per cent in each of cashew and arecanut. Plantation industry provides direct as well as indirect employment to many millions of people. For instance, tea industry offers direct employment to 10 lakh people and indirect employment to 10 lakh people, while-cashew processing factories alone provide employment to three lakh people besides two lakh farmers are employed in cashew cultivation. Plantation industry supports many by-product industries and also many rural industries. For example, coconut husk is used to produce coir fiber annually to a tune of 2,19,600 tonnes in India. These crops help to conserve the soil and ecosystem. Tea planted in hill slopes and cashew in barren and waste lands protect the land from soil erosion during the rainy season or due to heavy winds (Anon., 2014b).

Nutrition security is different from food security as it is about a community's access to essential nutrients, not just calories. Zinc, selenium, iron and all kind of vitamins should be present in human diets. A growing export market for cash crops is increasingly threatening such variety in diets, that is not adequately rich in essential nutrients, for local farmers and plantation workers in developing countries. Under-nutrition in regions that primarily produce cash crops is among the worst in most developing countries, resulting in poor productivity levels, low school performance of children and a poor health situation for farming families. Problems of under-nutrition are intergenerational; when under-nutrition is not adequately addressed, children will grow into undernourished adults and their children will experience the same problems. This affects not only families, but also local economic development. It is estimated that under-nutrition causes losses in GDP of 2 to 3%. Promoting nutrition security offers several opportunities to stimulate economic and pro-poor development in cost-efficient ways. This potential of nutrition security is stressed by the World Bank, the United Nations, the African Union and the World Economic Forum and in the Copenhagen Consensus, a reoccurring meeting of some of the world's most prominent economists. This recognition has attracted the attention of policymakers in governments all over the world and dramatically increased funding for programmes that focus on nutrition security. In regions that produce cash crops like tea, cocoa and coffee, nutrition security is rarely on the agenda of decision makers. However, nutrition is the key to trigger a positive effect on the health of farmers and plantation workers and on economic development and productivity (Klaas de Vries, 2014).

An effect on nutrition security can However, be created by diversifying the crops that farmers and plantation workers grow. Major plantation crops, production, nutritional value and health benefits, under-nutrition in major plantation crops growing areas, strategies to stimulate production of plantation crops and nutrition security are discussed in this paper.

2. NATIONAL PRODUCTION SCENARIO

2.1 Coconut

Coconuts (*Cocos nucifera* L.) are different from other fruits because they contain a large quantity of "water" and when immature they are known as tender-nuts or jelly-nuts and may be harvested for drinking. When mature, they still contain some water and can be used as seed nuts or processed to give oil from the kernel, charcoal from the hard shell and coir from the fibrous husk. The endosperm is initially in its nuclear phase suspended within the coconut water. As development continues, cellular layers of endosperm deposit along the walls of the coconut, becoming the edible coconut "flesh". When dried, the coconut flesh is called copra. The oil and milk derived from it are commonly used in cooking and frying; coconut oil is also widely used in soaps and cosmetics. The clear liquid coconut water within is potable. The husks and leaves can be used as material to make a variety of products for furnishing and decorating. It also has cultural and religious significance in many societies that use it. In India, the coconut palm is grown both under plantation and homestead management system. India produced 23billion nuts from two million hectare area with a per hectare productivity of 11,500 nuts/ha (CPCRI, 2015).

2.2 Cocoa

Cocoa (*Theobroma cacao* L.), a beverage crop having high commercial potential, is mostly grown in India as a mixed crop in arecanut and coconut gardens. It is mainly cultivated in four major southern States of Kerala, Karnataka, Tamil Nadu and Andhra Pradesh. The cocoa industry in the country had expanded to a considerable extent in recent years. The area under cocoa is 7,100 ha with a production is 15,100 tonnes of cocoa beans (CPCRI, 2015).

2.3 Oil Palm

Oil palm/red oil palm (*Elaeis guineensis* Jacq.) tree is known to be the highest edible oil yielding perennial crop. Palm oil is extracted from the pulpy portion (mesocarp) of the fruit of oil palm. It is the crop that has a greater advantage in terms of productivity and much higher than that of other major oil seed crops. Oil palm is the crop of the present and future vegetable oil economy of world as well as India. So far an area of 1.94 lakh ha was covered under oil palm but only 1.64 lakh ha exists at present as about 30,000 ha were uprooted due to various reasons. The production of palm oil in India continues to be at meager level with respect to its requirement. According to Majumdar (2014) the area under oil palm in India is 1, 89,624 ha and production is 79 lakh tonnes of fruits.

2.4 Palmyrah

Palmyrah palm or toddy palm or sugar palm or sea coconut or sea palm (*Borassus flabellifer*) is economically useful and widely cultivated in tropical regions. This is called in Tamil as *nungu* or *nongu.* The British named this as *ice apple,* as it resembled ice. *Nongu* cools many dried throats on a hot summer day, tastes best when it comes from arid fields. In South India, in summer the road side vendors pile up these young fruits, cut the shell and pop out the heart shaped icy *nongu* for their customers. It is nutritious when eaten in its natural form. This fruit is loaded with minerals and absolutely no fat and protein. The jelly part of the fruit is covered with a thin, yellowish-brown skin. These are known to contain watery fluid inside the fleshy white body. The pulp or the seed becomes harder as it grows older. Though the palmyrah palm has extensive distribution in drier parts of India, no reliable current statistics on its area and production are available. In India it is grown in Tamil Nadu andhra Pradesh, Odisha, Bihar, Madhya Pradesh, West Bengal, Kerala and Karnataka states. India has an estimated population of 102 million palms, with highest population in Tamil Nadu (51.9 million) (Kurian and Peter, 2007). The estimated area under palmyrah palm in India may be over 8-10 lakh ha.

2.5 Arecanut

Areca palm or areca nut palm or betel palm (*Areca catechu* L.) produces arecanut or betel nut. This palm is often called the betel tree because its fruit, the arecanut, is often chewed along with the betel leaf, a leaf from a vine of the Piperaceae family. The areca nut is the seed of the areca palm. It is commonly referred to as betel nut, as it is often chewed wrapped in betel leaves (*paan*). Areca palm is grown for its commercially important seed crop, the areca nut. The arecanut palm influences the economic life of people in rural India, providing a decent livelihood for more than three million people and assured employment of 10 million man days annually, besides its role in religions, social and cultural functions (Amudhan *et al.,* 2012).. India ranks first in both area and production of arecanut with annual production of 6.32 lakh tonnes of nuts from an area of 4.53 lakh hectares (CPCRI, 2015). India is the major producer and consumer of arecanut in the world. Production is concentrated in Karnataka, Kerala, Assam, Meghalaya, Tamil Nadu, West Bengal, Goa and the Andaman and Nicobar Islands. The ever increasing demand for areca products like *paan, supari* and *gutkha* has led to continuous increase in arecanut prices worldwide. The processing methods, maturity and consumer preferences influence the types of arecanut products prepared.

2.6 Cashew/Cashew Nut

Cashew (*Anacardium occidentale* L.) kernels are a popular snack and food source. Cashews, unlike other oily tree nuts, contain starch to about 10% of their weight. This makes them more effective than other nuts in thickening water-based dishes, such as, soups, meat stews and some Indian milk-based desserts. Many Southeast Asian cuisines use cashews for this unusual characteristic, rather than other nuts. Cashew nuts are commonly used in Indian cuisine, whole for garnishing sweets or curries or ground into a paste that forms a base of sauces for curries (*e.g.,* korma) or some sweets (*e.g.,* kaju barfi). It is also used in powdered form in the preparation of several Indian sweets and desserts. In Goan cuisine, both roasted and raw kernels are used whole for making curries and sweets. Cashew is presently grown in an area of 10.11 lakh hectares with a production of 7.53 lakh tonnes of raw cashew nuts in the country and the productivity is 0.74 t/ha (IHD, 2015). Most of the area under cashew is in East coast and West coast regions of the country.

2.7 Tea

Tea plant (*Camellia sinensis* L.O. Kuntze.) leaves and leaf buds are used to produce the popular beverage tea. Tea is one of the most popular and lowest cost beverages in the world and consumed by a large number of people. Owing to its increasing demand, tea is considered to be one of the major components of world beverage market. Currently grown in 35 countries, the tea industry provides a vital source of employment and export earnings, often in some of the world's poorest countries. And a vast chain or network of people have a relationship with tea – not just consumers, but growers, pickers, suppliers, traders and sellers – impacting on the lives and well-being of millions of people across the world. Tea cultivation is confined only to certain specific regions of the world due to specific requirements of climate and soil (Basu Majumder *et al.,* 2010). Next to water, tea is the world's most popular drink. More than three billion cups of tea are consumed every day across the world. Tea contains half the amount of caffeine found in coffee. Worldwide consumption of tea grew by 60% between 1993 and 2010. Life without tea is unthinkable for the billions of people who rely on their cherished cup of tea to get them through the day (Anderson, 2014). Tea growing regions in India are Anamallais, Nilgiri (Tamil Nadu), Wayanad, Travancore, Munnar (Kerala), Kangra (Himachal Pradesh), Dooars and Terai (Himalayan foot hills), Chikmagalur (Karnataka), Darjeeling (West Bengal) and Assam. India is one of the largest tea producers in the world. The area under tea in India is 5, 63,980 ha and the production is 1208.78 million kg (TB, 2014).

2.8 Coffee

Coffee (*Coffea canephora/Coffea Arabica*) seeds or beans are used to make coffee. Coffee ranks as one of the world's most valuable and widely traded commodity crops and is an important export product of several countries. Coffee is the second largest traded commodity in the world and is an extremely important foreign exchange earner. The coffee industry of India is one of the largest producer of coffee in the world. South India dominates the other states in coffee plantation. Hills of Western ghats receive plenty of rainfall during the monsoon season which makes these regions best places for coffee cultivation. The major coffee growing states are Karnataka (Coorg, Hassan, Chikkamagalur), Kerala (Waynad, Travancore, Nelliampathies), Tamil Nadu (Pulneys, Nilgiris, Shevroys/Salem, Annamalais/Coimbatore) andhra Pradesh (Araku, Chintapalli) orissa (Daringbadi), Himachal Pradesh and Assam. During the year 2013-14 India produced 3, 04,500 tonnes of coffee from an area of 3, 55,500 ha (Coffee Board, 2014).

3. NUTRITION AND HEALTH BENEFITS

3.1 Coconut

Coconut nutrition is vital for many people throughout the world. The people of India, Asia and throughout the tropics eat coconuts regularly. Also, the people in these cultures have special customs and ceremonies to honor them for all of their health giving benefits. Coconuts can be eaten in one of two stages: young or mature. The young ones are full of delicious water and soft, tasty meat. Mature coconuts contain less water but the meat is thicker and sweeter. Meat from mature coconuts also contains more fiber and oil. Without any question coconuts are one of the most nutritious foods on the planet and they possess powerful healing attributes that do wonders for our bodies (Anon., 2016b).

The following health benefits of coconut have been reported (Anon., 2016b):

1. **Coconuts are very high in vitamins, minerals and antioxidants**: Vitamins and minerals are always important and essential for a balanced system. Antioxidants help to reduce inflammation and fight stress in our bodies.
2. **Coconuts are high in fiber:** Coconut fiber is indigestible (insoluble) which means it contains no calories and the system doesn't break it down. It helps you to feel full and regulates bowel movements. Generally, it assists the digestive process by keeping the system clean and running smoothly. One of many great benefits of coconut nutrition!
3. **Coconut oil provides numerous healing benefits:** The oil can be used as a medicine to heal the body internally and externally. It's a powerful healing agent that has cleansing and restorative abilities.
4. **The oil is a natural digestive aid:** It's easy on the system because it naturally contains enzymes and therefore doesn't require any additional output from the pancreas. It's therapeutic for the digestive tract.
5. **Coconuts contain specific essential fatty acids that have potent *anti-microbial* properties:** These fatty acids are *lauric acid*, *caprylic acid* and *capric acid*. They control inflammation caused by any type of infection.
6. **Coconuts are healthy for you because they contain MCFA's:** These fatty acids are also referred to as medium-chain triglycerides. There are three types of fatty acids and they are: short-chain, medium-chain and long-chain. The body has an easier time breaking down short- and medium-chain fatty acids versus long- chain fatty acids. It all comes down to the size of the fat molecule. Larger fat molecules are harder for the body to metabolize. It doesn't matter whether you eat saturated or unsaturated fats in meat, eggs, dairy products or most oils, they all contain long-chain fatty acids (LCFA's). This type of fatty acid has been shown to increase cholesterol and is a risk factor for heart disease and other health conditions. In contrast, medium-chain fatty acids (MCFA's) actually help to normalize cholesterol and lower one's risk for heart disease (and other diseases).
7. **The MCFA's in coconuts promote absorption of nutrients from other foods:** By eating MCFA's with other foods you enable better absorption of nutrients from everything consumed. Coconut nutrition and its healthy fat profile is healing to the digestive system and because of this benefits the whole body. Nutrition from this fruit is extremely helpful for those suffering from chronic inflammation and pain. It can also provide relief from digestive issues like, constipation, diarrhea or cramping.
8. **Coconut nutrition increases metabolism and provides longer lasting energy between meals:** The type of fat in MCFA's bypass the liver and the pancreas. This

results in quick absorption into the bloodstream where the body converts them into energy. Long-chain fatty acids (LCFA's) don't do this. It takes a longer time for the body to process these types of fats. As a consequence, the LCFA's get stored more as fat which contributes to weight gain.

9. **Coconut nutrition is a great source of electrolyte replacement:** Coconut water is high in potassium but not as high in magnesium. The water contains other minerals as well which make it a good electrolyte replacement drink. It's great for everyone regardless of age or activity level. However, if you are someone who is active in sports, an outdoorsy type who sweats a lot or someone who works out often, then coconut water is especially helpful. It's completely natural with no added sugar and is a better choice than most other sports drinks or common beverages.
10. **No negative effects from heating coconut oil:** Unlike monounsaturated olive oil or all polyunsaturated oils that are susceptible to heat-induced damage, the oil from the coconut is highly recommended for cooking and baking because you can safely heat it up to 500 degrees.

3.2 Cacao

The cacao bean is full of antioxidants, fat, carbohydrates, protein, polyphenols like flavanoids that are antioxidants, minerals like calcium, magnesium, sulfur, copper, iron, zinc and potassium, oleic acid which is a heart-healthy essential monounsaturated fat, fiber and vitamins E, B2, B1, B5, B3 and B9. There are also small amounts of vitamin A and C in raw cacao. The maximum nutrients are found in raw cacao beans before any kind of processing. In a health hierarchy, raw cacao beans before processing rank first, followed by organic, cocoa powder (un-roasted) and organic dark chocolate which greater concentrations of cocoa powder and lower levels of processed sugar (Anon., 2016a).

Cacao has been dubbed the 'food of the gods' in the Americas where it was first cultivated. It is considered a 'superfood' due to the high density of essential nutrients, very few side effects and proven health benefits. The cacao beans are taken out and made into cocoa butter which is used to make toiletries, ointments and many pharmaceutical products. Inside each cacao bean is the cacao nib and this is used to make chocolate after they are dried, fermented and processed. All cacao products are said to be effective in treating many health conditions like lowering high blood pressure, high cholesterol, heart disease, liver, bladder and kidney disorders, asthma, diarrhea, preventing wrinkles, pregnancy stretch marks and more. Raw cacao nibs and beans offer the best health benefits since some of the nutrients are destroyed when they are processed into chocolate. They can be eaten raw, powdered and made into tea and included in various foods. There are also cacao extracts available (Anon., 2016a).

The following health benefits of cocoa have been reported (Anon., 2016a):

1. **Theobromine and antioxidants:** This is a mild, non-addictive stimulant found in raw cacao. It is believed to treat depression. Flavonoids which are antioxidants can help fight free radicals that can buildup due to exposure to environmental pollutants, sun exposure and cigarette smoking and this in turn reduces the risk of developing certain cancers. Both antioxidants and theobromine can improve blood flow, support healthy functioning of heart and reduce thickening of arterial walls

and plaque build-up. Theobromine can also improve mental and physical well-being by stimulating central nervous system, dilating blood vessels, increasing energy levels and relaxing tense muscles. Other benefits offer include reducing PMS and menopause symptoms. The levels of antioxidants in raw cacao are higher than that of green tea and red wine.

2. **Fiber content** in cacao can help digestion by increasing digestive enzyme production. Fiber content keeps you feeling full for longer. This controls appetite and helps promote weight loss.
3. **Essential fatty acids**, such as, oleic acid found in cacao can lower bad cholesterol and raise good cholesterol levels. Oleic acid also possesses antioxidant properties and can fight free radicals. It can strengthen cell membrane integrity, boost memory power, improve functioning of the heart, circulatory system and brain, increase energy levels, help with weight loss, reduce inflammatory conditions like arthritis, help to moisturize the skin and more. Minerals especially magnesium found in raw cacao promotes heart health and is essential for proper brain functioning. Tryptophan and anandamide are other brain-boosting minerals found in cacao.

Vitamins found in raw cacao benefit the cardiovascular system. They have antioxidant benefits and promote longevity.

Precautions: *Those who have chocolate allergies must avoid eating cacao nibs or beans. Polyphenols and other active substances found in them could trigger side effects like migraines and other allergic reactions.*

3.3 Oil Palm

Palm oil or palmolein is the oil produced from the oil palm/red oil palm tree. Palm oil is extracted from the pulpy portion (mesocarp) of the fruit of oil palm. The crude palm oil is deep orange red in colour and is semi solid at a temperature of 20 ^{0}C. Palm oil contains an equal proportion of saturated and unsaturated fatty acid containing about 40% oleic acid, 10% linoleic acid. 44% palmitic acid and 5% stearic acid. The unprocessed palm oil is used for cooking. Palm oil is a very rich source of beta carotene, an important source of vitamin A and it contains tocopherols and tocotrienols, a natural source of vitamin E. Vitamin A and vitamin E contents are the highest in palm oil in comparison with any other types of oil and hence, consumption of the same boosts health. By virtue of its high vitamins content, the palm oil is a nature's gift for the human beings. In view of the rich content of vitamins, palm oil can be utilized for the preparation of cosmetics as well (OPIL, 2014). Fats and oils account for 95% of a category of nutrients called lipids. More than 90% of the lipids in our body and food are simple lipids; the most common kind found in edible fats and oils. Palm oil is herbal supplement liquid that can be beneficial to our health and for the treatment of some medical ailments. Widely available in most local supermarkets, drug stores and online pharmacies, palm oil is considered safe enough for daily use due to the lack of documented harmful side effects. Palm oil supplements can serve as powerful antioxidants that are able to help rid body of harmful free radicals in the bloodstream. These free radicals, essentially harmful chemicals, can lead to an increased risk of developing heart disease and other cardiac conditions. Cardiac conditions like hardening of the arteries or atherosclerosis and other heart problems are preventable and treatable with herbal supplements like palm oil. Furthermore, this herbal supplement can lessen the sticky nature of blood platelets, thus lowering the risk of stroke or internal blood clots that can damage vital organs in the body.

One of the major benefits associated with palm oil is the fact that it is completely free of cholesterol. Palm oil supplements are considered part of the vegetable family and therefore do not contain cholesterol; this can prove to be beneficial to patients currently struggling with high blood cholesterol. As a result, our bodies are able to use this supplement more efficiently for the purpose of producing energy for everyday exercise and activity (Abraham, 2014).

3.4 Palmyrah

Palmyrah palm or sugar palm seeds or pulp are considered as a delicacy in India where the palmyrah trees are grown. During early summer, they are sold in markets and for a limited period of time. Summer holidays are enjoyed by preparing different types of drinks, pancakes and dishes prepared from the nutritious pulp. The tender fruit pulp is eaten raw during summer because it gives a cooling effect to the body. The ripened fribrous outer layer of the fruit can be eaten raw, boiled or roasted. The fresh ones are little crunchy and moderate sweet to taste. They are chopped to small pieces and used in desserts with other varieties of fruits, such as, pineapple, mango, papaya and offered along with some good flavoured ice cream. Matured palm fruit juice is used in making pancakes (Vivekananda, 2014).

The fruit contains vitamin A, B and also vitamin C in the form of ascorbic acid. It is rich in minerals too. Minerals, such as, potassium, iron, calcium, phosphorous and zinc are found in the fruit. If the palmyrah trees are grown in large scale commercial farms, it could help solve the world's malnutrition problem. Its medicinal uses are still unknown to people. The pulp from the mature fruit is used over the skin to cure dermatitis (Vivekananda, 2014).

The sap contains 0.23% protein, 12.6% total sugar and small amounts of fat, mineral water and vitamin C. A sugar content of more than 12.6% is observed and it ranges from 12 to 16%. The sap also contains most of the essential amino acids in the form of free amino acids. The sugar is mainly sucrose with traces of glucose and fructose, but during fermentation sucrose is hydrolyzed to glucose and fructose. 227 g or 8 oz of *neera* contains 28.8 g of total sugar, 35.4 g of calcium, 5.5 mg of iron, 32.4 g of phosphorous, 82.3 mg of thiamine, 44.4 mg of riboflavin, 12.2 mg of ascorbic acid, 674.1 mg of nicotinic acid and 49.7 mg of protein with a calorific value of 113.3 kcal (Kurian and Peter, 2007).

3.5 Arecanut

Chewing betel nut creates a short-lived feeling of euphoria, alertness and light-headedness when chewed. Many compare the feeling with ones gained from drinking coffee or smoking a cigarette. The degree of these feelings varies depending on how much we chew and also whether we spit or swallow the nut. Holding betel nut in our cheek like chewing tobacco also prolongs the effects. The sensations of betel nut may also be compounded with additives. Tobacco is the most notable one, but lime (calcium mineral, not the fruit) is another highly typical addition. Many additional spices, herbs and flavors get added to betel nut depending on the occasion and preference. Another noteworthy effect of the betel nut is its ability to stain the lips and teeth. Locals with exceptionally red teeth are a common sight throughout India and a sure-fire sign they have been indulging in *paan.*

Edible parts of betel nut per 100 g containing 76.4 g of water, 3 g protein, 14 g fat, 0.6 g alkaloid; its fatty oil containing 19.5% lauric acid, palmitic acid 12.7%, 1.6% stearic acid, oleic acid 6.2 %, 0.3% by acid, linoleic acid 5.4%, 0.3% dodecyl acid, acid 7.2% in the fourth carbon, *etc.*; betel nut contains free amino acids, more than 15% proline, tyrosine, more than 10% acid and arginine; In addition, it also contains betel nut red pigment, catechin and polymers, such as, flower IL (Anon., 2014d). The nuts also have medicinal and pharmacological properties. Betel nut chewing can increase gastrointestinal smooth muscle and enhance peristalsis and promote the secretion of gastric juice, it can help digestion, increase appetite. Peristalsis is a radially symmetrical contraction and relaxation of muscles which propagates in a wave down a muscular tube, in an anterograde fashion. Peristalsis is often found in the contraction of smooth muscle tissues to propel food chyme through a digestive tract, such as, the human gastrointestinal tract. The seed contains alkoids, such as, arecaidine and arecoline, which, when chewed, are intoxicating and slightly addictive. Areca palms are grown in India for their seeds. The seed also contains condensed tannins (procyanidins) called arecatannins which are carcinogenic. The extract of *Areca catechu* has been shown to have antidepressant properties in rodents, but it may be addictive.

3.6 Cashew

Mildly sweet yet crunchy, delicious cashew kernel is packed with energy, antioxidants, minerals and vitamins that are essential for robust health! Cashews are a popular ingredient in sweet as well savory dishes. Cashews are high in calories. 100 g of kernels provide 553 calories. They are packed with soluble dietary fiber, vitamins, minerals and numerous health-promoting phyto-chemicals that help protect from diseases and cancers. They are rich in "heart-friendly" monounsaturated fatty acids like oleic acid and palmitoleic acid that help lower LDL (low-density lipoprotein} cholesterol or bad LDL cholesterol and increase HDL (high-density lipoprotein) cholesterol or good HDL cholesterol. Research studies suggest that Mediterranean diet, which is rich in monounsaturated fatty acids helps to prevent coronary artery disease and strokes by favoring healthy blood lipid profile. Cashew kernels are very rich source of minerals. Minerals especially manganese, potassium, copper, iron, magnesium, zinc and selenium are concentrated in the kernels. A handful of cashew kernels a day in the diet would provide enough of these minerals and prevent deficiency diseases. Selenium is an important micronutrient, which functions as co-factor for antioxidant enzymes, such as, glutathione peroxidases, one of the most powerful antioxidant in the body. Copper is a cofactor for many vital enzymes, including cytochrome c-oxidase and superoxide dismutase (other minerals function as co−factors for this enzyme are manganese and zinc). Zinc is a co-factor in many enzymes that regulate growth and development, sperm generation, digestion and nucleic acid synthesis. Cashews are also rich in many essential vitamins, such as, pantothenic acid (vitamin B-5), pyridoxine (vitamin B-6), riboflavin and thiamin (vitamin B-1). These vitamins are essential in the sense that our body requires them from external sources to replenish and essential for metabolism of protein, fat and carbohydrates at cellular levels. In addition, the kernels are also containing small amount of zea-xanthin, an important flavonoid antioxidant, which selectively absorbed into the retinal macula lutea in the eyes. It is thought to provide antioxidant and protective UV ray filtering functions and helps prevent age related macular degeneration (ARMD) in the elderly (Anon., 2012a).

Cashew is a versatile tree nut. It is, in fact, a precious gift of nature to mankind. The cashew kernel is a unique combination of fats, proteins, carbohydrates, minerals and vitamins. Cashew kernel contains 47 *per cent* fat, but 82 per cent of this fat is unsaturated fatty acid. The unsaturated fat content of cashew kernel not only eliminates the possibility of the increase of cholesterol, but also balances or reduces the cholesterol level in the blood. Cashew kernel also contains 21 *per cent* proteins and 22 *per cent* carbohydrates and the right combination of amino acids, minerals and vitamins and therefore nutritionally, it stands on par with milk, eggs and meat. As cashew kernel has a very low content of carbohydrates, almost as low as one *per cent* soluble sugar, the consumer of cashew kernel is privileged to get a sweet taste without having to worry about excess calories. Consumption of cashew kernels does not lead to obesity and helps to control diabetes. In short, it is a good appetizer, an excellent nerve tonic, a stimulant and a body builder (Nayar, 1998).

3.7 Tea

Tea contains about 500 different organic compounds. Some of these organic compounds are nutrients that the body needs, such as, vitamins, protein, amino acids and lipid classes like carbohydrates and minerals. Many of these organic compounds have been found to have medicinal properties, such as, tea polyphenols, caffeine, lipopolysaccharide and more. Drinking tea could help reduce the risk of heart attack. Tea might also help protect against cardiovascular and degenerative diseases. The antioxidants in tea might help protect against a boatload of cancers, including breast, colon, colorectal, skin, lung, esophagus, stomach, small intestine, pancreas, liver, ovarian, prostate and oral cancers. But don't rely solely on tea to keep a healthy body — tea is not a miracle cure, after all. While more studies than not suggest that tea has cancer-fighting benefits, the current research is mixed. Tea helps fight free radicals. Tea is high in oxygen radical absorbance capacity ("ORAC" to its friends), which is a fancy way of saying that it helps destroy free radicals (which can damage DNA) in the body. While our bodies are designed to fight free radicals on their own, they are not 100 per cent effective — and since damage from these radical oxygen ninjas has been linked to cancer, heart disease and neurological degeneration, we will take all the help we can get. Drinking tea is linked with a lower risk of Parkinson's disease. When considered with other factors like smoking, physical activity, age and body mass index, regular tea drinking was associated with a lowered risk of Parkinson's disease in both men and women. Tea might provide protection from ultraviolet rays. The good news is that green tea may act as a back-up sunscreen. Regular tea drinking might also counteract some of the negative effects of smoking and might even lessen the risk of lung cancer. Tea could be beneficial to people with Type 2 diabetes. Studies suggest that compounds in green tea could help diabetics better process sugars. Tea can help the body recover from radiation. One study found that tea helped protect against cellular degeneration upon exposure to radiation, while another found that tea can help skin bounce back post-exposure. Green tea has been found to improve bone mineral density and strength. Tea might be an effective agent in the prevention and treatment of neurological diseases, especially degenerative diseases (like Alzheimer's). While many factors influence brain health, polyphenols in green tea may help maintain the parts of the brain that regulate learning and memory (Abraham, 2014). The tea leaves have been used in traditional Chinese medicine and other medical systems to treat asthma (functioning as a bronchodilator), angina pectoris, peripheral vascular disease and coronary artery disease. Recent medical research on tea (most of which has been on green tea) has revealed various health benefits, including anti-cancer potential, effects on cholesterol levels, antibacterial properties and positive effects for weight loss. It is considered to have many positive health benefits due to tea's high levels of catechins, a type of antioxidant. Among other interesting bioactivities, -catechin from *C. sinensis* was shown to act as agonist of PPAR-gamma, nuclear receptor that is current pharmacological target for the treatment of diabetes type 2. However, tea may have some negative impacts on health, such as, over-consumption of caffeine and the presence of fluoride and oxalates in tea (WIKI, 2014b).

3.8 Coffee

Coffee is not just black water. Many of the nutrients in the coffee beans do make it into the final drink, which actually contains a decent amount of vitamins and minerals. A cup of coffee contains, 6% of the RDA for pantothenic acid (vitamin B5), 11% of the RDA for riboflavin (vitamin B2), 2% of the recommended dietary allowance (RDA) for niacin (B3) and thiamine (B1) and 3% of the RDA for potassium and manganese (Gunnars, 2014). A stimulating aromatic drink is made from coffee seeds after drying, roasting and powdering. Coffee is moderate brown or dark greyish brown. In social gathering coffee is a refreshing drink. 100 million people are drinking coffee a day. Coffee is a beverage that will help fight depression. It is found to be good for our liver. Coffee is found to help lose weight. It is a legal performance boost for athletics. Caffeine, in particular, has been proven to increase the number of fatty acids circulating in the bloodstream. It may lower the risk of type II diabetes. It may lower the risk of Alzheimer's and Parkinson's disease (Abraham, 2014). Legend has it that coffee was discovered around 850 A.D. in Ethiopia by a goatherd who observed that his animals were unusually lively after eating bright-red berries. Inside those berries were the coffee beans that later went on a global journey. Now, with more than 400 billion cups consumed every year, coffee is the world's most popular beverage. Coffee has been studied for more than a century and claims a number of health benefits (as well as some risks). Your daily cup of coffee may reduce the risk of type 2 diabetes, Alzheimer's disease and more. It also packs a powerful punch of antioxidants. In fact, Americans get more antioxidants from coffee than from any other food or beverage. With just two calories per 8-ounce cup (no cream or sugar) and no fat, coffee is a pretty guilt-free way to boost your health. But don't overdo it. More than 2 or 3 cups daily may increase blood pressure, especially in those with borderline or high readings (Gabrick, 2009).

4. UNDER-NUTRITION IN MAJOR PLANTATION CROPS GROWING AREAS

Two important nutrition indicators are stunting and child mortality. Stunting is the most frequently used indicator for monitoring under-nutrition by measuring children under five years old and is a visible sign of chronic malnutrition (Source: WHO reference median: Percentage of children stunted is the percentage of children under five years who have a height-for-age below minus two standard deviations of the National Center for Health Statistics). When children are stunted, it means they are too short for their age as a result of a poor diet and poor health circumstances. Stunting levels of 25% or higher are alarming and were encountered in 18 out of 21 coffee producing regions analyzed in India, Vietnam, Indonesia, Kenya and Ethiopia. Stunting levels of that magnitude can only be addressed through interventions that are specifically aimed at reducing under-nutrition. In all countries but Vietnam, there are coffee-producing regions in which stunting levels top 40%. Under-nutrition also affects the immune system and decreases the defence of the human body to infectious diseases, making it an underlying cause of terminal infectious diseases. As many as 53% of children's deaths can be attributed to being under-weight, making child mortality an important indicator in measuring under-nutrition (Black *et al.*, 2003).

Child mortality refers to death of infants and children under the age of five. These rates are particularly high in Africa. In cocoa-producing regions in Ivory Coast, for example, 11-17% of children do not live to reach the age of five. Issues like stunting and child mortality can only be addressed during the first 1,000 days of pregnancy and in the first two years after birth. Furthermore, like physical consequences, if under-nutrition is not addressed within the first two years after birth, it leads to approximately 15% less cognitive capacity (Grantham-McGregor *et al.*, 1999). After two years, the damage done can no longer be 'repaired'. This is why interventions to stimulate nutrition intake always need to take into account young children and especially pregnant women.

Cocoa is cultivated in more than 40 countries across Africa, Asia and Latin America. Over 90% of global cocoa production is cultivated by an estimated 5.5 million smallholders and most live below the poverty line and lack access to nutritious food. It is reported that cocoa production statistics compared with nutritional data of the three major cocoa producing countries - Ivory Coast, Ghana and Indonesia demonstrates the consequences of under-nutrition for the sector. It outlines the possible interventions for combining Good Agricultural Practices with Good Nutritional Practices and shows the enormous potential for the cocoa sector. For the cocoa industry in particular, under-nutrition might lead to direct losses in cocoa productivity from,

1) reduced labor output and physical productivity due to illness, fatigue and other health related problems,
2) reduced cognitive development and educational performance due to malnutrition early in life, 3) losses in household resources from increased health care costs.

Additional productivity gains might be achievable in the cocoa sector if it invests in interventions to promote human nutrition security through a certification incentive (de Vries *et al.*, 2016).

5. CONTRIBUTION OF PLANTATION CROPS TO NUTRITION SECURITY

Nutritional security denotes the consumption and physiological use of adequate quantities of safe and nutritious food by every member of a family and encompasses the process of equitable distribution among members of households and communities. This calls for the consumption of a variety of food items which supply an entire range of essential macro and micronutrients. Fruits and vegetables are considered to be the richest sources of these essential and beneficial phyto-nutrients important for the human metabolism. Their large-scale production and consumption is, therefore, imperative for ensuring the nutritional security (Chadha, 2005). The first step any country should take to improve the micronutrient status of its population, is to "work on the food system" to ensure that it produces the right quantity and quality of foods to allow all people adequate access to micronutrient-rich foods (Ruel, 2014).

The plantation crops are cash crops which supplement and complement our nutrition security. By themselves, these cash crops and the related value chains have no direct effect on nutrition security for most of the plantation farmers and plantation workers. An effect on food and nutrition security for these people can however, be created by diversifying the crops that farmers and plantation workers grow. Qualitative and quantitative contribution made by these crops towards nutrition security is dealt with as under.

Food and nutrition security is an integral component of economic growth and development of the society. Usually the development will be measured in terms of economic and social indicators. Economic upliftment not only implies increase in income, but also the well being in terms of food and nutritional security of the communities. Household food security is a function of availability of food coupled with the purchasing power available with each household. Food security is defined as, "when all people at all times have both physical and economic access to sufficient food to meet their dietary needs for a productive and healthy life."

While formulating any research project to improve the food and nutritional security, it should encompass options of interventions for local production of food materials, additional income generation activities which can enhance the purchasing power for procuring other essential food items as well as the generation of knowledge for efficient utilization of available food items through proper processing and storage. The small and marginal coconut growers in most of the coconut growing areas are facing difficulties to sustain the livelihoods of their families from coconut-derived income alone. Apart from improving their income, it is equally important to enhance the food and nutritional security of their family. A concerted effort was undertaken with this purpose in strategically selected coconut-growing communities through a project network by the Central Plantation Crops Research Institute, Regional Station, Kayangulam, Kerala. The major objectives of the project were to improve food and nutritional security and income of the family members of the small and marginal coconut holdings through diverse interventions like nutrition gardening, livestock integration, product diversification and allied income generation activities along with nutrition education for proper utilization of the available food materials. Interventions like intercropping/ homestead gardening, livestock rearing and household level processing/off-farm activities along with nutrition education were proved as successful in improving food and nutritional security in terms of frequency and quantity of consumption as well as diet diversity in all the project areas. Despite all diverse interventions, the consumption of green leafy vegetables and milk and milk products were found to be below the Recommended Dietary Intake (RDI) levels. There is an urgent need to envisage an integrated nutrition and health program for all vulnerable groups, focusing on the role of gender. Community Based Organizations (CBOs) -wise analysis of the consumption patterns highlighted the need for location-specific interventions matching to the needs and preferences of the communities. This model strategy can be replicated in all coconut growing areas to test on how local communities can achieve higher on-farm income and food and nutritional security through capacity building and adoption of different cost-effective interventions as per the comprehensive strategic plans (Shanthakumari *et al.,* 2010).

In regions that produce cash crops like tea, cocoa and coffee, nutrition security is rarely on the agenda of decision makers. However, nutrition is key to triggering a positive effect on the health of farmers and plantation workers and on economic development and productivity. Nutrition security is different from food security as it is about a community's access to essential nutrients, not just calories. Zinc, selenium, iron and all kind of vitamins should be present in human diets. A growing export Market for cash crops is increasingly threatening such variety in diets for local farmers and plantation workers in developing countries. Under-nutrition in regions that primarily produce cash crops is among the worst in most developing countries, resulting in poor productivity levels, low school performance of children and a poor health situation for farming families. Problems of under-nutrition are intergenerational; when under-nutrition is not adequately addressed, children will grow into under-nourished adults and their children will experience the same problems. This affects not only families, but also local economic development. The World Bank estimates that under-nutrition causes losses in GDP of 2 to 3%. Promoting nutrition security offers several opportunities to stimulate economic and pro-poor development in cost-efficient ways. In practice nutrition security can be stimulated by supporting fruit and vegetable value chains; fortifying oil with micronutrients; creating awareness on the importance of eating healthy among consumers, *etc*. Diversification means engaging in the cultivation of fruits and vegetables, but introducing chickens, ducks or fish can also make a difference. This can be done on and around land where the cash crops are grown, as well as in home gardens (de Vries, 2014).

Research shows that a 1% increase in a person's height leads to 1.4% increase in that person's productivity, whereas, a 1% reduction in iron status (anaemia) leads to 1% reduction in productivity. In other words, there is a case to be made that investing in the nutrition status of the people who are at the beginning of the cash crop value chains will contribute to the increase of labour productivity and thus benefit the multinational companies that profit from the growing worldwide demand for cocoa, tea and coffee *etc*. And, more importantly, it benefits the farmers and their families who are the main beneficiaries. They can achieve a healthier diet that, among other things, increases their incomes and reduces the prevalence of anaemia. However, purchasing other goods (*e.g.*, staple crops like maize, wheat and rice) takes priority over consuming fruits and vegetables. This is because putting any kind of food on the table regularly is the first priority for low-income households and consuming staple crops has been a long-time habit. Eating staple crops is embedded in many cultures, as is growing them. Yet, farmers often grow staple crops without recognizing that markets for these commodities are regularly flooded, as the majority of farmers grow the same crops out of habit. This has a negative effect on farmer's income, but also limits the availability, for example, of fruits and vegetables. Limited availability and perishability of fruits and vegetables also negatively affects the accessibility of these commodities due to, for instance, the resulting relatively high prices.

By stimulating access to and increasing availability of micronutrient rich foods through diversification of value chains, the choice to eat healthier is promoted and also contributes to increased household income. Although this will not solve all the problems and constraints that farmers and their households face, the anticipated results would mean a major step forward in improving health and income. When more varied cultivation of vegetables and fruits is promoted along with nutritious food consumption, the empowerment of women and nutrition education, the health and income of farmers and plantation workers and their families will improve, as will productivity (de Vries, 2014).

The first intervention concerns the implementation of value chain diversification for local products. This can be achieved by first increasing local production and preserving and marketing of affordable nutrient dense foods (vegetables, fruits, animal-based food products) that are to be sold in local markets. The first intervention particularly aimed at plantations is encouraging 'backyard farming'. Plantation workers may not have as much land of their own as smallholder farmers, but they do have small plots surrounding their own houses. Encouraging plantation workers to produce vegetables and raise chickens in their backyard for their own consumption can make a difference in improving the household diet. The second intervention is increased nutritious food consumption through food diversification. This is done by promoting household food production that includes a diversified basket of nutrient rich crops including fruits and vegetables and animal-sourced foods, such as, eggs, chicken, fish and dairy products. The second intervention is strengthening the link with public health services for access to micronutrient supplementation. Promoting programmes in which plantation workers are provided with micronutrient supplements (vitamin A, iron, zinc and iodine) have already sparked positive results at plantation level for workers' health and productivity in countries like India. Plantation owners/management can cooperate with public health institutions. Diversification can be achieved with a minimal burden for the farmer. Trees to provide shade for cocoa trees can also produce fruits, such as, bananas and vegetables and can grow along fences or in otherwise unused corners of the farm. The third intervention is strengthening the role of women. Ensuring women are empowered to make household decisions is key to stimulating the spending of resources on improving nutrition for all family members at all stages of the lifecycle. Finally, the fourth intervention focuses on nutrition education and raising awareness about good nutrition, health and sanitation practices through education programmes. Together, these four interventions constitute a combination that benefits both good agricultural practices and good nutritional practices (Vries, 2014).

6. STRATEGIES TO STIMULATE PRODUCTION OF PLANTATION CROPS

The 21st century has presented humanity with an ominous reality. By 2050, we may face wide spread insufficiency of food, fiber and bio-energy to meet our needs. A rapidly growing global population, accelerating consumption, dietary shifts, climate change and other factors are driving unprecedented price volatility, resource shortages and other risks in supply chains. Assuming the projected population of India by the year 2050 as 1.62 billion, there is likely to be an increasing demand for plantation crops (coconut, cocoa, oil palm, palmyrah, arecanut, cashew, tea and coffee) products in the domestic and international market, owing to climate related production dips in many parts of the world. Further, cashew, coconut and cocoa demand in the domestic market is also predicted to increase due to life style changes coupled with increase in post harvest processing and product development activities.

The following Strategies are to be adopted for increasing the production:

The basic need for research on plantation crops has not decreased and challenges have increased rather than decreased in complexity given current consumer demands for affordability, safety and continuity of supply; increasing needs to achieve sustainable practices; and the need to deal with challenges arising from a more variable climate, the loss of productive soils through urban encroachment and the loss of low-cost labour.

There are, no doubt, many priorities that can be identified for plantation crops research.

The concepts of sustainability of plantation crops relate both to the use of resources, such as, water and nutrients, in a manner which considers future needs and the responsible use of compounds, such as, pesticides which will not compromise the quality of the environment (or the safety of the produce that is produced).

The most topical issue that is facing plantation crops production in almost all areas is the availability of water. Pressure on water resources for urban, industrial, recreational, conservation and other uses all appear to have higher priority within societies than the availability of water for plantation crop production. Current demands on water from both river and subterranean sources are outstripping availability resulting in attention being given to water harvesting options with on-farm dams, the use of waste water and to the use of more efficient application methods.

Research is urgently needed to resolve issues, such as, the development of drought tolerant crops, the management of crops under managed water deficits.

Integrated crop production systems have been shown to be effective on a number of crops for reducing the number of pesticide applications and for shifting to new control compounds that have less persistence in the environment and which are of lower overall toxicity.

India has produced 23 billion tonnes of coconuts from two million hectare area in 2013-14. Assuming the projected population of India by the year 2050 as 1.62 billion, the projected coconut demand for 2050 is predicted to be 45 billion nuts. As a matter of fact, coconut in future may experience a paradigm shift from the oil seed label, if promoted as food for nutrition, health care and environmental services to support the farming community. Moreover, the recent surge in export of coconut products and the rising demand for tender coconut in the country are noteworthy. In such a scenario, by 2050 the demand for coconut would be certainly more than the estimated figure. Therefore, it would be a challenge to meet the futuristic coconut demand, especially because of the scarce land, labour, water and energy resources at disposal. An appreciable growth in total factor productivity and appropriate capital substitution are the possible alternatives and to achieve this we must strengthen the traditional coconut based farming system through the use of modern research tools (CPCRI, 2015)

It is envisaged that the domestic cocoa sector will grow at a rapid pace due to increase in the cocoa consumption pattern in India due to growth of the chocolate industry. In India, cocoa is cultivated as an intercrop in coconut/arecanut gardens, mainly in the states of Kerala and Karnataka and its cultivation is gaining increasing popularity in the states of Tamil Nadu and Andhra Pradesh. At present, demand for cocoa beans far outstrips the local production, necessitating large scale imports to meet the national requirements for processing and confectionaries. Hence, there is wide scope for utilization of cocoa within the country and a scenario of domestic production surplus is not envisaged. India imports 40% of its needs and the demand is increasing @ 15% annually (Abraham, 2014).

At present, more than 15 industrial entrepreneurs and firms exist in the field and the present domestic cocoa production meets only about 40 per cent of the existing industrial demand of 30,000 tonnes of cocoa beans. Considering the market growth in the chocolate segment in India, which is about 20 per cent per annum, cocoa has a great potential to develop in future years. We can strengthen the transfer of technology activities to encourage cocoa intercropping in arecanut and coconut gardens to meet the future demand challenges regarding supply of cocoa. The area under cocoa is 7,100 ha with a production is 15,1000 tonnes of cocoa beans. The projected demand of cocoa during 2050 is 2.12 lakh tonnes. To achieve this target the production should grow at an annual growth rate of 7.68 per cent. Considering the market growth in the chocolate segment in India, which is about 20 per cent per annum, cocoa, has a great potential to develop in future years (CPCRI, 2015)

The country of 1.2 billion is currently the world's largest consumer of palm oil. India accounts for 16 per cent of global consumption of palm oil (50 million tonnes of palm oil) in 2014. The area under oil palm is 1,89,624 ha and production is 79 lakh tonnes of fruits (Majumdar, 2014). If concerted efforts are taken to bring one million hectare under oil palm in the potential areas, it will be possible to get 3-4 million tonnes of palm oil and 0.3-0.4 million tonnes of palm kernel oil by the year 2030 (Arulraj and Suresh, 2012).

India has an estimated population of 102 million palms, with highest population in Tamil Nadu (51.9 million) (Kurian and Peter, 2007). The estimated area may be over 8-10 lakh ha. Palmyrah as neera and jaggery along with its other products can fetch a very good demand in the country in the coming years. Therefore, the projected demand will be almost double the present production. Hence, the area under palmyrah should be increased to 20-25 lakh ha by 2050.

India ranks first in both area and production of arecanut with annual production of 6.32 lakh tonnes of nuts from an area of 4.53 lakh hectares. The estimated demand in the case of arecanut by 2050 is 1.214 million tonnes (CPCRI, 2015).

Cashew is presently grown in an area of 10.11 lakh hectares with a production of 7.53 lakh tonnes of raw cashewnuts. India requires about 1.3-1.4 million tonnes of raw cashewnuts to cater to the needs of large number of cashew processing units. In addition to the domestic production of about 0.75 million tonnes of raw cashewnuts annually, the balance of 0.60 to 0.70 milliion tonnes of raw cashewnuts is imported annually by India from African and South East Asian countries. The projected demand/ requirement of raw cashewnuts shall be at least 5.0 million tonnes by 2050. This target can be achieved by area expansion of cashew in many states, especially in non-traditional potential areas. Also the low yielding cashew plants of senile and non-descript origin need to be replaced by cashew grafts of high yielding varieties with adoption of improved production technologies. It is expected that India will be able to keep pace with the requirement of raw cashewnuts from cashew processing industries and will be self sufficient by 2030 in its raw cashewnuts requirement to meet both export and domestic requirement by the cashew processors.

Increasing productivity and expansion of area under cashew are the main strategies for enhancing raw cashewnuts production in the country. Increasing productivity is the main aim of research activities, such as, this Directorate of Cashew Research (DCR) and All India Coordinated Research Project on Cashew (AICRP-Cashew) Centers. More research effort is needed in order to generate cost effective cashew production technology. Various technological approaches, such as, high yielding varieties, high density planting, drip irrigation, optimum use of fertilizers *etc*., are being evaluated for their suitability in the various cashew growing tracts and are being recommended to enhance productivity (DCR, 2011).

With the efforts of Directorate of Cashew Research (DCR), Coordinating centers of All India Coordinated Research Project on Cashew (AICRP-Cashew) and State Agricultural Universities (SAUs), over 40 high yielding cashew varieties have been developed and released. Production of planting material of high yielding varieties through softwood grafting technique has been greatly successful. This has made tremendous impact in improving cashew productivity. Various improved cashew production technologies, such as, nutrient management, canopy management, intercropping, irrigation including drip irrigation, soil and water conservation techniques, high density planting system, pruning technique, rejuvenation of senile orchards, insect pest management *etc.,* have been developed and are available to farmers. Department of Horticulture / Agriculture of several cashew growing states are transferring the technologies through Central Sector Schemes including National Horticulture Mission (NHM). Due to the efforts of all the concerned both in research and development, the status of cashew has now been changed from a crop suitable for soil conservation and afforestation purpose into a commercial plantation crop (DCR, 2011).

Presently, India is one of the world's biggest tea producers (1,200 million kg). The world's population is set to grow by a third by 2050, increasing demand for food by up to 70%, according to the United Nations. And much of this population growth will be in rapidly-developing countries, such as, China and India - precisely those that produce the most tea (Anderson, 2014). Therefore, the expected demand for tea in India by 2050 will be about 2,000 to 2,100 million kg.

Presently coffee production in India is 3.04 lakh tonnes. Several studies had shown that global crop production needed to double by the middle of this century to meet demands from an increasing human population. Therefore, the expected production of coffee by 2050 will be 6.00 to 7.00 lakh tonnes.

7. STRATEGIES TO STIMULATE NUTRITION SECURITY IN PLANTATION CROPS GROWING AREAS

There are many opportunities, such as, supporting fruit and vegetable value chains; fortifying oil with micronutrients; creating awareness on the importance of eating healthy among consumers. There is, However, another option, *i.e.,* making cash crops nutrition-sensitive. Cash crops, like tea, coffee and cocoa, are sourced from developing countries and emerging economies and exported for consumption in high-income countries. By themselves, these cash crops and the related value chains have no direct effect on food or nutrition security, they only create an income that is very low for most of the farmers and plantation workers. An effect on food and nutrition security can however, be created by diversifying the crops that farmers and plantation workers grow. Diversification means engaging in the cultivation of fruits and vegetables, but introducing chickens, ducks or fish can also make a difference. This can be done on and around land where the cash crops are grown, as well as in home gardens. By introducing nutritious cash crops, they become so-called 'nutrition-sensitive cash crops'. One of the main reasons that nutrition-sensitive value chains are a remarkable window of opportunity relative to other interventions in food and nutrition security is that there are resources available within cash crop value chains through the involvement of large multinational companies. When these resources are partly invested in stimulating the nutrition security of farmers and plantation workers, the interests of different stakeholders in the chain can be served (de Vries, 2014).

Research shows that a 1% increase in a person's height leads to 1.4% increase in that person's productivity, whereas, a 1% reduction in iron status (anaemia) leads to 1% reduction in productivity (Behrman and Rozenzweig, 2001). In other words, there is a case to be made that investing in the nutrition status of the people who are at the beginning of the cash crop value chains will contribute to the increase of labour productivity and thus benefit the multinational companies that profit from the growing worldwide demand for cocoa, tea and coffee (Anon., 2010a, 2010b, 2012). And, more importantly, it benefits the farmers and their families who are the main beneficiaries. They can achieve a healthier diet that, among other things, increases their incomes and reduces the prevalence of anaemia.

A recent study in Nairobi showed that knowledge of the importance of eating healthy is increasingly present (Lans, *et al.,* 2012). However, purchasing other goods (*e.g.,* staple crops like maize, wheat and rice) takes priority over consuming fruits and vegetables (Ofwana, 2013). This is because putting any kind of food on the table regularly is the first priority for low-income households and consuming staple crops has been a long-time habit (Kamau *et al.,* 2010). Eating staple crops is embedded in many cultures, as is growing them. Yet farmers often grow staple crops without recognizing that markets for these commodities are regularly flooded, as the majority of farmers grow the same crops out of habit (Meridian Institute, 2012). This has a negative effect on farmer income, but also limits the availability, for example, of fruits and vegetables. Limited availability and perishability of fruits and vegetables also negatively affects the accessibility of these commodities due to, for instance, the resulting relatively high prices. By stimulating access to and increasing availability of micronutrient rich foods through diversification of value chains,' the choice to eat healthier is promoted and also contributes to increased household income. Although this will not solve all the problems and constraints that farmers and their households face, the anticipated results would mean a major step forward in improving health and income. When more varied cultivation of vegetables and fruits is promoted along with nutritious food consumption, the empowerment of women and nutrition education, the health and income of farmers and plantation workers and their families will improve, as will productivity. Results from programmes engaging in these activities are promising, as three recently published concept briefs by the Wageningen UR Centre for Development Innovation (CDI) and the Global Alliance for Improved Nutrition (GAIN) show.

7.1 Intervening in Food Systems

The first intervention concerns the implementation of value chain diversification for local products. This can be achieved by first increasing local production and preserving and marketing of affordable nutrient dense foods (vegetables, fruits, animal-based food products) that are to be sold in local markets. The second intervention is increased nutritious food consumption through food diversification. This is done by promoting household food production that includes a diversified basket of nutrient rich crops including fruits and vegetables and animal-sourced foods, such as, eggs, chicken, fish and dairy products. The third intervention is strengthening the role of women. Ensuring women are empowered to make household decisions is key to stimulating the spending of resources on improving nutrition for all family members at all stages of the lifecycle. Finally, the fourth intervention focuses on nutrition education and raising awareness about good nutrition, health and sanitation practices through education programmes. Together, these four interventions constitute a combination that benefits both good agricultural practices and good nutritional practices (Wiegers *et al.,* 2011).

As plantations supply a significant amount of coffee and tea in the world, two additional interventions that particularly target plantation workers are proposed. Addressing nutrition problems simultaneously increasing plantation workers' productivity will have a significant impact on raising worldwide production to meet growing demand. The first intervention particularly aimed at plantations is encouraging 'backyard farming'. Plantation workers may not have as much land of their own as smallholder farmers, but they do have small plots surrounding their own houses. Encouraging plantation workers to produce vegetables and raise chickens in their backyard for their own consumption can make a difference in improving the household diet. The second intervention is strengthening the link with public health services for access to micronutrient supplementation. Promoting programmes in which plantation workers are provided with micronutrient supplements (vitamin A, iron, zinc and iodine) have already sparked positive results at plantation level for workers' health and productivity in countries like India (Panhuysen and Van Reenen. 2012; Scholz *et al.*1997) Plantation owners/management can cooperate with public health institutions (Sarmishtha Biswas *et al.,* 2005). These interventions can be integrated in cash crop certification training programmes.

Organized farmer groups are the entry point for promoting household production of nutrient-rich crops and developing viable local value chains for animal-sourced food products. Farmer groups can also play a role in empowering women to make nutritious choices that benefit their families. In addition, farmer groups provide a platform for linking nutrition education to agricultural production (Dorp *et al.,* 2011).

The following four interrelated interventions are designed to achieve improved nutrition in the cocoa value chain. Together they constitute a smart combination of Good Agricultural Practices and Good Nutritional Practices:

1. **Nutrition sensitive value chains for local products:** building nutrition sensitive value chains beginning by increasing the local production, preservation and marketing of affordable nutritious, nutrient dense foods (vegetables, fruits, animal based food products) to be sold in local markets. Sustainable improvements in the food basket can best be brought about by commercially viable value chains with push and pull factors;
2. **Increased nutritious food consumption through food diversification:** promoting household production that includes: a diversified basket of nutrient rich crops, such as, fruits and vegetables and animal source foods, such as, eggs, chicken, fish and

dairy products;

3. **Strengthening the role of women:** ensuring women are empowered to make household decisions, especially to spend resources on improving nutrition for all family members at all stages of the lifecycle; and
4. **Nutrition education:** raising awareness on good nutrition, health and sanitation practices through education programmes (de Vries *et al.,* 2016).

The benefits of diversification of farm produce are well known in the agricultural context in developing countries. It, for example, serves as a safety net as farmers depend less on income from one single commodity. A recent study in Malawi shows that both the ecosystem and farmers' incomes benefit from crop diversification (Sieglinde S. Snapp *et al.,* 2010).

The proposition to make cash crops nutrition-sensitive through diversification is, however, still relatively new. Not much evidence is yet available of successful implementation. However, first indications are that the approach can indeed work out positively. It builds on the benefits of crop diversification in general with the extra benefit of having a positive effect on the availability and access to nutritious commodities for farmer households and their surrounding communities. In addition, the approach builds on earlier work of the promotion of backyard farming by Helen Keller International, including small livestock, which has proven successful in diversifying the dietary intake of household members in rural Bangladesh. In this project, results for women empowerment were particularly remarkable. Other positive results were increased incomes, an increase of vitamin A rich food sources and an improved food security situation for millions of people (Lora Iannotti *et al.,* 2009).

Recent farm visits showed that diversification can be achieved with a minimal burden for the farmer. Trees to provide shade for cocoa trees can also produce fruits, such as, bananas and vegetables and can grow along fences or in otherwise unused corners of the farm. Moreover, chickens and ducks fertilize the farms and catch mosquitoes and insects that could potentially harm crops. This provides farmers with sources of micronutrients, extra income in the low season for cash crops and a safety net in case of bad harvests. Increased labour productivity on plantations due to micronutrient supplementation. The key to success is making farmers and plantation workers aware of the benefits in combination with a low burden to implement changes. Of course, farmers and plantation workers cannot implement the described solutions on their own. A variety of challenges that frustrate access to healthy commodities as well as agriculture in general need to be addressed. They include limited access to quality seeds for vegetables, poor technology, high post-harvest losses and weak market linkages. Governments, both local and national, are key to creating an environment in which these issues can be addressed and for maintaining a positive attitude towards making the link between value chains and nutrition. For this reason, it is positive that governments in developing countries and emerging economies are increasingly acknowledging the importance of nutrition for the general development of their respective countries and regions. In the past few years, governments of countries like Ethiopia, Indonesia and Rwanda have committed to improving the nutrition status of their populations (de Vries, 2014).

On top of the above mentioned interventions, investments to produce better nutrition and achieve significant results in improved health and increased income are relatively low. This benevolent cost benefit ratio, in combination with an increase in productivity, makes it attractive for companies to get involved. The results of the analysis could encourage the cocoa, tea and coffee industries to invest in nutrition-sensitive cash crop chains in areas experiencing high rates of under-nutrition to prevent future generations of undernourished cash crop producers and plantation workers with decreased productivity (de Vries, 2014).

8. CONCLUSION

We cannot feed today's world with yesterday's agriculture. We certainly cannot feed the world of tomorrow with the agriculture of today. However, highly productive sustainable agriculture can feed a multiple of today's world population (Rudy Rabbinge, 2010). To meet the projected demand in 2050 there is an urgent need to develop appropriate production technologies with due emphasis on adaptation measures that are most likely to be effective in stabilizing yields grown under drought, flood and high temperature conditions in future climates.

Success in attaining higher production in the future scenario of worsening climatic conditions and severe constraints on the available natural resources base will depend on enhanced efficiencies, to attain higher yield per unit of input. This would require an eco-friendly knowledge-intensive agricultural revolution that will enable realization of dramatic yield increase commensurate with that realized as a result of the input intensive Green revolution of the 20th century. The emphasis therefore is to be on cultivation of diverse crop varieties that have enhanced efficiencies and hence, capable of producing 'more with less' in accordance with perspective of 'greener' crop production. This will of course require the re-orientation of many aspects of crop production systems with plant breeding and the cultivation of the resulting high yielding, well adapted, input use-efficient and resilient crop varieties constituting a major component of the interventions. Access to input support systems and infrastructure for processing and value addition and information and communication technologies for efficient market functioning are the constraints for functioning of the farmers' groups. Further, product diversification is to be encouraged in plantation crops to ensure competitiveness of these crops as compared to other crops (CPCRI, 2015). This would serve as a guide for developing technology led, demand driven and need based programmes. It will help in attaining sustainable cultivation of plantation crops for nutritional security and better livelihood option.

REFERENCES

Abraham, V.K. (2014). Role of Plantation Crops in Nutrition Security and Their Health Benefits. In: (Compiled by Prem Nath) The Basics of Human Civilization-Food, Agriculture and Humanity, Volume III: Agriculture. PNASF, Bengaluru and Scientific Publishers (India), Jodhpur. p. 235-252.

Amudhan, M.S., Begum, V.H. and Hebbar, K.B. (2012). A review on phytochemical and pharmacological potential of areca catechu L. seed. *Int. J. Pharm. Sci. Res.*, 3(11): 4151-4157.

Anderson, Richard. (2014). Tea industry's future depends on corporate collaboration. BBC News, 14 February 2014 . http://www.bbc.com/news/business-26123944.

Anon. (2012a). Cashew nut nutrition facts, medicinal properties and health benefits. http://www.nutrition-and-you.com/cashew_nut.html.

Anon. (2014b). Importance and Area of Plantation Crops. My Agriculture Information Bank. http://www.agriinfo.in/default.aspx?page=topic&superid=2&topicid=1338.

Anon. (2014d). All About Betel Nut in India. The Earth of India. 9 Oct 2012. http://theindianvegan.blogspot.in/2012/10/all-about-betel-nut-in-india.html.

Anon. (2016a). Cacao Nutritional Facts. The Superfoods, 2016. http://www.thesuperfoods.net/cacao/cacao-nutritional-facts.

Anon. (2016b). How Can Coconut Nutrition Benefit You. be-healthy-with-coconuts. http://www.be-healthy-with-coconuts.com/coconut-nutrition.html

Arulraj, S. and Suresh, K. (2012). DOPR Vision 2030. Directorate of Oil Palm Research, Pedavegi, Andhra Pradesh, India. p. 27.

Basu Majumder, A., Bera, B. and Rajan, A. (2010). Tea Statistics: Global Scenario. Ind. J. Tea Sci. **8** (1): 121-124

Behrman and Rozenzweig. (2001). The Returns to Increasing Body Weight. 2016 Social Science Electronic Publishing, Inc. PIER Working p. 1-52.

Black, R.E., Morris, S.S. and Bryce, J. (2003). Where and why are 10 million children dying every year?. *Lancet,* 361: 2226–34.

Coffee Board. (2014). Statistics on coffee. Coffee Board, Bengaluru. http://www.indiacoffee.org/contact-us.html.

Chadha, K.L. (2005). Horticultural Crop Production. The Horticultural Society of India, New Delhi. http://nutritionfoundationofindia.res.in/workshop_symposia/towards-national-nutrition-security-2005.pdf/Session%201_

CPCRI. (2015).Vision 2050 (Compiled and Edited by George V. Thomas, V. Niral, K.B. Hebbar, Kesavan Subaharan, Ravi Bhat and S. Jayasekhar). Central Plantation Crops Research Institute, Kasaragod, Kerala, India. p. 36.

DCR. (2011). Vision 2030. Directorate of Cashew Research, Puttur.

de Vries, Klaas. (2014). What about nutrition security ?. thebrokeronline Blogs. 09 January 2014.

http://www.thebrokeronline.eu/Blogs/Towards-a-food-secure-world/What-about-nutrition-security.

de Vries, K., McClafferty, B and van Dorp, M. (eds). (2016). Increasing cocoa productivity through nutrition. Global Alliance for Improved Nutrition. Eldis, 15-16 Sept 2016. http://www.eldis.org/go/home&id=65560&type=Document#.V8b3A5grLq8.

Dorp, M. van, Oenema, C.H. and Verdonk, I.D. (2011). Agriculture – Nutrition Linkages. Linking agriculture and food security to nutrition improvement. Wageningen Portals. November 2011.

Gabrick andrea. (2009).The humble coffee bean boasts plenty of health benefits -- and boosts the flavor of our chili recipe!. http://www.webmd.com/diet/features/does-coffee-have-nutritional-value-you-bet.

Grantham-McGregor, Sally, M.; Fernald, Lia C.; and Sethuraman, Kavita. (1999). Effects of health and nutrition on cognitive and behavioural development in children in the first three years of life. Part 1: Low birthweight, breastfeeding and protein-energy malnutrition. *Food Nutr. Bull.,* 20:76-99.

Gunnars, Kris. (2014). Why is Coffee Good For You? Here Are 7 Reasons. http://authoritynutrition.com/why-is-coffee-good-for-you/.

IHD. (2015). Indian Horticulture Database-2013 (eds. Mamta Saxena and Chander, P., Gandhi). Ministry of Agriculture, Government of India, Gurgaon.

Kamau, M., *et al.,* (2010). Consumption and Expenditures on Key Food Commodities and its Implications on Households' Food Security: the case of Nairobi. Expanding Kenya's Agricultural Competitiveness Market Access and Food Security: Research Findings and Policy Options. Nairobi, Kenya.

Kurian, A. and Peter, K.V. (2007). Palmyrah. In: Commercial crops technology. New India Publishing, New Delhi. p. 315-340.

Lans, C.J.M. van der, Snoek, H.M.; Boer, F.A. de, Elings, A. (2012). Vegetable chains in Kenya; Production and consumption of vegetables in the Nairobi metropolis. Research report. Wageningen UR centre for Development Innovation.

Lora Iannotti, Kenda Cunningham and Marie Ruel. (2009). Diversifying into healthy diets: Homestead food production in Bangladesh. https://books.google.ca/books?isbn=1136461469.

Majumdar, A. (2014). OPAE. DAC, MOA, GOI. *agricoop.nic.in/Kharif2011/JS%20Seed.ppt* · PPT file · Web view.

Meridian Institute. (2012). Science and innovation for African agriculture value chains: Maize value chain overview.

Nayar, K.G. (1998). Cashew : A versatile nut for health. In : *Proceedings of International Cashew and Coconut Conference. Trees for Life – the Key to Development* (Eds. CT Topper, PDS Caligari, AK Kulaya, SH Shomari, LJ Kasuga, PAL Masawe and AA Mpunami). BioHybrids International Ltd., Reading, UK. p. 195 – 199.

Ofwana. (2013). An Analysis of the Patterns of Food Consumption among Households in Kenya, *Journal of Emerging Trends in Economics and Management Sciences* 4(1): 111-113.

OPIL. (2014). Oil Palm India Ltd. http://kerala.gov.in/index.php?option=com_content&view=article&id=3467&Itemid=2531.

Panhuysen, S., Van Reenen, M. (2012). Coffee Barometer 2012, Tropical Commodity Coalition.

Rudy Rabbinge. (2010). Most crops perform 50% below their optimum In: Can we feed 9 billion people in 2050 in a sustainable way?.Nutreco. Feeding the future. A Nutreco Publication - March 2010. p. 26. http://www.nutreco.es/pdf/feeding_the_future_english.pdf.

Ruel. (2014). Micronutrients and improving nutrition through food systems. Jul 31, 2014 by IFPRI. http://www.ifpri.org/blog/micronutrients-and-improving-nutrition-through-food-systems.

Santhakumari Kalavathi, Vishuu Potti Krishuakumar, Regi Jacob Thomas, George V. Thomas and Maria Luz George. (2010). Improving food and nutritional security of small and marginal coconut growers through diversification of crops and enterprises. J.Agric. Rural Development in the Tropics and Subtropics, **3** (2):101-109.

Sarmishtha Biswas, Debasish Chokraborty, Sutay Berman, R.N. and Joshua Berman. (2005). Nutritional Survey of Tea Workers on Closed, Re-Opened and Open Tea Plantations of the Dooars Region, West Bengal, India. For Paschim Banga Khet Majoor Samity (West Bengal Agricultural Workers' Association), in association with the International Union of Foodworkers and the American Jewish World Service. AJWSnutritionreport.pdf.

Scholz, B.D. *et al.* (1997). Anaemia is associated with reduced productivity of women workers even in less-physically-strenuous tasks, *The British journal of nutrition* 1997;77:47-57.

Sieglinde, S. Snapp, Malcolm J. Blackie, Robert, A. Gilbert, Rachel Bezner-Kerr and George Y. Kanyama-Phiri. (2010). Biodiversity can support a greener revolution in Africa. Proc Natl Acad Sci U S A, 2010.

TB. (2014). Tea area as on 31-12-2013 & production in 2013-14. Tea Board, Kolkata. http://www.teaboard.gov.in/pdf/stat/Area.pdf.

Vivekananda. (2014). Sugar Palm fruit and its health benefits. http://vivekananda.hubpages.com/hub/Sugar_Palm_fruit_and_its_health_benefits.

Wiegers, E., Van Dorp, M., Torgerson, S. (2011). Improving nutrition through Agriculture.

WIKI. (2014b). Wikipedia 25.8.2014. *Camellia sinensis. http://en.wikipedia.org/wiki/Camellia_sinensis.*

ООО

24

Edible Mushrooms Towards Achieving Nutritional Security of Small and Marginal Farm Families

Meera Pandey and G. Senthil Kumaran

SYNOPSIS

Mushrooms are great recyclers and play a vital role in maintaining the health of our planet by returning the nutrients into the ecosystem. They also maintain human health by offering better nutrition and medicines. The scientific community in the developed world is beginning to understand this mycological power. However, in India, this inherent biological power embodied within the mycelial network of mushrooms remains a vast untapped resource. As the boundaries of scientific research shrink, ecologists, foresters, pharmacologists and mushroom growers can come together to do innovative research for the betterment of this planet. Mushroom biology has the inherent potential to address the three basic problems: shortage of food, diminishing quality of human health and pollution of the environment. At present, about 800 million people in the world are living in poverty. On the other hand, over 70% of agricultural and forest biomass has not been put to total productivity and have been discarded as waste. Mushroom technology is closely associated with three aspects of wellbeing- food, health and environment. It has the potential to convert lignocellulosic wastes into valuable human food and nutraceutical products, which have many health benefits. Further in the Indian context, mushroom cultivation can be integrated with social and rural upliftment policies of the Government like National rural employment guarantee scheme, mid-day meal schemes in schools, anganwadi schemes for rural women for creating rural employment, betterment of nutrition especially among the children and women. Hence, there is a need to diversify the Indian horticulture towards non green food resources and promote research in the areas of applied Mycology for a holistic horticultural/ agricultural growth. This article sums up in brief the multifaceted role mushrooms can play in the diversification of Indian horticulture. Although India is bestowed with all favorable conditions for successful mushroom cultivation (availability of agricultural waste, cheap labor and diverse climatic conditions), yet it has not been able to carve a niche for itself due to lack of awareness about this crop. The potential of mushrooms as a multifaceted crop can be utilized for food, medicine, diminishing environmental pollution, bridging the protein gap among vegetarians, mitigating malnutrition, creating self employment especially for women and develop rural – urban agri-network for rural employment, has largely gone unnoticed. Today when the Indian horticulture is looking for diversification, mushrooms stand tall as one of the most important biological components which can play a vital role in finding meaningful solutions to the problem of food, health and environment.

1. INTRODUCTION

India is an agrarian society where growing own food is a way of life. Culturally, agriculture has been considered by far the most respectable profession. This is evident by often quoted saying of the sage, Ghag Muni "Uttam Kheti, Madhyam Baan, Nishidh Chaakri, Bheekh Nidaan". However, in today's India, the very hands which till the soil often go hungry. According to 2011 census, agriculture is the largest private-sector enterprise in India, engaging nearly 119 million farmers ("cultivators") and another 144 million landless laborers. According to the National Sample Survey (NSS), small and marginal farmers had 2.5-5 acres and 0.05-2.5 acres respectively, as operational land or holdings in India. The proportion of marginal farmers has increased from 31% (1960-61) to 47% (2002-04) and the small scale farmers decreased from 16% (1960-61) to 11% (2002-03). Food and nutrition security is an important measure of development (Kalavathi *et al.,* 2010). Food and nutrition security can only be achieved by ensuring sufficient quantity and quality of food required for optimal physical and mental growth of individuals. Hence, food and nutrition security is a function of availability of sufficient amount of nutritious food through self production or having sufficient purchasing power to buy the same and can be effectively managed through the following:

(a) Creating diversified livelihoods options and income generation activities to enhance purchasing power

(b) Enhancing awareness and knowledge about nutrition and efficient utilization of available food items through proper processing and storage.

(c) Since women are the guardian of household nutrition; it is essential to emphatically include women at the forefront of technical, skill and knowledge empowerment for food and nutrition security.

(d) In this era of global climate change with low unpredictable yield of major edible grain crops, scarcity of water, land erosion and low productivity; there is a need to diversify agriculture and search for new sustainable technologies and novel food resources to fill the nutrition gap.

Edible mushroom can be an effective biological tool which can offer partial but meaningful solutions for nutritional security through (1) the generation of relatively cheap source of high quality food protein, (2) the provision of health-enhancing dietary supplements/mushroom nutraceutical and (3) the bioconversion/bioremediation of environmental adulterants for the maintenance of balanced ecosystems which will help in restoring and replenishing earths overburdened ecosphere.

2. WHAT IS MUSHROOM?

Mushrooms are amazing sculptors created by nature. Thomas carlyl rightly expressed this as "Nature alone is antique and the oldest art a mushroom". For thousands of years fungi in general and mushrooms in particular have evoked mixed response of adulation as expressed by John Ford "I am... a mushroom; On whom the dew of heaven drops now and then" to fear and loath. Culturally mushrooms have been an integral part of ancient Greek, Indian, Mesoamerican, Mayan and Egyptian cultures. Mushrooms have been used for food, medicine and euphoric experiences to poisoning tools through the usage of deadly

Amanita. Scientifically Mushrooms have been a part of fungal diversity for around 300 million years. The Western culture was largely ignorant about the usage of mushrooms and often termed as 'mycophobic'. Mushrooms were largely collected from wild for food and medicine, a practice still prevalent among the people living near forests. Wild mushrooms are an important component of forest products and global trade. The earliest record of mushroom cultivation dates back to A.D. 600, when the Jelly mushroom (*Auricularia auricula*) was first cultivated by Chinese on wooden logs. This was the stepping stone for the experiments to cultivate other wood inhabiting mushrooms like *Flammulina velutipes* (A.D.800); a delicacy in Japan till date and *Lentinula edodes* (A.D. 1000); the most sought after edible and medicinal mushroom of modern times.

Biologically the earlier classification of mushrooms was under lower plants due to their simple structure and some similarity to plants. As science and knowledge of fungi increased, their distinct features led them to be placed in a separate kingdom of their own called fungal kingdom. Mushroom is often defined as a macrofungus with a large fruitbody which can be above or below ground and large enough to be seen with naked eyes and to be picked by hand (Chang and Miles, 1992). Mushrooms like plant have varied features and uses. They can be delicious food (edible mushrooms), have healing properties (medicinal mushrooms), those having toxic effects (poisonous mushrooms), mushrooms producing hallucinations (Psychotropic mushrooms) and a large group of mushrooms whose properties have not yet been studied (Other mushrooms). Ecologically mushrooms can be grouped as saprophytes (growing on dead organic matter), mycorrhizal (in symbiotic association with trees) and parasitic (growing on other living organisms). Hawksworth (1991), estimated that about 1.5 million species of fungi inhabit earth out of which 1,40,000 species fall under the category of mushrooms (Hawksworth, 2001). Among the 14,000 known mushroom species, 7000 species are considered to have varying degree of edibility and more than 3000 species of mushrooms of 31 genera are known to be prim edible mushrooms. To date, 200 mushroom species are experimentally grown, 100 specie economically cultivated, 60 commercially cultivated and 10 have reached industrial status. Further, 2000 mushroom species are regarded as medicinal mushrooms with a variety of health attributes. The number of Poisonous mushrooms which is the major phobia for non consumption of mushrooms are very few (Approximately 1%) and some 30 species are considered to be lethal (Chang and Miles, 2004). The diversity of mushroom germplasm is enormous in tropical and subtropical regions and is least known. Macrofungi survive in diverse climate ranging from highly temperate arctic region (*Lycoperdon Molle,* Singh *et al.,* 2012) to highly tropical desert mushrooms found in Sahara and Kalahari (desert truffles, *Terfizia spp.*). Indian sub-continent with extremely diverse climate from temeperate– subtropical - tropical to arid harbours high variability of macrofungi.

The global mushroom production data (Table1) is 10.37 million tones with China contributing 73.57% of the world production. Although India's potential for mushroom production can be more than 7 million tonnes based on availability of diverse agricultural residue, cheap work force and suitable diverse climate; yet it contributes merely 0.27% of the global production.

Table 1. *Country wise mushroom annual production and its percentage contribution to the world production (2014)*

Country	Albania	Algeria	Armenia	Australia	Austria	Azer-baijan	Belarus	Belgium	Bosnia and Herze-govina	Brunei	Bulgaria	Canada	China	Croatia
A	100	2200	250	60023	1200	1500	6800	41754	1164	13	2239	102526	7634959	0
B	0.00	0.02	0.00	0.58	0.01	0.01	0.07	0.40	0.01	0.00	0.02	0.99	73.57	0.00
Country	**Cyprus**	**Czechia**	**North Korea**	**Denmark**	**Estonia**	**Finland**	**France**	**Germany**	**Greece**	**Hungary**	**Iceland**	**India**	**Indonesia**	**Iran**
A	694	531	5871	10113	130	1256	108540	59923	4480	22603	596	28000	37410	80239
B	0.01	0.01	0.06	0.10	0.00	0.01	1.05	0.58	0.04	0.22	0.01	0.27	0.36	0.77
Country	**Ireland**	**Israel**	**Italy**	**Japan**	**Jordan**	**Kaza-khstan**	**Kyrgyz-stan**	**Latvia**	**Lithuania**	**Luxem-bourg**	**Mad-agascar**	**Malta**	**Mongolia**	**Mon-tenegro**
A	69600	10000	600114	65811	833	513	227	598	12200	5	2171	1905	300	600
B	0.67	0.10	5.78	0.63	0.01	0.00	0.00	0.01	0.12	0.00	0.02	0.02	0.00	0.01
Country	**Morocco**	**Nether-lands**	**New Zealand**	**Philippines**	**Poland**	**Portugal**	**South Korea**	**Moldova**	**Réunion Island**	**Romania**	**Russia**	**Serbia**	**Singapore**	**Slovakia**
A	2050	310000	607	485	254224	1443	27130	2000	289	9758	7662	5005	80	2037
B	0.02	2.99	0.01	0.00	2.45	0.01	0.26	0.02	0.00	0.09	0.07	0.05	0.00	0.02
Country	**Slovenia**	**South Africa**	**Spain**	**Switzer-land**	**Thailand**	**Macedonia**	**Tunisia**	**Turkey**	**Ukraine**	**UK**	**USA**	**Uzbe-kistan**	**Viet Nam**	**Zim-babwe**
A	1081	17299	149854	8155	1000	2828	149	38767	6000	94857	432100	650	22000	663
B	0.01	0.17	1.44	0.08	0.01	0.03	0.00	0.37	0.06	0.91	4.16	0.01	0.21	0.01

WORLD TOTAL PRODUCTION , 2014 = 10378164 tonnes A = Annual production, Tonnes , B= Percentage contribution to world production ,

Source: *www.fao.org*

Among the global mushroom species, *Lentinula edodes* is the major species, contributing about 22% of the world's cultivated mushrooms. *Pleurotus* spp, including five to six cultivated species, contributes about 19% and *Auricularia* spp, including two to three species, contributes 17%, while *Agaricus bisporus* the common button mushroom accounts for 15% of the volume. The average global *per capita* mushroom consumption increased from 0.25kg to 5.0kg per annum in the last four decades. China has the highest consumption of 22kg *per capita* per annum and India's *per capita* consumption still remains very low at 0.08kg per annum (Fig. 1).

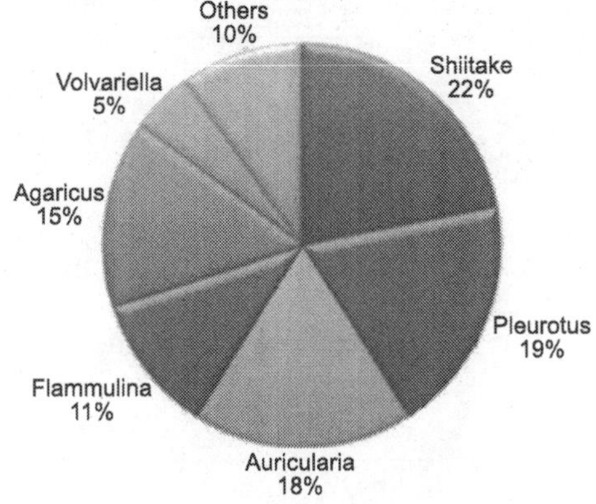

Source: *Royse et al., 2017*

Fig. 1. *World share of major cultivated mushroom species*

3. MUSHROOM TECHNOLOGY

Mushroom technology comprises of the following major components and its unit operations, which are required to make it suitable for entrepreneurship (Table 1).

Table 1. *Musuroom Technologies.*

Unit operations	Basic infrastructure	Suitability
1.Quality mushroom spawn production		
Grain cleaning	Permanent laboratory building and related equipments	Government agencies
Grain boiling		
Boiled grain and chalk power mixing and bagging		Entrepreneurs with investment capacity
Sterilization		
Inoculation		
Incubation		
2. Mushroom production		
Substrate preparation	(i) Structures with environmental conditions maintained naturally - low cost	Marginal Farmers
Substrate pasteurization/sterilization		
Bag filling and spawning	(ii) Permanent structures with artificially controlled environments, high cost	
Spawn running period		Entrepreneurs with investment capacity
Cropping		
Harvesting		

2a. Ready to fruit (RTF) bag production		
Substrate preparation	Basic equipments and space	Marginal Farmers Entrepreneurs with investment capacity
Substrate pasteurization / sterilization		
Bag filling and spawning		
3. Processing and Value addition		
Packaging, Dehydration, powdering, value added products and mushroom nutraceuticals	(i) Cottage type industries (ii) High tech Processing units	Marginal Farmers Entrepreneurs with investment capacity
4.Marketing as fresh and processed		
Local Market Export market	Required infrastructure or channel	Marginal farmers in a cooperative mode Entrepreneurs with investment capacity

Suitable varieties for Indian condition and cultivation method: There are numerous varieties of mushrooms which can be grown seasonally as per the natural conditions or under control climatic conditions (one variety throughout the year). The following table 2 shows the different varieties which can be grown in India.

Table 2. *Suitable mushroom varieties for Indian condition and their cultivation requirement*

	Oyster mushroom	Milky mushroom	Shiitake mushroom	Button mushroom	Paddy straw mushroom
Cultivation substrate	All agricultural wastes after pasteurization or sterilization	Paddy straw, wheat straw after pasteurization or sterilization	Sterilized and supplemented hardwood sawdust	Specially made compost	Paddy straw, cotton waste
Containers for growth	Polypropylene (PP) bags	Polypropylene (PP) bags	Polypropylene (PP) bags	Polybags, trays	Shelves
Seed rate	4-5% of wet substrate	4-5% of wet substrate	4-5% of wet substrate	2-3%	3-4%
Conditions for vegetative growth	Temperature 24-28°C Ambient humidity No light	Temperature 30-38°C Ambient humidity No light	Temperature 22-28°C Ambient humidity No light	Temperature 22-25°C Ambient humidity No light	Temperature 30-38°C Ambient humidity No light
Period of vegetative growth (days)	15-25 days (variety dependent)	30-40 days	70-80 days	30-40 days	10-15 days
Conditions for mushroom formation	Temperature 14-28°C (variety dependent) Humidity 80-85% 12 hours diffused natural or artificial light, Proper cross ventilation	Casing with pasteurized soil, Temperature 30-38°C Humidity 80-85% 12 hours diffused natural or artificial light, Proper cross ventilation	Cold water (12-15°C) shock treatment, temperature 20-28°C, Humidity 80-85% 12 hours diffused natural or artificial light, Proper cross ventilation	Temperature 15-18°C, Humidity 80-85% 12 hours diffused natural or artificial light, Proper cross ventilation	Temperature 30-28°C (variety dependent) Humidity 80-85% 12 hours diffused natural or artificial light, Proper cross ventilation

Period for pinhead initiation (days)	3-7 (variety dependent) after bag opening	5-10 days after casing	5-15 days after cold water shock treatment	5-7 days after casing	2-5 days after aeration
Period of mushroom harvest (days)	6-30 days (variety dependent) after bag opening	30-35	30-35	30-35 days	3-7 days
Total cropping period (seeding to last harvest) (days)	24-50 days (variety dependent)	60-70 days	100-110 days	60-70 days	12-15 days
Yield potential (biological efficiency, %)	50-100% of the dry substrate used (depends on growing and farm conditions)	50-100% of the dry substrate used (depends on growing and farm conditions)	50-100% of the dry substrate used (depends on growing and farm conditions)	15-30%	10-30%
Shelf life of fresh mushroom	2-3 days (depending on variety, harvest stage and packaging) at 25±2°C, 5-7 days in refrigerator and 10-15 days at 4°C.	2-3 days (depending on harvest stage and packaging) at 25±2°C, 4-6 days in refrigerator and 10-12 days at 18-20°C.	3-4 days (depending on harvest stage and packaging) at 25±2°C, 10-15 days in refrigerator and 18-20 days at 4°C.	2-3 days (depending on harvest stage and packaging) at 25±2°C, 7-10 days in refrigerator and 10-12 days at 4°C.	One day (at 25±2°C, can not be stored at low temperature
Post harvest management	Can be sold fresh or as dried mushroom or mushroom powder	Sold as fresh mushrooms	Can be sold fresh or as dried mushroom or mushroom powder	Fresh or canned	Fresh or canned
Spent mushroom substrate (SMS) usage	Excellent organic manure, can be used for vermi-composting, on farm multiplication of bioagent *Trichoderma* , as soil conditioner, as mulch	Excellent organic manure, can be used for vermi-composting, on farm multiplication of bioagent *Trichoderma* , as soil conditioner, as mulch	Excellent organic manure, can be used for vermi-composting, on farm multiplication of bioagent *Trichoderma*, as soil conditioner, as mulch	Excellent organic manure, can be used for vermicomposting, on farm multiplication of bioagent *Trichoderma* , as soil conditioner, as mulch	can be used for vermi-composting, on farm multiplication of bioagent *Trichoderma*, to grow oyster mushroom varieties

4. NUTRACEUTICAL POTENTIAL OF EDIBLE MUSHROOMS

Mushrooms are the vegetables that are full of nutrients and therefore can make a very valuable contribution to human nutrition especially in a country like India where the predominantly vegetarian population suffers from acute malnutrition. Edible mushrooms provide high quality protein that can be produced with greater biological efficiency than animal protein. Mushrooms are a rich source of proteins, (3-7%, Fresh weight basis and 25-40% dry weight basis). It can therefore be an important protein source among the vegetarians (Chang and Quimio, 1982). As a dietary source of proteins mushrooms are superior to most common fruits and vegetables with the exception of beans and peas (Chang and Miles, 1989). Mushrooms can be eaten fresh in salads or cooked unlike soya or yeast which have to be processed before they are accepted on the table. Mushrooms contain all the essential amino acids. Lysine is the most important amino acid in mushrooms which is low in cereals. Mushrooms rank very high in vitamin content, especially B vitamins as compared to most of the common vegetables eaten in the Indian diet. The riboflavin, thiamin, Vitamin B12 and niacin content of mushrooms is very high. Mushrooms contain high concentration of Ergosterol, Pro-vitamin D which can be converted to vitamin D with ultraviolet irradiation (Roberts *et al.,* 2008; Mau *et al.,* 1998; Koyyalamudi *et al.,* 2011; Holick 2010, Kalaras *et al.,* 2012). Mushroom is the only vegetable source of vitamin D. Exposure of fresh mushrooms to sun can also enhance the vitamin D content appreciably (Simon *et al.,* 2011). Mushrooms are a good source of minerals like potassium, calcium, phosphorus, iron and folic acid (Chang and Miles, 2004). Many varieties contain high amounts of antioxidants like ergothioneine too. Mushrooms are low in fat and carbohydrate and are low calorie food. They lack sugar and have glycemic index of less than 3 which makes them an ideal food for diabetics. At least 72% of the total fatty acids in mushrooms are unsaturated. This coupled with the presence of lovastatin and ergothioneine can help in reduction of cholesterol and become a healthy diet for heart patients (Schneider *et al.,* 2011). Due to its very low sodium content, it is recommended for lowering blood pressure and its high fiber content makes it an ideal food to reduce weight. (Mattila *et al.,* 2000; ISMS, 2008).

Mushrooms have been valued since millennia for their medicinal properties. Historically hot water soluble fractions (decoctions and essences containing mostly polysaccharides) from medicinal mushrooms were used as medicine in the Far East, where knowledge and practice of mushroom therapy primarily originated. Mushrooms, such as, *Ganoderma lucidum, Inonotus obliquus* had been historically used for medicinal purpose in Korea, Japan, China and Russia. About 700 species of mushrooms are known to possess significant pharmacological properties. Most of the mushrooms contain biologically active substances (BAS) contained in fruiting bodies, pure culture mycelia and culture filtrates. Polysaccharides represent the most important group of mushroom derived BASs. Soon after the report of anti tumor activity of polysaccharides derived from shiitake mushroom, three major drugs were developed from medicinal mushrooms. All three were polysaccharides specifically β glucans: Krestin from cultured mycelium of *Trametes versicolor*, Lentinan from fruiting bodies of *Lentinula edodes* and Schizophyllan from liquid medium broth of *Schizophyllum commune*. In 40 years since then, medicinal mushrooms have been intensively studied and many new anti-tumor and Immunomodulation substances have been identified and put to practical use (Mizuno *et al.,* 1995; Liu *et al.,* 1999). Apart from these, mushrooms also offer other potentially important therapeutic properties including antioxidants, anti-hypertensive, cholesterol lowering and liver protection, anti fibrotic, anti- inflammatory, anti- diabetic, anti- viral and anti- microbial compounds (Chihara and Maeda, 1969; Chihara *et al.,* 1970; Chihara *et al.,* 1987; Chiu *et al.,* 1998; Dennart and Tucker 1973; De Vere *et al.,* 2002).

A patent (PCT/NL2007/000086) was filed in March 2007 for the production of oral vaccine (immunogenic protein) through *Pleurotus* or S*chizophyllum* mushrooms against enterotoxin (LT) of *E. coli* or to a subunit of the cholera toxin (CT) of *Vibrio cholerae.* (Free Patent Online 2007). Due to their property of easy mass production on agricultural wastes, mushrooms are being studied for the production of vaccines and other therapeutic proteins to devise efficient and economical methods for their extraction and purification from mushroom tissue (PSU.edu., 2007).

Strategies for mass nutrition: Improving nutrition is widely regarded as the most effective form of aid. The methodology adopted is distribution of food or making the food available through cash vouchers. In any such strategy to tackle malnutrition, the mass availability of cheap source of protein and mineral rich healthy food is of utmost importance. Numerous grain based foods, food supplements, fortified foods etc., have been used depending on the type of malnutrition and the region and availability of food products. Mushrooms have long been consumed by man through collecting from forests. The development of commercial cultivation technology of many of the mushrooms renders it to become an easily available source of high protein. The most important aspect for its suitability under Indian conditions is the availability of colossal amounts of Agricultural residues which are burnt away. These agricultural residues can be profitably recycled through mushroom cultivation. Mass scale cultivation of mushrooms in the rural areas can become an effective tool to reduce malnutrition through direct intake of mushrooms in everyday diet or through enhanced income through sale of mushrooms to buy food. Although there are numerous varieties of mushrooms which can be cultivated, oyster mushrooms (*Pleurotus* and *Hypsizygus* species) are most suitable due to the wide range of agricultural residues on which they can be grown and under wide environmental conditions to which they can adapt. The technology of cultivation of oyster mushrooms is very simple, does not require very sophisticated facilities and can be grown from kitchen garden scale to large business scale. The yield of oyster mushrooms is high, they can be easily dried, stored and used as nutritionally rich supplements to fortify everyday diets of every region. Apart from consumption in the fresh form, oyster mushrooms can be easily sundried to make them vitamin D rich, can be easily powdered in the home blender and stored in airtight containers upto one year. The dried mushroom as such can be used in the form of vegetable after rehydration in warm water. The mushroom powder can be used for the production of mushroom fortified rasam owder, chutney powder etc. which are commonly consumed in every home. Addition of mushroom powder to these recipes enhances their nutrition manyfold and can be stored for a long period. These types of mushroom fortified recipes can be be easily adopted in mass nutrition programs. ICAR-IIHR has developed a series of products with dry mushrooms like 'Arka mushroom fortified rasam powder', five types of mushroom chutney powders, mushroom pulyogire powder which can be easily prepared at home.

India has numerous social schemes like mid day meals in schools, anganwadi schemes etc., through which mushrooms can be effectively brought into everyday diet of children. Dried oyster mushroom cubes and powder can be used to fortify any local meal without changing the original taste and flavor of the food. Both dried mushrooms and mushroom powder can be used with ease and in simplest methods of fortification for e.g. mushroom powder can be added to any vegetable while cooking so that the nutrition is enhanced.

Similarly mushroom powder can be added to *sambar, rasam* and other curries of everyday use. Fresh mushrooms contain 80-85% water. Hence, huge quantity of fresh mushrooms will have to be grown to get the dry product. This too can be solved at the village level itself by giving mushroom cultivation training to the women and the school children who can grow oyster mushrooms easily for their own requirement. Alternatively, it can be taken up as contractual farming at village level to supply the requisite quantity as required by the local schools in a particular area. This strategy will not only assure the constant supply of quality produce at reasonable price but also will create self employment at the village level and the agricultural wastes generated at the village level will be efficiently recycled. The Akshaya Patra Foundation runs the world's largest NGO-run midday meal program serving freshly cooked meals to over 1.2 million hungry school children in government and government-aided schools in India. Mushrooms can be made an integral supplement for fortification by such NGOs to mitigate malnutrition. Another advantage of supplementing with dry mushroom powder is that it does not change the aroma and flavor of the dishes to which it is added. Hence, it can easily be adopted by the people who have psychological reservation in eating fresh mushrooms (Table 3).

Table 3. *Statewise surplus Agrowaste available and estimated Oyster mushroom and protein production*

State	Ranking based on malnutrition index*	Surplus crop residue available (million tons /y)	Estimated mushroom production @ 50% conversion (million tons/y)	Estimated protein production through mushroom @ 2% fresh weight basis (million tons/y)	Estimated employment generation @ 150 mandays/ ton/y of mushroom production (million days/y)	Population (million)	Per cent of population whose daily protein requirement can be met per annum (%)
Haryana	5 (A)	11.34	5.67	0.11	850.72	25.35	24.86
Punjab	1(S)	24.84	12.42	0.25	1863.22	27.70	49.82
Rajasthan	7(A)	8.65	4.32	0.09	648.43	68.62	7.00
Gujarat	13(A)	9.09	4.54	0.09	681.41	60.38	8.36
Madhya Pradesh	17(EA)	10.33	5.16	0.10	774.70	72.60	7.90
Maharashtra	10(A)	14.79	7.39	0.15	1109.22	112.37	7.31
Karnataka	11(A)	9.03	4.51	0.09	677.04	61.13	8.20
Tamil Nadu	6(A)	8.90	0.45	0.01	66.75	72.14	0.69
Kerala	2(S)	6.35	3.18	0.06	476.41	33.39	10.57
Andhra Pradesh	3(S)	6.96	3.48	0.07	521.73	84.67	4.56
Chhattisgarh	14(A)	2.13	1.06	0.02	159.59	25.54	4.63
Orissa	12(A)	3.68	1.84	0.04	275.76	41.95	4.87
West Bengal	8(A)	4.30	2.15	0.04	322.61	91.35	2.62
Jharkhand	16(A)	0.89	0.45	0.01	66.75	32.97	1.50

Bihar	15(A)	5.15	2.57	0.05	386.04	103.80	2.75
Uttar Pradesh	7(A)	13.74	6.87	0.14	1030.34	199.58	3.82
Assam	4(S)	2.35	1.17	0.02	176.02	31.17	4.18

(**Source:** Menon *et al.,* 2009 and IISC, 2010)

(This table should go along with the recycling of waste – it has been given here in reference to protein malnutrition)

*Index explanation : EA: Extremely alarming A: Alarming S:Serious

5. MUSHROOM TECHNOLOGY FOR DIVERSIFIED LIVELIHOOD

Mushroom cultivation is a highly suitable diversified activity which can be easily integrated along with other agricultural activities. It can be taken up on any waste land, excess land not used for agriculture and courtyard of farm house by constructing a suitable shed using naturally available dry poles and covered with dry paddy straw and polythene sheet. The paddy straw or any dry agricultural residue derived from the field can be used as substrate. The quality spawn can be procured from nearby spawn laboratory. All the family members can be involved in mushroom cultivation in their free time/ leisure time in a day as it does not require any fixed time of operation. It is an unskilled activity which suits all type of labours. The farmer can acquire specialized skill in mushroom related activities and earn a livelihood by being employed in a mushroom farm.

A marginal farmer can grow mushroom in a make shift growing chamber of size 2m × 1m × 1m at an approximate cost of Rs. 10,000 with natural method of humidity maintenance. An approximate yield of 5 – 10 kg of mushroom can be obtained in a month leading to an additional income of Rs. 500 – 1000 per month. The selection of mushroom variety will depend on local weather conditions.

Potential of Agro residue utilization through mushroom cultivation: It is estimated that about 98.4 Mt (million tons) of crop residues are burnt per annum in India emitting major pollutants causing major health hazard causing respiratory diseases and other related ailments. Is there any alternative method to dispose farm residues without causing air pollution? Utilization of crop residues to cultivate mushrooms is a highly eco-friendly and viable alternative which can not only reduce air pollution and soil nutrient loss but also lead to the production of a highly nutritious and medicinal vegetable. Mushroom production through utilization of crop residues will also create employment especially in rural areas and for women. The spent mushroom substrate (SMS) generated after mushroom harvest can be utilized to produce high quality vermi-compost which will give organic carbon back to mother earth. The SMS can also form a very suitable raw material for industries involved in energy and alcohol production from biomass. The estimated potential benefits of utilizing just 10% of the surplus crop residues which are burnt; by growing mushrooms is as follows (Table 4).

Table 4. *Utilization of agro-residues.*

Surplus crop residues burnt in the country per annum	98.4 million tons per annum
10% of this crop residue	9.84 million tons per annum
Fresh oyster mushrooms produced through 9.84 million tons @50 % biological efficiency	4.92 million tons per annum
Employment generated @150 man-days per ton production	738 million man-days or 2.02 million people can be employed every year
Protein produced @ 2% of the fresh mushroom produced	98400 tons per annum
Spent mushroom substrate (SMS) available as organic manure @ 60% (minimum) of residue used	5.9 million tons per annum

Mushroom cultivation involves many activities like spawn production, substrate preparation, crop management, dehydration, product development and marketing.

All these activities require specific expertise, infrastructural investment and effective day-to-day management. The satellite model of mushroom production can help in sustainable production of large volume of quality and consistent mushroom through integration of small scale farms and creation of specialized mushroom production supporting services. Marginal and small scale farmers or women can become a part of satellite model of mushroom growing. Mushroom cultivation is better suitable to the rural areas due to

- Availability of cheap and ample raw material
- Availability of manpower
- Availability of space

Therefore, it is important to focus on the satellite-growing model where rural areas become the centre of production and urban areas are the hub for marketing and export. The satellite model requires the establishment of centralized energy and water efficient mechanized spawn production unit at district level which provides quality spawn and training to the. Spawn production can be the avenue for employment for many women and marginal farmers. At Taluk levels, units producing quality-pasteurized/sterilized substrate can be established which can be the avenue of employment as well as provide ready substrate to marginal farmers/women to cultivate mushrooms at a small scale in their homes. 'Alternatively Ready to fruit' (RTF) bag production units can also be established at taluk level to provide spawned (seeded) bags to women in villages to grow mushrooms to enhance household nutrition and supplement family income. The pasteurized casing medium producing unit can also be established at taluk level to provide ready pasteurized casing medium required for milky and button mushroom production. The produce from villages can be purchased from the farmer/women, pooled, packaged and marketed in the urban areas as fresh mushrooms or can be dehydrated either in sun or in dehydrators for the production of mushroom value added products like mushroom fortified *rasam* powder, *chutney* powder, mushroom *pappad* etc. Such processing units can be a source of employment especially for farm women. If done in an organized co-operative basis; fresh or dry mushrooms, mushroom value added products produced in the rural areas can be linked to the urban market by employing village youth who can be employed in such market linkage, transportation and supply management. In this way sustainable production can be made available to the market at all times. Since the individual grower will not be burdened with all the activities, there can be better and consistent production. By this system numerous small-scale growers can be developed and a consistent quality production and supply chain can be maintained. This kind of production can also be effective for export market where large quantity of qualitatively uniform produce at competitive cost is the pre-requisite.

Fig. 2. *Model for agro residue management*

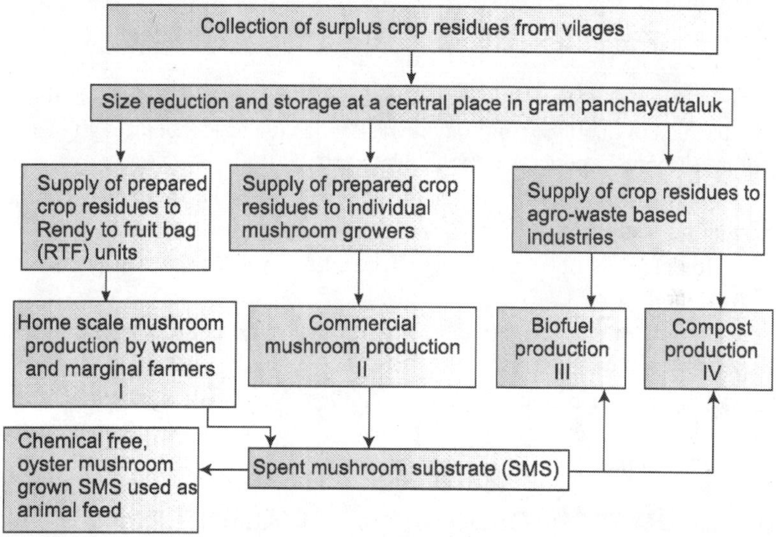

Fig. 3. *Benefits through the above model*

I	II	III	IV
1. Enhancement of household nutrition 2. Availability of fresh mushrooms for midday meal for village school 3. Supplementary income through sale of fresh or processed mushroom or mushroom products	1. Economic activity for self employment and generation of additional employment for marginal, small scale farmers, landless people, women, differentially abled people of the village 2. Enhancement of household nutrition 3. Availability of fresh mushrooms for midday meal for village school 4. Supplementary income through sale of fresh or processed mushroom or mushroom products	1. Economic activity 2. Employment generation 3. Betterment of environment through Prevention of pollution	1. Economic activity 2. Employment generation 3. Betterment of soil health 4. Promotion of organic farming

Net outcome of the above model

1. Efficient collection, storage and availability of biomass raw material for various economic activities like mushroom cultivation, compost preparation, biofuel etc.

2. Employment generation for landless and marginal farmers and unemployed youth through collection and transportation of the biomass.
3. Drastic reduction in environmental pollution which happens due to burning of such biomass wastes.
4. Villages will become clean (swatch gram).
5. Single window availability of large amount of raw materials for biomass based activities.
6. Sale of agricultural biomass like paddy straw etc., can bring in regular additional income to the farmer.
7. Creation of self employment in villages through mushroom cultivation and betterment of quality of mid-day meals of the village school through incorporation of mushrooms produced in the village itself.
8. Villages can have central composting yards to make their own compost in a scientific manner.
9. Surplus biomass not utilized at village/taluk level can be transported to district level warehouses.
10. Availability of biomass through a single window system for agricultural/industrial enterprises in peri-urban and urban areas.
11. Availability of nutritious and healthy food (mushrooms), employment, betterment of environment, energy, manure through efficient management of agricultural and forestry biomass.

6. MUSHROOM TECHNOLOGY AS A SOCIAL ENTERPRISE

Social entrepreneurship is a novel movement gaining momentum around the world. It is a novel concept of modern business model which can find sustainable solutions to social, economic and environmental issues. It is a concept which believes in engaging in profitable commercial activities for mutual social and community gain. It works on the principle "Together we win", Hence, community takes precedence over individuals. Social enterprises have a strong character of creating jobs and making a socially just and inclusive business model. Mushroom technology can be a very successful social enterprise in the Indian context where agricultural crop residues to the tune of 100 million tons/annum is burnt, where millions of youth are unemployed, where there is rampant undernourishment and ever increasing threat of climate change.

How Mushroom technology fits into the social Enterprise model: Mushroom technology with its multifaceted character has the potential of becoming a successful social enterprise especially in the Indian context due to the following reasons.

- **Multidimensional tool:** Mushroom production process involves numerous activities like spawn production, substrate production, crop production, packaging and processing, value addition, Marketing. Each of these activities has the potential to become a standalone commercial unit. When these individual units are integrated and networked, it can become a social enterprise which will be more sustainable as compared to individual units.

- **Environment and agro waste management:** As has been discussed earlier, mushrooms play a very important role in eco-friendly recycling of agricultural wastes leading to the production of highly nutritious and medicinal mushrooms and the spent mushroom substrate (SMS) after harvest can be used as animal feed or as organic manure.
- **Nutrition:** Mushrooms can play an important role in mitigation of protein and mineral (micronutrient) malnutrition.
- **Rural employment:** Flourishing rural areas are vital to regional and national development. Skills are central to improving employability and livelihood opportunities, reducing poverty, enhancing productivity and promoting environmentally sustainable development. India's 56% population is engaged in agriculture. Approximately two million people per annum migrate from rural to urban areas, of which about 1 million may be job seekers. Mushroom technology has the potential to create jobs in the rural areas which can reduce the migration to urban areas.
- **Women empowerment:** A sharp decline is being observed in the number of working women in rural areas. This is because of the dwindling number of jobs suitable for them. Women in rural areas, who are mostly burdened with responsibilities of running the home and taking care of the family, find it convenient to do part-time jobs near their homes. The most preferred form of employment would be farming related jobs. "However, the number of farming jobs has been shrinking, without a commensurate increase in other employment opportunities. Many researchers suggest that more than half of the decline in female labour force participation is due to the scarcity of suitable jobs at the local level. Mushroom technology being a women friendly soft technology can be a boon for rural women who can help by bringing additional income to the family without sacrificing their time for their family and children. When women take up mushroom cultivation, it has better chances of being used in the daily diet of their families which can also help in mitigating malnutrition.
- **Inclusive technology:** Mushroom technology is a simple technology with repetitive activities suitable for trainable physically challenged people especially in the rural sector where they are deprived of any skill or economical opportunity for equal participation. Mushroom technology has activities which with few physical, social and infrastructural modifications can be made suitable for special people and help in their socio-economic and psychological growth.

Fig. 4. *Mushroom technology as a social enterpreneurship model*

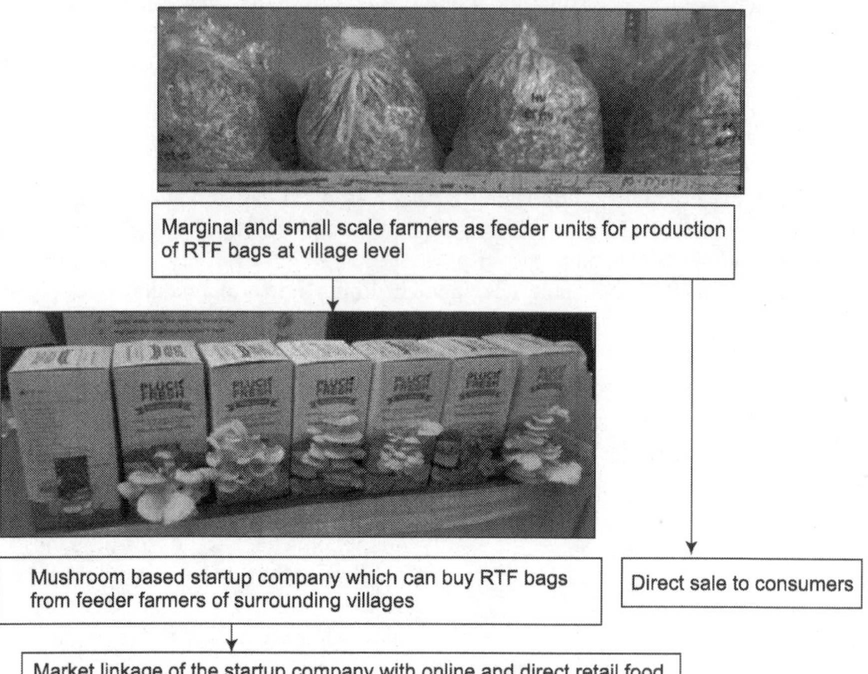

Net gain through the above model

1. Regular income to the marginal and small scale farmers
2. Employment generation
3. Regular availability of fresh, hygienic, chemical free, nutritious vegetable as mushrooms
4. Enhancement in mushroom consumption
5. Eco-friendly utilization of crop residues
6. Betterment of environment and cleanliness of villages.

7. WOMEN AS CATALYSTS TO IMPROVE HOUSEHOLD NUTRITION THROUGH MUSHROOMS

Women constitute 50% of the country's population and contribute significantly to 90% of the agriculture related activities at home or in industries. There is a natural multitasking demand from women at all levels for better functioning of family and society at large. Hence, it becomes important for women to choose and develop skills which are in harmony with their natural rhythm so that they can carry out their multiple tasks efficiently without stress. Mushroom cultivation is one such technology that has facets which can be harmoniously blended with socio-artistic and managerial skills of every rural and urban woman. Women are the guardians of family health which contributes towards a healthy nation. Mushrooms being highly nutritious can be incorporated in every family meal. This assumes special importance in Indian families which are predominantly vegetarian. Since mushroom is an indoor crop which can be grown with ease in any damp corner of the house or in small sheds, it can be taken up by women as kitchen garden activity. The concept of Ready to fruit (RTF) developed by ICAR-IIHR in 2014 has enhanced the household consumption of mushrooms both in urban and rural areas. RTF bags are ready seeded bags which are sold to housewives who can harvest mushrooms in 7-10 days after the purchase of such bags (Pandey, 2014; The Hindu, 2013). ICAR-IIHR in association with the NGO peoples trust has sold more than 5000 RTF bags during the past three years. This concept has helped in the inclusion of mushrooms in the diet of rural population of the villages of Sriramanahalli in Bangalore North Taluk. Efforts are on to disseminate this concept in the other taluks of Dodaballapur and Devenahalli in Karnataka. Growing oyster mushrooms at home and incorporating 100g oyster mushrooms per head per day in diet goes a long way in enhancing household nutrition.

Development of mushroom products and recipes is another highly suitable area for women to exploit their natural culinary talents and transform it into business. ICAR-IIHR, Bengaluru has been working on the development of healthy daily diet recipes from dry mushrooms which can also take care of the short shelf life of mushrooms. Mushroom fortified instant *rasam* powder is one such product which has been released by this institute for daily use. Products like mushroom powders, Mushroom rasam powder, Mushroom chutney powder, Mushroom pulyogire powder, Mushroom *pappad*, pickle etc., can be taken up as business venture. Production of such mushroom fortified products can become a home scale entrepreneurial activity for women SHGs who can make such products available for mass nutrition programs. In the present day era where people are conscious about what they eat, opening of exclusive mushroom eateries can be a novel business venture which can be taken up by women groups. Healthy recipes like mushroom sandwiches, mushroom rice, mushroom soups and curries can be marketed through such eateries. In today's fast life and a large population of upwardly moving middle class with women spending less time on elaborate cooking, simple home cooked healthy mushroom recipes will not only find many takers but will also aid in the perpetuation of healthy concept of eating.

8. CONCLUSION

Mushrooms are highly nutritious vegetables which can be grown on agricultural wastes in every household. The marginal and small scale farmers can be encouraged to grow mushrooms by utilizing their own farm waste and utilizing family labor for its cultivation. The other method could be to take up the technology on a community basis at village level which can make mushrooms available for consumption of every household in the village and also to create a business opportunity through urban market linkage to supplement income. It can be integrated into social schemes like mid-day meals for school children, NAREGA program for women and skill development program for the differentially abled people. Production of mushroom value added products like mushroom fortified *rasam* powder which can be a part of daily diet of every home can be an effective strategy for mass nutrition and avoid glut in the market. Mushroom technology gives diversified opportunities for enhancing nutrition and income. The challenge However, lies in formulating policies, strategies for agro residue management at national level with government, non government and private institutional participation to create awareness, skill empowerment, providing quality mushroom spawn (seed) at village level, providing processing facilities in rural areas and helping with domestic and export market linkage.

REFERENCES

Chang, S.T. and Miles, P.G. (2004). Mushrooms Cultivation, Nutritional Value, Medicinal Effect and Environmental Impact (2nd edition), CRC press, USA.

Chang, S.T. and Miles, P.G. (1989). Edible mushrooms and their cultivation, CRC press. Florida.

Chang, S.T. and Miles, P.G. (1992). Mushroom Biology – A new discipline, The Mycologist, 6, 64-65.

Chang, S.T. and Quimio, T.H. (1982). Tropical mushrooms-Biological nature and cultivation methods. Chinese University press, Hong Kong.

Chihara, G. and Maeda, Y. Y. (1969). Inhibition of mouse sarcoma 180 by polysaccharides from *Lentinus edodes*, Nature. 222: 687.

Chihara, G. Hamura, J. Maeda, Y. Y. (1987). Antitumor and metastasis-inhibitory activities of lentinan as an immunomodulator: An *overview*. Cancer Detection and Prevention. 1: 423–443.

Chihara, G. Hamuro, Maeda, J., Arai, Y.Y. and Fukuoka, F. (1970). Fractionation and purification of the polysaccharides with marked antitumor activity, especially lentinan, from *Lentinus edodes* (Berk.) Sing. (an edible mushroom). Cancer Research 30: 2776-278.

Chiu, J.H. Ju, C.H. Wu, L.H. Lui, W.Y. Wu, C.W. Shiao, M.S. and Hong, C.Y. (1998). *Cordyceps sinensis* increases the expression of major histocompatibility complex class II antigens on human hepatoma cell line HA22T/VGH cells. Am. J. Chin. Med. 26:159-70.

Dennart, G. and Tucker, D. (1973). Antirumor polysaccharide Lentinan, a T cell adjuvant. J. Natl. Cancer lnst. 51:1729.

DeVere White, Hackman, R.W., Soares, R.M., Beckett, S.E., L.A. and Sun, B. (2002). Effects of a mushroom mycelium extract on the treatment of prostate cancer. Urology. 60:640-644.

FAO. (1998). Mushroom cultivation for people with disabilities. FAO training manual, www.fao.org.

FAO. (2017). Statistical data on agriculture crop production. www.fao.org.

Free Patent Online. (2007). Immunization with transgenic mushrooms WIPO Patent Application WO/2007/111500 www.freepatentsonline.com.

Hawksworth, D.L. (1991). The fungal dimension of biodiversity: magnitude, significance and conservation. Mycological Research 95: 641–655.

Hawksworth, D.L. (2001). The magnitude of fungal diversity: the 1.5 million species estimate revisited. Mycological Research 105: 1422–1432.

Holick, M.F. (2010). The vitamin D deficiency pandemic: a forgotten hormone important for health. Public Health Reviews. 32: 267–283.

Indian Institute of Science (IISC). (2010). Biomass Resources Atlas of India 2000–2004. Available online: http://lab.cgpl.iisc.ernet.in/atlas/Default.aspx.

ISMS. (2008). Mushrooms and health global initiative bullettins. www.isms.biz.

Kalaras, M.D. Beelman, R.B. Holick, M.F. and Elias, R.J. (2012). Generation of potentially bioactive ergosterol-derived products following pulsed ultraviolet light exposure of mushrooms (*Agaricus bisporus*). Food Chemistry. 135: 396–401.

Koyyalamudi, S.R. Jeong, S.C. Pang, G. Teal, A. and Biggs, T. (2011). Concentration of vitamin D2 in white button mushrooms (*Agaricus bisporus*) exposed to pulsed UV light. Journal of Food Composition and Analysis. 24: 976–979.

Liu F. Ooi, V. E. C. and Fung, M. C. (1999). Analysis of immunomodulating cytokine mRNAs in the mouse induced by mushroom polysaccharides. Life Sci. 64, 12: 1005-11.

Mattila, P. Suonpaa, K. and Piironen, V. (2000). Functional properties of edible mushrooms. Nutrition. 16: 694-696.

Mau, J. L. Chen, P. R. and Yang, J. H. (1998). Ultraviolet irradiation increased vitamin D2 content in edible mushrooms. Journal of Agricultural and Food Chemistry. 46: 5269–5272.

Menon, P. Deolalikar, Anil. And Bhask, A. (2009). India state hunger index - Comparisons of Hunger across states http://motherchildnutrition.org/india/pdf/mcn-india-state-hunger-index.pdf accessed on 01-11-16.

Mizuno, T. Saito, H. Nishitoba, T. and Kawagishi, H. (1995). Antitumor active substances from mushrooms. Food Reviews International. 111: 23-61.

Pandey, M. (2014). Ready to Fruit (RTF) Bags- A Novel Concept to Enhance Household Nutrition through Mushrooms. Popular Kheti, 2:4, 7-8.

Pandey, M. and Veena, S.S. (2003). Mushroom cultivation-An appropriate technology for rural areas. Proc. International seminar on Downsizing technology for rural development. RRL, Bhubaneshwar 07-09 October. p. 202-209.

PSU.edu. (2007). Mushrooms may aid rapid vaccine response. *November 19, 2007 http://news.psu.edu.*

Roberts, J.S. Teichert, A. and McHugh, T.H. (2008). Vitamin D2 formation from post-harvest UV-B treatment of mushrooms (*Agaricus bisporus*) and retention during storage. Journal of Agricultural and Food Chemistry 56: 4541–4544.

Royse, D. J. Baars, J. & Tan, Q. (2017). Current overview of mushroom production in the world. In D. C. Zied (Ed.), *Edible and medicinal mushrooms: Technology and applications*. New York: John Wiley & Sons.

Santhakumari Kalavathi, Vishnu Potti Krishnakumar, Regi Jacob Thomas, George V. Thomas & Maria Luz George. (2010). Improving food and nutritional security of small and marginal coconut growers through diversification of crops and enterprises.

Journal of Agriculture and Rural Development in the Tropics and Subtropics, Vol. 111 No. 2, 101-109.

Schneider, I. Kressel, G. Meyer, A. Krings, U. Berger, R.G. and Hahn, A. (2011). Lipid lowering effects of oyster mushroom *(Pleurotus ostreatus)* in humans. Journal of Functional Foods 3: 17–24.

Simon, R.R. Phillips, K.M. Horst, R.L. and Munro, I.C. (2011). Vitamin D mushrooms: comparison of the composition of button mushrooms (*Agaricus bisporus*) treated postharvest with UVB light or sunlight. Journal of Agricultural and Food Chemistry 59: 8724–8732.

Singh, P., Singh, A., M. D'Souza, L., Roy, U. & Singh, M.S. (2012). Chemical constituents and antioxidant activity of the Arctic mushroom *Lycoperdon molle* Pers. Polar Research, Vol 31 (2012) incl supplements.

The Hindu, August 07, 2013. Buy it, water it and it's ready to cook.

25

Nutrition Awareness Through School Children: Experiences and Lessons Learnt

B. S. Prabhakar and Prem Nath

SYNOPSIS

The WHO and FAO of the UN have been striving through its theme 'Working together for health', to create awareness about nutrition and health care, especially among the low– and middle–income countries. Providing nutritional security through addition of fruits and vegetables with regular food is yet another effort to better health care. Dr. Prem Nath Agricultural Science Foundation (PNASF), Bengaluru has taken up a novel project on "Nutrition for and through School Children" with an objective of creating awareness about nutrition among school children, teachers and parents of the children and through this task force in the neighborhood.

The study conducted for three years from 2007-08 included six sets of students, parents and teachers, one who participated in the experiments and another control set. The study was conducted in a government school during the first year (2007-08) and it was studied in the second year (2008-09) in the same Government school and a private management school in addition for comparison. The level of nutrition knowledge improved significantly among the students who have participated in the experiment, but that of teachers, parents and others from the control group did not show any marked improvement.

Feedback from the schools indicates sincere appreciation of the nutrition education, but yet there is difficulty in incorporating such programmes in the routine syllabus. The PNASF, having committed to improve the nutrition and health security of the poor communities would like to work with National Council for Educational Research and Training (NCERT), New Delhi and similar institutions at the state level to develop lesson plans including Food and Nutrition in Biology for school children from 6th to 10th or 12th standards.

During the study in the first two years it was found that iron deficiency was one of the major dietary deficiencies. Therefore during the third year (2009-10) a study was taken up in a Government school to supplement iron nutrition through snacks using locally available inexpensive source of iron nutrient in *Shepu* (dill) leaves. Feeding children with iron rich food products proved more acceptable and economical compared to supplementing with tablets.

Keywords: nutrition security, nutrition awareness, school garden, nutrition garden, school children

1. INTRODUCTION

"*When diet is correct; medicine is of no need and when diet is wrong, medicine is of no use – AyurVeda*" The ancient knowledge from *Vedas* emphasizes the need for proper diet for healthy development of human beings. More than 10 million children die and more than 20 million children suffer from a handful of preventable and treatable disease conditions every year. Nearly all of these occur in low and middle-income countries and mostly among the poor, both in urban and rural areas. About two billion people, especially women and children, are suffering from nutrient and vitamin deficiencies. India's Ninth Plan document refers to the consumption of green leafy vegetables (Major inexpensive source of vitamins and minerals) that was one third of Recommended Dietary Allowance (RDA) among low-income group (LIG) in urban and rural areas. Vitamin A intake remained far below RDA because of poor bio-availability from Indian diets, as an effect, anemia continues to remain a public health problem. Energy consumption in laborers (Urban and rural low income groups) is low affecting their physical performance. National Nutrition Monitoring Bureau (NNMB) reported that overall prevalence of anemia is 67% among preschool children; 69% among 12-14 year old adolescent girls; 75% among pregnant women and 78% among lactating women (Nath *et al.*, 2009).

2. SCHOOL FEEDING PROGRAMMES

A study by UNESCO reported (Nath *et al.*, 2009) that iron deficiency (anemia) among school children is an impediment to learning which is likely to affect the level of alertness of children, which in turn affects attention and, therefore, learning. Children with low calorie intake over long periods of time reach a state of energy imbalance through reduction in activity level that represents a loss of significant opportunities for learning. A School Feeding Programme (SFP) and non - SFP in Haiti, covering 100 schools and 1936 primary school children was studied (Nath *et al.*, 2009). It was noted that there was strong correlation between home environment and attendance in both SFP and non-SFP schools. After a SFP operating in Dominican Republic with commodity support from the USA was suddenly terminated, a sample survey of the teachers revealed that enrolment had been adversely affected. The termination of the programme appeared to be much greater in rural schools and also on girl students.

School going child's nutritional status exerts significant influence on his/her academic performance; current diet is the single most significant predictor of class-room achievement; hunger cause inattentiveness, distractibility and SFPs that are successful either in reducing a child's feeling of hunger or improving his nutritional status are likely to facilitate cognitive development that is, (1) Mobilization and maintenance of attention, (2) Development of sensory - integrative capacity, (3) Exploratory, problem solving behaviors; memory.

School Feeding Programmes at primary schools have always been the most common way of using food aid for the food and nutrition security of school children. In 1992, these programmes accounted for more than two-third of all World Food Programme (WFP) assistance for human resource development, in many countries, school canteens have been operating with WFP support since 1960s (Nath *et al.*, 2009).

Children whose cognitive development during the first years of their life has been impaired by malnutrition and who live on one meal per day are more affected than better-nourished children. This is, unfortunately, true of many children in the developing world. A school meal can only address the immediate food needs of a child and would have to be accompanied by other interventions to have an effect on correcting malnutrition. The Chinese saying "Teach him how to fish, than give him a meal with fish" is opt to cite here as it is sustainable to provide the knowledge and skill of growing fruits and vegetables in schools and homesteads to supplement the School Feeding Programme. The awareness gained by this nutrition education is expected to trigger and encourage homestead gardens that provide clean ambience and nutritious food supplement.

Data on the influence of fruits and vegetables on providing nutrition and health care of poor children around the world is scanty. The available data had been more focused on calorie and protein nutrition. Average intake of fruit and vegetable in India is far below (3.5 servings or 280 g/person/day) the recommended dose of (5.0 servings or 400 g/person/day) and it is further low among students (2.94 servings or 235 g/person/day) (Singh, 2017) and this is true across the age groups and income groups studied (Figure 1). The situation is likely to be similar in the Least Developed Countries (LDCs) in Asia and Africa.

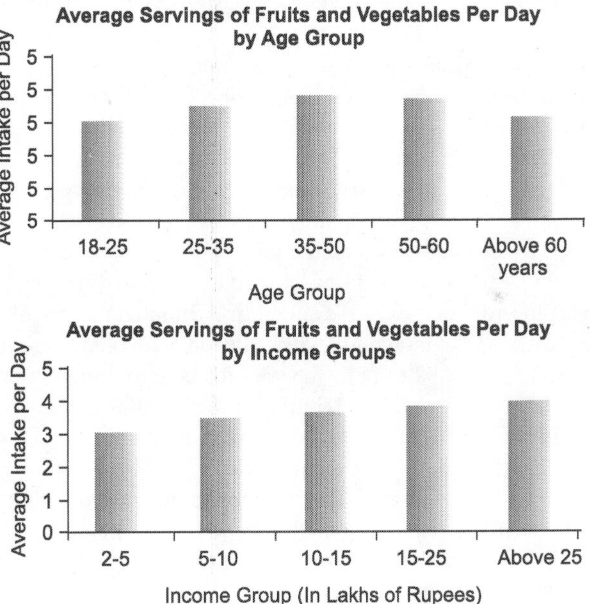

Fig. 1. *Average intake of fruits and vegetables in India (Pandey, 2016)*

Fruits and vegetables are important in children's diet because they provide vitamins, minerals, fibber and several phyto-chemicals necessary for growth, development and maintenance of health. In a country marked by malnutrition and under nutrition, fruits and vegetables have an important role as "Protective Food". School children need plenty of good food, as their energy and nutrient needs are particularly high in relation to their size. In their overall diet, it is often difficult to achieve adequate intake of energy, vitamin A, calcium, iron, zinc and iodine. Parents, teachers and school catering staff need to ensure that children receive plenty of nutrient-rich foods. Research supports the role of fruits and vegetables consumption in prevention of cancer and heart diseases. The protective role of fruits and vegetables in strokes, cataract, chronic obstructive pulmonary disease, diverticulosis and hyper tension is well documented (Mc Aleese and Rankin, 2007). Eating protective food from childhood takes them to a healthy adulthood (Eboh and Boye, 2006). A study reports that consumption of fruits and vegetables, as a habit in childhood, is an important predictor of higher fruit and vegetable consumption as adults and can help to prevent or delay chronic disease conditions (Heimendinge and Van Duyn, 1995).

Considering the facts in the foregoing, the PNASF, Bengaluru, has initiated a pilot project to promote nutrition awareness and gardening skills among school children in the first year 2007-08. Model Higher Primary School, Geddalahally, (Govt. school) and added during the second year (2008-09) Daffodils School (Private school), at Bengaluru, in the programme by providing financial assistance and expertise in horticultural practices.The study, particularly practical sessions on gardening, was carried out on vegetable crops to start with as the space available was limited as is the case in cities like Bengaluru. However, a garden in a school, depending on the space available, should include both fruits and vegetables so that a student gains basic knowledge about different plants that play important role in human nutrition.

The programme is learner-friendly and is applicable universally irrespective of nationality, language and wealth distribution among the communities. Teaching school children in their teen age ensures better reception, sincere commitment to the cause and also helps them to grow as responsible citizens contributing immensely to the welfare of the community. They belong to impressionable age group and good food habits can be easily be inculcated that they carry throughout their lives and also help to change the mind set of those in the families as well as community as a whole. Children are the potential ambassadors to bring change in the mindset of the families, bring in better awareness and concern about the role of protective foods like fruits and vegetables to correct the imbalances in the diets of the poor and less privileged. The hands-on-experience gained by the little soldiers is expected to pave the way for better consumption of these health foods and helps to improve the income security of the families as well as the community by triggering production, distribution and consumption in the community. This way, it helps to achieve the Millennium Development Goal (MDG) – improving the health of mothers and children and also helps to improve immunity and tolerance to diseases that are causing health hazards in the developing nations. Therefore, to start with the present curriculum in schools was examined to find out the details regarding the lessons on food and nutrition covered in the science syllabus.

3. LESSONS RELATED TO FOOD AND NUTRITION COVERED IN THE PRESENT CURRICULUM

Class room lessons related to food and nutrition of Central Board of Secondary Education (CBSE) and Karnataka State Syllabus are given here under:

3.1 Karnataka State Syllabus

8th Standard: Food and its constituents, general functions of food, constituents of food, energy giving foods- carbohydrates, lipids, proteins, vitamins, minerals, roughages, water, tests for major constituents of food- test for glucose, starch, biuret test for proteins, emission test for fats and oils.

9th Standard: Importance of food, nutrient composition, carbohydrates, proteins, lipids, vitamins, minerals, water and fibber, Nutrient deficiency disorders, causes and symptoms. Malnutrition, food spoilage, perishable and non-perishable food, food preservation, food adulteration.

10th Standard: Food plants: Vegetables, oil yielding plants, fruits, food grains, beverages, Food preservation, adulteration.

3.2 CBSE Syllabus

3rd Standard: Food for energy, growth and health, growing food, from field to market, milk, eggs, meat and fish, eating well.

4th Standard: Nutrients, carbohydrates, Fats, proteins, vitamins, minerals, water, healthy diet, healthy cooking, healthy habits.

5th Standard: Producing and storing food, - ploughing, improving fertility (manures and fertilizers), sowing seeds, irrigation, protecting the crops, harvesting, storing, canning.

6th Standard: Food plant and animal food, components of food – tests for sugars, fats and protein, functions of nutrients, balanced diet, deficiency diseases.

8th Standard: Crop production and animal husbandry.

9th Standard: Improvement in food resources – crops and animals while it starts from 3rd standard itself in CBSE syllabus. The subject covered is almost similar, including sources of food, constituents, nutrients.

The perusal of the above indicates that the mention about food and nutrition starts from 8th standard in the state syllabus ent deficiencies, health disorders, preservation, food adulteration and a brief mention about food production.

The coverage on the subject seems equally distributed among several other topics of science. Advances in science take place at a rapid rate as scientists are actively involved all over the world and more topics are to be covered in the school syllabus so that the student is well informed by the time he/she completes the school final (10th or 12 th Standard) examination .It is logical to touch upon all possible areas of science to impart a comprehensive knowhow of the vast subject like science.

But, the knowledge of food and nutrition is vital for proper growth and development of an individual. An individual's ability to perform and progress depends on one's physical and mental health. Intake of adequate quantities of nutritious food is must to maintain health of an individual. To achieve this goal it is important to know, what to eat and how much to eat, right from childhood.

India, the second most populous nation in the world, has a huge human capital and large part of it (About 65%) consists of young and energetic youth. This huge youth force needs nutritious food to perform optimally to contribute to the progress of the nation. The formal education at present is more oriented towards making the students employable so that the student can choose the profession of his or her choice soon after finishing the school studies. But the knowledge and practice of the science of food and nutrition is essential for everyone. This is the science for everyday life and to promote life skills.

Therefore, it should be a habit forming knowledge of food and nutrition to learn and practice in everyone's life.

A pre-intervention survey was carried out to assess the awareness about food and nutrition among the students, teachers and parents in the selected (Government) school. Based on the feedback, class room lessons, games and gardening sessions were designed to impart awareness about food and nutrition in the first year (2007-2008). During the second year (2008-2009) a private school was added for comparison. Since the initial survey indicated large scale prevalence of iron deficiency among the students a feeding intervention with iron enriched snacks was taken up in a government school during the third year (2009-2010). *Shepu* (Dill) leaves were used as a less expensive biological source of iron to fortify the snacks.

4. METHODOLOGY

The PNASF initiated a project on "Biological Science and Nutrition for school children" in 2007-2008 with the following objectives – 1) to create awareness about the nutrition requirement and the gaps, 2) to grow fruits and vegetables to learn their life cycles and 3) to prepare the young minds to create a healthy community.

4.1 Pre-Project Awareness Survey

Participating students, parents and teachers completed a questionnaire to assess the initial knowledge level about growing vegetables and fruits crops and nutritional aspects of these protective foods. Similar data was also generated for the control groups. Benchmark data included current knowledge level on nutrition, growing of vegetables, consumption pattern, Importance of fruits and vegetables in nutrition and health. The secondary data available from the school records and local Panchayat (Rural people's Self Governing Body) offices on the family household details collected were also used for the study purpose.

4.2 Project Implementation

In the first year (2007-2008) one hundred students were selected and grouped into 10 batches of 10 students each to take up pot cultivation. To start with, the participating students filled seedlings trays with steam sterilized coco peat. Seeds of 10 types of vegetables were sown and seedlings were raised in these trays. Students transplanted the healthy seedlings in 100 earthen pots filled with traditional pot mixture, red soil + sand + compost in equal proportion and placed on the terrace of the school building. Students cultivated the selected crops from seed to seed to understand the life cycle of the plants. The students were also introduced to different kinds of plants (Fruits and vegetables). The plants were classified based on the nutritional qualities like proteins, calories, minerals, dietary fiber and vitamins. They were also taught to use this knowledge in improving the environment in their households and community. The programme had three treatment groups and three control groups. The broad outline of the Non-Equivalent Control Group Design is described in Tables 1, 2 and 3. Data on Difference-in-Difference as detailed above was generated to account for the changes that affect both experimental and control groups. In addition, time-series data was collected to document changes in treatment and control groups.

Table 1. *Direct effect on school Children*

	Pre-response measure	Treatment	Post-response measure	Difference
Students (Experimental)	SX0	X	SX1	SX1-SX0
Students (Control)	S0	-	S1	S1-S0
Difference	SX0-S0		SX1-S1	

*T*able **2.** *Multiplier effect through Parents*

	Pre-response measure	Treatment	Post-response measure	Difference
Parents of Experimental students	PX0	X	PX1	PX1-PX0
Parents of Control Students	P0	-	P1	P1-P0
Difference	PX0-P0		PX1-P1	

Table 3. *Multiplier effect through Teachers*

	Pre-response measure	Treatment	Post-response measure	Difference
Teachers (Experimental)	TX0	X	TX1	TX1-TX0
Teachers (Control)	T0	-	T1	T1-T0
Difference	TX0-T0		TX1-T1	

The second year of study (2008-2009) was taken at the same Government school and another private school was added for comparison. Total 60 students belonging to age group of 10 - 13 years from each school made a control and experimental group for the study. Keeping in view the objectives of the study schedules were developed. Preliminary study was carried out to test the practicability of the schedule and the final schedule was developed after incorporating necessary modifications. Schedules were developed for relevant concepts, such as, attitude towards fruits and vegetables (with maximum score 20), Nutrition knowledge (with maximum score 20), snake and ladder game on fruits and vegetables (with maximum score 15). Classroom lessons including nutritional importance and health benefits of fruits and vegetables were taught to the children under experimental group for the period of eight weeks. Teaching aids like, charts, flash cards, flip charts and power point slides were used. And impact of the nutrition education was assessed with the nutrition knowledge questionnaire. Along with the classroom lessons children under experimental group also participated in practical growing of vegetables in school premises. The activities carried out by them involved, filling of potting mixture in the poly-bags, arranging them, sowing, watering, weeding and harvesting under the guidance of the scientist from PNASF. Study period included growing of vegetables in polybags by children for one season from August to January. Vegetables grown by children were lady's finger, beans, garden peas, spinach, tomato, coriander and brinjal. Snake and ladder game was developed to draw the attention of the children towards healthy food choices and deficiency disorders. Students in both control and experimental group were asked to sit in groups of four and allowed to play the game with dice for about 30 minutes. Impact of the game was evaluated on the basis of scores obtained before and after plying, using questionnaire.

In the third year (2009-10) a study was conducted for school children studying in government Higher Secondary School, Ganganagar, Bengaluru. Initially the haemoglobin level was estimated for 90 children and were screened into two groups, control and experimental, each group comprising thirty children of different class of anemia. Control group had more number of normal children, while experimental had high number of mild anemic children. General information, such as, family background, age, class in which studying, type of family, educational level of parents, anthropometric measurements which included height, weight, mid upper arm circumference and skin fold thickness, dietary intake, clinical symptoms and health status were collected from school children. Iron status of school children was assessed by haemoglobin estimation. Control and experimental groups were selected for intervention study. Before intervention deworming was carried out to enhance the absorption of iron through diet. The children getting iron supplements through Government, was stopped one month prior to intervention programme for both groups. The intervention programme was carried out for a period of 90 days. For control group/plain buns and rusk was fed and for experimental group children *shepu* greens (*Anethum graveolens*) incorporated bun and rusk were provided. After completion of feeding programme again blood was drawn from both groups for analyzing the hemoglobin level with pre level.

Shepu (Dill) greens are the one of the conventional greens available throughout the year in India. Generally the green leafy vegetables considered as less expensive but a good source of micronutrients. These green leafy vegetables are more perishable compared to other vegetables. The shelf life of the greens can be increased through dehydration. Therefore, both bakery and traditional products were developed incorporating dry and fresh shepu greens for evaluating the acceptability. Dry greens incorporated at 4, 5 and 6 *per cent* level and fresh were incorporated at 5, 10 and 15 *per cent* level. Bakery and traditional products were developed by incorporating these greens and they were masala bun, *Bombay khar, chilly biscuits, masala rusk, chapati,* rice *roti and nippattu*. All the products were acceptable at 5 *per cent* level in both dry and fresh *shepu* greens incorporated in the product. The acceptability values of the product along with method of preparation, iron content, feasibility for transportation and shelf life characters were considered for selection of products for intervention programme. *Masala* bun and *masala* rusk with 5 *per cent* incorporated dry *shepu* greens were selected for feeding trials. *Shepu* greens were analyzed for the nutrients and anti nutrient composition. The selected products were also analyzed for proximate and iron content to analyze the change in iron content on baking.

5. RESULTS AND DISCUSSION

5.1 Pre–project Awareness Survey

The pre–project test was conducted on 104 school children and four of their teachers in the school to know the current knowledge level on nutrition, growing of vegetables, consumption pattern and importance of fruits and vegetables in nutrition and health. The students surveyed were in the age group of 9 to 12 years and most of them were representing the Below Poverty Level (BPL) category. While their knowledge on vegetables was above average, the knowledge on nutritional aspects was below average. Many expressed keen interest in growing vegetables and to use them for healthy living. Parents had very limited knowledge about the vegetables and their nutrients content. Half of the parents had some knowledge of vegetable cultivation, but not of pot cultivation. Majority parents had moderate knowledge of clean kitchen practices and had good knowledge about vegetable consumption pattern including quantity and types of vegetable to be consumed. But the actual consumption pattern was low, because of non-availability and the cost factor of vegetables. The same could be improved by encouraging them to have kitchen gardens in their backyards, terraces or in front of their houses.

5.2 Post Project Implementation Test Results

Only 31 students took the post-test, where as 104 students had taken the pre-test. Out of the 31, only 11 students attended both the pre-test and post-test. So, the data of the 11 students who had taken both the pre and post tests has been compared to arrive at the preliminary impact of the project. Following are the results.

Overall, about 25.5 *percent* gain in knowledge was observed on various aspects of the project. In terms of various components of the project, gain in knowledge ranged from 33 *percent* gain in vegetable cultivation aspects to 12 *percent* gain in case of knowledge related to vegetable consumption pattern. The gain in knowledge of other components are (i) 28 percent on knowledge of vegetables (ii) 22 *percent* on the aspects related to pot cultivation at home and (iii) 23 *percent* gain in knowledge of clean kitchen.

The feedback from the students indicates that the students would like to have still more involvement in the project activities. In addition to the participation in practical (hands on experiences in vegetable cultivation), students needed more about theoretical aspects like nutrient status of vegetables, healthy consumption ways, clean kitchen parameters and pot cultivation at home. The parents expected that vegetables cultivated in the school could be simultaneously grown in one or two pots at their houses by their wards.

Table 4. *Impact of nutrition programme on attitude towards fruits and vegetables among control and experimental groups.*

		Response Scores			Paired '*t*' Value
	Status	**Mean (Max = 20)**	**Mean (%)**	**SD (%)**	
Control ($n = 60$)	Pre test	13.3	66.6	12.6	1.50^{NS}
	Post test	13.4	67.3	10.7	
	Enhancement	0.1	0.65	3.3	
Experimental ($n = 60$)	Pre test	13.1	65.5	13.6	7.10*
	Post test	15.1	75.5	11.9	
	Enhancement	2.0	10.0	10.9	

* Significant at 5% level, NS: Non significant

Nutrition programme has significantly ($t = 7.10^{*}$) improved mean scores (10%) of children's attitude towards fruits and vegetables among experimental group compared to control group (Table 4). This indicates positive impact of nutrition programme which includes nutrition education and participation in gardening activities on children's attitude towards fruits and vegetables. Similar findings were observed by Koon *et al.* (2006) in their study on evaluation of a school- based nutrition intervention conducted in Kuala Lumpur and Selangor, Malaysia, where nutrition education components had improved children's choice of food items and beverages during school breaks.

Table 5. *Impact of nutrition programme on knowledge among control and experimental groups*

Control (n = 60)	Status	Response Scores			Paired 't' Value
		Mean (Max = 20)	Mean (%)	SD (%)	
	Pre test	8.9	44.5	12.4	0.6^{NS}
	Post test	8.9	44.9	12.0	
	Enhancement	0.06	0.3	3.8	
Experimental (n = 60)	Pre test	10.3	51.7	11.8	6.54*
	Post test	15.4	77.1	13.4	
	Enhancement	2.6	13.4	15.9	

* Significant at 5% level, NS: Non significant

There was no significant difference observed between initial and final and also enhanced mean per cent scores of two experimental groups (Table 7). Nutrition programme has significantly (t=6.54*) increased the mean scores (13.1%) of children's nutrition knowledge among experimental group compared to control group (Table 5). which indicates that children's participation in gardening activity and nutrition education programme has a positive impact in increasing the nutrition knowledge mean scores (24.9%). These findings of the results are on par with the results of the study by Koon *et al.* (2006) in their study on evaluation of a school- based Nutrition intervention conducted in Kuala Lumpur and Selangor, Malaysia, where nutrition education component has significantly improved the nutritional knowledge of the children.

There was no significant difference observed between initial mean *per cent* scores of experimental groups of two schools, while significant difference was observed between final mean *per cent* (t = 2.93*) scores and also enhanced mean *per cent* scores (t = 3.85*; Table 7). Mean enhancement scores (31.5%) were significantly higher (t = 3.85*) in experimental group of government school compared to experiment group of private school (18.3%). This may be due to, private school children initially had some exposure to nutrition education in their syllabus and hence, they had higher initial scores than government school experimental group and also it depict the interest and involvement of the government school children which has lead to the higher final mean *per cent* scores than the private school experimental group. Imparting nutrition education through non formal methods is very important apart from formal methods of teaching. Nutrition games can be effectively utilized as a strategy for nutrition education and they can be effectively incorporated in school curriculum whereby children gain some knowledge as well as enjoy playing games (Suniti Punetha, 1995).

Table 6. *Impact of nutrition game on nutrition knowledge of Control and experimental groups*

	Status	**Response Scores**			**Paired '*t*' Value**
		Mean (Max = 15)	**Mean (%)**	**SD(%)**	
Control ($n = 60$)	Pre test	5.7	38.2	9.0	11.82*
	Post test	9.1	61.0	13.2	
	Enhancement	3.4	20.9		
Experimental ($n = 60$)	Pre test	9.3	62.0	9.4	12.63*
	Post test	11.7	78.0	9.6	
	Enhancement	2.4	16.0	9.8	

* Significant at 5% level, NS: Non significant

Nutrition game had a positive impact on nutrition knowledge of both control (t = 11.82*) and experimental groups (t = 12.63*; Table 6), which was indicated by significant difference between initial and final scores for knowledge (16% enhancement). Similar findings were observed by Suniti Punetha, (1995) in her study for assessing the efficacy of nutrition games by evaluating the gains in knowledge and awareness of children in pre and post test research design.

Initial scores were significantly higher among both the experimental groups (t = 14.50*, t = 11.46*) than their respective control groups. This may be due to the fact that initially experimental group children had nutrition knowledge as they had participated in nutrition education and garden programmes. Whereas, mean percent scores for enhancement of knowledge found to be significantly higher (t = 4.57*, t = 11.9*) in case of both the control groups than their respective experimental groups. This indicates effectiveness of games in imparting nutrition education (Table 8).

Table 7. *Comparison of mean scores (%) of attitude towards fruits and vegetables and nutrition knowledge between experimental groups.*

Parameter	Before				Student's 't' Value	After				Student's 't' Value	Enhancement				Student's 't' Value
	Private School ($n = 30$)		Government school ($n = 30$)			Private School ($n = 30$)		Government School ($n = 30$)			Private School ($n = 30$)		Government School ($n = 30$)		
	Mean %	SD %	Mean %	SD %		Mean %	SD %	Mean %	SD %		Mean %	SD %	Mean %	SD %	
X_1	64	13.0	66.5	13.2	0.79^{NS}	75.5	12.8	75.0	12.1	0.17^{NS}	11.5	12.8	8.0	9.2	1.31^{NS}
X_2	54	12.2	49.3	10.9	1.69^{NS}	72.5	12.9	81.5	12.7	2.93*	18.3	15.8	31.5	12.9	3.85*

* Significant at 5% level, NS : Non significant, X_1 = Attitude scores for fruits and vegetables, X_2 = Nutrition knowledge scores

Table 8. *Comparison of mean scores (%) of impact of nutrition game on control and experimental groups by schools*

School	Control ($n = 30$)		Experimental ($n = 30$)		Student's '*t*' value	Control ($n = 30$)		Experimental ($n = 30$)		Student's '*t*' value	Control ($n = 30$)		Experimental ($n = 30$)		Student's '*t*' value
	Mean %	SD %	Mean %	SD %		Mean %	SD %	Mean %	SD %		Mean %	SD %	Mean %	SD %	
Private	40.0	7.6	66.4	7.6	14.50*	67.3	14.13	82.4	5.5	5.83*	27.3	14.6	16.0	07.5	4.57*
Government	34.8	7.5	57.5	9.0	11.46*	55.3	08.2	74.0	8.2	9.49*	20.4	12.0	16.4	11.8	11.90*

* Significant at 5% level

The follow-up indicated that the health and nutritional status of school was inadequate, especially with anthropometric measurements, nutrient intake and hemoglobin level. Therefore, in the third phase (2009-10), the feeding of the *shepu* (Dill) greens incorporated products was undertaken and it was found to have significant influence on improving hemoglobin level after feeding. After screening 90 children were grouped into control and experimental (with and without intervention). The experimental groups were fed with the products with and without shepu (Dill) leaves for 90 days. The hemoglobin classification (indicator of anemia status) indicated that control group had higher percent of normal children (66.6%) while experimental group had more mild anemic children (33.3%)and moderately anemic children (13.3%; Table 9). Before feeding trials the mean hemoglobin level was higher in control group (11.0) than that in experimental group (9.8) because of inadequate intake of iron rich foods by the experimental group (Tab.10). Leela and Shanthapriya (2002) studied iron status and morbidity pattern among school children and reported 47% as mild anemic, 33% moderate anemic and 20% non-anemic. Seshagiri (2003) reported 45% children with severe anemia, 18% mild, 22% moderate anemia and 15% normal. Shalamli Naik (2010) found in her study of control and experimental groups 66% and 33.3% in normal. 33.3 % and 33-3% in mild anemic, 0.0 and 13.3% in moderate anemic group and none in severe anemic group (Table 9 and Table 10).

Table 9. *Hemoglobin level (Anemia status) in control and experimental groups (Naik, 2010)*

Classification	Hemoglobin* (g.dL)	Control		Experimental		X^2 value
		N	%	N	%	
Normal	5.11	20	66.6	10	33.3	
Mild	9.11	10	33.3	16	33.3	4234.68*
Moderate	7.9	0	0.0	4	13.3	
Total		30	100	30	100	

* Significant at 5% level, NS: Non significant

Table 10. *Mean Hemoglobin level (Anemia status) in control and experimental groups (Naik, 2010)*

Category	N	Hemoglobin, % Mean	SD
Control	30	11.0	2.21
Experimental	30	9.8	2.18

* Significant at 5% level, NS: Non significant

The variation in reports can be due to sample size and Government schemes distributing iron and folic acid tablets in some schools supplementing the iron nutrition. Feeding the children with the products incorporating *shepu* leaves for 90 days improved the hemoglobin level (11.8 %) in experimental group (Table 11).

Table 11. *Mean Hemoglobin level, percent (Anemia status) in control and experimental groups after feeding intervention (Naik, 2010)*

Category	Control		Experimental	
	PRE	POST	PRE	POST
Mean,%	11.0	11.2	9.8	11.8
SD	2.210	2.185	2.531	2.956
Paired 't' test	1.80^{NS}		9.18*	

* Significant at 5% level, NS: Non significant

Nawade *et al.* (2005) reported improved hemoglobin by feeding children with iron rich foods like roasted rice flakes (Chiwda) and sesame *chikki*. Sarwate *et al.* (1994) observed increased hemoglobin in children fed with laddos made of jaggery, sesame seeds, Bengal gram flour and dates. Feeding children with iron rich food products proved more acceptable and economical compared to supplementing with tablets.

6. SCHOOL GARDEN MAKES LEARNING ABOUT NUTRITION EASIER

Children spend much of their time in school. Therefore, schools have the opportunity to play a large role in adoption of healthy living habits. There are many ways teacher can present nutrition education. Gardening provides a fun and interactive way to teach and learn these habits. In many developing countries, school gardens have been used as a powerful tool to improve nutrition and education of school children and their families and community (Cederstrom, 2002; Wanasinghe, 2003; Childs Elizabeth, 2011). It is an innovative teaching tool and strategy that lets educators incorporate hands-on activities in a diversity of interdisciplinary and standards-based lessons. The garden engages students by providing a dynamic environment in which to observe, discover, experiment, nurture and learn. It is a living laboratory where lessons are drawn from real-life experiences rather than textbook examples, allowing students to become active participants in the learning process. Through the garden, students gain an understanding of ecosystems, an appreciation for food origins and nutrition and knowledge of plant life cycles (Morris *et al.*, 2002). It is expected that through school children their parents will also be motivated to grow horticultural crops in the available interspaces that in turn helps to provide nutrition security as well as income security in the households. It is proposed to encourage students' parents, especially women, to participate in lecture cum demonstration programmes to ensure finally that it will be a sustainable community activity in the neighborhood of the school and adjoining areas. Gardens are good for schools because they teach children the skills for survival (Ncube, 2004). Along with the World Bank sponsored Integrated Child Development Services (ICDS) programme, promoting nutrition gardens can make positive and significant impact on nutritional status of the children. Cultivation of vegetables and fruits can be taken up in the backyard, school, *anganawadi* centre, roof tops or in the community land, which is not difficult considering the benefits it provides of better nutrition and wholesome diet. Healthful food habits can be inculcated through garden-based programs in children by incorporating gardening, nutrition education and cooking through teaching interventions. Nutrition garden in school is emerging as a nutrition education tool in academic settings.

School Garden programmes improve students' attitude that helps them to do well in academic subjects. It also builds scientific temper and questioning attitude – "if so why so and if not why not". The changes will not happen instantly with a garden programme. Therefore long range programmes need to be established at a participating school to study the effects of an integrated garden programme. Garden based programme improves ability to tackle science related tasks and better academic performance in science. Students develop positive attitudes towards plants and gardening. The school garden improves peer and community relationships. Regular curriculum helps to promote nutrition awareness, performance in science and other academic subjects and attitude towards environment and community at large. School gardens have an important social purpose of creating awareness about nutrition and environment, improving academic achievement of children, promoting healthy life style and life skills and helping to build a healthy community.

School gardening offers children opportunities for outdoor exercise while teaching them a useful skill. Gardens containing fruit and vegetables help to revise attitudes about particular foods. There is mounting evidence that active learning in less structured, participatory spaces like nutrition gardens are more likely to transform children's food attitudes and habits and that school gardening, especially when combined with a healthy lunch programme and/or nutritional education, encourages more healthful food choices. Practical experience with fresh foods positively impacts dietary habits. When students take their preferences back to their families, they can help to improve family consumption choices also. Students are more likely to try eating vegetables they have grown themselves and to ask for them at home. The main actors themselves, the school children must be approached in a participatory way to insure their commitment to the programme. Nath *et al.* (2009) found that nutrition gardening and nutrition education including games have improved the nutrition knowledge of the students in general and especially those involved in the experiments. Nutrition gardens were useful tools used in several places in the world to improve nutrition knowledge and translate the knowledge into small scale economic activities of the communities.

7. SWOT ANALYSIS

7.1 Strengths

The concept is pertinent, timely and very effective means of promoting awareness about the nutrition security among the under and mal nourished communities. With the promotion of School garden Programmes a great number of malnourished children can be reached with relatively simple and cost effective means. School garden programs must be well adapted to the local habits, needs and to the specific climatic and environmental situation. Community involvement is a basic requirement for the success of these programmes.

School gardens are considered as catalyst to trigger change in food habits of school children, parents and community as a whole. School gardens can play pivotal role in encouraging community gardens that eventually support income, food and social security of the community. School children will also encourage both rural and urban youth to engage in community gardens and allied activities that helps to build a safe and healthy society.

7.2 Weaknesses

The schools sincerely appreciate the project objectives and methodology. However, the schools feel it difficult to continue such programme as a regular academic activity, because it does not fit into the existing curriculum of the schools. Sincere efforts are required to find out suitable slots in the existing curriculum to integrate the much needed knowledge capsules on food and nutrition.

7.3 Opportunities

One successful programme will have rippling effect and the benefits can be reaped by the BPL communities in length and breadth of the country and beyond the boundaries as well. School children are very enthusiastic to learn new things. School gardening provides outdoor activity that facilitates interactive learning and helps to promote overall learning process in schools. Seaman Knapp, known as father of extension wrote "What a man hears, he may doubt; what he sees, he may possibly doubt; but what he does himself, he cannot doubt." (Hinton, 2003). Learning by doing is a common descriptor of experimental learning. Learning by doing, learning in real life context, learning through projects and learning by solving problems are four pillars of experimental education (Knobloch, 2003). Therefore, studying and gardening at schools will be useful in promoting good eating habits in children.

7.4 Threats

Lack of commitment and responsibility on the part of stake holders will not yield the expected results. School garden programmes cannot be created in isolation but have to be linked with education, health and nutrition interventions. Therefore best program design seems to link school gardens with other programs related to nutrition, health and environment. Though national policies and interventions are aimed at tackling the problems of malnutrition and under- nutrition, the changes that take place at grass root level ensures social good.

8. CONCLUSION

Dr. Prem Nath Agricultural Science Foundation (PNASF), Bengaluru, India, has taken up the noble cause of spreading nutrition awareness through school children especially in urban areas where children are not exposed to gardening as is the case in rural areas. It is imperative to catch them young and inculcate in them good eating habits and awareness about nutrition and health benefits of consuming fruits and vegetables through structured lessons and hands on practice of gardening. These educated children are ambassadors of change of nutrition awareness among households and neighborhood that pave way for creating better informed and healthy society.

Most correlation studies have found associations between iron-deficiency anemia and poor cognitive and motor development and behavioral problems. Randomized controlled trials of iron treatment are warranted especially in younger children (Sally *et al.*, 2001) Therefore, the PNASF has launched yet another project on supplementing iron through bakery product to improve academic performance of the children in some selected schools.

This Project on 'Nutrition Education' for school children which aims at bringing awareness in communities below poverty line about nutrition and health care by educating school children about the role of fruits and vegetables in nutrition and health care and teaching skills of growing vegetables. This is a novel project to educate the urban and rural school children to use the knowledge of biological sciences to make a difference in improving environment and carry out experiments. The innovativeness of the project also lies in its *modus operandi*, wherein 'learning-by-doing' motivates students to learn the life cycle of plants and their growing habits by relating it to their own growth cycle. The unique nature of the project lies in its expected multiplier effect, first through 'student-parent-family-neighborhood-community' network and also through 'teacher-teacher-student' network.

Nutrition programme comprising nutrition gardening along with nutrition education using various teaching aids proved to be effective in improving children's attitude towards fruits and vegetables and nutrition knowledge. Healthy eating habits and improved nutrition knowledge can be achieved through effective Nutrition education and nutrition garden. Such nutrition programmes in schools are helpful in developing healthy eating habits among children, who are future wealth of the country. The study conducted for three years from 2007-08, included six sets, of students, parents and teachers who participated in the experiments and another control set. The level of nutrition knowledge has significantly improved among the students who have participated in the experiment, but that of teachers, parents and others from the control group did not show any marked improvement.

Feedback from the schools indicates sincere appreciation of the nutrition education, but yet there is difficulty in incorporating such programmes in the routine syllabus. The PNASF, having committed to improve the nutrition and health security of the poor communities would like to work with National Council for Educational Research and Training (NCERT), New Delhi and similar institutions a the state level to develop lesson plans on food and nitration in Biology for school children from 6th to 10th or 12th standards.

This Endeavour would ensure the student leaving the school is well equipped with knowledge of nutrition, role of fruits and vegetables in human nutrition, simple techniques of growing them in homesteads and community lung spaces and participate in community activities that help to improve the nutrition and health security of the community. This social activity is expected to involve the schools, State Departments of Horticulture and Social Forestry, local municipal bodies and responsible citizens to evolve favorable ambiance for healthy communities.

It is important to implant this concept of nutrition education into the regular school curriculum. The concept note may be sent to National Council for Educational Research and Training (NCERT), State Council for Educational Research and Training (SCERT) and Deputy director General (DDG; Education) Indian Council of Agricultural Research (ICAR) for funding the preparation of education material. Lesson plans from 6th to 10th standard to include nutrition and nutrition garden. It should be mandatory but not optional. The students especially the sample group chosen for this pilot study, 6th, 7th and 8th standard students are very receptive and can very well be the agents of change towards building a healthy nation through nutrition awareness.

The Government schools are generally starved of funds and therefore any intervention by the Govt. or non-Govt. agencies is seen as an opportunity to attract some more funds for the school rather than being focused on the deliverables of the mission, thus vitiating the results of the intervention. Private schools are better funded and well equipped to take up such programmers but they are more focused on the targets of achieving the academic excellence and therefore such interventions are looked upon as unnecessary interference and hurdle in the path of academic progress.

The concern of both Government and private schools is understandable as the education sector is not adequately funded and taken care of. It is disturbing that such an important human activity, which is fundamental to the development of good citizenry, is not given adequate attention and support it deserves. Therefore, a very important step to build much needed nutrition security of the children and society at large can only be achieved by making nutrition education as a part of regular curriculum. Class room lessons on food and nutrition and practical sessions on gardening have to be made mandatory so that the foundation for a healthy society is in place. There should be proactive debate at the national level and stake holders should commit to the cause of building a healthy, wealthy and wise nation.

Acknowledgement

The authors are thankful to Dr. Prem Nath Agricultural Science Foundation, Bengaluru, for providing fellowships and considering the paper for publication.

REFERENCES

Cederstrom, T. (2002). Learning to grow: How school gardens can provide food for education. Food Forum Online, Issue 61, 3rd quarter. www.foodaid.org/ff2002.htm.

Childs Elizabeth, (2011). Impact of school garden on student attitudes and beliefs. Graduate thesis and desertation. Iowa State Univ.

Eboh, L. O. and Boye, T. E. (2006). Nutrition knowledge and food choices of primary school children in the Niger- Delta region Nigeria. Pakistan journal of nutrition, 308-311.

Heimendinger, J. and M. Van Duyn. (1995). Dietary behavior change: the challenge of recasting the role of fruit and vegetables in the American diet. American Journal of Clinical Nutrition, 61:1397S-1401.) (Cited from: Research database: The collective school garden network. http://www. Csgn.org./research. 8 Jan., 2013).

Hinton, Karen. (2003). Celebrating 100 Years of Cooperative Extension. Communique. Retrieved from http://www.unce.unr.edu/news/newsletter/2003/Nov2003.pdf).

Knobloch. (2003). cited from Cornell garden based learning and highlights from journal articles. Blogs.cornell.edugarden.grow. you. Program. Research.

Koon, P. B., Karim, N. A., Talib, R. A. and Mohd Taib, M. N., (2006). Nutrition education for Malaysian children: Evaluation of a school-based nutrition intervention conducted in Kuala Lumpur and Selangor, Malaysia. Proc. Intl. Symp. Building leadership skills in food and nutrition essential for national development, CFTRI, Mysore (India): 23-29.

Leela,T.T. and Shanthapriya. (2002). Iron status and morbidity pattern among selected school children. The Ind.J Nutrition and diet: 39: 216-22.

Mc ALeese, J. D. and Rankin, L. L. (2007). Garden-based nutrition education affects fruits and vegetables consumption in sixth grade adolescents. Journal of American Dietetic Association, 662-665.

Morris, L.J., Briggs, M. and Zidenberg-Cherr, S. (2002). Development evaluation of a garden- enhanced nutrition education curriculum for elementary school children. The journal of child nutrition and management, (2) : 1-5.

Naik, Shalamli. (2010). Development of iron rich formulations from shipu greens (Peucedanum graveolens) and its impact on school children. M.Sc.Thesis, UAS, Bengaluru.

Nath, Prem, Prabhakar,B.S., Chandre Gowda, M.J., Mukund Sirgur, Patil, M. S., Yatnatti, Shilpa and Latha, S.V. (2009). Nutrition awareness Through school children Souvenir, International Conference on Horticulture: Horticulture for livelihood security and Economic Growth. Bangalore. Noember 9-12. p 97-101.

Nawade, V, Mahadevi, M, Vinaya, K and Varsha, Z. (2005). Nutritional intervention for ironand Vitamin A deficient school children The Ind. J Nutrition and diet 42: 465-474.

Neube, S. (2004). cited from FAO. (2012). Setting up and running a school nutrition garden project. http://www.fao.org/3/a-a0218e/A0218E01.html.

Pandey, Kundan. (2016). India's fruit and vegetable intake less than standard, finds study. Down to Earth.05.02.2016.

Punetha, Suniti. (1995). Nutrition education through games in primary schools. The Ind. J. Nutr. Dietet. 32: 270:273.

Sally, Grantham, McGregor and Cornelius, Ani. (2001). A Review of Studies on the Effect of Iron Deficiency on Cognitive Development in Children .J of Nutrition. 2001; 131:649-668.

Sarwate, N., Sharma, K., and Singh, S.D. (1994). Impact of supplementing indigenous sweet (Ladoo) on hemoglobin level of anemic children. The Ind.J Nutrition and diet 31:262-264.

Seshagiri, M. and Nath, K. M. (2003). Utilization of knol khol greens to improve iron and Vitamin A status in school children. M.Sc.Thesis, UAS, Bengaluru.

Singh, Jyotna. (2017). Consumption of fruits and vegetables short of requirement in Indian cities finds survey.Live Mint E Paper 12.05.2017.

Wanasinghe, A.D. (2003). From School Garden to Home Garden. Integrated Food Security Programme. Trincomalee, Sri-Lanka. Technical Paper 29.

❍❍❍

26

School Garden—The Harbinger of Nutrition Education

B.S. Prabhakar and Megha Katte

SYNOPSIS

Malnutrition in early childhood has serious and long-term consequences because it impedes motor, sensory, cognitive, social and emotional development. Malnourished children are less likely to perform well in school and more likely to grow into malnourished adults, with greater risk of disease and early death. In spite of impressive economic growth, India experiences pervasive and persistent child malnutrition. School children need plenty of good food, as their energy and nutrient needs are particularly high in relation to their size. In their overall diet, it is often difficult to achieve adequate intake of energy, vitamin A, calcium, iron, zinc and iodine. Parents, teachers and school catering staff need to ensure that children receive plenty of nutrient-rich foods. School gardens provide fun and interactive way of learning about food and nutrition. Nutrition garden is a precious tool to educate and improve awareness about healthy food and nutrition requirement in human life. It is a living laboratory where children cultivate, feel and grow with healthy habits. Gardens are good for schools because they teach children the skills for survival.

Keywords: Nutrition education, school garden, life style, community.

1. INTRODUCTION

Millennium Development Goal envisages halving of the proportion of hungry people between 1990 to 2015. But we are nowhere near that goal, about 250 m. Indians are still hungry (Bajaj, 2012). Despite economic progress and self-sufficiency in food grain production, hunger, poverty, lack of proper education and health services and want of required (employable) job skills are still failing the country's progress. Increasing population (1.25 billion in 2011) and continuing dependence on agriculture (65% of population) for livelihood, fragmented land holdings (65% of farm families depend on less than one hectare land for livelihood, Veeresh, 2013), depleted soil productivity, inefficient use of water resources and outdated production technologies, lack of credit and infrastructure for post harvest operations and marketing are responsible for low crop yields and income (Hegde, 2012).

Green revolution of 1960s has increased food grain production especially in Northern States of Punjab, Haryana, UP and Rajasthan, increasing the country's production from

82 MT in 1996 to 176.4 MT on 1990 to 241.56 MT in 2010. However, the growth of agricultural production declined from 10-12% in 1980 to 4.7% in 1990 to 2.3% during 2002-2007 (GOI, 2011). Stagnation in agricultural production suppressed the employment opportunities in rural areas bringing down the contribution of agriculture to national GDP from 50% in 1951 to 35% in 1991 to 17% in 2009 (Hegde, 2013).

While continuous efforts are on to increase food production, it is equally important to improve the awareness about the healthy food and nutrition to promote healthy lives. School garden is a potential tool to educate the children and through them the society at large about the healthy food and nutrition requirement in human life. School garden provides fun and interactive way of learning about food and nutrition. It is a living laboratory where children cultivate, feel and grow with healthy habits. Gardens are good for schools because they teach children the skills for survival.

2. MALNUTRTION A NATIONAL HELATH PROBLEM

Malnutrition is yet another compelling national problem besides hunger (Nath, 2013). It is imperative to tackle the problem of hunger on priority. Under nutrition and malnutrition are rampant among rural as well as urban poor while malnutrition is prevalent among rich families as well because of unhealthy dietary habits and life style. Among 1.25 billion Indians, more than 300 M are undernourished despite 50% increase in food production in the last two decades .India is ranked 124 in nutrition ladder even below African countries. Most of the children are suffering from ill health and dying every year. Anemia rate of 51 to 73 % in women and small children and of children less than 3 years of age 47% are under weight and 45% are stunted as per global standards. Every hour of every day 300 children die of malnutrition (Nath, 2013). National health survey (2002) reports that 65% of women and children in India are anemic and 50% world's anemic are in India according to WHO report. Malnutrition in early childhood has serious, long-term consequences because it impedes motor, sensory, cognitive, social and emotional development. Malnourished children are less likely to perform well in school and more likely to grow into malnourished adults, with greater risk of disease and early death. In spite of impressive economic growth, India experiences pervasive and persistent child malnutrition. The latest estimates suggest that around 48 *percent* of India's children aged 0-5 years are stunted making up almost half of world's stunted children. Children in low-income families are more malnourished than those in high-income families. Despite India's 50 *percent* increase in GDP since 1991, more than one third of the world's malnourished children live in India. Among these, half of them under the age of 3 years are underweight and a third of wealthiest children are over-weight. Vitamin and mineral deficiencies also affect children's survival and development. Anemia affects 74 *per cent* of children under the age of three years. Iodine deficiency, which reduces learning capacity by up to 13 *per cent*, is widespread because fewer than half of all households use iodized salt. Vitamin A deficiency, which causes blindness and increases morbidity and mortality among pre-school children, also remains a public-health problem (UNICEF, 2012).

The main theme of World Health Day 2006 was "Working together for health". FAO is struggling hard to promote food and nutrition security of the human kind. UNO has given a clarion call to all National Governments, private sector, NGOs and to all those who care for mitigating hunger, poverty and ill health to come together and act together to correct the malady at the earliest.

3. HORTICULTURAL REMEDY FOR MALNUTRTION

Agriculture continues to be the primary source of livelihood for the majority of nutritionally vulnerable households in India (Gillespie *et al.*, 2012 Indian Council of Medical Research's (ICMR)). Horticulture provides remedy for every nutrition malady. Recommendation is that every person should consume 300 g vegetables and 100 g fruits, 50 g greens (Pregnant women 100 g), 50 g roots and Rani (2012) reported that a healthy person should consume tubers and 200g other vegetables. Prathiba and 125 g greens, 100 g roots and tubers, 75 g other vegetables and 85 g fruits apart from 475 g cereals and 85 g pulses every day. Growing locally available fruits and vegetables helps to increase consumption at affordable costs (Chavasith, 2012). A well planned nutrition garden (Home garden) can meet daily requirement of a family round the year (Sheela *et al.*, 1998). Nutrition garden is helping the tribal community in Rajastahn to improve their food, nutritional and income security (Nandal and Mathur, 2012). Nutrition garden is essential for every household that can be easily managed by waste water and organic wastes from house hold (Indumathi *et al.*, 2010). Hydroponic garden of 20 sqm can supply 2 kg vegetables every day round the year (Bradley, 2009). Mitra (2014) demonstrated that a roof top unit that integrates vermiculture based on kitchen waste and vegetable growing together. Light portable poly houses can be useful for growing vegetables on roof tops (Mitra, 2004). Shukla *et al.*, (2012) reported that establishing nutrition gardens in Udam Singh Nager district in UP improved nutrition and income of the farm families.

United Nations Organization's (UNO) declaration on food security (1996) emphasizes safe and regular supply of food to mitigate hunger, ill health and poverty. Though India is the second largest producer of fruits and vegetables the consumption is low among rural and urban poor. At present more people in general are becoming health conscious; thereby the consumption and demand for fruits and vegetables is gradually increasing especially in urban areas.

4. SCHOOL GARDEN TO IMPART NUTRITION EDUCATION

School children need plenty of good food, as their energy and nutrient needs are particularly high in relation to their size. In their overall diet, it is often difficult to achieve adequate intake of energy, vitamin A, calcium, iron, zinc and iodine. Parents, teachers and school catering staff need to ensure that children receive plenty of nutrient-rich foods. Research supports the role of fruits and vegetables consumption in prevention of cancer and heart diseases. The protective role of fruits and vegetables in strokes, cataract, chronic obstructive pulmonary disease, diverticulosis and hypertension is well documented (Mc Aleese and Rankin, 2007). Eating protective food from childhood takes them to a healthy adulthood (Eboh and Boye, 2006). Consumption of fruits and vegetables, as a habit in childhood, is an important predictor of higher fruit and vegetable consumption as adults and can help to prevent or delay chronic disease conditions (Heimendinge and Van Duyn,

1995). Children spend most of their time in schools. Therefore, schools have opportunity to inculcate healthy eating habits in them.

Seaman Knapp wrote "what a man hears, he may doubt. What he sees he may also doubt. But what he does, he cannot doubt." Learning by doing is a common descriptor of experimental learning. Learning by doing, learning in real life context, learning through projects and learning by solving problems are four pillars of experimental education (Knobloch, 2003).

School gardens can have many different uses and have been seen in many different ways; some practical and some educational. School gardens provide fun and interactive way of learning about food and nutrition. Nutrition garden is a precious tool to educate and improve awareness about healthy food and nutrition requirement in human life. It is a learning laboratory where children cultivate, feel and grow with healthy habits. Gardens are good for schools because they teach children the skills for survival. (Ncube, 2004). Students gain knowledge about ecosystems and life cycles of different food and nonfood plants (Morris *et al.*, 2002) that provides nutrition, color and aroma therapy and together help them to grow into healthy adults and helpful citizen. Along with the Integrated Child Development Services (ICDS) programme, promoting nutrition gardens can make positive and significant impact on nutritional status of the children.

Cultivation of vegetables and fruits can be taken up in the backyard, school, *anganawadi* centre, roof tops or in the community land, which is not difficult considering the benefits it provides of better nutrition and wholesome diet. Healthful food habits can be inculcated through garden-based programs in children by incorporating gardening, nutrition education and cooking through teaching interventions. Nutrition garden in school is emerging as a nutrition education tool in academic settings. Initial survey indicated that the students of 5th, 6th and 7th standards form appropriate groups to act as ambassadors for change. They belong to impressionable age group and good food habits can be easily inculcated that they carry throughout their lives and also help to change the mind set of those in the families as well as community as a whole. Nath *et al.* (2009) found that nutrition gardening and nutrition education including games have improved the nutrition knowledge of the students in general and especially those involved in the experiments. Nutrition gardens were useful tools used in several places in the world to improve nutrition knowledge and translate the knowledge into small scale economic activities of the communities.

Sligoville, a farming community, was the first free village in Jamaica after slavery was abolished. The all-age school has been called the "most environmentally aware" in Jamaica. The head teacher encourages the garden strongly because she believes that children learn by doing. Children gain skills they can use, teachers find new ways to teach and everyone gets some delicious nutritious food. Most of the staff is experienced gardeners and there is an active parent-teacher association. Each grade manages a small plot with a few crops, such as, carrots, corn and thyme. The older children have an acre with a variety of vegetables and rows of plantain, banana and cocoa. There are chicken, rabbits and goats. The garden is completely organic. The children eat a lot of the garden food and take some home; some is cooked in the canteen and some is sold in the community. Each year a cross-curricular project focuses on one food. One year they researched corn, grew samples and produced songs, puppets and poems. Their book "Corn in the Classroom" was translated into 84 languages and featured on the radio. (Power, 2003; Bruce, 1998).

In northern Zambia's Mansa Township, the Seventh Day Adventist church opened a small orphanage for AIDS (acquired immune deficiency syndrome) orphans, with support from the Rotary Club and the Zambian Government. To provide the children's food, they established a half-hectare garden which the children cultivate themselves under the supervision of a qualified volunteer gardener. A small grant from Food and Agriculture Organization (FAO) provided seeds, tools and a small-scale drip irrigation system. Now, under the children's careful tending, the garden is flourishing, producing eggplants, peas, maize, green peppers, soybeans, sweet potatoes, tomatoes and chicken. "They are completely self-sufficient for vegetables, maize and poultry," reported Karel Callens, an FAO nutritionist. "They are producing enough surpluses to sell at market and are reinvesting the proceeds in the garden." The children are also learning teamwork and farming know-how, which will support them for the rest of their lives. "That's a pretty impressive return on a small investment" (Callens, 2004).

Nebiri School is in a game reserve in the Zambesi Valley in Zimbabwe. The garden has mango trees, paw-paw, a lemon tree and areas with tomatoes, cabbage and other vegetables. A small grant provided a solar-powered electric fence. Children bring in elephant and buffalo manure for the compost heap. The garden provides funds for the school for stationery and equipment: the school sells the produce in the local market and to families at half price. Older pupils market the food and keep the accounts. Under the responsibility of the head teacher, the garden work is organized by a "garden team" of four pupils, those change every month. Teachers use the garden as a learning resource for maths, biology and environmental science (Ncube and Chinanzavavana, 2004).

In Trincomalee in Sri Lanka, food insecurity was a severe problem after civil unrest and war. Agricultural clubs were formed in the schools, consisting of eight students and a teacher. Basic agricultural training was given by the Department of Agriculture. Vegetable plant nurseries and school gardens have been established on the school premises and the vegetable seedlings are sold to households. Knowledge and practice are transferred to the home: students grow their own seedlings at home and pass on information about how to grow them to friends and family. Students report that they have gained knowledge, practical skills and opportunities for self-employment (Wanasinghe, 2003).

Manorbier School in Wales, UK is in a depressed area with high unemployment. Although it is a rural area, many of the children have their first experience of growing plants at school. The school garden has herbs, flowers, a play area, a large apple tree, a small wood and a pond for studying wildlife. It grows beans, tomatoes, sunflower and leek. The whole school has created a pebble mosaic at the entrance, with a segment for each class. The idea is that the garden belongs to the children and they should run it. Children volunteer to be "garden monitors" for the week. Each class has a responsibility - *e.g.,* the pre-school class looks after flowers, older pupils see to the pond. A garden club meets once a week after school. The garden is used for learning by direct experience in science and environmental studies, mathematics, literature and art (Greenhouse and Carr, 2003).

School Garden programmes improve students' attitude that helps them to do well in academic subjects. It also builds scientific temper and questioning attitude – "if so why so and if not why not". The changes will not happen instantly with a garden programme. Therefore, long range programmes need to be established at a participating school to study the effects of an integrated garden programme. Garden based programme improves ability to tackle science related tasks and better academic performance in science. Students develop positive attitudes towards plants and gardening. The school garden improves peer and community relationships. Regular curriculum helps to promote nutrition awareness, performance in science and other academic subjects and attitude towards environment and community at large. School gardens have an important social purpose of creating awareness about nutrition and environment, improving academic achievement of children, promoting healthy life style and life skills and helping to build a healthy community.

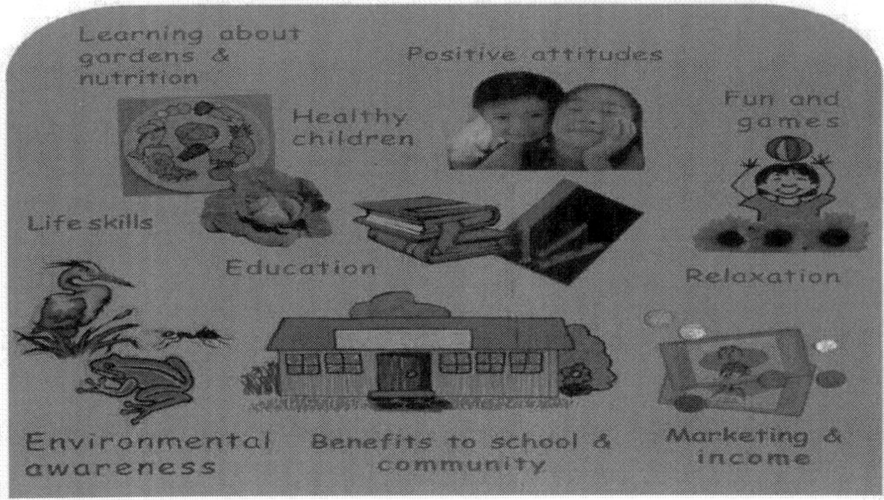

Source: *FAO – Nutrition Division*

Fig. 1. *School garden and its relationship with academics, health and community welfare*

4.1 Awareness About Nutrition

School garden based studies in Cornell University report how students (6 to 9 grade) change their dietary behavior during short term studies. Students were confident to answer questions in science and maintain plants in the garden. New students and outgoing students differed in their confidence in the activity of maintaining a garden. All outgoing students scored positive points towards consumption of fruits and vegetables. School gardens support the study of land through the production of local foods (fruits and vegetables), management of land and conservation of local resources. Students gain practical knowledge and lifelong skills by working through garden curriculum with their mentors and peers. Well developed garden activities help students faster positive attitudes towards science and build up their science self-efficacy (Childs, 2011).

Gardening improves opportunities for children, youth, adults and families to gain interest and eat fruits and vegetables. A study of middle school-aged students indicate that school gardening may affect children's vegetable consumption, including improved recognition of, attitudes toward, preferences for and willingness to taste vegetables. Gardening also increases the variety of vegetables eaten (Ratcliffe *et al.* (2011) School gardens can positively impact children's food choices by improving their preferences for vegetables and increasing their nutrition knowledge (Graham *et al.* 2005). A study in Tucson, Arizona, showed that children who participated in the garden learned to like healthy foods. The vegetables that the children grew had a high intrinsic value (Cavaliere, 1987). After gardening children have shown more positive attitudes toward fruit and vegetable snacks and an improvement in vegetable preference scores (Lineberger, 1999). It is reported that the adolescents who participated in the garden-based nutrition intervention increased their servings of fruits and vegetables more than students in the two other groups. Although further research is needed, the results of this study seem to indicate the efficacy of using garden-based nutrition education to increase adolescents' consumption of fruits and vegetables (McAleese and Ranklin, 2007).

In a project that involved integrating nutrition and gardening among children in grades one through four, the outcomes have gone well beyond an understanding of good nutrition and the origin of fresh food, to include enhancing the quality and meaningfulness of learning (Canaris, 1995). After participating in the nutritional program, children's knowledge about the benefits of eating fruit and vegetables significantly improved, but there were no significant differences found in participants' attitude scores toward fruit and vegetables. However, the participants did report eating healthier snacks after participation in the nutritional program (Koch *et al.,* 2006). Access to a garden also significantly increased the variety of fruits and vegetables in diets (Wright *et al.,* 2008). In a study to identify whether or not there are associations between frequency of eating homegrown produce among rural parents and their preschool children and overall intake in eight rural Southeast Missouri counties. Frequency of eating homegrown fruits and vegetables promoted a positive home environment with increased availability of produce, preschooler's preference for them and parental role modeling (Nanney *et al.*, 2007).

4.2 Academic Achievement

Academic achievement is the primary focus of educators throughout the world. Emphasis is placed on ensuring students perform at satisfactory levels guided by local, state and national academic standards. All lessons and activities must complement mandatory standards to merit the use of valuable classroom time. An eight-lesson gardening and nutrition curriculum with a hands-on gardening emphasis was taught as an after-school program to determine the effect it had on increasing children's nutrition knowledge, fruit and vegetable preference and improving children's self-efficacy and outcome expectations for gardening and for consuming fruit and vegetables. Seventeen fourth grade students participated in the experimental group as part of an after-school gardening club and 21 fourth grade students served as the control group. At baseline, measurements of gardening self-efficacy and outcome expectations were significantly different between the groups. The experimental group was able to maintain high self-efficacy and outcome expectations scores during the program, but the control group's scores increased significantly for gardening self-efficacy and outcome expectations at the end-program assessment. Further research to clarify aspects of gardening that have the greatest impact on influencing preference, self-efficacy and outcome expectations is needed (O'Brien *et al.*, 2006).

The study assessed school teachers' perceived attitudes and barriers associated with school gardens, as well as the purpose and use of gardens in schools, specifically in relation to the link between gardens and nutrition. Results indicate that the teachers perceived the garden to be somewhat to very effective at enhancing academic performance, physical activity, language arts and healthful eating habits. This research provides evidence for needed standards-based curricula materials and teacher training in relation to gardening and nutrition (Graham and Zidenberg-Cherr, 2005). Third, fourth and fifth grade students that participated in school gardening activities scored significantly higher on science achievement tests compared to students that did not experience any garden-based learning activities (Klemmer *et al.*, (2005). Several variables may have affected the outcome of the study, but the results show once weekly use of gardening activities and hands-on classroom activities help improve science achievement test scores (Smith and Motsenbocker, 2005).

Although educators widely use school gardens for experiential education, researchers have not systematically examined the evaluative literature on school-gardening outcomes. The author reviewed the U.S. literature on children's gardening, taking into account potential effects, school gardening outcomes, teacher evaluations of gardens as learning tools and methodological issues. Quantitative studies showed positive outcomes of school-gardening initiatives in the areas of science (Anon., 2013).

4.3 Healthy Life Style and Life Skills

Beyond academics, the garden provides broader life lessons including contributing to students' knowledge of how to maintain a healthy lifestyle. The number of overweight and obese youth is increasing causing concern to parents and society especially in urban areas. Results from survey respondents at the Tennessee Green Industries Field Day (McMinnville) and the Tennessee State Fair (Nashville) suggest that although gardeners select from a wide range of plant materials and activities in an individualistic manner, the interaction with nature in a nurturing environment provides a number of benefits important to them, including an opportunity for self-expression, physical and physiological benefit and restorative experiences (Catanzaro and Ekanem, 2004). Directed attention plays an important role in human information processing; its fatigue, in turn, has far-reaching consequences. Attention Restoration Theory provides an analysis of the kinds of experiences that lead to recovery from such fatigue. Natural environments turn out to be particularly rich in the characteristics necessary restorative experiences (Kaplan, 1995).

Results from a study reinforce findings of an earlier investigation of New Zealand mid-aged women gardeners that gardening serves a wide range of needs and benefits on many levels, including psychological, emotional, social and spiritual. The data also provide an understanding of the type of persons who engage in gardening and give an overall picture of their motivations for gardening. Finally, the study identifies which factors are apparent indicators for someone to find satisfaction and a sense of well-being from gardening

(Kidd and Brascamp, 2004). A study revealed important benefits of gardening on physical, emotional, social and spiritual well-being and highlighted a key role of gardening as a coping strategy for living with stressful life experiences. The prospective nature of the study revealed the personal and subjective ways in which interest in gardening might change in response to the person's own situation and needs (Unruh, 2004).

4.4 Awareness about Environment

A school garden is a powerful environmental education tool. Through gardening, students become responsible caretakers. They explore the web of interactions among the living and non-living players that sustain life. Elementary school and junior high school students gained more positive attitudes about environmental issues after participating in a school garden program (Waliczek and Zajicek, 1999). Gardens are often the most accessible places for children to learn about nature's beauty, interconnections, power, fragility and solace (Heffernan, 1994). Both passive and active interactions with plants during childhood are associated with positive adult values about trees. However, the strongest influence came from active gardening, such as, picking flowers or planting trees as a child (Lohr and Pearson-Mims, 2005). Gardening has been shown to increase scores on environmental attitude surveys of elementary school children (Skelly and Zajicek, 1998). A study discusses a strain of bacterium in soil, *Mycobacterium vaccae* that has been found to trigger the release of seratonin, which elevates mood and decreases anxiety. This bacterium has also been found to improve cognitive function (Lowry *et al.*, 2011).

5. BUILDING A HEALTHY COMMUNITY

Community and social development lessons do not receive the attention of academic achievement, but they are as crucial to our survival as reading and writing. Children must learn how to take responsibility for their environment and develop a strong sense of community to ensure the continuation of our society. There is ample evidence to show that gardening has positive effects on well being of the community. Acceding to Guy *et al.* (1996) you need to know only three things to run a successful School garden, *viz.*, 1) How to cultivate people, 2) How to cultivate plants and 3) Where to go for help. Divina Misericordia is a school for disabled children in Lima, Peru. The school has its own garden, which grows lettuce, beets, carrots and broccoli. Everyone was involved from the beginning: the director, the teachers, the students and some parents built the garden from scratch. When they started, there was nothing but sand and dirt. They cleaned it up, brought in soil and installed an irrigation system. The school director Elvira Pacherres says the children are passionate about the garden. "Gardening is now part of the curriculum and it works as therapy for these children. It shows them how easily they can have access to food if need be and it gives them responsibilities. Often, these kids are put aside in the family… here they learn to contribute to the household. Some grow their own little gardens at home too" (FAO Telefood, 2004).

Gangadhar Bidyaniketan is a secondary school surrounded by paddy fields in eastern India. During the rainy season students and teachers had to walk two kilometres in mud and water to reach the school. In summer there was no water at all and children had to carry water bottles to school. There was not a single tree in the area. The school decided to make changes. Each student agreed to plant and care for one tree. Choosing the locations carefully, they created a green belt around the school. With the help of a local NGO the villagers built an approach road and a water tank on the school campus. The school now has plenty of drinking water and can use the surplus water for growing vegetables and flowers the land that was barren and saline is now green and colourful (Patta Naik, 1998).

In Burkina Faso a school garden project has had a real influence on the community and vice versa. Women community workers helped schools set up gardens to grow foods rich in vitamin A. Some of these (*e.g.* carrots and orange sweet potato) were new to the area - they were first consumed by the children and then carried home and tried out there. Many home gardens have been started in imitation and now people are producing and eating far more vegetables. With the increase in vitamin A consumption, there are far fewer cases of night blindness. Home gardeners are even trying out foods (*e.g.* tomato leaves) which have not been promoted by the community workers. In northern Mexico a school garden project invited successful local gardeners to act as trainers and educators for the school garden. In Bangladesh a school garden project appointed a village resident to manage the garden and compensated her with a percentage of the proceeds (Cederstrom, 2002).

In Kenya, manual work is likened to punishment. But this stigma is changing because our school has made a success of the garden. The children love the garden. They eat the food we produce and are visibly healthier. Parents shake us by the hand because they can see the difference. The garden has given the school a name. The District Education Officer brings visitors to see it (Choday, 2004).

Gardening promotes relief from acute stress: Stress-relieving effects of gardening were hypothesized and tested in a field experiment. Thirty gardeners performed a stressful task and were then randomly assigned to 30 minutes of outdoor gardening in their own plot or indoor reading. Gardening and reading each led to decreases in cortisol during the recovery period, but decreases were significantly stronger in the gardening group. Positive mood was fully restored after gardening, but further deteriorated during reading. These findings provide the first experimental evidence that gardening can promote relief from acute stress (Van Den Berg *et al.,* 2011).

Greenery makes life more manageable: In a study conducted in a Chicago public housing development, women who lived in apartment buildings with trees and greenery immediately outside reported greater effectiveness and less procrastination in dealing with their major life issues than those living in barren but otherwise identical buildings. In addition, the women in greener surroundings found their problems to be less difficult and of shorter duration. Exposure to green surroundings refreshes the ability to concentrate, leading to greater effectiveness at coping with major life issues. And, in this study, even small amounts of greenery—a few trees and a patch of grass—helped inner city residents to feel and do better (Kuo, 2001). Community gardening has great potential to help alleviate some of the stress experienced by military spouses. It offers recreation, physical and mental health benefits and the potential to help cultivate community by fostering social networking and mentorship platform (Fairleigh, 2004).

Participation with nature enhances health and well-being: Participation with nature enhances mental health, reduces stress and can produce physiological benefits, such as, lower blood pressure and reduced muscle tension (Simson and Straus, 1998). Gardeners directly experience nearby nature by 'getting their hands dirty' and growing food. They enjoy the way vegetables taste and form emotional connections with the garden. The physical and social qualities of garden participation awaken the senses and stimulate a range of responses that influence interpersonal processes (learning, affirming and expressive experiences) and social relationships that are supportive of positive health-related behaviors and overall health. This research suggests that the relational nature of aesthetics, defined as the most fundamental connection between people and place, can help guide community designers and health planners when designing environment and policy approaches to improve health behaviors (James *et al.,* 2011).

Community gardens enhance nutrition and physical activity and promote the role of public health in improving quality of life. Opportunities to organize around other issues and build social capital also emerge through community gardens (Twiss *et al*, 2003). A study found that neighborhood aesthetics, social involvement and community garden participation were significantly associated with fruit and vegetable intake. Community gardeners consumed fruits and vegetables 5.7 times per day, compared with home gardeners and non-gardeners (3.9 times per day). Moreover, 56% of community gardeners met national recommendations to consume fruits and vegetables at least 5 times per day, compared with 37% of home gardeners and 25% of non-gardeners (Litt *et al*., 2011).

A study highlights the advantages of solar cookers and gardens to nutrition, health and the environment. Study results indicate that growing their own vegetables encouraged adults and children to try new foods, which in turn improved their eating habits. Free seeds were distributed for container gardens, which motivated new and experienced gardeners to make gardens. Both rural and urban participants gained enthusiasm for the fresh, flavorful, nutritious and economical vegetables they grew and for the varied foods they cooked in solar cookers. They realized that gardening helped to improve their nutrition, health, family, economics and the environment (Dow and Dow, 1999). In a study conducted at a Chicago public housing development, residents of buildings with more trees and grass reported that they knew their neighbors better; socialized with them more often; had stronger feelings of community; and felt safer and better adjusted than did residents of more barren, but otherwise identical, buildings. Such settings support frequent, friendly interaction among neighbors–the foundation of neighborhood social ties. These ties are the heart of a neighborhood's strength, encouraging neighbors to help and protect each other (Kuo *et al*., 1998).

A study on a youth gardening program in Detroit reports that after gardening, children have an increased interest in eating fruit and vegetables, possess an appreciation for working with neighborhood adults and have an increased interest for improvement of neighborhood appearance. In addition, they made new friends and showed increased knowledge about nutrition, plant ecology and gardening (Pothukuchi, 2004). Results from this study examining formal and informal community-based social networks and family adaptation in military communities suggest that communities can be important sources of tangible information and expressive support. Further, a sense of real belonging in a distinct place can help to balance the turmoil of the deployment cycle (Bowen *et al.*, 2003). This survey of community gardens in upstate NY indicated gardens in low-income neighborhoods (46%) were four times as likely as non low-income gardens to lead to other issues in the neighborhood being addressed; reportedly due to organizing facilitated through the community gardens (Armstrong, 2000).

6. CONCLUSION

School gardening offers children opportunities for outdoor exercise while teaching them a useful skill. Gardens containing fruit and vegetables help to revise attitudes about particular foods. There is mounting evidence that active learning in less structured, participatory spaces like nutrition gardens are more likely to transform children's food attitudes and habits and that school gardening, especially when combined with a healthy lunch programme and/or nutritional education, encourages more healthful food choices. Practical experience with fresh foods positively impacts dietary habits. When students take their preferences back to their families, they can help to improve family consumption choices also. Students are more likely to try eating vegetables they have grown themselves and to ask for them at home. The main actors themselves, the school children must be approached in a participatory way to insure their commitment to the programme.

With the promotion of School Garden Programmes a great number of malnourished children can be reached with relatively simple and cost effective means. School garden programmes cannot be created in isolation but have to be linked with education and health and nutrition interventions. Therefore best program design seems to link school gardens with other programs related to nutrition, health and environment. School garden programs must be well adapted to the local habits, needs and to the specific climatic and environmental situation. Community involvement is a basic requirement for the success of these programmes.

School gardens are considered as catalyst to trigger change in food habits of school children, parents and community a whole. School gardens can play pivotal role in encouraging community gardens that eventually support income, food and social security of the community. School children will also encourage both rural and urban youth to engage in community gardens and allied activities that helps to build a safe and healthy society. Though national policies and interventions are aimed at tackling the problems of malnutrition and under- nutrition, the changes that take place at grass root level ensures social good.

Learning about food by personally growing crops and livestock on small scale makes them familiar with ways of earning livelihood. Educated children can effectively communicate the ways to ensure nutritional security among households and community at large. Utilization of community lung spaces creates income generation activities in the community. Sharing responsibilities and benefits among the members brings harmony and social security in the communities. School children are potent agents of change that makes healthy and prosperous societies.

Acknowledgement

The authors are thankful to Dr. Prem Nath Agricultural Science Foundation, Bengaluru, for providing fellowships and considering the paper for publication.

REFERENCES

Anon. (2013). The child in the garden: An evaluative review of the benefits of school gardening. Abrt.csgn-master. (Cited from: Research database: The collective school garden network. http://www. Csgn.org./research. 8 Jan., 2013.)

Armstrong, D. (2000). A survey of community gardens in upstate New York: Implications for health promotion and community development. Health & Place. 6: 4: 319-327 (Cited from: Research database: The collective school garden network. http://www. Csgn.org./research. 8 Jan., 2013.)

Bajaj, Vikas, (2012). Hunger in the midst of abundant food. Panoraama. Deccan Herald. June 11, 2012.

Bowen, G. L. *et al.* (2003). Promoting the Adaptation of Military Families: An Empirical Test of a Community Practice Model. Family Relations. 52, p. 33–44. (Cited from: Research database: The collective school garden network. http://www. Csgn.org./ research. 8 Jan., 2013.)

Bruce, B. (1998). *Corn in the classroom.* Developing Countries Farm Radio Network Package 48, script 7. (a broadcast on Sligoville School, Jamaica).

Callens, Karel. (2004). cited from FAO (2012) Setting up and running a school nutrition garden project. http://www.fao.org/3/a-a0218e/A0218E01.html.

Canaris, Irene. (1995). Growing Foods for Growing Minds: Integrating Gardening and Nutrition Education into the Total Curriculum. Children's Environments, 12(2): 264-270. (Cited from: Research database: The collective school garden network. http://www. Csgn.org./research. 8 Jan., 2013.)

Catanzaro, C. and Ekanem, E. (2004). Home Gardeners Value Stress Reduction and Interaction with Nature. Acta Hort. (ISHS) 639:269-275.) (Cited from: Research database: The collective school garden network. http://www. Csgn.org./research. 8 Jan., 2013.)

Cavaliere, D. (1987). How Zucchini Won Fifth-Grade Hearts. Children Today, 16(3), 18-21.(Lineberger, S. (1999). The Effect of School Gardens on Children's Attitudes and Related Behaviors Regarding Fruits and Vegetables. Thesis, Texas A&M University. (Cited from: Research database: The collective school garden network. http://www. Csgn.org./research. 8 Jan., 2013.)

Cederstrom, T. (2002). Learning to grow: How school gardens can provide food for education. Food Forum Online, Issue 61, 3rd quarter. www.foodaid.org/ff2002.html.

Chavsith, V. (2012). Importance of vegetables to achieve food and nutrition security in South-East Asia. Proc.of regional Symp. On High value vegetables in SEA-Production, supply and demand.

Childs Elizabeth, (2011). Impact of school garden on student attitudes and beliefs. Graduate thesis and desertation. Iowa State Univ.

Choday, A. (2004). cited from FAO (2012) Setting up and running a school nutrition garden project. http://www.fao.org/3/a-a0218e/A0218E01.htm.

Dow,R.M. and Dow, C.R. (1999). Using solar cookers and gardens to improve health in urban and rural areas. Afallit internationalInc.99:9 (Cited from: Research database: The collective school garden network. http://www. Csgn.org./research. 8 Jan., 2013.)

Eboh, L.O. and Boye, T.E. (2006). Nutrition knowledgr and food choices of primary school childderen in in Nifer delta region in Nigeria. J. of Nutrtion.p.308-311.

Fairleigh, Megan E. (2004). "Gardens for the Green Machine: Investigating the Use of Community Gardening For Stress Treatment in Marine Corps Families." MLA Thesis. California State Polytechnic University, Pomona, (2004.) (Cited from: Research database: The collective school garden network. http://www. Csgn.org./ research. 8 Jan., 2013.)

FAO Telefood. (2004a). Zambia: helping AIDS orphans help themselves. Telefood website www.fao.org/food/english/stories/projects.

Gillespie, S., Harris, J. and Kadiyala, S. (2012). The agriculture - nutrition disconnect in India.IFPRI. http://www.ifpri.org/publication/agriculture-nutrition-disconnect-india.

Govt. of India. (2011). Faster ,sustainable nd more inclusive growth:An approach for the XII FYP. Planning Comission, New Delhi.

Graham, H. and S. Zidenberg-Cherr. (2005). California teachers perceive school gardens as an effective nutritional tool to promote healthful eating habits. Journal of the American Dietetic Association, 105:1797-1800. (Cited from: Research database: The collective school garden network. http://www. Csgn.org./research. 8 Jan., 2013.)

Graham, H. *et al.,* (2005). Use of School Gardens in Academic Instruction. Journal of Nutrition Education and Behavior. 37: 3: 147-151. (Cited from: Research database: The collective school garden network. http://www. Csgn.org./research. 8 Jan., 2013.)

Greenhouse, J. and Carr, L. (2003). cited from FAO (2012) Setting up and running a school nutrition garden project. http://www.fao.org/3/a-a0218e/A0218E01.htm.

Guy, L.A., Cromell, C. & Bradley, L.K. (1996). *Success with school gardens: how to create a learning oasis in the desert*. Phoenix, Arizona Master Gardeners Press (in cooperation with the University of Arizona).

Heffernan, M. (1994). The Children's Garden Project at River Farm. Children's Environments. 11(3): 221-231. (Cited from: Research database: The collective school garden network. http://www. Csgn.org./research. 8 Jan., 2013.)

Hegde, N.G. (2012). Water scarcity and security in India 99th Indian Science Congress, KIIT UNIv., Bhubaneswar. p19.

Hegde, N.G. (2013). Improved integrated farming to augment food and nutrition security of masses The bsics of Human civilization: Food, Agriculture and Humanity. Present Scenario.Ed.Prem Nath. New India publishing agency, New Delhi, India. p. 279-285

Heimendinger, J. and M. Van Duyn. (1995). Dietary behavior change: the challenge of recasting the role of fruit and vegetables in the American diet. American Journal of Clinical Nutrition, 61:1397S-1401.) (Cited from: Research database: The collective school garden network. http://www. Csgn.org./research. 8 Jan., 2013.).

Indumati, K., Shanmugam P.S. and Tamilselva, N. (2012). Nutrition garden as a valuable intervention to fight malnutrition in rural india. Absrt. Global conference on Hoerticulture for food, nutrition and livelihood options. Bhubaneswar. 25-31 May, 2012. p. 19.

James, H., *et al.* (2011). Connecting food environments and health through the relational nature of aesthetics: Gaining insight through the community gardening experience. Social Science Medicine, 72: 11: 1853-1863. (Cited from: Research database: The collective school garden network. http://www. Csgn.org./research. 8 Jan., 2013.)

Kaplan, Stephen. (1995). The restorative benefits of nature: Toward an integrative framework. Journal of Environmental Psychology. Volume 15: Issue 3: 169-182. (Cited from: Research database: The collective school garden network. http://www. Csgn.org./research. 8 Jan., 2013.)

Kidd, J.L. and Brascamp, W. (2004). Benefits of Gardening to the Well-Being of New Zealand Gardeners. Acta Hort. (ISHS) 639:103-112. (Cited from: Research database: The collective school garden network. http://www. Csgn.org./research. 8 Jan., 2013.)

Klemmer, C.D., Waliczek, T.M. & Zajicek, J.M. (2005). Growing Minds: The Effect of a School Gardening Program on the Science Achievement of Elementary Students. HortTechnology. 15(3): p. 448-452. (Cited from: Research database: The collective school garden network. http://www. Csgn.org./research. 8 Jan., 2013.)

Knobloch. (2003). cited from Cornell garden based learning and highlights from journal articles. Blogs.cornell.edugarden.grow. you. Program. research.

Koch, S., Waliczek, T.M. and Zajicek, J.M. (2006). The Effect of a Summer Garden Program on the Nutritional Knowledge, Attitudes and Behaviors of Children. Hort Technology. 16(4), p. 620-624. (Cited from: Research database: The collective school garden network. http://www. Csgn.org./research. 8 Jan., 2013.)

Kuo, F.E. (2001). Coping with poverty: Impacts of environment and attention in the inner city. Environment & Behavior 33.1: p. 5-34. (Cited from: Research database: The collective school garden network. http://www. Csgn.org./research. 8 Jan., 2013.)

Kuo, F.E., *et al.* (1998). Fertile ground for community: Inner-city neighborhood common spaces." American Journal of Community Psychology 26.6: p. 823-851. (Cited from: Research database: The collective school garden network. http://www. Csgn. org./research. 8 Jan., 2013.)

Lineberger, S. (1999). The effect of school gardens on children's attitudes and related behaviours regarding fruits and vegetablesThesis. Texas A&M univ (Cited from: Research database: The collective school garden network. http://www. Csgn.org./ research. 8 Jan., 2013.)

Litt, J. S. *et al.* (2011). The Influence of Social Involvement, Neighborhood Aesthetics and Community Garden Participation on Fruit and Vegetable Consumption. American Journal of Public Health, Vol. 101, No. 8, p. 1466-1473.) (Cited from: Research database: The collective school garden network. http://www. Csgn.org./research. 8 Jan., 2013.)

Lohr, V.I. and Pearson-Mims, C.H. (2005). Children's Active and Passive Interactions with Plants Influence Their Attitudes and Actions toward Trees and Gardening as Adults. Hort Technology. 15(3): p. 472-476. (Cited from: Research database: The collective school garden network. http://www. Csgn.org./research. 8 Jan., 2013.)

Lowry, C.A., *et al.* (2011). Identification of an immune-responsive mesolimbocortical serotonergic system: potential role in regulation of emotional behavior. Neuroscience 146.2: p. 756–772. (Cited from: Research database: The collective school garden network. http://www. Csgn.org./research. 8 Jan., 2013.)

McAleese, J. D. and L. L. Ranklin. (2007). Garden-based nutrition education affects fruit and vegetable consumption in sixth-grade adolescents. Journal of the American Dietetic Association, 107: p. 662-665. (Cited from: Research database: The collective school garden network. http://www. Csgn.org./research. 8 Jan., 2013.)

Mitra Pritvijit. (2015). India. City Farmer News. http:/www.cityfarmerinfo/category/india.

Mitra, Arjith. (2014). personal communication.

Morris, L.J., M. Briggs and S.Zidenberg –Cherr. (2002). Development evaluation of garden enhanced nutrition education curriculum for elementary school children. The J. of child nutrition amnd management. 2:1-5.

Nandal, V. and I.J. mathur. (2012). Backyard nutrition gardening; n effective solution for achieving food and nutrton security of tribal families in Sirohi (Rajasthan). Absrt. Global conference on Horticulture for food, nutrition and livelihood options. Bhubaneswar. 25-31 May, 2012. p.25.

Nanney, M.S. *et al.* (2007). Frequency of Eating Homegrown Produce Is Associated with Higher Intake among Parents and Their Preschool-Aged Children in Rural Missouri. Journal of the American Dietetic Association (2007) 107.4: p. 577-584. (Cited from: Research database: The collective school garden network. http://www. Csgn.org./research. 8 Jan., 2013.)

Nath, Prem, Prabhakar, B.S., Chandre Gowda, M.J., Mukund Sirgur, Patil, M. S., Yatnatti, Shilpa and Latha, S.V. (2009). Nutrition awareness through school childern Souvenir, International Conferene on Horticultur,: Horticulture for livelihood security and Economic Growth. Bengaluru. November 9-12. p. 97-101.

Nath, Prem. (2013). Challenges and opportunities in food, agriculture and humanity. The bsics of Human civilization: Food, Agriculture and Humanity. Present Scenario. Ed.Prem Nath. New India publishing agency, New Delhi, India. p. 565-667.

Neube, S. (2004). cited from FAO (2012) Setting up and running a school nutrition garden project. http://www.fao.org/3/a-a0218e/A0218E01.htm.

Nuebe, S. and Chinanzavavana, L. (2004). cited from FAO (2012) Setting up and running a school nutrition garden project. http://www.fao.org/3/a-a0218e/A0218E01.htm

O'Brien, S. A. *et al.* (2006). An After-school Gardening Club to Promote Fruit and Vegetable Consumption among Fourth Grade Students: The Assessment of Social Cognitive Theory Constructs. Hort Technology. 16(1), p. 24-29. (Cited from: Research database: The collective school garden network. http://www. Csgn.org./ research. 8 Jan., 2013.)

Pattanaik, B. (1998). "A school garden in India", Developing Countries Farm Radio Network, Package 50, Script 5.

Pothukuchi, K. (2004). Hortaliza: A Youth "Nutrition Garden" in Southwest Detroit. Children, Youth and Environments. 14(2): p. 124-155. (Cited from: Research database: The collective school garden network. http://www. Csgn.org./research. 8 Jan., 2013.)

Power, C. (2003). cited from FAO (2012) Settingup and running a garden project. http:// www.fao.org/3/a-a0218e/A0218E01.htm.

Pratibha and S. Rani. (2012). Nutrtion garden: A venture to mitigate malnutrition. Absrt. Global conference on Hoerticulture for food, nutrition and livelihood options. Bhubaneswar. 25-31 May, 2012. p. 11-12.

Ratcliffe, M.M. *et al.* (2011). The Effects of School Garden Experiences on Middle School-Aged Students' Knowledge, Attitudes and Behaviors Associated with Vegetable Consumption. Health Promotion Practice 12.1: p. 36-43.(Cited from: Research database: The collective school garden network. http://www. Csgn.org./research. 8 Jan., 2013.)

Sheela, K.B., Peter, K.V. and K.Krishnakumari (1998). Nutrition Garden. Directorate of Extension. KAU, Vllanikkara, trissur. p. 86.

Shukla, P., Rajkumari and R. Limbu. (2012). Nutrtion garden as a tool towards combating nutrient deficiencies. Absrt. Global conference on Horticulture for food, nutrition and livelihood options. Bhubaneswar. 25-31 May, 2012. p. 16-17.

Simson, S.P. and M.C. (1988). Straus.People-Plant Relationship. in Horticulture as Therapy. Ed. S.P. Simson and M.C. Straus. New York: The Food Products Press, 1988. p. 21-42. (Cited from: Research database: The collective school garden network. http://www. Csgn.org./research. 8 Jan., 2013.)

Skelly, S. and J. Zajicek. (1998). The Effect of an Interdisciplinary Garden Program on the Environmental Attitudes of Elementary School Students. Hort Technology, 8(4): p. 579-583. (Cited from: Research database: The collective school garden network. http://www. Csgn.org./research. 8 Jan., 2013.)

Smith, L. L. and Motsenbocker, C. E. (2005). Impact of Hands-on Science through School Gardening in Louisiana Public Elementary Schools. HortTechnology. 15(3), p. 439-443(Cited from: Research database: The collective school garden network. http://www. Csgn.org./research. 8 Jan., 2013.)

Twiss, J., *et al.* (2003). Community Gardens: Lessons Learned From California Healthy Cities and Communities. American Journal of Public Health 93.9: p. 1435-1438. (Cited from: Research database: The collective school garden network. http://www. Csgn.org./research. 8 Jan., 2013.)

UNICEF. (2012). http://unicef.in/whatwedo.

Unruh, A.M. (2004). The Meaning of Gardens and Gardening in Daily Life: A Comparison Between Gardeners With Serious Health Problems And Healthy Participants. Acta Hort. (ISHS) 639: p. 67-73. (Cited from: Research database: The collective school garden network. http://www. Csgn.org./research. 8 Jan., 2013.)

Van Den Berg, Agnes and Custers, Mariëtte H.G. (2011). Gardening Promotes Neuroendocrine and Affective Restoration from Stress. Journal of Health Psychology 16.1: 3-11. (Cited from: Research database: The collective school garden network. http://www. Csgn.org./research. 8 Jan., 2013.)

Veeresh, G.K. (2013). Making small farmers sustainable in India. The basics of Human civilization: Food, Agriculture and Humanity. Present Scenario.Ed.Prem Nath. New India publishing agency, New Delhi, India.p 155.

Waliczek, T.M. and Zajicek, J.M. (1999). School Gardening: Improving Environmental Attitudes of Children Through Hands-On Learning. Journal of Environmental Horticulture. 17(4): 180-184. (Cited from: Research database: The collective school garden network. http://www. Csgn.org./research. 8 Jan., 2013.)

Wanasinghe, A.D. (2003). From School Garden to Home Garden. Integrated Food Security Programme. Trincomalee, Sri-Lanka. Technical p. 29.

Wright M. *et al.* (2008). Accessing food resources: Rural and urban patterns of giving and getting food. Agriculture and Human Values. Volume 25, Number 1. (Cited from: Research database: The collective school garden network. http://www. Csgn.org./ research. 8 Jan., 2013.)

❍❍❍

VIII. Addressing Sustainable Horticulture Development in the New Millennium

27

Sustainable Horticultural Development in India: Growth and Pathways for Sustainability

S. Bisaliah

1. PREAMBLE TO THE THEME

With wide variability of climate, soil and topography, India is highly favorable for growing horticulture crops, such as, fruits, vegetables, flowers and aromatic, plantation crops and spices. It is the fastest growing subsector in agriculture, contributing over one-third of the agriculture GDP. Further, major contribution is expected to come from this sector for achieving a national goal of around 4% growth per annum in agriculture. In fact as of now, India is said to be the second largest producer of fruits and vegetables in the world, even though in terms of volume of production India produces less than 20% of what China produces. But the potentials for growing horticulture crops in terms of topography-soil-climate complex are said to be quite high.

A brief reference to reasons for delayed emphasis on horticulture development is in order. It was a historical necessity for India to place high emphasis on production of cereals after independence and Green Revolution of 1960s and 1970s was the outcome of this drive towards Cereals production. Hence, only in 1980s the country realized the importance of horticulture crops and much more so during the period of post – 1993 with increased intensity of globalization which in turn triggered the allocation of more financial outlay and provision of knowledge inputs through techno-generation and transfer. The arguments built up to place priority for the development of horticulture are many: For crop diversification, for increasing profitability of farming and for creating more employment opportunities at different stages of production, processing and marketing of horticultural crops, as well as for addressing the problems of nutrition as a part of total food security, beyond the notion of calories (Planning Commission, 2001). Further, many arguments have been advanced (Bisaliah, 2011) for considering horticulture as a candidate for addressing many socio-economic problems of agriculture sector: Nutritional contents of horticultural crops, change in food basket of consumers, proper use of diverse agro-climatic resources of the country, higher employment and net income, foreign exchange earnings and move towards more inclusive growth. Bisaliah (2011) has also argued that horticulture could support and lead the growth of crop production sector. Hence, horticulture-led growth of crop production sector is quite imperative, even to prevent /minimize growth deceleration in agriculture sector. Further, even though encouraging growth performance of horticulture has been responsible for preventing deceleration of productivity and production growth in crop production sector, production and productivity rates of horticulture sector have started decelerating since 1997/98 and this needs to be addressed through technological, institutional and policy interventions so as to place horticulture sector in high growth trajectory (Details in Bisaliah, 2011).

Yet, another way is to state the advantages claimed in support of farms diversification through horticulture crops (Planning Commission, 2001): Higher biomass, efficient use of natural resources, highly remunerative in rainfed, hilly arid and coastal belts, improvement of waste land, less water intensive, environment friendly, high potential for value addition and for foreign exchange earnings and more contribution to GDP from less cultivated area.

In view of all the reasons and arguments stated above, horticulture has moved from rural confines to becoming a commercial venture. In fact, the dawn of liberalization in international trade and the emergence of an integrated global market have offered new vistas for Indian horticulture. To support this drive, National Horticulture Mission (NHM) launched in 2005-06 has paved the way for India to exploit its potential in horticulture sector. NHM has a mandate to promote integrated development of horticulture, focusing on both production-oriented and market-oriented strategies. NHM has placed emphasis on the holistic growth of horticulture by adopting an area-based regionally differentiated cluster approach. The main objectives of the Mission are to: enhance productivity and production of horticulture crops, reduce post harvest losses, improve nutritional security, increase farmer income and to generate more employment opportunities at different stages of "farm-to-plate". It is beyond the scope of the present paper to conduct a performance appraisal of NHM against the mandated goals of the Mission. However, as a drive towards the development of horticulture in the country, establishment of NHM has provided a much needed institutional support for a end-to-end holistic development of horticulture, covering production, post-harvest management and processing, to enhance acreage and productivity growth, to promote crop diversification and to adopt a coordinated approach with public and private sector tie-up and to promote capacity building and human resource development as well at all levels. To cite a few statistics to suggest the advancement in the development of the sector, area under horticulture has increased from18.5 million hectares to 23.7 million hectare during 2005-06 and 2012-13 and production from 182.8 million tonnes to 268.8 million tonnes and productivity from 9.8 tonnne per hectare to 11.3 tonnne during the same period (Details later on the statistical profile of horticulture in India). As a result of this development drive, there has been a shift in cropping pattern from cereals to fruits and vegetables, leading to slow transition to sun-rise sector syndrome and to agro-diversification for the second Green Revolution in the country.

2. STATEMENT ON THE ISSUES ADDRESSED

Against the backdrop of the section preceded, the issues addressed in the present paper are as follows:

- Can we cast the sustainability of horticulture sector in the perspective of interaction between development and environment?
- How does the growth of horticulture look in the form of a statistical profile? What is the pattern of growth of the sector that has surfaced in India?
- Which ones would constitute the dimensions of sustainability in terms of production and trade orientation?
- What are the possible pathways for sustainable horticulture development and what lessons India can learn from the experience of some important players in foreign trade?

3. GLOBAL BACKDROP TO THE CONCEPT OF SUSTAINABLE HORTICULTURE DEVELOPMENT

Four key concerns have emerged from the collective concerns and aspirations of the global community: peace, freedom, development and environment. In this context, the nexus between development and environment has emerged as the theme drawing the attention of researchers, development agencies (both national and international) and policy makers. Further, it is in the same perspective that sustainable development, the most important concept of twentieth century, has become a global agenda and a movement and it is now central to the missions of international organizations, national institutions, corporate enterprises and locales. In a way the issue of sustainable development started evolving since 1972 and there have been many milestones of this concept (Details in Bisaliah, 2015). The concerns of these milestones are captured in Brundtland Commission Report (1987): our Common Future, where in the concept of sustainable development is defined as "one which meets the needs of the present generation without compromising the ability of the future generation to meet their needs". The Report articulates: Environment is where we live and development is what we all do in attempting to improve our lot within that abode. Both environment and development are inseparable. The need for protecting our planet and for securing our future is very well articulated by Rob Watson (In Fried Man, 2009 P.179): Mother Nature is just chemistry, biology and physics. Everything she does just the sum of these three things. You cannot negotiate with her, you cannot spin her around and you cannot evade her rules. All you can do is "fit-in" as species. And when a species does not learn to fit in with nature, it gets kicked out. If we fail to fit in as species, we end up being a bad biological experiment on this planet. Perhaps, we can learn from the Red Indians of America: we have not inherited the natural resources from our forefathers, but we have borrowed from our children. We are obligated to preserve the natural environment with its natural resources endowments when we are engaged in our development drive. This is absolutely required to protect our planet and secure our future and of the future generations.

3.1 Transition to Industrial Farming

Against the concept of sustainability discussed in the paragraphs preceded, one can examine the sustainability of agriculture/ horticulture with the adoption of unsustainable practices especially in 1960s and 1970s in India. With impending food crisis, attempts were made in 1960s and 1970s to search for technologies that would increase food production to overcome the crisis and search ended with choice of biochemical technology comprising of high yielding seeds coupled with chemical fertilizers, pesticides and other associated cultural practices, which were of great help for India to contain food crisis. But the farming practices under Green revolution technologies have proved to be unsustainable. Agriculture/ horticulture has become more dependent on external inputs produced in the industrial sector, leading to transition to industrial farming. For example, inorganic fertilizers have substituted for livestock manures, composts and nitrogen fixing crops. Pesticides have replaced biological, cultural and mechanical methods for controlling pests, weeds and diseases. Regenerative and resources conserving technologies have received less emphasis (Details in Niels R. *et al.,* Undated). All these unsustainable practices are found to have dampening effect on soil health and yield. The case in point is Gangetic plane, the adobe of "success" of Green Revolution Technology.

Yet, another example is the unsustainable practices followed for the development of horticulture in Himalayan region of Himachal Pradesh. In a case study by Ramesh Chand (1996), the positive impact of horticulture in the region in terms of diversification of agricultural sector through the development of high value cash crops, mainly fruits and vegetables, has been established. But the same study recognizes that horticultural production could tend to result in ecological degradation in the region due to damage to forest wealth (Details in Ramesh Chand, 1996). Obviously the main challenge in the region is whether economical and ecological sustainability of horticultural development is possible in the Himalayan region. Further, it is not merely sustainability of horticultural production without damaging the environment, but also sustainability of trade related issues to increase our share in the international trade by increasing our competitive advantage both terms of quality and price advantage in the global market.

4. HORTICULTURAL DEVELOPMENT MAP: A STATISTICAL PROFILE

To build up the horticulture development map of India, a statistical profile of parameters of the sector is used as an analytical tool. The parameters considered for building up a statistical profile are: Trends in area, production and productivity of total horticultural crops and crops groups, the share of horticulture output in the total agriculture output, India's share in production of fruits and vegetables vis-à-vis other major countries, export share in global trade, and geographical concentration of dominant horticultural crop in India. It is recalled from earlier sections that India has witnessed shift in area from food grains towards horticultural crops. In fact area under horticulture crops had increased from12.77 million hectares in 1991-95 to 23.4 million hectors, registering 83% increase in area during this period. The total production of horticulture crops increased by nearly 2.9 times during the same period and productivity 1.6 times.

- Fruits and vegetables occupy over two-third of the area under horticulture crops in 2014-15, contributing about 90% of total Production in the sector. The total value of output of the sector is estimated to be over 30% of all agriculture crops, and fruits and vegetables alone contribute about 22.4%. Hence, the present study focuses mainly on two dominant crop groups. Between 2009-10 and 2015-16, compound annual growth rates are estimated to be 6.5% in case of total horticulture crops, 7% in case of vegetables and 6% in case of fruits and nuts.
- Between 2001-02 and 2013-14, production of total food grains increased from about 21 3 million tonnes to 264 million tonnes in India, whereas, production of horticulture produce from about 146 million tonnes to 281 million tonnes. Hence, production of horticulture crops has outpaced production of food grains.
- Even though the percentage of value share of horticulture output in agriculture output is estimated to be one-third, but the area under horticulture crops in total cropped area is estimated to be around 9%.
- Horticulture was allocated 3.9% of total outlay for agri-allied sector during IX plans (1997-2002) and has been increased to 4.6% during XII plan (2012-17).
- Geographically, the total horticulture production is the highest in West Bengal, followed by Andhra Pradesh. The highest share of fruit production is in Andhra Pradesh, followed by Maharashtra. The highest production of vegetables is in West Bengal, followed by Uttar Pradesh. The share of flower production is the highest in Tamil Nadu and the share of production of spices is the highest in Uttar Pradesh.

- India's share in production of fruits and vegetables in the world production during the year 2012-13 is estimated to be about 14% and it is second only to China. But China produced 583 Million tonnes of major vegetables during the year and India only 121 Million tonnes, *i.e.,* about 21% of production in China and productivity in china is found to be close to two times the productivity in India.
- The share of horticulture products in total agriculture exports would come to 52%: India accounts for 1.7% of global trade in vegetables and 0.5% in fruits. India is a minor player in international trade in vegetables and fruits.

4.1 Trends in Area, Production and Productivity: Decadal Change

To build up dynamic change over a period of time between 1991-92 and 2014-15 in area, production and productivity of horticulture crops in India, and patterns of output growth that has been registered in India, let us examine results in Table 1. It is realized that difficulties in estimating horticulture area, production and productivity are to be very well recognized (Horticulture Statistics at a Glance, 2015).

Table 1. *Decadal percentage change in Area, Production and Productivity of Horticultural crops and Dominant Crop Groups*

Sl.No.	Decade/Crop Groups	Percentage Change		
		Area	Production	Productivity
1.	1991/92-2001/02			
	Total Hort. Crops	30.0	51.0	16.0
	Fruits	40.0	50.0	10.0
	Vegetables	10.0	51.0	38.0
2.	2001/02-2011/12			
	Total Hort. Crops	40.0	77.0	26.0
	Fruits	67.0	78.0	10.0
	Vegetables	46.0	76.0	21.0
3.	2011/12-2014/15			
	Total Hort. Crops	1.0	10.0	10.0
	Fruits	-1.0	16.0	22.0
	Vegetables	6.0	10.0	2.0
4.	Overall: 1991/92-2014/15			
	Total Hort. Crops	83.0	194.0	60.0
	Fruits	121.0	210.o	40.0
	Vegetables	71.0	188.o	69.0

Note:

1. Total horticultural Crops are defined to include fruits, vegetables, flowers and aromatic, plantation crops and spices.
2. Fruits and Vegetables are treated as dominant crop groups, since they occupy over 90% of area under hort. Crops.
3. The overall period pertains to over two decades: 1991/92 - 2014-15.

It is seen from Table 1that:

- Between 1991-92 and 2014-15, there has been 83% change in area under horticulture crops in India, but production increased by 194%, because of 60% increase in productivity. Hence, the production growth pattern has turned out to be area-cum-productivity-led with dominance of area expansion under the crop.

  With respect to crop groups, the increase in production of fruits during this period of over two decades has been 210%, dominantly- led by area expansion with some contribution from productivity increase. In case of vegetables, the output increase of 188% is found to be area-cum-productivity led. Hence, one could observe tangible increase in productivity of vegetables in addition to area expansion.
- In case of results from three sub-period analysis viz 1991/92 to 2001/02, 2001/02-2011/12 and 2011/12-2014/15,it could be inferred that:
  - Production of total horticulture crops during the first period (1991/92-2001/02) was dominantly area-led, and it was also area-led in case of fruits, but productivity-led in case of vegetables.
  - During the second period (2001/02-2011/12), the growth pattern of horticulture crops in total was both area-cum-productivity led, of fruits dominantly area-led and of vegetables area-cum-productivity led.
  - During the third period(2011/12-2014/15), horticulture crops in total are found to have experienced productivity-led and the same inference holds good in case of fruits and in incase of vegetables it has been dominantly area-led. However, the inferences from third period growth pattern need to be read with caution, because data pertain to a period of only four years.

Hence, the overall period analysis (1991/92-2014/15) of over two decades data suggests that growth of horticulture sector depends mainly on area expansion, excepting in the case of vegetables where both area and productivity have contributed for growth of output.

4.2 Analysis of Global Productivity Differentials

Two dominant factors that have a bearing on competitive advantage of a country in global trade are productivity of the crop and international trade-related strategies designed by a country. Given the importance of productivity as a determinant of competitive advantage in trade, Table 2 provides a broad glimpse of productivity differentials across major countries, growing fruits and vegetables.

Table 2. *Global Productivity Differentials of Major Fruits and Vegetables: 2013.*

Country	Productivity (Tonnes/Ha)
I. Major Fruits	
India	11.91
U.S.A	23.88
Indonesia	22.54
Brazil	16. 5
World average	11.35

II. Major Vegetables	
India	13.99
China	23.89
Spain	37.35
U.S.A	32.65
World average	19.50

Source: *Horticulture Statistics at A Glance: 2015.*

It is evident from Table 2 that:

- In case of fruits, productivity in India is found to be 11.91 tonnes per hectare (a little over world average), whereas in U.S.A, it is 23.88 tonnes and productivity of fruits in India is far below that of Indonesia and Brazil.
- With respect to productivity of vegetables, India (the second largest producer of fruits and vegetables in the world, next only to china) has turned out to be a country where productivity of vegetables is far below the world average and it is far below the productivity in countries, such as China, Spain and U.S.A.

The conclusion that emerges from this analysis of productivity differentials is how India will be able to emerge as a leading country in the productivity, among other things, to enable it to be sustainable as a player in world market. High productivity, among other factors, would enable India to be competitive in world market and to be competitive in domestic market as well under WTO trade liberalization drive.

5. SUSTAINABILITY: DIMENSIONS AND PATHWAYS

To place this section in the proper perspective, a few inferences are drawn from the earlier sections:

First, the importance of horticulture as a subsector in agriculture and allied sector group has been inferred in terms of crop diversification, profitability, employment generation at different stages of production, processing and trading, nutrition value, foreign exchange earnings, efficient use of diverse agro-climatic resources in the country, minimization of growth deceleration in agriculture sector, more contribution to agriculture GDP from less cultivated area and so on.

Second, given the rationale for exploiting the potential in horticulture, the National Horticulture Mission was launched in 2005-06 with mandate for holistic development of horticulture in the country.

Third, the theme of this paper has been placed in the conceptual framework of Sustainable Development (SD), the most important concept of twentieth century, where in the concept of SD has to be defined as "one which meets the needs of the present generation without compromising the ability of the future generation to meet their needs". In this context, the transition to industrial agriculture under which, the unsustainable practices of Green Revolution Technology are indicated.

Fourth, the percentage share of horticulture to total value of Agriculture GDP is one-third, and area under horticulture crops in total cropped area is around 9%. Further, to total horticulture production in the country, fruits and vegetables alone contribute 90%.

Fifth, a statistical profile of horticulture development in the country has been examined to infer expansion of area under horticulture production, increase in horticulture sector and marginal increase productivity.

Sixth, India is second only to China in total production of fruits and vegetables, but productivity in China is found to be close to two-times the productivity of India.

Seventh, the share of horticulture products in total agriculture exports is above 52%, but India accounts for 1.7% of global trade in vegetables and 0.5% in fruits. Hence, India continues to be a minor player in global trade.

Eighth, growth of horticulture sector depends mainly on area expansion, excepting in the case of vegetables where area-cum-productivity growth pattern has emerged.

Ninth, in global productivity of fruits and vegetables, India is a country of low productivity. This raises a pertinent question whether India will be able to be a competitive player in international market as well as in domestic market under free trade require of WTO.

This leads us to the theme of sustainability of horticulture in India and to the issue whether horticulture development in India can be a successful story. There are two dimensions of sustainability viz., sustainability in productivity and production, which could be called on-farm dimension of sustainability and trade-related sustainability with ability to compete with other countries under liberalized trade regime of WTO. For lack of proper concept, the second dimension of sustainability could be called off-farm dimension of sustainability.

5.1 Pathways for Productivity and Production Sustainability: On-Farm Sustainability

The challenge of productivity and production sustainability is addressed mainly through the pathways of production technologies and associated cultural practices (Newley *et al.,* 2006). In this context, the challenges facing the current and future generation are to maintain productivity and production sustainability by minimizing/eliminating their impact on the environment and to "repair" the impact of past inappropriate and unsustainable practices, which have been discussed under section 3:. The problems of salinity, soil erosion and degradation and residues of pesticides are the derivatives of unsustainable practices in horticulture. Hence, positive actions are needed to protect, restore and enhance natural resource base for horticulture as to ensure regular supply of safe, quality foodstuffs at reasonable cost to consumers, while preserving the productivity of the landscape.

The broad pathways for sustainability of horticulture at farm level are (Details in Newley *et al.,* 2006): Sustainable Vegetation management, water management and soil management, prevention of soil erosion, induction of organic matter, nutrition management, pest management, weed management, induction of cover crops, insect management, disease control, increase in skill levels of farmers in using pesticides, protection and enhancement of existing native vegetation for greater bio-diversities and security of rural environment, management of soil health and efficient use of non-renewable resources. It is understandable that research results on most of the aspects of on-farm management domain are available, but there are gaps in extension service delivery system as well as adaption gaps. The adaption gaps may be due to ineffectiveness of extension delivery system and or due to low profitability of adaption. The low profitability issue has to be addressed through productivity and trade-related pathways (More on this later). In fact, there are evidences to suggest that regenerative and resource conservation technologies and practices can bring both environmental and economic benefits to farmers. Further sustainability at farm level emanates when people individually/collectively are concerned with long-term productivity of the natural resources which is the main focus of on-farm sustainability domain.

5.2 Off-Farm Support System for Sustainability: Trade-Related Pathways

Efforts to increase productivity and production of horticulture produce are necessary to ensure the sustainability of the sector, but not sufficient. The sufficient condition for sustainability will have to emanate from off-farm pathways. Sustainable horticulture development needs both productivity-production pathways and trade-related pathways to make Indian horticulture to be competitive in the international market. According to a World Bank Report (Mattoo *et al.,* 2007), the biggest obstacles to the competiveness of horticulture exports lie outside the sector rather than inside it. The average price at a farm gate for a typical horticulture produce is estimated to be just 12-15 percent of the price at which it is retailed. So, a 20% improvement in productivity (yield) can translate into only 2.4-3 percentage points reduction in the final price, where as a 20% reduction in international transport costs can reduce final prices by 8-10 percentage points. Hence, drive is to be to-wards reduction in cost involved in taking the produce from farm to market destination. However, given the low productivity of Indian horticulture, India needs to use both productivity pathway and trade-related pathways (Ali, 2007) to be competitive in the international market destinations.

There are many directions of off-farm support needed for sustainability of horticulture development: Address the problems of inadequate infrastructure facilities like cold storage, market, roads, transport facilities, supply chain management, institutional buildings at farmers level to take advantage of expanding global markets, comprehensive R&D to focus on both productivity pathway and trade-related pathways; consumer education on processed horticultural products, needed Foreign Direct Investment in processing and marketing and some more fiscal incentives to processing industry.

To elaborate some of the needed off-farm support for sustainability of horticulture sector:

- Seasonality governs the arrivals and prices of agricultural commodities, but horticultural commodities register fluctuations of prices from month to month. This makes it imperative to release the produce in an orderly manner to the market. For this, storage plays an important role, because of perishable nature of horticulture produce. This requires a chain of cold storage and shelf-life management, which are far from adequate at present. In addition to cold storage, other infrastructural facilities like roads, markets and transportation facilities are needed.
- The supply chain management is yet to gather momentum, involving all stakeholders for vertical and horizontal integration of domestic and international markets.

- There have been studies (Eg. Surbhi Mittal, 2007) identifying potential states for fruits and vegetables which could be globally competitive with comparative advantage in production. It is true that these are the states to be targeted for the enhancing export potential of country. But still the pertinent issue is whether India is competitive in quality of products and in trade. Unless quality issue is addressed and trade-related constrains are released it is difficult for India to be competitive based only on on-farm productivity advantages, unless improvements are effected in supply chain efficiency and management, post-harvest management, market reforms and forward and back-ward linkages.
- North East Region of India offers a favorable set of climate conditions for cultivation of various kinds of fruits, vegetables and other horticulture crops. For example, for small regions like Manipur, horticulture can be a lifeline. But ecological sustainability should not be disregarded.
- A research project conducted in Vijayawada, Andhra Pradesh (India) to support Vijaya Fruits and Vegetables Grower's Association provides two major lessons, among others, on the linkage between technical and institutional factors for sustainable horticulture development (Details in Andrew Hall *et al.*, 2001):
  - ♦ While technology development and transfer is certainly key to sustainable development of horticulture, it will succeed if institutional innovations are introduced to gell with technological innovations. This gelling of two broad types of innovations could provide a strong foundation for sustainable development drive.
  - ♦ Success in production, marketing and export has to be seen as a total system activity.

  Yet another study (Hegde, 2017) suggests the need for enrolling small and marginal farmers as partners in the development of agri-horti pastures on marginally productive and waste land, and that would increase the institution building capacity at the local level. In a partnership mode between farmer's associations and corporate, these pastures could be developed. This could be an experiment in mobilizing smallholder to enter into business cluster like in Kenya. The success depends on the creation of positive environment for development alliance between farmers and corporate. If organic production of fruits and vegetables is organized in these business clusters, it provides the opportunity for capturing market in Europe.
- Doubts have been expressed (Sidhu, 2005) about building heavy weight fruit and processing industry in India. In fact, in commercial processing of fruits and vegetables, India is far behind countries like Malaysia and Philippines, inspite of several fiscal incentives and other policy initiatives such permitting foreign equity investment. But due to inadequate domestic demand for processed fruits and vegetables and due to less competitive in international market, the industry is yet to experience growth. In this situation, it is different to expect private sector to invest in fruits and vegetable processing industry. Perhaps consumer education campaign and of on-farm and off-farm initiatives like the ones discussed earlier could provide foundation of sustainability of Indian horticulture sector.

6. EXPERIENCES OF AND LESSONS FROM OTHER COUNTRIES

This section is to provide a brief on experiences of and lessons from four horticulture produce export countries viz., Kenya, Thailand, Israel and China. Some of these lessons may be relevant for us to adopt and some for fine tunning our own horticulture development drive.

- **Kenya:**
  - The growth in Kenya horticulture exports is very much associated with increasing involvement of small holders with support from Government and corporate sector. Linking small holders to high-value urban and export market is an important strategy for raising rural incomes and reducing poverty.
  - Government is directly involved in the provision of extension services and promotion through R&D.
  - There are institutional arrangements to provide inputs on credit, to obtain reliable high quality supplies and to ensure repayment of loans.
  - In processing operations, there are joint ventures with foreign companies. This kind of ventures with long-term investments is facilitated by macroeconomic stability.
  - To sum up, the factors which have contributed to the success of horticulture sector in Kenya are: A realistic exchange rate, stable policies, a good investment climate, competitive international transport connections, institutional and social links with market in Europe and market institutions to link farmers with exporters.
- **Thailand:** Thailand is one of the leading fresh fruit and vegetable producers in Asia and Pacific region for export. There are three major supportive development initiatives that have sustained the horticulture sector.
  - The on-farm management system is emphasized on Integrated Pest Management (IPM) and Integrated Crop Management (ICM).
  - Emphasis on safety and quality.
  - Business clusters of growers and collaboration between growers and traders.
- **Israel:** In case of Israel, two strategies stand out:
  - Emphasis on R&D to support the growth of modern horticulture sector under the conditions of water scarcity.
  - To improve efficiency at all stages and opening of the markets to foreign competition.
- **China:** China, a dominant producer of fruits and vegetables globally, has placed emphasis on following key strategies:
  - Raise quality standards through "Green Food" and "Pollution-Free" production standards to be competitive quality-wise.
  - China's competitive advantage also lies in its low production costs.
  - Processing industries have grown rapidly due to the combination of China's low costs, growing market and Government policy of encouraging processing as a means of helping farmers.

- Marketing efficiency has improved due to emergence of supermarket chains with good procurement system. Added to these, whole sale and distribution sectors have been opened up for foreign competitors.

7. RECAPITULATIONS AND SUBSTANTIVE INFERENCES

The focus of the present section is to recapitulate the main results of the study and to derive inferences for sustainable development of horticulture:

First, India is endowed with wide variability of soil, climate, topography and rainfall. As a result, the potentials for development of horticulture are quite high. Further, the importance of horticulture as a sub–sector of agriculture and allied sector could be gauged in terms of crop diversification, nutrition, efficient use of diverse ago-climatic resources, employment, profitability and its contribution to Agriculture GDP from its 9% of total cultivated area.

Second, the concept of sustainable development, the most important concept of twentieth century, provides a very strong foundation for sustainable development of horticulture with an emphasis on meeting the needs of the present generation as well as of future generations. This necessitates the imperatives of maintaining the ecological security of our land-scape. We are obligated to preserve the natural environment with its natural resource endowments, while our drive for development of horticulture is accelerated. But under Green Revolution Technology, a historical necessity, there has been a transition to industrial agriculture/horticulture with more dependence on external inputs produced in industrial sector. The unsustainable practices of this new technology has led to damaging effect on our soil health and ecological security. The ecological degradation has become much more in Himalayan region raising the issue of sustainability of horticulture across the country.

Third, a statistical profile of horticulture development has been examined, leading to the inference of area and production expansion, with marginal increase in productivity. Area under horticulture crops in the country had increased from 12.77 million hectare in 1991-92 to 23.41 million hectare in 2014-15, total production of horticultural crops had increased nearly 2.9 times during the same period and productivity 1.6 times. As a result, India is second only to China in total production of fruit and vegetables, but productivity in China is found to be close to two-times the productivity in India. This is reinforced by the result that growth of horticulture sector as such depends mainly on area expansion, with low contribution from productivity excepting in the case of vegetables where both area and productivity have contributed for growth of output in this subsector of horticulture. Added to this, the global productivity differentials analysis leading to the inference that India is country of low productivity in major fruits and vegetables. Following this is the question how India will emerge as a leading country to be competitive in global market. That explains why India is a minor player in international market, accounting for 1.7% of global trade in vegetables and 0.5% in fruits. It is true that with dawn of liberalization in international trade under WTO, new vistas have been opened to Indian Horticulture sector. To support this drive of India to exploit its horticulture development potential, the National Horticulture Mission was launched in 2005-06 to promote holistic development of horticulture in the country.

Fourth, doubts have been expressed (Sidhu, 2005) about building heavy weight fruit and vegetable processing industry in India. Infact, in commercial processing of fruits and vegetables, India is far behind countries like Malaysia and Philippines, inspite of several fiscal relief and other policy initiatives, such as, permitting foreign equity investment (Details in Sidhu, 2005) for the growth of fruits and vegetables processing industry. But due to inadequate domestic demand for processed fruits and vegetables and due to our minor role as players in international market, it is difficult to expect private sector to invest in fruits and vegetable processing industry. To enlarge the domestic demand for processed produce, consumer education campaign is required.

Fifth, how India will be able to emerge as a leading country to sustain horticulture sector both in domestic and international market. It is recalled that India accounts for 1.7% of global trade in vegetables and 0.5% in fruits. Given the minor role of India in global trade, sustainability of the sector depends on on-farm sustainability through productivity enhancement and off-farm strategies related to trade are required to make Indian horticulture competitive in the market-both domestic and international.

The study has indentified pathways for sustainability of horticulture at farm level: sustainable vegetation management, water management and soil management, prevention of soil erosion, induction of organic matter, protection and enhancement of existing native vegetation for biodiversity so on to protect the on-farm environment both for present and future generations.

The study has also identified pathways for releasing constraints on trade so as to make horticulture sector competitive. On-farm efforts could increase productivity and production which are only necessary, but not sufficient. The sufficient conditions will have to come from off-farm pathways: Address the probles of inadequate infrastructure facilities like cold storage, market, roads and transport facilities, supply chain management, promoting development alliance between farmers' associations and corporate, forgoing the linkage between technical and institutional factors and fruit and vegetables processing industry in the country.

Sixth, what are the lessons which India can learn from other countries like Kenya, Thailand, Israel and China for enlarging its role in international trade in horticulture products: Mobilize small holders (who are dominant in India) to be partners in the development of horticulture and in trading horticulture products with support from Government and Corporate sector and encourage joint ventures with foreign companies in processing operations, develop business clusters in partnership mode between growers and traders, promote quality standards to be competitive and open up wholesale and distribution sectors for foreign competitors. Hence, use of pathways on both on-farm and off-farm domain could make Indian Horticulture Sector sustainable to face competition both in domestic and international markets.

REFERENCES

Ali, M. (2009). Developing Sustainable Horticultural Production Systems For Socio-Economic And Nutritional Development in Asia Vol II- Edited by Ali, M.

Bisaliah, S. (2011). Growth of Agriculture In Horticulture Sectors and Livelihood Security. Horticulture and Livelihood Security. P.N. Agri. Science Foundation (PNASF), Bengaluru.

Bisaliah, S. (2015). Climate Change: Drivers, Impacts and a Response Road Map for Coping with. In Shantha Kumar,: Global Climate Change- Issues, challenges and Policy Implications. Excel India Publishers, New Delhi 2015.

Brundtland Commission, World Commission on Environment and Development: Our Common Future. (1987). Oxford University Press, New York.

Chand, Ramesh. (1996). Ecological and Economic Impact of Horticultural Development in the Himalayas: Evidences from Himachal Pradesh, Economic and Political Weekly, Vol. 31, No.26, June 29, 1996.

Choudhary, S.N. (2013). Contribution of National Horticulture Mission in Agriculture Development. International Journal of Advanced Research in Management and Social Sciences, Vol.2. No.6, June, 2013.

Friedman, T.L. (2009). Hot, Flat and Crowded world. Penguin Books, London.

Hall Andrew. *et al.* (2001): International Learning Through Technical Projects: Horticulture Technology R&D Systems in India. AgREN, Agriculture Research and Extension Network.

Hand Book on Horticulture statistics. (2014). Ministry of Agriculture. Government of India. New Delhi.

Hegde, N.G. (2017): New Technologies to Enhance Agricultural Production and Sustainable Rural Livelihood, International communication.

Horticulture statistics at a Glance. (2015). Horticulture statistics Division, Ministry of Agriculture and Farmers welfare, Government of India, New Delhi.

Kamei, P. (2013). The Overview of Horticultural Growth, The Role of Horticulture in Tamenglong District: A Survey Account of Ten Villages in the Tamenglong District, Manipur State, International Journal of Scientific and Research Publications. Vol. 3, Issue 3, March, 2013.

Matto, Aditya, Mishra, Deepak and Narain Ashish. (2007): From Competition at Home to Competition Abroad: A Case study of India's Horticulture. The World Bank and Oxford Press.

Mittal, Surbhi. (2007). Can Horticulture Be A Success Story for India. Indian Council for Research on International Economic Relations, New Delhi.

Newley, P.D. and Treverrow, N. L. (2006). Sustainable Horticulture. Prime Facts, profitable and Sustainable Primary Industries, Prime Fact, 144.

Planning Commission. (2001). Report of the Working Group on Horticulture Development, Government of India, New Delhi.

Roling, Niels and Pretty, Jules N. (1997). Extension's Role in Sustainable Agricultural Development. FAO corporate Documentary Depository. Rome. Edited by Burton E. Swanson, Robert P. Bentz andrew J. Sofranko, Prepared under the guidance of the Extension, Education and Communication Service Research, Extension and

Training Division, Sustainable Development Department, Food and Agriculture Organization of the United Nations, Rome, 1997.

Sidhu, M.S. (2005). Fruit and Vegetable Processing Industry in India: An Appraisal of the post Reform Period, Economic and Political Weekly. Vol 40. No.28 July 9-15, 2005.

Singh, A.K. *et al.* (2015). Role of Horticulture in Agriculture Development and Food Security in India. National Seminar on Sustainable Agriculture and Food Security, New Delhi.

Singh, Mahendra and Mathur, V.C. (2008). Structural Changes in Horticulture Sector In India: Retrospect and Prospect for XIth Five Year Plan. Ind. Jn. of Agri. Econ. Vol 63. No.3. July – September 2008.

❍❍❍

28

Pathways to Addressing Research for Sustainable Horticulture Development

H. Ravishankar

1. INTRODUCTION

Many sources define sustainability in agriculture-horticulture only in terms of local environmental effects. The definition of a sustainable system is one that serves people, sustains or improves the environment and enhances on the scale of the entire planet and over the next hundred or several hundred years. Sustainable horticulture may broadly revolve around sustainability of environment, social milieu and economics. Thus these production systems need to efficiently use natural resources *viz.*, land, water, energy and with care; minimize generation of waste, reuse, recover through recycling and finally dispose-off sustainably without causing pollution in order to conserve the natural systems and protect biodiversity. Horticultural production systems encompassing fruits, vegetables ornamentals, plantation, spices and medicinal and aromatic crops are spread across the subcontinent under the diverse agro-ecologies and are driven by changing food baskets, economic development and market dynamics.

Sustainable development recognizes principally three tenets: 1) Environmental sustainability; 2) Social sustainability and 3) Economic sustainability. A sustainable community anywhere often seeks to : 1) Protect and enhance the environment and 2) Promote economic success, as furnished below:

1.1 Protect and Enhance the Environment

- Use energy, water and other natural resources efficiently and with care.
- Minimize waste, reuse or recover it through recycling, composting or energy recovery and finally sustainably dispose of what is left.
- Limit pollution to levels that do not damage natural systems.
- Value and protect the diversity of nature.
- Protect human health and amenity through safe, clean and pleasant environments
- Emphasize on health service; prevention as well as cure of diseases.
- Ensure access to good food, water, housing and fuel at reasonable cost. Meet local needs locally wherever possible.
- Maximize everyone's access to the skills and knowledge needed to play a full part in society.

- Empower all sections of the community to participate in decision making and consider the social and community impacts of decisions.

1.2 Promote Economic Success

- Create a vibrant local economy that gives access to satisfying and rewarding work without damaging the local, national or global environment.
- Value unpaid work
- Encourage necessary access to facilities, services, goods and other people in ways which make less use of the car and minimize impacts on the environment.
- Make opportunities for culture, leisure and recreation readily available for all.

2. CHALLENGES AND OPPORTUNITIES FOR HORTICULTURE

The projected population of India is likely to range between 1.64 and 1.74 billion by 2050, when the world population is likely to reach nine billion. Already with a global share of 2.3% land, 4.2% of water, the *per capita* availability of resources in India is 4-6 times less than the world average. The pressure on limited natural resources will increase further and it is feared that food demand of the country may exceed production from 2021 onwards. Our burgeoning population and the rise in income level will lead to increase in demand of high- value agriculture (HVA) produce (fruits, vegetables, meat, eggs, milk, fish) and value-added food products.

The transformation in favour of HVA is driven mainly by changes in food consumption patterns and income elasticity. Increased demands of value-added processed products and imported food items in big city markets in recent years signal the nature of domestic markets that are likely to emerge (Ghosh, 2012).

The scenario of horticultural crops in India has presently become very encouraging. The percentage share of horticulture output in agriculture has become more than 30%. Under the purview of agriculture and allied activities, the share of plan outlay for horticulture which was 3.9% during Ninth Five Year Plan, increased to 4.6% during the Twelfth Plan. India has witnessed voluminous increase in horticulture production over the last few years. Significant progress has been made in area expansion resulting in higher production. Over the last decade, the area under horticulture grew by about 2.7% per annum and annual production increased by 7.0%. During 2013-14, the production of horticultural crops was about 283.5 million tonnes from an area of 24.2 million hectares. Out of the six categories, that is, fruits, vegetables, flowers, aromatic, spices and plantation crops, the highest annual growth of 9.5% is seen in fruit production during 2013-14. The production of vegetables has increased from 58,532 thousand tons to 168,300 thousand tons since 1991-92 to 2014-15. During 2016-17, horticulture production has surpassed food production in the country indicating the scope which However, needs not only to be sustained but also boosted in the years to come for meeting the surging demands and ensure nutritional security of the growing population.

By 2051, 52% of India's population is expected to be living in urban areas with higher incomes. This scenario will exert competing demands for environment market for farm space *viz.,* green space, buffer strip and carbon sequestration concomitantly demanding more, diversified, better quality and safe foods. It is estimated that the *per capita* household food demand for fruits (kg/year) will appreciate from 15.4 (2009-10) to an upside of 115.5 by 2051. This is projected to boost the total domestic demand of fruits from 71.2 million tons in 2009-10 to 183.4 million tons by 2050. These requirements are to be met from the continued shrinking of land, water, energy and labour resources. The challenges perceived in this background include, climate change manifested by rise in temperature, erratic precipitation, rise in sea water, enhanced greenhouse gas (GHG) emissions, deteriorated ecosystem, reduced biodiversity besides trade and competitiveness. The potential impacts of climate change on Indian agriculture are expected to result in productivity decreases of most cereals due to increase in temperature, CO_2 and decrease in water availability; a projected loss of 10-40% in crop production by 2100. It is estimated that 1°C increase in temperature may reduce yields of major food crops by 3-7%. Much greater losses at higher temperatures with longer duration are expected in *rabi.* It is also expected that the length of the growing period in rain fed areas is likely to reduce, especially in the peninsular regions of the country. Meanwhile, it is also expected that increase in CO_2 to 550 ppm increases yields of rice, wheat, legumes and oilseeds by 10-20%. In this scenario, as majority of the tropical and subtropical fruits are produced during *rabi*-post *rabi* periods, their production, productivity and quality are going to be impaired by truncated degree days impacting phenological stages. *Kharif* crops may get impacted more by rainfall variability while the *rabi* crops by minimum temperature with duration of cold adversely impacting the phenological stages, specifically the reproductive ones besides affecting yield through reduced number and activity of pollinators.

The emerging scenario though presents more opportunities for rain water harvesting due to high intensity rainfall but greater loss of top soil due to erosion imposes challenges of soil conservation. In 2050, 1.6 billion people need to have adequate nutritious, safe and healthy food and fiber within the limits of natural system. This would necessarily entail four-fold increase in land productivity; a three-fold increase in water productivity; doubling of energy use efficiency; six-folds increase in labour productivity, about half through labour-capital substitution and low carbon emissions. The journey to 2050, thus would have to tread the pathway of scientific innovations-led growth in sustainable productivity, characterized by knowledge intensive rather than input intensive strategies, improved and new crop varieties, crop husbandry focusing on good horticulture practices, new cost effective technology including biotechnology, without associated increase in resources aimed at improving the profitability to the grower, infusing competitiveness, thus ushering in '*Golden revolution 2*'. Such a realization could be harnessed through search for new genes / edited genes, address challenges of climate change paradigms, conserve biodiversity; prioritized attention on phenomics and genomics, molecular breeding, precision farming, diagnostics for nutrients, pests and diseases, value addition, high value and secondary horticulture for converting waste to wealth, Nano-biotechnological applications. Dynamic private sector, strong institutions and vast potential with strategic partnerships of public, private international and farming community undoubtedly would be the game changer in this vision.

2.1 World Scenario

India is the second largest producer of vegetables and fruits and our presence in global market is significant. Grapes occupy the premier position in exports with 107.3 thousand tons valued Rs 108,649 lakh. Other fruits which have attained significant position in exports include banana and mango. Exports of fresh vegetables (such as peas, potatoes) have been on the rise. However, export of fresh onion has declined from 17 to 12.4 lakh tons over the last three years. The developments achieved in the horticulture sector are indicative of the fact that there is growing demand for horticulture produce both for domestic consumption as well as exports. The past experience have been rewarding for enhanced output from the investment. Availability of timely robust information in this sector will certainly improve the socioeconomic conditions of Indian farmers by providing self-reliance besides environmental protection. The Government has identified horticultural crops as a means of diversification for making agriculture more profitable through efficient soil management by introducing 'Soil Health Card' Scheme, efficient land and water use through 'Per Drop More Crop' Scheme and creating skilled employment for rural masses. Recent efforts have been rewarding in terms of opening new vistas for farmers' welfare through 'e- Kisan Mandi' portal and easing export of value added horticultural products (Anon., 2015a).

2.2 Emerging Scenario of Horticulture Sub-sector in India

The income elasticity for fruits and vegetables is reported to be 0.42% and 0.35% respectively, against only 0.05% for rice and –0.06% for wheat (Ghosh, 2012). The annual growth rate in domestic demands for fruits and vegetables is estimated at 3.34% and 3.03% respectively. About 127.2 Mt of vegetables and 86.2 Mt of fresh fruits will be required by 2020–21 (Ghosh, 2012). During 2005, area-wise vegetables covered 59.73%, fruits covered 29.03% and plantation crops covered 7.75% of the total horticultural area.

The National Horticulture Mission (NHM) launched during 2005 in 18 states and 3 Union Territories (UTs) (371 districts covered) aimed to record a production of 300 Mt by 2012. Up to 2009–10, 16.57 lakh ha additional area had been brought under different horticultural crops under NHM. For fruits, maximum area coverage had been under mango, aonla, citrus, guava, sapota, banana and pineapple.

Table 1. *Arrival of horticulture produce and number of markets involved in its trade (Quantity in '000 tons)*

	2004-05		2009-10	
Commodity	**Quantity**	**Number**	**Quantity**	**Number**
Fruits	11,907	227	28,683	600
Vegetables	12,147	367	35,582	897
Spices	4,457	334	7,671	828
Flowers	2,172	9	54,462	57
Plantation	2,372	40	8,396	123

(**Source:** *Ghosh, 2012*)

The enhanced market arrivals of horticultural produce and larger involvement of markets in horticultural trade (Table 1) during implementation period of NHM (2004–05 to 2009–10) has been highly impressive (Jose, 2010). In the horticultural sub-sector, impressive growth in fruits and vegetables helped in improving nutritional security and increased export earnings. Share of fruits, vegetables and flowers helped in increasing agricultural export basket considerably. Further, growth in the horticulture sector in the years ahead is most likely to be through enhancement of productivity, quality assurance and reduction of post-harvest losses and value-addition.

2.3 The Present Growth Pattern and its Sustainability

The present growth rate in the horticulture sub-sector is reasonably high. However, reduction in regional disparities across eco-regions, careful treatment of ecologically fragile areas and water management in rain fed areas and concerted group actions for small and marginal farmers appear crucial for sustenance of gains achieved. The *Fertile Crescent area* (Punjab, Haryana and western Uttar Pradesh), the mainstay of India's food security system, is presently facing serious ecological and economic problems associated with soil health, groundwater depletion and unfavorable cost-risk return structure. It pointed towards the need to change the very approach for sustainable growth by combining crops, including horticulture, animal husbandry, fisheries, agro-forestry to advance productivity in perpetuity, without the associated ecological harm.

Horticulture should receive greater attention in the hills after taking care of proper soil and water-conservation measures. Climate-change adaptations, particularly in temperate fruits and nuts requiring specific chilling hours for flowering, irrigation management (drip-fertigation) and drought management (water harvesting, hardy rootstock use) are the immediate research and extension issues. Transfer of proven technologies has already brought significant improvements in productivity of horticultural crops, such as, banana, grapes, potato, *etc.* To meet the annual growth rate in domestic demands for fruits (3.34%) and vegetables (3.03%), a partial shift in agricultural policy and practices will be essential. Expansion of area in higher-productivity zones, increase in fertilizer and irrigation water-use efficiency in the crop sector and reinforcement of package of technology will be possible options. The low (191) and very low (66) land productivity districts offer immense opportunity for production of hardy horticultural crops with minimum infrastructure support. Small-holder horticulture through group (SHGs)/ contract farming (horticulture estates as suggested by the National Commission of Farmers (NCF)) and cluster approach of the NHM should be attempted. In most of the '*very high*' and '*high*' productive districts, horticultural crops have a predominant presence.

It is estimated that by 2050, about 22% of the geographical area and 17% of the Indian population would face acute water scarcity. Water supply for horticultural crops and livestock will face more intense competition among the multiple users of water and therefore, investment in micro-irrigation for horticultural crops will need attention to mitigate the problem. Enhanced use of agro-bio inputs (bio fertilizers, bio pesticides, botanicals, *etc*.,), Integrated Pest Management (IPM) technology and use of hardy rootstocks are certain other options for greater ecological balance and sustenance. For sustainable development and realization of high productivity levels, crop planning based on agro-ecological considerations and efficient use of inputs has been attempted, only with partial success.

Another partially tapped region for horticultural growth is the NE hill region. In the North East, out of 2.2 million ha of shifting cultivation (*Jhum*) land, only 17.5% is cultivated at any point of time. Excellent quality apple and Kiwi fruit in the rain-shadow, temperate zone belts of Arunachal Pradesh; mandarin oranges and pineapple in the mid-hills of almost all the hill states; ginger and turmeric in Mizoram, Meghalaya, Sikkim, Nagaland; large cardamom in Sikkim; flowers (*Anthurium*, *Gerbera*, rose, *Lillium*, *Cymbidium*) in most of the N-E states, small-holder tea in Assam and Nagaland and rubber cultivation in Tripura are some examples in support of the great potential remaining untapped.

2.4 Futuristic View for High Value Horticulture to Meet the Demands of 2050

In the area of R&D, a strong research base has been already created and with proper technology transfer and fast- track development of established crops should be possible (Ghosh, 2008).The importance of horticultural biodiversity has been well recognized among the mitigation strategies for climate-change adaptation. For example, in order to mitigate temperature rise in temperate zones, introduction of newer varieties adapted to a wide range of chilling hours and use of hardy rootstocks to overcome lower / delayed rain-fall appear possible options. Use of insect-pest-tolerant crop varieties and soil-borne disease-resistant rootstocks needs to be encouraged. In future, during crop and land-use planning, possible climate change dimensions must be kept in mind. The indigenous genetic resources provide a buffer in times of drought, heat and biotic stresses and genetic resources with bio-alternatives may reduce the vulnerability of production systems to climate change. In planning *rabi* vegetables cultivation, selection of suitable varieties with lesser thermo-sensitivity should not be difficult even from within the available gene pool / variety base. Developing cultivars tolerant to heat, drought and agronomic manipulations, including water management so as to match adverse climate change impacts should receive research priority. Therefore, institutional reforms, such as, strengthening of market, credit and pricing, accelerated goal-oriented research and technology transfer for horticultural development need greater focus (Ghosh, 2012).

3. ADVANCING HORTICULTURE

3.1 Revamping of Horticulture Education

Horticultural systems are quite diverse across the world and are highly dynamic, driven by the economic development and market globalization parameters. All these changes enormously challenge the role of education, which needs to articulate on issues of providing information, knowledge and practice change. The focus on technical information gives way to the need to strengthen the socio-economic component *e.g.* ecological, ethical and social dimensions of horticulture. The excess of information is handled by emphasizing critical thinking skills, problem-solving attitudes and decision-making and by creating innovative learning environments. Horticulturists are well prepared through continuous updating so as to become perpetual long learners. Modern horticultural education requires dedicated teachers combining the mastery of their subject matters together with good communication skills. "Capacity building through imparting contextual education in horticulture though is quite difficult and complex but it appears to be the best long-term investment that could happen for sustainable horticulture development anywhere".

Realizing the importance of agricultural growth, the Govt. of India through successive Plan periods supported development of agricultural education in the country. The human resources developed by the agricultural education system have been primarily responsible for green, white, yellow and blue revolutions signifying respectively increase in food grain, horticulture products, milk and oil seed and fish production. While on one hand the country can be justifiably proud of its achievements on the other hand, concerns have been expressed time and again on declining total factor productivity, rising unemployment, fast degrading natural resources, declining agricultural growth in the recent years. This has largely been attributed to decline in quality of instructions imparted to a great extent in institutions of higher learning. There have been several factors attributable to this, the chief among them are declining financial resources, lack of faculty competence in new and emerging frontier areas, extensive inbreeding and lack of adequate skills, entrepreneurship and experiential learning. The graduates passing out, today it is expressed do not match the requirements of emerging market. Globalization and demand-led markets necessitate a relook at the content and delivery of *curricula* such that the graduates coming out not only meet the expectations of different stakeholders but also be propellers of agricultural growth. For India to occupy pre-eminent position in the international market, it is imperative that the graduates also remain in the forefront of developing new technologies and disseminate them to the farming community (Anon., 2006).

Department of Agriculture Research and Education (DARE), Ministry of Agriculture, Government of India and Indian Council of Agricultural Research (ICAR) have been vested with the mandate of regulating higher agricultural education in the country. This responsibility is being discharged by the Education Division of ICAR in partnership with the State Agricultural Universities (SAUs) . From the beginning, the ICAR has been providing support for developing and strengthening higher agricultural education in the country. In addition, ICAR has put in motion, a number of initiatives and reforms which include establishment of Accreditation Board for quality assurance, faculty competence improvement through training, library strengthening, institution of scholarships and fellowships, measures for reducing inbreeding, infrastructure support for library, hostels and procurement of state of the art equipment *etc*., so as to have quality assurance in agricultural education.

Different recommendations have been made to reorient the course *curricula* in order to develop much needed skills and entrepreneurial mind-set among the graduates so as to take up self-employment, contribute to enhanced rural sustainable livelihoods and food security, sustainability of agriculture and be the propellers of agricultural transformation.

3.2 Restructuring of Under Graduate Programmes for Increased Practical and Practice Contents

* General changes/ modifications have been recommended by the Fourth Dean's Committee across the faculty:

- Experiential learning programmes have been recommended in all disciplines. Flexibility in the course curriculum has been recommended.
- Courses on entrepreneurship development, communication skills, computer knowledge, agri-business, environment science, biotechnology *etc*., have been included in all the disciplines.
- Practical content has been increased substantially and presently is 50-60% of the total course load.

3.3 Central Assistance for Strengthening Higher Agricultural Education

The ICAR support for education has increased exponentially from 8.9% in VIII plan to about 14% in Xth Plan. The review committee constituted strongly recommended increasing development assistance to SAUs and Deemed Universities by ICAR to the extent of 20% of its budget. Increased investment was particularly felt necessary because of the major revision in course curriculum, which necessitated investment in infrastructure, faculty training and procurement of the state of the art equipments for promoting entrepreneurial skills and training under the real-time situation. The committee recommended starting non-formal education programmes by SAUs especially for need assessed vocational modules for generating Para-professionals for promoting technological and economic empowerment of the rural youth.

Recognizing the need for quality improvement in agricultural education, the World Bank supported ICAR through Agricultural Human Resource Development Programme with an outlay of US $ 74.2 million for bringing in much needed reforms in agricultural education. The major reforms brought (Anon., 2006) in through this project included:

(i) ***Establishment of Accreditation Board:*** The new accreditation procedure has been institutionalized which involves preparation of Self Study Report by the institutions and its validation by a Peer Team and finally decision by the Accreditation Board on accreditation of the institutions. The Board also approved three sectoral committees on Accreditation Norms and New Institutions / Programmes, *Curricula* and Equivalence, Governance and Personnel / Financial Policies.

(ii) ***Norms and Standards for improving education:*** Uniform academic regulations, examination and evaluation system including grading system have been made in SAUs. Rural Awareness Work Experience (RAWE) is included in the programme, courses in new emerging areas, such as, Computer applications, Bio-statistics, Biodiversity, Biotechnology, Agri-business management, Marketing and Environment were incorporated in all the degree programmes.

(iii) ***Capacity building for human resource development (HRD):*** The HRD programmes were strengthened by supporting modernization of class-rooms, library strengthening, establishment of students' laboratories at college level, students counseling and placement cell, supporting construction of International student hostels, girls hostels, education technology cells and providing support for modernization of UG and PG practical and research laboratories.

(iv) ***Faculty competence improvement:*** In order to improve the faculty competence, large number of training programmes are being organized through Centers of Advance Studies, Summer and Winter Schools in which 9,000 scientists benefited. Besides, nearly 1000 scientists were sent for training in the best of the institutions overseas. Sabbatical leave rules as well as Visiting Scientists scheme were operationalized. In order to promote excellence in teaching, five Best Teacher Awards were instituted both at the university and national level. Faculty was made computer literate and necessary support was provided for participating in seminars, symposia both nationally and internationally.

(v) ***Reducing inbreeding:*** Extensive inbreeding was identified as the single factor contributing largely for decrease in quality of education imparted. From 1995-2000 major steps were taken for reducing inbreeding.

It was emphasized that the new *curricula* should align with the new developments taking place nationally and internationally for making agriculture education contextual, sustainable and productive. The paradigm shift from production to cost competitiveness and consumer quality acceptance was articulated by the committee.

Additional emphasis in these meetings of the committee was on how to add value to the time spent by the graduates, to provide skills in the latest cutting edge technologies and build confidence to undertake farming or other vocations profitably. Reorientation recommended took care of developing needed knowledge, skills, entrepreneurship and agri-business acumen and communication skills, so that graduates developed confidence to undertake self employment and become job providers instead of job seekers. Additionally, they get empowered to contribute to sustainability of agriculture, economic and technological empowerment and ultimately to the sustainable livelihoods, food and nutritional security of the country. Another cardinal principle was to provide adequate coverage with regard to new technologies, such as, biotechnology, information technology, bio-informatics, GIS, remote sensing, precision agriculture, for new emerging global scenario, coverage on organic farming, international agriculture, processing and value addition, agri-business management, communication skills *etc*., across different disciplines.

Horticulture education until 1970 was imparted as a part of the agriculture *curriculum*. Realizing the importance of horticulture in promoting livelihood security, economic empowerment and nutritional security, horticulture emerged as a separate discipline. Separate colleges of Horticulture were established with the first college of Horticulture coming up in Kerala in 1972. Since then, 15 more colleges of Horticulture have been established in the country and with one of the Universities specializing in Horticulture namely Dr.Y.S.Parmar University of Horticulture and Forestry, Nauni (H.P.) latest being University of Horticultural Sciences, Bagalkot in Karnataka.

The country is fortunate to have different agro climatic regions and can produce tropical, sub-tropical and temperate fruits and vegetables. This is a unique position that could be advantageously exploited internationally provided due attention is given to develop human resource with skills to take up different aspects of horticultural development including processing and value addition.

There is a need to adopt mission mode approach for capacity building in order to produce quality horticultural produce at reasonably low cost so as to compete in the global market and to promote self-employment among the horticultural graduates. This entails to equip the graduates with current trends and requirements of horticulture industry which underscores, enabling them to help and guide the farmers or involve themselves in achieving cost and quality competitiveness of obtaining horticultural produce; enable market intelligence and participate in the global trade of horticultural produce.

3.3.1 Hands– on training / experiential learning

Final year B.Sc. (Hort.) students are required to select any two areas of the following to undergo specialized training (Anon., 2006) *viz.,* protected cultivation of high value crops ; post- harvest technology and value addition ; nursery production and management and floriculture and landscape gardening. Features of the new curriculum focused on experiential learning for one year in two specific areas , increased practical skills through experiential learning , entrepreneurship development , inclusion of new courses on organic farming, farming systems, IPR international treaties, communication skill development, horticulture business management, biotechnology *etc.* and infusing flexibility in the curriculum in the final year.

It was felt by the committee that the graduates coming out of institutions must have adequate hands on experience on different aspects of horticulture for which it was decided that the final year programme be restructured to provide for six months of experiential learning and six months of attachment with industry/modern horticulture farm. In order to provide experiential learning it is necessary to have adequate infrastructure, therefore, the budget requirement for experiential learning have been worked out. It is important to instill entrepreneurship and confidence in taking horticulture as an avocation. The experiential learning need to cover different aspects of horticulture and therefore four areas were identified (Anon., 2006) but institutions would have freedom to add more areas relevant to their region. Following four areas for experiential learning have been detailed as a model with different activities for learning and evaluation. The identification of areas needs to be based on present day needs, *horticulture development scenario in the region and University, industry, private partnership potential.* There would also be flexibility in choosing experiential learning areas. For this purpose, the students would be required to prepare a work plan in the area selected with an end-to-end approach *i.e.,* from purchasing the inputs to producing a product and marketing.

Era of individual research of the past has now given way to identification of challenges and addressing them by harmonizing teamwork. Globally, institutions are partnering for problem resolution and institutions of National Agricultural Research and Education system (NARES) could in no way be exception. In order to capitalize on the strength, which exists among private Research and Development (R&D) institutions, Indian Institute of Technology (IITs) and general universities, it is necessary to open dialogue with national institutions and private R&D institutions, which have proven track records (Anon., 2006). In fact, it was expressed by the committee that the quality of our education programmes will improve substantially with such a linkage. The major question put forth However, was how to operationalize? In this regard (SAUs), must change their policies and aggressively initiate dialogue with institutions, outside NARES, which have proven national and international standings. For SAUs, which open such a dialogue and sign MOU for effective quality improvement, a provision of Rs. 10 lakh has been recommended under development grants.

3.3.2 Overseas universities

There is a need to have tie ups and linkage for major rejuvenation of our agricultural education taking advantage of Indo-US Knowledge Initiative and National Agricultural Innovation Project. ICAR need to help in operationalisaiton of this which the Committee considered most crucial (Anon., 2006). To begin with, the collaboration could be in the specialized areas *viz.,* biotechnology, processing and value addition, water management, integrated natural resource management and development of *curriculum* and delivery. The collaboration should not only be in respect of joint research work but also exploring the possibility of sandwich post graduate programmes leading to PhD degree as also post-doctoral research.

In order to have the major impact of the initiatives taken under NARES system it is necessary to develop a mechanism of continuous dialogue and synergy among SAUs, ICAR institutes and line departments of the states. It is recommended that the mechanism be developed for regularly holding at least two meetings in a year at state level for bringing together SAUs, ICAR institutes and line-departments in the state. In these meetings the representation of ICAR should not only be of the institute level but also from ICAR headquarters. With this there will be synergy and complementation in the efforts of the three major institutions (Anon., 2006).

3.4 Revitalizing Horticultural Research

3.4.1 Research orientation

The country's horticultural research orientation has not seen any major changes except, articulating on issues of climate changes, soil health, increasing nutrient use efficiency and water prudence in isolation but not holistically. However, since 1990, some improved crop varieties and production technologies have been developed and transferred. Current research focus remains demand-driven and participatory research. Major reasons for this could be '*farmer-centric*' so as to obtain funding and to enhance the adoption as the country's economy is predominantly agrarian. There is a need to align the goals in tune with the emerging global perspectives of emerging demands and markets for sustainability.

Keeping in view of the strength and weaknesses of the research system and priority areas of horticulture development the following strategies and programmes are suggested for horticultural research:

3.4.2 Rationalization of research

There is a need to shift from commodity/discipline oriented research to system based research and to establish stronger Inter-institutional linkages with SAUs, Central Scientific and Industrial Research (CSIR), Department of Biotechnology (DBT), Indian Space Research Organization (ISRO) and Bombay Atomic Research Centre (BARC). Greater private sector partnership for diversification, value addition and export promotional research and seed and planting materials production programmes will be required to modernize horticulture industry. Introduction of project based budgeting in the Institutes / National Research Cente (NRCs) will bring better accountability and one time catch up grant to modernize old Institutes will be essential. Development of database on technologies evolved, market intelligence, export projections/removal of quantitative restrictions, R&D scenario in horticulture should receive priority attention (Anon., 2006).

3.4.3 Safeguard of intellectual property rights (IPR)

A fool-proof description of varieties, characterization (including DNA finger printing) and their registration and finalization of material transfer agreement and channelization of germplasm exchange need to be institutionalized. Similarly, specification of quality / codex standards for export of indigenous fruits and vegetables need to be developed. Phytosanitary regulations for importing vegetatively propagated materials need a relook / revision and rigorous enforcement.

3.4.4 Genetic resource management

Greater emphasis need to be given on *in situ* conservation of endangered species, *ex situ* conservation of base collections and *in vitro* storage and cryopreservation of important germplasm. Clonal repository of vegetatively propagated crops and germplasm screening for processing and diversified use will be certain other areas of priority research.

3.4.5 Horticultural biotechnology

Biotechnology as a tool for rapid multiplication of quality planting materials, virus cleaning, genetic transformation *etc.*, will be of considerable. The priority research programmes included, development of micro-propagation protocols in selected crops; molecular characterization of germplasm and marker assisted selection; value addition to products and preservation of post-harvest losses through control of metabolic process.

3.4.6 Integrated production system

Production related technologies can bring quick improvement in production and productivity on different regions. Short or medium range programmes on horticulture based cropping systems; water management including micro irrigation and fertigation, greenhouse cultivation of vegetables and flowers; integrated nutrient and pest managements; environment pollution and pesticide residue problems have already received due research attention. Further refinement of the technologies and their transfer will bring perceptible change / improvement in production of different crop commodities.

3.4.7 Export -promotional research

There is a need for development of bulk handling system of tropical fruits, including pre-cooling and controlled atmosphere / modified atmosphere storage and post harvest protocols for sea transport of major fruits like banana, mango, litchi, sapota, Kinnow and pomegranate. Disinfestations technology including vapor heat treatment (VHT) for export of fresh fruits and extension of shelf life by preventing desiccation of vegetables should help in further export promotion. Organic farming for vegetable and spice crops and residue– free IPM technology are other important areas of research.

3.4.8 Quality seeds and planting materials production

For resolving the long standing problems of supply of good quality seeds and planting materials of different horticultural crops, need to articulate on such research areas as system of certification and standardization of planting materials; micro-propagation protocols for mango, litchi, coconut, walnut, date palm, oil palm and apple rootstocks; refinement of TPS and micro-tuber production technology in potato; improvement in STG techniques and cross-protection in citrus for virus elimination and control; and seed production systems for hybrid vegetables and commercialization of micro-propagation in floriculture.

3.4.9 Post-harvest management

In order to reduce post-harvest losses at production centers, low cost ecofriendly on-farm storage structures can play a crucial role. Significant advancements have been made in that direction and some small and medium sized cool chambers on the principles of evaporative cooling have been devised. Further refinement of the technology will go on a long way in reducing post-harvest losses. Also, standardization of packing line operations and proper packaging of different commodities are of urgent need. Pesticide residue management and newer product developments will add values to the produce.

3.4.10 Human resource development (HRD)

Advanced training in research methodologies and instrumentation, biotechnology, micro-irrigation, fertigation, Integrated Pest Management (IPM), Integrated Nematode Management (INM), bio-fertilizer, biopesticides, pesticide residue and Post Harvest Technology (PHT) and product development need priority attention for increasing research capabilities of the scientists. Skill development for state level development functionaries through in-service training at different R&D institutions will enhance capabilities of extension staff. Post Graduate programmes in fruit, vegetable, floriculture, plantation crops and post harvest management of horticultural crops will help in providing trained manpower in specialized areas.

3.4.11 Challenges and research gaps

Non-existence of centralized facilities for conducting advanced biotechnology research; non-existence of controlled environment phenotyping centers and biotic and abiotic stress tolerance.

3.4.12 Future pathways

Germplasm mining for traits and genes; controlled environment phenotyping centers; identifying new traits for future research; resistance to biotic stresses: more basic research to understand basis of resistance, broad spectrum resistance/ non-host resistance, more attention to partial resistance and plant secondary metabolism for insect resistance; tolerance to abiotic stresses: breeding objectives to be dissected into component traits, identification of individual component traits and genes from crops, molecular basis of stress– induced alterations in phenology, development of non-destructive high throughput phenotyping methods and understand cross-talk between multiple stresses.

3.4.13 Horticultural research

Challenges and research gaps:

These include, changing climate; biotic and abiotic stresses; growing demand of organic food; lack of quality seed and planting materials to end users; old and unproductive orchards; huge post-harvest losses and lack of trait-specific germplasm ready for utilization in breeding programme.

Future pathways:

(A) Genetic enhancement: Identification / breeding of climate resilient crops / genotypes; trait specific breeding, pre-breeding and genetic enhancement; development of core and mini-core of the fruit genetic resources; potential of indigenous ornamentals and Medicinal and Aromatic plants (MAP) to be explored phyto-medicine rich varietal development; research on plant based insulin, edible vaccines.

(B) Production system: Foolproof technology for organic horticulture; improved production technology to reduce production cost, meeting the quality and safety standard, grafting technology in vegetables; explore the possibilities to deploy nanotechnology, such as, nano-fertilizer, nano-pesticides, precision farming using Nano-sensor, *etc.,* aeroponic and hydroponics; collaboration with medical science to harness the medicinal value of horticultural crops.

(C) Post-harvest research:

Challenges and research gaps: Continuous availability of high quality fruits and vegetables; improving nutritional quality of fruits and vegetables after harvest.

Future pathways: To develop novel foods (fermented/ unfermented) with functional ingredients; develop evidences of food fermentation processes with functional benefits accrued along with added phytochemicals; food processing with appropriate techniques.

Production systems and NRM: Research on comprehensive organic production systems and safe integrated production systems; pollinators and pollination; reducing the cost of production; rootstock breeding and technology for grafting vegetable crops especially to overcome soil-borne diseases; Improving the resource use efficiency including water productivity, residues management, Plant Growth Promoters (PGP) microbial consortia, efficient delivery systems like Nano composites; plant architecture engineering including canopy management, solar energy harvesting, management of flowering physiology, exploiting dwarfing genes and rootstocks in fruit crops for ultra high density cropping systems in fruit trees and polyhouse cultivation of vegetable crops.

Plant health management: There are several national challenges like *Tospo* viruses in vegetables, pomegranate bacterial blight, banana bunchy top, guava wilt, phytoplasma in perennial crops like coconut, *etc.* Therefore, there is a need to have a programme on surveillance; field based diagnostic kits, developing forecasting models, decision support systems and eco-genomics for understanding host-pathogen-vector interactions at gene expression level.

Post-harvest technology and value addition: Stable natural pigments to replace synthetic colors; nutrition in terms of bioavailability; fermentation technology and bioreactors; nutraceutical and specialty foods; waste utilization; food and environmental safety.

'Horticulture and Post-Harvest Technology' emphasized on integration of horticultural research with cereal crop research. The major emphasizing points include germplasm management of horticultural crops, pyramiding multiple traits, breeding for development of large scale population, stripping in breeding cycles, generating information for plant architecture and developing forecasting model and biodegradation management schemes. There is a need to emphasize on the techniques like Marker Assisted Selection (MAS) and harnessing the latest advances like *Cenh* histone based haploidy. Improving the resource use efficiency including water productivity, residues management, PGP microbial consortia and efficient delivery systems like nano-composites need special emphasis in horticultural crops too. The post-harvest technology emphasized on stable natural pigments to replace synthetic colors, nutrition in terms of bioavailability, fermentation technology and bioreactors and nutraceutical and specialty foods.

3.4.14 Farmers' perspective

Some of the critical challenges today's farmers are facing include increasing input cost and low profits due to low quality produce as a result of unpredictable weather and abrupt market fluctuations for farm produce. To deal with situation, the farmers are in dire need of timely and farmer friendly information on inputs, weather conditions and technologies suited to specific needs of the farmers.

In order to address these issues in a better way, the national and international institutes need to work in harmony with local people for a comprehensive solution to the challenges that farmers are facing. One such initiative was taken and a society named "*Society for Conservation of Natural Resources and Empowering Rural Youth*" was launched. Through this society, farmers are exposed to new farm machinery and approaches of conservation agriculture. But, the machines are not accessible to all the farmers because of high cost and they are not economically feasible.

There is urgency for farmer-led research and extension programmes through graduate students that involve problem identification to methodology development, result assessment, evaluation and final recommendations of the best evolved technology to other farmers. Due acknowledgments or rewards for the involvement of farmers should be given from time to time in publications and meetings (Saharawat *et al.*, 2013).

3.4.15 Strengthening extension activities

Small and marginal farmers account for a major proportion of holdings in horticulture too. Indeed, they constitute the backbone of the agriculture sector in majority of the developing countries of the Asia-Pacific region. In well-endowed regions with irrigation facilities, small farmers have been quick to adopt the seed-fertilizer package and make their small farms economically viable. In fact, they have benefited from the first generation effects of the Green Revolution and are now moving on to diversification of agriculture into high value crops i.e., horticulture. Their problems are no longer related to production but rather to marketing of surpluses and value addition on the one hand and natural resource management for sustainable production on the other.

In the vast rainfed areas of the region small and marginal farmers still face major constraints in the production process. They suffer from low productivity and instability in yields and are highly risk averse. They are constrained in the production process by lack of access to inputs, technology, credit and markets. In addition, they are prone to deal with land degradation and poor and inequitable irrigation resources. Their organizational capacity is also poor which further weakened their bargaining position *vis-à-vis* the better organized service providers and purchasers.

3.4.16 Reforms in technology dissemination

Reforms in the agricultural extension system encompass a restructured technology dissemination system, based on a multi-agency system, comprising of the public as well as the private sectors. Innovative, decentralized institutional arrangements need to be promoted to make the extension system farmers' responsive and farmers' accountable. The thrust of the reorganized technology dissemination mechanism need to articulate to, ensure greater farmer participation in determining research and extension agendas; strengthen research-extension linkages; allow greater location-specificity in both technology generation and dissemination; emphasize more sharply the provision of knowledge and skill-based technologies as opposed to material-based technologies this can be increasingly left to the private sector; forge new partnerships between public institutions, technology users, private sector and NGOs; mainstreaming gender issues in technology generation and dissemination; harness information technologies more effectively for technology transfer; improve financial sustainability of the technology dissemination systems; and create an enabling policy environment at the State level that allows for institutional arrangements wherein the comparative advantage of the public and private sectors can be optimally harnessed for rapid and cost effective technology dissemination.

3.4.17 Innovations in technology dissemination

The Ministry of Agriculture, Government of India, implemented the National Agricultural Technology Project (NATP). Innovations in technology dissemination (ITD) were pilot-tested through new organizational arrangements and operational procedures for extension delivery by decentralized decision-making at district level and below. Agricultural Technology Management Agency in 24 project districts spread over six States in the country was established as an autonomous agency with farmers as major stakeholders. Main objectives of the ITD component included: increase the effectiveness of technology dissemination system by making it demand-driven and responsive to location-specific issues of farmers; bring in bottom-up planning procedure for technology dissemination; develop and strengthen institutional linkages that would result in shared objectives and ownership of system by key stakeholders; empower farmers through farmers' organizations/associations to play an active role by making the technology dissemination farmer-driven and farmer accountable; promote model farmers; increase financial sustainability of technology dissemination and develop private sector partnerships.

In order to increase the knowledge and awareness among farmers to enhance the production and productivity of various crops, the Government of India had initiated different programmes, such as, Front Line Demonstrations and Extension through network of Krishi Vigyan Kendra's (KVKs), National Mission of Agricultural Extension and Technology (NMAET), National Food Security Mission (NFSM), Soil Health Management Scheme, Mission for Integrated Development of Horticulture (MIDH), Bringing a Green Revolution to Eastern India (BGREI), Rashtriya Krishi Vikas Yojana (RKVY) and recently '*Farmers First*'.

The *Kisan Call Centre* (KCC) scheme in the past was also implemented across the country for the benefits of farmers. The main aim of the scheme is to answer queries of farmers on telephone calls in their own dialect. A countrywide common eleven digit toll free number 1800-180-1551 has been allotted for Kisan Call Centre for the purpose.

Â Sub-Mission on Seeds & Planting Material (SMSP) covers the entire gamut of seed chain from nucleus seed supply to farmers for sowing and also to the major stakeholders in the seed chain. It promotes adoption of quality seeds and planting materials for increasing agricultural production and productivity.

Considering the growing importance of horticulture in the agrarian economy and in ensuring sustainable livelihoods of farmers; nutritional security of the population and the prevailing gaps in technology dissemination and assimilation in to the real time farming situations, it is high time that the NARES seriously considers mooting the idea of developing Para-horticulturists on the existing lines of Para-medics in the National health care system. This could go a long way in stabilizing and sustaining small holder horticulture that is largely prevalent in the country.

3.4.18 Role of Indian Council of Agricultural Research (ICAR)

ICAR has established a network of 642 Krishi Vigyan Kendra's (KVKs) in the country aiming at assessment and demonstration of technologies/ products and its dissemination through number of extension programmes including training of farmers so as to update their knowledge and awareness to increase the productivity of crops.

Due to the training programmes conducted by KVKs on improved technologies related to agriculture and allied fields, benefited the farmers in terms of increased crop production and improved farm income. The successful technological interventions identified by KVKs through On-farm testing (OFTs) and frontline demonstrations (FLDs) are up-scaled with the help of Department of Agriculture and other line departments for the spread of the technologies to large number of farmers in the district by the respective KVKs and also through organizing the extension activities like technology week in the demonstration farm of KVKs for exposure of farmers to the improved technologies of agriculture.

3.4.19 Strengthening public extension services

Though there is influx of multiple players (private agencies, corporate houses, Non Government Organization, NGOs), trust and authenticity among the clientele group continues with the public extension services, simply because their concern is free of profit motive. Also the biggest advantage is the researches back up to maintain credibility. In this background following aspects merit due attention:

- It is high time to revitalize and strengthen the extension machineries of SAUs and State Department of Horticulture, KVKs, Extension Education Institutes (EEI), State Agricultural Management Extension Training Institutes (SAMETIs) .
- Provision of vehicle for mobility; incentives; decentralized management; creation of single window delivery outlets; intranet connectivity among the different departments of essential services; commodity programme development; implementation, monitoring and evaluation unit; mobile exhibition and video show vans, etc must become integral part.
- The public extension system needs recasting as learning organization.

3.4.20 Promoting pluralistic extension system in public–private partnership mode

Farmers groups, corporate houses, NGOs, Voluntary Organizations, input dealers and other private agencies should be encouraged to disseminate routine information and provide extension services. If farmers are willing to have paid extension service in specialized areas, provision need to be made for such export- oriented farmers.

Capacity building of extension functionaries to infuse pro-activeness, professionalism, responsiveness, accountability, missionary zeal and sound managerial abilities to handle the World Trade Organization (WTO) related farming constraints and challenges, the entire functionaries require to be trained in newer and relevant areas of behavioral dynamics like motivation and attitudinal orientation, entrepreneurship development, group formation and functioning, skills in conflict management among growers; use of media mix and computer aided communication aids (internet); video-conferencing; help-line; enterprise management; agri-business management skills; WTO and implications to horticulture;

participatory need assessment technique; innovative techniques in production, protection and post-harvest management; quality standards for exports *etc.*

3.4.21 Information management

- Mechanism for information support to farmers needs to be streamlined. Organization of horticultural fairs and exhibitions at state, regional, national and international levels, production and relay of audio-video films and development of extension literature in local language need to be emphasized.
- Exploitation of booming mass media channels to reach the unreached and improve their Dissemination Index (DI).
- Face-to-face contact programme has its own importance and incomparable impact in technology dissemination. They must be used for follow-up and convincing campaign for major changes in production system of farmers.
- Farm School on All India Radio (AIR) can be replicated effectively in horticulture at wider scale.
- Community radio could be promoted for information sharing among the horticulture farmers in cluster villages.

3.4.22 Information technology (IT) enabled extension services

Development of commodity specific decision support system (DSS) and Expert system, on- line market info-portal could bring about improvement in precision decision-making by the growers and lead to betterment in production and trade of horticultural produce.

3.4.23 Intensification of field extension programmes

Besides conventional methods of demonstrations, exhibition, excursions, growers-scientists interface, field-visits, commodity shows and competitions at block, district, state and national levels and innovative approaches of *Farmers' Field School*, Farmer-to-Farmer communication ('*model farmers*') and rewarding the pioneer adopters in public gatherings may be effective in dissemination of knowledge and infusion of competitive spirit among the growers to some extent.

3.4.24 Strengthening contract farming institutions / mechanisms

Dependency upon private sectors, contracting firms and companies is becoming inevitable for small-scale farmers in order to facilitate accelerated capital inflow, technology and essential service delivery and market integration for quality crop production. However, appropriate structural arrangements should be initiated to develop an institutionalized mechanism for designing organizing and operationalizing the contract system with the watchful eyes of the government and public institutions keeping in view of the long-term implications. Structural arrangements should ensure fair deal in arbitration and quick dispute resolution and for justice it should be the domain of the government. At the same time, it is highly essential to educate the participating farmers about the dynamics of contract and corporate world.

In fact, 'Niti Ayog', Government of India presently is seriously contemplating to bring about an act in the parliament on 'contract farming' in order to infuse dynamism and profitability in futuristic agriculture for higher sustainability and profitability.

3.4.25 Distance learning programme in horticulture

Capacity building of farmers, entrepreneurs and other potential end users could be facilitated with education and training through distance learning system for practicing horticulture with latest technical know-how and nitty-gritties of do-how.

3.4.26 Marketing extension

Besides the task of promoting quality production through organic means, marketing has assumed considerable significance in the light of new international trading regime under WTO. Public extension system will need to address the issues related to marketing of produce in order to project horticulture, as an economically viable and remunerative avocation. Market surveillance; export-quality standards, terms of global trade, export ventures in partnership with exporters, linkage with APEDA, NHB and export houses should be strengthened (Anon., 2011b).

3.5 Urban and Peri-Urban Horticulture (UPH)

Urbanization is taking place at a faster rate in India. Population residing in urban areas in India, according to 1901 census, was 11.4%. This count increased to 28.53% according to 2001 census and crossing 30% as per 2011 census, presently standing at 31.16%. According to a survey by the Unite Nations (UN) State of the World Population report published in 2007, by 2030, 40.76% of country's population is expected to reside in the urban areas. In a developing and highly populous country like India, resources are even scarcer. The recent world food price crisis has rendered the importance of understanding and confronting the causes of food and nutritional insecurity of the urban poor. Poor urban dwellers, to a great extent net food buyers and depend upon mostly on markets for their food supplies, are particularly vulnerable to adverse food price shocks and are consistently the group in the society that suffers most from ever increasing food prices. Urban horticulture, defined as plant production activities that are conducted in a city or suburb that produce horticultural plants that are wholly or partially edible, economically viable and has the potential to reduce CO_2 emissions. Urban horticulture perhaps, is one of the solutions that are perceived globally in order to meet the demand of nutrient dense-foods of the urban population and insulate them against malnutrition and hunger besides combating the challenges of Green House Gas emissions.

There are a number of ways through which urban horticulture could, in principle, have an impact on the food and nutritional security system. At the household level, it can be a source of income; can provide direct access to a larger number of nutritionally rich foods (vegetables, fruits) and a more diversified food basket, can increase the stability of household food consumption against seasonality or other temporary shortages and can increase the time, mothers spend caring for their children, as opposed to non-agricultural activities that are more likely to be located farther away from home (Awasthi, 2013).

Urban and Periurban Horticulture could envision planners to establish greener cities by contributing to food and nutritional security; generate employment, waste management and community well-being. Moreover, in order to increase productivity in limited resource areas, closed or semi closed systems (*i.e.,* greenhouses) are considered more advantageous than open systems (*i.e.,* fields). The production and daily delivery to market of perishable horticultural crops from smallholder farmers close to large urban centers (peri-urban horticulture; market gardening) have been important for the millennia. The issues of unsorted vegetables, fruits, flowers and also medicinal and aromatic plants delivered directly to urban consumers with little or no official oversight for quality or safety needs to be addressed through education and extension services infrastructure. However, the surge in world population from two to seven billion during the past 70 years and especially rural-to-urban migration, have profoundly impacted both the practicality of traditional peri-urban horticulture as well as the numbers of the urban poor. Competition for water resources can be fierce and modern food quality and safety expectations increasingly demand better compliance and oversight. Developments in horticultural sciences could educate on issues about what, how and where to grow with '*crop per drop*' sensitivities besides food safety. This sector knows about marketing, delivering safe and attractive products to consumers. Perhaps most importantly, horticulturists can credibly address issues about land, labour, water and the crop production / protection inputs needed for successful grounding.

Food and Agriculture Organization (FAO) is assisting governments and cities to put in place policy, institutional framework and provide technical support to develop urban and peri-urban horticulture with adapted good agricultural practices (GAPs), irrigated gardens, hydroponic micro-gardens and green rooftops in densely populated cities.

In cities, environmental benefits and synergies could be achieved when UPH is planned as a part of the urban landscape including safe recycling of solid waste and waste water (Sharholy *et al.,* 2008). Further, trees and other ornamental plants are crucial for carbon sequestration of from atmosphere and play an important role in reducing carbon footprints. Moreover, flowering / foliage plants in the garden not only add aesthetic value but also help to improve the ecosystem services. India has responded though fairly well to the needs for effective urban and peri-urban horticulture with emphasis on green space, green building, development of parks and gardens and promotion of peri-urban vegetables production; however, the integration appears to be poor. The design of UPH must include a large component of awareness creation aimed at improving access to food and nutrition while enhancing the livelihoods of people living in and around cities besides nurturing the environment. UPH is now a dire necessity rather than a luxury. The government's initiative of peri-urban vegetables production alone is not enough for the huge challenges to be met. This necessitates of holistic approach having vertical and horizontal integration of the efforts of all the stakeholders, which should concurrently link all components of UPH. Among the various gardening options, terrace gardening, a raised ground space around a dwelling house or on the sides of a hill, forms a link between the house and the rest of the outdoor living space and must, therefore, be designed in harmony with the plan of the house. Roof garden is one of popular alternatives in urban and peri-urban areas, because of the lack of available space on the grounds of a house. Particularly, in the big cities and towns, the only space left for garden enthusiasts is the roof of the house and the balcony. In order to ensure the success of roof gardening, technical and developmental support is critical.

Emphasis is on increased and quality production under protected cultivation for year round supply of fruits, vegetables and flowers and home scale production of horticultural crops; popularize nutrition gardening / terrace gardening / container gardening / turf gardening / roof top gardening, with due emphasis on safe use of pesticides, recycled sewage water etc.

The use of waste water for irrigation without careful treatment and monitoring could result in the spread of diseases and an issue of public health. Besides, cultivation on contaminated land also represents a health hazard for the consumers. Therefore, public health safety is paramount to ensure that any of the activities of UPH do not predispose unacceptable risks to public health and negate its potential economic, nutritional and environmental services. Utilization of available water and pre-treatment measures for waste water recycling and utilization for growing plants in the cities is an important aspect. On the other hand pesticide residues in the urban and peri-urban horticulture may become high, besides polluted water with heavy metals contamination and post-harvest cleaning of vegetables in water contaminated with bacteria and viruses is also a major issue. Use of environment–friendly microbial formulations in converting recyclable agri-horticulture residues including those arising in vegetable / fruit mandis in to nutrient rich manure is a viable and cost-effective option and needs consideration. Building professional skills among the urban youth for growing indoor plants, interior and exterior landscape, avenue plantations, landscape planning, turf grass management, mushroom production, post-harvest management, including packaging storage and transport and overall entrepreneurial and agribusiness skills need due consideration.

Science-based packages of technologies and authentic literature on UPH modules However, presently are not available. In particular, user-friendly technologies for protected cultivation suiting to urban and peri-urban horticulture are needed.

UPH in the emerging scenario should drive the agenda of greener cities. Need to enforce laws for safe handling and utilization of urban wastes, effluents, sewage sludge, waste water and urban compost in UPH. Urban development authorities would do well to take appropriate measures for recycling and making safe water and other inputs available for use in urban and peri-urban horticulture.

Singh (2013) suggested the following approaches to strengthen and consolidate the UPH initiatives:

3.5.1 Research

Inventorise, validate, refine and develop (if not available), UPH technologies on input supply (seeds, seedlings, manures, pesticides), growing vegetables ornamental plants and herbs in pots, shade, trays and hydroponics and on small scale protected cultivation structures for both roof top gardening as well as in peri urban production of fruits, vegetables and flowers.

3.5.2 Development and coordination

Integrate biodiversity conservation with urban landscaping in order to make the urban areas biologically and aesthetically rich.

3.5.3 Strengthen research on risk assessment and management and on development of safety standards

Inputs like water, manures and pesticides as well as for products like fruits, vegetables, milk, meat, egg and fish; and also assess extent of heavy metal removal from city waste by vermi-composting and earthworm activity need standards for regulation. National

Seeds Corporation, State Seed Farms Corporation and National Horticulture Board should jointly strengthen production and cost-effective distribution of quality seeds and planting materials of selected varieties tolerant to diseases and insect pests and suitable for UPH.

Strengthen value addition in the different value chains of horticultural, livestock, poultry, fisheries and aquaculture products. The success stories of UPH need to be documented while organizing pilot demonstrations and upscale and popularize the selected ones should receive priority.

3.5.4 Women empowerment

According to FAO, half the world's farmers are women and in South Asia women are involved in almost all stages of the agricultural process. Steady feminization of the agricultural labour force has created a major role for women in the agriculture process making women's time a crucial factor. Women are also an absolute importance when it comes to child nutrition and household well-being. Ironically, South Asia is home to a large population of undernourished children and this begs the question: Will empowering the women force in agriculture better her nutrition? Will she have more time for child care and will this lead to better child nutrition (Swaminathan, 2017)?

Horticulture is one of the major disciplines of agriculture where women play a crucial role in production of horticultural crops, particularly, in upland areas of the country. The extent of total area with poor soil with low water retention capacity poor vegetation, very frequently occurrence of drought, famine and frost, difficult to execute agro-techniques, poor transportation, communication and marketing facilities.

3.5.4.1 Women in horticulture

Knowledge and research gaps among women farmers were identified under the Network project on Assessment of gender issues–identification and refinement of selected women-specific technologies in horticultural crops being implemented at six centers. Interventions taken up to address the gaps were: (*i*) raising vegetable seedlings in portrays; (*ii*) planting on raised beds for better survival; (*iii*) seed replacement with high-yielding brinjal, bitter gourd, cucumber and pumpkin; (*iv*) cultivation of offseason and high-value vegetables under protected structures; and (*v*) proper staking/training of vegetables.

3.5.4.2 Capacity building of farm women

Training modules on bee-keeping, value-addition to tomatoes, fruit preservation, vermi-compost, dairy management, mushroom cultivation and production of bio fertilizer were developed for establishment of micro-enterprises for economic empowerment of women.

For drudgery reduction of women in agriculture, improved agricultural tools/technologies, namely, fertilizer trolley, manual seed drill, mat nursery, vegetable plucker, vegetable bag, water bag, face protector, dung collector, fodder chopper, fodder collector, long handle fork, manual maize sheller, mango harvester, potato picker and revolving stool, were field validated. Trainings/demonstrations were organized for capacity development of women in agriculture. Nutrition gardens were established in *rabi* and *Kharif* seasons to improve the daily diet of rural families to reduce micro-nutrient deficiencies. For promotion of vocational skills among adolescent girls and young mothers, need-based skill oriented training programmes were organized for their capacity development benefiting adolescent girls/mothers and anganwadi workers.

Value-addition and income-generating technologies have to be disseminated in the adopted villages for empowerment of women Protective garments have to be designed and developed to protect farm women from the health hazards during pesticide application. During pesticide application, the farm workers are exposed to pesticide vapours leading to various health hazards and it enter body via dermal absorption, inhalation and ingestion due to improper clothing. Hence, farm workers suffer from short term advantages of protective clothing among pesticide applicators (Anon., 2013).

Gender sensitivity and systematic approach are the two important factors for rural development programme targeted either at women or both women and men. Empowerment is a complicated process, which comprises short-term and long-term strategies. Under each strategy, several projects need to be formulated and carried out seriously. The process of farm women's empowerment is conceptualized in terms of personal assertions of confidence and ability to project themselves as women, attaining economic independence, ownership of productive assets, ability to handle capital and assets and provide leadership in both women and community related issues at all levels (Mishra and Mishra, 2012).

4. INCREASING HORTICULTURAL PRODUCTION

Innovative horticulture production in risk-prone marginal environment may involve the following:

4.1 Saline Horticulture

Among the options for augmenting freshwater resources is the desalination of salty groundwater, brackish drainage water and seawater. However, progress on modern desalination was made during the 1960s and plants have been developed since the 1970s, starting with some countries of the Persian Gulf because of their ready availability of energy and relevant scarcity of freshwater resources. Intensive research for large-scale commercial desalinating technologies began in the United States of America in the early 1960s (Buros, 1999).

The first desalination technology to be developed was thermal distillation. In this process, saline water is distilled into steam, which in turn is condensed into pure water. Later, membrane processes, such as, electro-dialysis (ED) and reverse osmosis (RO), were developed. In an ED process, salts are separated from water by means of an electric load application. In RO, pressure is applied to the intake water to force it to flow through a semi-permeable membrane that prevents most of the salts from passing through. The higher the salt concentration of the intake water is, the higher the pressure that is required. The pressure required for desalinating brackish waters range from 100 to 270 *kPa*, while that required for seawater ranges from 550 to 1 000 *kPa* (FAO, 2003a). Solar energy may also be used to produce water vapour, which is then condensed on a cooler surface to form the desalinated water. However, this process produces only a small amount of water and is generally ignored as a technological solution.

4.1.1 Use of desalinized water for agriculture

In the past, the high cost of desalinating and the energy required have been major constraints on large-scale production of freshwater from brackish waters and seawater. However, desalinated water is becoming more competitive for urban uses because desalinating costs are declining and the costs of surface water and groundwater are increasing. In spite of this development, the costs of desalinated water are still too high for the full use of this resource in irrigated agriculture, with the exception of intensive horticulture for high-value cash crops, such as, vegetables and flowers (mainly in greenhouses), grown in coastal areas (where safe disposal is easier than in inland areas).

For agricultural uses, RO is the preferred desalination technology because of the cost reductions driven by improvements in membranes in recent years. Spain provides a significant example of the application of desalinated water in irrigation. Spain has more than 300 treatment plants (about 40 *percent* of the total number of existing plants) and 22.4 *percent* of the total desalinated water is used for agriculture. Most of these plants process brackish water (only 10 *percent* of the total desalinated water for agriculture originates from seawater) and are located in coastal areas or within 60 km of the sea (FAO, 2003b). Moderately saline waters can be used for irrigation where control of soil salinity in the crop root zone is by means of leaching and drainage of dissolved salts. However, where brackish water is the only resource available, prior desalination is needed. Desalinated brackish waters and seawater are not used worldwide for irrigated agriculture because of the costs involved. However, in some countries, they are used for high-value horticultural cash crops. As irrigated agriculture does not require the strict standards that apply for drinking-water requirements, opportunities appear to exist for blending high-quality desalinated water with lower-quality waters. In this way, the final cost of a cubic metre of irrigation water can be reduced. Moreover, the desalination of saline waters for urban supply will also have a considerable impact on the production of low salt-content wastewater to be treated subsequently, with lower costs for use in irrigated agriculture.

Seawater farming addresses the severe lack of freshwater and undesirable soil conditions for agricultural activities in coastal regions. Saltwater, instead of freshwater, can be used to directly support a wide range of sustainable agricultural activities and enrich the soils in the coastal regions (Martinez eltrán and Kooshima, 2013).

Seawater integrated farm indirectly reduces hunger by creatively increasing means of productions, creating jobs and eradicating the poverty problem. Mission 2014 study has indicated that the world agricultural system is currently producing enough food to feed everyone but the poorest do not have the means to access them. Thus, when the people are employed through seawater integrated farming, they can use the money to purchase the food that they need. Currently there is no vegetable or grain crop that has high saline tolerance.

A British company has created an irrigation system that can grow crops using salt water. The dRHS irrigation system consists of a network of sub-surface pipes, which can be filled with almost any water, whether pure, brackish, salted or polluted. The system can even take most industrial waste-water and use it without the need for a purification process. The pipes are made from a plastic that retains virtually all contaminants while letting clean water through to the plants' roots.

4.2 Multiple Uses of Sea and Waste-water

Irrigation of horticultural plants using sea water directly is possible, according to a study by the Aqua Maris Foundation, as reported by *Hortoinfo*. Given the lack of the necessary fresh water to obtain the nutrients to feed humanity in the near future, Aqua Maris and other organizations are pushing for sea water, the most widely available, to be granted its rightful position, not only for use as drinking water (which is or can be) but also as a great source for minerals.

4.3 Alternative Production Technologies in Risk-Prone Environments

Water-saving solutions *e.g.*, rain water-harvesting, protected cultivation, use of desalinized water using solar energy, drip irrigation, mulching *etc.*, should be considered. In arid and semi-arid regions, where precipitation is low or infrequent during the dry season, it is necessary to store the maximum amount of rainwater during the wet season for use at a later time, especially for agricultural and domestic water supply. One of the methods frequently used in rainwater harvesting is the storage of rainwater *in situ.* Topographically, low areas are ideal sites for *in situ* harvesting of rainfall. This technique has been used in the arid and semi-arid regions of northeastern Brazil, Argentina and Paraguay, primarily for irrigation purposes. The *in situ* technology consists of making storage available in areas where the water is going to be utilized. All rainfall harvesting systems have three components: a collection area, a conveyance system and a storage area. In this application, collection and storage is provided within the landscape. Topographic depressions represent ideal collection and storage areas. In many situations, such areas are impermeable, being underlain by clay soils that minimize infiltration.

4.4 Use of Topographic Depressions as Rainfall Harvesting Areas

This technology increases water supply for irrigation purposes in arid and semi-arid regions. It promotes improved management practices in the cultivation of corn, cotton, sorghum and many other crops. It also provides additional water supply for livestock watering and domestic consumption. This technology is applicable to low topographic areas in arid or semi-arid climates. This technology requires minimal additional labor. It offers flexibility of implementation; furrows can be constructed before or after planting.

Rainwater harvesting allows better utilization of rainwater for irrigation purposes, particularly in the case of inclined raised beds. Rainwater harvesting is compatible with agricultural best management practices, including crop rotation. It provides additional flexibility in land utilization. Permeable *in situ* rainwater harvesting areas can be used as a method of artificially recharging groundwater aquifers. Disadvantages include:

in situ rainwater harvesting cannot be implemented where the slope of the land is greater than 5%; it is difficult to implement in rocky soils. Areas covered with stones and / or trees need to be cleared before implementation; additional costs incurred in implementing this technology could be a factor for some farmers; it requires impermeable soils and low topographic relief in order to be effective; the effectiveness of the storage area can be limited by evaporation that tends to occur between rains.

Rainwater harvesting system in some geographical regions improves the ability to capture rain water during wet periods and has the potential to reduce pressure on extracting groundwater resources during dry periods. Rainwater harvesting structures can recharge aquifers and improve water storage in the soil. These structures include ponds, percolation tanks, check dams, stream bunds and gully plugs. In many states in India, large areas of rain water harvesting have been successfully implemented (Kahlon *et al.,* 2012).

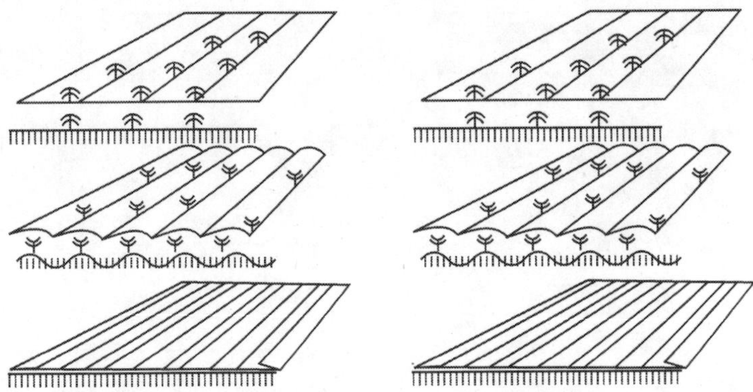

Fig. 1. *Site preparation methods for in situ rainwater harvesting*

Furrows may be used as an *in situ* means of storing harvested rainwater. They are built prior to or after planting to store water for future use by the plants. A variation on the use of topographic depressions to store rainfall, this method uses flattened trenches between the rows of crops to store water (Figure1.A-C). Furrows may have mud dams or barriers every 2 m to 3 m along the row in order to retain water for longer periods of time and avoid excessive surface runoff and erosion (Figure D). Raised beds may also be used to trap the water in the furrows or uncultivated areas may be left between rows, spaced at 1 m apart, to assist in capturing rainwater falling on the land surface between furrows (Fig. 1).

4.5 Soil Moisture Conservation

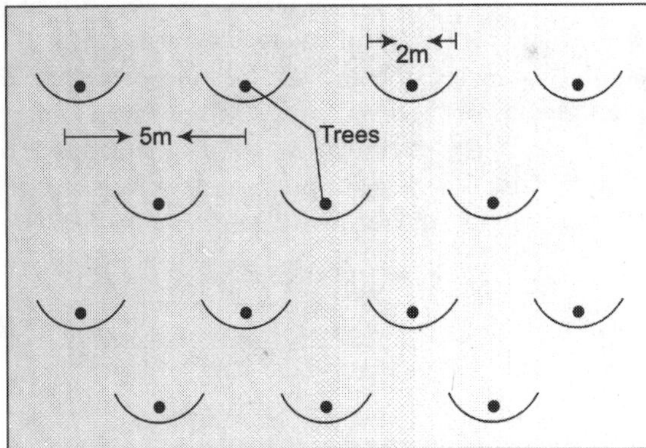

Fig. 2. *Type of bunds are forms*

- To conserve *in situ* moisture and reduce soil loss, circular basin of one meter diameter for level lands depending upon infiltration and rainfall; 'V' ditches of size 5m × 5m with trees planted centre and height according to the rainfall and slope of slopy lands; saucer basins / semi-circular bunds with 2m diameter to a height of 15-20cm across the slope are to be prepared (Fig. 2).

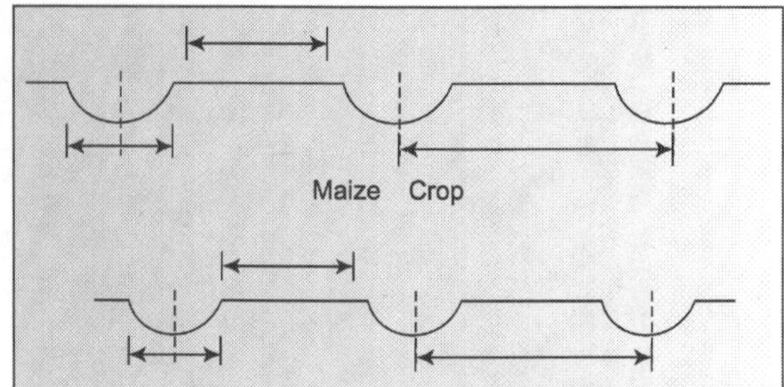

Fig. 3. *Broads beds and furrows*

Broad beds and furrows: This is laid within the field boundaries. The land level is taken and it is laid using either animal or tractor drawn ridges. Approximate cost for laying beds and furrows is Rs. 10000/- per ha. This has advantages in conserving soil moisture under dry land farming situations, besides controlling soil erosion. It also serves as drainage channel during heavy rains (Fig. 3).

4.6 Farm Mechanization

Over the last few years, there has been considerable progress in agriculture mechanization. It is generally believed that the benefits of modern technology have been restricted to farmers with large land-holdings. Yet the fact remains that even small farmers are adopting and utilizing selected farm equipments for efficient farm management through custom hiring. Mechanical equipments for various farm operations like tillage, sowing, irrigation, plant protection and threshing ~~etc~~ are generally being used by the farming community.

Moreover, reliance on animate power for day to day management of farm operations is showing a continuous decline owing to higher gains in form of improved land and labour productivity resulting from use of mechanical power. Improved implements, such as, M B plough, puddler, disc harrow, peg tooth harrow, spring line harrow being more efficient have been adopted. Further, use of sowing or planting devices is also registering a higher growth due to their impact on seed and fertilizer use (Table 2).

Table 2. *Level of Farm Mechanization in India*

Equipment manufacturers	No. of units
• Agricultural tractors	13
• Power tillers	2
• Irrigation pumps	600
• Plant protection equipment	300
• Combine Harvester	48
• Reapers	60
• Threshers	6000
• Seed Drills and planters	2500
• Diesel oil engines	200
• Plough, cultivators, harrows	5000
• Chaff cutter	50
•Rural artisans (hand tools)	>1 million

Table 3. *Women-friendly farm machinery*

Crop	Women friendly farm machinery
Maize	Hand operated maize dehusker-cum-sheller Gender-friendly on-farm model capable of dehusking and shelling maize cobs
Flowers	Finger guard for plucking flowers ; Fruit harvester
Paddy	Direct paddy seeder, Paddle operated paddy thresher, Two row and four row Rice planter
Sugarcane	Sugarcane stripper
Groundnut	Groundnut stripper
Cotton	Cotton stalk puller

Some concerted action is required from the state / centre on the following:

- Increasing the reach of farm mechanization to small and marginal farmers and to the regions where availability of farm power is low; promoting '*Custom Hiring Centers*' in order to offset the adverse economies of scale arising due to small landholding and high cost of individual ownership; creating hubs for hi-tech and high value farm equipments; creating awareness among stakeholders through demonstration and capacity building activities; Ensuring performance testing and certification at designated testing centers located all over the country.

India is the global farm equipment market. Indian market is expected to contribute nearly 10% in 2006 to the worldwide farm equipment market estimated at $ 66 Billion. Over the last few years, Indian agricultural equipment market has been widely exposed to international trade.

Global majors like John Deere, New Holland, Carraro and Same Deutz have already made an entry in to Indian market, whereas, Indian players are looking for alliances with foreign partners overseas.

Many factors affecting agricultural equipment sales in India include the monsoon, government declared support prices for crops, commodity prices, crop production expenses (including fuel, fertilizer, pesticides and other costs) and the credit policy announced by banks.

4.7 Status of Farm Mechanization in India

Even though farm mechanization shows an increasing trend, there are wide ranging disparities in the levels of mechanization across states. Northern States, such as, Punjab, Haryana, Uttar Pradesh (particularly Western and Terai belt) have achieved a faster growth in mechanization over various Plans. The sale of other implements and machines like combine harvesters, threshers and other power-operated equipment have been increasing almost throughout the country. The pace of mechanization in North-Eastern States has not been satisfactory due to constraints, such as, hilly topography, socio-economic conditions, high cost of transport and lack of institutional financing and lack of farm machinery manufacturing industries. Mechanization in Western and Southern states of the country *viz.,* Gujarat, Maharashtra, Rajasthan and certain areas of Tamil Nadu, Andhra Pradesh *etc*., has increased with the increase in area under irrigation and also with the growing awareness among farmer.

The market for farm equipments in India is on a growth phase post a decline period that lasted during period 1999 to 2002. The reasons behind the revival include easy availability of finance schemes and reduction in interest rates. The Government focus on building and improving infrastructure also had a positive effect on the overall performance of the farm equipment sector. Contract farming, which is being encouraged by the Government since 2005, will also have a favorable impact on the farm equipment demand. Farmers will now be hedge their crop risks through technical advice provided to them by the corporate organizations, besides assurance of upfront prices and market outlets would increase the farmer affinity for mechanized Agri technologies.

In the absence of standardization of parts and components farmers are compelled to carry their machines to the manufactures for repair and replacement of parts and components. Due to this, their market size is limited to their proximity and they are not able to develop their businesses. The village artisans on the other hand are located in the villages and therefore provide immediate attention to the needs of the farmers in their immediate neighborhoods. Therefore, the tools and implements, *etc*., made by them are against specific requirements of individual customers (Table 3).

4.8 Popularization of Agricultural Machinery

The assimilation of R & D requires an effective technological infrastructure of institutions and services to develop and test prototypes, to set up pilot plants for intensive evaluation and extensive demonstrations besides, training and credit support. New technology also requires network for transfer of technology to the manufactures. Popularization of agricultural machinery in the country is undertaken by the Provincial Governments through Department of Agriculture or Department of Agricultural Engineering. The activities are coordinated by the Department of Agriculture in Cooperation with the Ministry of Agriculture, Government of India. The Ministry of Food Processing promotes technology related to agro-processing. The extension system deals with the first-line extension projects with a view to: i) demonstrating the latest technologies to the farmers as well as the extension agencies; (ii) testing and verifying the technologies on the farmers field (iii) providing opportunities to get firsthand scientific feed-back; (iv) developing extension or technological models for the state extension systems; (v) providing training and communication support; and (vi) promoting research in transfer of technologies (Singh, Joginder, 2006).

4.9 Cropping Systems in Horticultural Crops

4.9.1 Cover crops

Cover crops in many horticultural production systems especially, in perennial crops are raised mainly in the beginning or advanced rainy season so that these may provide necessary protection to the soil when damage by soil erosion is the greatest. Retention of cover crops during summer is undesirable as it will compete with the main crop of fruit trees for soil moisture. In general, cover crops need to be complementary to the optimal land and natural resources, seldom becoming competitive to the main crops; Hence, choice of crops carefully looking to morphological frames, rooting patterns, water and nutrient mining capacities, pests and disease reactions *etc.,* is essential. A site-specific multi-storied cropping model for efficient harnessing of natural resources need to be explored. During monsoon, cover crops help in drawing off excess moisture. Experiments on cover cropping in citrus orchard showed that when berseem and fenugreek were grown as cover crops, there was a significant increase in nitrogen content of the soil.

The different cultural operations adopted, generally aim at: i) reducing the prevalence of weeds and ii) conservation of moisture.

In order to meet these objectives, use of artificial mulches like saw dust, alkathene black film, dried leaves are becoming popular. The experiments conducted at Pune on effect of different mulching materials on yield and growth of grapes has shown that the yield of grapes was significantly higher under black alkathene film and sawdust mulches. Similarly, giving a mulch of dried leaves and grass is beneficial for Cashew nut under the Konkan conditions of Maharashtra.

4.9.2 Intercrops

The crops which are growing between the rows of fruit trees in pre-bearing stage with a view to get some income are known as intercrops. The commonly grown inter crops are vegetables, groundnut, mung and black gram. The following factors need to be kept in mind while selecting the intercrops to be grown in an orchard:

- The short duration intercrops should not occupy the space for more the five to nine months; the water requirement of the inter crop should be minimum; inter crops should not be taller than the main crop; vegetables like ground vines should not be taken as inter crops as the creepers are likely to climb on the trees. Such intercrops may be good for mandarin orchards. In the first and second year, papaya may be taken as intercrop. The mandarin trees are spaced 20' and if papayas are planted at a distance of 10', an acre of mandarin orchard would accommodate about 261 papaya plants which would yield an additional income of Rs. 40,000 to 50,000/- per acre; after papaya, until the mandarin attains three to four years old, vegetables could be profitably grown in between the rows; when the mandarin trees are four to five years old, instead of growing vegetables, an inter crop of groundnut could be taken, where ever there is no trouble of pigs. In areas where there is such trouble an inter crop of peas may be taken.

4.9.3 Mulching

Mulching is a system of soil management in which hay, straw or cut grass is placed on the soil between the tree rows to conserve the soil moisture by checking surface evaporation. Mulching checks germination of weeds. However, the demerits include, mulching could favour fire outbreaks and also facilitates the hibernation of rodents and pests.

4.10 Horticulture Production Through C_4 and CAM (Crassulacean Acid Metabolism) Pathways

There are three distinct biochemical variants or types of photosynthesis based on the mechanism that plants employ by which carbohydrate is formed from CO_2: C_3 photosynthesis, C_4 photosynthesis and CAM photosynthesis. The Table 4 shows their respective features.

Table 4. *Summary comparison of C_3, C_4 and CAM photosynthesis in crop plants*

Feature	C_3 Photosynthesis	C_4 Photosynthesis	CAM Photosynthesis
Pathway of CO_2 fixation	*via* C_3 cycle only	*via* C_3 and C_4 cycles, spatially (C_4 in the mesophyll cells followed by C_3 in the bundle sheath cells	*via* C_3 and C_4 cycles, both spatially (in different parts of same cells) and temporally (C_4 at night, C_3 at daytime)
Normal diurnal occurrence	Daytime	Daytime	Day and night
Initial CO_2 acceptor	Ribulose-1,5-bisphosphate (*Rubisco*)	Phosphoenolpyruvate (*PEP*)	PEP at night, *RuBP* in the day
CO_2-fixing enzyme	Ribulose-1,5-bisphosphate carboxylase / oxygenase (*Rubisco*)	Phosphoenolpyruvate carboxylase (*PEP case*) then *Rubisco*	*PEP case* at night, *Rubisco* at daytime
First stable product of CO_2 fixation	3-phosphoglycerate (3-PGA)	Oxaloacetate (OAA) in C_4 cycle	OAA at night, 3-PGA at daytime
Cells Involved	Mesophyll cells	C_4- mesophyll cells, C3- bundle sheath cells	Both C_3 and C_4 in same mesophyll cells
Energy needed for complete reduction of one molecule of CO_2	3 ATP, 2 NADPH	5 ATP, 2 NADPH	
Environmental conditions favoring most efficient photosynthesis	moderate conditions; temperature 15 °C-25°C	hot, dry conditions; temperature 30°C-47°C	extremely dry or xeric conditions; temperature ~35°C
Amount of water needed to produce 1 g dry matter	450-950 g	250-350 g	50-55 g
CO_2 compensation point	30-70 ppm	0-10 ppm	0-5 ppm
Photorespiration	Yes	Absent or suppressed	Absent or suppressed
Annual dry matter production per hectare	~20-25 tons	~35-40 tons	Usually low and variable

Photosynthesis converts solar energy to chemical energy, allowing plants to build the tissues that sustain life on Earth. Pineapple makes use of a special type of photosynthesis, called crassulacean acid metabolism or CAM, which has evolved independently in more than 10,000 plant species. Pineapple is the most economically valuable horticulture plant among those 10,000 species.

Most crop plants use a different type of photosynthesis, called C_3. CAM plants use only 20 percent of the water used by typical C_3 crop plants and CAM plants can grow in dry, marginal lands that are unsuited for most crop plants. A closer look at the pineapple genome revealed that some genes that contribute to CAM photosynthesis are regulated by the plant's circadian clock genes, which allow plants to differentiate day and night and adjust their metabolism accordingly.

CAM photosynthesis allows plants to close the pores in their leaves during the day and open them at night. This contributes to pineapple's resilience in hot, arid climates, as the plant loses very little moisture through its leaves during the day. CAM photosynthesis allows the plant to absorb and "fix carbon dioxide into molecules during the night, concentrate it in its leaves and release it the next day for photosynthesis.

CAM and C_4 photosynthesis, which is common among grasses, use many of the same enzymes to concentrate carbon dioxide in plant leaves, the researchers report. Understanding the evolution of these different types of photosynthesis will help scientists in their efforts to develop more productive, drought-tolerant varieties of essential crops. For example, the U.S. Department of Energy has funded a project to explore the genetic mechanisms that enable CAM photosynthesis and drought tolerance in desert-adapted plants - with the aim of introducing those traits to potential biofuels crops. Adapting food crops to be more tolerant of drought will also help humans adapt to climate change, "Higher water-use efficiency is a highly desirable trait, given the need to double food production by 2050 in the context of a changing climate."

C_3 plants use the Calvin cycle to fix carbon. C_4 plants use a modified Calvin cycle in which they separate Ribulose-1,5-bisphosphate carboxylase oxygenase (*Rubisco*) from atmospheric oxygen, fixing carbon in their mesophyll cells and using oxaloacetate and malate to ferry the fixed carbon to *RuBisCO* and the rest of the Calvin cycle enzymes isolated in the bundle-sheath cells. The intermediate compounds both contain four carbon atoms, which give C_4. In Crassulacean acid metabolism (CAM), time isolates functioning *RuBisCo* (and the other Calvin cycle enzymes) from high oxygen concentrations produced by photosynthesis, in that O_2 is evolved during the day and allowed to dissipate then, while at night atmospheric CO_2 is taken up and stored as malic or other acids. During the day, CAM plants close stomata and use stored acids as carbon sources for sugar, *etc.*, production.

The C_3 pathway requires 18 ATP for the synthesis of one molecule of glucose while the C_4 pathway requires 20 ATP. Despite this reduced ATP efficiency, C_4 is an evolutionary advancement, adapted to areas of high levels of light, where the reduced ATP efficiency is more than offset by the use of increased light. The ability to thrive despite restricted water availability maximizes the ability to use available light. The simpler C_3 cycle which operates in most plants is adapted to wetter darker environments, such as, many northern latitudes. Corn, sugar cane and sorghum are C_4 plants. These plants are economically important in part because of their relatively high photosynthetic efficiencies compared to many other crops.

4.11 Indigenous Vegetables

There is an urgent global need to increase agricultural systems diversification in order to improve human nutrition through better balanced diets and to achieve more resilient, profitable and sustainable small-holder production and marketing systems. Indigenous vegetables are in many cases highly nutrient-dense both in vitamins and minerals. Thus they are powerful tools in the present battle against malnutrition and non-communicable diseases worldwide and especially in the Pacific Island communities and other locations in Africa and Asia where poor current dietary choices have led to critically high human health costs to society. In addition, owing to the comparative lack of research to date on their improvement throughout all aspects of the value chain from field to fork, investment in this area can bring much higher rates of return than in conventional staples.

A large number of indigenous vegetables crop species are used, particularly, by the tribal population. Wide range of *Solanum* species is found in various parts of the region (Table 5). The local tribals of Arunachal Pradesh grow a vegetable having red tomato like fruits slightly bitter in taste belonging to the genus *Solanum*. In Manipur, another kind of brinjal, having round fruit and intermediate in appearance between tomato and brinjal, is grown. In the hilly areas, tree tomato (*Cyphomandra betacca*), a perennial shrub producing red tomato like vegetables is also grown and used as such. Tree tomato is consumed as delicious chutney when raw or after roasting and peeling off the skin.

Tree bean (*Parkia roxburghii* G. Don.) is one of the most common of multipurpose tree species in the Manipur and Mizoram (Kumar *et al.*, 2002). Another vegetable tree growing in the lower altitude zones and popular among the people is drum stick or horse radish, locally called Sajina (*Moringa oleifera*). One of the legume species, *Vigna vexillata,* is grown by the tribals of Tripura, rich in carbohydrates and minerals.

Many species of cucurbits are found. The wild species *Cucumis hardwickii*, the likely progenitor of cultivated cucumber, is found growing in natural habitats in the foothills of Himalayas and NE region, particularly Meghalaya. Cho-Cho (*Sechium edule*) a native of tropical America is a very popular vegetable in the region. Cho-Cho produces large starchy edible roots in addition to fruits.

Table 5. *Diversity of Solanum species in North East India*

Cultivated species	Remarks
Solanum macrocarpon L	Introduced in NE region
Solanum xanthcarpum Schard & Wendl	Used as vegetable and for medicinal purposes
Solanum indicum L.	Domesticated, used as vegetable and medicine
Solanum mammosum L.	Possibly introduced ornamental with high solasodine percentage
Solanum khasianum Clarke	Wild-cultivated for solasodine alkaloid
Solanum torvum Swartz	Wild - sold in the market in Mizoram
Solanum berbisetum Nees	Edible ripe fruits
Solanum ferox L.	Wild - leaves are used for medicinal purpose
Solanum spirale Roxb.	Wild but domesticated for medicinal use in Arunachal Pradesh
Solanum sisymbrifolium Lam	Native of Africa, wildly grown in Meghalaya
Solanum kurzii Br.	Endemic to Garo hills, Meghalaya
Solanum gilo Raddi.	Introduced in NE region as a Vegetable

(**Source:** Rai *et.al.,* 2004)

Keeping in view, the regional demand for vegetable crops, underutilized species needs to be identified for collection, particularly, for high yield, quality, resistance to diseases and pests, tolerance to frost and acidity.

Unfortunately, traditional diets and indigenous crops that have reliably fed communities for centuries across the world are rapidly being replaced by unhealthy foods. FAO observed that approximately 75 *percent* of the Earth's genetic resources are now extinct and another third of plant biodiversity is predicted to disappear by the year 2050.

But initiatives like the World Vegetable Center (AVRDC)'s Vegetable Genetic Resources System and Slow Food International's Ark of Taste are working to catalogue indigenous species of fruits and vegetables all over the world. With the help of these incredible resources, Food Tank has compiled a list of indigenous vegetables in different continents.

***Bitter Melon*:** Despite its distinctive appearance and bitter taste, this vegetable originally from the Indian subcontinent, is popular in a number of Asian countries. The gourd has cancer-fighting properties; helps cure diabetes and can help cleanse the body of toxins.

***Okra*:** The edible green seed pods of this plant are a common ingredient in soups and sauces and popular in Indian and Pakistani cuisine. The vegetable is a rich source of vitamins and minerals and the seeds provide quality oil and protein.

***Mungbean*:** The mung bean is important in Asian diets and valuable for its easily digestible protein.

High levels of iron in the vegetable help improve the diets of the most vulnerable women and children and mung bean production offers an opportunity for increased income for small-scale farmers. In addition, the vegetable can fix nitrogen in the soil, making it valuable for crop rotations.

***Lemongrass*:** This valuable herb is used in many Asian countries, including India, Sri Lanka, Thailand, Vietnamese, Malaysia, Cambodia and Indonesia. Lemongrass possesses many minerals and essential vitamins, which help control blood pressure and prevent heart disease. It is also valuable for its essential oils, which are known to possess anti-microbial, anti-fungal, antioxidant and anti-septic properties.

Indigenous vegetables have enormous potential to contribute in food and nutritional security in tribal or geographical challenged regions like Andaman and Nicobar Islands, India. These are highly preferred and frequently consumed vegetables by tribal and rural communities as compared to the exotic vegetables. Their easy accessibility, cost effectiveness, low pesticide or chemical uses and abundance in and around habitat make them preferred choice for the rural and tribal population. They are rich in micronutrients iron, calcium, zinc and copper and phosphorus and antioxidants like phenolics, carotenoids, ascorbic acid, anthocyanin, flavonoids and tannins (Singh *et al.,* 2013).

4.12 Vertical Farming

It is estimated that by the year 2050, the world's population would increase to ten billion and close to 80% of human settlements would be concentrated in and around urban locations. To feed this urban-centric population, huge tracts of land would be required. In this context, well-designed farms that utilize space vertically rather than horizontally, can eliminate the pressure on an already overburdened earth and help create a cleaner environment.

Vertical farming can be considered as an extension of indoor farming that evolved in the 1700s with the advent of greenhouses, with the primary objective of harnessing the off-season crop cultivation potential during unfavorable seasons in the Indian context too (Aswath and Selvakumar, 2016). Vertical farming facilitates year-round cultivation in multi-storied structures using artificial illumination. Most commercial vertical farm ventures operate out of existing warehouses or abandoned factories that have been suitably converted to hydroponic/ aeroponic facilities with LED-based illumination systems.

Though vertical farming has been in vogue since ancient times as evidenced by the Hanging Gardens of Babylon, the modern concept of vertical farming involves the union of plant biology and engineering know-how. The advantages of vertical farming includes: year-round crop production, elimination of agricultural run-off, significant reduction in the use of fossil fuels for operation of farm machines and transport of crops, utilization of abandoned or unused properties, reduction in weather-related crop failures, sustainable food supply for all urban centers, conversion of black and grey-water to drinking water, addition of energy back to the grid via methane generation or plasma arc gasification, creation of new urban employment opportunities, reduction of the risk of infection from agents transmitted at the human–agricultural interface and return of farmland to nature, thereby helping restore ecosystem functions and services.

Over the last few years, multi-storied vertical farms have been established in Japan, Singapore, Taiwan, China, Korea and the United States of America. Some vertical farms have been constructed by adding on to the existing buildings, while few are green field edifices. But this practice is yet to gain momentum in India although it has captured the idea of some visionaries and few enterprising urbanites. Perhaps time is ripe for the country to gear up and introduce this space saving practice, especially in the urban and peri-urban clusters in order to meet the nutritional needs of a burgeoning population. An urban vertical aeroponic farm of 400 sq ft area has the distinct advantage of producing the same quantum of better-quality produce than a 1600 sq ft soil-based farm. With the right vertical aeroponic technology, good quality fresh food can be grown year round in a profitable manner. The future of vertical aeroponic farming in India and worldwide depends on sound economics of scale that takes advantage of the latest technologies to deliver fresh produce and offers good profit margins for investment ventures. The quantum of water saving in vertical farming is reported to be 90% of the water used in normal horticulture besides improving the photosynthetic efficiency to the tune of 15–18% as against 9% which is normally achieved under outdoor conditions.

The country is blessed with varied agro-climatic zones, multitude of crops and dietary preferences. The choice of crops suited for vertical farming under Indian conditions was also debated. Leafy vegetables, especially the high-value ones like lettuce, celery and cucurbits with export potential, tomato, capsicum, bottle brinjal, fruits like strawberry and flowers like gerbera, carnation and structures in a seamless manner. Therefore, we need to harness solar energy that is available in abundance in order to minimize the energy requirements of illuminating multi-storied growing structures. Approaches to harness the solar radiation in frugal vertical growing structures, which could be easily integrated into existing terraces and balconies, need focus. The limitations of vertical farming in the Indian context could be as follows:

- Limited vegetable crops have been identified for the purpose. However, vegetable crops in vogue are being grown without any scientific validation. No varieties/ hybrids have been exclusively bred for the purpose of vertical farming. As a result, yield and quality of the produce cannot be guaranteed ; production technologies and Good Agricultural Practices for these crops have not been standardized ; presently, the high-rise inclined buildings are not built to suit vertical farming. There is a need to reengineer the existing buildings by providing additional structures to suit the purpose.

The above limitations could be taken up as researchable issues. It is essential that all the high-rise buildings coming up in the smart / green cities and tier-two cities in the near future should be engineered in such a way that these buildings could be used for vertical farming (Aswath and Selvakumar, 2016).

5. ESSENTIAL FARM INPUTS

5.1 Improved Seed Quality

The Indian Seed Improvement Programme is backed up by a strong crop improvement programme in both the public and private sectors. At the moment, the industry is highly vibrant and energetic and is well recognized in the international seed arena. Several developing and neighbouring countries have benefited from quality seed imports from India. India's Seed Programme has a strong seed production base in terms of diverse and ideal agro-climates spread throughout the country for producing high quality seeds of several tropical, temperate and sub-tropical plant varieties in enough quantities at competitive prices. Over the years, several seed crop zones have evolved with extreme levels of specialization.

In terms of seed marketing and distribution, more than about 20000 seed dealers and distributors are in the business.

Over the years, seed quality specifications comparable to international standards have been evolved and are adopted by the Indian Seed Programme in both the public and private sectors. The country has a strong rigorous mechanism for seed quality control through voluntary seed certificate and compulsory labelling monitored by provincial level Seed Law Enforcement Agencies.

Thus, the Indian Seed Programme is now occupying a pivotal place in Indian agriculture and is well poised for continued growth in the years to come. National Seeds Corporation, which is the largest single seed organization in the country with such a wide product range, pioneered the growth and development of a sound industry in India. National Seed Corporation (NSC), SFCI, States Seeds Corporations and other seed producing agencies are continuously and gradually expanding all its activities especially in terms of its product range, volume and value of seed handled level of seed distribution to the un-reached areas, *etc*. Over the past four decades, these seed producing agencies have built up a hard core of competent and experienced seed producers and seed dealers in various parts of the country and have adequate level of specialization and competence in handling and managing various segments of seed improvement on scientifically sound and commercially viable terms.

5.2 Varietal Protection in India

India is signatory of World Trade Organization (WTO). WTO has at least half a dozen intergovernmental agreements that directly affect agriculture. Under the TRIPS Agreement Article 27(3) (b), which resulted from the negotiations of the Uruguay Round, requires members of protect plant varieties either by patents or by an effective '*sui generis*' system of protection or by a combination of both these systems. In compliance to the TRIPS Agreement India established Protection of Plant Varieties and Farmers Rights (PPV&FR) Authority, under the Protection of Plant Varieties and Farmers Rights Act, 2001. PPV & FR Authority has become operational since 11th November, 2005. PPV & FR Authority shall maintain a National Register of Plant Varieties. The certificate of Registration shall be valid for nine years in the case of trees and vines and six years in case of other crops. The total period of variety shall not exceed 18 years for trees and vines and 15 years for extant varieties notified under Seeds Act and for other crops. Farmer who is engaged in the conservation of genetic resources of land basis and wild relatives of economic plans and their improvement and preservation shall be entitled to recognition and reward from the Gene Fund provided the material so selected and preserved has been used as a donor of genes in varieties register able under the PPV & FR Act. Any person or group of persons (whether actively engaged in farming or not) or any other Governmental or Non-governmental organization may stake a claim on behalf of the village or local community. There is a provision for compulsory licensing to meet the reasonable requirement of the public for seed or other propagating material. (http://seednet.gov.in/Material/IndianSeedSector.htm).

5.2.1 Indian seed market: growth over years and growth drivers

Issues to be addressed are: Enactment of suitable legislations; Market liberalization to increase availability of high-quality seeds to Indian farmers. Relaxation of norms for export to overseas country; regulatory framework ; sharing of germplasm and breeder seeds of public-bred varieties. Minimum restrictions and proper safeguard mechanism for import of germplasm and hybrids/varieties; acceptance and commercialization of new seed technology, GM traits, use of biotechnology; Investments by public/private players.

5.2.2 Research and technology

Technical and financial assistance in the early stage of development from foreign aid agencies, such as, USAID, Ford Foundation and Rockefeller Foundations, World Bank to strengthen the breeding, seed production, distribution, processing and quality control infrastructure in public and private sectors. Entry of multinational corporations into the Indian seed business through equity participation.

5.2.3 Foreign investment

- Public-private partnership.
- Special schemes for upgrading quality of farm saved seed and improving Seed Replacement Rate.
- Extension programs for awareness generation and easy accessibility of quality inputs.
- Introduction of the Seed Technology subject at graduate and Post graduate level.

5.3 Easy Credit Facilities

Credit helps farmers to invest in creating assets to generate outputs and income through deploying science, technology and modern methods of business management. Investments in farm can generate farm income sustainably when credit is simultaneously adequately supported by backward and forward linkages, *viz.,* inputs of production, technology and services.

However, according to Amrit Patel (2014), the following pressing issues are still continuing:

- impact on improving crop productivity and output has been not so satisfactory, leave alone expected; significant disparities in the flow of credit among States, districts, villages and even within the village; significant imbalance between short-term and term loans, between credit disbursed to agriculture and allied activities as also across components of agricultural term loans, *viz.,* irrigation development, farm mechanization, land development, plantation and horticulture, hi-tech agriculture, *etc.,* difficulties experienced to easy and reliable access of institutional credit by small, marginal and tenant farmers, share croppers oral lessees, landless laborers, households residing in hilly, tribal, desert, drought prone and most backward and vulnerable areas in particular[v]problem of loan repayments leading to building up huge amount of NPA has become serious and pernicious.

In order to ensure that credit in particular and financial services in general achieve their intended objectives, now the need to create enabling policy environment and effective credit planning is greater than before. Government, RBI and the legislators need to demonstrate a serious concern and commitment to address these issues. Following policy intervention, ***inter alia,*** need immediate attention along with formulating a strategic action plan to resolve most issues by the end of the 12thFYP [2016-17] (Amrit Patel, 2014).

5.4 Enabling Environment

Policy should address to create enabling environment that can significantly enhance credit absorption capacity of the geographical areas and farmers in each village. For this, Amrit Patel (2014) opined that the policy should focus on critical areas, *viz.,* investing significantly in rural infrastructure to improve farm fuel/energy, transport, communication, storage, processing and marketing facilities; establishing State of Art Agri-meteorology in all agro-ecological regions; expanding irrigation and reclamation of wastelands; integrating agricultural education, research and extension services and strengthening capacity-building of farmers to bridge the huge yield gap between the potential yields and actual yields at field level in rain-fed and irrigated farming systems; development and use of genetically engineered seeds, micro-irrigation system, greenhouse technology, integrated nutrient and pest management technique, computer-based modeling to track disease and pest incidence; expanding farm mechanization program; promotion of remote sensing technology.

Policy and programs involving timely and adequate investments in agriculture would facilitate farmers' access to frontier technologies for production, food processing and preservation, farm-to-market/fork linkages, agricultural research and extension, weather and crop forecasting, large-scale development of bio-diesel, modernization, mechanization and commercialization of agriculture. Appropriate proportion of public, private and foreign investments should remedy the situation of investment shortage in agriculture and help transform a 'negative subsidy regime' into a 'capital-intensive positive Agricultural Marketing Service regime' and stimulate India's agricultural producers to access global markets (Amrit Patel, 2014).

5.5 Improving Farmers' Income

Role of credit to agriculture cannot be viewed just as a support to food-producing activity but it should focus "need to improve the overall income and economic well- being of the farmers" as agriculture has been the basic requisite for national sovereignty. The extensive literature analyzing the relationship between agricultural and non-agricultural growth in India confirms that farm & non-farm sector in rural areas are complimentary to each other and risks mitigating. Rural credit policy and programs need to focus on farm & rural non-farm sector development to alleviate rural poverty, deprivation and suffering (Amrit Patel, 2014).

5.6 Reaching the Unreached

ICT has been fast changing the shape of the world and as newer technologies are ushered in, banking services become more affordable and accessible to all. Harnessing the power of technology presents significant opportunities to financial institutions to reach dispersed and hitherto hard-to-reach clients for providing them financial services in rural hinterland of developing economies.

More importantly, the resourceful commercial banks in the developing economies can profitably make use of these technologies to automate agricultural, rural & micro-finance (ARMF) systems to reach a large number of hitherto unreached rural households.

The technology can yield significant benefits to ARMF service providers and their customers. Banks may consider implementing the technology in a customer-friendly and cost-efficient manner so that automation can certainly facilitate ARMF systems to reach dispersed and hitherto hard-to-reach clients, work smoothly and efficiently to deliver new financial products and services. ARMF service providers to increase the outreach and improve the competitiveness can adopt ways and means of cost savings associated with automatic ARMF operations, such as:

- Personal Digital Assistants (PDAs) relieve loan officers from carrying bulky client files to the field, improve data quality and reduce transaction time by automating transactions and record keeping. All of these help loan officers serve more customers; credit scoring efficiently and reliably categorizes customers according to risk that helps loan officers exclude ineligible & risky applicants from borrowing. This directly saves costs and improves allocation of loan loss provisions ; installation of ATM that can handle multiple transactions can reduce the cost of opening branch offices and providing staff in a remote rural area; technologies, such as, Smart Cards, ATMs & *m-banking services* enhance security and flexibility, which customers value much and help increase customer loyalty. Automating processes enable banks to conduct transactions faster, at lower cost per transaction and in more dispersed locations than branch staff can do. Linking *Aadhar* based applications in this sector also need to emerge so as to reach the unreached and the needy.

5.7 Storage in Vicinity and Transport Facilities

Storage facilities require adequate ventilation in order to help extend shelf life and maintain produce quality. The following are three types of fans found in common use. Ventilation in storage structures is improved if air inlets are located at the bottom of the store, while air outlets are at the top. A simple, light-proof exhaust vent is a pressure-relief flap. Any type of building or facility used for storage of horticultural crops should be insulated for maximum effectiveness. A well– insulated refrigerated building will require less electricity to keep produce cool. If the structure is to be cooled by evaporative or night air ventilation, a well-insulated building will hold the cooled air longer. For storage facilities that are refrigerated, using outside air for ventilation is wasteful of energy. For these systems, a simple recirculation system can be designed by adding a fan below floor level and providing a free space at one end of the storeroom for cool air to return to the inlet vents. In cooler regions, suitable storage temperatures can be maintained by bringing outside air into the storage facility. For best results, air vents should be located at the base of the storage structure. An exhaust fan located at the top of the structure pulls the cool air through the storeroom. Vents should be closed at sunrise and remain closed during the heat of the day.

5.8 Controlled Atmosphere (CA) Storage

Controlled or modified atmosphere storage should be used as a supplement to and not as a substitute for, proper temperature and relative humidity management. Some simple methods for modifying the composition of air in the storage environment are listed below (Kader, 1993). Air coming into the storeroom or being recirculated within the room must pass through a monitoring and control system.

5.9 Relative Perishability and Storage Life of Fresh Horticultural Crops

Precise data on post-harvest losses of horticultural commodities though is still lacking in the country, it is reported that the system may experience somewhere between 28-30 per cent losses which is unsustainable for a developing country like India. The relative perishability and storage life of fresh horticultural crops could vary greatly depending upon wide range of factors including both pre-harvest protocols integrated with post-harvest handling is presented in Table 6.

Table 6. *Classification of fresh horticultural crops according to their relative perishability and potential storage life in air at near optimum temperature and relative humidity*

Relative Perishability	Potential Storage Life (Weeks)	Commodities
Very high	<2	Apricot, blackberry, blueberry, cherry, fig, raspberry, strawberry; asparagus, bean sprouts, broccoli, cauliflower, green onion, leaf lettuce, mushroom, muskmelon, pea, spinach, sweet corn, tomato (ripe); most cut flowers and foliage; minimally processed fruits and vegetables.
High	2-4	Avocado, banana, grape (without SO_2 treatment), guava, loquat, mandarin, mango, melons (honeydew, Crenshaw, Persian), nectarine, papaya, peach, plum; artichoke, green beans, Brussels sprouts, cabbage, celery, eggplant, head lettuce, okra, pepper, summer squash, tomato (partially ripe).
Moderate	4- 8	Apple and pear (some cultivars), grape (SO_2-treated) orange, grapefruit, lime, kiwifruit, persimmon, pomegranate; table beet, carrot, radish, potato (immature).
Low	8-16	Apple and pear (some cultivars), lemon; potato (mature), dry onion, garlic, pumpkin, winter squash, sweet potato, taro, yam; bulbs and other propagules of ornamental plants.
Very low	>16	Tree nuts, dried fruits and vegetables

(***Source:*** *Kader, 1993*).

Temperature management is critical during long distance transport, so loads must be stacked so as to enable proper air circulation to carry away heat from the produce itself as well as incoming heat from the atmosphere and off the road. Transport vehicles should be well insulated to maintain cool environments for pre-cooled commodities and well ventilated to allow air movement through the produce. During transport, produce must be stacked in ways that minimize damage and then be braced and secured. An open air vehicle can be loaded in such a way that air can pass through the load and provide some cooling of the produce as the vehicle moves. Transport during the night and early morning can reduce the heat load on a vehicle that is transporting produce.

6. LINKING HORTICULTURE PRODUCE TO MARKETING

6.1 Electronic Marketing of Perishable Produce

Horticultural commodities are highly perishable. Their markets are thin and sparsely spread. Markets are inefficient, which is evidenced by price gap between producer's price and consumer rupee. Till recently, horticultural commodities are routed through mandis, established under APMC. They add to marketing costs of the produce. These markets are controlled by a few big traders of the respective regions; the operations of them are non-transparent. The net result of all these is much lower realization of income by the farmer.

Majority of the horticultural commodity markets in India generally operate under the normal forces of demand and supply. The buying and selling of horticulture produce takes place in the market yards where numbers of market functionaries are involved. Government of India (2001) reported that as per survey conducted, the farmer is getting only one rupee out of every Rs. 3.50 paid by the consumer, the retailers is getting Rs. 0.75, the wholesaler is getting Rs. 0.50 and rest of the amount Rs. 1.25 is going to commission agents and traders. The share of producer in consumer's price depends upon the types of marketing channels followed in sales transactions by the farmers. Studies indicated that the share of producers varies from 56 to 83 per cent in food grains and 79 to 95 *per cent* in pulses, 6596 *per cent* in oilseeds and 33 to 75 *per cent* in case of fruit and vegetables. In order to provide the remunerative prices to the farmers, there is a need to eliminate the chain of middlemen *etc.,* by introducing innovative marketing channels *viz.,* direct marketing and contract farming. India amended APMC act and allowed cooperatives and private entrepreneurs to set up special markets for fruits, vegetables and flowers. India opened up post –1990, a number of new liberalized public markets for F&V in the WTO regime. The main functions of these farmers markets are empowering the farmers to participate effectively in the open market to get a remunerative price for their produce; to avoid the exploitation of both the farmers and the consumers by the middlemen; enhancing the distributional efficiency of the marketing system; price stabilization and risk management. In India organized retailing for fruits and vegetables is gaining a lot of impetus with huge investments by leading Indian corporations. Traditional markets are making way for new formats, such as, departmental stores, hypermarkets, supermarkets and specialty stores and western-style malls. Their number and scale of operation have been under- going a phenomenal growth in the past few years. In horticulture, the gap between prices received by the farmers and those paid by urban consumers is large, reflecting inefficient marketing arrangements. The study concluded that the marketing model adopted by private markets particularly *Reliance* and *Subhikshais* are more efficient than government markets like *Rytha bazaar*. Therefore, there is an immediate need to replicate such models in a much larger scale in order to cover not only the cities but also the interior villages in the country. The study further suggested that it is necessary to amend outdated laws restricting the establishment of markets so as to allow the cooperatives and private entrepreneurs to set up modern markets. Both public and private retail markets have to adopt the new marketing models in order to enhance the distributional efficiency of the marketing system.

6.2 Price Rise

Globally, people were going hungry because they could not afford food. With higher prices, the numbers of people suffering from extreme hunger has increased even further and the 1st UN Millennium Development Goals ('*Eradicate extreme poverty and hunger*') has become an even greater challenge. Immediate and substantial aid is required, wherever appropriate in the form of cash or vouchers so as to strengthen, rather than undermine the domestic markets in the recipient countries.

In the medium term, there is a real need to improve the purchasing power of poor food buyers so that they could acquire enough food even at the higher prices, relative to past averages, that are expected to prevail in the future. Fundamentally that requires to foster growth and development in poor countries. In some of the poorest countries, investment in agriculture, including in agricultural research, extension and education, may be the best way to cut poverty and stimulate economic activity. In other words, a tailored approach is needed, one that could build upon the capacity and potential of individual countries, rather than a generalized rush to develop agriculture.

Studies have indicated that, the current hike in food prices is an issue of a truly global nature. It has complex causes and impacts and requires a complex response at the international level. Current developments in global food markets are having dramatic implications for food security among poor people. At the same time, speculative factors and inward looking policy actions contribute to the nervousness and volatility of markets. What is needed perhaps now is an objective, effective and coherent global response so as to avoid rendering a difficult situation worse.

6.3 Indian PM's Proposal

Pradhan Mantri Krishi Sinchayee Yojana is a central scheme that aims at providing irrigation facilities to every village in the country by converging ongoing irrigation schemes implemented by various ministries. It will focus on 'end-to-end solution' in irrigation supply chain by implementing new programme in a "project mode" with decentralized state-level planning and execution. This scheme has amalgamated three ongoing programmes of three different ministries viz. Accelerated Irrigation Benefit Programme of the Ministry of Water Resources; Integrated Watershed Management Programme of the Ministry of Rural Development; and Farm water management component of the National Mission on Sustainable Agriculture. For 2015-16, Rs.1, 000 crores are allotted for the scheme. The funding pattern between centre and states is 75:25 percent. In case of north-eastern region and hilly states it will be 90:10. (Ref: http://www.gktoday.in/blog/2016-agriculture-schemes-terms-and-updates/#Pradhan_Mantri_Krishi_Sinchayee_Yojana)

Soil Health Card: The scheme has been launched to provide every farmer a Soil Health Card in a mission mode. The card will carry crop wise recommendations of nutrients / fertilizers required for farms, making it possible for farmers to improve productivity by using appropriate inputs. Since the scheme is being implemented by the State Governments; it indicates that states need to increase their manpower requirement and funding. Paramparagat Krishi Vikas Yojana Paramparagat Krishi Vikas Yojana (Traditional Farming Improvement Programme) has been launched by Government of India to support and promote organic farming and thereby improving soil health. This will encourage farmers to adopt eco-friendly concept of cultivation and reduce their dependence on fertilizers and agricultural chemicals to improve yields. Government has made budgetary allocation of Rs. 300 Crores for the same in the Union Budget 2015-16 (http://www.gktoday.in/blog/2016-agriculture-schemes-terms-and-updates/#Pradhan_Mantri_Krishi_Sinchayee_Yojana)

7. TECHNOLOGICAL INNOVATIONS IN HORTICULTURE

7.1 Use of Wireless and Digital Technology

Wireless Sensor Technologies (WST) is entering a new phase. Recent advances offer vast opportunities for research and development. This is the consequence of the decreasing costs of ownership, the engineering of increasingly smaller sensing devices and the achievements in radio frequency technology and digital circuits. The mobile app allows easy collection of field data. It also provides a web-based program where farmers and their crop consultants can assess the whole farm in a user-friendly, consolidated view. Integration of the web-based and mobile apps allows nutrient management records to be stored online in a "*cloud*" database, accessible from any location or device via Internet access.

Information and communication technology in agriculture (ICT in agriculture), also known as e-agriculture, is developing and applying innovative ways to use ICTs in the rural domain, with primary focus on agriculture. ICT in agriculture offers wide range of solutions to some real time agricultural challenges. It is seen as an emerging field focusing on the enhancement of agricultural and rural development through improved information and communication processes. In this context, ICT is used as an umbrella term encompassing all information and communication technologies including devices, networks, mobiles, services and applications; these range from innovative Internet-era technologies and sensors to other pre-existing aids, such as, fixed telephones, televisions, radios and satellites. *E-agriculture* continues to evolve in scope as new ICT applications continue to be harnessed in the agriculture sector. More specifically, *e-agriculture* involves the conceptualization, design, development, evaluation and application of innovative ways to use ICTs in the rural domain, with a primary focus on agriculture. Provisions of standards, norms, methodologies and tools as well as development of individual and institutional capacities and policy support are all key components of *e-agriculture.*

Yet it may be wrong to assume that wireless applications have not penetrated the agricultural sector at all. 2-way radios and "*push-to-talk*" cell phones are two examples. These are wireless tools that are relatively cheap, reliable and very simple to use. For several generations, farmers in countries with large farms have used 2-way radios to communicate with employees. Because farmers already understood the benefits of wireless communications, they were some of the earliest adopters of cell phone technology, especially "*push-to-talk*" cell phones. These devices gave them the mobility to contact their employees, farm supplier, equipment dealer, extension agent, buying point or spouse from anywhere at any time. Today, the cell phone is indispensable to farming (Kvien, 2005).

In addition to being cheap, reliable and simple to use, cell phones have one more important attribute – all brands are compatible. Unfortunately, the same cannot be said for the wireless sensor and control networks which turn switches on and off, check fuel levels or stream video. At the moment, these technologies all work differently, require proprietary software, include components which are expensive and often not compatible and are frequently cumbersome to use. However, that is likely to change because of the rapid pace of development in certain sectors of internet communications (Pocknee, 2005).

7.2 Mobile Apps for Soil and Crop Nutrient Management

ICAR-Indian Institute of Soil Science, Bhopal has developed a mini lab named '*Mrida parikshak*'. It is a digital mobile quantitative mini lab to provide soil testing service at farmers' doorsteps. It determines all the important soil parameters i.e. soil pH, EC organic carbon, available nitrogen, phosphorus, potassium, sulphur and micronutrients *viz.*, zinc, boron and iron. It also provides crop and soil specific fertilizer recommendations directly to farmers on his mobilc through SMS.

The following are the latest agriculture apps for the Android, *iPhone*, *iPad* and other mobile electronic devices that could be adopted:

Simplot Spray Guide: Developed in cooperation with Precision Laboratories and Simplot, the Spray Guide app quickly and accurately identifies the ideal mixing order of crop protection products. The app streamlines the process of mixing, spraying and record keeping. Users can instantly document and share data on products sprayed, location and weather conditions.

Mobile Farm Manager: This app from *John Deere* gives producers easier access to important farm and field information through their mobile device. The app also connects to customers' Apex farm management software. Features include field maps, historical reports, GPS tracking, field navigation and soil sampling grids.

Spray Select: The Tee Jet Technologies Spray Select app enables users to quickly and easily choose the proper spray tip for a given application. Enter speed, tip spacing and target rate, then select the droplet size category and a list of recommended tips is provided. Specific gravity input is also included for use when applying liquid fertilizers.

Connected Farm: This app from Trimble allows farmers and agronomists to map field boundaries , enter scouting attributes for pests (weeds, insects, diseases), take geo-referenced photos and manage collected data online through Trimble's Connected Farm. The app is flexible to use with any crop, including corn, wheat, soybeans and cotton.

AgStudio MAP: MapShots AgStudio Mobile Application Platform (MAP) is optimized for mapping field boundaries and recording data associated with precision ag soil sampling programs. Designed to seamlessly integrate with MapShots *AgStudio* precision agronomy application, the app wirelessly transmits soil sample work order information and completed job files.

Onsite: Onsite is a cloud-based, mobile and desktop app that assists with file management and communications to and from the field by socially connecting people. Developed by *AgIntegrated*, the program allows users to directly access equipment telematics systems, such as, Raven Slingshot build a network of connections to send and share files to other colleagues and locate connections (individuals or equipment) on their desktop or smartphone.

Maximum Return To Nitrogen (MRTN) Calculator: Developed by the University of Illinois Extension and the Illinois Council on Best Management Practices, the app helps farmers and crop advisors determine the optimum nitrogen rate for corn and plan for split applications of nitrogen (fall, pre-plant and post-applied). The MRTN calculator also enables users to choose from various sources of nitrogen, add in stabilizers and calculate the corresponding application costs.

Weed Manager PLUS: Weed Manager PLUS is the latest tool added to Monsanto's Roundup Ready PLUS Weed Management Solutions platform. The app provides weed management recommendations by region and crop, calculates potential incentives for farmers who use endorsed residual herbicide products and delivers a tank mixing tool and measurement conversion calculator.

Ag Seed Select: This is Monsanto's first attempt at building an app for its seed brands: Asgrow, DEKALB and Delta pine. *ag Seed Select* lets users create, store, e-mail and print a custom seed guide tailored to their specific geography and crops. Featuring videos by territory agronomists, the app provides detailed information on products for corn, soybeans, cotton and other crops.

Agrowdata: This mobile portal contains the latest global commodity price data. The app offers easy-to-view screens, which include graphs of historical and futures prices and comparison graphs of commodity prices over the same period allowing users to identify trends and correlations. Other features include handy conversion calculators, futures and swaps contract specifications and recent ag-related news.

AgFleet: ZedX recently launched a new app to complement its Web-based Ag Fleet service. Functionality includes AgFleet account synchronization; boundary and attribute logging; soil sampling; option to provide driving directions to a field; and in-field navigation to target soil sample points.

Sprayer Calibration (in development): Mississippi State University Extension is in the process of developing a new app to help farmers properly calibrate sprayers, an involved process with the potential for making mistakes. This app will require the end user to enter a variety of inputs, such as, orifice size on the spray tips, spray pressure, speed of the tractor, chemical quantity and tank size. The app will use these figures to calculate proper calibration for the sprayer.

Mix Tank 2.0 (update): Precision Laboratories has released an update to its Mix Tank app (along with an Android version). Mix Tank 2.0 allows applicators to quickly create, save and share spray logs with GPS location information and integrated weather data. New spray logs feature a stopwatch to record start and stop times, as well as total spraying time. Spray details, including location, time sprayed and weather data, are integrated into the new spray logs for easy sharing and recordkeeping.

Optimizer 2.0 (update) : Developed by Advanced Ag Solutions, the newly updated Optimizer 2.0 allows users to upload GPS soil sampling data, including phosphorus, potassium, pH and organic matter. Other apps use the standard national soil data set, but Optimizer 2.0 projects relevant information that is related to the user's actual management practices.

AGCOMMAND: AGCO's new app for its AGCOMMAND telemetry system gives the user real-time access to valuable machinery information. It features the ability to show turn-by-turn directions, the history of machine status and machine data, ability to compare the performance of multiple machines and user-defined alerts from vehicles and machines. Plus, a unique integrated radar shares weather conditions where the machine is located ***(Web source).***

Although the wireless farm of the future offers great potential for improving efficiencies, it will not materialize without an equal quantum leap in the ability of farmers to understand electronics. Even the most robust wireless systems will require regular maintenance, repair and upgrades. As learnt from a decade of implementing precision agriculture, adding technology to the farm may reduce the number of people required but dramatically increases the needed educational and competence level of the remaining workers. There will be a market for network and sensor specialists and other technologists in rural areas which may reverse the century-long trend of the brightest minds migrating from the countryside to the city.

In the coming decade, wireless networks will offer the same type of quantum leap forward for farming that GPS provided during the past decade. Internet networks will eclipse all previous forms of distance communication and will provide the communications backbone for farms in the future (Bareja, 2011).

8. ATTAINING FOOD AND NUTRITION SECURITY

8.1 Horticulture-Major Contributor to Nutrition Security

Right now, India is self sufficient in food grains. In general, sustainability in food security stands on three basic pillars: namely, (i) sufficient quantities of food available on a consistent basis; (ii) having sufficient resources to obtain appropriate foods for a nutritious diet; (iii) appropriate use based on knowledge of basic nutrition and care, as well as adequate water and sanitation. Once the country fulfils all these targets, feed the population efficiently. According to the International Food Policy Research Institute's (IFPRI) 2011 around 60 million children in India are underweight and malnourished and nearly 21 per cent of the population is malnourished. The introduction of the National Food Security Bill threatening to continue market inefficiencies in food supply and extend the problem of malnutrition far into the future. In addition, the report of World Bank says productivity losses in India due to stunted growth, iodine deficiencies and iron deficiencies are equivalent to almost 3 percent of GDP. India has taken many efforts to build up a food security system to ensure that the threat of famine like, Investment in agriculture and the green revolution have ensured that the food production has kept pace with the population growth; establishment of adequate buffer stocks has ensured availability of food stuffs within affordable cost even during the times of drought. Moreover, India is currently undergoing demographic, economic, social, educational and agricultural and health transition. These factors individually and collectively can bring about substantial alteration in health and nutritional status of the population. If through effective planning and inter-sectoral coordination, appropriate synergy is brought about, it will be possible to achieve substantial improvement in nutritional and health status of the population. Infrastructure and capital flow play critical role in the food production as well growth of economy. Adequate infrastructure facilities, such as, transport, communications, storage and processing facilities and other factors take part in food security. Policies that encourage holistic approaches including the engagement of the private sector should feature in any national and international approach to address food security. The private sector can invest in clean new technologies and develop innovative market mechanisms to combat climate change on agriculture, livestock and fishery (Prasanna Kumar and Harish Kumar, 2013).

There is an urgent need for better nutrition monitoring arrangements in India. There are serious gaps in India's nutrition statistics and even the most basic nutrition trends are far from clear. The NFHS surveys, for their part, are conducted at wide intervals (six to seven years) and the reports focus mainly on state-level indicators. Effective action in this field requires regular and reliable large-scale surveys that would make it possible to monitor the nutrition situation at the district level at intervals of, say, two to three years at most. Revamping the District Level Health Surveys (DLHS) and conducting NFHS surveys at more regular intervals, is two possible steps in this direction. Another useful option would be to "*link*" these surveys with NSS surveys from time-to-time, making it possible to combine detailed nutrition data with household expenditure data (Angus Deaton and Jean Drèze, 2009).

8.2 Proper Utilization of Horticultural Produce

India is the world largest producer of many fruits and vegetables but there still exist huge gap between *per capita* demand and supply due to enormous waste during post-harvest storage and handling caused by improper bagging without crating, lack of temperature controlled vehicles, unavailability of cold chain facilities in various parts of country for preserving the produce, along with significant processing of the agricultural produce which results in immense losses to the nation. Hence, a proper supply chain management in fruits and vegetables has to be improved in all the stages of the supply by adopting best global practices in storage, packaging, handling, transportation, value added service *etc* to meet the country's demand of fruits and vegetables. As per this paper important drawbacks of the current supply chain are high level of wastage, quality degradation, poor infrastructural facilities and high cost. Government and private operators have to join hands to improve the physical infrastructure, information sharing and the service required for quality improvement of the supply chain (Sheoran, 2015).

There is a colossal waste during the post-harvest storage and handling due to improper bagging without crating, lack of temperature–controlled vehicles, no cold chain facilities for preserving the produce, coupled with significant processing of the agricultural produce resulting in enormous losses to the nation. Given the characteristics of fruits and vegetables, such as, perishability, seasonality, bulkiness and delicate nature of the products coupled with inadequate storage and transport facilities, the supply chain can be made efficient by reducing the length of the chain improving cold chain facilities. The supply chain management in vegetables has to be improved in all the stages of the supply by adopting global best practices in storage, packaging, handling, transportation, value added service *etc*. Several initiatives can be taken by the Indian government to solve the problems existing in the Indian food supply chain which includes allowing more organized retailers to enter into retail market, setting up of basic production factors, an optimal crop management system, developing a post-harvest infrastructure, entrepreneurial management and expertise and logistical infrastructure and finally by improving post-harvest operations related to handling, storage and marketing of produce. All these initiatives will ensure consistency in supply and provide recorded and demonstrated traceability of products will benefit the farmers with good remunerative price and consumers with good quality produce and also will reduce the losses incurred due to poor post-harvest management and thereby ensuring adequate supply to the consumers. Creation of proper fruit and vegetable supply chain system will link together farmers, vegetable vendors, farmer's self-help groups and women vegetable/fruits vendors ensuring the availability of quality produce at competitive price to consumers at a convenient time and place.

9. ADDRESSING SUSTAINABLE HORTICULTURE DEVELOPMENT IN THE NEW MILLENNIUM

9.1 Future Research Priorities

Following research priorities for the tropical and subtropical fruits sub-sector are flagged here below (Anon., 2011 and Anon., 2015):

Priority I: Enhancing Productivity and Quality

Germplasm resources, conservation and utilization

Genetic erosion and vulnerability of biological diversity / genetic resources over the past decade in many countries due to impacts of pests and diseases outbreaks or to absence of abiotic stresses, such as, heat, drought or frost looms large based on country reports.

These losses are expected grow further under the adverse manifestations of global climate change having consequences for the adaptation and geographic distribution of crops, specific varieties and crop wild relatives (CWR). A case in point with reference to grapes underscores the importance. In this context, it may be crucial for the respective countries to maintain their own traditional varieties in national or regional ampelographic collections, conserve clonal variability and CWR as far as possible for posterity. Climate change has been impacting tropical and subtropical fruit production scenario in significant manner.

- Development of cultivars to ensure vertical growth and sustain climate change while, meeting the requirements of high yield, nutritive and nutraceutical values, value addition and offer export competitiveness. This strategy needs to harness the power of biotechnology that could generate trait - specific gene products, genomic resources and molecular breeding; and nanotechnology.
- Standards for quality planting materials' production that would encompass such issues as rootstocks, biosafety, quarantine, diagnostics, traceability for quality assurance and plant architecture models for optimizing resource use are highlighted .

Priority II: Production System Management for Augmented Input Efficiency and Addressing Issues of Abiotic Stress

- Rainfed, arid, hills and coastal fruit culture focusing on identification of suitable crops, genotypes with traits for resilience to abiotic stresses and weather dynamics
- Rejuvenation of old and unproductive orchards and canopy engineering for efficient use of resources; HDP systems
- Designing and integration of systems for efficient management of land, water and nutrients; good horticulture practices; site-specific management strategies; harnessing the power of microbial dynamics for enhancing soil health including bioremediation .
- Precision and organic farming modules syntheses to ensure food and environmental safety.
- Develop climate resilient tropical and subtropical fruit culture; frost fighting mechanism through exploring use of anti-transpirants, metabolic inhibitors, osmo-protectants, antibiotics/antifungal agents; genetic engineering to protect candidate crops from the adverse effects of epiphytic INA Bacteria (Ice Nucleating Active Bacteria) and increasing cold hardiness.

Priority III: Addressing the Contextual Issues of Biotic Stresses and Their Management

- Breeding varieties for multiple resistance / tolerance to major pests and diseases .
- Develop systems for management of biotic stresses-rapid diagnostics for detection of pathogens and pests, pests and diseases forecasting models, integrated bio-control systems for IPM and effective management strategies.
- Surveillance for prevailing and emerging threats including their diversity analysis; identify PFA and internal quarantine.
- Residues and environmental safety are critical .

Priority IV: Mechanization for Small Holder Fruit Growers Through Development of Low Cost User-Friendly Tools to Reduce Drudgery

Priority V: Post Harvest Management Addressing Issues of Profitability of Orcharding and Utilization of Produce

- Integration of pre and post harvest protocols for enhanced productivity, quality and minimization of losses; establishment of value chains.
- Develop systems of off-season production for extended markets.
- Secondary horticulture for value addition, product diversification, byproduct utilization, residues and wastes recycling.
- Supply chain management.

Priority VI: Transfer of Technology and Horticulture Education

- Impact assessment and economic appraisal of profitability of orcharding and supply chain analysis.
- Issues of researchers- extension-grower linkages; technologies commercialization; re-innovation; innovative extension methods for promoting adoption of technologies.
- Harnessing of ICT (Concepts of information technology) and space technology modules
- Capacity building and human resource development through fine tuning of horticultural education rendered relevant and contextual.

Priority VII: Credits to Growers and Market Integration

- Policy interventions for smooth flow of micro-finance at lower interest rates; institutional reforms .
- Market research, market intelligence, creation of markets and market integration.
- Risk insulation-crop insurance to fruit growers .

9.2 Constraints and the Path Forward

Despite the research achievements and generous support for the developmental activities by the Government of India to the NARES, Indian horticulture is confronted with several constraints which need to be tackled effectively (Anon., 2011).

Low productivity: One of the worrisome features of the Indian fruits sector is the low productivity of many components. Among the fruits, with the exception of grapes and banana, the average productivity is much lower (11 t/ha) in India as compared to many leading fruit producing countries (21 t/ha in USA, Brazil). Fruits and vegetables together registered a growth of 4.14 per cent in production, 3.2 per cent in area, but a meager 0.91 per cent in productivity in India during 1993-94 to 2007-08. Low productivity levels decrease the profitability of growers, especially the small and marginal and can adversely impact sustainable livelihoods. Large tracts of perennial fruit culture in the country are constrained by the prevalence of old and unproductive orchards that could be as short and medium term measures need rejuvenation, central opening and canopy management to optimize productivity. In order to address the widespread nutrient deficiencies prevalent across the fruit crops and the agro-ecologies, site-specific nutrient management strategies are emphasized.

Deteriorating production environment: Indian fruit culture is predominantly a small holders' horticulture as more than 82 *per cent* of the holdings are small and marginal. The average size of the landholding declined from 2.30 ha in 1970-71 to 1.32 ha in 2000-01 with a concurrent increase in the absolute number of operational holdings from about 70 million to 121 million. If this trend continues, the average size of holding in India would be at an unsatisfactory 0.32 ha by 2030.The contribution of agriculture in the GDP has declined from about 55 per cent during 1950s to about 14 per cent in 2011-12 yet 55 per cent population still engaged in agriculture rendering large proportion of the rural populace to sustain on dwindling resources. This situation has lead to widespread poverty, hunger, malnutrition and widened rural-urban income disparities. The holding size for the fruit sector is no different, with the added constraint that in several regions it is mostly the marginal lands that is available for expansion. Moisture and salinity stress, poor soil fertility *etc* are characteristic features of such marginal lands to which fruit culture has been extended recently may continue in future. Thus, while striving for increasing the productivity of small and fragmented arable lands, it is also imperative to use the degraded and marginal lands as well as places with abiotic stresses for extending the cultivation of fruits. Even in good arable lands, the problems of land and water quality are to be addressed to sustain productivity. In the emerging scenario, due focus on water productivity of different fruits is inevitable. Consolidation of land holdings, contract farming, mechanization, off-farm employment opportunities by promoting secondary horticulture through primary processing and value addition, linking with markets, access to credit *etc*. need due focus. With the exodus of males to urban locales and feminization of fruit culture occurring, gender mainstreaming demands attention. In order to tide over the acute labour shortages prevailing and minimize drudgery of orchard related operations, there are critical requirements of low cost and efficient tree pruners, harvesters *etc*. Further adoption of compatible fruit based cropping / farming systems in different agro-ecologies to augment factor productivity merits due consideration.

Post-harvest losses and its impact on *per capita* availability: Despite phenomenal increases in the production of fruits across the country, the nutritional status of the population has not improved much, as the *per capita* availability of fruits is around 104 g day^{-1} respectively, far less than the standard recommended levels of 120 g day^{-1}. One of the main reasons for low *per capita* availability is the enormous post-harvest losses of fruits, which occur at different stages of handling, transport, storage, processing and distribution. India loses fruits worth Rs.12700-15876 crores (20-25% loss) annually. The country continues to lag in the arena of value addition and horticultural waste utilization.

Dynamics of quality awareness and global competition: As the consumers are becoming more quality conscious around the globe, improving the quality of fruit produce is imperative in order to render Indian fruit culture globally competitive and harness the advantage of the global trade liberalization platform. Quality here denotes physical appearance, chemical composition with reference to taste and flavour as well as hygiene and health factors, especially pesticide residues and heavy metal contamination and the *Codex Alimentarius* standards are highly dynamic. Developing good management practices (GMP) are crucial to promotion of both domestic as well as export trade opportunities.

Climate change and fruit culture: Global temperatures of future is predicted to vary gradually leading to more frequent hot extremes, floods, droughts, cyclones, cold periods, recession of glaciers and possibilities of occurrence of a range of other abiotic and biotic stresses loom large. Soil organic carbon stocks would be depleting fast and pests and diseases dynamics and pollination mechanism are likely to be impacted adversely resulting in greater instability in fruit production and sustainable livelihoods. There is a need to critically profile germplasm resilience of different fruits and develop adaptation strategies.

Market linkages, price stabilization and crop insurance: Absence of organized market infrastructure for obtaining information on fluctuating prices poses a major threat for the fruit producers in realizing optimum returns. Putting in place efficient market linkages, modalities of market regulation, price support and crop insurance on the pattern of commodities sector are emphasized for rendering fruit culture, a profitable avocation. In this regard, adoption of innovations available in IT and ICT sectors will provide fillip to the fruit production sector.

Technology adoption pattern cannot be uniform throughout the country and would vary from crop to crop and even from region to region. Certain degree of flexibility in research planning and research strategy is therefore obvious. Also, with the opening of global markets and removal of quantitative restrictions under the WTO, export-import scenario is likely to change at much faster pace. Market forces will play a more dominant role and demands for modern and cost-effective technologies will increase. Research systems in horticulture will have to be very alert, dynamic and should resilient. Development of both short term and long term strategies for modernizing Indian horticulture will depend largely upon the research support and strength of the research system.

10. HORTICULTURE AND HUMANITY

Ethics refers to moral considerations on a human level. Ethics is not an absolute concept but rather a balancing act of many points of view and issues. It includes standards of what is considered inherently "*right or wrong*" in terms of rights, obligations and benefits to individuals and to society. Further, ethical considerations involve the precepts of specific virtues, such as, honesty and fairness and impose obligations on conduct, including the obligation to avoid harm to others and to abstain from misrepresentation or selectively using known information. Ethical standards in horticulture include growth, management and use of fruits, vegetables ornamentals, culinary spices and medicinal plants (Aitken *et al.,* 2012). Two overriding guidelines in this regard: "*do no harm*" and "*do good*" for producers, workers, consumers and the environment. Many people are directly or indirectly involved with horticultural activities along the entire supply chain and that a balance must be maintained between conflicting and competing issues and beliefs (Jules Janick and Ian Warrington, 2015). Pesticide residues and food safety are the two important considerations in the emerging trends.

Equally, the globalization of food production has resulted in monumental shifts in sources of supply as consumers expect year-round supply of affordable fruits, vegetables and ornamental plants. Consequently, movements of horticultural industries cannot be regarded as being unethical *per se,*. unless it can be demonstrated that the shift in locations involves exploitative practices. The product, often but not always a food, must be assured to "*produce no harm*" and the product must be safe. This involves efforts to eliminate any contamination from poor sanitation, freedom from known and excessive harmful residues, both natural and chemical and avoidance of misrepresentation of any kind. To ensure product safety, the producer and consumer need to be linked. Thus, methods to preserve the chain of identity of products, including the marketplace, are underscored. Increasingly, producers around the world must conform to the use of good agricultural practices (GAPs) where, many of these issues are addressed in a systematic way and can be subjected to close scrutiny through independent audits (FAO, 2003).

10.1 Producers

Horticultural activities involve growers of plants who are selling either a product or service as a means of livelihood, a fair return or profit is necessary for the efforts and investments that they expend. Ethical producers whether single individuals or corporations do not exploit workers and must respect and reward their contributions in a fair and equitable manner. Producers must certainly not exploit children and must avoid all forms of discrimination in the workforce. Too often, the margins of return available to producers, even in the developed world, are so small that workers employed in the production and processing sectors are usually paid only at or even below, the minimum wage specified by society. Any exploitive employment practices that deviate from the norms expected by society should be regarded as being unethical, if not illegal (Wasley, 2011). Unfortunately, employment in many horticultural industries is now highly unattractive. Therefore, producers increasingly shift to labor-reducing methods, such as, mechanization and rely on migrant labor.

Modern horticultural producers are themselves consumers of inputs required for the production of different horticulture commodities. Access to germplasm (improved planting materials) is one such critical input that can lead to various ethical considerations. For example, newly bred elite cultivars of fruit trees in modern production systems can be legally protected for a limited period, usually about 20 years. Clonal plants in seed-propagated plants are protected by Plant Variety Protection globally including India. In the European Union and other countries, plant breeder's rights allow the breeder to recover a royalty through their unique intellectual contribution in developing that particular cultivar. During this period of protection, it is unethical to propagate plants or plant parts without paying royalties based either on tree number or, if the license requires, on the amount of product produced. To propagate such materials without a license or where production is subject to royalties, to grow the product without declaring the amount produced, is both an unethical and often illegal practice. In emerging economies where, farmers are often poor, there is much controversy regarding plant variety rights but it should be noted that there is usually an exemption for farmer-saved seed. Nonetheless, commercial seed companies often protect their cultivars of many vegetables, such as, eggplant and onion by producing F_1 hybrids. In addition to providing hybrid vigor and uniformity, the off-spring do not grow true-to-type and thus the grower must continually purchase seed. Producers must make the decision to assess the return on the extra cost of the seed through the enhanced performance and uniformity of product.

10.2 Marketers

Those who market and handle horticultural products must strive to avoid misrepresentation including cultivar identity, source and origin of product, treatments imposed on the product and sanitary handling. Unfortunately, except for apples, pears and some ornamentals, such as, roses, most horticultural products on the market do not have cultivar labeling and this vital piece of identification is lost. Modern breeding is highly sophisticated and expensive and the cultivars that emerge from such programs can be superior in terms of size, flavor or appearance, as well as having improved traits, such as, disease resistance. Accordingly, they usually command a premium. Nonetheless, to a consumer, there is often little visual difference between an improved cultivar and a "standard" one, leading some producers and marketers to unethically misrepresent or obscure cultivar designation.

The treatment of fresh fruit and vegetables with ionizing radiation, such as, *gamma-rays*, is a controversial issue that has ethical implications. There is consumer resistance to such practices because of concerns of the chemical changes that might occur with such treatments. The practice is widely used in imported dried spices yet labeling is usually not required. The irradiation of fruits may be necessary because of import restrictions in certain countries that prohibit the presence of specific invasive insects, such as, spotted winged fruit fly (*Drosophila suzukii*), marmorated stink bug (*Halyomorpha halys*) and the Asian citrus psyllid (*Diaphorina citri*) that is responsible for spreading citrus yellows (*Huanglongbing*) caused by the bacterium (*Candidatus Liberibacter asiaticus*).

10.3 Processors

Many horticultural products are preserved or processed in various ways (canned, frozen). Processors have an ethical obligation to truthfully represent the treatments that have been imposed and to precisely state the amount of each component that is in a processed product. This requirement falls within the jurisdiction of product labeling, which is fraught with controversy. Many consumer advocate organizations are demanding more complete information on the product label. Labeling of processed food is a tremendously challenging issue that involves serious ethical considerations.

A number of plant introductions, especially landscape ornamentals, have turned out to be invasive and have become a serious danger to the environment in various locations. Thus, introduction of exotic plant species involves a serious ethical situation for the ornamentals industry (Niemiera and Phillips, 2006). In academia, there are important ethical guidelines (codes of ethics) for research, including the appropriate use of human and animal subjects, intellectual honesty, falsification of experiments and results and especially plagiarism and the appropriation without attribution of another scholar's language, thoughts and ideas in the preparation of publications. Plagiarism, However, can often be a complex and contentious area since all scientists appropriate and build on the ideas and language of others. Modern soft-ware packages are now frequently used in academia and by scientific journals to detect incidences of plagiarism in a range of outputs including students' assignments and scholarly contributions.

10.4 Malnutrition

Addressing malnutrition should be the foremost priority of any government. Malnutrition, defined as any nutritional disorder whether due to dietary deficiency or excess, is a major challenge to the nutritionists, agriculturalists, horticulturists, food scientists and social scientists which needs to be tackled holistically. Malnutrition impacts negatively on the health and wellbeing of the most vulnerable in the society, primarily pregnant women, lactating mothers and children plagued with nutritional anemia – low dietary iron, calcium, malaria and intestinal parasites which, leads to poor nutrition consequently resulting in morbidity, mortality and poor growth in children and in the long run could cripple the health of any nation (Table 7), Hence, needs due focus and accordingly nations would do well and initiate appropriate strategies to address MDG 4 and 5 in the right perspective. On the contrary, excess or imbalanced nutrition may also result in non-communicable health problems, such as, obesity, high blood pressure, heart diseases, diabetes and certain cancers as could be seen from the problems of the developed world. In this scenario, horticulture has a major role to play. Of course, the future horticulture has to address issues of sustainable soil health as produces raised on impoverished soils cannot supplement the population with critical vitamins, minerals, bioactive compounds and nutracetical.

Table 7. *Nutrition recommendation for pregnant women, lactating mothers and school children is based on FAO recommendations.*

Nutrient	Amount per		
	100	**100 Kcal**	**100 KJ**
Protein	15	5.2	1.2
Fat (g)	10	2.6	0.6
Crude fiber (g)	5	1.3	0.3
Vitamin A (as retinol)	400	100	24
Vitamin D (as cholesterol	10	2.5	0.6
Vitamin E (mg)	5	1.25	0.3
Ascorbic acid (mg)	20	5.2	0.12
Thiamine (mg)	0.5	0.12	0.03
Riboflavin (mg)	0.8	0.2	0.05
Niacin (mg)	9	2.2	0.57
Vitamin B6	0.9	0.22	0.06
Folic acid	100	27	6
Vitamin B12 (mg)	2	0.52	0.12
Calcium (mg)	800	200	48
Phosphorous (mg)	800	200	48
Iron (mg)	10	2.7	0.6
Iodine	70	18	4.5

(**Source:** The processed Cereal based Foods and Baby foods for infants and young children. *Codex Alimentarius Commission Amendment Regulations*, 2000)

Sustainable Horticulture' is the buzzword every stakeholder is using today. The present and the emerging horticulture globally has to thrive in the scenario of shrinking resources of land, labor, water, energy besides the ever looming threat of climate change paradigms. Food and nutritional security globally also needs address the issues of food and environment safety and access and affordability to safe and residue-free food in the right perspectives. Horticulture commodities need to be rendered free from a range of contaminants so as to make them safe for consumption. This situation obviously demands ensuring horticulture not only *'sustainable'* but also *'smart'* in the days ahead. Empowering all the stakeholders with knowledge and information on producing safe food sustainably though appears a formidable challenge, is certainly inevitable.

Countries need to enforce stringent laws to support issues of regulating availability of safe horticulture commodities for consumption. Researchers would do well by contributing, conservation of pollinators and ecosystem – services, trait-specific varieties tolerant to range of abiotic and biotic stresses, water prudent, nutrient use efficient and develop value chains that could fit into integrated farming systems for a viable and profitable horticulture avocation in the long run through system and mission approach.

REFERENCES

Aitken, A., Hewett, E.W., Warrington, I.J. and Hale, C. (2012). Harvesting the Sun: A Profile of World Horticulture. *Scripta Horticulturae* 14. International Society for Horticultural Science.

Amrit Patel. (2014). Issues facing Agricultural Credit in India, Agribusiness, *Web source.*

Anon. (2006). Report of 4th Dean's committee on agricultural education, Education division, ICAR, New Delhi.

Anon. (2011 b). Planning Commission: Report of the working group on Horticulture and Plantation crops . XII Five Year Plan. p. 264.

Anon. (2012). Vision, 2030, Central Institute for Subtropical Horticulture, ICAR, New Delhi.p.25.

Anon. (2013). DARE / ICAR ANNUAL REPORT 2012–13, ICAR, New Delhi.

Anon. (2013). *Web source*, 13 New Mobile Agriculture Apps for 2013.

Anon. (2015). Vision, 2050, Central Institute for Subtropical Horticulture, ICAR, New Delhi. p.15.

Anon. (2015a). Horticultural statistics at a glance 2015, Horticulture Statistics Division, Department of Agriculture, Cooperation & Farmers Welfare, Ministry of Agriculture & Farmers Welfare, Government of India.

Angus Deaton and Jean Drèze, (2009). *Food and Nutrition in India*: Facts and Interpretations, Economic & Political Weekly EPW February 14, Vol. xliv No 7.

Aswath and G. Selvakumar, (2016). Utilization of vertical spaces for horticultural crop production in urban and peri-urban areas. *Current Science*, 10 (11), p. 2048-9.

Ayyappan, H.S. and Paroda, R.S. (2013). *Foresight and Future Pathways of Agricultural Research through Youth Proceedings and Recommendations*, NASC Complex, Pusa, New Delhi 1-2 March, 2013.

Awasthi, P. (2013). Development and Monitoring *International Journal of Environmental Science*: (IJESDM) ISSN No. 2231-1289, Volume 4 No. 2.

Bareja, B.G. (2011). "Climatic factors promote or inhibit plant growth and development" CropsReview@.com: Towards an Informed Application in Agriculture. Retrieved from http://www.cropsreview.com/climatic-factors.html.

Chadha, K.L. (2013). Horticulture Research in India-Infrastructure, Achievements, Impact, Needs and Expectations, Web source.

Chand, R. (2008). High value agriculture in India. *In* Souvenir-*3rd Indian Horticulture Congress*, The Horticultural Society of India, New Delhi, 2008.

Dastagiri, M.B., Kumar, B.G. and Diana, S. (2009). Innovative models in horticulture marketing in India. Indian Journal of Agricultural Marketing, 23(3): 83-94.

Ghosh, S.P. (2008). High value crops of India: Prospects and policies. *NAAS News*, January–March 2008, **8**.

Ghosh, S.P. (2012). Carrying capacity of Indian horticulture, *Current Science*, 102 (6): 889-893.

Jules Janick and Ian Warrington. (2015). Ethics and horticulture, *Chronica Horticulturae*, 55(2): 10-14.

Kader, A.A. (1993). Post harvest Handling. *In*: Preece, J.E. and Read, P.E., *The Biology of Horticulture- An Introductory Textbook.* New York: John Wiley & Sons. p. 353-377.

Kvien, C. (2006). Don't sit back and wait. *The Precision Ag. Guide*, The Precision Ag. Guide – insert to the November issues of *The Peanut Grower, Cotton Farming and Dealer & Applicator*, www.nespal.org/precisionag/paguide 2005.asp. pp. 3

Mishra, A. and A. Mishra, 2012, Empowering the Farm Women for Greening the Upland through Horticultural Crops, *Indian Research Journal of Extension Education* Special Issue, I : 291-295.

Marzella, Wüstefeld. (2013). Food and Nutrition Security UNSCN Meeting of the Minds Nutrition impact of food systems In UN standing committee on Nutrition.

Pocknee, S. (2005). Applications of wireless networks on the farm. *The Precision Ag. Guide*, p.4-5, The Precision Ag. Guide –

Prasanna Kumar, H. M. and Harish Kumar, H. V. (2013). Food Security and Nutritional scenario of India- an overview, IOSR *Journal of Agriculture and Veterinary Science* (IOSR-JAVS), 2 (5): 28-39.

Rai, N. Asati, B. S. and. Yadav, D. S. (2004). Conservation and genetic enhancement of underutilized vegetable crop species in North Eastern region of India, Leisa India.

Saharawat, Y.S, Dhillon, M.K., Ranjan Bhattacharyya, Jat, M.L., Malavika Dadlani, Gupta, Jose, C. S. (2010). Heralding golden revolution through NHM. *In: Indian Horticulture*, DIPA-ICAR, New Delhi.

Sharholy. (2008). Municipal solid waste management in Indian cities–A Review, Waste management 28: 459 –467.

Sheoran A Rais M. (2015). Scope of supply chain management in fruits and vegetables in India. *J Food Process Technol.,* 6:427. doi:10.4172/2157-7110.1000427.

Shiva and Crompton. (2011). Capital Study on "*Sowing the seeds of success*", Department of Agriculture and Cooperation; New Delhi.

Singh, Shrawan, D.R. Singh, L.B. Singh, Subhash Chand and S. Dam Roy. (2013). Indigenous Vegetables for Food and Nutritional Security in Andaman and Nicobar Islands, India, *International Journal of Agriculture and Food Science Technology*. ISSN 2249-3050, 4 (5): 503-512.

Singh, Joginder. (2006). *Scope, Progress and Constraints of Farm Mechanization in India;* published in Status of Farm Mechanization in India; IARI, Agricultural Machinery Sector in India, FICCI.

Vellidis, G., Garrick, V., Pocknee, S., Perry, C., Kvien, C., Tucker, M. (2007). How wireless will change agriculture. *In*: Stafford, J.V. (Ed.), *Precision Agriculture* '07 – Proceedings of the Sixth European Conference on Precision Agriculture (6ECPA), Skiathos, Greece, p. 57-67.

29

Micro-Financing Towards Sustainable Horticulture Development

Bindu Ananth

1. INTRODUCTION

Improving access to financial services, particularly agricultural credit, has been a long-standing goal of the Government and the Reserve Bank of India (RBI). Various approaches have been attempted, including creation of the cooperative credit system and the Regional Rural Banks, nationalization of the banking system in 1969 and the Priority Sector Lending (PSL) policy that requires all commercial banks to lend up to 40 percent of their Annual Net Bank Credit (ANBC) to designated sectors, 18 percent of ANBC or Credit Equivalent Amount of Off-Balance Sheet Exposure, whichever is higher. Within the 18 *percent* target for agriculture, a target of eight percent of ANBC or Credit Equivalent Amount of Off-Balance Sheet Exposure, whichever is higher is prescribed for Small and Marginal Farmers. In recent years, there has been a renewed emphasis on financial inclusion. The current Government has leveraged what it refers to as "the JAM strategy" (Jan Dhan Accounts – Aadhaar – Mobile) to ensure that all individuals have access to a bank account and convenient transactions.

2. STATUS OF AGRICULTURAL CREDIT AND ROLE OF MICROFINANCE

As on September 30, 2016, the total bank credit outstanding was approximately Rs. 70 trillion. Of that, agricultural credit accounted for Rs. 9.5 trillion. Overall, institutional sources account for approximately 60% of all farmer borrowings although this percentage is lower as the farm size decreases. Specific to horticulture, 24 percent of fruit growers, compared to only 10.4 percent of vegetable growers and 11.4 percent of non-growers availed themselves of institutional credit. The average loan outstanding is Rs 47, 000 ranging from Rs. 31,100/- for landless farmers and Rs 2,90,300/- for farmers having 10 ha or above. There is a general decline in interest rates with a decline of average rate from 18.91 percent in 2002 to 17.00 percent in 2012. The decline is higher in case of institutional loans by 264 basis points from 14.31 percent in 2002. Interest rate on non-institutional loans declined but by 114 basis points from 26.34 percent in 2002.

Microfinance as an approach for increasing the penetration of rural credit has a vintage of about twenty years. India has two distinct approaches within microfinance – one is the Self-Help Group Bank Linkage (SHG-BL) model where a commercial bank lends to groups of typically 15-20 women with support and oversight provided by a Non-Government

Organisation (NGO). The second is the Micro Finance Institution (MFI) model where a Non-Bank Finance Company (NBFC) or a Section eight company lends to groups of five women. Both approaches do not rely on collateral from the borrower as is typical of the banking sector but on the principle of joint liability, *i.e.,* each member is responsible for the repayment of every other member of her group. Through these two models, over 100 million individuals have been reached as on date, making it one of the most scaleable approaches to delivering financial services to the rural and urban poor. Unfortunately, farmers and agriculture have been at the margin of this innovation.

One of the features of microfinance is that the average loan amounts are small (approximately Rs. 25,000) and they are structured as term loans with weekly, fortnightly or monthly repayments. As a result of these features of the loan and the group methodology, the typical microfinance Average loan tenors (maturity) are 18 months, if the borrower is a landless labourer or a small-holder farmer (including livestock) in the rural context or a vegetable vendor or low-income salaried worker in the urban context. Most of the borrowings are used for consumption smoothing purposes and working capital. The loan features and the rigidity associated with it do not make it suitable for individuals with more seasonality and volatility associated with their cash flows. In order for micro finance to be availed by a broader segment of farmers, including horticulture farmers, we will need significant innovations in product design and delivery channels that I discuss next.

3. INNOVATIONS IN AGRICULTURE AND HORTICULTURE FINANCE

3.1 Product Design

Farmer finance has seen little to no product innovation. The key issue is that the cash flows of a farming household have a high degree of seasonality. Managing liquidity in the face of this seasonality is a challenge. This problem is vastly exacerbated when it comes to horticulture, particularly fruit crops. For example, pomegranate while being a highly remunerative crop for export markets is a three year crop. Fruits require more capital compared to vegetables and staple crops, meaning that farmers seeking to cultivate fruit crops must often supplement their own resources by borrowing from institutional and non-institutional sources. Uncertainty is intrinsic to farming, longer tenors (maturity) compound this issue. What are some implications of this liquidity challenge for farmers?

A unique study by Anandi Mani *et al.* (2013) among sugarcane farmers in Tamil Nadu attempted to measure changes in farmers' financial situation and cognitive capacities before and after harvest. Their data show that farmers indeed faced greater financial pressures at pre-harvest as compared with post-harvest: They pawned items at a higher rate (78 versus 4%, $P < 0.001$, $n = 462$ participants) and were more likely to have loans (99 versus 13%, $P < 0.001$, $n = 461$ participants). On average, farmers had 1.97 more loans before harvest than they did after it. They were also more likely to answer "Yes" to the question, "Did you have trouble coping with ordinary bills in the last fifteen days?" before harvest than after (1.62 and 1.76, respectively, on a 3-point scale, where 1 corresponded to low ability and 3 to high ability to cope; $P < 0.001$, $n = 462$ participants). The farmers in the study showed lower cognitive performance before harvest as compared to after harvest, highlighting the role of financial stress on decision making and cognition. On Raven's test, the farmers scored an average of 5.45 items correct post-harvest but only 4.35 items correct pre-harvest ($P < 0.001$, $n = 460$ participants). On the Stroop task, they took an average of 131 seconds to respond to all items post-harvest, as compared with 146 seconds pre-harvest ($P < .001$, $n = 452$). In addition, the average number of errors the farmers committed was higher before harvest than after harvest (5.93 versus 5.16 errors; $P < .001$, $n = 453$).

This study alludes to one of the reasons why, despite horticulture being far more remunerative than staple crops, farmers are slow to adapt to growing these crops. The liquidity challenge associated with long-cycle crops as is typical of horticulture can be easily solved through well-structured credit and savings products. For example, if the farmer can be given periodic disbursements at pre-harvest stage that allow him smooth consumption, it takes considerable pressure off the cash flows of the household when the crop is in the field. Interestingly, this structure is often used in the context of infrastructure finance where disbursements prior to the project being operational are used to service interest obligations; this is referred to as interest during construction. There is an analogous need for horticulture farmers.

There is an unfortunate policy mind-set that consumption loans are not appropriate and are to be discouraged. For example, MFIs are required to cap the amount of their lending that can be used for consumption purposes. For households with income seasonality and where human capital or labour is the most critical factor of production, consumption finance is critical. A small farmer who allocates land to a commercial crop often has to depend on market purchases to meet food requirements, resulting in an additional source of risk.

3.2 Value Chain Financing

This is an approach where financing is not provided to the farmer on a stand-alone basis but in the context of buyer and supplier relationships. The appeal of this approach is that given the information challenges of assessing the credit worthiness of a farmer, it may be easier to lend against specific value chain contracts. One of the celebrated examples of this approach comes from Kenya where small growers of fruit and vegetables are able to access bank finance for agro-chemicals against their export contract. The exporter pays the farmers through the bank, which deducts the scheduled loan payments before releasing the net proceeds to the farmer group (Kenneth Marangu, 2007).

In the Indian context, there have been several experiments in contract farming – where farmers are contracted *ex-ante* to supply their produce to a specific buyer. In this case, the buyer has typically facilitated access to other inputs such as seed, fertilizer and credit. Some recent examples of contract farming in India include INI Farms. They work with small-holder farmers to grow pomegranate and bananas for the domestic and export market. Through their linkages, they are able to assist the farmers to adopt better practices for irrigation and crop management. They are one of the few horticultural companies that have been able to secure funding from venture capital companies. Similarly, Siddhi Vinayak Agri connects with over 6000 potato farmers in seven states of India. They provide door-step access to high quality farm inputs, including organic fertilisers. They also facilitate access to credit and insurance for farmers in their network. One of the reasons for their success has been the assurances of continuous purchase of produce from the farmers. Samunati Finance is an example of a recent value chain financier. They work with input suppliers, farmers, traders and aggregators to take a holistic view of the financial health of the value chain, not just that of the farmer. Through this approach, they are able to provide working capital to the value chain participants and also enable equipment leasing and other interventions aimed at increasing productivity.

This is certainly an easier environment as far as financing is concerned but there are some concerns around the repudiation risk by buyers and farmers in an adverse market environment. Particularly, horticultural crops that have less domestic market but rely entirely on export markets are vulnerable to down cycles and that risk cannot be borne by the farmer.

3.3 Presence of Local Financial Institutions

One of the key challenges of the banking sector in India has been limited outreach. We have a little over one lakh branches for a country that has over 6,00,000 villages and hamlets. The credit depth of the country (credit as a percentage of GDP) is also fairly low as compared to peer countries, at around 60%. One solution here could be significantly increasing the number of banks and banking outlets in the country to serve the needs of all segments. However, this is not ideal for a number of reasons. 1) Since banks are deposit taking entities, the regulator needs to be very prudent about licensing only such promoters who fulfill the "fit and proper" criteria to undertake banking business. 2) The cost structure of commercial banks makes them unsuitable to profitably expand small-value lending. Deepti George (2013) reported that the cost-to-serve for small loans for national full-service banks exceeds 30 *per cent* while they are required to lend money to this sector at spreads ranging from three *percent* below their base rate to two *percent* above them.

For these reasons, we cannot overly rely on banks to get the task of agricultural credit delivery done. There is a need for more institutional diversity. In particular, financial institutions that are not deposit taking but have a good understanding and physical proximity to customers hold significant promise to expand access to credit. One such institutional innovation I have been closely involved with is the Kshetriya Gramin Financial Services (KGFS) model. KGFS is an NBFC that has a physical presence at the Panchayat level in over 4000 villages. It hires local youth and trains them to be "wealth managers". Their ability to go beyond standard micro-loans to crop finance and small business loans is attributable to their local understanding and customer-centricity. India needs hundreds of such institutions that will meet the financial service needs at a micro-level.

3.4 Risk Management Innovations

One of the principal challenges for horticulture farmers and lenders is the fact that these crops are more susceptible to pest attacks and price volatility than staple crops. In a market like India where there are limited risk management options even for corporates, these risks tend to get pushed to the farmer. This perhaps explains the take-up issues in horticulture to a large extent. Once again, there are several attempts underway aimed at improving risk management.

Earlier this year, the Pradhan Mantri Fasal Bima Yojana (PMFBY) was launched under which a premium of 5% has been announced for commercial/horticultural crops. The difference between this premium and the premium offered by the insurance company will be subsidized equally by the Central and State Governments. This scheme promises to address earlier shortcomings of crop insurance schemes including capped payouts and slow claims processing and targets bringing at least 50% of the cultivated area under insurance coverage.

Crop insurance has traditionally relied on assessment of losses in a farm, thus making the costs unsustainable. Innovations in index-based insurance made it possible to link the insurance payout not to farm-specific losses but an observed weather parameter such as rainfall or frost, thus reducing the administrative expenses of the insurance programme significantly. While rainfall insurance was first launched in India about ten years ago, scale-up has been low on account of high premia costs and basis risk – arising from the absence of a Panchayat level network of automatic weather stations.

As far as price risk management is concerned, there are no insurance mechanisms currently available for this. Various Government supported organisations do tend to provide some support price environment. These are regarded as being distortive in nature. More recently, commodity exchanges have been created in India with the goal of providing exchange-traded derivatives like futures and options. Currently, the National Commodity and Derivatives Exchange (NCDEX) offers futures trading in certain crops, including turmeric, pepper, *jeera* (cumin) coriander, chilli and potato (Kenneth Marangu, 2007). While the number of contracts need to be expanded to cover more crops, there is also a need for innovations in facilitating access to such risk management mechanisms among farmers. NCDEX currently works with producer organisations to appoint them as aggregators. Aggregators pool price risk on behalf of a defined group of farmers and undertake futures trades on their behalf. They also provide other services such as commodity testing and facilitating physical delivery. Deepening of spot and futures markets also brings about significantly improved price transparency for farmers.

4. POLICY ASPECTS AND CONCLUSION

In conclusion, while there have been important developments in access to basic financial services (savings accounts, micro-loans), a lot more innovation in products and channels is required to extend these gains to agriculture, especially commercial crops and horticulture. Such financial innovation will need to account for the unique liquidity requirements due to long tenor of horticulture crops of the fruit and tree varieties and the risk management needs on account of price volatility and susceptibility to pests.

While Priority Sector Lending (PSL) policy is an enabler as far as allocation of bank credit to the agricultural sector is concerned, this is not equitable across different regions of the country and farmer segments. One specific reform in PSL that could be very beneficial is to provide higher weightage to those districts of the country that have a low credit-depth but high potential/growth elasticity so that we improve spatial distribution of agricultural credit. The current practice is creating concentrated availability in some pockets.

NABARD has significant resources available to it under the Rural Infrastructure Development Fund (RIDF) that is to be administered through state governments. A focused effort to assess high-potential horticulture regions and creating project templates for infrastructural investments in horticulture will also go a long way in alleviating public investment shortfalls (Satyasai, 2015).

REFERENCES

Anandi Mani, Sendhil Mullainathan, Eldar Shafir, and Jiaying Zhao. (2013). Poverty Impedes Cognitive Function. *Science,* **341**(6149): 976-980.

Deepti George. (2013). Cost of Delivering Rural Credit in India. IFMR finance Foundation. http://www.ifmr.co.in/blog/2013/04/23/cost-of-delivering-rural-credit-in-india/.

Kenneth Marangu. (2007). Kenya BDS Program Experience in Value Chain Facilitation. presentation at the AFRACA Agribanks Forum. http://slideplayer.com/slide/4165939/.

Satyasai, K. J. S. (2015). How Indian farmers borrow, produce and earn? Evidence from recent NSSP surveys, *Rural Pulse*, Issue VIII, March-April, NABARD, Mumbai.

❍❍❍

30

Sustainable Development of Horticulture Under Declining Resource Base in India

Prem Nath and B.S.Prabhakar

SYNOPSIS

India being the most populous country in the world (next only to China) and continued addition of people, the pressure on natural resources like land, water, labour *etc.* is increasingly demanding immediate attention and correction for ensuring well being of its population. Agriculture is still the major employer for majority of the population and a significant partner in building national economy. Horticulture has emerged as a potential partner in progress by providing much needed food, nutrition and income security to the farmers and others involved in farm related activities. Therefore, it is pertinent to discuss the impact of changing resources on horticultural production and look for strategy alternatives to sustain production, productivity and income from horticultural crops.

Continuing fragmentation of land, small and uneconomical holdings, depleting land availability for agriculture, challenges due to climate change, gradually increasing urban and periurban agriculture, changing preference for jobs by the rural youth and growing need for skill development are some of the compelling reasons to reorient our horticulture development Strategies.

Despite challenges, India has witnessed voluminous increase in horticulture production over the last few years. Significant progress has been made in area expansion resulting in higher production. Over the last decade, the area under horticulture grew by about 2.7% per annum and annual production increased by 7.0%. During 2013-14, the production of horticultural crops was about 283.5 million tonnes from an area of 24.2 million hectares.

Keywords: Sustainable development, resource base, horticulture.

1. INTRODUCTION

Horticultural produce and processed products from the developing world are becoming increasingly popular both in domestic and international markets. Global production and exports are rising steadily. However, yield increases have been smaller than area growth and have been negligible or even negative in some of the developing countries. While experience shows that horticulture can offer good opportunities for poverty reduction as it increases income and generates employment, care must be taken that small and poor farmers are not excluded from the opportunities in these market sectors (Weinberger and Lumpkin, 2007).

Owing to a number of economic and political compulsions, the Indian strategy for agricultural growth remained preoccupied with the goal of achieving quick increases in food-grains production by concentrating the resources and efforts on the relatively better-endowed areas and strata of cultivators. A consequence of this approach has been the less-than- optimum allocation of the critically scarce inputs like water and fertilisers across crops and a group of farmers. An underdeveloped agricultural sector on account of low productivity is the main reason for low levels of diversification of the economy. A buoyant agricultural sector has been the basis for industrial development in the developed countries.

Productivity is a function of natural inputs, coupled with the skill and knowledge of farmers. Often, the output is compared in absolute figurative terms and decided as low/high productivity. All the more, as the productivity is now linked with demand, the concerns rose about production per acre. Given this definition, Indian productivity is on par with the available level of natural resources. Indian farmer does production against many odds, including deteriorating soil and water conditions (Sahasrabuddhe, 2012). If there is a comparison on productivity, one should include all these parameters. In poor quality soils, if a farmer produces more per acre, merely because of using more chemical fertilisers and other chemicals he cannot be glorified and a farmer who produces despite poor soil condition need not be condemned, as inefficient. All parameters have to be considered. A US farm may produce triple per acre, but it is all based on fossil fuel inputs, a luxury which a marginal Indian farmer does not have access to (Narasimha Reddy, 2014).

India being the most populous country in the world (next only to China) and continued addition of people, the pressure on natural resources like land, water, labour *etc.* is increasing demanding immediate attention and correction for ensuring well being of its population. Agriculture is still the major employer for majority of the population and a significant partner in our national economy. Horticulture has emerged as a potential partner in progress by providing much needed food, nutrition and income security to the farmers and others involved in farm related activates. Therefore, it is pertinent to discuss the impact of changing resources on horticultural production and look for strategy alternatives to sustain production, productivity and income from horticultural crops.

2. LAND

In the aftermath of the Government of India (GOI) promulgating an ordinance to amend the Land Acquisition Act, it is feared that the farm land will be taken away for other purposes and the farm land will come down drastically. Data from World Bank show that around 60.3 *percent* of India's land area is agricultural land. The bank defines agricultural land as share of land area that is arable and that under permanent crops and pastures. India has the second largest agricultural land in the world (157.35 million hectares) next only to USA. India has enough land dedicated to agriculture and even if some of it is taken away for other purposes there will still be enough land left for agriculture. Nevertheless, there are bigger problems when it comes to Indian agriculture. India has more (34%) arable land than China, but China produces 40% more rice and wheat than India does. India is also the second largest producer of fruits and vegetables in the world after China, but China's fruit production is three times India's production.

India's agricultural productivity is low compared to that of China and many other countries in the world. FAO (2013) report points out that India produces 106.19 million tonnes of rice a year from 44 million hectares of land (2.4 tonnes per hectare) placing India at 27th place out of 47 countries. China and Brazil have yield rates of 4.7 and 3.6 tonnes per hectare, respectively. In case of wheat the productivity is better than that of rice, with 93.51 million tonnes of wheat from 29.65 million hectares (3.15 tonnes per hectare) placing India 19th out of 41 countries. Here, we do better than Brazil's yield rate of 2.73 tonnes per hectare, but lag behind South Africa (3.4 t/ha) and China (4.9 t/ha). Maize yield in India is 2 t/ha while world average is 5 t/ha. Apples yield in India is 7 t/ha while the world average is 14 t/ha.

There are multiple reasons for this low productivity. The average holding size of land has come down over the decades. As per Agriculture Census 2010-11 (Table 1) small and marginal holdings of less than two hectares account for 85 percent of the total operational holdings and 44 *percent* of the total operated area. The average size of holdings for all operational classes (small and marginal, medium and large) have declined over the years and for all classes put together it has come down to 1.16 hectare in 2010-11 from 2.82 hectare in 1970-71.

Table 1. *Distribution of operational land holdings in India*

Year	Total			Distribution of holdings (%)				
	Holdings (no. in '000)	Area ('000 ha)	Ave size (ha)	Marginal	Small	Semi-Medium	Medium	Large
1970-71	91,011	162,178	2.3	51.0	18.9	15.0	11.2	3.9
1980-81	88,883	163,797	1.8	56.4	18.1	14.0	9.1	2.4
1990-91	106.638	165.507	1.6	59.4	18.8	13.1	7.1	1.6
2000-01	119.931	159.435	1.3	62.9	18.9	11.7	5.5	1.0
2005-06	129.222	158,323	1.2	64.8	18.5	10.9	4.9	0.8
2010-11	137,757	159,181	1.2	67.0	17.9	10.0	4.3	0.7

Note: Marginal (<1 ha); Small (1-2 ha); Semi-medium (2-4 ha); Medium (4-10 ha); Large (10 and above ha); ha-hectare; Operational holding-All land which is used wholly or partly for agricultural production and is operated as one technical unit by one person alone or with others without regard to the title, legal form, size or location; Operated area-This includes both cultivated and uncultivated area, provided part of it is put to agricultural production during the reference period. Figures may not add up exactly to 100 due to rounding off.

Sources: IDFC's India Rural Development Report 2012-13; 'All India Report on Number and Area of Operational Holdings, Agriculture Census, 2010-11', Ministry of Agriculture, 2012, website (http://agricoop.nic.in/imagedefault/whatsnew/Agriculture%20Census.pdf).

Over 70% of farmers are small holders (less than one ha of land). This means that one can rarely apply technological solutions that work best on the large scale. Small farm sizes inhibit mechanization .This creates difficulties in application of modern inputs, adoption of scientific land improvement, water conservation and plant protection measures. The tardy progress of land reforms in most states has compounded this problem.

Consolidation of land can help to improve productivity. On the other hand, substantial chunks of scarce land remain untilled because of landowners' reluctance to lease out land for fear of losing its ownership. The problems of absentee landlordism also act as an impediment towards consolidation of fragmented land holdings (Figure1 and Figure 2).

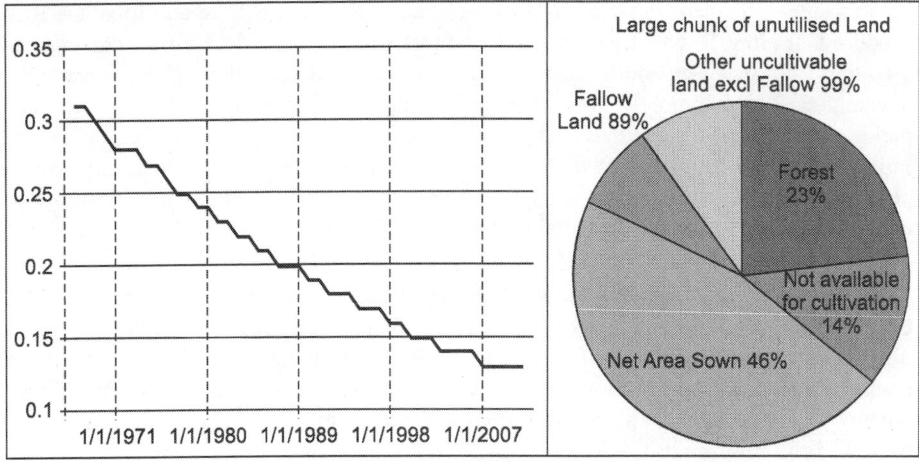

Fig. 1. *Temporal changes in land holding*

Fig. 2. *Unutilized land in hectares per person in India*

However, we cannot be complacent about seemingly large portion of land available for farming. Land is limited and has almost reached the level where more expansion in cultivated area is not possible. The growth in population creates immense pressure on land. It causes further fragmentation of the already meagre *per capita* landholdings. Therefore, every effort has to be in, place to preserve the precious farm land for the posterity. Even though the land-human ratio in India is better compared to some of the developed countries like Japan, the Netherlands, Belgium and even China, other factors like very low yields and low levels of industrialisation in India compound the problem of population pressure on agricultural land. It is wise to save the scarce farm land and optimise food and other agricultural crops production per unit of land and time with innovative technologies.

The shrinking size of the average land holding of an Indian farmer has held back agricultural productivity and it is opined that not much that can be done about this. But there are other areas which can be worked upon. To enhance productivity, easy and reliable access to inputs, such as, quality seeds, fertilizers, pesticides, access to suitable technology tailored for specific needs, the presence of support infrastructure and innovative marketing systems to aggregate and market the output from large number of small holdings efficiently are required.

It is clear from foregoing that India has enough land for farming. The problem is that it is not productive enough. The other huge issue is that for nearly 58 *percent* of India's population, agriculture is a primary source of livelihood. If we keep on taking away land from agriculture in the name of development, won't we end up in a situation where we will not produce enough food and will have to import it? As countries have rapidly urbanized in the past, land has been taken away from agriculture. Take the case of the British industry. Prohibiting the use of agricultural land for industries is ultimately self-defeating. The locations of great industry, be it Manchester or Lancashire, these were all on heavily fertile land. Industry has always competed against agriculture because the shared land was convenient for industry, trade and transportation (Amartya Sen, 2007). This has often led to a situation where as land has been taken away from agriculture less food has been produced, leading to countries having to import food in order to feed its citizens. As a country industrialises and modernises, cropland is used for industrial and residential development. As automobile ownership spreads, the construction of roads, highways and parking lots, takes valuable land away from agriculture. This phenomenon was observed in Japan as well. Japan was essentially self-sufficient in grain when its grain harvested area peaked in 1955. Since then, the grain land area has shrunk by half. With grain consumption climbing and production falling, grain imports soared. A similar analysis for South Korea and Taiwan shows a pattern almost identical with that of Japan.

Government of India trying to make land acquisition easier by trying to push through the *The Right to Fair Compensation and Transparency in Land Acquisition, Rehabilitation and Resettlement (Amendment) Ordinance, 2014*, is there a danger of India getting into a similar situation like Japan, South Korea and Taiwan have in the past? India has a large amount of arable land. India's arable area as a proportion of its land area is at 48 *per cent* (159 million ha/329 million ha) which places India as the country with the highest arable land area in the world. Hence, even if some amount of this land is taken away from agriculture for urbanization as well as industrialization, it won't create much of a problem, as it has in other countries like Japan and South Korea in the past. Even if the most optimistic scenarios of India's urbanization were to play out over the next decade or so, the total urban land area will be less than six *per cent* of its land mass, compared to the 48 per cent that is currently under agriculture. Given this, in the coming years an adequate amount of land will continue to remain dedicated to agriculture. Further, despite a little under half of the land area of the country being dedicated to agriculture, India lags behind the rest of the world when it comes to agricultural productivity (Vivek Kaul, 2015). The poor in rural India are located mostly among the landless agricultural labourers, marginal and small farmers. Where efficiency of small and large farms is found more or less equal (Table 2), it is necessary to allow size of holding which can generate the investable capital in agriculture sector. It is possible to achieve this through pooling the smaller holdings to form formal/ informal groups of producers and contract out their produce for the purpose of markets. In a recent study addressed to the investigation of the causes of suicides it was found that a good number of farmers who committed suicide had rented in land and could not borrow from the regular financial institutions as the land was not legally tenanted. They borrowed from moneylenders and could not pay back due to heavy interest rates (Deshpande, 2012). Legalizing tenancy brings the small and marginal tenants in the ambit of the institutional credit. If the status of the tenant is conferred on the tenant it may not leave any room for the land market imperfections. It is expected that opening of the lease market will bring in the private investment in agriculture that has been hitherto a major problem. It is possible to keep out the possible imperfections by restricting the upper limit to the 25 hectares and imposing time limit on the lease period say for three years.

Among the immediate policy interventions the drive to correct and computerize the land records assumes importance. Legalizing tenancy and relaxing the ceilings on land holding assumes next priority. The analysis of the emergence of land policy in India went through different phases. These included land reforms in its two stages, bringing in the lagging area under main stream growth through drought prone area programme, desert development programme. This was followed by Wasteland Development Programme that was environmentally designed to reclaim the degraded land. Land development

Table 2. *Productivity of different size farm holdings*

Agricultural output per household, per hectare and per capita on different farm-size categories

Farm size class: hectare	Household size: number	Per capita land: hectare	Output value (₹)		
			Per household	Per capita	Per hectare
0.01-0.4	5	0.04	4,783	965	25,173
0.4-1.00	5	0.12	12,563	2,364	18,921
1.01-2.00	6	0.24	23,292	3,801	16,780
2.01-4.00	6	0.43	40,403	6,734	15,091
4.01-10.00	7	0.82	77,120	10,588	13,565
>10	8	2.20	137,473	16,782	7,722
All	6	0.22	18,585	3,143	15,426

Source: Situation Assessment Survey, NSSO 59th Round, Report No. 49.
DARE/ICAR Annual Report 2011-12

and treatment of degraded forestland and wastelands is another important component. This is being taken up by the Waste Land Development agency but the achievements are not significant. All these policy interventions have a significant influence on poverty and overall development process of the country. In addition to these, other land policy instruments are to be used for the purpose of sharpening the policy. Among the key areas for action the land requires, legalising the tenancy market, contract farming and wasteland development assume greater significance. The implementation issues relate to the political and economic aspects of these reform measures and setting right the institutions for the purpose of achieving these goals. It is strongly felt that research should be directed towards locating the institutional framework in order to get these issues into a broader policy frame (Deshpande, 2012). The National Institution for Transforming India (NITI,) Aayog, has drafted a model land-leasing law that would serve as guide for those states that wish to amend their existing legislation. Such a measure would lead to uncultivated land owned by absentee landowners becoming productive and would allow small and marginal farmers to expand their holdings to a viable size by leasing in additional land.

2.1 Soil Quality

There are about 1.5 billion hectares of arable land in the world of which 1/3 is being destroyed by topsoil thinning, over tilling, salt deposits, water logging or lack of water causing gradual desertification (Houston and Griffiths, 2008). For an agricultural country like India, soil is a precious resource and degradation of soil is a serious problem, which leads to depletion of soil fertility. Soil erosion is the main form of degradation which occurs because of deforestation and unscientific agricultural practices like shifting cultivation. Increasing salinity, alkalinity and aridity because of mismanagement and repeated use of chemical fertilizers indiscriminately are other reasons for loss of soil fertility. Also the excess unscientific methods of irrigation further harm the fertility levels of the soil. Barring a few regions the soil is not really suitable for sustainable agriculture. Added to these woes is the fact that over use of irrigation and chemical fertilisers (Punjab, Haryana, western Maharashtra) has lead to increased salinity of soil making it even more unsuitable for agriculture. Apart from inherently poor soil quality there is no area where intervention won't help, what is required is the political will to do it.

Since the transfer of land to non-farm activities continues with the growth of developmental activities like urbanisation, development of infrastructure, residential and industrial areas, it is imperative to have a feel good factor like adequate monetary compensation or making the land owners as stake holders in the new venture for which the land his being acquired. The State Government of Gujarat plans 'equity model' of land acquisition. The farmers whose land is acquired will be made partners in this new company apart from Gujarat Industrial Development Corporation (GIDC). The farmers will be given shares in the company, proportionate to the value of their land or provided with a regular-flow of rental income for the plots acquired by the state government in the future. This sort of an equity model has been gainfully implemented in Magarpatta City near Pune. The biggest advantage of such a model is that the farmers or the land owners continue to remain invested and is part of the potential upside to the development that happens on the piece of acquired land. They do not feel left out. In the Margarpatta model, the farmers who pooled in their land and set up a private limited company that developed a commercial-cum-residential project, owns shares of the company, have got plush homes and earn dividends on the shares they hold.

3. WATER

About three billion people live in areas where water demand outstrips supply. Up to five billion people may live in water stressed areas by 2050. In northern India, nearly 50 million people live on Indus and Ganges Rivers which are largely fed by glacier melt water. India receives 4000 cubic kilometres of water through rain alone received over a period of four months that means about one million gallons of fresh water per person per year. This water has to be conserved and used judiciously for the rest of the year. The three reasons for water insecurity are, threefold increase in population since independence (more water for personal use),expansion of industry and manufacturing using more water, intensive agriculture requiring more water. In the name of the development we already ripped off the watersheds and mining, housing and industry have come up indiscriminately resulting in encroachment of catchment areas drying the tanks. In addition we are exhausting the ground water as well adding to the water scarcity (Chandrasekhar, 2016).

Water saving technologies and more efficient irrigation methods are must to promote judicious use of scarce water. Finance, technology and infrastructure are serious limitations of water management in developing nations. It is observed that in India 40 to 90 *percent* withdrawal of fresh water per annum putting pressure on fresh water resources. Inter-basin transfer of water was first proposed by Sir Arthur Cotton. Connecting all Indian rivers, both rain and snow fed, was proposed by Mr. K.L.Rao early in 1972. Time has come now to revisit the idea to make use of water, the elixir of life, which otherwise would drain into the seas. Some actions that need immediate attention are population control, restoring the tanks and watersheds, crop planning based on water available, efficient use of water – *more crop per drop*, judicious use of ground water, reforestation and watershed management, use of desalinated water by industry in coastal areas and fine tuning of industrial policy by locating and relocating the industries that need relatively less water like garments, software and electronic (Assembly) industry *etc.* in water scarce regions. Water scarcity is bound to continue owing to continuing population pressure and increasing needs of development. Though the problem is over population, the solution also lies with the civilized population to find ways and means to address these problems.

Water stress is estimated from the percent of run-off to be captured. Countries using less than 5% have few problems, 10 – 20% has increasing difficulties and more than 20% are in trouble. India uses 50% of run-off and rest from ground water. By 2025, 92% will be needed to be captured? About 150L/head/day of water is used in UK, 300 L/head/day in USA. Around 100L can be recycled from water used for bathing, washing and cleaning. There are about 2500 desalinisation plants in the world; one in Jabel Ahni in UAE converts 300 million m3 water /year and one in Riyadh 2-3 million litters per day.

With the shifting of ocean climatic profiles, over pumped aquifers, saline pollution of the soil, erosion of the top soil and increasing desertification, the benefits of green revolution is fading. Consumption of water has quadrupled in the last 50 years; up to 40% of food production is from irrigated farms. Global cropped land is about 1.5 billion hectares and most of it is currently under production. One billion hectares are needed to make up the balances as 40% of food is produced from irrigated agriculture, the loss of output will be felt in countries with large populations, Asia, Africa and the Middle East (Houston and Griffiths, 2008).

Because of climate change a 10-15% increase in monsoon precipitation in many regions, a simultaneous precipitation decline of 5-25% in drought-prone central India and a sharp decline in winter rainfall in northern India are also projected. This implies changes in output of winter wheat and mustard crops in north-western India. A decrease in number of rainy days (5-15 days on an average) is expected over much of India, along with an increase in heavy rainfall days in the monsoon season (Indian Institute of Tropical Meteorology, Ministry of Earth Sciences, Government of India). These changes are expected to increase the vulnerability of Indian agriculture. This is particularly important in India, where agriculture is highly sensitive to monsoon variability as 65% of the cropped area is rain-fed (Chang, 1987; Zickfeld, 2005). Changes in temperature and precipitation could have a significant impact on more than 350 million people who are dependent on rain-fed agriculture.

It is also argued that watershed development holds the potential for enhanced livelihood security even in geo-climatic conditions where the watershed cannot bring direct irrigation benefits on a large scale. In such fragile environments, However, watershed development is a necessary but not a sufficient condition for sustaining rural livelihoods. While the focus of watershed development is primarily on strengthening the ecological base, such as, water bodies (including traditional tanks), grazing lands and wastelands, it should be complemented with other programmes which focus on landless poor households in order to make it pro-poor. In the context of low rainfall regions where improvement in irrigation facilities is slow, agriculture alone cannot support the communities. Policies and programmes should aim at creating an environment for diverse livelihood activities, which are the choice of the household rather than distress activities (Ratna Reddy, 2004).

4. LABOUR

The shrinking farm land and poor productivity are driving the rural population to urban areas in search of livelihood. There are nearly nine million fewer farmers in 2011 than there were in 2001(Figure 3, Times of India, 2013).

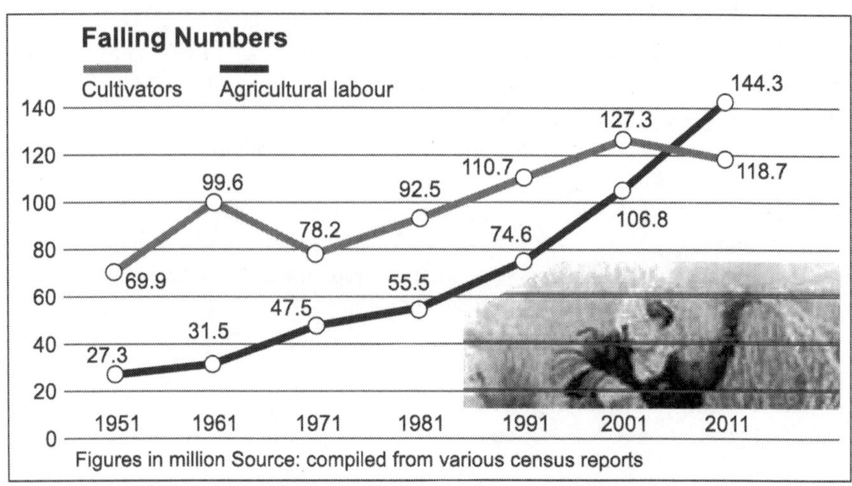

Fig. 3. *Falling number of farmers*

Continuous migration from rural to urban areas in search of livelihood is equalising the population and also increasing poverty in both urban rural areas. Rural areas are losing youth power that would have been engaged in farming and urban areas are getting choked with unemployed and unskilled youth. Increasing unemployable youth due to lack of required skills both in urban and rural areas are posing a serious problem resulting in social unrest and disturbed peace in society. Therefore, skill development among youth both in farm and non-farm sectors is the major problem to be addressed to day. Majority of Indian population (60 – 65 %) are still dependent on agriculture and related activities for livelihood, making agriculture the largest employer. Agriculture and horticulture in particular, should assume similar status as industry with sustainable wages, skill training and employment opportunities. Many farmers in rural areas do not have the most updated information on how to grow food effectively and economically. Improving their knowledge about techniques and technologies in addition to providing them physical and financial resources necessary for implementation would dramatically increase the productivity of farmers (Rosegrant and Clime, 2003).

As an alternative we have to move people away from agriculture into other areas. This can be done if enough industry and jobs are created. For that land is required and farmers have that land. The worrying point is that in the past when governments have taken away land from farmers, they have not been adequately compensated. Neither have they been re-skilled so that they can take on other professions. And too many times in the past, the government has taken land from farmers at cheap prices and sold it on to industrialists at a higher price. The industrialists have then sold it on further and made still higher profits. Hence, the trust system that is required for acquiring land for public purposes and building industry has broken down completely and building this trust will not be easy.

Further, the industry has had a totally lackadaisical attitude towards the entire issue. They seem to be just interested in getting the land and profiting from it. They can pay but they do not want to take responsibility for resettlement. It is opined that it should be left to a government agency. The question to ask is why can't the industry be responsible for resettlement? The issue of land acquisition is a complicated one. If India has to develop, land is required. But the entire system of compensating and re-skilling farmers so that they can move away from agriculture needs to be thought through as well (First Post, 2017).

Horticulture is now turning to be a high technology activity. The gardeners (Malis) who actually work in the fields and their immediate supervisors have also to be trained in the advanced methods of cultivation and use of various technically superior and efficient equipment and implements. The present courses being offered to the gardeners are not only numerically few but also highly inadequate content-wise. Consequently, age old practices continue to be adopted, resulting in low productivity. Presently there is a wide gap between the demand for trained managers and supervisors as also malis and their supply. This gap is only likely to widen with further developments in this field.

A number of Agricultural Universities and Training Institutions in the country are turning out a large number of graduates with specialization in horticulture. But the courses offered by these institutions tend to be theoretical with very little practical input. The students receive little or no hands-on experience. Nor are they trained adequately in the management aspects of horticultural plantations and farms. As such there is absence of campus recruitment as has been the case with Management and I.T. trained students. Prospective employers look for people trained in the practical aspects of horticulture, capable of performing supervisory and managerial functions. Horticulture is now turning to be a high technology activity. The gardeners (Malis) who actually work in the fields and their immediate supervisors have also to be trained in the advanced methods of cultivation and use of various technically superior and efficient equipment and implements. The present courses being offered to the gardeners are not only numerically few but also highly inadequate content-wise. Consequently, age old practices continue to be adopted, resulting in low productivity. Presently there is a wide gap between the demand for trained managers and supervisors as also malis and their supply. This gap is only likely to widen with further developments in this field.

5. IMPACT OF CLIMATE CHANGE ON NATURAL RESOURCES

Global warming and climate change is the greatest concern of mankind in 21st Century. The impact of climate change has serious consequences for peoples' lives and livelihoods. Climate change dilators displace about 20 million people every year. The most vulnerable people are those depending on agriculture, water and natural resources for their livelihoods. Preventing erosion results in less silting of rivers and reservoirs and clean water is available for use. Several strategies, those for agriculture, coastal zones, energy, health, water and urban infrastructure require planning at all levels. Migration due to melting of permafrost, coastal erosion and sea level rise may be unavoidable; However, planning ahead can reduce the cost of replacing infrastructure, such as, roads and pylons restrict major developments in less dangerous zones and diversify economic activities. Planners need to balance the costs – benefits between hard investment and major imaginary losses and soft investments that build flexibility to manage a wide range of possible impacts (Chandrasekhar, 2016). It is possible that climate change may force the pace of rural-urban migration (rurbanisation) over the next few decades (Revi, 2008). The ongoing agrarian crisis in rural India could be catalyzed by climate change into a migratory rout, driven by greater monsoon variability, endemic drought and flooding and resource conflict.

Climate related major changes like submergence of coastal lands cannot be controlled. Such events drive the vulnerable population inlands increasing pressure on land, food, water and energy *etc.* The farm land which is already stressed beyond endurance has to be managed to produce more food for the increasing population. It is imperative, therefore, to utilize the technological interventions developed in the past and continue to develop innovative technologies to use the scarce inputs to meet the increasing demand for food and nutrition of the growing populations. The consequences of climate change are - global warming, change of seasonal pattern, excessive rain, melting of ice cap, flood, rising sea level, drought, *etc.* leading to extremity of all kinds. Decrease in potential yields is likely to be caused by shortening of the growing period, decrease in water availability and poor vernalization. Western Ghats and surrounding regions may be deprived of normal precipitation due to abnormal monsoon. Vulnerability, rarity and rapid extinction of plant species will be among the other consequences. Plains of India will face similar kind of problems. Nobel Laureate R. Pachauri said, total agricultural land will shrink and the available land may not remain suitable for the present crops for too long. Farmers have to explore options of changing crops suitable to weather. He also pointed out that climatic changes could lead to major food security issues for a country like India. Climate change is predicted to cause an increase in average air temperature of between 1.4°C and 5.8°C, increases in atmospheric CO_2 concentration and significant changes in rainfall pattern (Houghton *et al.*, 2001).

Impact of climate change on four sectors of the economy, namely, Agriculture, Water, Natural Ecosystems and Biodiversity and Health in four climate sensitive regions of India, namely, the Himalayan region, the Western Ghats, the Coastal Area and the North-East Region. The present challenges like global climate change, water and soil pollution, less water availability, urbanization *etc* adds up to the situation. In combination with elevated temperatures, decreased precipitation could cause reduction in availability of irrigation water and increase in evapotranspiration, leading to severe crop water-stress conditions.

6. HORTICULTURE DEVELOPMENT IN INDIA

For decades we have wallowed in socialistic principles of growing our own food. Food sufficiency does not mean a household being able to grow its food but being able to access food it needs. In many cases it would be wiser for a small holder to focus on cash crops and buy cereals needed from the income. But that is not the way a small holder has been asked to think. Horticulture offers ample opportunities to raise production and income per hectare. The need for diversification to horticulture sector was acknowledged by the Government of India in the mid-1980s, by focusing its attention on Investment in this sector. Presently, horticulture has established its credibility in improving income through increased productivity, generating employment and in enhancing exports. Resultantly, horticulture has moved from rural confines to becoming a commercial venture. The recent emphasis on horticulture in our country consequent to the recognition of the need for attaining nutrition security and for more profitable land use has brought about a significant change in the outlook of the growers. The need for greater utilization of available wastelands against the background of dwindling water and energy resources has focused attention to dry land, to arid and semi-arid tracts and to horticultural crops which have lesser demands on water and other inputs besides being 3 to 4 times more remunerative than field crops. The scenario of horticultural crops in India has become very encouraging. Our farming comrades have busted the myth that India produces more food grains than fruits and vegetables.

The fact is that fruits and vegetables are grown more than food grains in India. In 2012-13, food grain production was 257 million tonnes but the total horticulture production was 268.9 Million Tonnes. Also, India now produces and consumes more fruits and vegetables than rice and wheat. In recent times, horticulture has become the growth driver of India's agriculture with India's emergence as second largest producer of fruits and vegetables in the world after China. The percentage share of horticulture output in agriculture has become more than 30% (Figure 4).

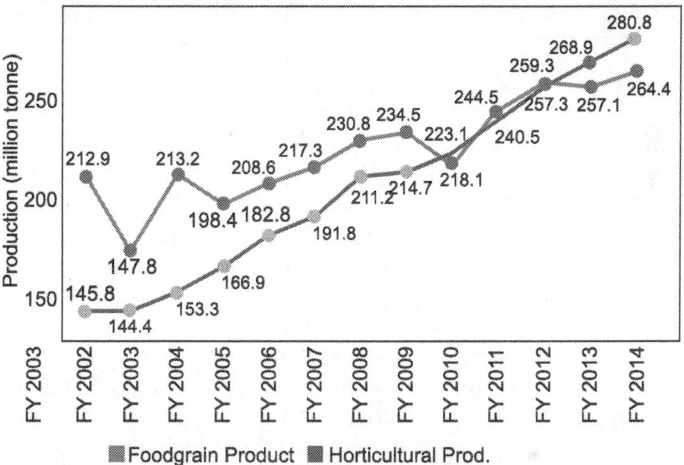

Source: *Ministry of Agriculture, FY 2014 figures are provisional.*

Fig. 4. *Horticulture versus Food grain production*

The role of horticulture in Indian agriculture growth is increasing steadily due to the following reasons. 1) Different horticultural crops can be grown in different regions because of the variance of climatic conditions in India, 2) Development in irrigation facilities and availability of technical information improves the scope for growing horticultural crops, 3) Due to increase in communication and transport facilities, the market potential of horticultural crops has been high, 4) There is a greater demand of fresh and processed products, including products, in local as well as international markets and 5) Government aid, development of financial institutions and co-operatives in rural areas help farmers in replacing their traditional crops with horticultural crops by increasing their returns from these crops (Anon., 2015).

Under the purview of agriculture and allied activities, the share of plan outlay for horticulture which was 3.9% during Ninth Five Year Plan, has increased to 4.6% during the Twelfth Plan (Table 3).

Table 3. *Plan wise share of agriculture and horticulture*

(Rs in Crores)

	Plan Outlay	
	Agriculture & Allied activities	**Horticulture**
IX plan (1997-2002)	37546	1453 (3.9%)
X Plan (2002-07)	58933	5025 (8.5%)
XI Plan (2007-12)	136381	15800 (11.6%)
Annual Plan (2007-08)	17971	1941
Annual Plan (2008-09)	27270	2049
Annual Plan (2009-10)	28772	1876
Annual Plan (2010-11)	36983	2700
Annual Plan (2011-12)	46225	3088
XII Plan (2012-2017)	363273	16840 (4.6%)
Annual Plan (2012-13)	54748	3062
Annual Plan (2013-14)	18781	4128
Annual Plan (2014-15)	11531	2922

Source: (1) Agriculture & Allied Activities: Agricultural Statistics at a Glance 2013 and 2014, DES. (2) Horticulture: Mission for Integrated Development of Horticulture (MIDH), DAC & FW.

India has witnessed voluminous increase in horticulture production over the last few years. Significant progress has been made in area expansion resulting in higher production. Over the last decade, the area under horticulture grew by about 2.7% per annum and annual production increased by 7.0%. During 2013-14, the production of horticultural crops was about 283.5 million tonnes from an area of 24.2 million hectares (Figure5).

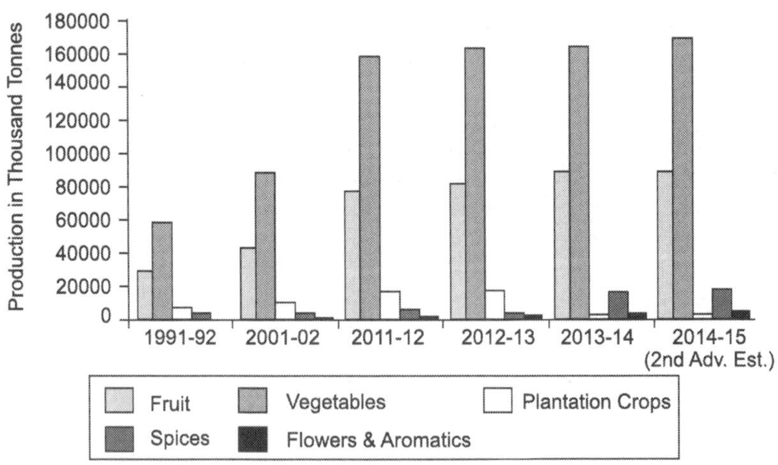

Fig. 5. *Production of horticultural crops during 1991-92 to 2014-15*

Out of the six categories, that is, fruits, vegetables, flowers, aromatic, spices and plantation crops, the highest annual growth of 9.5% is seen in fruit production during

2013-14. The production of vegetables has increased from 58,532 thousand tonnes to 168,300 thousand tonnes since 1991-92 to 2014-15. Apart from the health improvements, the production of vegetables improves the economy of a country as these are very good source of income and employment. The contribution of vegetables remains highest (59–61%) in horticulture crop productions over the last five years. During 2013-14, the area under vegetables is estimated at 9.4 million ha with a production of 162.9 million tonnes in India. For this period the total vegetable production was highest in West Bengal (23,045 thousand tonnes) followed by Uttar Pradesh (18,545 thousand tonnes).

Great potentialities exist for cultivation of flowering plants. Increasing trends in area and production of flowers has been observed since 2003-04. In addition to the beautification of the local landscape, great scope exists for export of flowers; and floriculture is important for bee-keeping industry which too provides an alternate source of income to the Indian farmers. The highest production of flowers (loose) was recorded in Tamil Nadu (343.65 thousand tonnes) followed by Karnataka (211.50 thousand tonnes). India witnessed the shift in area from food grains towards horticultural crops over last five years (2010-11 to 2014-15). The area under horticultural crops has been increased about 18% but expansion of area under food grains is only 5% during the stipulated period. The production of horticultural crops has outpaced the production of food grains since 2012-13.

The initiatives taken by the GOI and positive interventions by the agencies like APEDA, Horticulture Mission and NABARD, the production is increasing and is projected to increase further (Table 4).

Table 4. *Horticulture production in India: Present and future projection*

Commodity	Production, mt		
	1998-99	2011-12	2020-21
Fruits	44-04	81.00	98.00
Vegetables	87.53	185.00	220.00
Spices	2.91	5.50	6.50
Coconut	10.27	20.00	24.00
Cashew nut	0.46	1.70	2.00
Cocoa& others	3.00	6.8	9.5
Total	146.82	300.00	360.00

Source: (Ahlavat *et al.*, 2010)

7. IMPACT OF CLIMATE CHANGE ON HORTICULTURE

Climate change poses serious challenges to human and places unprecedented pressure on sustainability of horticulture industry. Two major parameters of climate change that has far reaching implications on agriculture in general and horticulture in particular are more erratic rainfall patterns and unpredictable high temperature spells that will consequently reduce crop productivity. Latitudinal and altitudinal shifts in ecological and agro-economic zones, land degradation, extreme geophysical events, reduced water availability, rise in sea level and salinization are postulated (FAO 2004). Vegetable production is threatened by increasing soil salinity particularly in irrigated croplands which provide 40% of the world's food. Fruits, vegetables, flowers, medicinal plants and tubers are grown from tropical to temperate, some horticultural crops like spices and plantation crops are location specific. In order to sustain our horticultural production with present day challenges we have to have packages to manage abiotic stresses. The nature and magnitudes of stress vary. The established commercial varieties of fruits, vegetables and flowers will perform poorly in an unpredictable manner due to aberration of climate. Melting of ice cap in the Himalayan regions will reduce chilling effect required for the flowering of many of the horticultural crops like Apple, Saffron, Rhododendron orchid, *etc*. Commercial production of horticultural plants particularly grown under open field conditions will be severely affected. Due to high temperature physiological disorder of horticultural crops will be more pronounced like Spongy tissue of mango, fruit cracking of litchi, flower and fruit abscission in solanaceous fruit vegetables, *etc*. Air pollution also significantly decreases the yield of several horticultural crops and increases the intensity of certain physiological disorder like black tip of mango. Hence, there is a need to protect these valuable crops for sustainability against the climate change scenario. The most effective way is to adopt conservation agriculture, using renewable energy, forest and water conservation, reforestation *etc*. To sustain the productivity, modification of present horticultural practices and greater use of greenhouse technology are some of the solutions to minimize the effect of climate change. Development of new cultivars of horticultural crops tolerant to high temperature, resistant to pests and diseases, short duration and producing good yield under stress conditions, as well as adoption of hi–tech horticulture and judicious management of natural resources will be the main strategies to meet this challenge (Dutta, 2013).

The climate change will have many impacts on horticulture and a few examples are: 1) A study conducted at IISR, Calicut using GIS models have shown that many areas presently suitable for spices would become unsuitable in another 25 years. There would be new areas which are presently unsuitable, become highly suitable for cultivation of spices. This will be applicable in other horticultural crops. 2) Production timing will change due to rise in temperature. Due to rise in temperature, photoperiods may not show much variation. As a result, photosensitive crop will mature faster. 3) The winter regime and chilling duration will reduce in temperate regions affecting the temperate crops. 4) Pollination will be affected adversely because of higher temperature. Floral abortions flower and fruit drop will be occurred frequently. 5) The requirement of annual irrigation will increase and heat unit requirement will be achieved in much lesser time. 6) Higher temperatures will reduce tuber initiation process in potato, reduced quality in tomatoes and pollination in many crops. In case of crucifers, it may lead to bolting; anthocyanin production may be affected in apples and capsicum. Tip burn and blossom end rot will be the common phenomenon in tomatoes. 7) Coastal regions can expect much faster percolation of sea water in inland water tables causing more salinity. Ahlawat *et al*. (2010) reported the following to be some of the possible effects of climate change on horticulture production; 1) suitability and adaptability of current cultivars may change, 2) shifting of major potential suitable zones with raise in temperature, 3) increased incidence of physiological disorders, such as, tip burn, 4) increase in pollination failures due to heat stress, 5) early maturity of citrus, grapes, melons and onion, 6) strawberry may produce more runners than fruits, 7) poor quality of potato tubers, 6) bolting in crucifers, 8) advancement of bud burst in almond, 9) delayed panicle emergence in mango due to low temp, 10) low perfect flowers 11) low set in citrus due to warm temp, 12) increased risk of spread and proliferation of soil borne diseases as a result of more intense rainfall events coupled with warmer temperatures and 13) changes in the distribution of existing pests, diseases and weeds and an increased threat of new incursions. Therefore, the development of horticultural crops that can withstand stress will be the single most important step we may take to adapt the changes we have faced today and will face in the future.

7.1 Fruit Crops

A large variety of fruits are grown in India, of which mango, banana, citrus, guava, grape, pineapple and apple are the major ones. Due to rise in temperature, crops will develop more rapidly and mature earlier. For example, Citrus, grapes, melons *etc.* will mature earlier by about 15 days. Strawberries will produce more runners at the expense of fruits. Specific chilling requirements of pome and stone fruits will be affected and hence, dormancy breaking will be earlier. Delay in monsoon, dry spells of rains and untimely rains during water stress period, supra-optimal temperatures during flowering and fruit growth, hailstorms are some of the most commonly encountered climatic conditions experienced by the citrus growers over the past decade or so. The climate change increases the atmospheric temperature and change of rainfall pattern, as a result, banana cultivation may suffer from high temperature, soil moisture stress or flooding / water logging. High temperature and moisture stress also increase sunburn and cracking in apples, apricot and cherries and increase in temperature at maturity will lead to fruit cracking and burning in litchi (Kumar and Kumar, 2007). Air pollution also significantly reduced the yield of several horticultural crops and increases the intensity of certain physiological disorders like black tip of mango which is induced by coal fume gases, sulphur dioxide, ethylene, carbon monooxide and fluoride. Leaf production increases by one leaf per month for every 3.3 to 3.7°C rise in minimum or mean temperature from 10-20 °C or 13.5 to 25 °C respectively. Higher temperature (31-32°C), in general, increases the rate of plant maturity in annual species, thus shortening the growth stages, during which developing fruits and suckers absorb photosynthetic products. The temperature below 10°C leads to impedance of inflorescence and malformations of bunches. Chilling symptoms on leaves are not seen immediately but it may take 2 to 4 days to appear. The production of apple has gradually increased but the productivity has fallen from 10.8 to 5.8 t/ha (Awasthi *et al.*, 2001). The reasons attributed to it are climate variability, soil, crop improvement *etc.* Among all the productivity reducing factors, climate is difficult to manage. The changes in climate in the form of erratic precipitation, increase in temperature, lesser days serving as the chilling period have started affecting the mountain agricultural production systems and ultimately the food security of the people. Immediately after the frost period, the survey was conducted in the Bikaner and adjoining areas in order to assess the effect of frost on the survival and severity of damage on arid fruit crops. Study revealed that the crops could be classified on the basis of severity of damage into four groups *viz.* severely affected which included crops, such as, aonla, gonad, phalsa, moringa, ber, Ficus sp. *etc.* It was also observed that few crops, such as, pomegranate was moderately affected, sapota and bael less affected and crops, such as, date palm was unaffected by the frost. Lists of some abiotic stress varieties are given in Table 5.

Table 5. *Fruit crop varieties tolerant to abiotic stresses*

S.No.	Crop	Variety	Tolerant
1.	Pomegranate	Ruby	Drought tolerant
2.	Annona	Arka Sahan	Drought tolerant
3.	Fig	Deanna and Excel	Drought tolerant
4.	Grape (rootstock)	Dogridge	Salinity tolerant.
5.	Mango	Bappakai	Salinity tolerant
6.	Lime	Rangpur lime and Cleopatra mandarin	Salinity tolerant

Source: Bose and Mitra (1996)

7.2 Vegetables Crops

India is the second largest producer of vegetables in the world (ranks next to China) and accounts for about 15% of the world's production of vegetables. The current production level is over 90 MT and the total area under vegetable cultivation is around 6.2 million hectares which is about 3% of the total area under cultivation in the country. Environmental stress is the primary cause of crop losses worldwide, reducing average yields for most major crops by more than 50% (Bray *et al.*, 2000). Climatic changes will influence the severity of environmental stress imposed on vegetable crops. The response of plants to environmental stresses depends on the plant developmental stage and the length and severity of the stress (Bray, 2002). Plants may respond similarly to avoid one or more stresses through morphological or biochemical mechanisms (Capiati *et al.*, 2006). Environmental interactions may make the stress response of plants more complex or influence the degree of impact of climate change. Lists of some abiotic stress vegetables and their varieties are given in Table-6 and 7 respectively.

High temperatures can cause significant losses in tomato productivity due to reduced fruit set and smaller and lower quality fruits. Pre-anthesis temperature stress is associated with developmental changes in the anthers, particularly irregularities in the epidermis and endothesium, lack of opening of the stromium and poor pollen formation (Sato and Thomas, 2002). Hazra *et al.* (2007) reported that symptoms causing fruit set failure at high temperatures in tomatoes includes bud drop, abnormal flower development, poor pollen production, dehiscence and viability, ovule abortion and poor viability, reduced carbohydrate availability and other reproductive abnormalities. In pepper, high temperature exposure at the pre-anthesis stage did not affect pistil or stamen viability, but high post-pollination temperatures inhibited fruit set, suggesting that fertilization is sensitive to high temperature stress (Erickson and Markhart, 2002). Plant sensitivity to salt stress is reflected in loss of turgor, growth reduction, wilting, leaf curling and epinasty, leaf abscission, decreased photosynthesis, respiratory changes, loss of cellular integrity, tissue necrosis and potentially death of the plant. Most of the vegetable crops are highly sensitive to flooding and genetic variation with respect to this character is limited. Flooded crops especially in tomato plants accumulate endogenous ethylene that causes damage to the plants (Drew, 1979). Under low oxygen levels stimulate an increased production of an ethylene precursor, 1-aminocyclopropane-1- carboxylic acid (ACC), in the roots. The severity of flooding symptoms increases with rising temperatures; rapid wilting and death of tomato plants is usually observed following a short period of flooding at high temperatures (Kuo *et al.*, 1982). During the last 40-50 years air pollution level increasing at an alarming rate in the developing countries and causing potential threat to the crop production. Sulphar dioxide, nitrogen oxide, hydrofluride, ozone and acid rain are the primary air pollutant. Ozone has adverse effect on vegetable production in terms of reducing growth, yield and quality. Risk of the air pollution is more when vegetable crops grown close to the densely populated areas. A recent study indicated that the ambient air pollution significantly decreased the yield upto more than 50 percent in case of Brassica oleracia, Lactuca sativa and Raphanus sativus. Many vegetable crops namely tomato, water melon, potato, squash, soyabeans, cantaloupe, peas, carrot, beet, turnip, *etc* are more susceptible to air pollution damage. Yield of vegetable can be reduced by 5-15 percent when daily ozone concentrations reach to greater than 50 ppb (Raj Narayan, 2009) (Table 6 & 7).

Table 6. *Vegetable varieties tolerant to abiotic stresses.*

S.No.	Tolerant	Crop
1.	Drought tolerant	Chilli, melons, tomato, onion
2.	Heat tolerant	Peas, tomato, beans, capsicum
3.	Salinity tolerant	Melons, peas, onion
4.	Floowind/excess moisture tolerant	Tomato, onion, chilli

Source: *Rai & Yadav (2005)*

Table 7. *Vegetable varieties and improved lines tolerant to abiotic stresses*

S.No.	Tolerant	Crop	Variety	Advanced Line
1.	Drought/rainfed	Tomato	Arka Vikas	RF-4A
		Onion	Arka Kalyan	MST-42 and MST-46
		Chilli	Arka Lohit	IIHR Sel.-132
2.	Photo insensitive	Dolichos	Arka Jay, Arka Vijay, Aarka Sambram, Arka Amogh, Arka Soumya	IIHR-16-2
		Cow pea	Arka graima, Arka Suman, Arka Samrudhi	
3.	High temperature	Capsicum		IIHR Sel.-3
		French bean		IIHR-19-1
		Peas		IIHR-1 and IIHR-8
		Cauliflower		IIHR 316-1 and IIHR-371-1

Source: Hazra & Som (1999) and Rai & Yadav (2005).

7.3 Plantation Crops

Consecutive drought here reduced the coconut production by about three lakh nuts/year for four years. Productivity loss was to the tune of about 3500nuts/hectare/year in India. Apart from drought other natural calamities like cyclone *etc* have impacted the crop production and productivity. In coconut, arecanut and cocoa increased CO_2 led to higher biomass production. But a slight decline in biomass production was apparent at elevated air temperature. All three crops responded differentially under elevated air temperature. In coconut, net photosynthesis rate has reduced but increased that of arecanut and cocoa. However, TDM was slightly decreased in all three crops.

Studies conducted on "Impact of climate change in cashew" at Directorate of Cashew Research, Puttur, India indicated that the rainfed cashew crop is highly sensitive to changes in climate and weather vagaries, particularly during reproductive phase. Cashew requires relatively dry atmosphere and mild winter (15-20°C) coupled with moderate dew during night for profuse flowering, but high temperature (>34.4°C) and low relative humidity (<20%) during afternoons cause drying of flowers subsequently resulting in yield reduction. Unseasonal rains and heavy dew during flowering and fruiting period aggravated the incidence of pests and diseases. All these situations resulted in reduction yield up to 50 to 65%.

7.4 Spice Crops

In general, due to increase in maximum and minimum day temperature and decreasing the annual rainfall the productivity showed decreasing trend in most of the black pepper growing areas of India. In black pepper, Accs 1380(IC 316801), 1387(IC 316803), 1410(IC 316817), 1423(IC 316825) and 1430(IC 316832) were identified as relatively tolerant to drought. In cardamom, RR1(IC 349591), CL-893 (IC 349537), Green Gold (IC 349550) were found relatively tolerant under Kerala, India condition. Kashmir's prized saffron crops have suffered a 40% drop in production, one of the three places in world - besides Iran and Spain - most famous for its saffron, water shortages are straining the crops. Some of the saffron farmers who traditionally relied on rainwater are now looking at irrigation measures to save their rare and labour intensive crop Seed Spices are winter season crops and commonly grown in arid and semi arid track of Rajasthan and Gujarat requiring certain period of low temperature for optimum vegetative growth. Heavy losses have been observed due to combined effect of chilling and frost injury. Cumin, coriander, nigella, ajowan are the crops which are very sensitive to frost. Incidence of frost causing serious loss in yield almost reaches up to zero. Fennel and fenugreek are also affected by frost but growth stage plays an important role. So far no efforts have been made to identify the source of resistance against low temperature injury in available germplasm of seed spices crops.

7.5 Flower Crops

Melting of ice cap in the Himalayan regions will reduce the chilling requirement for the flowering of many of the ornamental plants like Rhododendron orchid, Tulip, Alstromerea, Magnolia, Saussurea, Impatiens, Narcissus *etc.* Some of them will fail to bloom or flower with less abundance while others will be threatened. Plant species requiring high humidity and water may find them under difficult conditions for survival. Plains of India will also have similar kind of problems and will be affected either by drought or excessive rains, floods and seasonal variations. Commercial production of flowers particularly grown under open field conditions will be severely affected leading to poor flowering, improper floral development and colour. Chrysanthemum is a short day plant. So flowering round the year in open field condition is not possible. Low temperatures shut down flowering in Jasmine (<19°C) and lead to reduction in flower size. Flowers do not open up fully in tropical orchids wherever temperatures below 15°C. High temperature leads to flower bud drop and unmarketable spikes in tropical orchids when temperature remains > 35°C.

Climate change poses serious challenges to human and places unprecedented pressure on the vegetables has also exhibited rising trend. Horticultural crops play a unique role in India's sustainability of horticulture industry. Therefore, the development of horticultural crops that can withstand stress will be the single most important step we may take to adapt the changes we have faced today and will face in the future. It is a largely considered opinion in India that the sophisticated high-tech greenhouse technology is meant for elite and resource rich farmers and private investors. However, simultaneously Indian farmers have adopted by improvising this proven technology to suit their individual investment potential and technical knowledge. It is seen that the technology is being followed in varying degrees in contrast to the standard format seen in developed nations. It is imperative, therefore, to take all stake holders at different levels into confidence while promoting protected horticulture. In India, most of the farmers are not economically sound and high initial cost of green house structure restricts the farmers to establish this structures. It is pertinent, therefore, to develop greenhouse technology in participatory mode especially involving small farmers. It is important to note that greenhouse technology is amenable for adaptation in small and otherwise uneconomical land holdings, those lands unsuitable for traditional farming and hilly terrains. Small land holders are in majority holding the precious and scarce farm land 'Brown Gold'. Most of this land is less productive due to impoverished soils, lack of investment potential and are not available for large scale land development activities. Indian farmers are at the receiving end being considered as resource poor, ill informed and hard to change. The large extension network serves the technical know-how on platter and expects the farmers to follow in total. Financing institutions supply finances if they are credit worthy or Government schemes allow liberal loans. Input suppliers are also keen to market their products. But there is no guarantee that these efforts fructify in proper dividends as there is no coordination of these well meant efforts at all levels and especially at grassroots level. Rural youth are also looking for smart jobs compared to less attractive farming activates. Therefore, it is important to integrate these efforts to succeed as is the case in engineering industry. Suboptimal production from scarce farm land is a national loss and farm land is our strength and our national natural asset to feed our growing millions and produce exportable surplus.

Protected horticulture offers a proven opportunity to utilize the scarce land resource to produce more from limited area to meet nutrition and income security of our farming community which comprises largely of small land holders with poor investment potential. It is important; therefore, to make individual grower or grower groups as partners in progress. Both public and private investment is to be promoted to make the venture profitable. It should be an organised activity coordinating money, men, material and management.

1. **Money:** Money is available with public and private investors, but this has to be properly channelized to meet the growing demand.
2. **Men:** Skilled and unskilled work force is available but there is an urgent need to develop skills and contractual services for different production and maintenance operations as the case in an engineering industry that employ the workforce from ITI trained trade personnel.

3. **Material:** Timely supply of inputs is to be ensured in the vicinity of the greenhouse facility and
4. **Management:** India is producing a large number of efficient managers and IT professional who are ably contributing to a great extent to the economic progress all over the world. This knowledge potential is to be utilised to integrate different aspects of horticulture technologies including greenhouse technology and coordinate various activities to make the mission successful. The small farmer forms a proud partner in progress and contributes his might for inclusive economic growth of the nation at large. It is required to work together towards the envisaged economic progress.

8. TECHNOLOGICAL INTERVENTIONS DEVELOPED TO MEET THE CHALLENGES OF DEPLETING RESOURCES

There have been continuous efforts to cope up with the ill effects of climate change, such as, developing varieties/hybrids tolerant or resistant to both biotic and abiotic stresses and production and protection techniques to avoid or minimise the adverse effects of climate change.

Majority of the farm land in India is rain fed and therefore agriculture is considered as gambling with monsoon. Climate change is likely to change the rainfall pattern and agricultural production can be more unpredictable. There are many horticultural crops (fruits and plantation crops and medicinal crops), energy plantations and economic forest tress that require relatively less water (compared to the annuals like vegetables and flowers) that can be grown on the marginalised lands.

Excess precipitation and heat are deterrent to commercial production of good quality vegetables. Excess moisture in soil and air predisposes the crops to fungal and bacterial diseases and excess heat retards plant growth and affects flowering and fruit development and it also increases the activity of insect vectors that transmit viral diseases. Yield per unit area from horticultural crops (Vegetables) is far more higher compared to that from other food crops. They also provide direct and indirect employment opportunities, provide food, nutrition and livelihood opportunities and pave way for national economic progress (Swamy and Nath, 2012). Under the garden revolution new crops are being introduced, extending the supply of vegetables by staggered planting, accommodating vegetables in non traditional crop sequences, increasing producvity through intercropping, mixed cropping, relay cropping, multilayered cropping and alley cropping systems, thus providing much needed livelihood security to the farmers who were hitherto growing traditional crops like rice and millets. In Karnataka the dry lands where the staple food grain *Ragi* (Finger millet) is rapidly being replaced by mango plantations which require relatively less water. Ragi can also be grown with adequate rains in the interspaces during the pre-bearing period and till the crop canopy covers the interspaces completely.

8.1 Development of Varieties and Hybrids

Heat tolerant cabbage and cauliflower hybrids (Cool season crops) are now grown in summer months also in tropical parts of the country. Disease and pest resistant varieties/ hybrids are being developed from time to time all over the world. Liberalisation of seed policy in 1988 in India facilitates importing seeds from different parts of the world to meet the demand for tolerant /resistant verities. Photoperiod and thermo insensitive verities have broken the seasonality barrier for successful round the year production. This genetic engineering of the crops is breaking the regional and seasonal barriers for crop production.

8.2 Production Technologies

Apart from genetic improvement, best agronomical practices like adjusting the sowing dates by providing need based protection to the seedlings revolutionised the production of melons in northern states (UP and Bihar), silver beet and other vegetables in Ladakh. Protected nurseries and grafting in melons and growing tomato and cauliflower under rain shelters, tomato and melons under insect proof screens have indeed made it possible to boost vegetable production.

Adjusting sowing /planting dates had been a proven technology to avoid adverse weather conditions in the given location. Season wise selection of crop and varieties was the first step to decide the cropping plans. However, due to increasing demand, production has to be stretched to less suitable locations and seasons. Varietal adaptation trials at different locations and periods of time proved to be useful in finding additional temporal and spatial intervals that could be utilised to grow offseason and non traditional crops. Rice is the traditional crop of eastern and southern parts of the country but gradually rice cultivation moved to north-western parts. Similarly wheat has moved from north-west to north and eaten parts of the country, from saline soils of north-west to acid late rites of the north-east. Cole crops typical of temperate and sub-tropical climate moved southwards to tropical plains.

Vegetables were grown on small plots with lift irrigation during rainy free months in length and breadth of the country enabling round the year production of vegetables. The majority of the farm land is in tropical zones providing long growing periods suitable for vegetable production. Sub-tropical and temperate regions though limited are also being used to grow vegetables to meet the growing demand for vegetables. As demand increased growing vegetables has become an important economic activity and profitable proposition especially in the hinterlands of urban markets due to Increase in purchasing power and knowledge about need for consumption of nutritious vegetables for healthy living. Therefore, efforts have begun to utilize both spatial and temporal vacancies in the farm land to produce vegetables to enhance income and nutrition security of both urban and rural people. Thus growing vegetables in areas where they are not grown commercially help to supply vegetables to the markets round the year.

Production of vegetables is spread over the entire country. Bihar, Odisha, UP (Plains) and West Bengal are the major producers followed by AP, Assam, Karnataka, Kerala, Tamil Nadu and Maharashtra and other states also started producing vegetables as they are proved more remunerative compared to the income from staple food grains. There had been gradual rise in area, production and productivity of vegetables in India, but yet there is scope to increase yields per acre by balanced management of inputs to realise potential yields in different agro-climates in India. In Himachal Pradesh, the elevation ranges from 350 m to 700 m, providing divergent agro-climate to facilitate production of different sub-tropical and temperate vegetables at different elevations (Kanwar, 2008). Apart from commonly grown vegetables like tomato, capsicum, French bean, cucumber at lower elevations exotic vegetables like Swiss chards, red cabbage, Chinese cabbage, kale, celery, parsley, are also grown at different elevations though on small scale. Protected cultivation in walk-in poly tunnel and poly green houses are used to grow the vegetables that cannot be grown in open (Kanwar, 2008). Cool season vegetables like beet, carrot, cabbage, cauliflower, capsicum and broccoli are being grown in tropical coastal plains of Kerala. The area under cool season vegetables increased from 100 ha in 2011 to 600 ha in 2014 (Rajiv, 2014).

The area under food crops (Cereals, grain pulses *etc.*) is not replaced by vegetables except in urban and periurban areas to some extent. Vegetables in these areas are grown in place of cereals/millets as vegetables excel in per acre yield and income. Large numbers of farmers (1.45 lakh) in Kerala have given up paddy cultivation to grow more profitable vegetables like bitter gourd, cucumber, cowpea and okra (Nidhi, 2011). Since they are perishable they are limited to the proximity of the urban markets. Most of the food grain production is during monsoon rains and vegetables are grown after the food crop on residual moisture in rain free months as vegetables cannot stand the standing water. Short duration crops like beans, greens and radish are grown in rice fallows on residual moisture after the main rice crop. Growing melons on river beds in Gangetic plains on receding moisture level is yet another example of offseason production. Summer fallows are being utilised to grow vegetables with supplementary irrigation from tube wells. Irrigation had been the major factor for expanding area under vegetables. In Himalayan region (HP, UK, J&K) where vegetables like capsicum, tomato, cabbage cauliflower and peas, that prefer cool climate, have changed hill agriculture with increasing returns to the farmers. In southern region, Ooty (TN), Wynad (Kerala) and Coorg and Chikkamagalur (Karnataka), cool season vegetables like capsicum and garden peas are grown and supplied to the plain areas during lean season. Retail market chains are trying to supply vegetables round the year all over the country there by increasing consumption and demand for vegetables round the year. This is triggering production and supply of vegetables from all parts of the country both during on and off seasons from traditional and non traditional areas. Inclusion of vegetables tomato, cabbage, peas and watermelon in irrigated sequential cropping systems including cereals and oil seeds proved to be remunerative (Prabhakar *et al.*, 1985). Vegetables like dolichos bean, cluster bean and cowpea grown as inter crops in rain fed grain millets like upland raagi (Finger millet) improved nutrition and income (Rao *et al.*, 1983, Prabhakar and Shukla, 1988). Intercropping in widely spaced water melon with short duration beans, greens and radish helps to green the brown triangle (Spatial and temporal gaps in the growth of the main crop – watermelon) and increased the income. In an intensive vegetable production system, sequential intercropping of vegetables increased the crop land use efficiency up to 500 % and doubled the income (Prabhakar and Shukla, 1989). Cover cropping in fruit orchards during prehearing period with vegetables increased cropping efficiency and income (Prabhakar *et al.*, 1989). Vegetable based cropping systems were found to be more remunerative in different parts of India with net returns ranging from Rs.15,000 to Rs.1,00,000 per acre (Vanitha *et al.*, 2013). Therefore, the vacancies in land and time during the initial growth stages of both perennials and annuals can be beneficially utilised to grow vegetables.

In hill states of UK, HP, J&K the vegetables seedlings are grown under cover to transplant early in the season to realise better prices (Anon., 2013a; Parmar and Choudhary, 2011). In cold deserts of Ladakh, Lahul and Spiti, the climate is harsh except during the period from May to September. Therefore, off–season production of vegetables is under protected cover. Underground poly houses (poly sheet covered trenches) protect both soil heat and solar energy for crop production. Lean to type poly houses on northern, western and eastern side with unbaked brick walls and UV stabilised poly sheet fastened to poplar tree poles, sunken greenhouses with poly tunnels covering the trenches three feet deep where vegetables are grown, poly sheets covering poplar tree pole frame on terraced benches are popularly used for growing vegetables during off season from October to April, supplying fresh and nutritious vegetables to local population and defence personnel. Producing seedlings of cucurbits prior to regular season of May to September has made it possible to grow them and supply to early market that makes it remunerative.

8.3 Protection Technologies

Providing need based protection to the crops is now a proven technology to beak the seasonality barrier and other limitations for crop production. Within the same location and prevailing weather conditions protected cultivation increased the crop yields manifolds compared to that from the open field cultivation. Thee mean fruit yield of okra in open field was 15 t/ha while that in a naturally ventilated greenhouse it was 28 t/ha registering an increase of 87 *per cent* during off -season (October to January). The mean fruit yield of muskmelon in the open field was 29 t/ha compared to that in naturally ventilated greenhouse 60 t/ha registering an increase of 107 *per cent* during off- season (October to January). Garden pea variety Magadi Local recorded the maximum green pod yield of 10.29 t/ha in a naturally ventilated greenhouse compared to that in open field (4.13 t/ha) during summer (Anon., 2005).

In trials conducted under subtropical climate in India, where winter temperature does not favour fruit set in tomato, poly house yields have been found even better (88–157 tonnes/ha. An average yield of 100 tonnes/ha has been obtained from tomato hybrid Naveen grown in a medium-cost greenhouse at IARI, New Delhi. At Jorhat, tomato planted in rainy season (July) under plastic rain shelter gave 60–70% higher yield in off-season with the maintenance of single stem and reduction of the intra-row spacing by half (50cm × 25cm) (NCPAH, 2001). Tomato grown during summer months the yield has been increased to about 180 tonnes per hectare per crop of six months at Bengaluru (Anon., 2004).

Capsicum is a crop of temperate and subtropical climate. In the subtropics in certain areas, the low temperature does not allow proper plant growth and fruit set during winters; Hence, it is grown only during spring summer. After the first one or two flushes, the fruit size in this crop remains smaller. A yield of 105.9 tonnes/ha could be obtained from mid-September transplanted crop of California Wonder in a low-cost poly house. Whereas, crops planted in open field or plastic low tunnels do not prove successful. In medium-cost poly house, the yields are still better (149.4 tonnes/ha) (NCPAH, 2001). Capsicum recorded 100t/ha in 6 months compared to 30 to 40t/ha in open field during winter at Bengaluru (Anon., 2004). Cucumber seeds fail to germinate if sown in winter in north Indian plains. Cucumber variety Poinsette gives a fruit yield of about 200 tonnes/ha in a poly house. October is ideal time for sowing of cucumber Poinsette in a low-cost poly house (NCPAH, 2001).

The soil is saturated in certain areas due to heavy rains during rainy season and it becomes difficult to raise successful crop of early varieties of cauliflower. The yield of cauliflower Pusa Deepali, Pusa Katki, Early Chinese Prince, Heavy Sliver Plate and Selected Early Dawn transplanted on 24th August in a plastic greenhouse-cum-rain shelter is 12.7–16.6 tonnes/ha. Similarly cauliflower Pant Gobhi four gives good yield in a low-cost poly house (NCPAH, 2001).

Heavy precipitation inundates the crop fields and damages the crops. Both surface and sub-surface drainage systems carry away the excess water that can be collected in the water harvesting structures. Rain water harvesting both in rural and urban areas and storing in ponds is an important technology to save water in rainfall deficit areas.

Cropping plans need to be tailored to use the available water in the farm site. Water from supplemental irrigation systems has to be judiciously used to meet the moisture needs throughout the crop growth. Annual horticultural crops are sensitive to moisture stress and cropping need to be planned according to the availability of water during the cropping period. Indians farmers have adapted swell for such eventualities by following flexible cropping systems like sequential, inter and sequential-inter cropping systems, alley cropping and relay cropping systems. Production of melons on river beds on receding moisture is one such technique adapted by the farmers in northern India along the rivers like Ganges. Vegetable production in rice fallows in eastern and southern parts of India is yet another proven technology being adapted by the farmers. Advancing planting of vegetables like melons by covering the early stages of seedlings, vegetable seedling production under cover are the popular technologies to meet the growing demand for vegetables. Short duration crops like greens, beans and radish are grown in rice fallows on receding moisture and other temporal and spatial gaps in the existing cropping systems, in crops like corn, pigeon pea, cotton sugarcane *etc.* Melon seeds are sown during late winter in northern states of UP and Bhar and the seedlings are covered with straw caps to protect from them cold damage. This facilitates to harvest an early crop and reap better returns by supplying melons to the early markets.

Mulching the crop lands provides protection from moisture and temperature loss from the soil and movement of moisture in the root zone, controls weeds effectively and prevents movement of salts to the top of the soil caused by evaporation. Reflective mulches reflect the excess sunlight falling on the ground and prevent heating of the soil that affects soil temperature and moisture.

8.4 Polytunnels and Polyhouses

Providing need based protection facilitates off-season cultivation, efficient insect vector management to combat viral diseases and integrated disease management by adjusting green house environment and growing medium which finally helps to modify ambient microclimate required for expression of intrinsic genetic potential of the crops to the fullest extent.

Producing vegetable seedlings in polytunnels or polyhouses helped to protect the seedling from cold in Northern states and protected the seedlings from insect vectors in Southern states. Incidence of spotted wilt virus in tomato could be delayed by covering te nursery with 40 mesh insect proof net. Similarly bud necrosis virus in watermelon could also be delayed by covering the crop up to fruit set with agri-net.

Underground polyhouses (polysheet covered trenches) protect soil heat and permit solar energy for crop production. Lean to type poly houses have on northern, western and eastern side unbaked brick walls and UV stabilised poly sheet fastened to poplar tree poles on southern side. These sunken greenhouses are popularly used for growing vegetables during off season from October to April, supplying fresh and nutritious vegetables to local population and defence personnel. Producing seedlings of cucurbits prior to regular season of May to September has made it possible to grow them and supply to early market that makes it remunerative.

Protected cultivation is the best alternative to use the land and other resources efficiently. Plant protective structures include hot beds, cold frames, polytunnels, green houses, glass houses, polyhouses, screen houses and poly cum net houses; Different protective structures are being used to grow vegetables that fetch better returns to the growers.

Vegetable nurseries are raised in northern plains for early transplanting during spring to catch up early market that fetch better prices Evaporative cooled greenhouses are used to raise expensive flower and vegetable hybrids, hybrid seed production and vegetative propagation of chrysanthemum, carnation, rose and grapes. Strawberry is cultivated both in naturally ventilated and evaporative cooled greenhouses. Production of exotic vegetables, culinary, medicinal and aromatic herbs are also grown under cover.

Protected cultivation has been found beneficial to produce hybrid seeds and seedlings of vegetables and flower crops, off season cultivation of vegetables in tropical plains and breaking the seasonality barrier in vegetables in temperate Himalayas, export oriented production of superior quality vegetables and flowers, combating biotic and abiotic stresses in vegetable production organic production of vegetables, propagation of saplings of fruits, planting material of ornamentals, culinary herbs, aromatic and medicinal crops and production high value plantation crops like Vanilla.

Different structures like low tunnels for seedling production, Cole crops and salad vegetables, walk-in tunnels or net houses for vegetables like tomato, capsicum and melons with cladding material like insect proof nets, UVS polyfilms, cost effective evaporative cooling and retractable shading systems can be used to develop a economically viable and sustainable greenhouse structures to produce vegetables and flowers and others free from both biotic and abiotic constraints both in regular and off-seasons.

9. THE ROAD AHEAD

Climate related major changes like submergence of coastal lands cannot be controlled. Such events drive the vulnerable population inlands increasing pressure on land, food, water and energy *etc*. The farm land which is already stressed beyond endurance has to be managed to produce more food for increasing population. It is imperative, therefore, to utilize the technological interventions developed in the past and that continue to be developed in future to use the scarce inputs to meet the increasing demand for food and nutrition of the growing populations.

Horticulturists will have to play a significant role in envisaging proper technological interventions to combat the challenges of declining natural resources due to climate change. The most effective way is to adopt conservation agricultural practices; using renewable energy, forest and water conservation, reforestation *etc.* to sustain productivity. Modification of present horticultural practices, development of new cultivars tolerant to high temperature, resistant to pests and diseases, short duration and producing good yield even under stress conditions, protected horticulture and judicious management of land use resources are some of the possible solutions to minimize the effect of climate change.

There had been several developments in horticulture technologies to reorient coping strategies and reduce the adverse effects of climate change and changing resource base and provide much needed food and livelihood security. It is a dynamic process to mitigate the adverse effects of climate change. The challenge continues making us better equipped to evolve into wiser and safer world. Horticulture is comparable to any other conventional industry as there is adequate scope for both organized operations and investment opportunities. Horticultural produce processes and products are akin to an assembly line of any engineering industry and there is ample potential for employment generation. Similar to industry with innovative and risk averting technologies horticulture is the best alternative for rural youth to stay back comfortably in villages and contribute to national economic progress. On the other hand, the government has to commit and invest in spreading evenly the civic amenities like safe drinking water, sanitation, health services, education and communication in length and breadth of the country as the population is being equally distributed in both urban and rural areas.

Horticulture is amenable to skill development in areas like farm mechanization, land development, water harvesting, micro- irrigation systems, crop growth regulation, plant protection, protected nurseries, greenhouse technology, post harvest handling, processing, seed production and processing. Besides contractual services for regular garden operations, such as, seed bed preparation, planting, irrigation, interculture, plant protection, harvesting, handling and transport. Equity models similar to those discussed earlier in Gujarat for setting up industry and in Maharashtra for building residential cum commercial buildings can be tried in horticulture as well, wherein the farmers would continue to own the shares proportionate to the extent of land pooled for horticulture development. With an appropriate legal frame work the hurdles and fears in pooling the fragmented lands from absentee landowners and farmers having large chunk of untilled land can be eliminated. Public-Private participation models can well be tried to augment productivity of innumerable figments of land that have not yet seen the light of development.

10. CONCLUSION

Continuing fragmentation of land, small and uneconomical holdings, depleting land availability for agriculture, adverse effects of climate change, gradually increasing urban and periurban agriculture, changing preference for jobs by the rural youth and growing need for skill development are some of the compelling reasons to reorient our agriculture/ horticulture development Strategies. As cultivable land area is decreasing day by day due to rapid urbanisation and industrialisation, protected cultivation is the best alternative to use the limited land and other resources efficiently.

Indian exports of horticultural produce increased gradually over years. The export basket comprises mainly of fresh fruits and vegetables, cut flowers, bulbs and planting material and significant quantities of processed fruits, vegetables and dry flowers. The USA is the single largest importer of Indian floriculture produce, dry flowers and pot purees. The major destination of cut flowers from India is the Netherlands, Germany and Japan. With diverse agro-climates India has great potential to produce high quality vegetables and flowers in protected structures to cater to the domestic as well as export markets.

India, it turns out, is the world's second-largest producer of fruit. China leads the pack, producing 154.4 million tonnes of horticultural produce, exceeding India's 82.6 million tonnes by 87%. India is also a place where a major part of the produce is lost or damaged: precisely how much is anyone's guess. The point is that the country has great scope to increase output and productivity in horticultural produce and add value besides, ranging from basic washing, sorting and climate-controlled storage to food processing.

India stands poised to make a breakthrough in distributed agro-processing, thanks to the spread of electricity to rural India and extensive road connectivity even in the hinterland. Granted, power lines drawn to a village does not mean that power is supplied to it. But if the problem of electricity boards not being able to collect realistic tariffs for the power they supply can be fixed, as it has been in some states, power availability will no longer be a constraint on the creation of modern food processing in rural areas. India has diverse agro-climatic conditions, making it possible to grow an enormous variety of fruit and vegetables in the country. Once such output is freed from the clutches of institutionalized middlemen in the Agricultural Produce Marketing Committee and farmers are linked to final consumers with a minimum number of intermediaries, the combination of power to run cold storages and roads for swift transportation via refrigerated trucks will see the production and productivity of fruit and vegetables shoot up. Freeze-drying of fruit and vegetables, processing into pulp and puree and ready-to-cook meals, conversion into juice, jam and other kinds of preserved foods, the scope to add value at low cost is huge. The point, However, is to realize the potential (GK Today, 2015).

Further, National Agricultural Research System and Extension net work are striving to serve the technical know -how on a platter, input suppliers are eager to sell and financial institutions are liberalizing the credit lending even to the small and marginal farmer who is ill-informed and ill-equipped to manage the resources effectively and efficiently on his fragmented, impoverished and uneconomical land holding. Is it not a top heavy administrative decision to thrust the entire responsibility upon the poor peasant? We can no more neglect the fact that the precious farm land, an important and critical national asset, being underutilized or wasted. The farmer alone cannot be the only arbiter in the mission of augmenting productivity and income from farm land. The equity models discussed earlier show promise if planned and implemented properly. With rapid changing resources' position it is imperative to collectively participate to reap the best possible dividends from farming, especially from horticulture.

It is comforting to understand that there have been several innovative technologies developed to enhance productivity of horticultural crops under favorable as well as less -favorable growing conditions that makes us geared up to meet the possible challenges of resource crunch in times to come.

REFERENCES

Ahlawat, V.K., Sehrawat, S.K. and Sharma, Rupali. (2010). Horticulture production in present climate change scenario. Department of Horticulture CCS Haryana University, Hissar.

Sen, Amarthya. (2007). Cited fromVivek Kaul (2015) Land Acquisition: India has enough land for agriculture. The Daily Reckoning. 26 Feb. 2015.

Anon. (2004). Annual Report, Indian Institute of Horticultural Research, Bengaluru.

Anon. (2005). Annual Report, Indian Institute of Horticultural Research, Bengaluru.

Anon. (2009). India's land availability shrinks to 0.3 hectare per farmer. Indianexpress.com Nov 06 2009.

Anon. (2015). 5 Factors Boosting The Scope Of Horticulture In India. Cooling India, Sep 15, 2015.

Anon. (2017). How much land is actually cultivated in India? Live Mint E-Paper, 19 January 2017.

Awasthi, R.P., Verma, H.S., Sharma, R.D., Bhardwaj, S.P. and Bhardwaj, S.V. (2001). Causes of low productivity in apple orchards and suggested remedial measures: In Productivity of temperate fruits. Jindal, K.K. and Gautam, D.R. (Eds), p. 1-8.

Bose, T.K. and Mitra, S.K. (1996). Fruits: Tropical and Subtropical. Nayaprakash, Kolkata, India.

Bray, E.A. (2002). Abscisic acid regulation of gene expression during water-deficit stress in the era of the Arabidopsis genome. Plant Cell Environ 25: 153-161. Cited from Datta,S. (2013) Impact of Climate Change in Indian Horticulture: A review. International Journal of Science and Technology, 2 (4): p. 661–671.

Bray, E.A., Bailey-Serres J. and Weretilnyk, E. (2000). Responses to abiotic stresses: Biochemistry and molecular biology of plants. Gruissem, W., Buchannan, B. and Jones, R. (eds). ASPP, Rockville, MD 1158-1249. Cited from Datta,S. (2013) Impact of Climate Change in Indian Horticulture: A review. International Journal of Science and Technology, 2 (4):p 661– 671.

Capiati, D.A., País, S.M. and Téllez-Iñón, M.T. (2006). Wounding increases salt tolerance in tomato plants: evidence on the participation of calmodulin-like activities in cross tolerance signalling. Journal of Experimental Botany 57: 2391-2400. Cited from Datta,S. (2013) Impact of Climate Change in Indian Horticulture: A review. International Journal of Science and Technology, 2 (4): p. 661– 671.

Chang, J. H. (1987). The Indian summer monsoon- Geographical Review 57(3):373-396. (Cited from www.What is Climate Change? - Climate Action Network Canada.

Chandrasekhar, Krishnappa. (2016). Drought...the most dreaded word in Indian economy Krishi Tantrajna 10 (1): p. 1, 6-7.

Datta, S. (2013). Impact of Climate Change in Indian Horticulture: A review. International Journal of Science and Technology, 2 (4): p. 661– 671.

Despande, R.S. (2012). Progressive changes in land reforms and agriculture policy. In The Basics of Human Civilisation – Food, Agriculture and Humanity – Present Scenario.Prem Nath (Ed) Dr. Prem Nath Agricultural Science Foundation (PNASF), Bengaluru. New India Publishing Agency, New Delhi. p. 157–173.

Drew, M.C. (1979). Plant responses to anaerobic conditions in soil and solution culture. Curr. Adv. Plant Sci 36: 1-14. Cited from Datta,S. (2013) Impact of Climate Change in Indian Horticulture: A review. International Journal of Science and Technology, 2 (4): p. 661– 671 Economic Times. (2016). January 19, 2016.

Erickson, A.N. and Markhart, A.H. (2002). Flower developmental stage and organ sensitivity of bell pepper (Capsicum annuum L) to elevated temperature. Plant Cell Environ 25:123-130. Cited from Datta, S. (2013) Impact of Climate Change in Indian Horticulture: A review. International Journal of Science and Technology, 2 (4): p. 661–671.

FAO. (2004). Impact of climate change on agriculture in Asia and the Pacific. Twentyseventh FAO Regional Conference for Asia and the Pacific. Beijing, China, 17-21 May 2004.

FAO. (2013). Annual Report.

First Post. (2017). Saturday, January 21, 2017Latest E-book.

GK Today. (2015). September 22, 2015.

Hazra, P. and Som, M.G. (1999). Technology for Vegetable Production and Improvement. Naya Prokash, Kolkata, India.

Hazra, P., Samsul, H.A., Sikder, D. and Peter, K.V. (2007). Breeding tomato (Lycopersicon Esculentum Mill) resistant to high temperature stress. International Journal of Plant Breeding, 1(1).

Houston, William and Griffiths, Robin. (2008). Water – the finite resource – how the politics of water will impact on the world. m/s Harriman house Ltd., Hampshire, UK.

Houghton J., Ding, Y., Griggs, D. Noguer, M. and Van der Linden, P. (2001). Climate Change 2001: The Scientific Basis. Published for the Intergovernmental Panel on Climate Change. Cambridge University Press, Cambridge, UK and New York. 881p.

Kanwar, J.S. (2008). Avenues for vegetable production in hilly regions. http:ecohort.org/blog.

Kumar, R. and Kumar, K.K. (2007). Managing physiological disorders in litchi. Indian Horticulture 52 (1): 22-24.

Kuo, D.G., Tsay J.S., Chen, B.W. and Lin, P.Y. (1982). Screening for flooding tolerance in the genus Lycopersicon. Hort Science 17(1): 6-78. *Cited from* Datta,S. (2013) Impact of Climate Change in Indian Horticulture: A review. International Journal of Science and Technology, 2 (4): p. 661–671.

Parmar, Manu and Choudhary, H. (2011). Protected cultivation of offseason vegetables. The Tribune May21.

Reddy, Narasimha Donti. (2014). Quora, Jul.26 2014.

NCPAH. (2001). Greenhouse designs and Environmental Control, IARI, New Delhi.

Nidhi Nath Srinivas. (2011). Farmers dump paddy for more profitable vegetables. ET Bureau. Nov.14.2011.

Prabhakar, B.S. Srinivas, K., Mohan Doss, V., Subraminian, T.R. Rao, J.V., Shukla,V. and Shrma, C.B. (1985) Yield and price variability in the analysis of multiple cropping systems. Singapore J.Pri.Ind.13 p. 57-63.

Prabakar, B.S. and Shukla, V. (1988). Vegetables in rainfed cropping systems. Proc. Ntl,eminar Dry land horticulture, CRIDA, Hyderabad. p. 25.

Prabhakar, B.S. and Shukla, V. (1989). Sequential intercropping systems: a clue to increase crop land use efficiency proc.Ntl.seminar Natural resource management for sustainable agriculture. ISA, p.21. Rai, N and Yadav, D. S. (2005) Advances in Vegetable production. Research Co. Book centre, New Delhi, India.

Raj Narayan. (2009). Air pollution–A threat in vegetable production. International Conference on Horticulture (ICH-2009) Horticulture for Livelihood Security and Economic Growth, 158-159.

Rao, J.V., Srinivas, K. and Siubramainian, T.R. (1983). Interropping legume vegetable with cereals in dry lands Indian J.agric. Sci. 53; p. 889-892.

Rajiv, K.R. (2014). Kerala farmers warm up to cool season.:/timesofindia.indiatimes..com.

Ratna Reddy, V., M. Gopinath Reddy, S. Galab, John Soussan and Oliver Springate-Baginski. (2004). Participatory Watershed Development in India: Can it Sustain Rural Livelihoods? Development and Change, Institute of Social Studies The Hague Edited By: Ashwani Saith, Murat Arsel, Kees Biekart, Amrita Chhachhi andrew M. Fischer, Bridget O'Laughlin, Maha Abdelrahman, Kate Meagher and Servaas Storm.

Revi, C. A. (2008). Climate change risk: an adaptation and mitigation agenda for Indian cities. Environment and Urbanization 20(1):207-229.

Rosegrant, M. And S. Clime. (2003). Global food security challenges and policies. Farm Education and Extension Mission, 2004.

Sahasrabuddhe, Makarand. (2012). Why does India have such low agricultural productivity? Quora 31 May 2012.

Sato Peet, M.M. and Thomas, J.F. (2002). Determining critical pre- and post-anthesis periods and physiological process in Lycopersicon esculentum Mill. exposed to moderately elevated temperatures. Journal of Experimental Botany 53: 1187-1195. Cited from Datta,S. (2013) Impact of Climate Change in Indian Horticulture: A review. International Journal of Science and Technology, 2 (4):p 661– 671 Swamy,K.R.M. and Prem Nath.(2012) Messages from international horticultural conferences and symposia in the new millennium. The basics of human civilisation – Food, Agriculture and Humanity – Present Scenario.Premnath (Ed) Dr. Prem Nath Agricultural Science Foundation (PNASF), Bengaluru. New India Publishing Agency, New Delhi, p. 157 – 173.gricultural science Foundation.(PNASF), Bengaluru. New India publishing Agency, New Delhi. p. 199 – 224. Times of India (2013) Farmer population falls by 9 million in 10 years - Times of India May 1, 2013 timesofindia.indiatimes.com.

Vanitha, S.M., Chaurasia, S.N.S., Singh P.M. and Prakash, S. Naik. (2013). Vegetable Statistics. Technical Bulletin No. 51, IIVR, Varanasi, p. 250.

Kaul, Vivek. (2015). Land Acquisition: India has enough land for agriculture. The Daily Reckoning. 26 Feb. 2015.

Weinberger, Katinka and Thomas A. Lumpkin. (2007). Diversification into Horticulture and Poverty Reduction: A Research Agenda World Development, 35, (8) August 2007, p 1464–1480 Zickfeld, K *et al*. (2005), "Is the Indian summer monsoon stable against global change?" Geophysical Research Letters 32, L 15707. (Cited from www.What is Climate Change? - Climate Action Network Canada.

❍❍❍

31

Extension Education for Promotion of Food and Nutrition on 'Antyodaya' Model

C. Prasad

1. PREAMBLE

The basic purpose of agriculture is sustaining the life and living of humans, animals and other living genera over the world. Food and Nutrition are the ultimate products on which we live and survive. So its scope and importance is unquestionable. There are several dozens of disciplines and sub-subjects comprising agriculture, but all of them lead to "food & nutrition", so dear to Dr. Prem Nath – a horticultural scientist of international eminence. Even at his fag end of life, he is so devoted to food and nutrition – the foundation of all human and animals – in fact, all sources of life and living. Great credit goes to Dr. Prem Nath, who inspires all humans and works on their livelihoods and sustenance. I am proud of him being one of his classmates at the graduation level as well as we studied at the highest academic levels in the USA.

If you look at the international scenarios, 139 developing nations are in the Globe, where the measure weakness is relatively poor agriculture and therefore, the shortage of food and proper balanced nutrition. Again, these ills are due to human prolific weaknesses and that is, we do not work hard to make the living rich and better. Most of the Countries learn from "Germans" how to make our good living by hard-work – they are known inspirations for all the nations – developed or developing including America – the USA. Since many Countries rushed to this new found land called America, the Germans who worked so hard for developing agriculture – an inspiration for others who came from different Countries there. So it is not only agriculture – food and nutrition, but the real good habits of people who make you rich or poor. Agriculture will change and contribute to the health of people who are alive to the reality and devoted workers in their respective areas – agriculture or non-agriculture.

India is a big country comprising 121 crores population (2011 Census) with a decadal growth of 17.64 per cent. The male population being 62.37 crores and female 58.64 crores. The literacy is 74 per cent: male 82.10 and female 58.69 per cent; the male female ratio is 1000/940. There are 29 major States, 7 Union Territories, 640 Districts, 5767 Tehsils, 7933 towns and more than 6.0 lakh villages. There are 18 languages and 8 different religions: (i) Hinduism – 84.1 per cent, (ii) Islam – 9.8 per cent, (iii) Christianity – 2.3 per cent, (iv) Sikhism – 1.79 per cent, (v) Buddhism – 0.74 per cent, (vi) Jainism – 0.46 per cent, (vii) Zoroastrianism – 0.13 per cent, (viii) other religions – 0.43 per cent. India has 2.2 per cent surface of the World and 17.5 per cent of the population. All these factors influence food and nutrition directly or indirectly.

Mahatma Gandhi after independence (1947) promulgated the concept of development as Antyodaya – approaching the poorest of the poor first! On the same line, Shri S.K. Dey duly supported by Pandit Jawaharlal Nehru, the then Hon'ble Prime Minister of India, advocated to follow the 'Socialistic' pattern of development *vis-à-vis* the 'materialistic' approach of the West – the developed nations. But unfortunately, we ignored them and followed the materialistic mode, which made rich people richer, the poor people struggling economically and socially and the power loving people more powerful. And this trend continues till date, can you beat it? The poverty continues unabated. We must change our development strategy drastically for the better in the tune and philosophy of Mahatma Gandhi. Unless we adopt this extension approach, we are bound to maintain the big gulf between have and have not's!

Let us look at the present scenario after nearly 70 years of Independence, where are we? The Census of 2011 points out that still 60 per cent rural households and 35 per cent of the urban households are poor. This is a startling different than what we have been preaching – today only about 30 per cent poor in the Country. The total households of the Country have the Scheduled Castes (SCs) – 19.59 per cent, Scheduled Tribes (STs) – 8.63 per cent, Other Backward Classes (OBCs) – 40.94 per cent and landless agricultural labourers (56% rural household) – all told almost 90 per cent of the population. This also includes 1,80,657 manual scavengers. These poor groups are struggling for their livelihoods – the big & basic question of *Roti, Kapda* and *Makan!* Obviously they got since past only the flip side extension services and, therefore, we are repeating the same figures of achievement in a round about manner – not clean and clear, for we are guilty of not bringing the real facts to fore – their hand to mouth and precarious living.

By now, we have enough of technologies – farm and non-farm, but they are not reaching where they are needed. We must change our development strategies. For coming 15-20 years the Country's 60 per cent resources should go for the poor and weaker. The rich people have enough to sustain themselves and in fact they can also contribute for the poorer sections of the society. This requires a very strong and committed Government.

Added to that, our field extension efforts, where the interrelationships between extension workers/professionals and the subject-matter scientists must be ensured and continuity maintained. Unfortunately, the extension education and the trainings in the rural set-up *vis-à-vis* working in the laboratories and the air-conditioned labs, are two different arenas – the farmers and farming being harsh and hard to the scientists. Thus their basic interest in the extension work and activities are affected; they naturally wished to be posted in the towns and cities. To overcome this natural barrier to extension work and activities, it would be essential to develop a CADRE of Extension Professionals and the Subject-Matter Specialists (Agronomist) – a lifelong process where some rural special allowances need to be given to them. Thus, this becomes their life-time profession and so the natural attachment to the extension system and areas. Cadre ensures the promotion and allied privileges of the good workers at the right time and letting not the extension professionals run away from the field works and the natural hardship. This is not only necessary, rather imperative for the successful work and transfer of technology (TOT) on a sustained basis – a permanent cure!

2. CHILDHOOD MALNUTRITION

The Asian Development Bank (ADB) and UNICEF had taken an initiative to establish a Regional Technical Assistance (RETA) programme to undertake an in-depth situation analysis in eight Asian Countries aimed at reducing Childhood Malnutrition based on experiences in the region. It was aimed at creating awareness among the *Policy Makers* that the child nutrition was an important policy issue which needed at most attention. In India, this exercise was undertaken under the stewardship of the Department of Women and Child Development, Ministry of Human Resource Development, Government of India and the Administrative Staff College of India, Hyderabad (1997). The studies, broadly speaking, recommended more of food-based nutrition interventions. They also recommended to encourage for non-food determinants of childhood malnutrition.

Despite considerable improvement in the quality of life and health status, the levels of child malnutrition in India was still high – more than a half of the young children showed moderate to severe degrees of malnutrition. Chromic energy deficiency in similar proportion of adults suggests perpetuation of childhood malnutrition into adulthood. High prevalence of nutritional anaemia (57%), affecting cognitive development is another major cause for concern. Protein energy malnutrition and anaemia adversely affect the productivity and consequently reduce the national GDP. Other deficiencies of micro-nutrients, such as, Vitamin A and Iodine are also important public health problems in India. About 0.7 per cent pre-school children exhibited frank deficiency signs of Vitamin A which is incidentally one of the most limiting nutrients in Indian diet. Out of 200 million people residing in Goitre endemic areas of India, over 50 million have Goitre and about nine million suffer from different degrees of mental and motor disabilities.

3. FOOD AND NUTRITION FOR WEAK AND POOR

The developed Countries are further developing in all respects; the developing nations are also hastening their progress socially, economically and politically, but the weaker and poor, almost 40 per cent, have been mostly neglected and somehow progressing at the snail scale. Often the Governments have shown their interest in the poor and their policy on papers have shown concerns, have also ear-marked resources, but unfortunately, the so called resources do not reach them and they do not have this basic resources to build on. Their voices are unheard unfortunately.

The International Food Policy Research Institute (IFRPI, USA) has been concerned about this state of affairs and they have several studies reflecting on the Food & Nutrition especially for the weaker and poor. In 2006, they observed 'some 800 million Asians still live in poverty and 90 million preschool Children remain malnourished'. Even today (2015), they are no better. The first edition of the *Global Nutrition Report (1914),* puts a spot light on worldwide progress by the 193 Member Countries of the U.N. The Report observed: *'Malnutrition is truly a global challenge; of the 122 Countries studied, experience at least one of the three forms of malnutrition – under 5 stunting, anemia among women of reproductive age and most Countries experience multiple forms of malnutrition'.*

Most of the developing nations, since the Green Revolution in 1960s, are in a position to feed their population except for the lopsided policies and development. The resource ear-marked for the weaker and poor do not reach them. A time has come when over

40 per cent weak & poor should have at least 60 per cent resources of the Governments for 15-20 years to bring them above the poverty-line. All we need to create a holistic and committed extension education system. It has not been easy, but otherwise there is no other mechanisms. 'In spite of various nutrition intervention programmes being operational in India, it still accounts for 40 per cent of the world's malnourished children and nearly 60 per cent of Indian women are anemic. It has been realized that *agricultural interventions* are more beneficial and sustainable in addressing malnutrition among the rural poor than the direct nutritional intervention programmes' (Mruthyunjaya, 2003).

Incorporating nutritional objectives in agriculture will be a win-win situation for the poverty alleviation, income generation, nutritional security and ultimately achieving the goals of India's agricultural development. In this context, this study has thrown light on the issues of agricultural production systems *viz.* rice, vegetables and sugarcane influence on nutritional security of rural households in those systems. The study also gives a view about scope for inter-institutional linkages among developmental departments to promote nutritional security.

The study further depicts the potential of vegetable systems to supply healthy food especially rich in vitamins and micronutrients to the weaker sections of the society and expose the weaknesses of non-vegetable production systems. Agricultural based kind of wages and women income management in the households seem to be favourable for enhanced household nutritional security. Location specific agriculture-based interventions would be necessary to safe guard regional nutritional security balance. The existing weaknesses in institutional linkages amongst development departments suggest immediate reorientation in their approaches. The study underlines the need for redefining of agricultural extension to address nutritional security concerns.

4. HORTICULTURAL CROPS AND NUTRITIONCE

Human health and agriculture are synonymous. One cannot visualize good health without nutritious food and balanced diets. Varied foods provide varying amounts of food energy, proteins, fats, vitamins and minerals. Foods are grouped into the following heads: beverages, dairy products, eggs, fats and oils, fish and other sea foods, vegetables and vegetable products, fruits and fruit products, grain products, legumes, nuts and seeds, meat and meat products, mixed dishes and fast foods, poultry and poultry products, soups, sauces and gravies, sugars and sweets and miscellaneous foods. For promoting vitamin A rich foods, one must select food items which are rich in it, such as, lettuce, mustard greens, pumpkin, mango, melons, milk, liver. Some facts about foods rich in vitamin A are given in Table 1 and 2. Grains (cereals) and legumes, by and large, do not have vitamin A unless their food products are fortified e.g. bread (dry) 910 IU or 273 RE[1], corn flakes 1250 IU or 375 RE[2].

There are differences in programmes of production and promotion of foods between the developed and the developing Countries. Affluent population of industrialized Countries are scientific, selective and balanced in their dietary intakes and food habits. But in most of the developing nations, people are striving hard to get two meals a day, except for a handful of people who are rich and who can afford all the luxuries of the

1 IU - International units of vitamin A; RE[2] – Retinol equivalent

West. Under such circumstances, promoting specific food items for production as well as consumption becomes not only arduous, but at times impossible. People are conditioned by circumstances. Their first effort is to fill their bellies somehow; quality and selectivity in food, normally would follow suit after this basic need is fully met.

If one looks at the growth in all the sectors of agriculture, the development programmes in India have also been lopsided. The Green Revolution in the Country has been brought about by producing more cereals, especially wheat and rice. Horticultural crops, livestock production, fisheries, plantation crops, hill agriculture etc. could not get the same attention. This development also followed the traditional food habit of the nation, which by and large is cereal based. In lighter vein, one rightly say that Indians eat vegetables or meat for eating more cereals – the staple food, may that be rice, wheat, maize, pearl, millet or barley. One may also say that, if balanced food habits and integrated agricultural production programmes were followed in the Country, self-sufficiency in food could have been reached long before the Green revolution with salutary effects on health of the people – nutritionally rich.

Research in agriculture, till lately, also insisted on increasing productivity per se, rather than productivity and quality both. The researchers now have started focusing on qualitative aspects of cereals and other foods. Integrated agriculture is being emphasized both in research and development including agro-forestry. Further, from the research in cropping sequence, we have started shifting to the *'farming system research and development'*. It is high-time that we emphasize in all our research, education and development programmes, the qualitative aspects of food production and consumption. Foods rich in *vitamin-A* is only one such area.

The important sources of vitamins A are of animal origin viz. liver, egg yolk, butter, cheese, milk, fish *etc.,* but they are relatively expensive and are not consumed in sufficient quantities in developing Countries like India. The major plant sources are *fresh dark green leafy vegetables*, such as, spinach (palak), amaranth, methi, cabbage *etc.*; *deep yellow and orange coloured vegetables and fruits* viz. carrot, pumpkin, mango, papaya. Plant foods contain precursors of vitamin A – carotenoids of which β-carotene has the highest vitamin activity. A daily intake of 100 g of leafy vegetables, which is well within the means of common people, is a simple and practical way of meeting the vitamin-A requirements.

The production figure of fruits during 1990-91 was estimated at 27.83 million tons and that of vegetables about 70.00 million tons including potato and onion. Taking the rate of post-harvest loss to be 30%, the net availability during 1990-91 was 19.48 million tons of fruits and 49.00 million tons of vegetables. Thus, with a population of 844 million in 1990-91, there was a short fall of about 6.70 million tons of fruits and 28.02 million tons of vegetables, considering the fact that the minimum daily *per capita* requirement is 85 g of fruits and 250 g of vegetables.

Strategies for promoting vitamin-A rich foods are: (i) Demonstrations and education, (ii) Special training programmes, (iii) Research on food production and consumption, (iv) In-service training and (v) use of Mass media on vitamin-A rich foods.

Longback Runyan had said, "The Twentieth Century has seen an impressive amount of growth and development in the young science of nutrition... Many of the most difficult problems in the past lay in the areas of science and technology. In contrast, most of the obstacles which now must be overcome before the gap can be closed between the actual and the ideal are not primarily scientific and technical. Rather, they are social, political, economic, educational and administrative problems, many of which urgently require international co-operation of a scope never previously known in the history of humanity". The situation is no better today.

After we have nearly reached self-sufficiency in foods, basically in terms of cereals, the time is ripe to promote quality in food, which should get reflected in the research, education, training and extension programmes and planning. The nutritional aspect of agricultural production is so vital that it should find a mention in the agricultural policy document.

There are many ways to promote production and consumption of nutritious foods including vitamin-A rich foods. As a policy matter, all community lands including Wastelands and Social Forestry should be utilized for cultivation of nutritious crops and plants with community help and participation. These sites could also be utilized for demonstrations and training purposes. Thus, a social nutrition consciousness will emerge together with environmental improvement.

"The world is entering a period when the relationship between food availability in low-income Countries and rising population is increasingly perilous and, for some Countries has already reached the crisis proportions. The technology symbolized by the term 'Green Revolution' brought hope and sharply increased cereal yields of several of the most populous under-developed Countries, but it has become clear that without an integrated nutrition and population policy, it will not, in its present form, provide more than temporary relief", said Scrimshaw.

Table 1. *Nutritive Value of Some Foodstuffs*

Foods	Approx. quantity per serving (g)	Energy (Kcal.)	Protein (g)	Fat (g)	Vitamin A	
					IU	RE
Dairy Products and Eggs						
Milk(1 cup)	246	430	28	32	1210	330
Cheese,	11	445	28	37	1200	342
Ice-cream (1 cup)	133	270	5	14	540	133
Yoghurt (1 cup)	227	145	12	4	150	36
Butter (1 spoon)	14	100	Tr.	11	430	106
Egg (One)	50	80	6	6	260	78
Fruits						
Apple	138	80	Tr.	Tr.	70	7
Banana	114	105	1	1	90	9
Grape	50	35	Tr.	Tr.	40	4
Lemon	58	15	1	Tr.	20	2
Mango	207	135	1	1	8060	806
Melons	267	95	2	1	8610	161

Orange	131	60	1	Tr.	270	27
Papaya	140	65	1	Tr.	400	40
Vegetables						
Beans	170	170	10	1	320	32
Carrots	72	30	1	Tr.	20250	2025
Lettuce	163	20	2	Tr.	1580	158
Mustard greens	140	20	1	Tr.	4240	424
Peas greens	170	115	8	1	1310	131
Pumpkin	245	50	2	Tr.	2650	265
Spinach	136	220	11	18	3460	675
Tomatoes	123	25	1	Tr.	1390	139
Meat and Fish Products						
Beef	85	325	22	26	Tr.	Tr.
Beef liver	85	185	23	7	30690	9120
Pork	19	110	6	9	0	0
Chicken	140	365	35	18	90	28
Chicken liver	20	30	5	1	3270	983
Fish	85	65	11	1	90	26
Oysters	240	160	20	4	740	223

Table 2. *Carotene Content of Some Fruits and Vegetables*

Fruits	Carotene (µg)	Vegetables	Carotene (µg)	Leafy Vegetables	Carotene (µg)
Apricot (fresh)	2160	Carrot	1890	Amaranth, tender	5520
Banana, ripe	78	Yam	260	Beetgreens	5862
Cape gooseberry	1428	Bittergourd	126	Carrot leaves	5700
Jackfruit	175	Celery stalk	520	Celery leaves	3990
Mango ripe	2743	Cluster bean	198	Colocasia leaves	10278
Orange	1104	Cowpea pods	564	Coriander leaves	6918
Papaya ripe	666	Drumstick	110	Curry leaves	7560
Passion fruit juice	1968	Field beans	187	Drumstick leaves	6780
Parsimmon	2368	Capsicum	427	Fenugreek leaves	2340
Rasberry	1248	Onion stalks	595	Mustard leaves	2622
Sapota	97	Parwal	153	Raddish leaves	5295
Tomato ripe	391			Spinach	5580

Next to cereal crops, horticulture is a very potential discipline covering fruits, vegetables and flowers as the major sub-areas; their nutritional and economic importance for the Country is well known. They are grown in almost all kinds of climate and condition. Besides, there are *under-utilized crops especially vegetables* which are a potential source of nutrition and more so for the poorer sections of the society. They are grown almost in all kinds of farm or forest conditions. And for the poor, they are the boon; under-utilized vegetables can be grown in the vast areas of uncultivable waste lands, undulating uplands, gullied ravines, water logged, marshy areas, degraded postures, sandy and saline soils as explained by Chakrabarti, Munshian & Kumar (2008) in Social Science Perspectives in Agriculture by Prasad *et al.* Their paper entitled, *'under – utilized vegetables – a potential source of nutrition'* gives a vivid account and scope of such crops in all sorts of agro-climatic conditions of the Country suiting the poorer sections in the society – marginal, small, SCs, STs, landless labourers, *etc.* (Table 3) the nutritional composition of under-utilized and rare vegetables per 100g edible portion.

5. SOME BASIC NUTRITIONAL RULES

Rujuta Diwekar, an outstanding Food and Nutrition Expert, in her book – *Don't Lose Your Mind Lose Your Weight*, gives some practical guidance worth noting. Her recommends are:

- Body is made of two major parts – fat weight of bones and the muscles.
- Eat properly and do regular exercise; no extreme dieting; everything herbal, the better.
- Don't take chips, if at all use sparingly made at home.
- Five nutrients for the body – carbohydrate, protein, fat, vitamins, minerals plus water.
- Carry your salad materials – tomato & cucumber etc. in your pocket, if eating outside, for normal salad take away the skin parts of vegetables & fruits – they have minerals & vitamins; only wash them.
- Eat fresh food before three hours of cooking; don't use *fridge* for storing them.
- Peanuts have good fatty acids; improves/increases HDL – good cholesterol, protects heart, protein, vitamin B-3, folic acid, fibre, amino acid, improves skin and digestion & give lasting energy.
- Eat only low fat; avoid Paneer Parathas.
- No Pizza please.
- Eat coconut chutney – very rich.
- Take full meal between 7.00 AM to 10.00 AM.
- Carbohydrates give energy for day-to-day working and many processes in the body – available in all plant foods (not in meat); carbohydrate: (i) Glycemic Index (GI) & (ii) Glycemic Load (GL) – GI increases blood, glucose and insulin levels.
- In absence of carbohydrates, fat cannot be utilized for energy.
- Pulses and Vegetables add fibre are important.
- In most of our foods – carbohydrates may not be processed in order to retain their fibre, vitamin-B and minerals.
- *Tadka* (oil+mustard+Jeera) add fibre, vitamins, minerals and essential fatty acids.
- Cut low fibre – biscuits, cakes, pastries, pizza, cold drinks etc. are no good; Sweets, deserts high on the GI (Glycemic Index) – Sugar eat once in a while and not with the meals – *high GI.*

- *Macro* nutrients – carbohydrates, protein & fat; *micro* – vitamins & minerals,
- Take carbohydrates after exercise, any stressful activity, hard work/activities.
- Proteins – primary build up and repair our body.

Table 4. *Important Issues – Health of Mother and Child – Problems & Protection – the Top Ten*

S.No.	Issues/Problems	Protection/Cure
1.	Spacing pregnancy Ideal Pregnancy	At least two years a part Two Children After 18 years
2.	Pregnant women	Regular Check-ups Prenatal care by health worker Birth to be assisted by a trained person
3.	First Month of Children is very delicate	Save them from infections Five cleans during delivery Keep baby warm & on breast feeding
4.	First 4-5 Months	Breast Feeding Need some food (meshed vegetables, little fat/oil)
5.	Diarrhoea Common	Give plenty of liquids – dal, water, rice water, butter, milk etc. Help of health worker needed
6.	All Immunizations in fist year (Pregnant) – Tetanus Toxoid	For measles, tetanus, whooping cough, Polio
7.	Cough & Cold + breathing	Take to health Centre – give Plenty of liquids.
8.	Many illness by Germs etc.	Wash hands with soap frequently Give boiled water Extra meal everyday
9.	Family planning services	Visit Health Centre regularly
10.	During pregnancy	More quality food

Five Cleans: (i) clean hands of birth attendant, (ii) delivery on clean bed/surface, (iii) tie the cord by clean thread, (iv) new blade for cutting the cord, & (v) handle child by one or two persons.

Table 5. *Immunization Schedules*

S.No.	Issues/Problems	Protection/Cure
1.	One month of Pregnant Women	TT-1 injection TT-2/TT booster injection Children
2.	For infant	
	1 ½ months	BCG injection DPT-1 injection OPV-1 (dose)

	2 ½ months	DPT-2 injection OPV-2 dose
	3 ½ months	DPT-3 injection OPV-3 injection
3.	If Infant delivered in a hospital/clinic	BCG injection at Birth Zero dose of Polio vaccine

First Prime Messages: (i) Keep Baby warm, (ii) Prevent infections, (iii) help in breathing, (iv) if danger signals, rush to the hospital & (v) breast feeding.

6. FIVE PILLARS OF DEVELOPMENT

The USA system of research and development introduced triple functions of development in 1914 – Teaching, Research and Extension, the integrated functions for agriculture R&D. Prasad C. advocates five pillars of development: (i) Teaching, (ii) Research, (iii) Extension, (iv) Training and (v) Management, they must be employed in an integrated manner for reaching the farmers most effectively (Figure 1). Here it may be reminded that initially the USA genre, who came from different Countries, developed the habits of working together – sincerely, cooperatively and wholeheartedly – these included the training and retraining as well as the management practices and principles. The Americans in their humbleness talked loudly about the triple functions, perhaps education/teaching included training and the hard-work became the part of management. But that is becoming costly for the developing nations, for without insisting on the training and management functions which have become so important today. Therefore, the five pillars of development must be utilized to achieve the hundred *per cent* development. This also be noted that all these five pillars are social sciences applied to agriculture, medical, health, industries etc. of-course the research works both ways – either it could be social science researches or researches in the basic or the natural sciences. These understandings will help develop better perspectives of appropriate use of these two broad areas – social science or natural/ basic sciences.

Fig. 1. *The Five Pillars of R&D Promotion for Food & Nutrition for all*

Let us see the measurement of our progress by the annual GDP – the Gross Domestic Product, which does not include any social sciences components, whereas, a tiny Country like Bhutan measures its progress by Gross National Happiness (GNH), which includes the measures of social inputs like happiness and more. This important understanding is necessary to look after the social angles of the society, not only the material angles! We need to change our measurement of progress.

7. EXTENSION SCHEMES WHERE NUTRITION IS FOCUSED

"Nutrition" should be a forerunner in every agricultural R&D; that is what we ultimately want to achieve – healthy mind in a healthy body! The thinking of Dr. Prem Nath Agricultural Science Foundation, to put nutrition on the fore-joint in agriculture is an innovation, for that is the ultimate goal of our life and living. In this respect, the Home Science Colleges – home science education came as a gift of the USA, especially by Mrs. Alice M. Fisher of USDA and Dr. Elda Roble of University of Tennessee, USA in 1961, who started the Home Science College in Pantnagar, U.P., India first. Thus almost all of the Agricultural Universities (70), have Home Science Colleges by design of the Government – all have very strong food & nutrition division devoted to teaching, research, extension, training and management.

In India, Dr. Rajamal Devdas started the Home Sciences Education from school to college level (1959) in Coimbatore, Tamil Nadu and ultimately raising the Home Science education to the level of a Home Science University – an extremely successful venture.

The genesis of Home Science Education and colleges came from the USA as mentioned earlier especially when *Ellen Swallow Richards,* the first women who attended the Massachusetts Institute of Technology (USA) and who became the first female Instructor. She was the student of Chemistry, she talked about water quality which was related to the domestic situations. At World's Fair in Chicago (1893), she designed the Rumford Kitchen – a tiny kitchen which served nutritious food to thousands of fair goers plus they were given a healthy dose of nutrition education. On this initial thought and initiative, she was invited to the famous Women's Building – a Leadership Honour in the USA.

Richards further promoted how the domestic science could improve the quality of life of families. Thereafter, Melvil Deway of New York joined hands with them. After over 10 years of hard-work, this discipline saw the light of the day as *Home Economics* – the Official terms used in 1899. Then Richards founded the American Home Economics Association in 1909. In the very next year, by Government legislation, this subject become the research and teaching part of the higher institutions. Thus it became a movement to train women to become more scientific and efficient as the household managers and home-makers where food and nutrition was a prominent subject.

This subject gradually spread to European Countries during 1940; the Iowa State College, later the University, started Master in Home Science degree. This subject thus travelled from Schools to Colleges to Universities. Besides home related all sciences and disciplines, food and nutrition was a basic and important subject everywhere. In India, besides home sciences colleges in the Agricultural Universities, the College of Home Science at Baroda, Gujarat State and the similar College at Delhi contributed a lot by way of promoting home science discipline.

The USA agriculture and rural development system is based on three basic programmes: agriculture, Home Science and farm-youth development. These triple functions were never stopped, rather they have basic source of both rural as well as urban development – an extension design adopted by many developing nations.

8. INTEGRATED CHILD DEVELOPMENT SERVICES (ICDS)

The Ministry of Women & Child Development, promotes ICDS Scheme for the Children upto six years and also the pregnant and lactating mothers. This was launched in 1975. Under this scheme, the Government's welfare programme, provides food, pre-school education and primary health care for under six years of age group and their mother. A study in 2006, indicated that this programme was not effective in reducing the malnutrition, because the poorer States did not get enough Central support. These Services are provided through Anganwadi Centres established mostly in the rural areas. During 2012-13, the Central Government spent Rs. 159 billion (US $2.5 billion) on ICDS activities. The focus of the programme has been on the weaker sections of the society.

The services are to attain the following objects: (i) immunization, (ii) supplementary nutrition, (iii) health check-up, (iv) referral services, (v) formal-non-formal education and (vi) nutrition and health information. It is reported that by 2010, they had reached to 80.6 lakh expectant and lactating mothers and 3.93 crore children. The programme runs in 1,241,749 operational Anganwadi centres. Madam Mina Swaminathan, an expert on Children education, health and allied services has been contributing a lot in strengthen this movement.

The NABARD, State and Central Governments and international agencies, such as, UNICEF and World Bank are supporting this programme.

9. ATARIs AND KVKs ON VOCATIONAL TRAINING

At the instance of the National Commission on Education (1964-65), the ICAR agreed to establish Krishi Vigyan Kendras (KVKs) in the Country for organizing vocational trainings for the young farmers and farm women. Earlier there were eight zones in the Country which served as the Zonal Coordination for the KVKs being established in each District – starting with one, as per policy of the ICAR and as recommended by the National Commission on Agriculture (1974-75) and ultimately three KVKs per District in view of larger Districts and heavily populated. The Country has over 600 Districts and till date we have 664 KVKs. Each KVK has a Home Scientist who organizes vocational trainings including food & nutrition. In view of the importance being given to food, nutrition and health in the Country, the scope for trainings on food & nutrition should be enlarged.

Lately, the Zonal Coordination Centres have been changed to Agricultural Technology Application Research Institute (ATARI) where, besides vocational training applied researches are also being promoted. For food and nutrition, the KVKs may contribute immensely in training the trainers for those who promote education at the lowest levels for health and nutritional services for children, pregnant women and mothers.

Table 3. *Nutrient Composition of Under-Utilized and Rare Vegetables. Per 100 g Edible Portion*

Name of the vegetable	Calories (k cal)	Moisture content (g)	Carbo-hydrates (g)	Protein (g)	Fat (g)	0-carotono (Ng)	Thiamine (mg)	Riboflavin (mg)	Vitamin C (mg)	Calcium (mg)	Iron (mg)
Amaranthus tender	45	85.7	6.1	4.0	0.5	5520	0.03	0.30	99	395	25.5
Basela	32	90.8	4.2	2.8	0.4	7366	0.03	0.16	87	200	10.0
Bathua leaves	30	89.6	2.9	3.7	0.4	1740	0.01	0.14	35	150	4.2
Brussels sprout	45	85.2	8.3	4.9	0.4	330	0.10	0.16	102	36	1.5
Broad bean	48	85.4	7.2	4.5	0.1	9	0.08	-	12	50	1.4
Cassava (Tapoica)	157	59.4	38.1	0.7	0.2	0	0.05	0.10	25	50	0.9
Celery	17	94.1	3.9	0.9	0.1	3990	0.03	0.03	9	39	0.3
Chicory leaf	13	92.0	1.1	1.7	0.3	2400	0.06	0.10	24	100	0.9
Chinese cabbage	14	95.0	3.0	1.2	0.1	90	0.1	0.04	25	43	0.6
Cho-cho (immature fruit)	27	92.5	5.7	0.7	0.1	30	0.00	0.04	4	140	0.6
Cluster bean	16	81.0	10.8	3.2	0.4	198	0.09	0.03	49	130	4.5
Colocasia stems	18	94.0	3.6	0.3	0.3	104	0.07	0.07	3	60	0.5
Coriander leaves	44	86.3	6.3	3.3	0.6	6918	0.05	0.06	135	184	18.5
Colocasia	97	73.1	21.1	3.1	0.1	24	0.09	0.03	0	40	1.7
Colocassia leaves	56	82.7	6.8	3.9	1.5	10278	0.22	0.26	12	227	10.0
Dolichos bean	48	86.1	6.7	3.8	0.7	187	0.10	0.06	9	210	1.7
Drumstick	26	86.9	3.7	2.5	0.1	110	0.05	0.07	120	30	5.3
Drumstick leaves	92	75.9	12.5	6.7	1.7	6780	0.06	0.05	220	440	7.0
Elephant-foot yam	79	78.7	18.4	1.2	0.1	258	0.06	0.07	0	50	0.6
Endive	11	95.0	1.2	1.3	0.2	1500	0.07	0.08	8	42	2.0
Hyacinth bean	48	86.1	6.7	3.8	0.7	185	0.10	0.06	9	210	1.7

Name of the vegetable	Calories (k cal)	Moisture content (g)	Carbo-hydrates (g)	Protein (g)	Fat (g)	0-carotono (Ng)	Thiamine (mg)	Riboflavin (mg)	Vitamin C (mg)	Calcium (mg)	Iron (mg)
Ivy gourd	18	93.5	3.1	1.2	0.1	155	0.07	0.08	15	40	1.4
Kakrol	52	84.1	7.7	3.1	1.0	1592	0.05	0.18	N.A.	33	4.6
Kale	53	82.7	9.0	6.0	0.8	6000	0.16	0.26	186	249	2.9
Korai (Coccinia sp.)	18	93.5	3.1	1.2	0.1	156	0.07	0.08	15	40	1.4
Leek	77	78.9	17.2	1.8	0.1	18	0.23	-	11	50	2.3
Lettuce	21	93.4	2.5	2.1	0.3	990	0.09	0.13	10	50	2.4
Lotus root	53	85.9	11.3	1.7	0.1	-	0.1	-	22	21	0.4
Parsley	16	90.0	1.3	2.2	0.3	3120	0.08	0.11	90	125	2.0
Parsnip	53	81.0	11.6	1.2	0.3	0	0.09	0.05	17	40	0.7
Paruppu keerai (Portulaca)	27	90.5	2.7	2.4	0.6	2292	0.10	0.22	29	111	14.8
Sprouting broccoli	32	89.1	5.9	3.6	0.3	1500	0.10	0.23	113	103	1.1
Taro	97	73.1	21.1	3.0	0.1	24	0.09	0.03	0	40	1.7
Taro leaf stalks	24	93.0	6.0	0.5	0.2	357	0.02	0.04	13	49	0.9
Winged bean	25	92.0	4.0	2.1	0.3	-	-	-	-	-	-
Yam	102	74.0	24.0	1.5	0.2	15	0.1	0.01	15	12	0.8

REFERENCES

Adhiguru, P. & Ramasamy, C. (2003). Agricultural Based Interventions for Sustainable Nutritional Security Policy Paper, 17, National Centre for Agricultural Economics and Policy Research, New Delhi, 2003.

Chakrbortim A.K. Munshi, A.D. & Kaur, Ravinder. (2008). Under-Utilized Vegetables – a Potential Source of Nutrition, in Social Science Perspectives in Agriculture – A thrust for Integration, VARDAN Trust, New Delhi, India and IFPRI, Washington, D.C., USA, 2008.

Diwekar. Rujuta. (2009). Don't Lose Your Mind Lose Your Weight, Random House Group Limited, London SW1V 2SA, UK. 2009.

Final Report. (1997). National Strategy to Reduce Childhood Malnutrtion, RETA initiative by ADB & UNICEF, Department of Women & Child Development, Government of India & administrative Staff College, Hyderabad, India, 1997.

Gohardt, S.E. and Mathews, R.H. (1989). Nutritive value of Foods. Washington State University Co-operative Extension and USDA, College of Agriculture and Home Economics, Pullman, USA, 1989.

International Food Policy Research Institute (IFPRI). (2006). Asia, 2006, New Delhi & IFPRI, Washington D.C., USA

International Food Policy Research Institute (IFPRI). (2014). Global Nutrition Report, 2014, Washington, D.C., USA.

Prasad, C. & Babu, Suresh. (2008). Social Science Perspectives in Agriculture – A thrust for Integration, VARDAN Trust, New Delhi, India and IFPRI, Washington, D.C., USA, 2008.

Prasad, C. & Lal, Chhotey. (2016). Dynamics of Agricultural Education and Training, Studium Press LLC, Houstan, Texas, USA, 2016

Prasad, C. (1994). Promoting Production and Consumption of Vitamin A Rich Foods – Role of Krishi Vigyan Kendras, Vitamin A from Plants Foods, Dietary Promotion of Vitamin A Rich Foods through Agriculture, Horticulture and Social/Farm Forestry – A Report, Institute of Home Economics, University of Delhi, 1994.

Prasad, C. (2015). A Holistic A-Z Extension Strategies for Promoting Exportable Rice Varieties, International Conference on Extension-Research Interface Promoting Exportable Rice Varieties and Evolving a Sustainable Development Model, New Delhi, 2015.

Runyan Thora, J. (1976). Nutrition For Today, Harper and Row Publishers, New York, USA, 1976.

Scrimshaw, Nevin. (1974). Protein and Nutrition Policy in Low-income Countries by Francis Aylward and Mogens Jul, Charles Knight and Company Ltd., London, UK, 1974.

Websites:

http://indianexpress.com/article/india/india-others/landlessness-high-in-kerala-west-bengal-despite-land-reforms/.

http://timesofindia.indiatimes.com/india/OBCs-form-41-of-population-Survey/articleshow/ 2328117.cms.

IX. Horticulture and Humanity

32. Bondage Between Horticulture and Humanity

32

Bondage Between Horticulture and Humanity

Prem Nath

1. BACKGROUND

Among the natural resources available, such as, human beings, plants, animals, earth, sun, air and water the most important natural resource considered is the human resource and India inhabit one of the two the largest population of humans on earth. For all purposes the human resource has tried to prove as the leader among all natural resources. But it is true that human resource is unable to control birth/life of other natural resource but can manage after birth the natural resources generated by mother nature (Nath, 2016).

To be an efficient manager of all natural resources, the human resource with the nomination of human being qualifies for inheritance of necessary virtues. Here comes the pivotal role of human being to do good to fellow beings without damaging other natural resources. Hence, the humanity understood and defined by different people of different lands and has been sharpened from time to tune through the passage of decades and centuries. Each human being has a heart for sustenance which treasures humanity and two arenas- food and agriculture for growth.

In reality, the human behavior is the composite result of heart, mind, attitude and functions which defines humanity (Nath, 2016).

2. HUMANITY

Humanity is the entire reflection of human kind. To understand better it's the qualification of humans, the reflection is illustrated in behavior and action which is expressed by a person and received by another one. It may be expressed in person or through writing or phone or mass communication. In other words, it is the cause and effect of behavior of a human being. The behavior expressed by one and received by other is felt good or bad depending on the thinking or consideration of other person concerned. This feeling coupled with action designates a person good or bad. Further, we- the humans, designate a person good or bad. For all purposes, the human conception understood most probably by all is good behavior and action by human, hence, humanity has all plus points and anything against good is bad or against humanity (Nath, 2016).

Human Involvement

In addition to above mentioned there are other feelings and actions related to food and agriculture where the combination of two supplement or complement each other resulting in good human action (Nath, 2016).

- Agriculture and human being.
- Agriculture scientists and farmer.
- Economic accessibility and food accessibility.
- Policy maker, developer and farmer.
- Food, agriculture and humanity.
- Farm land and agriculture.
- Agriculture and food.
- Food and health.
- Vegetarian and non-vegetarian.
- Human and agriculture.
- Human and food.
- Diet and health.

On examination of the above the suitable combinations give rise to positive results in the interest of positive outcome.

3. EVIDENCES OF DOMESTICATION AND ASSOCIATION–HORTICULTURE PLANTS AND HUMANS

Scientists at the Royal Botanic Gardens at Kew have estimated that the number of species of plants alive on the earth today is probably in excess of 400,000. Of these it is thought that many more than half of them could be considered edible to humans. It is entirely possible that we could eat an amazing 300,000 plant species. However, the reality is that we only consume a tiny fraction of what is possible. *Homo sapiens*, which is the most cosmopolitan of all Species and one that thrives by virtue of being a supreme generalist, survives, by routinely eating only about 200 plant species (Warren, 2015).

Humans have been domesticating crops for around 10,000 years. But not until 1900 and the rediscovery of Gregor Mendel's experiments with breeding peas, did we have any real understanding of the laws of inheritance. In fact, it was as late as 1676 when plant anatomist, Nehemiah Grew addressed the Royal Society that we started to appreciate that plants actually indulge in sex. In other words, for the vast majority of agricultural history we have not really had much of a clue about what we have been doing. The process of domestication has been one of simply identifying the most desirable or just unusual plants and propagating them. This duplication has occurred either by seed or vegetable means, from a cutting or simply dividing the plant in to two. Although the sexual production of seeds has the potential to generate lots of variation amongst the next generation in comparison to time, vegetative propagations, given enough time, both of these processes are capable of delivering the genetic variation required for crops to become differentiated from wild plants (Warren, 2015).

The second most important family of crop plants has to be the legumes. There are more than 23,500 recorded species related to beans and peas which equates to approximately eight per cent of all known plants. As with the cereals, leguminous crops were independently domesticated by all the early agricultural civilisations. Wherever you go, they have already bean (Warren, 2015)!

The broad bean was first cultivated in the Near East during the Neolithic period and was well known to the ancient Greeks and Romans. They offered bean cakes to their gods and used beans as a form of ballot paper in elections, with black beans signifying a no vote and white beans a yes vote. Broad beans were originally small and black in colour, while more modern varieties have larger, green or white seeds. The writings of the Greek scientist (Warren, 2015).

When trying to understand our ultra-conservatism in the limited selection of plants that we deem fit to cultivate and consume, we find a paradox. Throughout human history, so often we have been willing to rapidly assimilate the new and exotic. In some ways this is the most unusual and un-natural of all aspects of our modern diet. All other species consume the things that they find locally. In contrast, modern humans have transplanted their preferred crops into almost every patch of ground in which they would flourish. While many human immigrants experience prejudice from resident indigenous populations, the same is less likely to be true of alien crops. Thus, today wherever we travel we find familiar fruit and vegetables. Surprisingly many of these crops are the primary ingredients in national dishes and regional speciality foods (Warren, 2015).

Tomato: An obvious example of this is the tomato. It is virtually impossible to conceive of Italian cuisine deprived of the intensely flavoured, blood red flesh of the tomato. But of course, tomatoes are natives of Southern and Central America and therefore, the Roman Emperors whose images adorn so many pizza palaces, could never have experienced the delights of munching on a margarita! In the UK, trying to identify any food crops that are genuinely native is an almost impossible task. The British diet and the British themselves comprise of components assembled from almost every corner of the globe. In many ways our food crops are more diverse than ever before, but at the same time we still limit ourselves to a fraction of what is edible (Warren, 2015).

The exact date of the introduction of the tomato into Europe is uncertain. It seems probable they first arrived in Spain following Cortes' capturing of the Aztec capital city, Tenochtitlan in 1521. Inspite of its toxic associations, within 23 years tomatoes were being described by the Italian botanist Mattioli as being a form of eggplant that can be cooked with salt, black pepper and oil. Even so, the majority of tomatoes grown in Italy at that time seem to have been for ornamental purposes and were used as table decorations. It took the best part of a hundred years before they became a regular component of the Italian diet. However, in comparison with the typically reserved English, this was rapid assimilation. John Gerard's famous herbal, published in 1597 considered that, "The whole plant is of a ranke and stinking savour", although he mentioned that they were eaten in Spain and other hot regions, boiled with pepper, salt and oil, "But they yield very little nourishment to the body and the same naught and corrupt" (Warren, 2015).

Kiwi Fruit: It generally comes as something of a surprise to learn that the kiwifruit is an undomesticated crop. It just seems a little bit odd to think that people still pick kiwifruit from the wild in its native China, as you might collect blackberries from the wayside in Europe. Given this, it seems even stranger to discover that the Chinese have never really eaten them, except to recommend them as a tonic for children or for women following childbirth. It was New Zealanders who first exploited the delights of this fruit, establishing the first commercial planting in 1937 (Warren, 2015).

The possible reason that the Chinese failed to appreciate the potential of the kiwifruit (or yang tao as they call it) is that many people are allergic to its fuzzy brown skin, which is covered in hairs, which act as an irritant. Not only does the kiwifruit contain allergenic factors that are commonplace in wild plants, there are in fact several different related edible species that are all eaten under the name of kiwifruit. This phenomenon is generally restricted to recently domesticated species, since species with longer histories of cultivation are more likely to be clearly differentiated from similar but undomesticated relatives (Warren, 2015).

The most commonly eaten species of kiwifruit has the readily identifiable fuzzy brown skin and is known in Latin as Actinidia deliciosa. The smoother skinned golden kiwifruit goes by the name Actinidia chinensis that gives away its Chinese, rather than New Zealand nationality (Warren, 2015).

Cabbage: Ever resourceful and ingenious, mankind always seems able to find uses for every plant species taken into cultivation. In many crops varieties have been selected especially suited to their particular applications. A simple example of this phenomenon is the apple with its cooking, dessert and cider varieties. More extreme cases include: fibre and high resin forms of hemp and linseed grown for oil or flax for fibre. The undisputed master of this art is the humble cabbage. This single species occurs as spring, summer and autumn varieties, in shades of red, green and white, with Savoy or smooth round heads. But this is just the beginning, because kale, collards, kohlrabi, calabrese, broccoli, Brussels, sprouts and cauliflower are still all the same species not to mention the more obscure ornamental kales, the palm tree cabbage and walking-stick cabbage. What is even more amazing about this long list of apparently different vegetables is that many of them appear to have been invented not just once but twice. This appears to have been relatively easy since plants, which resemble many of these different types, can still be found growing wild (Warren, 2015).

No one is entirely certain if the wild cabbage is truly native to Britain or not. Non-cultivated plants can be found groving around much of our coastline, but many of these populations are probably the descendants of plants that made it over the garden fence, rather than being relic ancient Brits. These coastal cabbages are highly variable with some plants looking more like broccoli, while in others the family resemblance is closer to kale. Whatever their origins, the cabbages that live in the wild in Britain are impressive plants. Some botanists claim that they are able to age plants by counting annual whorls of leaf scars and that individuals cabbages as old as thirty years been reported. In contrast, around the Mediterranean, plants of what is arguably the same species are unable to survive through the dry hot summers. They have adapted to the conditions by growing rapidly during the mild wetter months of winter and flowering just the nee before dying. The same switch in life-style from long-lived northern types to ephemeral Mediterranean winter annuals occurs in many familiar species including the common daisy. As we shall see later in crops, such as, and runner beans, there are many examples of crops which are similar to the perennial cabbage in that although they have the ability to survive for many years, . we tend only to grow them as annuals (Warren, 2015).

The earliest records of cultivated cabbages are to be found in the writings of ancient Greece and Rome and date from around 600 B.C., although they were probably grown much earlier than this. These classical cabbages must have been derived from the short-lived annual Brassicas native to the region. By the first century A.D., there were already accounts of varieties, which resembled leafy kales, heading cabbages, kohlrabi and something like a cauliflower or broccoli. Following the decline of the Roman Empire these early vegetables slipped into obscurity as Europe entered the dark ages.

The modern cabbage emerged in medieval Germany. Botanical genealogists have traced its ancestry to the long-lived northern forms. Unable to claim descent from the noble cabbages of ancient Rome, they have remained sour kraut ever since. As the name suggests, kohlrabi in its current incarnation is another German invention and dates from the fifteenth century. The cauliflower and its progenitor, the broccolis, are thought to have been reinvented around the Mediterranean with the island of Cyprus usually being identified as their particular Garden of Eden. This theory is supported by the fact that they are frequently fast growing annuals. However, perennial broccolis are also known, suggesting that northern Europeans have long enjoyed holiday romances in the sun. Compared with other manifestations of the cabbage, thee sprout is a new-fangled invention. It was first recorded in Belgium in 1750, near Brussels, where else? From there it took about 50 years for the crop to spread to France and Britain. This particular form of the species appears to have been completely unknown to the Emperors of Rome; the Belgians can therefore take full credit for the invention of *Brassica oleracea* var. *gemmifera* (Warren, 2015).

Recent genetic analysis has revealed that there are a number of crops, including; rice, beans, peas and bottle gourds that have all been domesticated independently on a number of occasions. However, the roman diversity of forms of cabbage remains somewhat unusual in that coming back to they seem to have returned to the wild before becoming coming back to the fold. In doing so they blur our understanding of domestication, although sometimes, it can occur instantaneously through a hybridization event, often it is best considered to be a spectrum of slow genetic change (Warren, 2015).

Beans: In Central and Southern America four closely related beans were domesticated. In the cool uplands, runner beans; in warmer temperate regions, common or French beans; the Lima or butter bean was grown in sub-tropical regions and the tepary bean was cultivated in semi-arid zones. In the high Andes water boils at well below 100 degrees centigrade. To cope with this, the Incas developed beans that did not require boiling, but could be consumed after rapid frying. Although all four of these beans are now usually cultivated as annuals, actually only the tepary bean is truly so. In its native seasonally arid habitat, the tepary bean has evolved unable to survive for more than a single year. In contrast, French beans, runner beans and butter beans can all live for several years. But being intolerant of frosts this is impractical when cultivated in temperate regions.

These American beans arrived in Europe via different routes. The runner bean was introduced to Britain in the early seventeenth century by the plant collector, John Tradescant the younger, son of the famous plants collector John Tradescant the elder, in whose honour the Tradescanti is, named. For about 100 years, runner beans were cultivated purely as ornamentals for their attractive red flowers which resemble sweet until Philip Miller of the Physic Garden in Chelsea rediscovered that the pods were good to eat (Warren, 2015).

The bright, scarlet-red flowers of runner beans are not only attractive to gardeners. In their native South America they are visited by humming birds. This fact probably partly explains why many runner bean flowers fail to produce beans when grown in European gardens and mutant white flowered varieties are considered to set more pods. Ignorant of this fact, many gardeners attribute the failure of pod production to erratic water supply. The other three cultivated American beans are self-fertile and to not require the visit of a bird or a bee to produce seeds. The flower of the broad bean is a complex trip trap, which when triggered showers visiting bees with pollen, although some .modern varieties are able to produce beans in the absence of a pollinating insect. Careful inspection of a few broad bean flowers will frequently reveal that instead of fighting the trip mechanism to gain access to the flower, bumblebees often bite holes at the base of the flowers to plunder their contents (Warren, 2015).

Fig: There is another crop plant that has even more highly specialized flowers than the orchids and even more specialized pollinators. Their flowers are so complex that manual pollination is unfeasible, so humans have had to resort to another trick to ensure that fruit are produced in the absence of insects. This crop is the fig, which like vanilla boasts lots of close relatives. There are about 850 species of mostly tropical tree in the genus *Ficus*; all of which are edible, but only one is truly domesticated (Warren, 2015).

The fig has an association with mankind as ancient as the Garden of Eden, where Adam and Eve first used its leaves to cover their nakedness. Archaeological records show that figs have been consumed rather than used as clothing since Neolithic times (Warren, 2015).

Figs are thought to have been first cultivated in Western Asia, but spread rapidly to the Mediterranean, where they were to become a staple food in ancient Greece. The Spartans fed their athletes almost exclusively on figs, believing it would increase their strength and speed — a phenomenon now termed as having the runs! The Greeks prized their figs so much that laws were passed forbidding the export of the best fruit. The word sycophant has been linked to this legislation, as its original usage was to describe loathsome individuals who informed the authorities about fig smuggling. Sycophant literally means 'shower of figs'. The fruit was also important to the ancient Romans who regarded it as sacred because, according to legend, Romulus and Remus were suckled by a wolf in the shade of a fig tree (Warren, 2015).

Over the years the fig has not only screened the sexuality of Adam and Eve and many classical statues, it has also been successful in concealing its Own. With the weirdest sex-life of any crop, it is easy to imagine why the fig. may have wished to keep its habits private. Only in the twentieth century were the fig's sexual secrets finally exposed. The fig tree has been able to achieve this because of the unique structure of the fig itself (Warren, 2015).

Avocado: Wild plants don't just have weird florals that complicate their structures like orchids and figs that complicate love lives and make them problematic to domesticate; they can also do remarkably complex tricks with very simple flowers. A beautiful example of this is the avocado (Warren, 2015).

Throughout the developed world, the avocado is regarded as a luxury item. As one of very few fruits that are high in fat, being up to 30 per cent oil, it is often considered an extravagance. However, in much of central and southern America the avocado has been a staple of the rural poor for thousands of years. Its everyday usage has resulted in it being termed 'the butter of the poor', vegetable butter and midshipman's butter (Warren, 2015).

The avocado is a tropical or subtropical tree, which grows to about ten metres tall. It exists as three distinct types, sometimes considered as sub-species. These are: the Mexican, the Guatemalan and the West Indian (which is originally from Colombia in northern South America). These different types may well be the descendants of independent cultivation events from the wild. Archaeological records indicate that humans have been consuming avocados for as long as 9,000 years and have been cultivating them for as much as 7,000. These three types are adapted to the different climatic zones of their origins, with the Mexican being the most cold tolerant and the West Indian being truly tropical. The majority of avocados finding their way to the tables of Europe and North America are of the Mexican type which is smaller and richer in fat than the tropical form. Over time, however, hybridization between these three traditional types is resulting in a blurring of these distinctions (Warren, 2015).

We have already seen that the complex sexual that occur within the flowers of orchids and figs are the result of their unusual morphologies. In contrast, the avocado, mangoes to spice up its sex-life by using very simple flower forms but effectively changing their gender over time. Just imagine how weird your sex-life might be it you were to combine all of these possibilities within one species. This is what exactly happens in the papaw. The papaw or papaya not only has two well know names, but it is said to have 31 different sexes and presumably a highly complex social life as a result (Warren, 2015).

Carrots: White the exotic sex change antics of the papaw are well known to all who try to grow them, some crop plants are a lot more discreet about their love lives. Indeed some of the tricks that are employed by crops to attract pollinators are seductively subtle. The humble carrot is such a beast. The wild carrot which is a member of a family of plants known affectionately by botanists as the 'Umbels' is a master of deception. The Umbelliferae, to give the family its formal title, contains many cultivated species, such as, celery and parsnip and the herbs fennel, parsley and aniseed, but also contains many toxic species including hemlock and the blistering giant hogweed. The Umbels are so called because their flowers are grouped into clusters like miniature umbrellas, which in turn are grouped into larger umbrellas to produce a mass of flowers. The wild carrot, which is not uncommon in Britain, can be distinguished from the other Umbels because the central flower of its umbrella is not white like the rest, but is dark Purple or pink. This solitary pigmented flower is thought to act as a decoy beetle, tricking passing insects into visiting the mass of flowers and hopefully bringing about pollination. The carrot pulls off this amazing trick by activating an anthocyanin gene that produces a strong red colour only in the central flower. Not surprisingly traditional herbalists spotted this curiosity and regarded these pigmented flowers as possessing special healing properties. Given anthocyanin have antioxidant properties, they may have genuine health benefits, but there are probably more efficient ways of incorporating them in your diet (Warren, 2015).

The majority of edible bananas and plantains are descended from just two wild species; *Musa acuminata* and *Musa balbisiana*, unfortunately neither of these really have common names. The first bananas to be cultivated were probably plants of *Musa acuminata* growing in the humid forests of Malaysia. In the wild, most of these plants produce fruits full of large seeds, but if unpollinated (and pollination is usually by bats) some plants develop fruit without seeds. Mankind by choosing plants that are, female sterile which do not produce seeds but yet still developed fruit, created the first edible bananas. Banana cultivation then spread north into the seasonally drier monsoon areas, where it came into contact with *Musa balbisiana* and because very occasionally bananas can produce seeds these two species were able to hybridize for the first time (Warren, 2015).

In spite of the fact that we routinely eat so few of the plant species that are available to us, we have found a seemingly endless list of ingenious uses for every plant product on offer. We dig up roots and underground storage organs, while fruits and seeds are devoured with relish.

Vavilovian Centres: Following is a list of the six primary Vavilovian centres of crop domestication; associated with each of these are their locally domesticated crops in each of the following four plant families; the cereals, the legumes, the cucurbits or melons and finally the Amaranthaceae or spinach family. Each of these ancient agricultural civilizations has developed traditional local recipes that incorporate their endemic crops (Warren, 2015).

Mexico and Central America

Cereals: Maize.

Legumes: Runner bean, small seeded common bean, lima bean and tepary bean.

Cucurbits: Squash, chayote, marrow, courgette and green striped cushaw. Amaranthaceae: Quinoa, huauzontle and quelite.

Classic dishes: Frijoles Refritos (refried bean) and tortillas, Frijoles de la Olla and Torrtillas de Maiz (stewed beans and corn tortillas) Black Bean Corn Casserole.

South America

Cereals: Maize.

Legumes: Lima bean, large seeded common bean and peanut.

Cucurbits: Winter pumpkin and fig-leaved gourd.

Amaranthaceae: Quinoa and qainiwa or canahua.

Classic dishes: The national dish of Argentina, Bolivia, Chile and Ecuador is a thick stew called Locro, made of corn, beans and gourds. There are many local variants which may include: potatoes, various meats and chillies.

Mediterranean and Middle East

Cereals: Wheat, oats, barley and rye. lentil and lupin.

Legumes: Pea, broad bean.

Cucurbits: Melon.

Amaranthaceae: Beetroot/ sugar beet/ mangelwurzel, chard and goosefoot.

Classic dishes: Hummus and pitta, Chobra frik (Algerian crushed wheat and chickpea soup), Chakhchoukha (semolina flat bread and chickpea stew).

Africa

Cereals: Millet, sorghum and teff (*Eragrostis teD*, a cereal of the Ethiopian uplands which is also knows as Williams Lovegrass.

Legumes: Cowpea, Bambara bean, hyacinth bean and locust bean. Cucurbits: Muskmelon, white flowered gourd, oysternut and watermelon.

Amaranthaceae: Doodo (Amaranthus dubius) and celosia.

Classic dishes: Food varies across the continent of Africa, but the importance of these staples ensures there are some recurring themes. In Nigeria, Moin moin (steamed bean pudding) is eaten with Ogi porridge). This is not dissimilar to Shiro, an Ethiopian paste of bread-beans or chickpeas which are eaten served on Injera (teff flat bread). The Yoruba people of Nigeria consume Ira (fermented locust beans) with Egusi soup which is made of roasted pumpkin seeds and celosia or spinach leaves. In much of the south of Africa maize porridge which has many local names, such as, Sadza, Pap and Putu is eaten with bean relish called Cankalaka.

China

Cereals: Rice and millet.

Legumes: Soya bean, adzuki bean and velvet bean.

Cucurbits: Winter melon (Benincasa hispida) or Chinese Preserving melon.

Amaranthaceae: Sand-rice (Agriophyllum squarrosum) and Chinese Spinach (Amaranthus tricolor).

Classic dishes: Bean tempura, tofu and rice and stir-fried bean shoots and rice.

India

Cereals: Rice.

Legumes: Chick pea, mung bean, pigeon pea and urad bean.

Cucurbits: Cucumber, bitter melon and luffa.

Amaranthaceae: Spinach.

Classic dishes: The vegetarian cuisine of India contains many combinations of these crops, either cooked or served together. Examples of this are: Bisi Bele Bath (hot lentil rice). Dal Puri Roti (flat bread containing split peas), Mujaddar (cooked lentil and grots).

Pumpkins: The utility of the four plant families that have been so universally domesticated is reflected in the fact that they were all amongst the first crops that humans chose to cultivate. However, the archaeological evidence of ancient use of the third of these families is rather limited because melons, cucumbers, marrows and pumpkins *etc.,* are fleshy and tend to leave very little trace of their existence. The cucurbits (members of the melon family) make such versatile crops that not only have many different species of them been domesticated, but recent genetic analysis has confirmed that this has occurred in *Cucurbita pepo* on several different occasions across South America. These different events have given rise to pumpkins, squashes and marrows. Even more remarkable is the bottle gourd that is thought to have originated in Africa. More than 5000 years ago the ancient Egyptians were using bottle gourds as water containers. Prior to this it appears to have floated across the Atlantic to South America, where there is evidence of it being grown as early as 9000 BP, well before Columbus made the trip. On both sides of the Atlantic, many different shaped hard-shelled fruits were selected for use as bottles, drinking vessels, ladles, floats for fishing nets and even musical instruments. Only the immature bottle gourd fruit are actually edible because when ripe (like many wild members of the family) they become incredibly bitter (Warren, 2015).

It can be rather confusing discussing the different members of the melon family, because the same common names, such as, pumpkin, squash and melon are frequently used to refer to several different species. Generally speaking fruits with hard skins that can be stored are termed pumpkins, whereas, soft skinned types are called squashes. The word gourd is even more ambiguously used and is applied to similar looking fruit from completely unrelated plants that are used ornamentally or as containers. Perhaps the most bizarre example of this is the penis gourd. In the highlands of New Guinea, the male indigenous peoples wear the dried shells of the bottle gourd to cover their genitals.

As crops, the cucurbits are remarkably different from both cereals and legumes. But perhaps they did not start that way. It has been suggested that they were first grown for their large oil rich seeds. Only later were mutant plants with non-bitter fruits identified. The cucurbits also differ from cereals and legumes nutritionally, in that they provide us with very little other than a few vitamins and fibre. In these terms the lexicographer Samuel Johnson may have had a point when he said "a cucumber should be well sliced and dressed with pepper and vinegar and then thrown out, as good for nothing". However, in regions with prolonged dry periods the cucurbits can be invaluable in providing a store of fresh vegetable matter. In arid places, such as, the Kalahari Desert, which is home to the watermelon, these fruit are a great source of clean safe water (Warren, 2015).

Spinach: Long before marketing departiments had thought of selling us anti-oxidant rich fruits as 'Super-foods' there was spinach. This annual vegetable is a member of the last of the four crop families that have been universally domesticated. In many ways the spinach family make the ultimate crops. They appear to have evolved the more efficient C4 form of photosynthesis on 15 separate occasions. Many members of the family have long growing seasons, during which their leaves can withstand being harvested repeatedly. Nutritionally they contain all nine essential amino acids, making them the perfect protein sources and as Elzie Crisler Segar's cartoon character, Popeye the Sailor Man knew, spinach is an amazing source of iron. Or is it (Warren, 2015).

Popeye first appeared as a cartoon strip character in 1929. Initially he acquired his super human strength by rubbing himself with a magic hen, but this fowl habit was soon replaced by an addiction to spinach! Although it has sometimes been claimed that Popeye was an invention of the spinach industry, this is not the case, although several important spinach growing towns in the USA have built monuments to the cartoon sailor as a mark of their appreciation. There is another more extreme myth associated with Popeye's fondness for spinach. For many years it was thought that spinach was selected by Segar because of its incredibly high iron content. This was in turn based on chemical analysis carried out in the nineteenth century by Professor von Wolff, who accidentally placed a decimal point in the wrong place and thus, overestimated the iron content of spinach by tenfold. More recently However, this story has been questioned by Mike Sutton of Nottingham Trent University. It appears that several other early estimates of the iron content of spinach were all dramatically higher than the true value. This was not because of mass multiple misplacing of decimal points, but related to iron contamination of the samples due to bad experimental practice. The French military also seem to have been impressed by spinach's high iron content, because during World War-I, spinach juice was added to wine given to soldiers suffering from haemorrhages (Warren, 2015).

Warren (2015) stated that what limits the number of species that we currently grow and consume, is our own imaginations, prejudices, traditions and vested interests. If this is true, in the future we may enjoy a whole myriad of new fruits and vegetables that are better for our health and less demanding of the world's limited resources.

4. HORTICULTURE AND HUMANS

Horticulture is intimately connected to humankind in various critical and essential ways – through provision of essential minerals and nutrients in diets, the use of ornamental and amenity plants to enhance leisure and pleasure pursuits, through to enhancing livelihoods through the use of horticultural subjects in art and literature. Horticulture also enhances the environment, especially in major urban centers where trees, shrubs and ornamental plantings in parks, roadside avenues and rooftop gardens can provide shade and shelter, reduce the impacts of heat islands caused by buildings, roads and infrastructure and even enhance road safety and the security of residents. The presence of and interaction with green spaces has also been shown to enhance physical and mental well-being and the rate of recovery of patients in health care. Horticulture truly feeds us, sustains and cures us and makes life worth living. It provides food for body and soul (Janick and Warrington, 2016).

Although grain crops, such as, wheat, rice, sorghum and maize are principle suppliers of calories, horticultural crops are prized for their flavors, nutrients, minerals and phytochemical contents, such as, vitamins, polyphenols, terpenoids organo-sulphurs and alkaloids (Desjardins, 2014). Horticultural crops,principally legumes, are important sources of protein, while fruits and tuber crops are rich sources of carbohydrates, vitamins and fiber and leafy vegetables are an important source of minerals, nutraceuticals and vitamins; some fruits, such as, olive and avocado are sources of lipids. The advantage of diets rich in plant over animal sources is production efficiency since animals must be fed on crops that could feed humans. For example, it has been calculated that it takes approximately 15,000 liters of water to produce a kilogram of animal protein (dry weight) as compared to approximately 240 liters for an equivalent weight of lettuce or 125 liters for a 150 g apple or for one glass of wine (Mekonnen and Hoekstra, 2010; 2011). Consequently, a shift to a horticulturally rich diet may be essential to feed the planet as world population continues to increase.

In conclusion, horticulture is so basic to human existence that it is no wonder that horticultural plants and practices have permeated so deeply into our cultures. The universal appeal of its beloved plants with their delightful fragrances and delicious flavors and beauty, their health and medicinal values and their importance in the environment, underscores the close relationship of horticulture to humans and humanity. Horticulture truly feeds us, sustains and cures us and makes life worth living. It is provides food for body and soul.

5. EVIDENCES OF HUMAN CONTRIBUTIONS

- Rules work for people, but what works even better, are rules tempered with an equal amount of compassion and love (Mehta, 2013).
- In a democracy, the poor will have more power than the rich, because there are more of them and the will of the majority is supreme (Aristotle, 2013).
- A good head and a good heart are always a formidable combination (Mandela, 2013).
- **Well-groomed** (Anon., 2014a): The owner of a farm asked his child to work every day at the farm. His friend said to him, "You do not have to make your son work so hard. The crops would grow just as good." The farm owner replied, "I am not cultivating my crops, but my child." A simple way to groom a child is to let him experience some hardships. If not cut, jade would not turn into useful ware.
- **Case Dismissed** (Nasruddin, 2014): Judge Nasruddin was presiding over a case. The plaintiff went first and exclaimed, "The defendant refuses to pay his restaurant bill!" "I would have," the defendant countered, "but he charged me $200 for three

hardboiled eggs!" "Is this true?" Nasruddin asked the plaintiff. "Well," the other replied, "as I explained to him earlier, those eggs could have been hatched into chickens, which in turn would have produced more eggs, which in turn would have hatched into more chickens and so on and so forth. The way I see it, his three eggs would have yielded me hundred of dollars worth of chickens and eggs." "Okay," Nasruddin replied. "Wait here while I go plant some boiled peas in my garden." "But Judge," the plaintiff said, "boiled peas will not grow into anything." "In that case," Nasruddin replied, "this case is dismissed!"

- **Ubuntu** (Anon., 2014c): An anthropologist proposed a game to some African tribal children. He placed a basket of fruits near a tree and made the children stand 100 meters away. He now announced that whoever reached the fruits in the basket. When he said 'Ready, steady, go', do you know what these small children did? They all held each other's hands and ran towards the tree together and divided the fruits among themselves, thus eating the fruits and enjoying them. When the anthropologist asked them why they did so, they said 'Ubuntu' in chorus, which meant: 'How can one be happy when all the others are sad?' Ubuntu in their language means: 'I am, because we are!'
- **Fruit of Heaven** (Anon., 2014e): There was once a woman who had heard of the fruit of heaven. She coveted it. She asked a certain dervish, whom we shall call Sabar, "How can I find this fruit, so that I may attain to immediate knowledge?" "You would best be advised to study with me," said the dervish. "But if you will not do so, you will have to travel resolutely and at times restlessly throughout the world." She left him and sought another, Arif, the wise one and then found Hakim, the sage, then Majzup, the madman, then Alim, the scientist and many more. This way, she spent 30 years searching for the fruit. Finally, she came to a garden. There stood the tree of heaven and from its branches hung the bright fruit of heaven. Standing beside the tree was Sabar, the first dervish. "Why didn't you tell me when we first met that you were the custodian of the fruit of heaven?" She asked him. "Because you would not have believed me. Besides, the tree produces fruit only once in 30 years and 30 days."
- **Positive Attitude** (Anon., 2014i)
  - Your smile is a signature of God on your face. Do not allow it to be washed away by your tears or erased by your anger. Life is best for those who are enjoying it and difficult for those who are comparing it and worst for those who are criticizing it. Your own attitude defines your life.
  - Once you are on the journey of self-discovery, your consciousness evolves to unfold the steps necessary for growth. (Bhardwaj, 2014).
  - Gandhiji said (Anon., 2014j), the connection between human rights, peace and ecological balance seems different but fundamentally are the same. They are part of the same reality. We evolved over a thousand years from nature. We are a part of nature, so we should function as part of nature and take care of it as we take care of ourselves. So, to honour animals and plants is honouring whatever it takes to be human. Our brain has evolved as far we can do all the things we do, but we haven't learnt how to live at peace and to take care of the environment. We think we can treat everything as though we were God but we are not God. We must

address peace as a systemic problem, work for systems that do no harm, work to end nuclear weapons and to prevent the worst effects of climate change. We must create laws and ways that do not destroy life. Too much dependence on gadgets and increasing consumerism has depleted our resources. If we follow the ideals of simplicity as adopted by Gandhi, we could restore our ecosystem.

- The landscape belongs to the person who looks at it (Emerson, 2015).

Hope and Despair Crippling Shortage of Drinking Water in North Karnataka as Borewells Run Dry

Water Woes - 52-Year-Old Woman Digs 60ft Well To Nourish Her Garden (Anon., 2017)

Faced with the bleak prospect of her small plantation drying up, Gouri Naik, a 52-year-old agricultural labourer in Ganesh Nagar, a small town in Sirsi (Uttara Kannada district) single-handedly dug a 60-foot well in her backyard.

Gouri, who peels arecanuts for a living, planted around 150 arecanut plants, 15 coconut trees, in addition to some banana plants in the vacant space in her backyard. A mother of two children, Gouri lives with her 20-year-old son, while her daughter is married. Failure of the monsoon last year resulted in the two wells in her com pound drying up. Former Prime Minister Manmohan Songh once said that it was crisis that concentrated the mind and Gouri showed that she had the resilience to confront the adversity. "The plants in my garden would have withered and dried without water. So, I thought of digging a well exclusively for my plantation", said Gouri.

A ritual steeped in beliefs (Mogata, 2016): In Yellapur taluk of Uttara Kannada district, a special ritual is observed during Deepavali. The practice that is followed by the local people since ancient times is called 'Holehuliyappana Puje'.

Farmers of the region worship Lord Holehuliyappa (the name translates to 'river tiger' in Kannada), situated on the banks of River Bedthi and pray to him to protect the cattle. During Deepavali, they offer coconuts to Holehuliyappa. As per the norms set by people themselves, they offer one coconut per one ox.

It is said that in the olden days, tigers that lived in the thick forest around used to hunt cattle, thus affecting the farmers. To protect the cattle from the tiger, the farmers decided to worship Lord Holehuliyappa. It has continued to this day and along with coconut, farmers also offer fruits, flowers, butter and milk to the god.

The festival starts for them only after this ritual is performed. Approximately, 15,000 coconuts are offered to the god during Deepavali.

From farm to home (Kurian, 2016): Madhusudhan and his wife were travelling back from Tamil Nadu when they saw a vegetable vendor washing carrots with drainage water. Shocked at the sight, they stopped to find out why he was doing so. Upon enquiring, Madhusudhan understood that the farm this vegetable came from was also being watered with drainage water. This shocking revelation led to a strong decision — to grow vegetables and eat only at places they knew where the produce was sourced from. Thus began the journey of 'Back2Basics', an organic farm that offers home delivery services.

While many vegetable markets and vendors are available in the city, the consumers aren't aware of where their produce comes from. Even by changing one's diet to a vegetarian one, many are falling sick because the produce isn't fresh.

Jyothi Rajesh, an entrepreneur and blogger orders from an online organic vendor called 'FirstArgo' twice a month. She says, "I'm happy to place my order and buy from an online store like this because they assure good produce. I can also get my hands on items like baby carrots and cherry tomatoes without paying an exorbitant amount. I know that the produce is fresh because it tastes better and when you cut into it, the colours seem lush."

Amar, J. a co-producer at 'Farm Direct Express', is happy to be a part of the venture. They deliver vegetables throughout Bengaluru and have their farms within 50 km of the city. He says, "We have a total of 31 farms in areas like Singapura, Kanakapura Road, Nelamangala, Hosakote and more. Our main aim is to deliver fresh produce and bring back the joy of eating fresh vegetables."

Bhairavi Madhusudhan, the idea cultivator of 'Back2Basics', says, "We cultivate about 90 varieties of fruits, vegetables and greens in the four commercial farms we have in rural Bengaluru and close to the Tamil Nadu border. We control the entire supply chain and grow all our produce ourselves. To give one an experience of how our farm works, we also have an experiential farm that anyone can visit."

Jasmines from Bhatkal bring in moolah, luck for Shankarapura (Pinto, 2016): Mangaluru: In the early 1950s, the then parish priest of St John's Evangelist Church in Pangala, Father Basil Salvadore Peris, gave a bunch of jasmine creepers to 10 families and predicted that they would transform their lives.

Today, the robust economy of Shankarapura village is testimony to Father Peris' prescient words. The jasmine plants the priest, who later went on to become the Bishop of Mangalore Diocese (1955-58), not only helped those families out of economic distress but have come to determine the very identity of the village. The flower, called Shankarapura Mallige, was accorded geographical indication (GI) status in 2008.

Mallige in Every Region

- Jasmine from Mysuru, popular as Mysuru Mallige, grow in sandy loam with a comparatively high pH; they bloom from March to July, are known for rich fragrance.
- Hadagali Mallige, which derives its name from Hoovina Hadagali in Bellari district, grows in sandy red soil; dry weather and continuous water supply offer a conductive environment for cultivation.
- Udupi Mallige are the least fragrant of the three, but boast a longer shelf life; grown in laterite soil, they depend on heavy rainfall and thrive in humid conditions.

9 `sacred' lemons fetch Rs 68,000 in annual auction in TN temple (Dominique, 2017): A lemon usually costs less than Rs 10. But not the ones spiked on the spear of the presiding deity at a temple at Tiruvanainallur in Villupuram district of Tamil Nadu. Nine lemons that were auctioned by the temple management at the end of the 11day Panguni Uthiram festival fetched Rs 68,000, of which one went for Rs 27,000.

It is the custom of the Balathandayuthapani temple management to auction the lemons spiked on the spear of the presiding deity, Muruga. One lemon is spiked every day during the first nine days of the 11-day festival. Villagers of Tiruvanainallur and surrounding region firmly believe that possessing these lemons will bring them prosperity. Moreover, they believe that childless couples will be blessed if the family possesses the sacred lemons.

4. SHARING OF HUMANITY

- Mango tree in my garden had started flowering. I was very happy wishing for a good crop of hybrid Mallika (Alphonso x Neelam) as there was very little harvest last year. Here came my next door neighbor and complained that my mango tree had crossed the fence and heavy mango flowers were dropping in his area and making his parked car dirty. I replied that in any religion or culture the flowering fruit trees are never cut but one has to wait for harvest; you should understand more being Hindu. He walked away sadly and sent a message next day to get his car cleaned by my gardener because it was covered with mango flowers.
- On harvest, my family was delighted greatly to see lovely Mallika mangoes looking so beautiful and weighing 500 to 800 grams each. I felt so happy to take care of the tree and didn't forget to send a basket of mango to my neighbor who had complained.
- I sent lovely large size Mallika mangoes to my neighbours and friends, for them to enjoy delicious mangoes. Some of them phoned and thanked me the day they received, some of them sent e-mail, few of them thanked me when they met and some of them did not acknowledge at all, may be they did not like the gift.
- Sometime, I and my family were driving from Rome, Italy to Vienna, Austria. In the morning, we left Rome and by evening, we crossed through the check-post and decided to spend night somewhere there. Finally, we selected a farm guest house at the top of the hill. On arrival, it gave a beautiful view of lights in the residences of the hill and down in the valley the light was glittering.
- When we got up in the morning, we were served by fresh tea or coffee, of course, by the farm owner. At the breakfast table it was a beautiful sight of fresh eggs, apples, grapes, bread, cheese, butter and milk, all harvested from the farm. Everything was in plenty to consume depending on the individual. It was excellent treat indeed.
- On departure, the host farmer took us round the farm to witness fruiting in vining apple plants, lovely grapes, attractive vegetables and poultry. We were presented with basket of fruits to consume on the way ahead.
- In the country side, about 100 mile from Vienna, Austria, we stayed at another farm guest house. During day time we visited Vienna and enjoyed the architecture, people, life and orchestra. We spent a couple of nights at the farm home with such a lovely host; lovely green area to go around and get lost. The old farm house built of solid wood completely had beautiful dining hall with ancient lights and pillars. The breakfast was sumptuous with home preparations and taste with fresh farm products. We enjoyed bread, egg, cheese, tomatoes or other vegetables freshly harvested from own farm, cooked in typical Viennese taste and served with smile and love. Great enjoyment!
- On having breakfast, we needed to buy some boiled eggs and breads to consume on way further to our journey. The host lady smiled and said wait a bit. No sooner she came with beautiful basket full of boiled eggs, bread and cheese, of course a gift to us (Nath, 2016).

5. IN PERSPECTIVE

- In the developing world, millions eat cereals only, millions eat both cereals and vegetables only and less people can afford to fruits with meals unless it is locally grown and economically affordable. There are a few developing nations, they do not produce fruit or cow milk, hence, both requirement and affordability are restricted. But in our country, Pooja Prashad must have locally grown fruits.
- The question is how many could afford to have two meals a day; How many could afford to have vegetables with cereals? How many could afford to have a healthy squash meat healthy diet of cereals, vegetables and fruits together for a healthy diet?
- Our main aim is to have healthy diet comprising of cereals, legumes, vegetables and fruits, for all. In this context, our focus should be to make vegetables and fruits available and affordable to one and all.

6. CONCLUSION

Without the essential inputs of humans, there is no flower, fruit and vegetables production and to achieve these for all the inputs of humanity are essential to further ensure sustainable nutrition security. With humanity in mind, we can enjoy fresh air under tree shade, enjoy flowers of the garden, pluck vegetables from homegarden, look fresh vegetables, eat fresh fruits and enjoy juices of fruits and vegetables to be human and healthy. Horticulture serves humans and promotes humanity.

REFERENCES

Anon. (2013b). The National Food Security Act 2013, No. 20 of 2013, The Gazette of India No. 29, New Delhi, September 10, 2013.

Anon. (2014a). Well- groomed, The Times of India, Bengaluru, March 02, 2014.

Anon. (2014c). Ubuntu, The Times of India, Bengaluru, March 16, 2014.

Anon. (2014e). Fruit of Heaven, The Times of India, Bengaluru, March 16, 2014.

Anon. (2014i). Positive Attitude, The Times of India, Bengaluru, July 27, 2014.

Anon. (2014j). The Times of India, Bengaluru, November 16, 2014.

Anon. (2017). Water Woes - 52-Year-Old Woman Digs 60ft Well To Nourish Her Garden, The Times of India (Bengaluru), Karwar, Times Network, Apr 16 2017.

Aristotle. (2013). Evidences of Human Contributions, The Times of India, Bengaluru, December 01, 2013.

Desjardins, Y. (2014). Fruit and vegetables and health: An overview. Chapt. 28. p.965-1000. In: G.R. Dixon and D.E. Aldous (eds.), Horticulture: Plants for People and Places, Vol. 3. Springer. Dordrecht, Heidelberg, New York, London.

Dominique, Bosco. (2017). 9 'sacred' lemons fetch Rs 68,000 in annual auction in TN temple, The Times of India (Bengaluru), Villupuram, Apr 15 2017.

Emerson, Ralph W. (2015). The Times of India, Bengaluru, January 11, 2015.

Janick, Jules and Warrington, Ian. (2016). The Human Connection to Horticulture: Humanity in Humans, by Humans and for Humans in Securing Food for All

(The Basics of Human Civilization – Food, Agriculture and Humanity, Vol. IV – Humanity), PNASF, Bengaluru.

Kurian, Anila. (2016). From farm to home, Deccan Herald, DHNS, Bengaluru, July 15, 2016.

Mandela, Nelson. (2013). The Times of India, Bengaluru, December 07, 2013

Mehta, N. (2013). Evidences of Human Contributions, The Times of India, Bengaluru, December 01, 2013.

Mekonnen, M.M. and A.Y. Hoekstra. (2010).The green, blue and grey water footprint of crops and derived crop products, value of water. Research Report Series No.47, UNESCO-IHE.

Mekonnen, M.M. and A.Y. Hoekstra. (2011).The green, blue and grey water footprint of crops and derived crop products. Hydrology and Earth System Sciences15(5): 1577-1600.

Mogata, Beeranna Nayak. (2016). A ritual steeped in beliefs, Deccan Herald, November 01, 2016.

Nasruddin, Mullah. (2014). Case Dismissed, The Times of India, Bengaluru, March 02, 2014.

Nath, Prem. (2016). Positive Human Forces in Support of Food and Nutrition Security, Humanity in Humans, by Humans and for Humans in Securing Food for All (The Basics of Human Civilization – Food, Agriculture and Humanity, Vol. IV – Humanity), PNASF, Bengaluru.

Nath, Prem. (2016). Some Human Perceptions Related to Food and Humanity, Family Farming and Food Insecurity, Humanity in Humans, by Humans and for Humans in Securing Food for All (The Basics of Human Civilization – Food, Agriculture and Humanity, Vol. IV – Humanity), PNASF, Bengaluru. p. 103-104.

Pinto, Stanley. (2016). Jasmines from Bhatkal bring in moolah, luck for Shankarapura, Times of India, TNN, Jun 29, 2016.

Warren, John. (2015). The Nature of Crops: How We Came to Eat the Plants We Do, CABI Publisher, Oxfordshire, U.K. 2015. p. 192.

Dr. P.N. Agricultural Science Foundation (PNASF)

To promote agricultural education, research and sustainable development with a focus on food and nutrition security

PNASF is a Charitable Trust registered with the Government of Karnataka, India. It is established to assist and encourage Indian agricultural scientists to address India's agricultural problems and at the same time use their skills for the benefit of developing countries. It facilitates interaction of knowledge and ideas among leading professionals in agriculture, development and planning, on issues of national and international importance related to agriculture and food.

PNASF Supports

1. Assistance, organization and/or participation in national, regional or international conferences, workshops, expert consultations, seminars etc;
2. Assistance to small field projects through grant received;
3. Assistance in the dissemination of technical messages through technical publications;
4. Assistance to young agricultural scientists in research through modest grants;

5. Assistance to post-graduate students in agriculture through awards, scholarships etc and
6. Development of linkages or partnership with national, regional and international agriculture and related institutions and societies

For further enquires please contact

Dr. Prem Nath Agricultural Science Foundation

9, 1st Cross, 1st Main, 1st Block,
Rajmahal Vilas Extension 2nd Stage,
Bangalore – 560 094, Karnataka State, INDIA
Tel: + 91 – 80 – 2341 5188, Fax: + 91 – 80 – 2351 1555
E-mail: drpremnath@vsnl.net ; Website: www.pnasf.org

Vegetable Science International Network

To strengthen and inter-institutional co-operations among the members among the vegetable sectors worldwide towards improved production and utilization of vegetables

1. **The main objectives of the VEGINET are to:**
   - Facilitate inter-and intra-regional and inter-institutional cooperation through exchange of scientific, technological and related economic information.
   - Identifying priority areas of cooperative research and development.
   - Facilitate development of human resource through better education and training.
   - Promote improved and sustainable production and utilization of vegetables to achieve:
     - o Food security Nutritional security, and Economic empowerment of the farming community.

2. **The overall approach of the VEGINET will be to:**
   - Help in identifying priority areas for research and development.
   - Help in the development of competent human resource for better production, protection, processing, marketing, distribution and utilization of vegetables.
   - Help in formulating collaborative programmes of action.
   - Harness knowledge and competence to achieve the objectives.
   - Help in exchange of information through interactive internet portal and organization of international workshops, symposia and conferences.

3. **Some of the areas identified for this purpose are:**
   - Development of policies, strategies and integrated programmes of action to increase availability of vegetables.
   - Production technology and productivity
   - Development and use of value added germplasm
   - Hybrid seed development and production
   - Genetic improvement of vegetables
   - Biotechnological applications
   - Biosafety of GM vegetables
   - Crop protection technology in the field and at post-harvest level
   - Post-harvest management of fresh produce and processed products
   - Organic vegetable production and means of its natural protection
   - Development of vegetables for better nutrition
   - Under-utilized vegetables and by-products for food and feed
   - Marketing and Trade Infrastructure
   - Technical co-operation among the countries having similar climate and ecological protection.

For further enquiries please contact

VEGINET Secretariat-Dr. Prem Nath Agricultural Science Foundation (PNASF)

#9, 1st Cross, 1st Main, 1st Block, Rajmahal Vilas (RMV) Extension 2nd Stage, Bangalore - 560 094, Karnataka, India; Tel: +91-80-23415188; Fax: +91-80-23511555

Email: drpremnath@vsnl.net, pnasf@rediffmail.com, pnasfoundation@gmail.com

Website: www.pnasf.org